T0206010

Metals and Non-Metals

Five-Membered *N*-Heterocycle Synthesis

Navjeet Kaur

Department of Chemistry
Banasthali Vidyapith
Rajasthan, India

CRC Press
Taylor & Francis Group
Boca Raton London New York

CRC Press is an imprint of the
Taylor & Francis Group, an **informa** business

A SCIENCE PUBLISHERS BOOK

CRC Press
Taylor & Francis Group
6000 Broken Sound Parkway NW, Suite 300
Boca Raton, FL 33487-2742

First issued in paperback 2021

ISBN-13: 978-0-367-32157-4 (hbk)
ISBN-13: 978-1-03-217556-0 (pbk)
DOI: 10.1201/9780429321580

Publisher's Note

The publisher has gone to great lengths to ensure the quality of this reprint but points out that some imperfections in the original copies may be apparent.

Library of Congress Cataloging-in-Publication Data
Names: Kaur, Navjeet, author.
Title: Metals and non-metals : five-membered N-heterocyle synthesis / Navjeet Kaur.
Description: First.	Boca Raton : CRC Press, Taylor & Francis Group, [2020]	"A science publishers book."	Includes bibliographical references and index.
Identifiers: LCCN 2019042096	ISBN 9780367321574 (hardcover)
Subjects: LCSH: Heterocyclic compounds--Synthesis.	Heterocyclic chemistry.	Chemistry, Organic.	Organic cyclic compounds.
Classification: LCC QD400.5.S95 K377 2020	DDC 547/.593--dc23
LC record available at https://lccn.loc.gov/2019042096

Visit the Taylor & Francis Web site at
http://www.taylorandfrancis.com

and the CRC Press Web site at
http://www.crcpress.com

Preface

Due to significant biological activity associated with N-, O-, and S-heterocycles, a number of reports on their synthesis have appeared in recent decades. Traditional approaches either require expensive or highly specialized equipment, or would be of limited use to a synthetic organic chemist due to their highly inconvenient approaches.

The largest classical divisions of organic chemistry are formed by heterocycles, which are important biologically and industrially, as well as for the functioning of any developed human society—their role in various areas cannot be undermined. Heterocyclic natural products such as anti-biotics, pesticides, cardiac glycosides, and alkaloids are of great importance to human and animal health.

Nowadays, new approaches that employ efficient pathways are being developed; researchers are following natural models to design and synthesize heterocycles. In doing so, transition metal-catalyzed protocols have attracted their attention more than other synthetic methodologies due to their easy availability as substrates for building and multiplying substituted complicated molecules directly under mild conditions. In organic synthesis, one of the most powerful and useful category of protocols is built upon transition metal-catalyzed coupling transformations. N-Heterocycles are synthesized by this convenient and useful tool.

The key area of research in organic synthesis these days is the development of rapid, efficient and versatile tools for the preparation of heterocycles and, consequently, protocols involving transition metal catalysis have gained much prominence of late. The traditional strategies that involve long reaction times, harsh conditions and limited substrate scopes are overshadowed by these catalytic practices.

In recent decades, the use of metal and non-metal complexes for heterocyclic construction has become common. In this book, synthesis of five-membered N-heterocycles using metals and non-metals has been focused.

Contents

Abbreviations

AAC	:	azide-alkyne cycloaddition
ABCB1	:	ATP binding cassette subfamily B member 1
AD	:	asymmetric dihydroxylation
AIBN	:	azobisisobutyronitrile
ARCM	:	asymmetric ring-closing metathesis
BINAP	:	2,2'-bis(diphenylphosphino)-1,1'-binaphthyl
BINOL	:	1-(2-hydroxynaphthalen-1-yl)naphthalen-2-ol
BIPHEPHOS	:	2,2'-bis[(1,1'-biphenyl-2,2'-diyl)phosphite]-3,3'-di-*tert*-butyl-5,5'-dimethoxy-1,1'-biphenyl
BMIM	:	1-butyl-3-methylimidazolium
BSA	:	bis(trimethylsilyl)acetamide
BTF	:	benzotrifluoride
BTI	:	[bis(trifluoroacetoxy)iodo]benzene
BQ	:	benzoquinone
CAN	:	ceric ammonium nitrate
CB1	:	cannabinoid type-1
Cbz	:	carboxybenzyl
3CC	:	three-component condensation
CDC	:	cell division cycle
CDK	:	cyclin dependent kinase
CNBr	:	cyanogen bromide
CNS	:	central nervous system
COD	:	cyclooctadiene
COX-2	:	cyclooxygenase-2
CPT	:	camptothecin
CR	:	component reaction
CSA	:	camphorsulfonic acid
DBU	:	1,8-diazabicyclo[5.4.0]undec-7-ene
DCB	:	dichlorobenzene
DCC	:	*N,N'*-dicyclohexylcarbodiimide
DCE	:	dichloroethane
DCM	:	dichloromethane
DDQ	:	2,3-dichloro-5,6-dicyanobenzoquinone
DHQD	:	3-dehydroquinate dehydratase
(DHQD)$_2$PYR	:	hydroquinidine-2,5-diphenyl-4,6-pyrimidinediyl diether

DIEA	:	*N,N*-diisopropylethylamine
DIEPA	:	*N,N*-diisopropylethylamine
DIPEA	:	*N,N*-diisopropylethylamine
DMA	:	dimethylaniline
DMAC	:	dimethylacetamide
DMAD	:	dimethylacetalenedicarboxylate
DMAP	:	4-dimethylaminopyridine
DMBA	:	7,12-dimethylbenz[*a*]anthracene
DMC	:	dimethyl carbonate
DME	:	1,2-dimethoxyethane
DMEDA	:	1,2-dimethylethylenediamine
DMF	:	dimethylformamide
DMPU	:	*N,N'*-dimethylpropyleneurea
DMSO	:	dimethylsulphoxide
DNA	:	deoxyribonucleic acid
Dpe-phos	:	bis-[2-(diphenylphosphino)phenyl]ether
DPPA	:	diphenylphosphorylazide
DPPE	:	1,2-bis-(diphenylphosphino)ethane
DPPF	:	1,1'-bis(diphenylphosphino)ferrocene
DPPM	:	1,2-bis(diphenylphosphino)methane
DPPP	:	1,3-bis(diphenylphosphino)propane
DTBP	:	di-*tert*-butyl peroxide
EAS	:	electrophilic aromatic substitution
EDC	:	1-ethyl-3-(3-dimethylaminopropyl)carbodiimide
EWG	:	electron withdrawing group
FDA	:	food and drug administration
GCMS	:	gas chromatography mass spectrometry
HATU	:	hexafluorophosphate azabenzotriazole tetramethyl uronium
HCV	:	hepatitis C virus
HIV-1	:	human immunodeficiency virus-1
HMG-CoA	:	3-hydroxy-3-methyl-glutaryl-coenzyme A
HMPA	:	hexamethylphosphoramide
5-HT	:	5-hydroxytryptamine
HTIB	:	[hydroxyl(tosyloxy)iodo]benzene
IBX	:	*o*-iodoxybenzoic acid
IDA	:	intramolecular dehydrogenative aminooxygenation
IL	:	ionic liquid
IMCR	:	isocyanide-based multi-component reactions
IMDAF	:	intramolecular Diels-Alder furan
LAH	:	lithium aluminium hydride
LDA	:	lithium diisopropylamide
LiHMDS	:	lithium hexamethyldisilamide
LTMP	:	lithium tetramethylpiperidide
MAP	:	mitogen-activated protein

MBA	:	4-mercaptobenzoic acid
MCAP	:	multi-component assembly process
m-CPBA	:	*meta*-chloroperoxybenzoic acid
MEK	:	methyl ethyl ketone
MIRC	:	Michael initiated ring closure
(*R*)-MNEA	:	*N*-methyl-bis[(*R*)-1-(1-naphthyl)ethyl]amine
MOM	:	methoxymethyl
MW	:	microwave
MWI	:	microwave irradiation
NBS	:	*N*-bromosuccinimide
NCS	:	*N*-chlorosuccinimide
NHC	:	*N*-heterocyclic carbene
NIS	:	*N*-iodosuccinimide
NMM	:	*N*-methyl-maleimide
NMO	:	*N*-methylmorpholine-*N*-oxide
NMP	:	1-methyl-2-pyrrolidinone
NMR	:	nuclear magnetic resonance
NOE	:	nuclear Overhauser effect
OTf	:	trifluoromethanesulfonate
OTMS	:	octadecyltrimethoxysilane
PANI-BC	:	polyaniline-bismoclite
PEG	:	poly(ethylene glycol)
PHAL	:	1,4-phthalazinediyl
PIFA	:	phenyliodine bis(trifluoroacetate)
PMB	:	*p*-methoxybenzyl
PMP	:	polymethylpentene
PPA	:	polyphosphoric acid
PS-BEMP	:	polystyrene 2-*tert*-butylimino-2-diethylamino-1,3-dimethylperhydro-1,3,2-diazaphosphorine
PTP	:	protein tyrosine phosphatase
RCM	:	ring closing metathesis
SEAr	:	electrophilic aromatic substitution
TBAB	:	tetrabutylammonium bromide
TBAF	:	tetrabutylammonium fluoride
TBDPS	:	*tert*-butyldiphenylsilyl
TBHP	:	*tert*-butylhydroperoxide
TBS	:	*tert*-butyldimethylsilyl
TBTH	:	tributyltin hydride
TBTU	:	tetramethyluronium tetrafluoroborate
TCD	:	thiocarbonyldiimidazole
TC-PTP	:	T cell protein tyrosine phosphatase
TEA	:	triethylamine
TEG	:	triethylene glycol
TEMPO	:	(2,2,6,6-tetramethylpiperidin-1-yl)oxyl or (2,2,6,6-tetramethylpiperidin-1-yl)oxidanyl

TES	:	triethylsilyl ether
TFA	:	trifluoroacetic acid
TFP	:	tri(2-furyl)phosphine
THF	:	tetrahydrofuran
TLC	:	thin layer chromatography
TMEDA	:	tetramethylethylenediamine
TMSA	:	tetramethylsilane
TMSCl	:	trimethylsilyl chloride
TMSCN	:	trimethylsilyl cyanide
TMSOTf	:	trimethylsilyl trifluoromethanesulfonate
TosMIC	:	toluenesulfonylmethyl isocyanide
TPPTS	:	triphenylphosphinetrisulfonate salt
p-TSA	:	*p*-toluenesulfonic acid
TTMSS	:	tris(trimethylsilyl)silane

1

Five-Membered *N*-Heterocycles

1.1 Introduction

Nitrogen-containing heterocyclic compounds, because of their presence in biologically active compounds and natural products, are by far the most explored heterocycles. Five-membered nitrogen-containing heterocycles like indoles, pyrroles, and carbazoles are present in a number of biologically active compounds. Due to this, synthetic chemists are increasingly interested in the functionalization and generation of these heterocyclic compounds. Saturated five-membered *N*-heterocycles are significant not only for the preparation of pigments, drugs, and pharmaceuticals, but also for the development of organic functional materials [1a–d]. Substituted pyrroles are vital compounds showing remarkable biological properties including anti-viral, anti-bacterial, anti-tumor, anti-inflammatory, and anti-oxidant activities [2]. Therefore, a variety of methods have been investigated for the synthesis of pyrrolic rings. Among these procedures, Paal-Knorr reaction [3a–g], in which 1,4-dicarbonyl substrates are reacted with primary amines or NH_3 in the presence of many promoting agents, is one of the most utilized methods for the construction of pyrroles [4]. Therefore, the preparation of these compounds has gained a longstanding interest. Various protocols for the formation of these heterocyclic compounds involve carbon-nitrogen bond-forming reactions like reductive amination, nucleophilic substitution, or dipolar cycloaddition for ring-closure [5–16]. This chapter describes the approaches for the synthesis of five-membered nitrogen heterocycles with the aid of metals and non-metals.

1.2 Metal- and non-metal-assisted synthesis of five-membered *N*-heterocycles

Aluminium-assisted synthesis

1,3-Bis(silyl)propenes and *N*-Ts-α-amino aldehydes undergo [3+2]-annulation to afford an efficient, stereoselective preparation of densely functionalized pyrrolidines **(Scheme 1)** [17].
N-Tosyl iodopyrrolidines are obtained in good yields by iodocyclization of unsaturated tosylamides promoted by oxone oxidation of potassium iodide. Alcohols are converted into tosylamides by this new facile method **(Scheme 2)** [18].

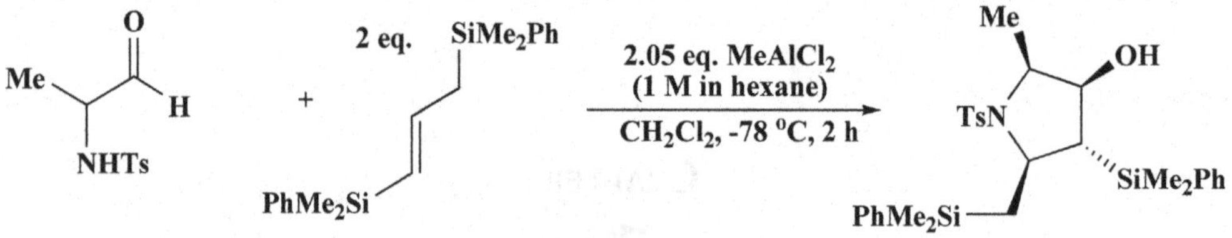

Scheme-1

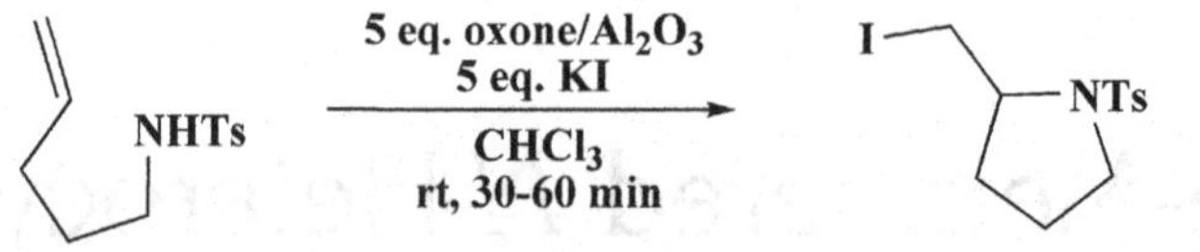

Scheme-2

Bismuth-assisted synthesis

Pyrroles are an important class of heterocycles and are widely used in material science and synthetic organic chemistry. Extensive investigations have been made to develop procedures for the synthesis of substituted pyrroles for their distinctive properties. The 1,2,3,4-tetrasubstituted pyrroles are synthesized by Knorr reaction [19], Hantzsch pyrrole synthesis [20], or 1,3-dipole addition of azomethyne ylides with alkynes [21–25]. Classical methods to obtain pyrrole derivatives also involve condensation reactions of 1,4-dicarbonyl reactants [26–27]. This procedure of synthesizing pyrrole derivatives is adapted using ionic liquids [bmim]BF_4 [28] for the condensation reaction of primary amines and 1,4-dicarbonyl reactants. The ionic liquid is utilized in a molar ratio of 1:16 (reactant/IL) and is recovered and reused four times with only a gradual decrease in activity (87%, 85%, 81%, and 76% yields). The reaction occurs smoothly in ionic liquids as well as in refluxing toluene in the presence of 5 mol% Bi(OTf)$_3$. For these condensation reactions, Bi(OTf)$_3$/[bmim]BF_4 has been found to be the ideal catalytic system. Moreover, the recovery and reuse of Bi(OTf)$_3$ is especially easy in ionic liquids than in toluene. Although the reaction in ionic liquid includes a catalyst, Bi(OTf)$_3$, it is considered faster and more convenient than I$_2$-catalyzed reactions using molecular solvents like CH$_2$Cl$_2$ or tetrahydrofuran (20 h at rt) **(Scheme 3)** [29, 30].

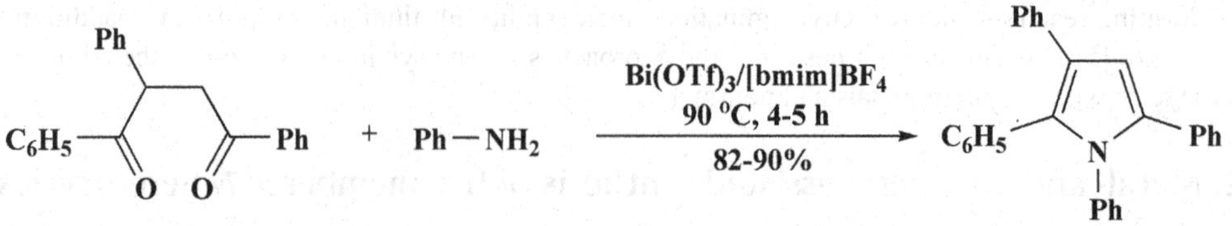

Scheme-3

Hayashi and Cook [31] reported the preparation of pyrrolidine from bis-allylamine containing an allyl bromide moiety *via* 5-*exo-trig* cyclization in the presence of halophilic Bi(OTf)$_3$ catalyst involving the activation of allyl bromide **(Scheme 4)** [32].

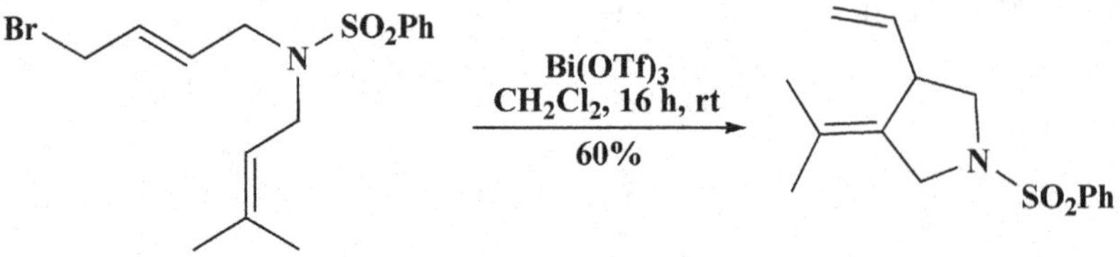

Scheme-4

Pyrrole can also synthesized from 1,4-dicarbonyl compounds under acid catalysis (Paal-Knorr reaction). Various pyrrole derivatives are prepared from aryl amines and 1,4-diketones in $Bi(OTf)_3 \cdot xH_2O$ immobilized in 1-butyl-3-methylimidazolium tetrafluoroborate, $Bi(OTf)_3 \cdot xH_2O/[bmim]BF_4$ as catalyst **(Scheme 5)**. In such cases, high yields of products are obtained and the catalytic system is recycled and reused. Bismuth chloride does not however show satisfactory performance in this reaction. In another instance, the reaction of 2-aminopyridine with hexane-2,5-dione in the presence of $Bi(NO_3)_3 \cdot 5H_2O$ catalyst provides pyridine-pyrrole derivative with 70% yield **(Scheme 6)** [33, 34].

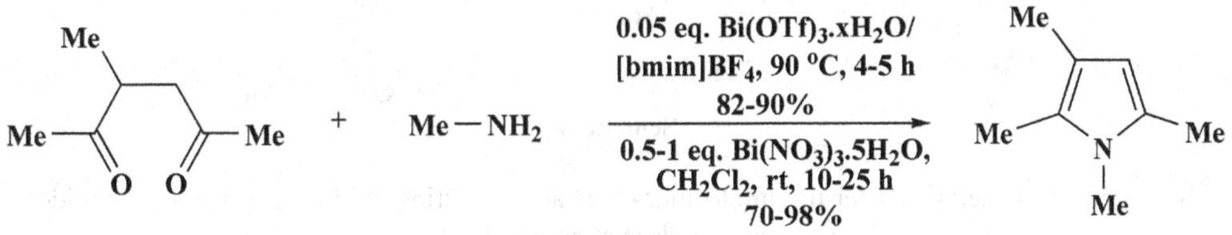

Scheme-5

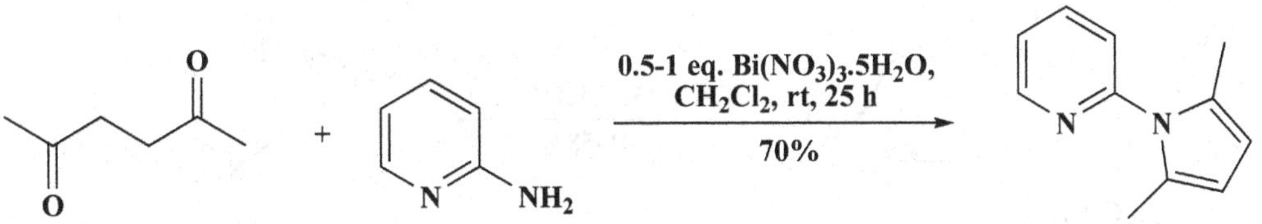

Scheme-6

Calcium-assisted synthesis

The chemistry of organometallic alkaline earth metal complexes closely resembles that of rare-earth elements [35–36]. Therefore, it is not beyond belief that alkaline earth metal complexes are active in the hydroamination/cyclization of aminoalkynes and aminoalkenes, as reported by Hill et al. [37], employing a diketiminato calcium amide complex **(Scheme 7)**. For the synthesis of pyrrolidines, the catalyst activity of calcium amide complex is comparable to rare-earth metal-based catalysts. However, a limiting factor is observed in the form of a facile Schlenk-type ligand redistribution reaction under catalytic conditions, resulting in the formation of diamido and homoleptic bis(diketiminato) species and catalyst deactivation. Initially, asymmetric hydroamination is performed using a chiral bis-oxazolinato calcium complex to afford only low *ee* (up to 10% *ee*) as a result of a facile Schlenk equilibrium [38]. Datta et al. [39–40] reported the catalytic activity of aminotroponiminato strontium and calcium complexes applied in catalyst loadings as low as 2 mol%. With increasing ionic radius of the metal, the catalytic activity of alkaline-earth metal catalysts decreases, which contrasts with the trends for rare earth and alkali metal-based catalysts [41].

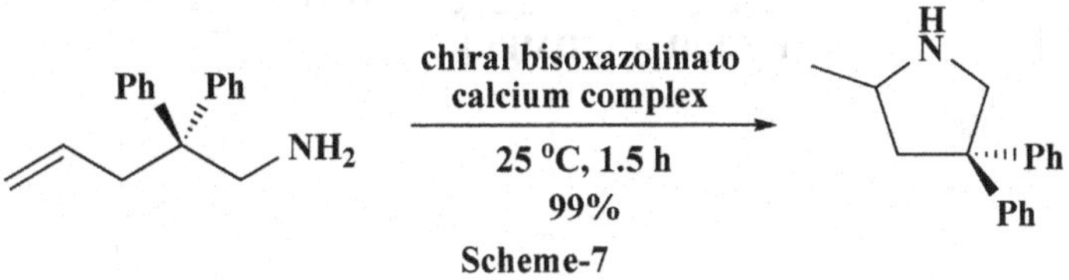

Scheme-7

Cerium-assisted synthesis

Many *N*-heterocycles are synthesized using enamines substituted with an electron-withdrawing group in *β*-position as useful building blocks. Surprisingly, pyrrole derivatives can also be prepared by Lewis acid-supported reactions of unsaturated aldehydes bearing an alkyl group in α-position. The variability of both the carbonyl and the enamine component affords a useful approach for the synthesis of many simple and condensed pyrroles. The new *N*-heterocycles have been found to be transformed in further investigations **(Scheme 8)** [42, 43].

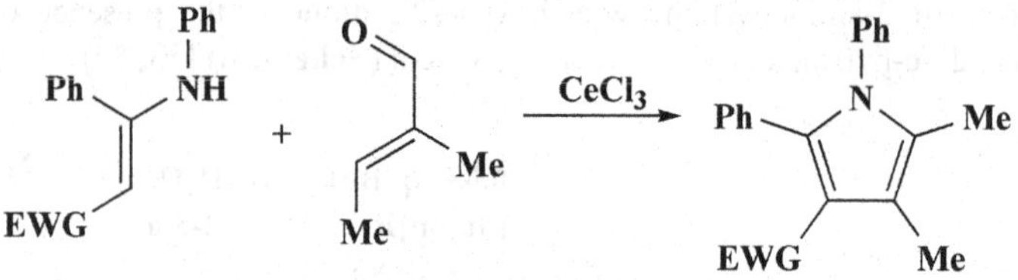

Scheme-8

Nair et al. [44] used CAN for intramolecular cyclization reactions of bis(cinnamyl)tosylamides for the preparation of pyrrolidines in moderate yields **(Scheme 9)** [32].

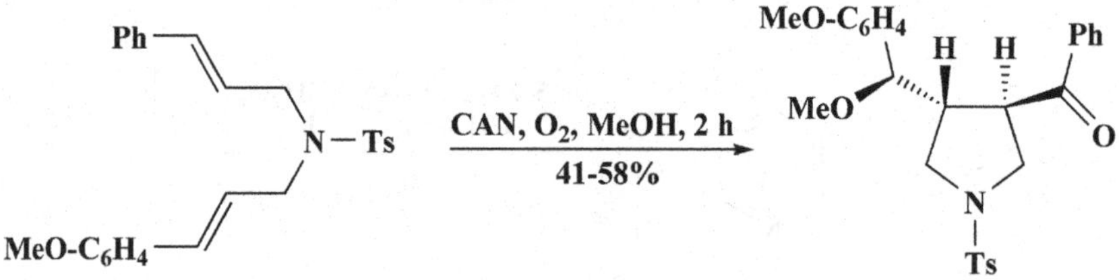

Scheme-9

Chromium-assisted synthesis

Watanuki et al. [45–47] synthesized *N*-heterocycles using chromium Fischer carbenes to promote the ring-closing ene-yne metathesis. The metathesis products were formed in 30% combined yield when substituted enol ether was subjected to ring-closing ene-yne metathesis employing chromium Fischer carbene in substoichiometric amounts **(Scheme 10)** [48].

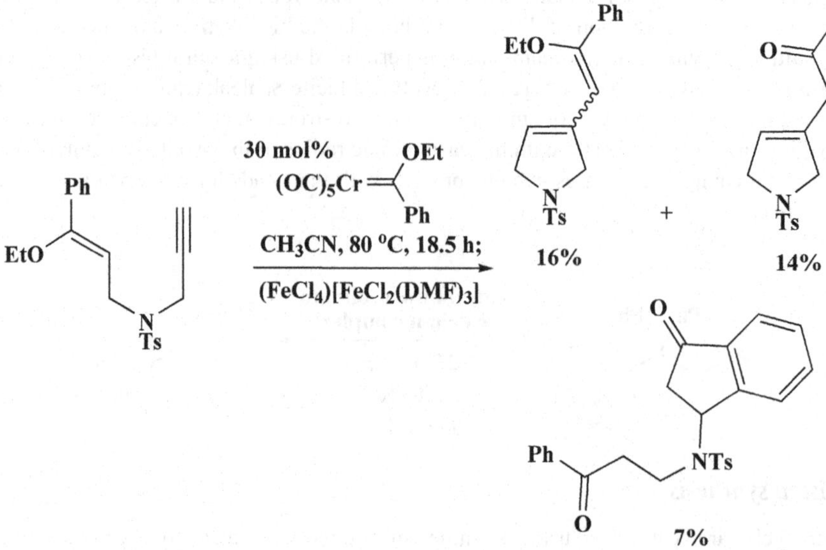

Scheme-10

Results observed with isomeric enol ethers are however different (**Scheme 11–12**). Whereas enone is formed from *E*-isomer through clean ene-yne metathesis and hydrolysis of the isomeric enol ethers, the metathesis product is formed in only 45% conversion and a rearranged enone is also obtained along with the hydrolysis product of unreacted enol ether from *Z*-isomer [48, 49].

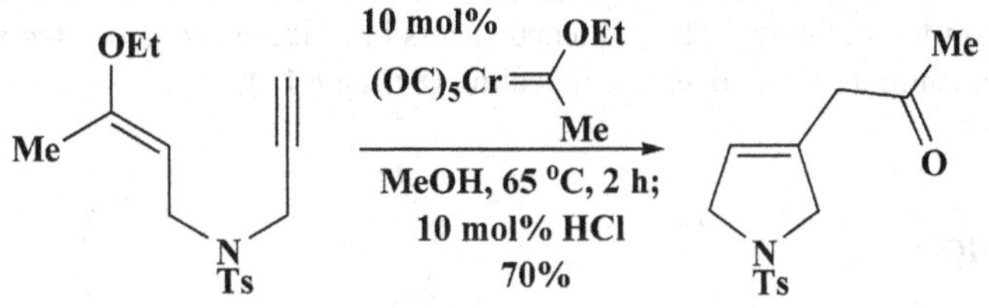

Scheme-11

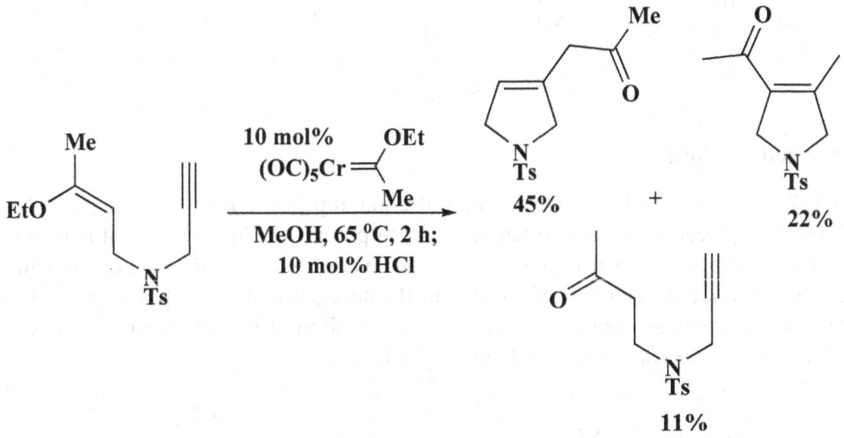

Scheme-12

Korkowski et al. [50] reported cyclopropanes in carbocyclic systems. It has been noted that the stoichiometric (1.2 eq.) Fischer carbene complex is needed for cyclopropanation. Moreover, the metathesis product is obtained in good yield when the alkene is substituted with a phenyl group (**Scheme 13**). The phenyl-subsituted alkene results in greater stability of the chromium-(0) carbene formed by the fragmentation of chromacyclobutane and which favors the metathesis. Moderate yields of cyclopropyl piperidines are obtained from chain-extended enynes [48, 51–55].

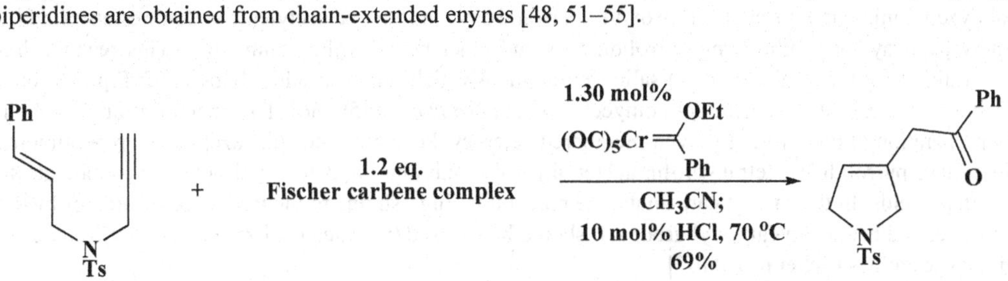

Scheme-13

Copper-assisted synthesis

Lindsay et al. [56] developed a tandem aza-Cope rearrangement-Mannich cyclization for the preparation of acylpyrrolidines (*cis* and *trans*) from aldehydes and amino alcohols with varied moieties. The reaction was reported to occur smoothly at 60–90 °C under MW heating for 5–150 min to afford acylpyrrolidines in 22%–84% yields. Higher diastereoselectivities have been reported in some cases when the reactions are performed at lower temperatures. The sequence occurs in a single synthetic step with reaction times significantly reduced compared to conventional heating (**Scheme 14**).

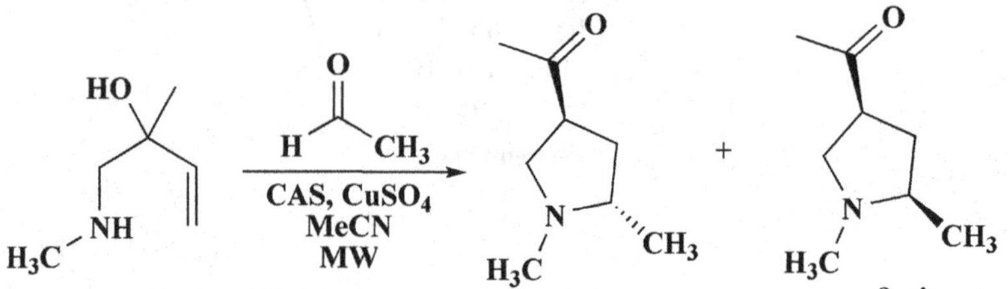

Scheme-14

Gallium-assisted synthesis

Araki et al. [57] reported a cycloaddition reaction of cyclopropenes to nitriles in the presence of Lewis acids. This reaction proceeds at elevated temperatures to provide moderate yields of pyrroles. Initially, zwitterionic complex is formed when cyclopropene double bond is activated with a Lewis acid, followed by a nucleophilic attack at the carbocationic center by the nitrogen atom of the nitrile group. Zwitterionic intermediate undergoes ring-expansion to provide imine, followed by subsequent de-protonation and protonolysis to ultimately yield pyrrole (**Scheme 15**) [58].

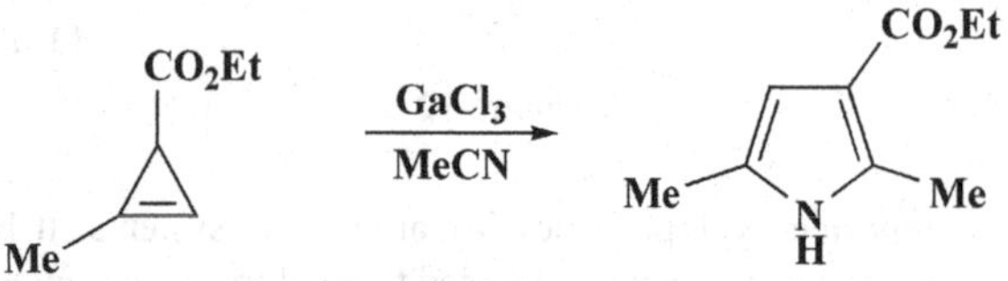

Scheme-15

Indium-assisted synthesis

Catalyzed Conia-ene-type cyclization of nitrogen- and oxygen-tethered acetylenic malonic esters provides a new pathway for synthesizing pyrrolidinones and other heterocyclic compounds. This reaction has been utilized for the synthesis of (–)-salinosporamide A which serves as a highly potent 20S proteasome inhibitor synthesized by marine actinomycete *Salinispora tropica* [59–66]. This method is applicable in the preparation of other five- to seven-membered heterocyclic compounds like azepanone, piperidinone, piperidine, pyrrolidines, tetrahydrofuran, tetrahydroisoquinoline, and tetrahydropyran in moderate to excellent yields. In the case of carbamate, the reaction is sluggish due to the tight coordination of In(III) with ester and benzyloxycarbonyl groups. It should be stressed that such cyclization occurs cleanly even with basic amines (**Scheme 16**) [67].

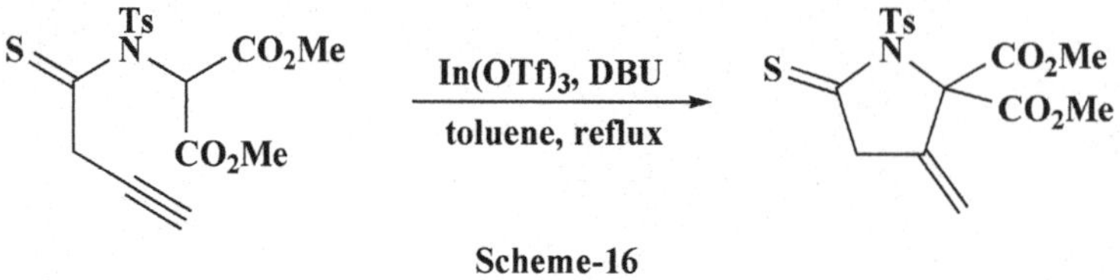

Scheme-16

Au- [68], Ni- [69], and In- [70–71] catalyzed reactions of amide have also been studied and reported **(Scheme 17)**. Although satisfying results are not observed from nickel and gold-catalyzed conditions, In(OTf)$_3$ effectively catalyzes cyclization to provide nearly considerable yield of pyrrolidinone. The synthesis of pyrrolidinone suggests that this In(OTf)$_3$-catalyzed reaction also occurs with racemizable chiral alkynes and non-terminal alkynes. The cyclization occurs without serious racemization and with complete *E*-selectivity even at higher temperatures. Also, the reaction is accelerated and provides better yields in cases of non-terminal alkynes upon addition of an equimolecular amount of 1,8-diazabicyclo[5.4.0]undec-7-ene to In(OTf)$_3$. It is important to note that no *endo*-cyclization and no isomerization of the olefinic double bond from β,γ- to α,β-position are observed in this reaction [72].

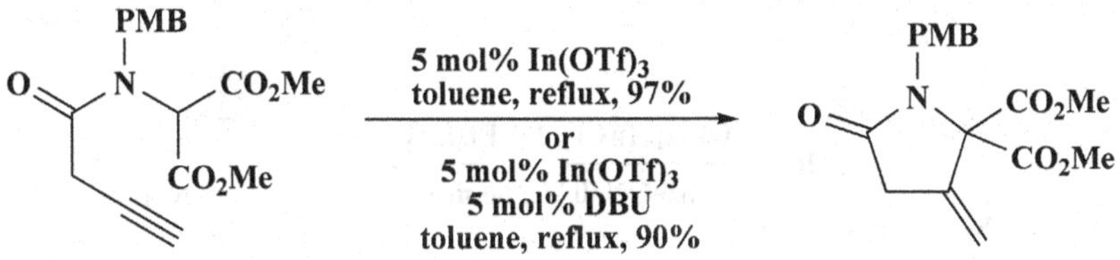

5 mol% In(OTf)$_3$
toluene, reflux, 97%

or

5 mol% In(OTf)$_3$
5 mol% DBU
toluene, reflux, 90%

Scheme-17

Another general and convenient one-pot method reported for the synthesis of substituted pyrroles involves silyl enol ethers, propargylic acetates, and primary amines in the presence of InCl$_3$ catalyst. Many pyrrole derivatives are formed in high yields through this route **(Scheme 18)** [73].

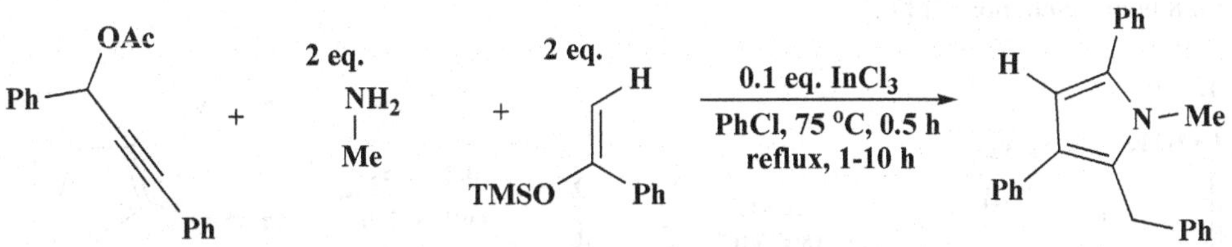

2 eq.

2 eq.

0.1 eq. InCl$_3$
PhCl, 75 °C, 0.5 h
reflux, 1-10 h

Scheme-18

Polindara-García and Miranda [74] synthesized 2,3-dihydropyrroles in two steps, through base-promoted cyclization of allenamide intermediates **(Scheme 19)**. The propargyl amide was prepared by

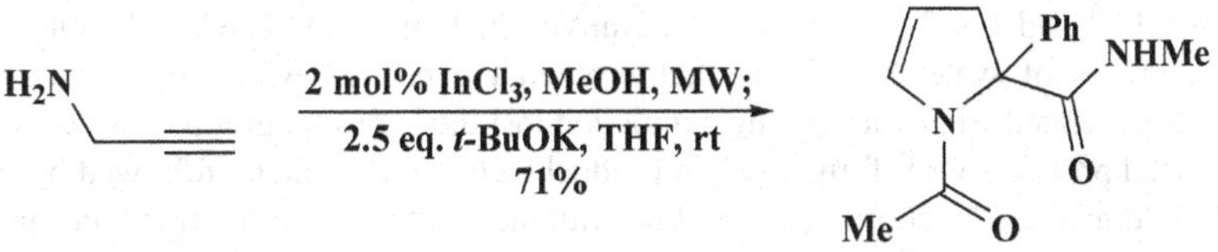

2 mol% InCl$_3$, MeOH, MW;
2.5 eq. *t*-BuOK, THF, rt
71%

Scheme-19

indium(III)-catalyzed Ugi four-component reaction (Ugi 4-CR) [75a–b]. Subsequently, base-promoted isomerization occurred to produce allenamide intermediates, which further underwent 5-*endo-trig* cyclization to afford 2,3-dihydropyrroles.

Readily available triethylsilane and indium chloride produce indium hydride under low toxicity and mild conditions, and are therefore a promising alternative to Bu$_3$SnH **(Scheme 20)** [76].

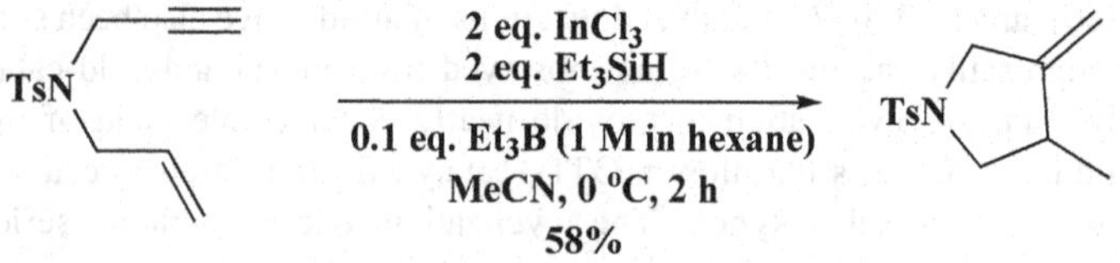

Scheme-20

Organic azides have been reported to easily and chemoselectively reduce to amines when reacted with dichloroindium hydride under very mild conditions. Following this, γ-azidonitriles can be cyclized to form pyrrolidin-2-imines **(Scheme 21)** [77].

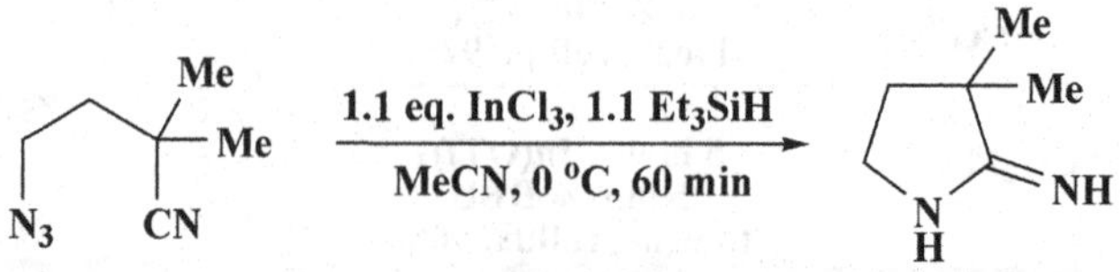

Scheme-21

Iodine-assisted synthesis

A cost-effective, efficient, simple, and metal-free four-component coupling reaction of amines, aldehydes, nitromethane, and dialkyl acetylenedicarboxylates has been explored for the production of 1,2,3,4-tetrasubstituted pyrroles in high yields in the presence of molecular I_2 as a catalyst under reflux within 8 hours **(Scheme 22)** [78].

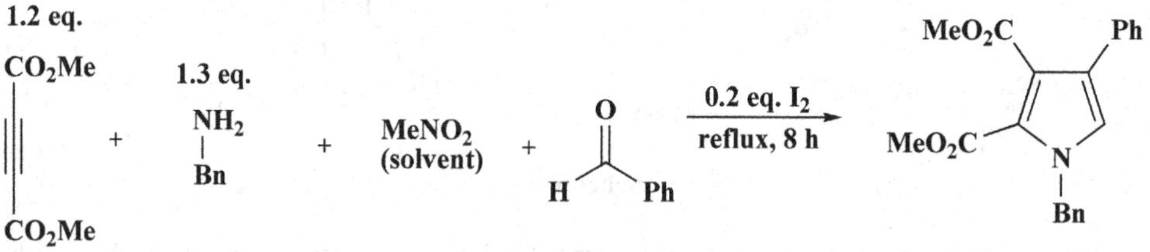

Scheme-22

Ji et al. [79] synthesized 3,4-diiodoheterocyclic compounds by I_2-induced tandem cyclization of 4-aminobut-2-yn-1-ol and but-2-yne-1,4-diol derivatives **(Scheme 23)**. The I^-, H^+, OH^-, and I^+ were produced when traces of water present in dichloromethane reacted with iodine. The allene cation intermediate was generated when the starting compound lost the hydroxyl group as water in the presence of H^+. The cyclized products were formed when I^- attacked the intermediate, followed by the formation of iodoiranium intermediate by coordinating iodine with the resultant iodo intermediate, and subsequent intramolecular attack of the heteroatom in an *endo-trig* fashion. Among the various solvents tested, moist dichloromethane provided optimum yields. The reaction also worked well with aromatic substituents. Better results were reported with electron rich aryl groups, over electron-withdrawing aryl groups. The product was also obtained with aliphatic substituents. Moreover, spiro products were also yielded in significant quantities [80].

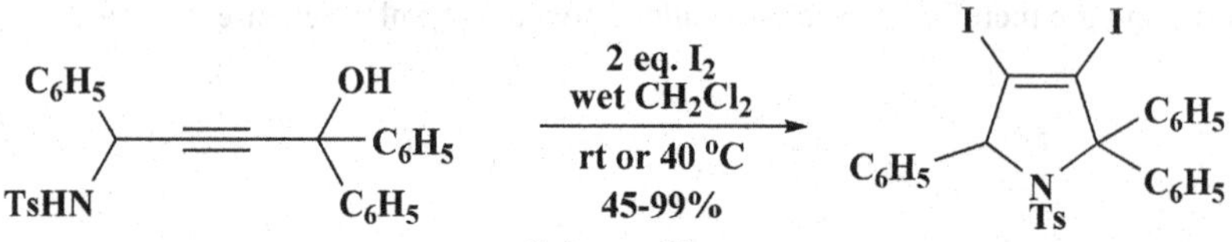

Scheme-23

Diaba et al. [81] prepared azaspirane ring system (found in some natural products) through 5-*endo-trig* iodoamino cyclization reaction (**Scheme 24**), wherein I_2 activation of carbon-carbon double bond afforded iodoiranium intermediate which then underwent 5-*endo-trig* cyclization to afford azaspiro compound [80].

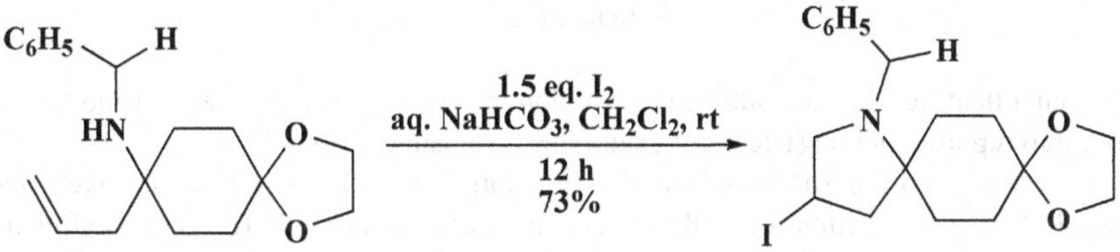

Scheme-24

Yu et al. [82] synthesized several 8-methyleneazaspiro[4.5] trienes by intramolecular electrophilic *ipso*-cyclization (**Scheme 25**) [83–84]. The iodoirenium intermediate was generated when I_2 coordinated with the carbon-carbon triple bond of the compound. The spiro-compounds were produced by intramolecular Friedel-Crafts cyclization followed by loss of proton. The yield of the product was reduced in the presence of a base (sodium bicarbonate). Moreover, analogous amides with methyl group replaced with an acyl group or hydrogen were found unsuitable as substrates. The reaction also failed with terminal acetylenes. Nevertheless, this reaction has also been studied with some examples of additional substituents (such as chloro, methyl, and bromo) on the phenyl ring [80].

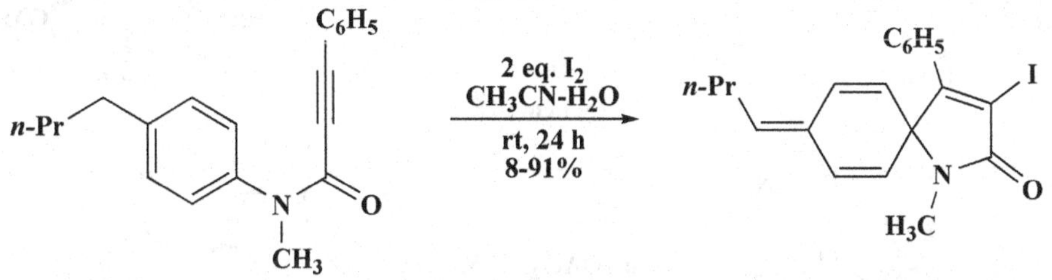

Scheme-25

Mai et al. [85] used an aminocyclization reaction to synthesize 3-amino-2,2-dimethyl-8-thia-1-azaspiro [4.5] decane (**Scheme 26**). The iodoiranium intermediate produced by the coordination of I_2 with the carbon-carbon double bond of the compound underwent 5-*endo-trig* cyclization to form a spiro compound. Then, the iodo group was transformed into amino moiety in the usual manner [80].

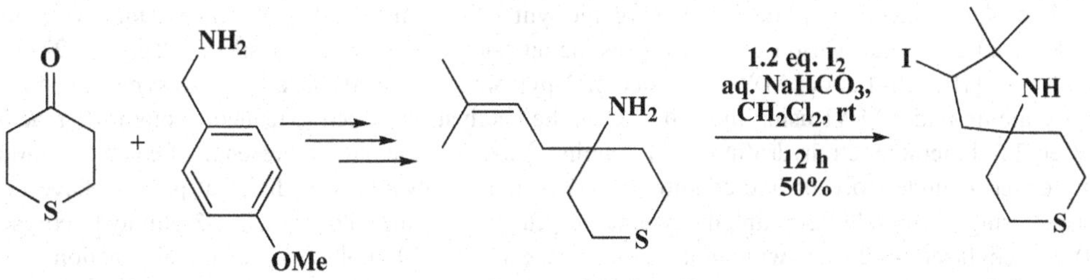

Scheme-26

Yoshida et al. [86] reported an I_2-promoted electrophilic cyclization of propargylic aziridines for the synthesis of 2,5-disubstituted 3-iodopyrroles (**Scheme 27**). The iodoirenium intermediate was produced by the coordination of I_2 with the carbon-carbon triple bond of the compound. The intermediate, formed by intramolecular attack of azidine nitrogen, readily eliminated the proton to yield aromatized products in

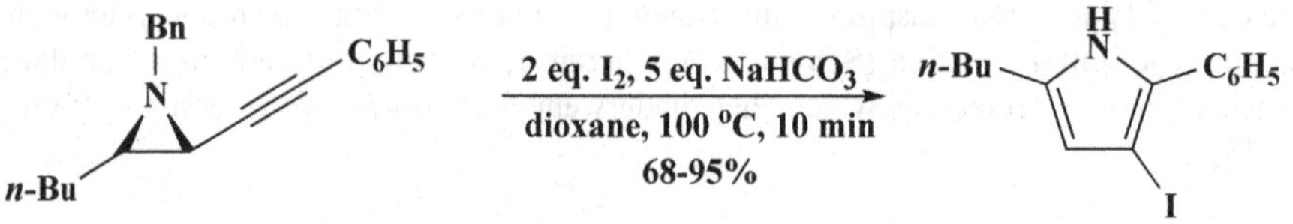

Scheme-27

good to excellent quantities. Various substituted 3-iodopyrroles were also synthesized and the importance of the iodo group reported in one successful example of Negishi coupling [80].

Studies have reported that (n+3) and (n+4) ring enlarged lactones as well as spiroketolactones can be synthesized from cycloalkanones by alkoxy radical fragmentation with (diacetoxyiodo)benzene and I_2 [87]. The appropriate amides when reacted with (diacetoxyiodo)benzene in the presence of I_2 generate nitrogen-centered radicals which undergo cyclization or rearrangement [88]. Specific demonstrated examples include the preparation of bicyclic spiro lactams from amides **(Scheme 28)** [89] and the synthesis of oxa-azabicyclic systems by intramolecular hydrogen atom transfer reaction promoted by phosphoramidyl and carbamoyl radicals produced from appropriately substituted carbohydrates **(Scheme 29)** [90].

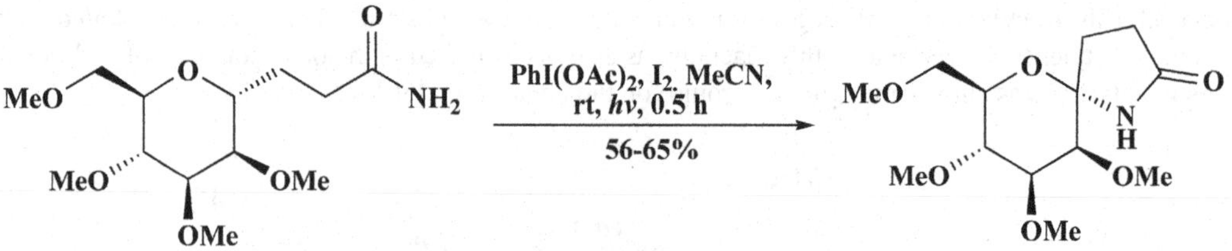

Scheme-28

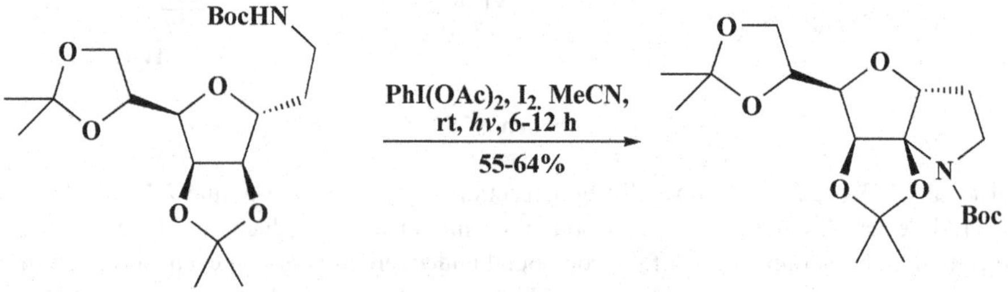

Scheme-29

Molecular I_2 plays an important role in organic synthesis because of its powerful and unique features [91–98] as a Lewis acid. Various new protocols include stereoselective synthesis of β-lactams [99–101] and pyrroles [102–104] under MWI. The idea of 3-pyrrole-substituted 2-azetidinones synthesized from primary amine under MWI using molecular I_2 as the catalyst, has been extended through the reaction between 2,5-dimethoxytetrahydrofuran and 3-amino-2-azetidinones in the presence of catalytic amount of molecular I_2 under solvent-free condition. Earlier, many solvents with different polarity have been screened using 1.2 mmol of 2,5-dimethoxytetrahydrofuran and 1 mmol of (±)-*trans*-3-amino-1-(chrysen-6-yl)-4-phenylazetidin-2-one, with molecular I_2 as catalyst (20 mol%), as a model reaction under automated CEM MWI conditions (90 °C, 300 W, 3 min). The results suggested that this reaction was convenient without a solvent. The same reaction also resulted in 21% yield in three minutes, without any catalyst (only MWI under solvent-free condition). The yield of 2-azetidinone has been reported to increase to 32% if the reaction is conducted for 30 minutes. The same reaction, when used to optimize the amount of catalyst to identify the best conditions, has shown that 10 mol% molecular I_2 is needed to complete the reaction within three minutes. Considering these observations, a series of reactions have been performed using 2,5-dimethoxytetrhydrofuran and various 3-aminosubstituted 2-azetidinones in catalytic amounts of molecular I$_2$ (10 mol%) under solvent-free conditions following an automated microwave-assisted procedure (90 °C, 300 W, 24–50 psi), and stirring the reaction mixture at room temperature. The efficacy of the newly developed reaction is evident from the formation of pyrrole derivatives in good to excellent yields when optically pure 3-amino-2-azetidinones are subjected to the reaction conditions, both in the microwave-induced procedure as well as at room temperature **(Scheme 30)**.

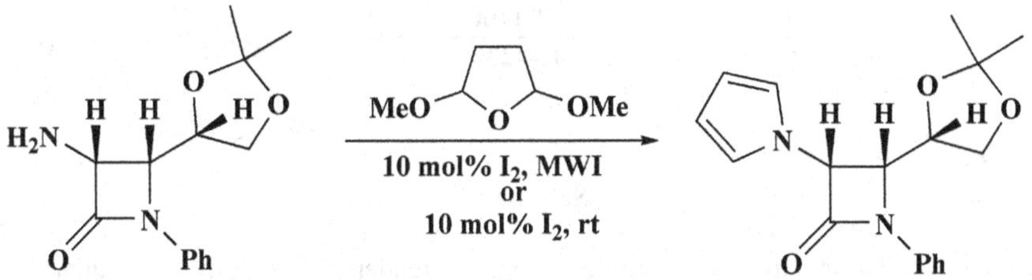

Scheme-30

Paal-Knorr method utilizes 1,4-diketones for the preparation of pyrroles. Bandyopadhyay et al. [105–106] synthesized many pyrroles from 2,5-dimethoxytetrahydrofurans under MW conditions using 5 mol% I$_2$ as catalyst. β-Lactam fused pyrroles can also be generated from 1,4-diketo compounds and 3-amino-β-lactams **(Scheme 31)** [80].

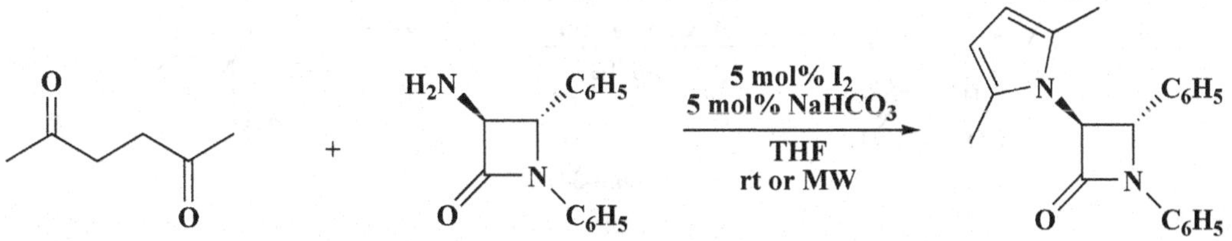

Scheme-31

Pyrroles and 2-azetidinones are two highly important classes of molecules in medicinal and organic chemistry. A practical and green protocol for the synthesis of 3-pyrrole-substituted 2-azetidinones with a variety of substituents at C-4 and at *N*-1, using molecular I$_2$ in catalytic amounts under MWI, has been found equally effective for mono- as well as polyaromatic groups at the *N*-1 position of the 2-azetidinone ring. The C-4 substituent has no effect on the rate of reaction or the yield. This protocol has also been found useful for the synthesis of optically pure 3-pyrrole-substituted 2-azetidinones. Even with highly acid sensitive group possessing substrates, no de-protection/rearrangement has yet been identified in this method. A synergistic effect of MWI and Lewis acid catalyst (molecular I$_2$) results in an extreme rapidity with excellent reaction yields. Recently, the focus of interest of researchers has shifted to modification and preparation of β-lactam ring for producing compounds with diverse biological properties. As a part of ongoing research on preparation of anti-cancer agents [107–110], some *trans*-acetoxy β-lactams [111–113] have demonstrated selective anti-cancer activity against a number of human cancer cell lines *in vivo* and *in vitro*. This finding justifies efforts towards the development of an efficient method for the preparation of pyrrole-substituted β-lactams as displayed in **Scheme 32** [114–115].

BTI or (diacetoxyiodo)benzene-induced intramolecular phenolic oxidation has been found to be an efficient pathway to synthetically useful polycyclic products. [Bis(acyloxy)iodo]arenes-promoted

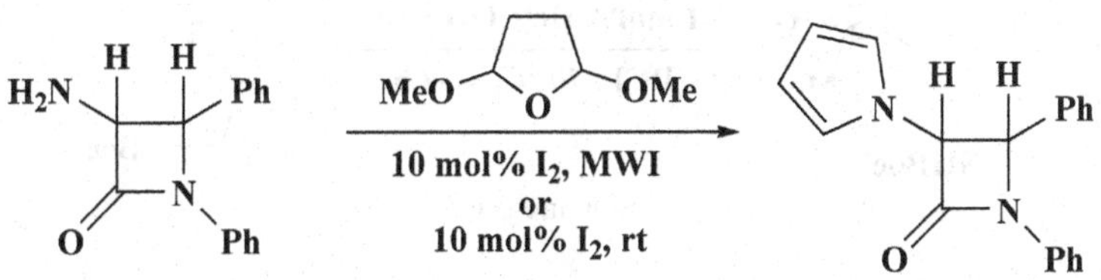

Scheme-32

oxidative phenolic cyclization has been shown in **Scheme 33**. It has been observed that phenolic oxazolines specifically undergo oxidative cyclization to afford synthetically useful spirolactams [90, 116].

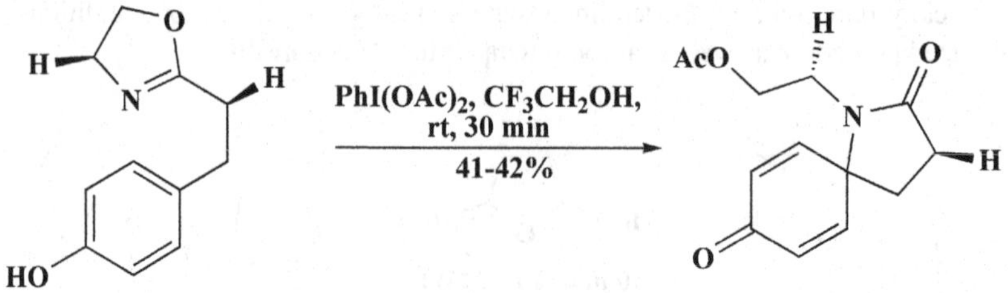

Scheme-33

N-Methoxy-3-(4-halophenyl)propanamides have a tendency to form spirodienones having 1-azaspiro[4.5] decane ring system *via* intramolecular *ipso*-cyclization of nitrenium ion generated with HTIB in trifluoroethanol **(Scheme 34)** [116]. 2,1-Benzothiazine derivatives are also synthesized from sulfonamides [117] and (–)-lapatin B *via* oxidative cyclization of *N,N*-diacetylglyantrypine by HTIB-promoted cyclization of appropriate amides [90, 118].

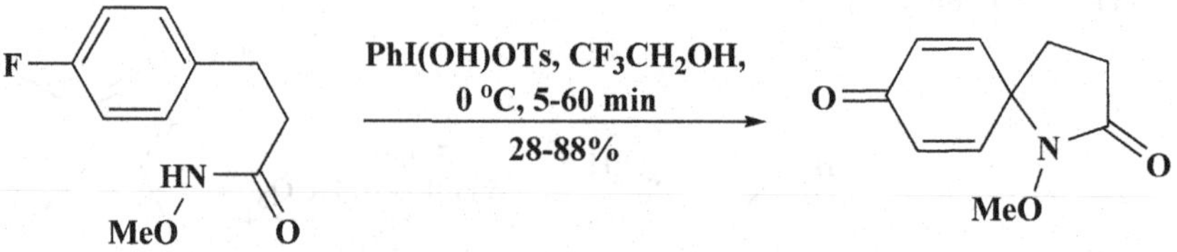

Scheme-34

Iridium-assisted synthesis

Various un-activated alkenes with pendant secondary amines undergo an effective intramolecular hydroamination with [Ir(COD)Cl]$_2$ pre-catalyst at relatively low loadings, without any need of other co-catalysts or ligands **(Scheme 35)** [119].

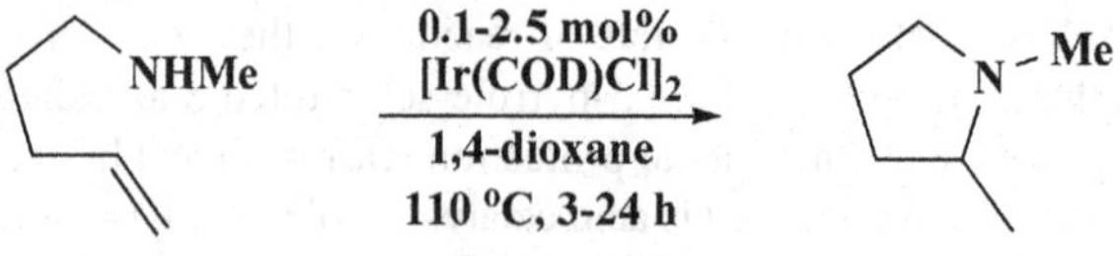

Scheme-35

Sulfoxonium ylides can be used as a carbene source in the presence of simple, commercially available iridium catalyst for various inter- and intramolecular X-H bond insertions, including ring-expansion for lactams. They are used as preferable surrogates to traditional diazo esters and ketones due to their stability and safety attributes **(Scheme 36)** [120].

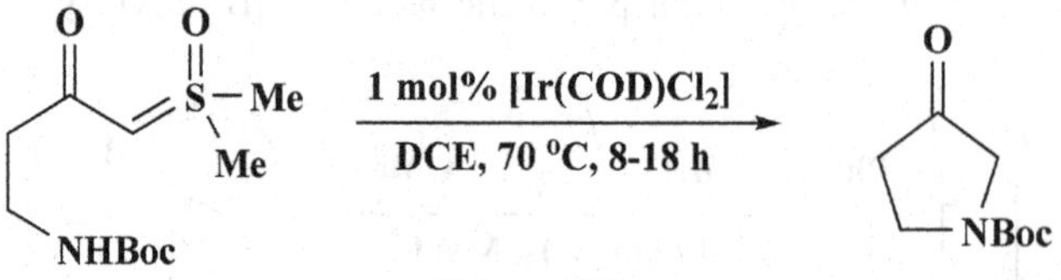

Scheme-36

Fujita et al. [121] synthesized 5-, 6-, and 7-membered cyclic amines when diols were cyclized with primary amines (**Scheme 37**). This reaction was compatible with alkyl-, benzylamines and anilines. The scope of diol was earlier limited to simple primary and secondary alkyl alcohols and benzyl alcohols. In this study N-heterocyclization of primary amines with diols was reported through a new and efficient method, in the presence of Cp*Ir complex catalyst. Consequently, many five-, six-, and seven-membered cyclic amines were prepared in good to excellent yields.

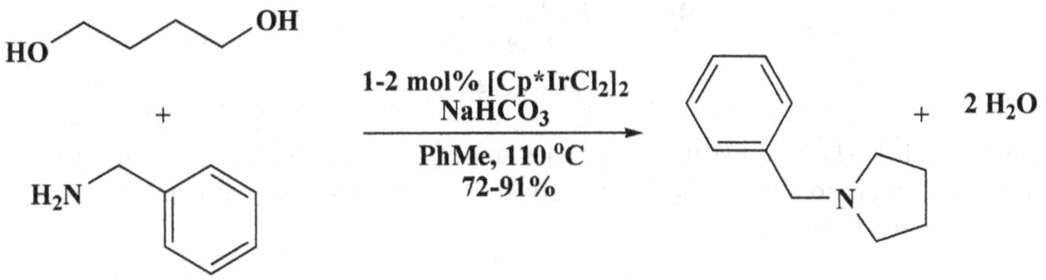

Scheme-37

Additional versatility and complexity have also been introduced to the hydroamination reaction by one-pot multistep reaction sequences [122–126]. Alkyne hydroamination results in imines which act as good targets for further functionalization as they are prone to hydrolysis. In fact, the imines are isolated and characterized after hydrolysis with hydrogen chloride or SiO_2. Synthetically useful reductions employ hydrogen/palladium/carbon, lithium aluminium hydride, or $NaBH_3CN$/zinc chloride. Yet, sequential hydroamination/hydrosilylation reactions, utilizing metal as catalysts in both processes, are more elegant. This reaction is performed using Ir [127], Ti [128] and rare-earth metal catalysts (**Scheme 38**) [41].

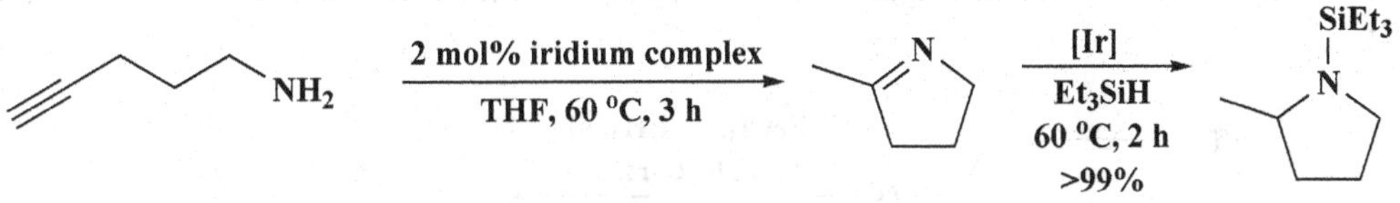

Scheme-38

Lactams are industrially useful N-heterocycles. The β-lactam family comprises anti-biotics [129–131] and caprolactam, the latter acting as an intermediate for the synthesis of nylon [132–133]. Lactams are prepared from amino acids [134–135], by Beckmann rearrangement [136], or by the Schmidt reaction [137–138]. They are synthesized from an amine (or ammonia) and a lactone in the presence of iridium which acts as a catalyst. The reaction occurs *via* aminolytic ring-opening, oxidation of alcohol, ring formation, and reduction (**Scheme 39**). Although amides are poor nucleophiles [139], the intramolecular nature of the reaction is favorable.

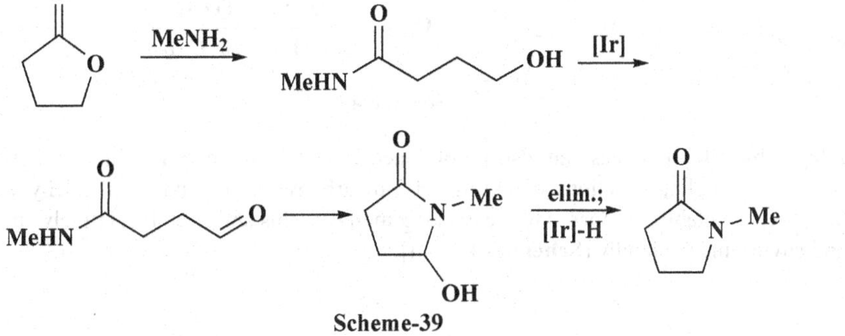

Scheme-39

Iron-assisted synthesis

Un-activated olefins undergo a simple iron-catalyzed intramolecular hydroamination reaction under mild conditions and tolerated halide functionalities bearing aminoolefins **(Scheme 40)** [140].

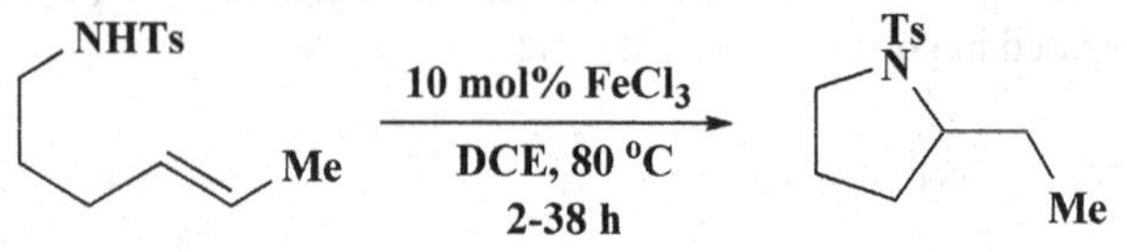

Scheme-40

Taniguchi et al. [141] reported that 1,6-dienes undergo radical cyclization with Fe(Pc) or ferric chloride in the presence of sodium borohydride and oxygen or air to provide five-membered functionalized pyrrolidines **(Scheme 41)** [32].

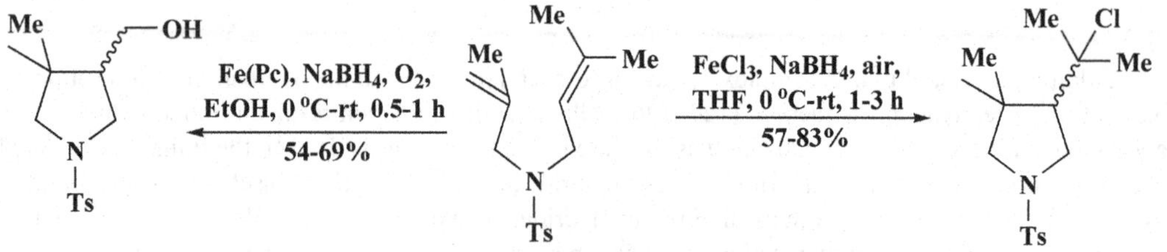

Scheme-41

3-Methylenepyrrolidines are prepared from 1,6-enynes under reaction conditions **(Scheme 42)** [32, 142].

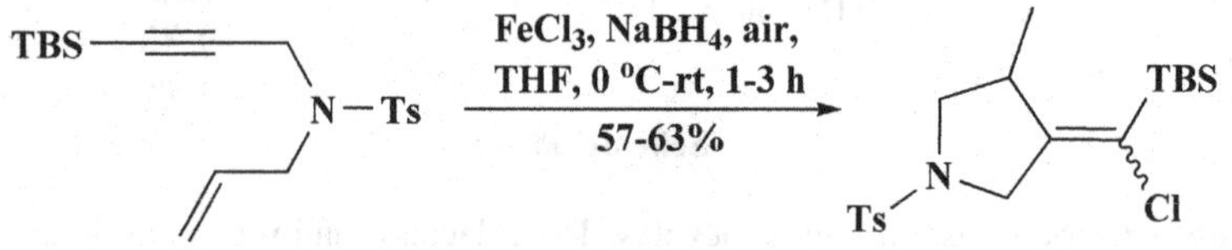

Scheme-42

Another operationally practical, simple, and economical method of synthesizing *N*-substituted pyrroles in good to excellent yields includes Paal-Knorr pyrrole condensation of 2,5-dimethoxytetrahydrofuran with various sulfonamines and amines in the presence of catalytic amounts of $FeCl_3$ in water under very mild reaction conditions **(Scheme 43)** [143].

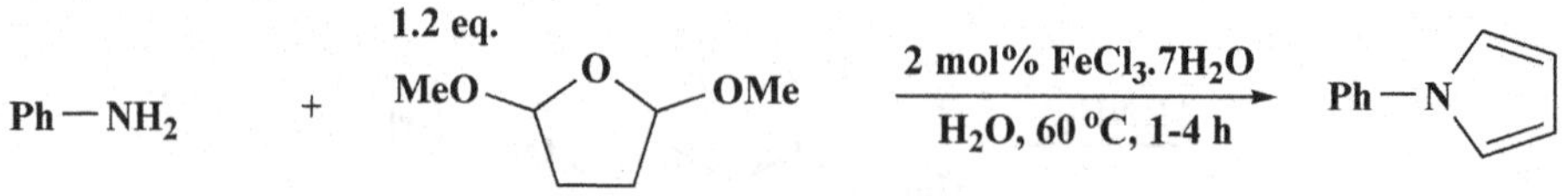

Scheme-43

Highly substituted pyrroles can also be obtained in considerable yields through Fe(III)-catalyzed four-component coupling reactions of amines, 1,3-dicarbonyl compounds and aldehydes without an inert atmosphere. Notably, compared to the existing methods, this approach is relatively straightforward, cheap and environment friendly **(Scheme 44)** [144].

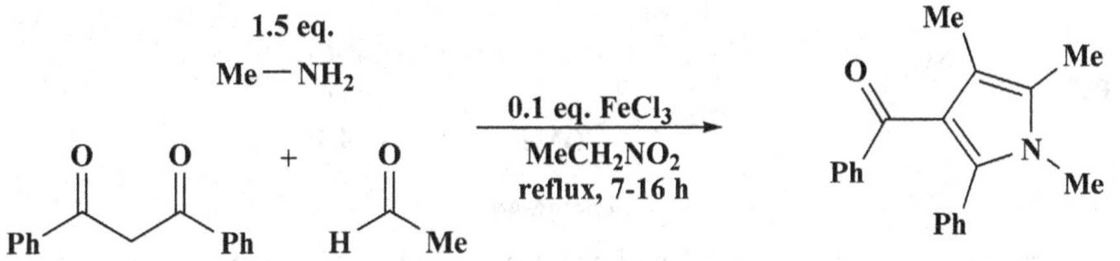

Scheme-44

A three-component reaction of amines, arylacyl bromides, and dialkyl acetylenedicarboxylate results in high yields of poly-substituted pyrroles at room temperature, in the presence of $FeCl_3$ as catalyst **(Scheme 45)** [145].

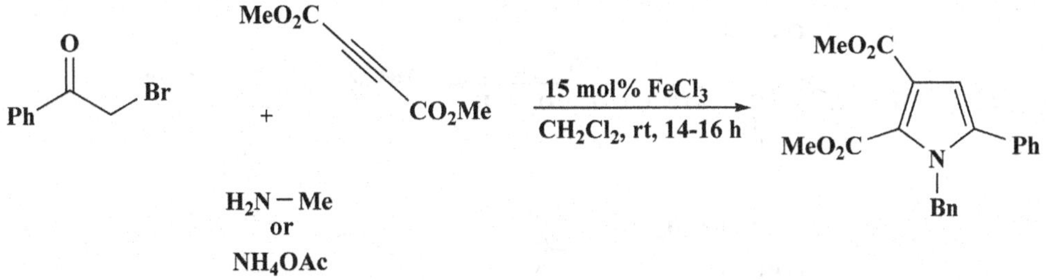

Scheme-45

Imhof et al. [146] reported an instance of hetero-Pauson-Khand-type [2+2+1]-cycloaddition reaction of CO, ketimine, and ethylene **(Scheme 46)** [147].

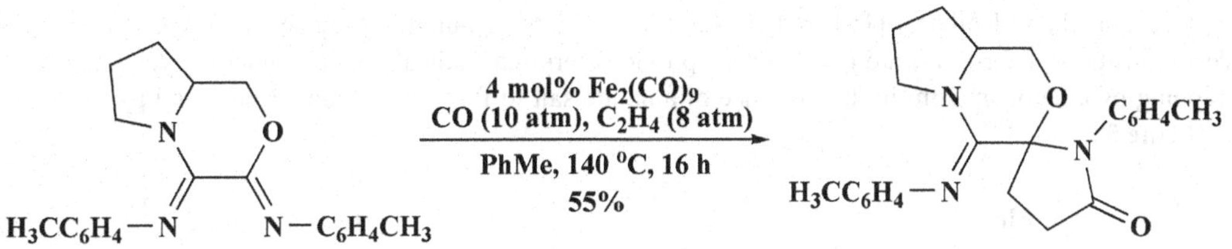

Scheme-46

Alkylidenebutenolides and alkylidenepyrrolinones can be synthesized from allenyl ketones (as well as aldehydes) [147–149] and allenyl imines [150], with iron as the catalyst, under fluorescent light **(Scheme 47)**.

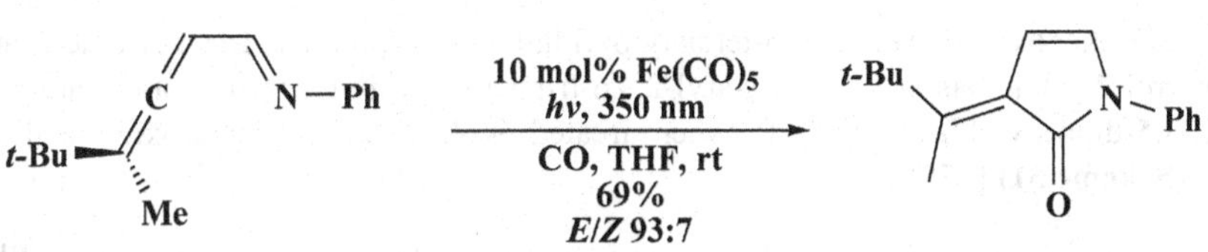

Scheme-47

Tetra-substituted NH pyrroles obtained in good yields through ferric chloride-catalyzed addition and cyclization of enamino esters with nitroolefins show considerable tolerance towards various functionalities. Further, substituted furans can also be synthesized through this efficient potassium acetate-promoted addition and cyclization method **(Scheme 48)** [151].

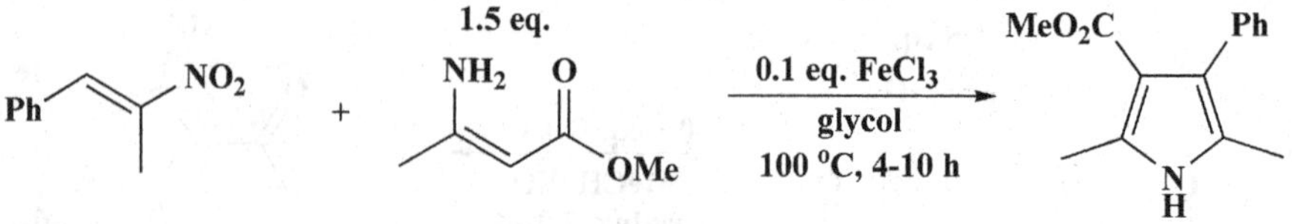

Scheme-48

Amino derivatives prepared by reduction of nitrogen atom represent the most obvious transformation of the nitro group. A chemoselective reduction of the nitro group can produce pyrrolidin-2-one derivatives directly when an ester function is also present in the substrate at the γ-position. Nitroalkanes when added to Baylis-Hillman products afford unsaturated nitro-esters which can be chemoselectively reduced by a 'one-pot' method to synthesize 2-alkylidenepyrrolidinones using iron in boiling acetic acid **(Scheme 49)** [152–153].

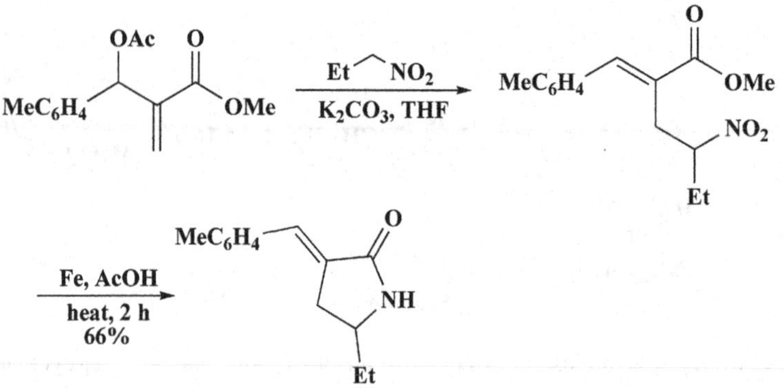

Scheme-49

Taniguchi and Ishibashi [154] radical addition of a nitro group to 1,6-dienes, promoted by thermal decomposition of ferric nitrate non-hydrate, provides terminal radicals which undergo cyclization and trapping by a halogen atom in the presence of a halide salt to form pyrrolidine-2-ones and pyrrolidines **(Scheme 50)** [30, 32].

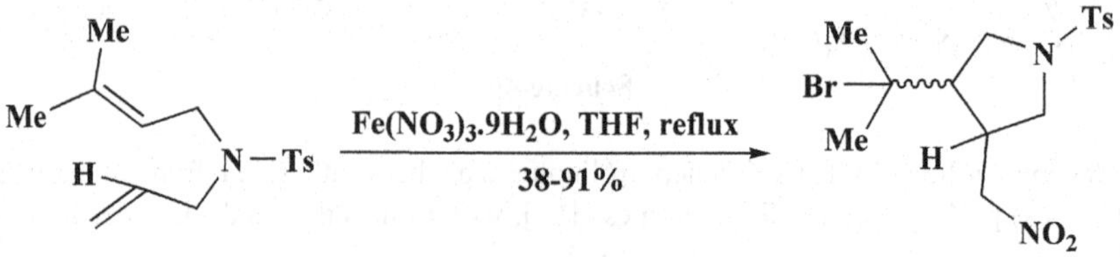

Scheme-50

Extended Decker oxidation of 1,2,4,6-tetraarylpyridinium salts produces a tropoisomeric aryl-(1,3,5-triaryl-1*H*-pyrrol-2-yl)methanones. Also, 1-aryl-2,4,6-triphenylpyridinium-perchlorates are transformed into (1-aryl-3,5-diphenyl-1*H*-pyrrol-2-yl) when treated with potassium hydroxide and potassium ferricyanide **(Scheme 51)** [155].

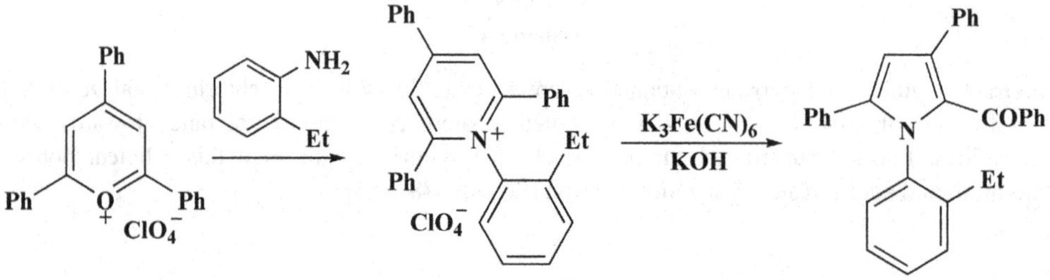

Scheme-51

3-Alkyl-1-tosyl pyrrolidines can be obtained in sufficient yields by an efficient alkene aza-Cope-Mannich cyclization of aldehydes and 2-hydroxy homoallyl tosylamine in the presence of Fe(III) salts *via* a γ-unsaturated iminium ion, 2-azonia-[3,3]-sigmatropic rearrangement, and intramolecular Mannich reaction. Furthermore, dihydro-1*H*-pyrroles can be synthesized by cyclization of 2-hydroxy homopropargyl tosylamines **(Scheme 52)** [156].

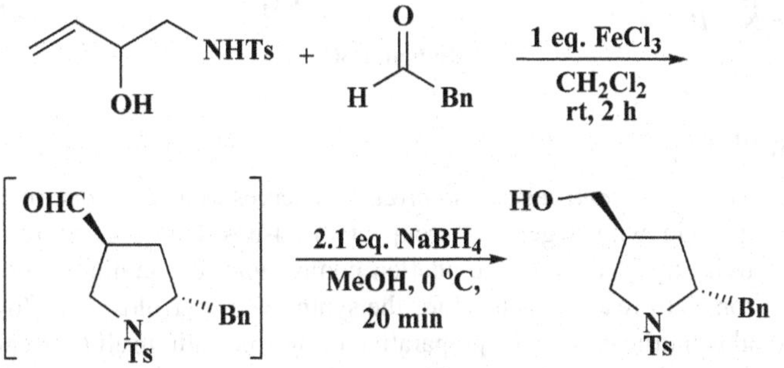

Scheme-52

Lanthanum-assisted synthesis

N-Heterocycles are found in naturally occurring alkaloids. These *N*-heterocycles are generated by organolanthanide-catalyzed intramolecular hydroamination of aminoalkenes **(Scheme 53)** [13, 157–163].

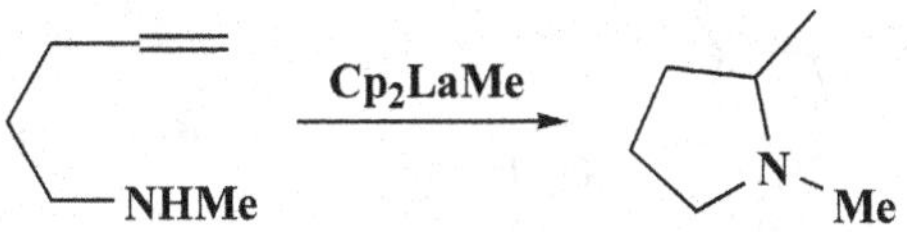

Scheme-53

Intramolecular hydroamination of terminal aminoalkenes occurs in the presence of rare-earth metal catalysts (found to be the most versatile and most active catalysts) for the synthesis of pyrrolidines **(Scheme 54)** [164–167]. However, with increasing ring size, the rate of cyclization decreases [41, 168–169].

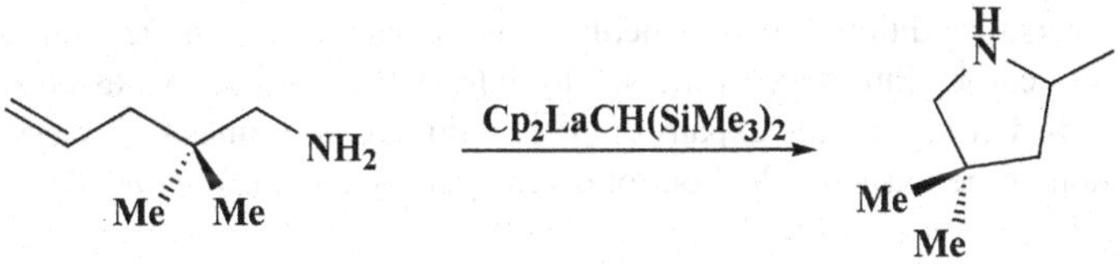

Scheme-54

Aminoallenes undergo hydroamination/cyclization at a rate faster compared to the reaction of terminal aminoalkenes, but slower (by a factor of 5–20) than aminoalkynes [170–171]. The cyclization follows two different pathways to afford two possible regioisomers. Mono-substituted, terminal allenes provide a mixture of products; however, the synthesis of cyclic imine is generally favored through the endocyclic pathway **(Scheme 55)** [41].

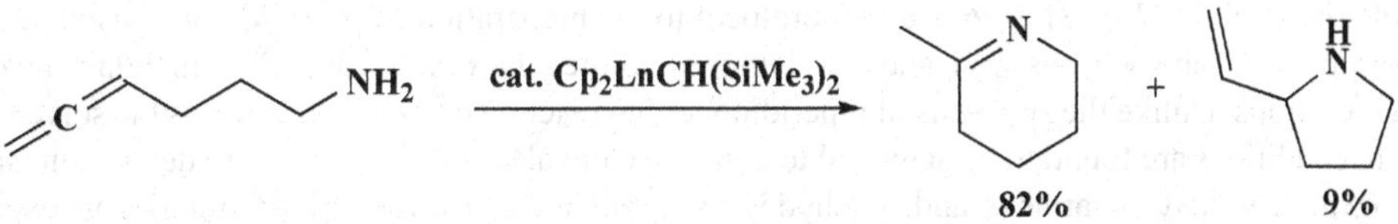

82%   9%

Scheme-55

Nitrogen-containing heterocyclic compounds are synthesized by organolanthanide-catalyzed hydroamination of the aminoalkynes [172]. The reaction of secondary amines results in cyclic enamines while the reaction of primary amines provides cyclic imines (**Scheme 56**) [13].

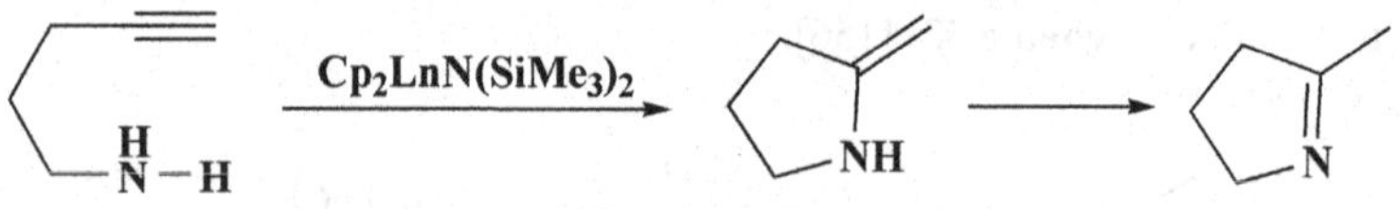

Scheme-56

Lithium-assisted synthesis

Pyrrolostatin isolated from *Streptomyces chrestomyceticus* serves as a peroxidation inhibitor. It consists of pyrrole-2-carboxylic acid with a geranyl group at the 4-position. This pyrrole derivative shows *in vitro* inhibitory activity against lipid peroxidation comparable to vitamin E (tocopherol), acting as anti-oxidant. The Barton-Zard reaction is used for the synthesis of pyrrolostatin. However, with merely 13% yield, this method is inefficient for the preparation of pyrrolostatin itself (**Scheme 57**) [173–174].

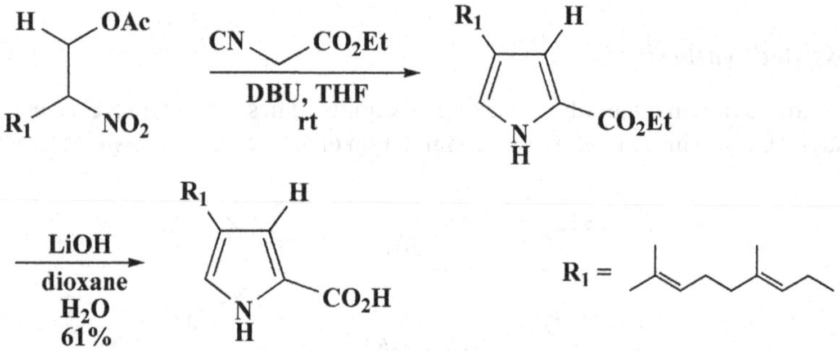

Scheme-57

2-Cyano azetidines serve as building blocks for the synthesis of functionalized heterocyclic compounds. On the contrary, hydrolysis of cyano derivatives into carboxylic acids needs drastic conditions and prolonged heating in concentrated acid to complete the hydrolysis of amide intermediate [175a–b]. Although harsh conditions are used, neither ring-opening nor epimerization have been reported yet. These derivatives can be introduced successfully into peptidic sequences as constrained mimetics of natural amino acids. Ultimately, the expansion of enantiopure azetidines to 3-mesyloxy pyrrolidines occurs upon reduction of the nitrile to alcohol, followed by mesylation (**Scheme 58**) [176].

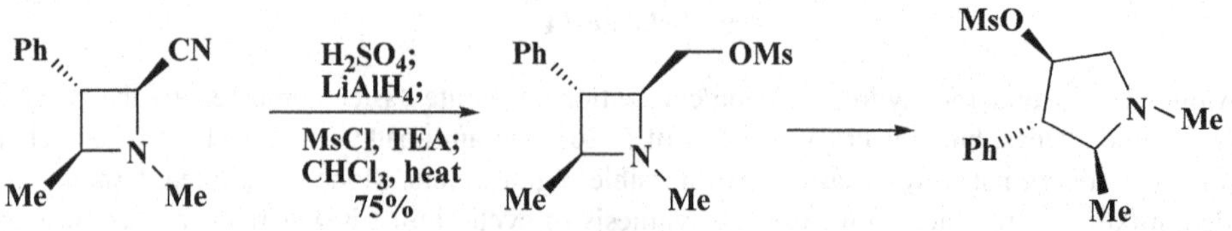

Scheme-58

Pelletier et al. [177–178] extended this protocol to the preparation of pyrrolidinones using methyl-3-nitropropanoate and various aryl and alkyl imines formed *in situ*, in nitro-Mannich/lactamization cascade reactions. Unlike the synthesis of piperidinones, no reaction occurs with water as the solvent. The optimum conditions are found to be degassed toluene, an equivalent of benzoic acid under an atmosphere of nitrogen. A variety of amines and aldehydes were tolerated to afford pyrrolidinones in excellent diastereoselectivity and moderate to good yields (**Scheme 59**). The reaction also occurs with cyclic

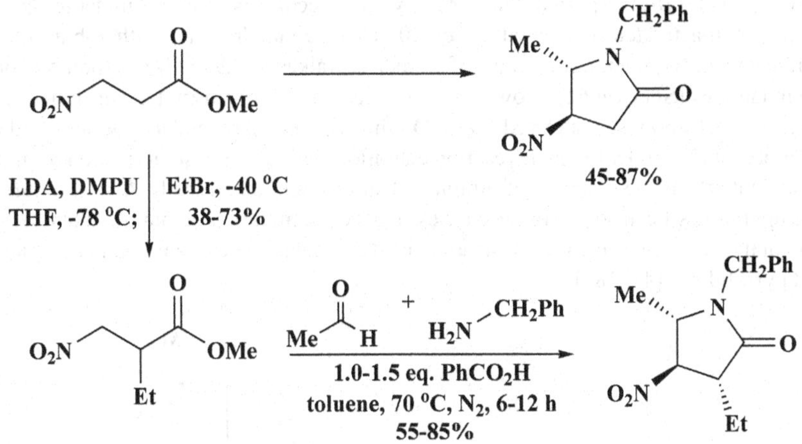

Scheme-59

imines but lower stereoselectivities have been reported. Alkylation prior to nitro-Mannich/lactamization is necessary to introduce moiety α to the carbonyl as all attempts to alkylate fail.

Sakai et al. [179] reported that linear 3-aminoalkanoates when treated with *t*-butyllithium produce five- and seven-membered lactams stereoselectively *via* an initial cyclization of 3-aminoalkanoates to result in azetidin-2-one and a subsequent aza-[1,2] or -[2,3] rearrangement reaction **(Scheme 60)**.

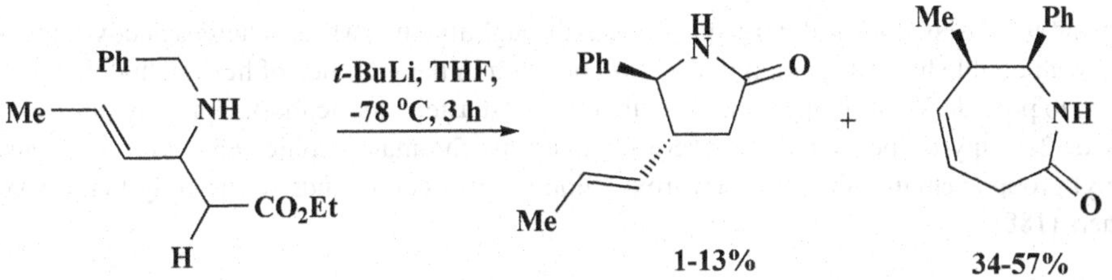

Scheme-60

Ketoesters are transformed into natural products which exhibit pharmacological properties. Keto methyl ester is produced when ketoester is hydrolyzed with lithium hydroxide followed by esterification with diazomethane [180a–b]. The stereochemistry of Si-containing chiral centre, by comparing the magnitude and the sign of the optical rotation value of methyl ester ([α]D24 = –0.8, *c* 0.8, chloroform) with the reported value ([α]D21 = –0.8, *c* 0.79, chloroform), has been concluded to be of (*S*) configuration. The absolute stereochemistry of ketoester has also been confirmed. Methyl ester can be converted into pyrrolidine easily and its conversion within a few steps into (+)-preussin, a hydroxylated pyrrolidine natural product, has already been reported **(Scheme 61)** [181].

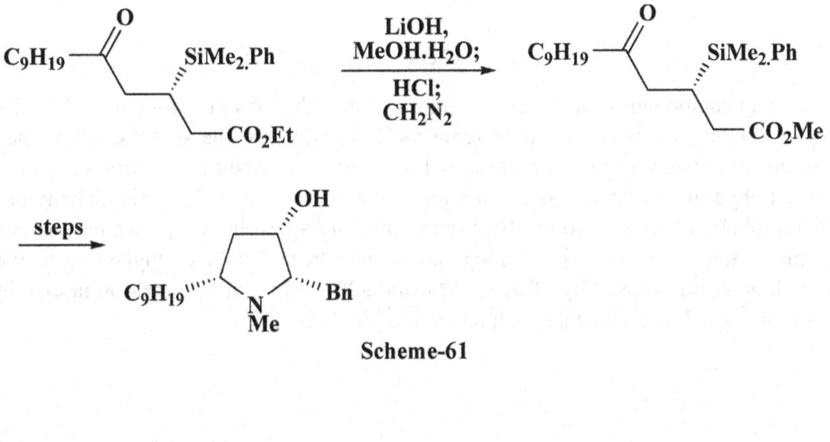

Scheme-61

Ates and Quinet [182] reported that primary and secondary amines undergo intramolecular cyclization onto non-activated alkenes at 50 or 20 °C in tetrahydrofuran with *n*-butyllithium (5–16 mol%) as pre-catalyst. In general, a mixture of isomerized alkene and cyclized product is obtained for primary alkenylamines **(Scheme 62)**. However, in the case of 2,2-disubstituted substrates (*gem*-dialkyl effect), alkene isomerization is suppressed [183]. On the other hand, secondary amines yield the desired pyrrolidines more rapidly under the same reaction conditions while olefin isomerization occurs to a lesser extent. The solvent affects the structure of lithium-substrate complex critically. A 1:1 mixture of toluene and tetrahydropyran has been found to be the best for hydroamination, avoiding alkene isomerization [184]. BuLi-catalyzed stereoselective cyclization of 1,5-disubstituted 5-aminoalkenes yields *cis*-2,5-disubstituted pyrrolidines [41, 185].

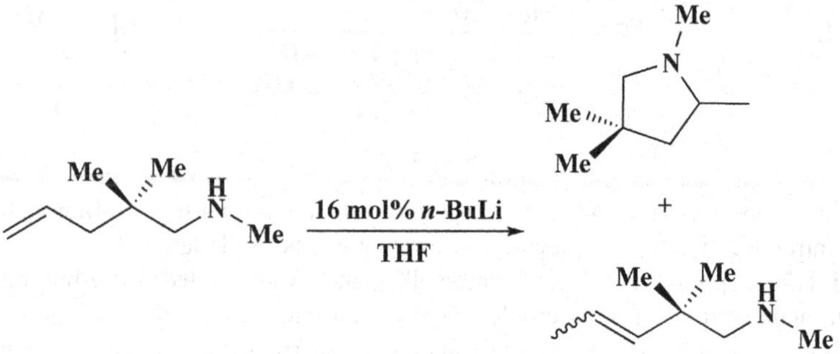

Scheme-62

Davis et al. [186] synthesized chiral sulfinimines through a method wherein stereoselective [3+2]-MIRC occurred with double bond migration to the *endo* position in the presence of hexamethylphosphoramide at −100 °C to provide *N*-sulfinylpyrrolines in up to 88:12 diastereomeric ratio. Optically pure substituted pyrroles or 2-arylpyrroline derivatives were obtained by chromatographic separation and removal of the sufinyl group **(Scheme 63)** [187]. Hydroxylation of such compounds is currently being studied by researchers [188].

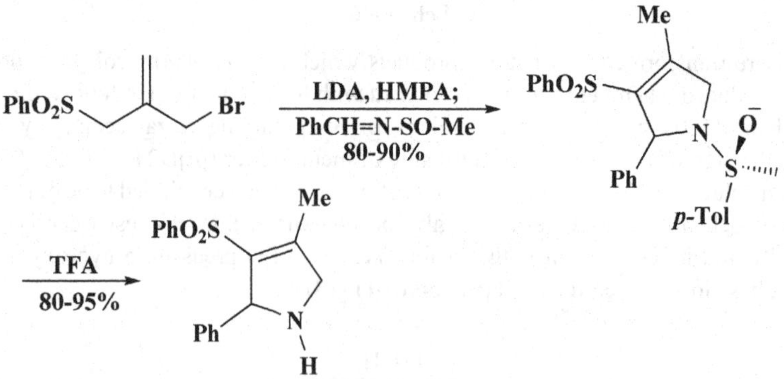

Scheme-63

The potency of compounds is increased by optimization of the substitution-pattern [189]. Pyrrolidine functionality is synthesized as shown in **Scheme 64**. Subsequent to the synthesis of imine, 2 eq. of lithium bromide are utilized for the promotion of 1,3-dipolar cycloaddition reaction of methyl acrylate and imine in tetrahydrofuran under basic conditions. The chiral acid, *R*-2,2'- bis(diphenylphosphino)-1,1'-binaphthyl phosphate is used to resolve the racemic *endo*-pyrrolidine species into its enantiomers by diastereomeric salt formation. The formed salt is then treated with triethylamine to yield chiral pyrrolidine with an enantioselectivity of over 95%, and 82% yield. Further transformations furnish the desired pyrrolidine in a 7-step sequence with an overall yield of > 25%.

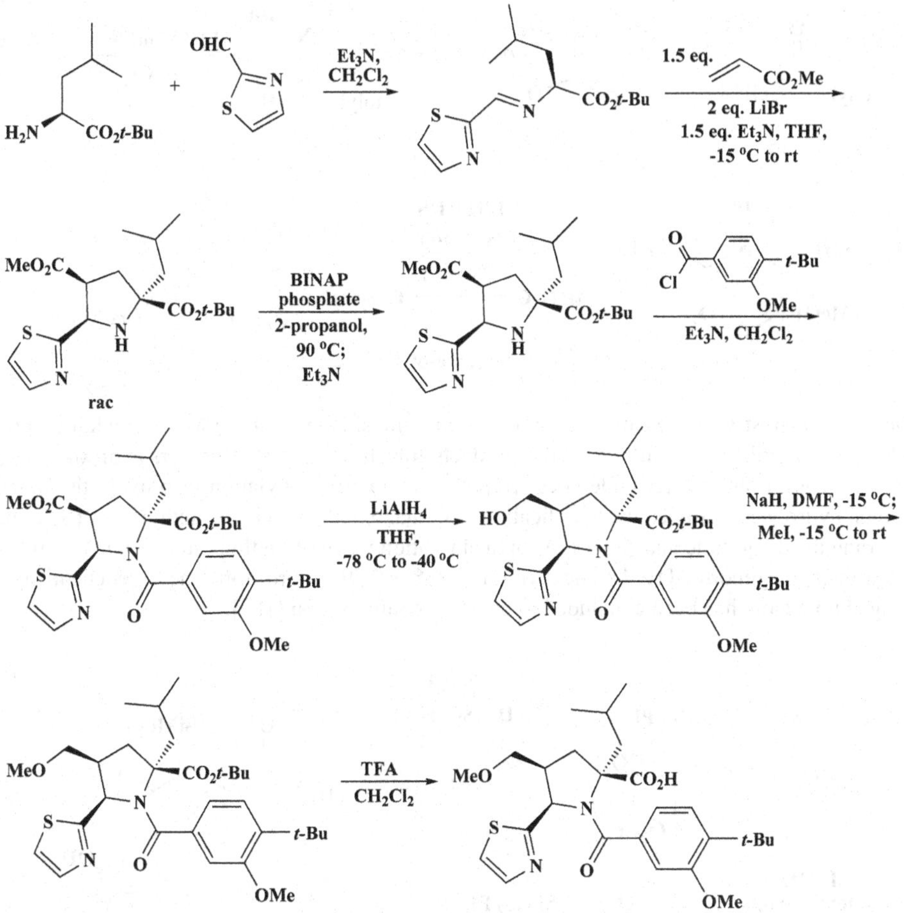

Scheme-64

Commercial trimethylamine *N*-oxide produces an un-stabilized azomethine ylide which undergoes 1,3-dipolar cycloaddition with unpolarized and electron rich olefins to produce good yields of 3,4-di-substituted pyrrolidines. A wide range of substituents can be tolerated on alkenes provided they are compatible with excess lithium diisopropylamide **(Scheme 65)** [190].

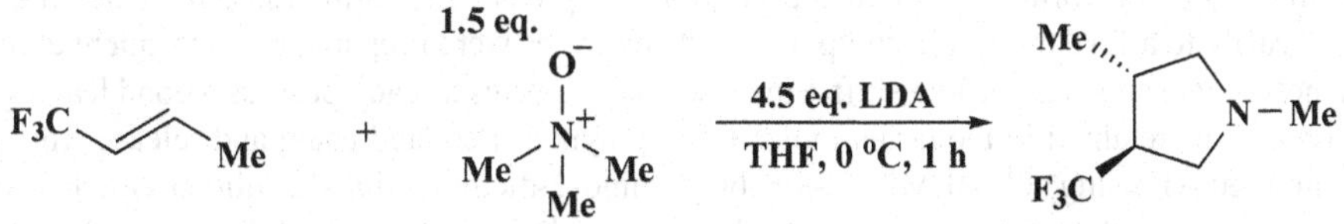

Scheme-65

Alternative routes have been created to produce these dipoles by Arbuzov intermediates in this reaction. Mironov et al. [191] synthesized amido-substituted phosphonates when imines were reacted with benzoate-substituted phosphonites. This reaction was carried out with a catechyl-substituted phosphorus reagent to synthesize the intermediate in a single step from commercial (*o*-catechyl)PCl, imine and *p*-toluic acid **(Scheme 66)**. This method avoids both the use of sensitive acid chloride (and iminium salt) precursors and the need to generate (*o*-catechyl) P(OTMS). Pyrroles were finally synthesized when this synthesis was coupled with subsequent one-pot cycloaddition from imines, carboxylic acids and alkynes [192].

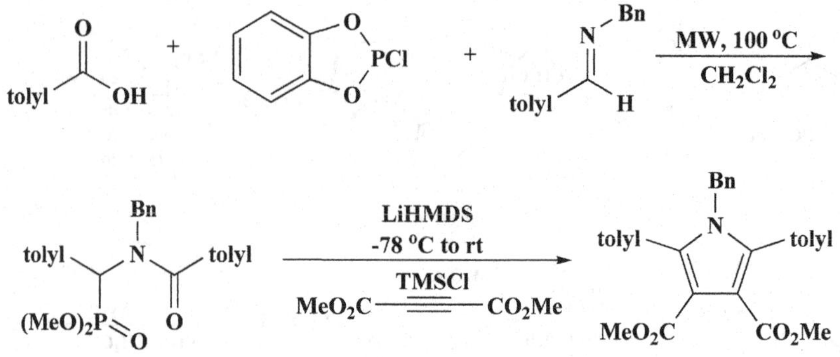

Scheme-66

The stereochemistry of the carbon atom containing the silicon group in Michael addition products has also been determined. When one of the products transforms into the precursor of known natural product (+)-preussin, the diester undergoes Krapcho deethoxycarbonylation to provide the monoester which, upon hydrolysis followed by esterification with diazomethane, yields methyl ester (**Scheme 67**). By comparing the magnitude and sign of the optical rotation value of methyl ester ($[\alpha]D24 = -0.8$, c 0.8, chloroform) with the reported [193] value ($[\alpha]D21 = -0.8$, c 0.79, chloroform), the stereochemistry of the Si bearing chiral centre has been concluded to be of (S) configuration [180].

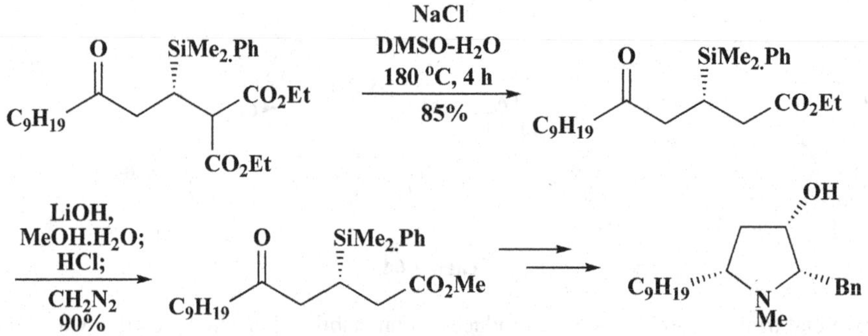

Scheme-67

Baxendale et al. [194a] synthesized 4-nitroindoles by an interesting hybrid of classical van Leusen and Barton-Zard pyrrole syntheses. The isocyanoacetate intermediate was generated when TosMIC was treated with ethyl chloroformate in the presence of a strong base; the formed intermediate attacked the nitroalkene easily to afford the cyclic compound. Nitropyrroles were prepared by subsequent elimination of p-toluenesulfinate and [1,5] proton shift. Since isocyanoacetates already possess a good leaving group, the NO_2 group was retained in the target molecule. A polymer mediated catch and release work-up and purification method using PS-BEMP (2-*tert*-butylimino-2-diethylamino-1,3-dimethylperhydro-1,3,2-diazaphosphorine on polystyrene) provided products in good yields. In general, the yields do not exceed 40% under usual work-up and column chromatography conditions (**Scheme 68**) [174, 194b].

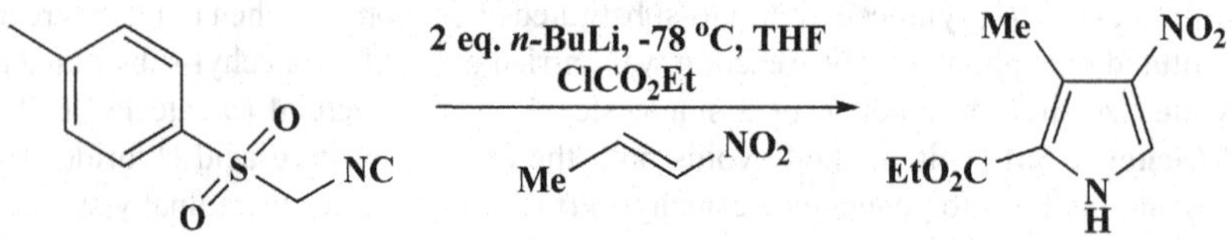

Scheme-68

In another study, Jahn et al. [195] reported that de-protonation of *N*-allylic *β*-alanine esters generates enolates which undergo ferrocenium hexafluorophosphate (C-29) mediated 5-*exo*-cyclization to result in the formation of pyrrolidine **(Scheme 69)** [32].

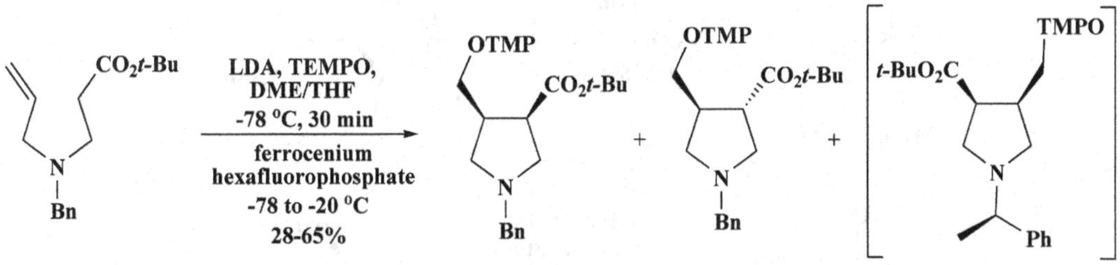

Scheme-69

Magnesium-assisted synthesis

An intermolecular cyclization of alkylidenecyclopropyl ketones with amines results in an efficient synthesis of 2,3,4-trisubstituted pyrroles. The reaction proceeds through a distal cleavage of carbon-carbon bond of the cyclopropane ring **(Scheme 70)** [196].

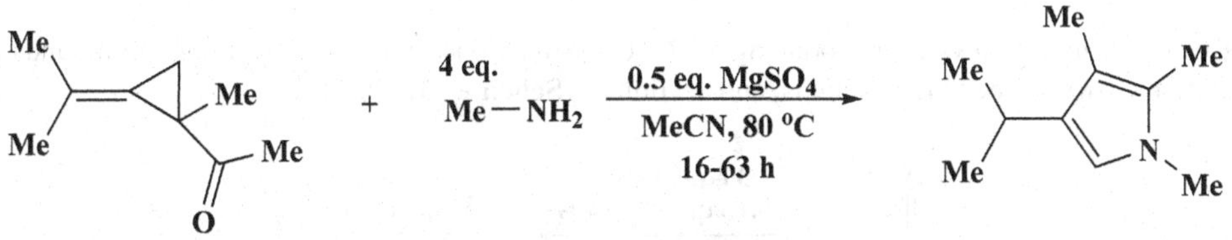

Scheme-70

Diene undergoes cyclization-reduction producing both pyrrolidinones and azocanones **(Scheme 71)**. The pyrrolidinones produced by 5-*exo*-cyclization consist of at least three isomers (pyrrolidinones as the major isomers), whereas 8-*endo*-cyclization provides azocanones as single stereoisomers. Although 8-*endo*-cyclization is slightly preferred over 5-*exo*-cyclization, the yield of 8-*endo*-cyclization products decreases upon increasing the bulkiness of the substituent present in the starting compound [32, 197–198].

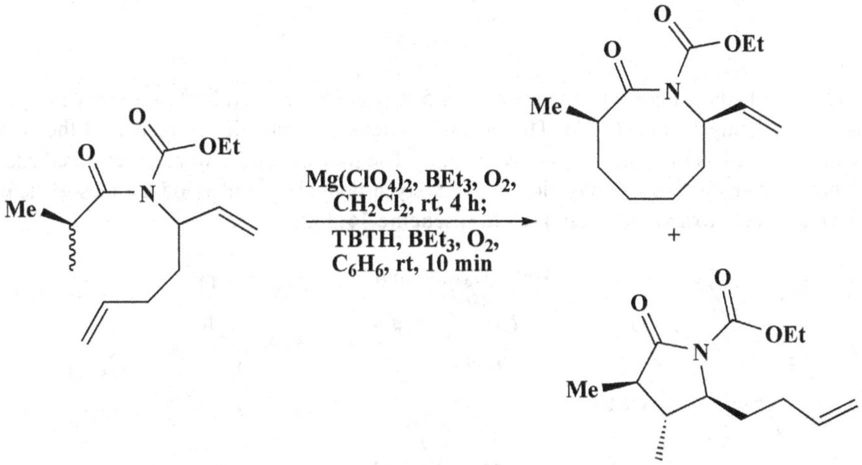

Scheme-71

Kim et al. [199] reported a functional transformation of substituted allylamine *via* an oxazolidinone intermediate for stereoselective synthesis of bioactive (2*S*,3*S*,4*S*)-3,4-dihydroxyglutamic acid hydrochloride salt **(Scheme 72)** [32].

Zirconium-catalyzed reaction of EtMgCl with imines under selected conditions delivers C,N-dimagnesiated compounds which are further trapped with electrophiles. The overall transformation

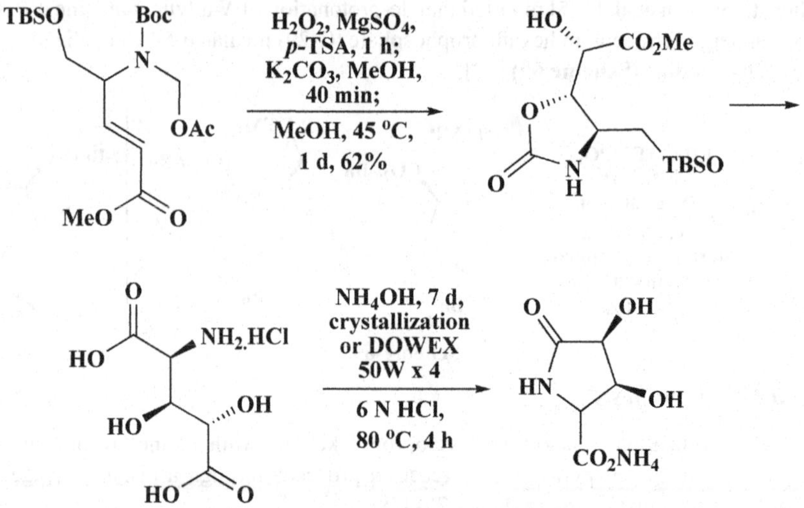

Scheme-72

gives way to a new pathway for producing cyclic or bi-functional nitrogen-containing compounds like pyrrolidines, 1-azaspirocyclic γ-lactams, and azetidines (**Scheme 73**) [200].

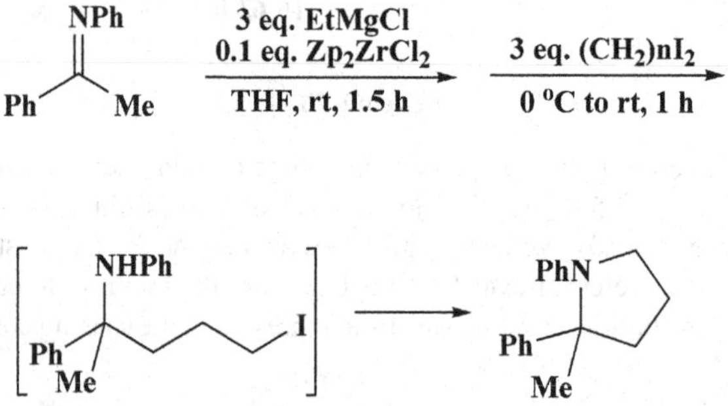

Scheme-73

Fang et al. [201] developed an I₂-atom-transfer 5-*exo* and 8-*endo*-cyclization of carbamoyl radicals with bidentate chelating ligand (C-26). The bidentate chelation dramatically increased the stereo- and regioselectivity as well as the efficiency of cyclization. The pyrrolidinones were synthesized exclusively as single stereoisomers by 5-*exo*-cyclization of *N*-ethoxycarbonyl substituted iodoamides in the presence of $Mg(ClO_4)_2$ and a bis(oxazoline) ligand (C-26) (**Scheme 74**) [32].

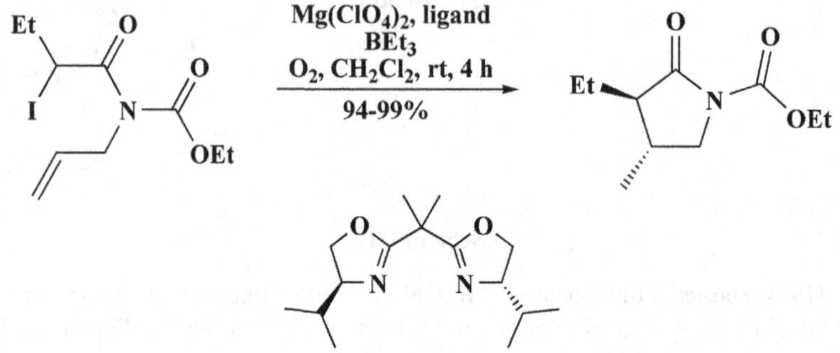

Ligand

Scheme-74

Spiro-pyrrolidine ring has been investigated broadly for its role in constituting the core structure of many biologically important 2-oxindoles and its presence in various natural products. It has been reported that a p53-MDM2 interaction inhibitor MI-219 bearing spiropyrrolidine-oxindole backbone is about to enter phase I clinical trials as an anti-cancer agent [202]. Other studies show that MgI$_2$-catalyzed ring-expansion of a cyclopropane ring in *N*-benzylspiro[2-oxindole-3,1'-cyclopropane] synthesizes spiro-pyrrolidine-oxindoles in the presence of aldimines **(Scheme 75)** [203–204].

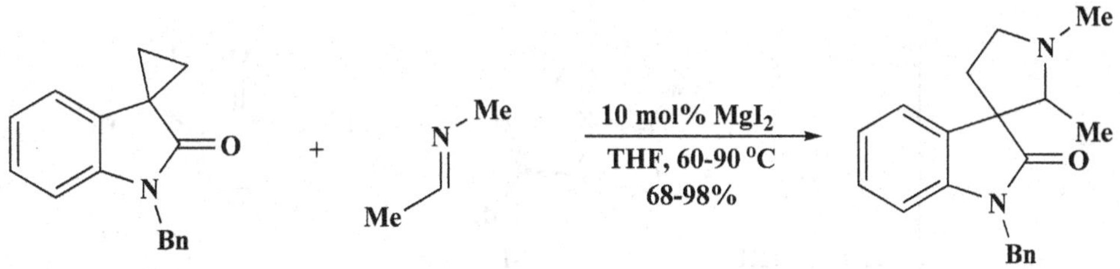

Scheme-75

Manganese-assisted synthesis

A wide range of substituents bearing substituted NH pyrroles have been synthesized by manganese(III)-catalyzed formal [3+2]-annulation of *β*-keto acids and vinyl azides **(Scheme 76)** [205].

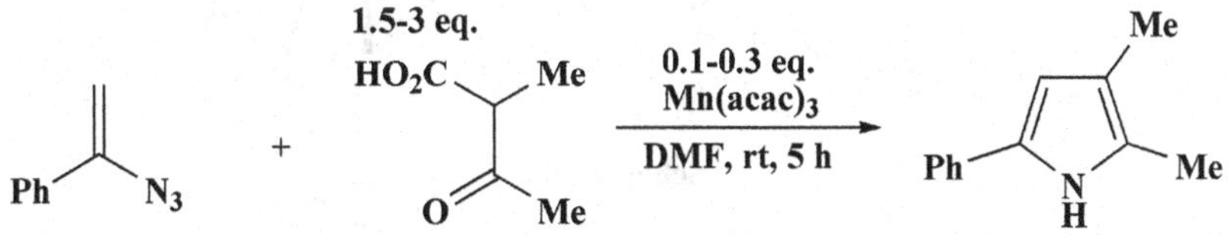

Scheme-76

Good yields of numerous poly-substituted N-H pyrroles through Mn(III)-catalyzed reaction of vinyl azides with 1,3-dicarbonyl compounds have also been reported **(Scheme 77)** [206].

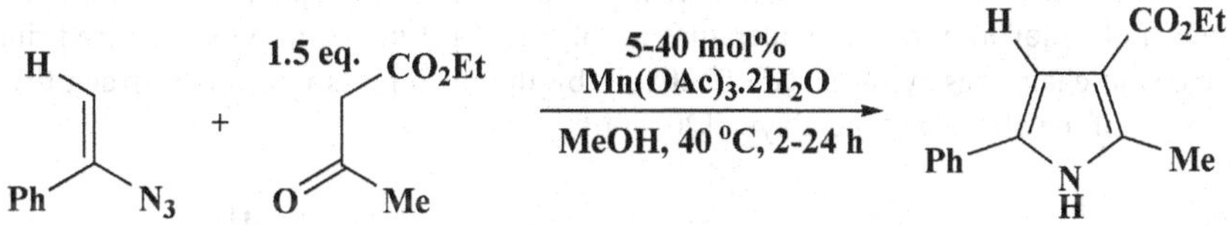

Scheme-77

Asahi and Nishino [207] investigated the use of Cu(OAc)$_2$ and Mn(OAc)$_3$ as radical initiators for the effective synthesis of 3-azabicyclo [3.1.0] hexan-2-ones by intramolecular oxidative cyclization of *N*-propenyl-3-oxobutanamides. Additionally, other cyclized products were also isolated **(Scheme 78)** [32].

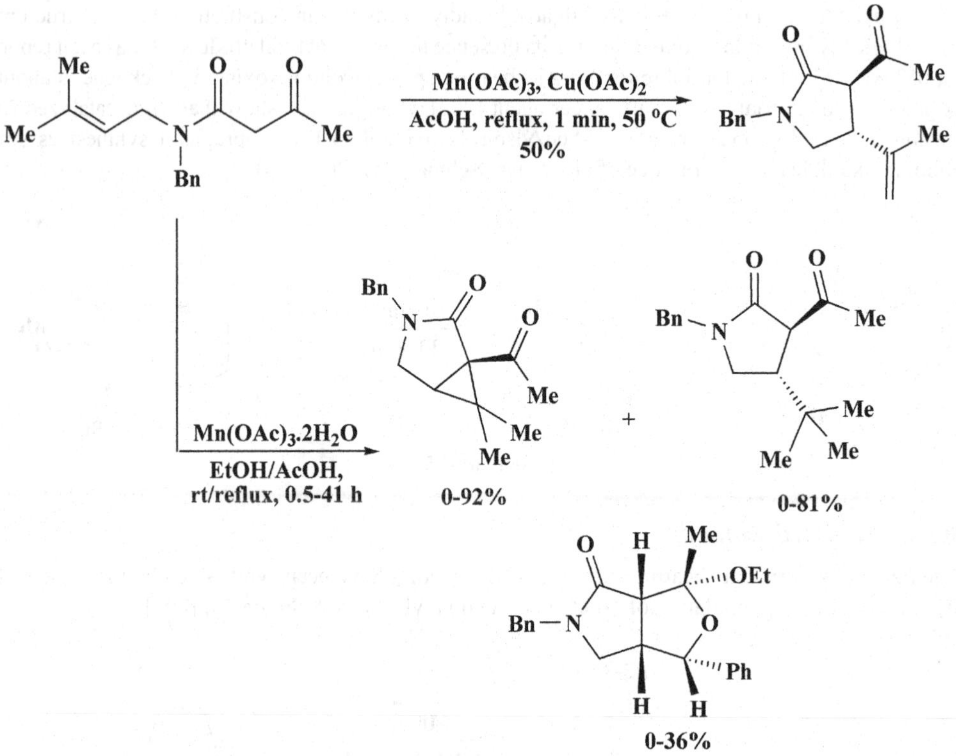

Scheme-78

Molybdenum-assisted synthesis

Murakami et al. [208] reported that allenynes underwent first ring-closing metathesis reaction in the presence of molybdenum complex at room temperature to provide closed ring vinyl allenes. This showed that the vinylallene framework was generated by a metathesis-type reaction between the proximal C-C double bond of the allene and the alkyne functionality (**Scheme 79**) [56]. Allenynes were treated with Schrock catalyst (15 mol%) at room temperature in toluene for 3 hours; a five-membered ring product with an allene side-chain was synthesized effectively by the metathesis reaction between the proximal C-C double bond of the allene and the alkyne [209a–b].

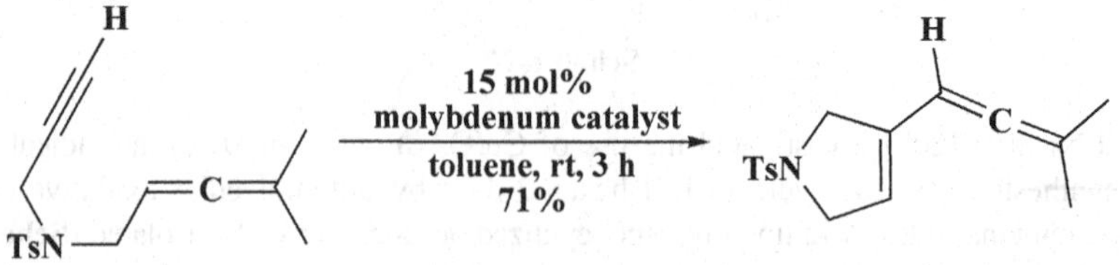

Scheme-79

Another study by Kaneta et al. [210] reported alkyne co-trimerization reactions for the synthesis of isoindoline in moderate yields using Mortreux's catalyst system derived from *p*-chlorophenol and Mo(CO)$_6$. Mono-alkynes possessing *o*-hydroxyphenyl group provided trimerized products, while alkynes bearing *m*- or *p*-hydroxyphenyl group resulted in cross-alkyne metathesis products under these conditions. Under similar reaction conditions, the diyne underwent co-trimerization with many mono-alkynes to provide the cyclized product along with the by-product (**Scheme 80**) [211]. The isolation of the by-product indicated that this catalytic reaction occurred using molybdenacyclopentadiene complex. The yields of desired trimerized products improved when 15 eq. of monoalkyne was used. Absolute intramolecular [2+2+2]-cyclization reactions were also performed using similar catalyst systems [212].

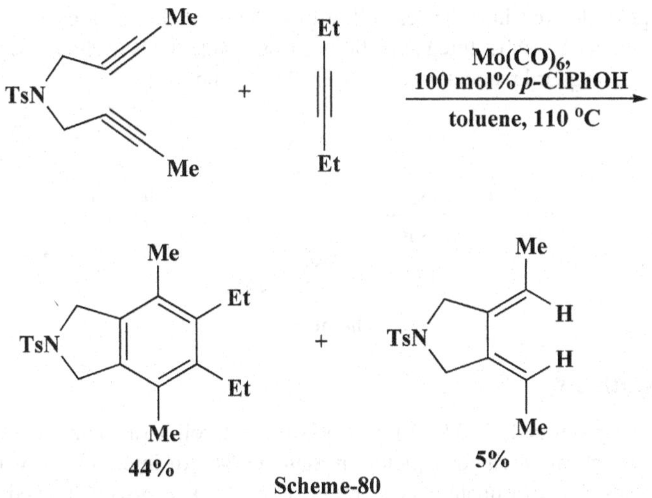

Scheme-80

Nickel-assisted synthesis

In an attempt towards further elaboration with tertiary nitroalkane, mixtures of lactam and amino ester have been obtained under 400 psi of hydrogen with Raney nickel. To complete the lactamization, the mixture was separated from the nickel catalyst and subjected to mild heating. Moderate yields of lactam were obtained after the complete consumption of the amino ester (**Scheme 81**) [213–216].

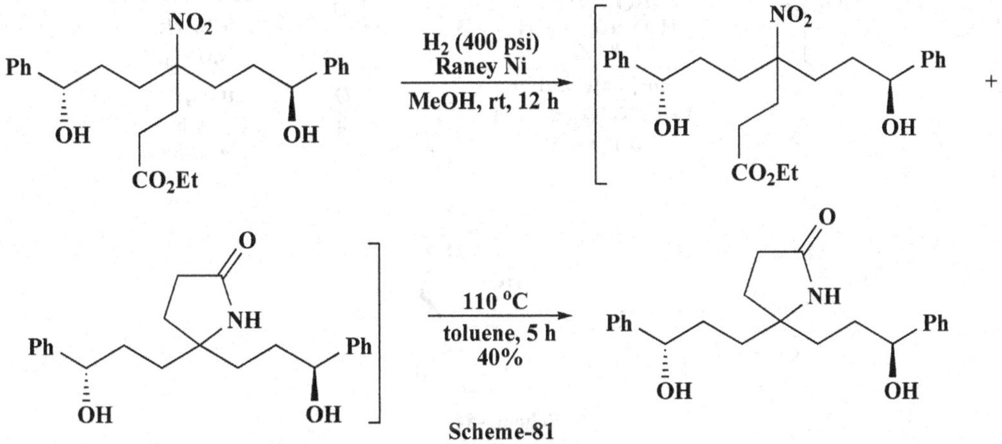

Scheme-81

Seebach and Konchel [217] studied the use of nitroalkene containing a pivalate as the leaving group, wherein nitropropene was used as a powerful nitroallylating reagent which coupled with two nucleophiles consecutively. This reagent assisted the synthesis of γ-lactams (**Scheme 82**).

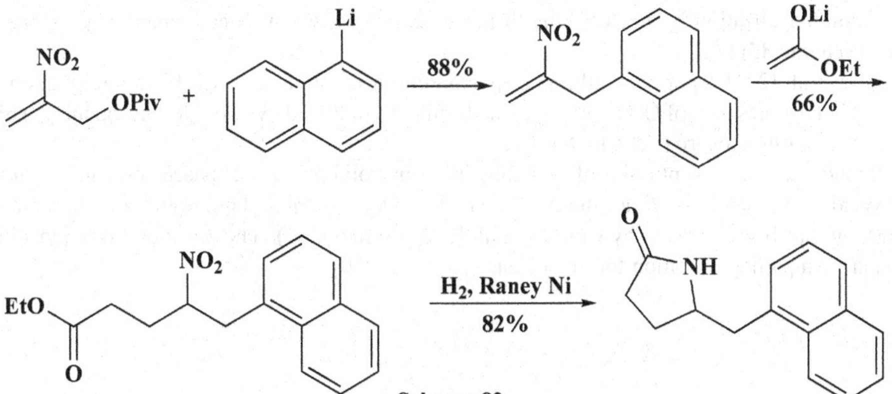

Scheme-82

Uemura et al. [218] looked into the feasibility of utilizing nickel as a catalyst in Heck reaction. Consequently, they produced 3-substituted pyrrolidines from allylamines **(Scheme 83)**. The reaction was conducted with an aryl Grignard reagent with Cp*CH₂PPh₂ ligand [32].

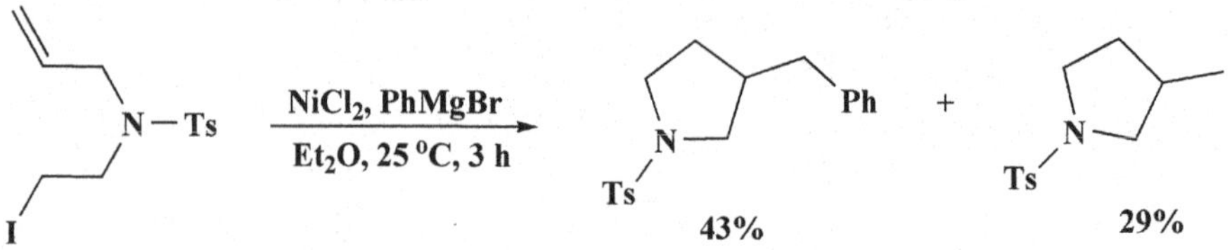

Scheme-83

Osmium-assisted synthesis

Yoshitomi et al. [219] synthesized (2S,3R)-3-hydroxy-3-methylproline (a component of polyoxypeptins) in large quantities through the method depicted in **Scheme 84**. An intermediate with two continuous asymmetric carbons possessing a quaternary stereogenic centre was also generated by intramolecular aldol reaction of the aminoacetaldehyde derivative in the presence of (2S,3R)-3-hydroxy-3-methylproline organocatalyst. Besides, preparation of aminoacetaldehyde derivative *via* dihydroxylation followed by oxidation of the double bond of the substituted allylamine has been found to be general, it provides various analogues of the proline compound [32].

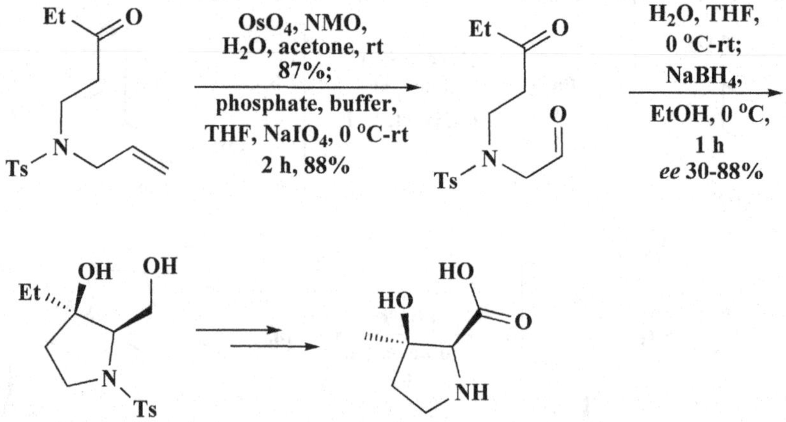

Scheme-84

Reddy and Rao [220] synthesized anti-biotic (−)-codonopsinine from an intermediate obtained from a chiral allylamine. The key steps involved were asymmetric dihydroxylation of the allylic double bond in chiral allylamine *via* a modified Sharpless reaction, and a highly stereoselective intramolecular acid-catalyzed amidocyclization by nucleophilic displacement of acetate with carbamate to yield the target molecule **(Scheme 85)** [32].

Vanlaer et al. [221] reported synthesis of pyrrolidine from allylamine *via* bis-hydroxylation of the double bond in the presence of OsO₄, and epoxide formation followed by epoxide ring-opening with base and concomitant ring-closure **(Scheme 86)** [32].

A method for the synthesis of 3,4-dihydroxy-pyrrolidin-2-ones, based on the asymmetric dihydroxylation of γ-amino-α,β-unsaturated esters followed by triethylamine-induced intramolecular lactamization, has been reported by Coutrot et al. [222] **(Scheme 87)**. The 3,4-dihydroxy-pyrrolidin-2-ones displayed a partial inhibition for glucosidase [32, 223–230].

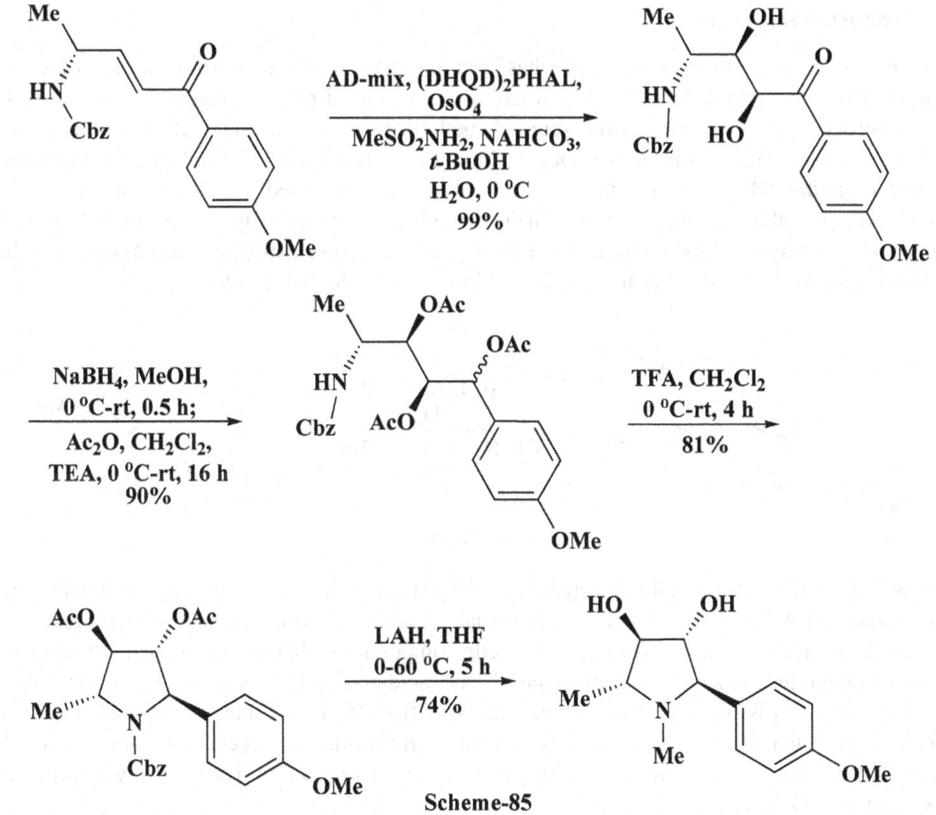

Scheme-85

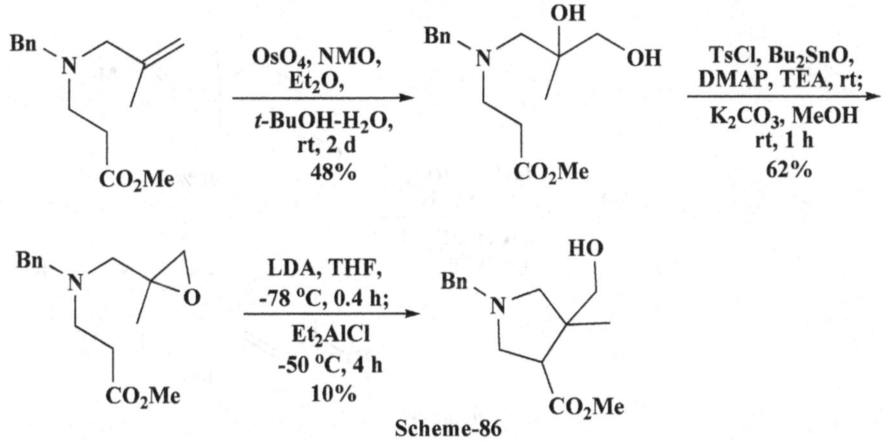

Scheme-86

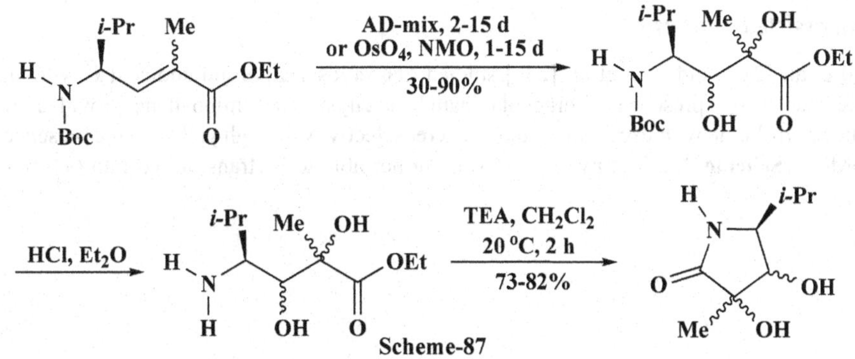

Scheme-87

Rhenium-assisted synthesis

Catalytic Beckmann rearrangement of oximes has been reported to occur with *p*-toluenesulfonic acid and tetrabutylammonium perrhenate [231–237], wherein cyclization of phenethylketone oximes occurs on the oxime nitrogen atom, with the phenyl group. Phenethylketone oximes are transformed into aza-spiro compounds when reacted in catalytic amounts of trifluoromethanesulfonic acid and tetrabutylammonium perrhenate (**Scheme 88**) [238]. Furthermore, quinolines are delivered by phenethylketone oximes substituted with *p*-methoxycarbonylamino, with the two substituents originally present at 1 and 4 positions shifted to 1 and 3 positions. Moreover, when *p*-hydroxyl or methoxy-substituted phenethylketone oximes are utilized, spirocyclic compounds are synthesized instead of quinolines [239–240].

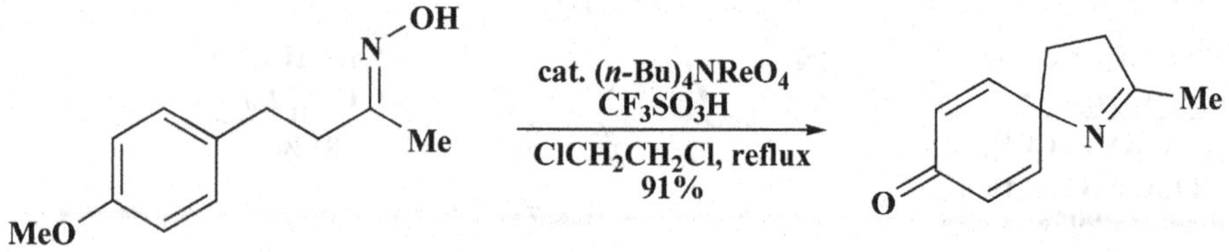

Scheme-88

Phenethylketone oxime containing a hydroxyl group at the *para*-position can also react with various acids (**Scheme 89**). When polyphosphoric acid (usually used in Beckmann rearrangement) reacts with *p*-OH oximes, it yields amides. However, spirocyclic compound is obtained as the main product, with the Beckmann product when trifluoromethanesulfonic acid (CF_3SO_3H) is used. However, only the spirocyclic compound is obtained when oxime reacts with *n*-Bu_4NReO_4 and trifluoromethanesulfonic acid. This suggests that either cyclization or Beckmann rearrangement may occur based on a very subtle difference in the reaction conditions. The starting material for these reactions is an oxime with protonated hydroxyl group [241–242].

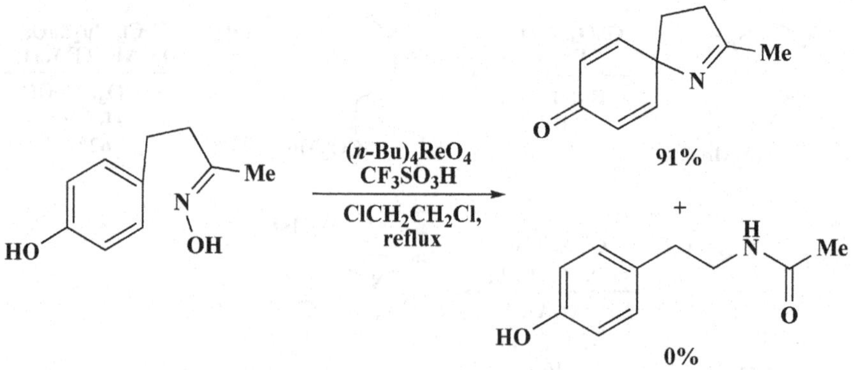

Scheme-89

Samarium-assisted synthesis

Arredondo et al. [243] and Tian et al. [244] studied diastereoselective intramolecular hydroamination of aminoallenes in the presence of organolanthanide catalyst. The aminoallene provided *trans*-2,5-disubstituted pyrrolidine with excellent *Z*- and diastereoselectivity, in high yields, in the presence of Cp_2-$SmCH(SiMe_3)_2$ (**Scheme 90**). The pyrrolidine was further able to be transformed into (+)-pyrrolidine-197B [13].

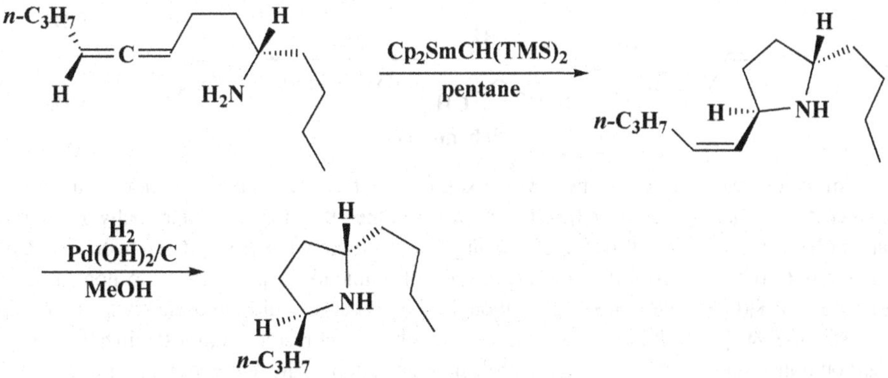

Scheme-90

In another study, it was found that aza-sugar scaffold is obtained as a single stereoisomer when initial *syn*-enol ether is hydroborated (**Scheme 91**) [245]. For example, for the cleavage of N-O oxazine bond, samarium(II) iodide was required, which resulted in the synthesis of *N*-benzylaminosugar with an overall yield of 73%. This material could ultimately be cyclized to a highly functionalized pyrrolidine.

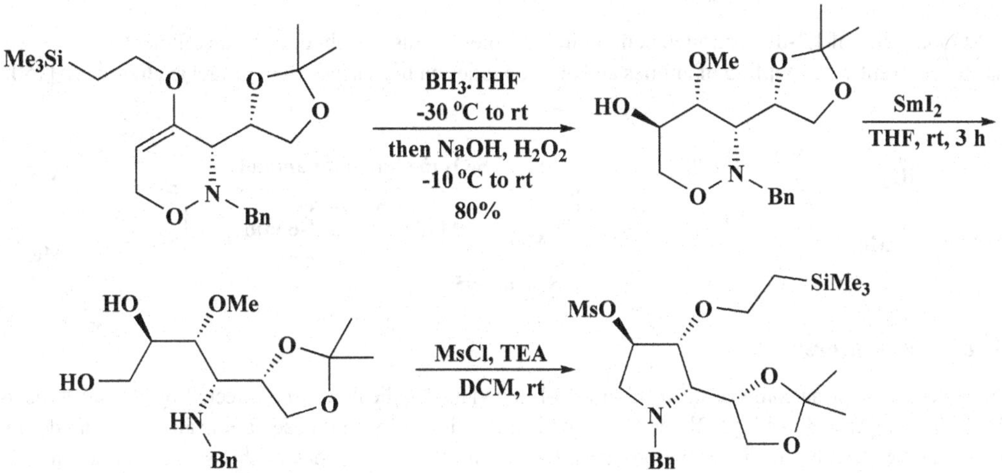

Scheme-91

Scandium-assisted synthesis

The highly diastereoselective synthesis of multi-substituted pyrrolidines has also been known to occur through a tandem ring-opening cyclization reaction of cyclopropanes with imines and scandium triflate (5 mol%) (**Scheme 92**) [246].

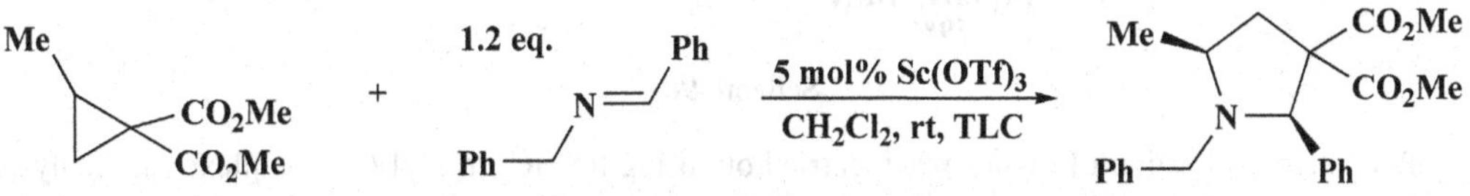

Scheme-92

Silicon-assisted synthesis

Aydogan et al. [247] performed reactions between *cis*-1,4-dichloro-2-butene and various amino alcohols, amine compounds and amino acid esters under MWI without solvent for 2–4 minutes on silica gel. Consequently, good yields (49–69%) of *N*-substituted homochiral pyrrole were obtained instead of 3-pyrrolines (**Scheme 93**) [248].

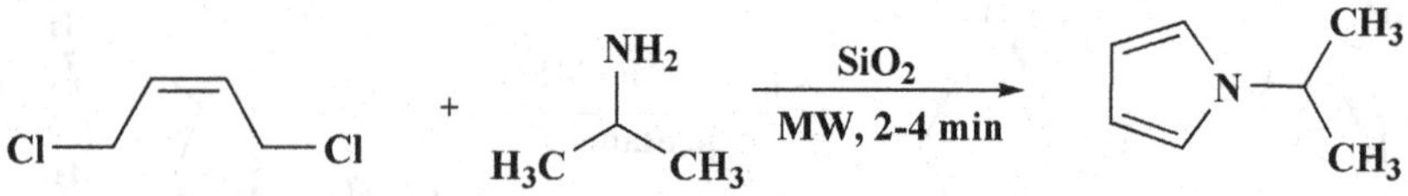

Scheme-93

Hydroamination reactions performed with oxophilic lanthanide catalysts in non-polar aromatic or aliphatic solvents have been observed to be efficient, whereas their rate decreases marginally in polar solvents like tetrahydrofuran [166]. *Ansa*-yttrocene containing an ether moiety tethered to the silicon bridge displays increased reactivity (up to five-fold) in the cyclization of aminoalkenes, compared to non-functionalized *ansa*-yttrocene Me$_2$Si(C$_5$Me$_4$)$_2$YCH(SiMe$_3$)$_2$, though the effect on diastereoselectivity is minimal to negligible (**Scheme 94**) [249]. It has been suggested that this effect results from a stabilization of the polar olefin insertion transition state through an intramolecular chelation of the tethered donor group [41].

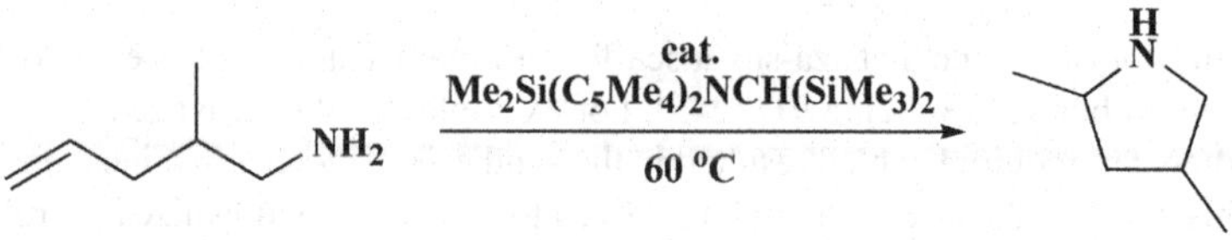

Scheme-94

Good yields of 1,2-disubstituted homochiral pyrroles can also be obtained by an efficient, microwave-assisted, solvent-free coupling of amines and chloroenones on the surface of silica gel (**Scheme 95**) [250].

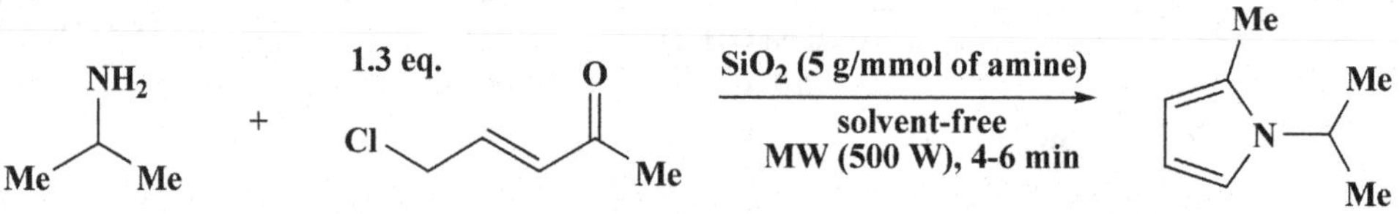

Scheme-95

Tin-assisted synthesis

The stereochemical outcome of cyclization of 5-hexenylalkyl radicals can be affected by the substituents [251]. Spellmeyer and Houk [252] and Beckwith and Scheisser [253] proposed a transition state model for predicting the stereochemistry of the products. Despite fruitful applications of this stereoselective method to synthesize cyclic structures employing carbon-centered radicals, only scattered examples of analogous cyclization of aminyl radicals have been reported [254–257]. The products are obtained in low to moderate levels of diastereoselectivity by cyclization of 1-substituted aminium cation radicals [258–259] (**Scheme 96**).

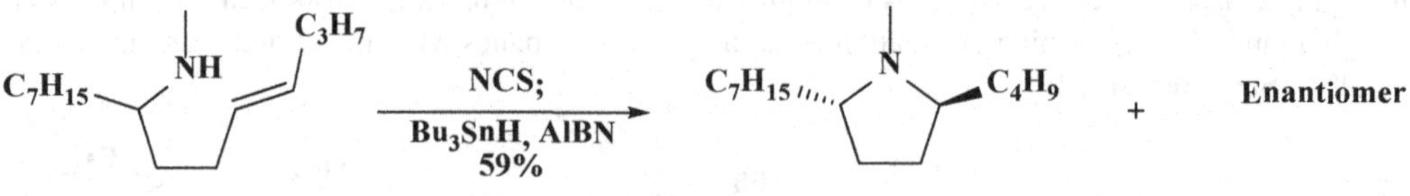

Scheme-96

Reductive cyclization of diynes when carried out using [(NHC)Pt(allyl)Cl] complexes as catalysts results in the formation of 2,5-dihydrofurans, -cyclopentenes, and -pyrroles in fair to high yields under hydrogen atmosphere [260]. Enynes can also be utilized as starting substrates to provide the same reaction products, although the temperature needs to be increased to 70 °C. A rare four-hydrogen addition occurs to the substrate when the reaction is performed under D$_2$ atmosphere. Furthermore, under a mixed D$_2$/H$_2$ atmosphere, crossover products are also formed, pointing towards the heterolytic activation of hydrogen in this method (**Scheme 97**) [225]. However, no conversion method without *N*-heterocyclic carbene or upon phosphine coordination at the platinum center has yet been reported.

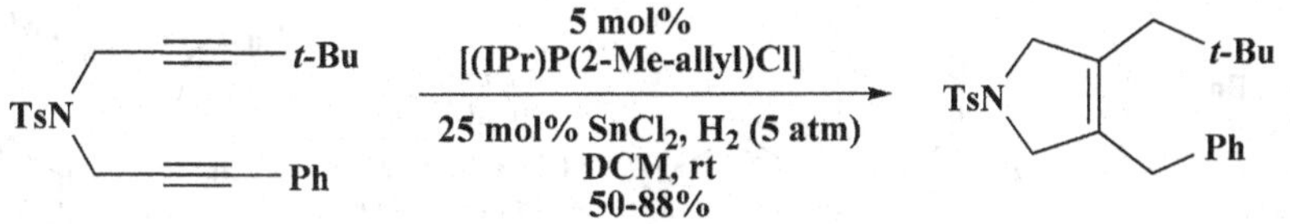

Scheme-97

Titanium-assisted synthesis

Secondary amines are formed in moderate to good yields upon subsequent nucleophilic addition of phenyllithium or *n*-butyllithium [261], while reaction with methyllithium is sluggish (< 20% yield). Although diminished yields are obtained with reduced regioselectivity, the separation of undesired Markovnikov hydroamination by-products is simple due to the significantly reduced rates of nucleophilic addition to the ketimines. Secondary amines are also directly available through hydroamination of internal alkynes, followed by reduction to amine. However, regioselectivity is more difficult to control in these reactions, potentially providing the wrong regiosiomer **(Scheme 98)** [41].

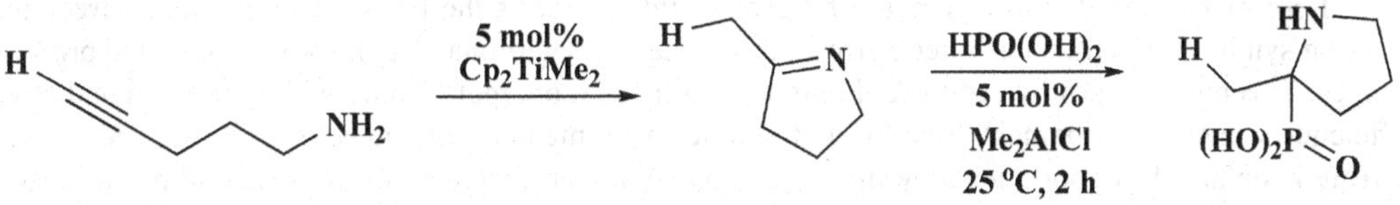

Scheme-98

Ytterbium-assisted synthesis

Mixtures of azapyranose and azafuranose are obtained by Yb(OTf)$_3$-catalyzed ring-opening of bis-aziridine with water and allyl alcohol, followed by intramolecular heterocyclization **(Scheme 99)** [262]. The ratio of products is dictated by the Lewis acid used in the reaction. For instance, poor selectivity is reported in BF$_3$.OEt$_2$-mediated reaction of allyl alcohol [263].

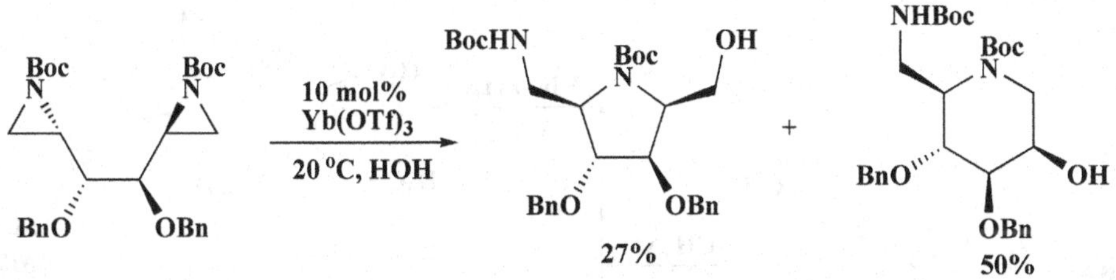

Scheme-99

When a nitrone undergoes catalytic asymmetric 1,3-dipolar cycloaddition with a dipolarophile using chiral Yb (derived from (*R*)-(+)-BINOL, Yb(OTf)$_3$ and *cis*-1,2,6-trimethylpiperidine) as the catalyst [264], the desired isoxazolidine is obtained in 78% enantiomeric excess with perfect diastereoselectivity (*endo/ exo* 99/1) through the reaction of benzylideneamine *N*-oxide with 3-(2-butenoyl)-1,3-oxazolidin-2-one **(Scheme 100)**. On the contrary, reverse enantioselection is observed with achiral scandium catalyst. The chiral Yb catalyst can be improved for asymmetric 1,3-dipolar cycloaddition by addition of *N*-methyl-bis-[(*R*)-1-(1-naphthyl)ethyl]amine [(*R*)-MNEA], a chiral amine [265]. Thus, 1,3-dipolar adducts are obtained with good to excellent *endo*- and enantioselectivities using a combination of Yb(OTf)$_3$, (*R*)-MNEA, and (*S*)-BINOL. A reversed selectivity results from the use of the same chiral catalyst by mere replacement of the molecular sieves with benzylidenebenzylamine *N*-oxide or *N*-methylmorpholine *N*-oxide [263, 266].

Rare earth metal-catalyzed cyclizations of aminoalkynes, aminoalkenes, and aminodienes synthesized exocyclic hydroamination products exclusively. However, homopropargylamines were cyclized to the

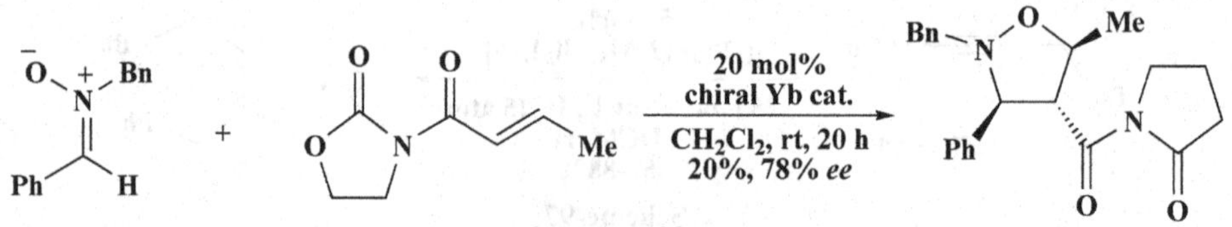

Scheme-100

endocyclic enamine product *via* a 5-*endo-dig* hydroamination/cyclization **(Scheme 101)** [267], due to steric strain associated with four-membered ring exocyclic hydroamination product. Interestingly, the 5-*endo-dig* cyclization was preferred even with an olefin group which led to a 6-*exo*-hydroamination product [41].

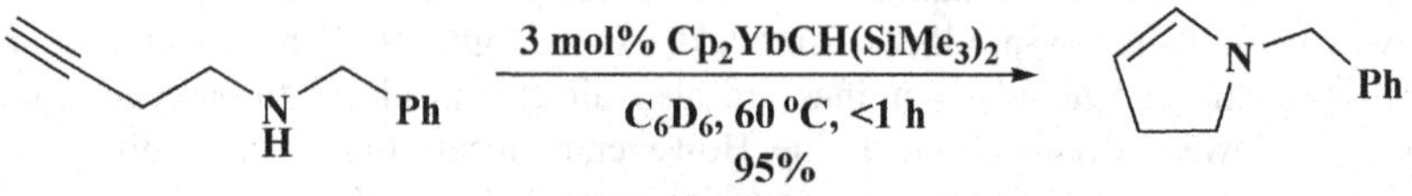

Scheme-101

Microwave-assisted rearrangement of 1,3-oxazolidines forms the basis for direct, metal-free, and modular synthesis of tetra-substituted pyrroles from aldehydes, terminal conjugated alkynes, and primary amines. This method uses two coupled domino reactions in a one-pot manner with both bond and atom efficiencies under environment friendly and simple experimental conditions. 1,2,3,4-Tetrasubstituted pyrrole is obtained in good yields by microwave-assisted rearrangement of 1,3-oxazolidines absorbed on silica gel for 5 minutes. The amine and the conjugated alkynoate are absorbed on 1 g of silica gel and irradiated for 8 minutes at 900 W to yield 1,2,3,4-tetrasubstituted pyrroles in 38–77% yields. The reaction is general for amines and tolerates a variety of moieties in the aldehyde. Several poly-substituted pyrroles have been prepared by this one-pot protocol. While the reaction takes hours under conventional heating conditions, and months at room temperature, it occurs in 8 minutes under microwave-assisted conditions. This method is ideal for diversity-oriented synthesis as a variety of substituted products with a pyrrole backbone can be obtained in short reaction times, under environmentally benign conditions **(Scheme 102)** [268–269].

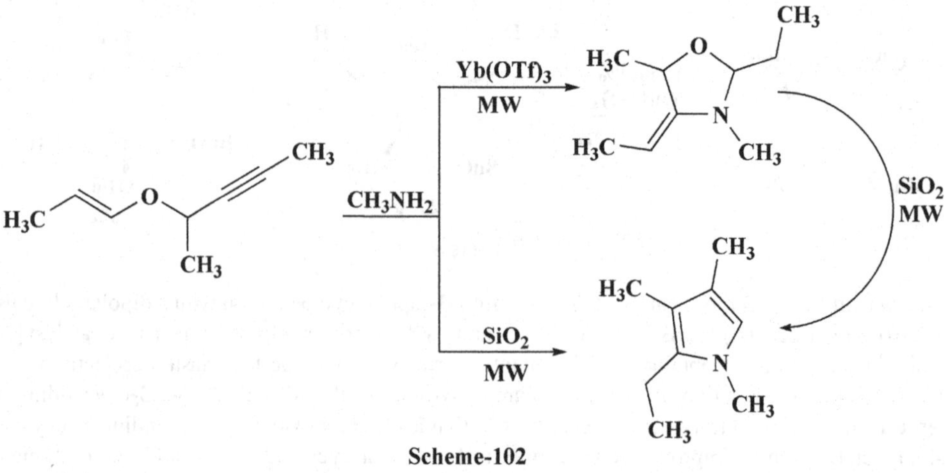

Scheme-102

Zinc-assisted synthesis

The aldol condensation of a 2-indolinone fragment and pyrrole aldehyde produces a complex pyrrole substituted 2-indolinone core. The oxime moiety in the compound, derived by nitrosation of *tert*-butyl acetoacetate [270], is reduced with zinc to provide an unstable aminoketone intermediate. The fully substituted pyrrole ring is generated by subsequent enamine formation and ring-closure. The formylation of ethyl 2,4-dimethylpyrrole-3-carboxylate (synthesized by selective de-protection of *tert*-butyl ester with concomitant decarboxylation) at the free ring position by trimethyl-*o*-formate yields fully pyrrole aldehyde. The aforementioned aldol condensation unites both the key fragments of ester hydrolysis and amide formation which complete the synthesis **(Scheme 103)** [271].

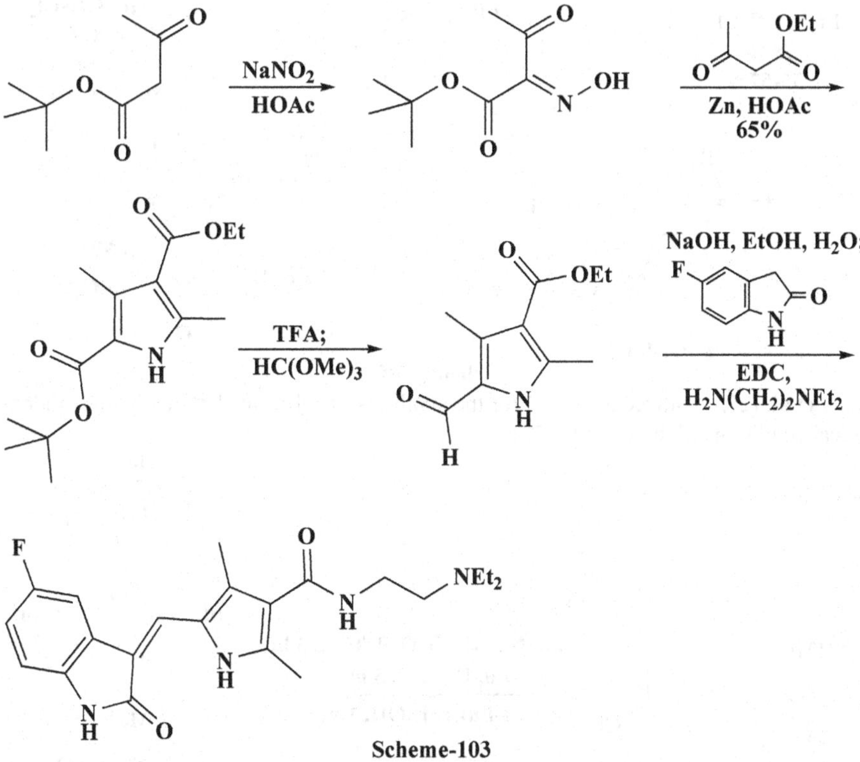

Scheme-103

Murakami et al. [272–273] synthesized α-isocyano-δ-ketoesters in high yields by Michael reaction of α,β-unsaturated carbonyl compounds and α-isocyanoesters in tetrabutylammonium fluoride. Another Michael reaction using the same reactants, accelerated by *N,O*-bis(trimethylsilyl)acetamide (BSA), produces silyl enolates under mild conditions. The isocyanide group is activated through coordination by a Lewis acid which promotes different types of nucleophilic cyclizations, following which silylated Michael adducts are cyclized using Zn(OAc)$_2$ to form pyrroline-2-carboxylic acids. This reaction occurs through the cyclization of the complex after coordination of zinc(II) with the isocyano functionality **(Scheme 104)** [174].

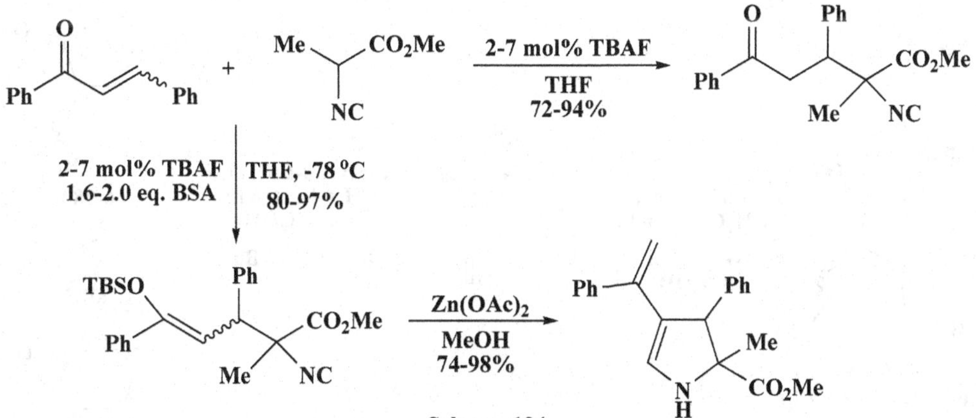

Scheme-104

Miyabe et al. [274] studied the enantioselective radical addition-cyclization-trapping of allylamides for the synthesis of chiral γ-lactams **(Scheme 105)**. Therein, enantioselectivity was observed with Lewis acid and C-28 ligand [32].

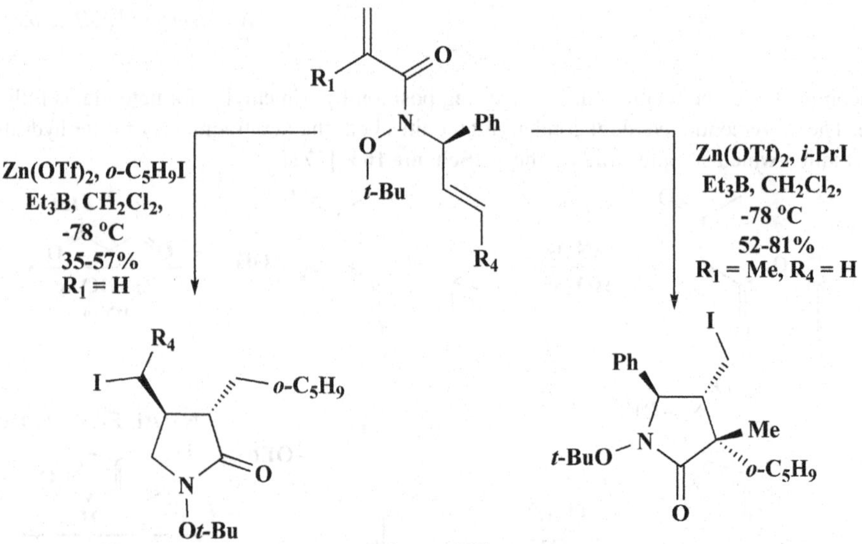

Scheme-105

Aminov et al. [275] realized a method for the synthesis of substituted *N*-methyl pyrrolidones under sonochemical conditions (**Scheme 106**) [32].

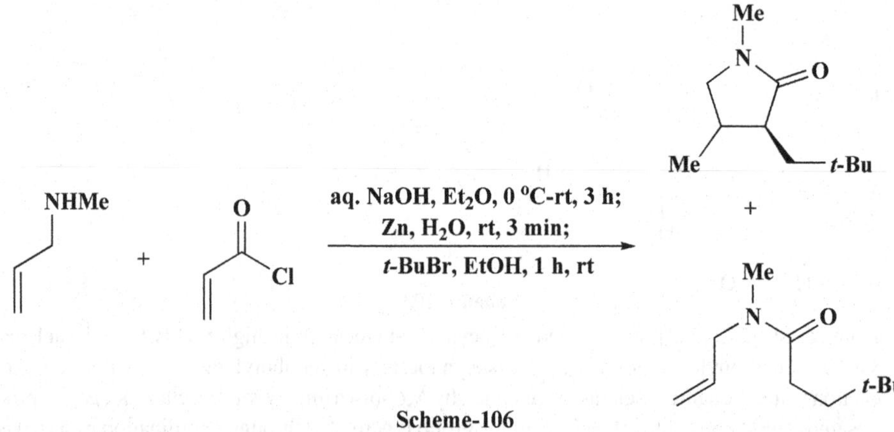

Scheme-106

Feray and Bertrand [276] successfully synthesized α-alkylidene-γ-lactams by dialkylzinc-assisted alkylative cycloisomerization of *N,N*-diallylpropiolamide in an aerobic medium (**Scheme 107**) [32].

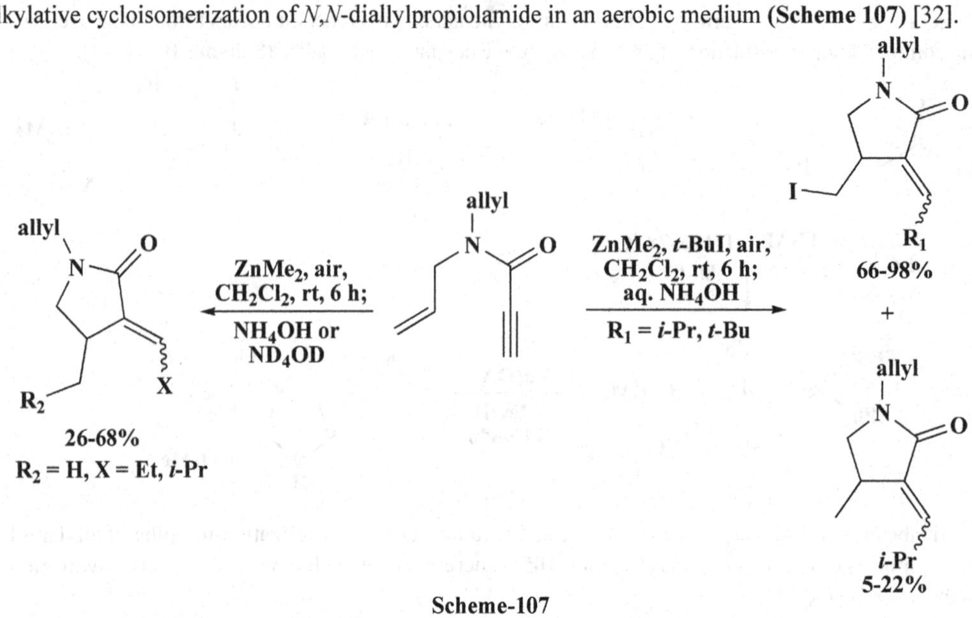

Scheme-107

Adducts are obtained as a single diastereomer in quantitative yields when functionalized nitrocyclopentane reacts with methyl acrylate due to the stereo-directing effect of the vicinal phenyl group (**Scheme 108**) [277]. Reduction of the nitro group produces spirolactam using zinc metal in ethanolic hydrogen chloride, without any migration of the exocyclic double bond. The lactam acts as a key intermediate in the synthesis of alkaloid cephalotaxine [132].

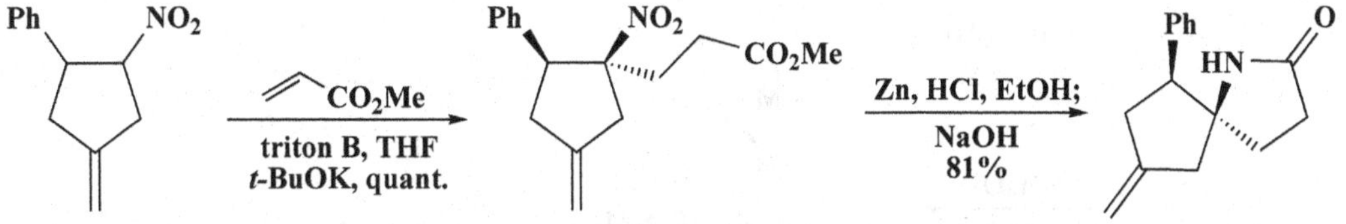

Scheme-108

Various multi-component reactions have explored and increased the diversity of compounds and reaction speeds. For macrocyclization of linear peptides, aziridine aldehyde dimers have been utilized in the multi-component Ugi reaction. Zinc acetate-catalyzed multi-component aldehyde-amine-alkyne couplings with aziridine aldehydes have been investigated for the synthesis of densely functionalized propargyl amino aziridines in good yields and with upto 20:1 diastereoselectivity (**Scheme 109**) [277].

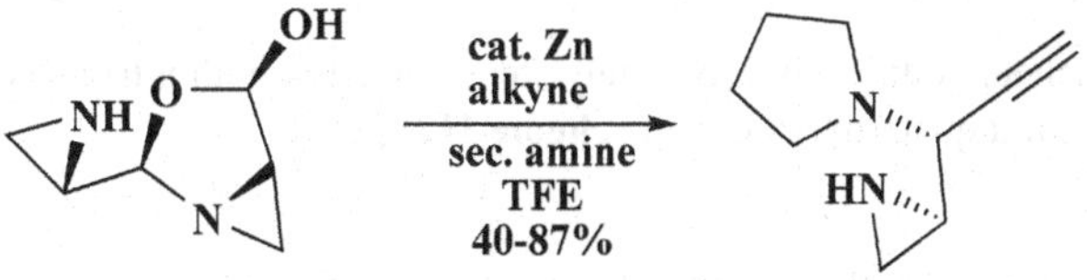

Scheme-109

Various transition metal salts catalyze pyrrole synthesis against indole formation (**Scheme 110**) [278]. Apart from rhodium(II) carboxylates, copper(II) triflate and zinc iodide have also been found to be competent catalysts for these reactions. Pyrrole is easily formed at room temperature in the presence of 5 mol% zinc iodide. The conversion rates of these reactions reveal different substrate preferences: dienylazides containing electron-deficient aryl groups provide higher yields of the desired products using rhodium perfluorobutyrate, while zinc iodide is preferable for substrates with electron-donating dienyl azide substituents. Many 2,4,5-trisubstituted and 2,5-disubstituted pyrroles can also be prepared from dienyl azides using $Rh_2(O_2CC_3F_7)_4$ or ZnI_2 as catalysts at room temperature [279].

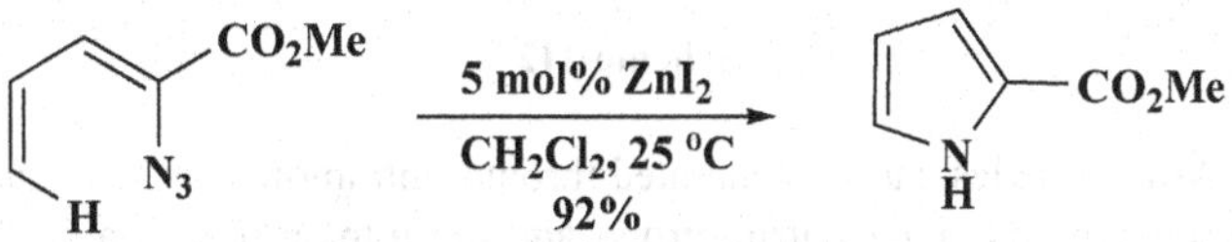

Scheme-110

Denes et al. [280] reported a highly diastereocontrolled carbometalation reaction involving a C-centred zinc-enolate for the conversion of substituted allylamines into enantioenriched 3,4-disubstituted β-prolines (**Scheme 111**) [32].

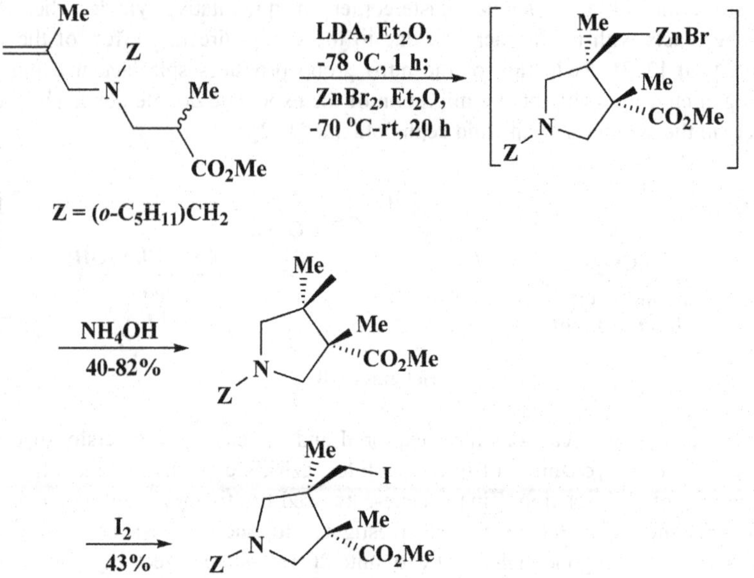

$$Z = (o\text{-}C_5H_{11})CH_2$$

Scheme-111

Intramolecular Michael addition is completely stereoselective with a tri-substituted alkene as well, leading to the synthesis of a quaternary center **(Scheme 112)** [281].

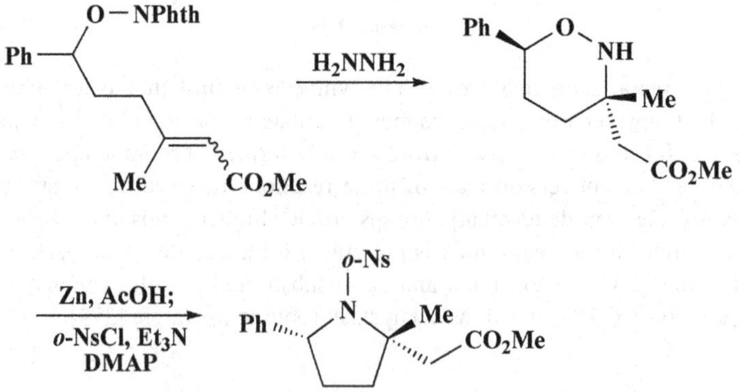

Scheme-112

Five-membered *N*-heterocyles can be generated through intramolecular cyclization by 1,4-relation between the carbonyl group and/or the nitrile group and the nitro group in a conjugate adduct. This strategy has been applied for the preparation of monocyclic compounds like pyrroline, nitrone and pyrrolidine efficiently **(Scheme 113)** [282].

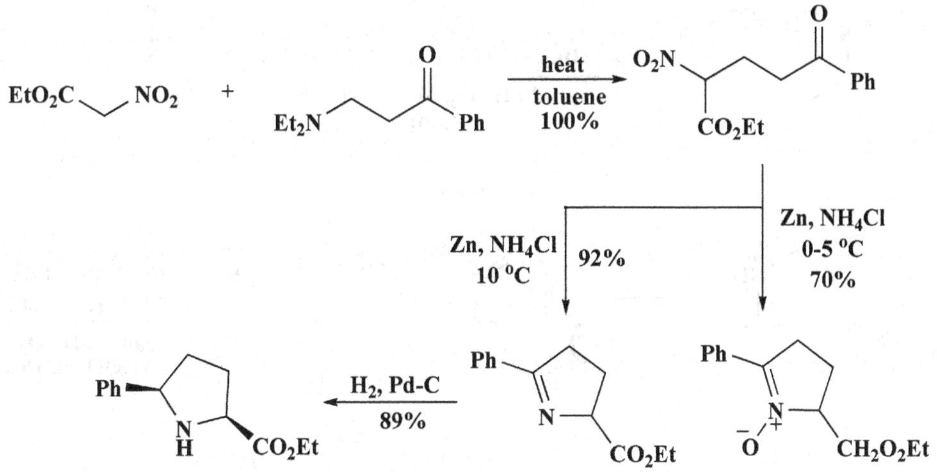

Scheme-113

Nitro compounds are reduced to amines through the formation of hydroxylamine intermediates. Adducts are synthesized through the reaction of nitroalkanes and enals or enones and are subsequently reduced using various metals in acidic conditions (**Scheme 114**) [283–284]. Cyclic nitrones can be synthesized directly from the hydroxylamine intermediate by an intramolecular reaction with the carbonyl group [132].

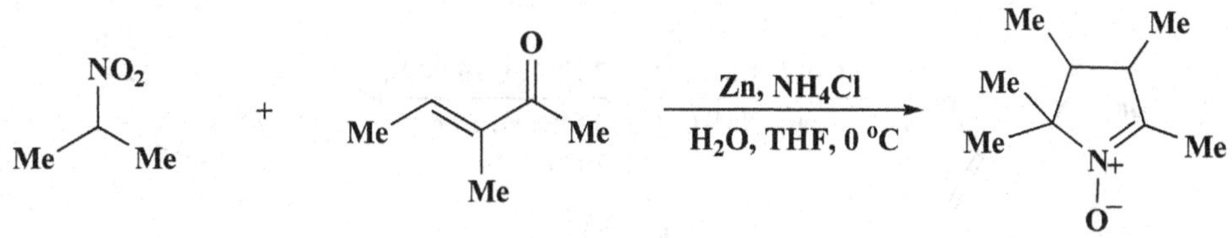

Scheme-114

The method illustrated in **Scheme 115** provides fully substituted pyrrole rings by zinc chloride-catalyzed condensation of a simple benzoin and functionalized enamine. However, this protocol is less successful in the synthesis of more complex pyrroles such as atorvastatin [271].

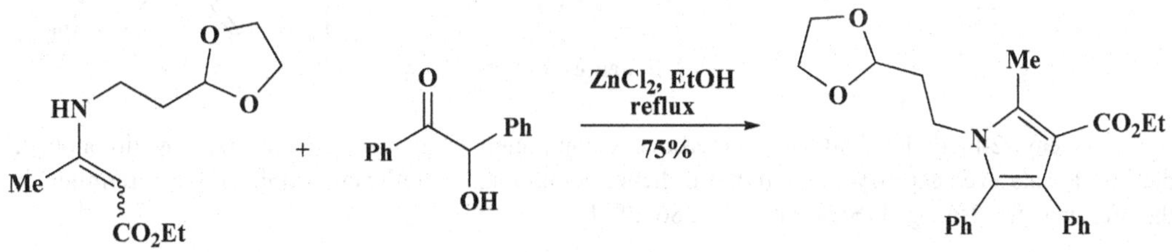

Scheme-115

Zirconium-assisted synthesis

Hunter et al. [285] transformed bis-allylamines into 3-benzyl-4-methylpyrrolidines through Zr-assisted (Negishi's reagent) cyclization both in solution (**Scheme 116**) as well as in solid-phase (**Scheme 117**). Higher overall yields were observed in the latter protocol. Under similar reaction conditions, allylamines form 3-arylpyrrolidines (**Scheme 118**), whereas allylamines produce 3-benzylidenepyrrolidines (**Scheme 119**) [32].

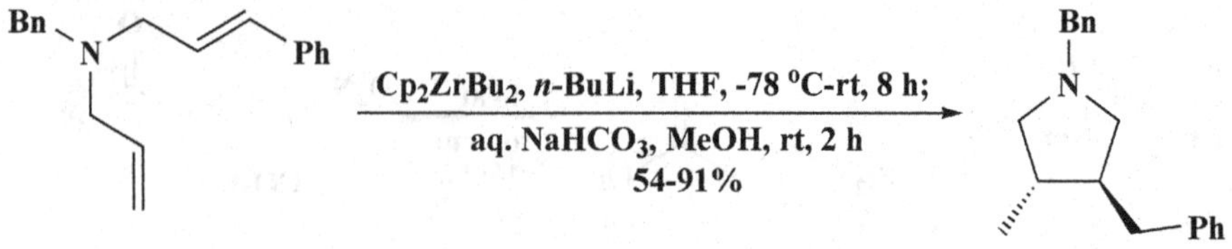

Scheme-116

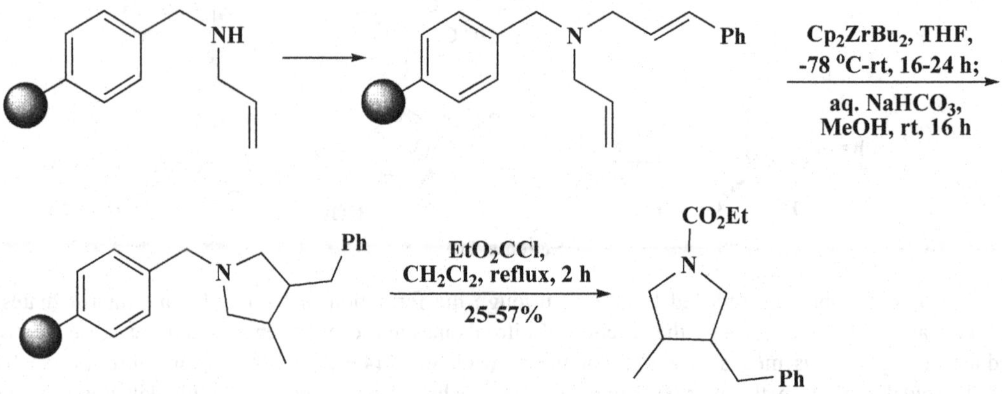

Scheme-117

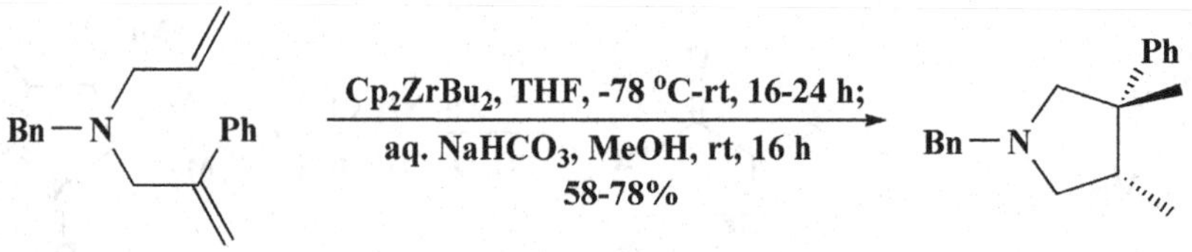

Scheme-118

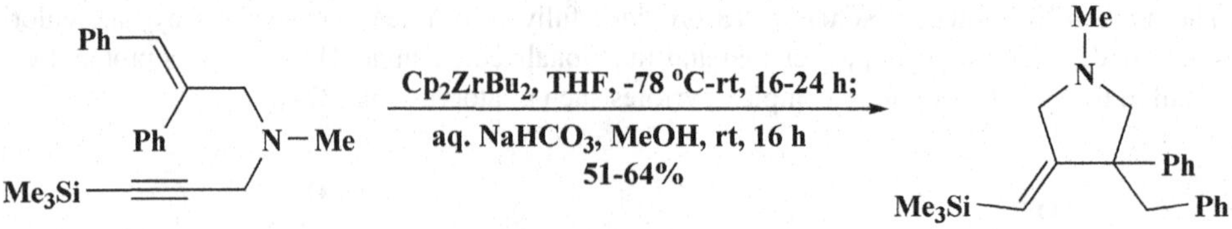

Scheme-119

1-Ethenyl-2-methylcycloalkanes (heterocyclic compounds) can also be synthesized by zirconocene dichloride-catalyzed carbocyclization of 1,6-dienes containing an allylic ether linkage at the terminus in the presence of *n*-BuMgCl (**Scheme 120**) [286–288].

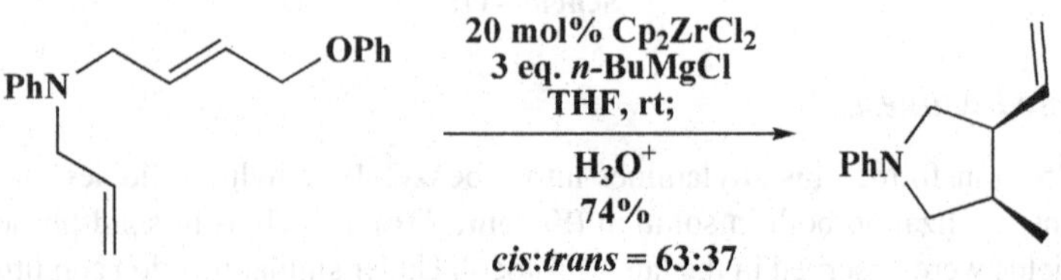

Scheme-120

Neutral group metal catalysts do not catalyze hydroamination reactions of secondary aminoalkenes as metal imido species are not formed in this case. However, it has been observed that dipyrrolylmethane zirconium complexes and constrained geometry zirconium catalyst, $Zr(NMe_2)_4$, are capable of cyclizing secondary aminoalkenes and it has been argued that a lanthanide-like mechanism, involving insertion of the olefin into a metal-amide σ-bond, is operational therein **(Scheme 121)** [41, 289–290].

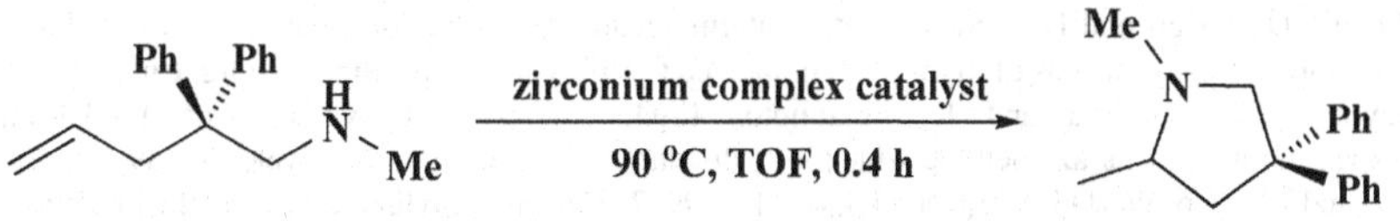

Scheme-121

References

(1)   (a) A. Fuerstner. 2003. Chemistry and biology of roseophilin and the prodigiosin alkaloids: a survey of the last 2500 years. Angew. Chem. Int. Ed. 42: 3582–3603. (b) N. Kaur. 2019. Palladium acetate and phosphine assisted synthesis of five-membered *N*-heterocycles. Synth. Commun. 49: 483–514. (c) N. Kaur. 2019. Applications of palladium dibenzylideneacetone as catalyst in the synthesis of five-membered *N*-heterocycles. Synth. Commun. 49: 1205–1230. (d) N. Kaur. 2019. Palladium acetate and phosphine assisted synthesis of five-membered *N*-heterocycles. Synth. Commun. 49: 483–514.

(2)   J.S. Russel, E.T. Pelkey and S.J.P. Yoon-Miller. 2009. Five-membered ring systems: pyrroles and benzo analogs. In Prog. Heterocycl. Chem. Elsevier Ltd. 21: 145–178.

(3)   (a) N. Kaur. 2015. Benign approaches for the microwave-assisted synthesis of five-membered 1,2-*N,N*-heterocycles. J. Heterocycl. Chem. 52: 953–973. (b) N. Kaur. 2017. Methods for metal and non-metal-catalyzed synthesis of six-membered oxygen-containing polyheterocycles. Curr. Org. Synth. 14: 531–556. (c) N. Kaur. 2017. Photochemical reactions: synthesis of six-membered *N*-heterocycles. Curr. Org. Synth. 14: 972–998. (d) N. Kaur. 2017. Ionic liquids: promising but challenging solvents for the synthesis of *N*-heterocycles. Mini Rev. Org. Chem. 14: 3–23. (e) N. Kaur. 2016. Metal catalysts for the formation of six-membered *N*-polyheterocycles. Synth. React. Inorg. Metal-Org. Nano-Metal Chem. 46: 983–1020. (f) N. Kaur. 2017. Applications of gold catalysts for the synthesis of five-membered *O*-heterocycles. Inorg. Nano-Metal Chem. 47: 163–187. (g) A. Domling. 2006. Recent developments in isocyanide-based multi-component reactions in applied chemistry. Chem. Rev. 106: 17–89.

(4)   X. Jing, X. Pan, Z. Li, X. Bi, C. Yan and H. Zhu. 2009. Organic catalytic multi-component one-pot synthesis of highly substituted pyrroles. Synth. Commun. 39: 3833–3844.

(5)   N. Kaur. 2014. Palladium-catalyzed approach to the synthesis of five-membered *O*-heterocycles. Inorg. Chem. Commun. 49: 86–119.

(6)   N. Kaur. 2015. Metal catalysts: applications in higher-membered *N*-heterocycles synthesis. J. Iran. Chem. Soc. 12: 9–45.

(7)   N. Kaur. 2015. Microwave-assisted synthesis: fused five-membered *N*-heterocycles. Synth. Commun. 45: 789–823.

(8)   N. Kaur. 2015. Environmentally benign synthesis of five-membered 1,3-*N,N*-heterocycles by microwave irradiation. Synth. Commun. 45: 909–943.

(9)   E.J. Kang and E. Lee. 2005. Total synthesis of oxacyclic macrodiolide natural products. Chem. Rev. 105: 4348–4378.

(10) N. Kaur. 2015. Role of microwaves in the synthesis of fused five-membered heterocycles with three *N*-heteroatoms. Synth. Commun. 45: 403–431.

(11) M. Saleem, H.J. Kim, M.S. Ali and Y.S. Lee. 2005. An update on bioactive plant lignans. Nat. Prod. Rep. 22: 696–716.

(12) A. Bermejo, B. Figadere, M.-C. Zafra-Polo, I. Barrachina, E. Estornell and D. Cortes. 2005. Acetogenins from *Annonaceae*: recent progress in isolation, synthesis and mechanisms of action. Nat. Prod. Rep. 22: 269–309.

(13) N. Kaur. 2015. Palladium catalysts: synthesis of five-membered *N*-heterocycles fused with other heterocycles. Catal. Rev. 57: 1–78.

(14) J.W. Daly, T.F. Spande and H.M. Garraffo. 2005. Alkaloids from amphibian skin: a tabulation of over eight-hundred compounds. J. Nat. Prod. 68: 1556–1575.

(15) A.E. Hackling and H. Stark. 2002. Dopamine D3 receptor ligands with antagonist properties. Chem. Bio. Chem. 3: 946–961.

(16) J.R. Lewis. 2001. *Amaryllidaceae*, *Sceletium*, imidazole, oxazole, thiazole, peptide and miscellaneous alkaloids. Nat. Prod. Rep. 18: 95–128.

(17)   P. Restorp, A. Fischer and P. Somfai. 2006. Stereoselective synthesis of functionalized pyrrolidines *via* a [3+2]-annulation of *N*-Ts-α-amino aldehydes and 1,3-bis(silyl)propenes. J. Am. Chem. Soc. 128: 12646–12647.

(18)   M.C. Marcotullio, V. Campagna, S. Sternativo, F. Costantino and M. Curini. 2006. A new, simple synthesis of *N*-tosyl pyrrolidines and piperidines. Synthesis 16: 2760–2766.

(19)   L. Knorr. 1884. Synthese von pyrrolderivaten. Ber. 17: 1635–1642.

(20)   M.W. Roomi and S.F. MacDonald. 1970. The Hantzsch pyrrole synthesis. Can. J. Chem. 48: 1689–1697.

(21)   T. Uchida. 1978. Synthetic approaches to fused heteroaromatic compounds by the condensation reactions of functional pyrroles. J. Heterocycl. Chem. 15: 241–248.

(22)   P. DeShong, D.A. Kell and D.R. Sidler. 1985. Intermolecular and intramolecular azomethine ylide [3+2]-dipolar cycloadditions for the synthesis of highly functionalized pyrroles and pyrrolidines. J. Org. Chem. 50: 2309–2315.

(23)   M. Toyota, Y. Nishikawa and K. Fukumoto. 1994. An efficient synthesis of 1,2,3,4-tetrasubstituted pyrroles *via* intramolecular azomethine ylide [3+2]-dipolar cycloaddition. Heterocycles 39: 39–42.

(24)   A.R. Katritzky, J. Yao, W. Bao, M. Qi and P.J. Steel. 1999. 2-Benzotriazolylaziridines and their reactions with diethyl acetylenedicarboxylate. J. Org. Chem. 64: 346–350.

(25)   A.S. Konev, M.S. Novikov and A.F. Khlebnikov. 2005. The first example of the generation of azomethine ylides from a fluorocarbene: 1,3-cyclization and 1,3-dipolar cycloaddition. Tetrahedron Lett. 46: 8337–8340.

(26)   G. Minetto, L.F. Raveglia and M. Taddei. 2004. Microwave-assisted Paal-Knorr reaction. A rapid approach to substituted pyrroles and furans. Org. Lett. 6: 389–392.

(27)   F. Freeman, D.S.H.L. Kim and E. Rodriguez. 1992. Preparation of 1,4-diketones and their reactions with bis(trialkyltin) or bis(triphenyltin) sulfide-boron trichloride. J. Org. Chem. 57: 1722–1727.

(28)   J.S. Yadav, B.V.S. Reddy, B. Eeshwaraiah and M.K. Gupta. 2004. Bi(OTf)$_3$/[bmim]BF$_4$ as novel and reusable catalytic system for the synthesis of furan, pyrrole and thiophene derivatives. Tetrahedron Lett. 45: 5873–5876.

(29)   B.K. Banik, S. Samajdar and I. Banik. 2004. Simple synthesis of substituted pyrroles. J. Org. Chem. 69: 213–216.

(30)   M.A.P. Martins, C.P. Frizzo, D.N. Moreira, N. Zanatta and H.G. Bonacorso. 2008. Ionic liquids in heterocyclic synthesis. Chem. Rev. 108: 2015–2050.

(31)   R. Hayashi and G.R. Cook. 2008. Bi(OTf)$_3$-catalyzed 5-*exo-trig* cyclization *via* halide activation. Tetrahedron Lett. 49: 3888–3890.

(32)   S. Nag and S. Batra. 2011. Applications of allylamines for the syntheses of aza-heterocycles. Tetrahedron 67: 8959–9061.

(33)   B.K. Banik, I. Banik, M. Renteria and S.K. Dasgupta. 2005. A straightforward highly efficient Paal-Knorr synthesis of pyrroles. Tetrahedron Lett. 46: 2643–2645.

(34)   J.A.R. Salvador, R.M.A. Pinto and S.M. Silvestre. 2009. Recent advances of bismuth(III) salts in organic chemistry: application to the synthesis of heterocycles of pharmaceutical interest. Curr. Org. Synth. 6: 426–470.

(35)   S. Harder. 2004. The chemistry of CaII and YbII: astoundingly similar but not equal! Angew. Chem. Int. Ed. 43: 2714–2718.

(36)   F. Buch, J. Brettar and S. Harder. 2006. Hydrosilylation of alkenes with early main-group metal catalysts. Angew. Chem. Int. Ed. 45: 2741–2745.

(37)   M.R. Crimmin, I.J. Casely and M.S. Hill. 2005. Calcium-mediated intramolecular hydroamination catalysis. J. Am. Chem. Soc. 127: 2042–2043.

(38)   F. Buch and S. Harder. 2008. A study on chiral organocalcium complexes: attempts in enantioselective catalytic hydrosilylation and intramolecular hydroamination of alkenes. Z. Naturforsch. 63b: 169–177.

(39)   S. Datta, P.W. Roesky and S. Blechert. 2007. Aminotroponate and aminotroponiminate calcium amides as catalysts for the hydroamination/cyclization catalysis. Organometallics 26: 4392–4394.

(40)   S. Datta, M.T. Gamer and P.W. Roesky. 2008. Aminotroponiminate complexes of the heavy alkaline earth and the divalent lanthanide metals as catalysts for the hydroamination/cyclization reaction. Organometallics 27: 1207–1213.

(41)   T.E. Muller, K.C. Hultzsch, M. Yus, F. Foubelo and M. Tada. 2008. Hydroamination: direct addition of amines to alkenes and alkynes. Chem. Rev. 108: 3795–3892.

(42)   P. Nemes, B. Balázs, G. Tóth and P. Scheiber. 1999. Synthesis of fused heterocycles from β-enaminonitrile and carbonyl compounds. Synlett 2: 222–224.

(43)   P. Nemes, Z. Vincze, B. Balázs, G. Tóth and P. Scheiber. 2003. Novel microwave-assisted [3+3]-cyclocondensations with 1-cyanomethylene tetrahydroisoquinoline. Synlett 2: 250–252.

(44)   V. Nair, K. Mohanan, T.D. Suja and E. Suresh. 2006. Stereoselective synthesis of 3,4-*trans*-disubstituted pyrrolidines and cyclopentanes *via* intramolecular radical cyclizations mediated by CAN. Tetrahedron Lett. 47: 2803–2806.

(45)   S. Watanuki and M. Mori. 1995. Enyne metathesis using chromium carbene complexes. Synthesis of heterocycles from enynes using chromium carbene complex. Organometallics 14: 5054–5061.

(46)   S. Watanuki, N. Ochifuji and M. Mori. 1994. Chromium-catalyzed intramolecular enyne metathesis. Organometallics 13: 4129–4130.

(47)   S. Watanuki, N. Ochifuji and M. Mori. 1995. Chromium-catalyzed intramolecular enyne metathesis. Organometallics 14: 5062–5067.

(48)   S.T. Diver and A.J. Giessert. 2004. Enyne metathesis (enyne bond reorganization). Chem. Rev. 104: 1317–1382.

(49)   T.J. Katz and T.M. Sivavec. 1985. Metal-catalyzed rearrangement of alkene-alkynes and the stereochemistry of metallacyclobutene ring-opening. J. Am. Chem. Soc. 107: 737–738.

(50) P.F. Korkowski, T.R. Hoye and D.B. Rydberg. 1988. Fischer carbene-mediated conversions of enynes to bi- and tricyclic cyclopropane containing carbon skeletons. J. Am. Chem. Soc. 110: 2676–2678.

(51) T. Niwa, H. Yorimitsu and K. Oshima. 2007. Palladium-catalyzed 2-pyridylmethyl transfer from 2-(2-pyridyl)ethanol derivatives to organic halides by chelation-assisted cleavage of unstrained $C(sp^3)$-$C(sp^3)$ bonds. Angew. Chem. Int. Ed. 46: 2643–2645.

(52) Y. Koganemaru, M. Kitamura and K. Narasaka. 2002. Synthesis of dihydropyrrole derivatives by copper-catalyzed cyclization of γ,δ-unsaturated ketone O-methoxycarbonyloximes. Chem. Lett. 31: 784–785.

(53) K. Tanaka, M. Kitamura and K. Narasaka. 2005. Synthesis of α-carbolines by copper-catalyzed radical cyclization of β-(3-indolyl) ketone O-pentafluorobenzoyloximes. Bull. Chem. Soc. Jpn. 78: 1659–1664.

(54) M. Wasa, K.M. Engle and J.-Q. Yu. 2010. Pd(II)-catalyzed olefination of sp^3 C-H bonds. J. Am. Chem. Soc. 132: 3680–3681.

(55) N. Jeong, S.D. Seo and J.Y. Shin. 2000. One-pot preparation of bicyclopentenones from propargyl malonates (and propargylsulfonamides) and allylic acetates by a tandem action of catalysts. J. Am. Chem. Soc. 122: 10220–10221.

(56) H. Lindsay, B. Johnson, E. Marrero and W. Turley. 2007. Controlling diastereoselectivity in the tandem microwave-assisted aza-Cope rearrangement-Mannich cyclization. Synlett 6: 893–896.

(57) S. Araki, T. Tanaka, S. Toumatsu and T. Hirashita. 2003. A new synthesis of pyrroles by the condensation of cyclopropenes and nitriles mediated by gallium(III) and indium(III) salts. Org. Biomol. Chem. 1: 4025–4029.

(58) M. Rubin and P.G. Ryabchuk. 2012. Rearrangements of cyclopropenes into five-membered aromatic heterocycles: mechanistic aspect. Chem. Heterocycl. Compd. 48: 126–138.

(59) R.H. Feling, G.O. Buchanan, T.J. Mincer, C.A. Kauffman, P.R. Jensen and W. Fenical. 2003. Salinosporamide A: a highly cytotoxic proteasome inhibitor from a novel microbial source, a marine bacterium of the new genus *Salinospora*. Angew. Chem. Int. Ed. 42: 355–357.

(60) L.R. Reddy, P. Saravanan and E.J. Corey. 2004. A simple stereocontrolled synthesis of salinosporamide A. J. Am. Chem. Soc. 126: 6230–6231.

(61) L.R. Reddy, J.-F. Fournier, B.V.S. Reddy and E.J. Corey. 2005. New synthetic route for the enantioselective total synthesis of salinosporamide A and biologically active analogues. Org. Lett. 7: 2699–2701.

(62) A. Endo and S.J. Danishefsky. 2005. Total synthesis of salinosporamide A. J. Am. Chem. Soc. 127: 8298–8299.

(63) T. Ling, V.R. Macherla, R.R. Manam, K.A. McArthur and B.C.M. Potts. 2007. Enantioselective total synthesis of (–)-salinosporamide A (NPI-0052). Org. Lett. 9: 2289–2292.

(64) N.P. Mulholland, G. Pattenden and I.A.S. Walters. 2006. A concise total synthesis of salinosporamide A. Org. Biomol. Chem. 4: 2845–2846.

(65) G. Ma, H. Nguyen and D. Romo. 2007. Concise total synthesis of (±)-salinosporamide A, (±)-cinnabaramide A, and derivatives *via* a bis-cyclization process: implications for a biosynthetic pathway? Org. Lett. 9: 2143–2146.

(66) V. Caubert, J. Massé, P. Retailleau and N. Langlois. 2007. Stereoselective formal synthesis of the potent proteasome inhibitor: salinosporamide A. Tetrahedron Lett. 48: 381–384.

(67) K. Takahashi, M. Midori, K. Kawano, J. Ishihara and S. Hatakeyama. 2008. Entry to heterocycles based on indium-catalyzed Conia-ene reactions: asymmetric synthesis of (–)-salinosporamide A. Angew. Chem. Int. Ed. 47: 6244–6246.

(68) J.J. Kennedy-Smith, S.T. Staben and F.D. Toste. 2004. Gold(I)-catalyzed Conia-ene reaction of β-ketoesters with alkynes. J. Am. Chem. Soc. 126: 4526–4527.

(69) Q. Gao, B.-F. Zheng, J.H. Li and D. Yang. 2005. Ni(II)-catalyzed Conia-ene reaction of 1,3-dicarbonyl compounds with alkynes. Org. Lett. 7: 2185–2188.

(70) H. Tsuji, K. Yamagata, Y. Itoh, K. Endo, M. Nakamura and E. Nakamura. 2007. Indium-catalyzed cycloisomerization of ω-alkynyl-β-ketoesters into six- to fifteen-membered rings. Angew. Chem. Int. Ed. 46: 8060–8062.

(71) K. Endo, T. Hatakeyama, M. Nakamura and E. Nakamura. 2007. Indium-catalyzed 2-alkenylation of 1,3-dicarbonyl compounds with un-activated alkynes. J. Am. Chem. Soc. 129: 5264–5271.

(72) S. Hatakeyama. 2009. Indium-catalyzed Conia-ene reaction for alkaloid synthesis. Pure Appl. Chem. 81: 217–226.

(73) M. Lin, L. Hao, R.-D. Ma and Z.-P. Zhan. 2010. A novel indium-catalyzed three-component reaction: general and efficient one-pot synthesis of substituted pyrroles. Synlett 15: 2345–2351.

(74) L.A. Polindara-García and L.D. Miranda. 2012. Two-step synthesis of 2,3-dihydropyrroles *via* a formal 5-*endo*-cycloisomerization of Ugi 4-CR/propargyl adducts. Org. Lett. 14: 5408–5411.

(75) (a) A. Dömling and I. Ugi. 2000. Multi-component reactions with isocyanides. Angew. Chem. Int. Ed. 39: 3168–3210. (b) T. Lu, Z. Lu, Z.-X. Ma, Y. Zhang and R.P. Hsung. 2013. Allenamides: a powerful and versatile building block in organic synthesis. Chem. Rev. 113: 4862–4904.

(76) N. Hayashi, I. Shibata and A. Baba. 2004. Triethylsilane-indium(III) chloride system as a radical reagent. Org. Lett. 6: 4981–4983.

(77) L. Benati, G. Bencivenni, R. Leardini, D. Nanni, M. Minozzi, P. Spagnolo, R. Scialpi and G. Zanardi. 2006. Reaction of azides with dichloroindium hydride: very mild production of amines and pyrrolidin-2-imines through possible indium-aminyl radicals. Org. Lett. 8: 2499–2502.

(78) B. Das, N. Bhunia and M. Lingaiah. 2011. A simple and efficient metal-free synthesis of tetra-substituted pyrroles by iodine-catalyzed four-component coupling reaction of aldehydes, amines, dialkyl acetylenedicarboxylates, and nitromethane. Synthesis 21: 3471–3474.

(79)   K.-G. Ji, H.-T. Zhu, F. Yang, X.-Z. Shu, S.-C. Zhao, X.-Y. Liu, A. Shaukat and Y.-M. Liang. 2010. A novel iodine-promoted tandem cyclization: an efficient synthesis of substituted 3,4-diiodoheterocyclic compounds. Chem. Eur. J. 16: 6151–6154.

(80)   P.T. Parvatkar, P.S. Parameswaran and S.G. Tilve. 2012. Recent developments in the synthesis of five- and six-membered heterocycles using molecular iodine. Chem. Eur. J. 18: 5460–5489.

(81)   F. Diaba, G. Puigbó and J. Bonjoch. 2007. Synthesis of enantiopure 1-azaspiro[4.5]decanes by iodoaminocyclization of allylaminocyclohexanes. Eur. J. Org. Chem. 18: 3038–3044.

(82)   Q.-F. Yu, Y.-H. Zhang, Q. Yin, B.-X. Tang, R.-Y. Tang, P. Zhong and J.-H. Li. 2008. Electrophilic *ipso*-iodocyclization of N-(4-methylphenyl)propiolamides: selective synthesis of 8-methyleneazaspiro[4,5]trienes. J. Org. Chem. 73: 3658–3661.

(83)   D.J. Wardrop and M.S. Burge. 2005. Nitrenium ion azaspirocyclization-spirodienone cleavage: a new synthetic strategy for the stereocontrolled preparation of highly substituted lactams and N-hydroxy lactams. J. Org. Chem. 70: 10271–10284.

(84)   T. Dohi, A. Maruyama, Y. Minamitsuji, N. Takenaga and Y. Kita. 2007. First hypervalent iodine(III)-catalyzed C-N bond forming reaction: catalytic spirocyclization of amides to N-fused spirolactams. Chem Commun. 12: 1224–1226.

(85)   W. Mai, S.A. Green, D.K. Bates and S. Fang. 2010. Synthesis of 3-amino-2,2-dimethyl-8-thia-1-azaspiro[4.5]decane. Synth. Commun. 40: 2571–2577.

(86)   M. Yoshida, M. Al-Amin and K. Shishido. 2009. Synthesis of substituted 3-iodopyrroles by electrophilic cyclization of propargylic aziridines. Tetrahedron Lett. 50: 6268–6270.

(87)   T. Pradhan and A. Hassner. 2007. A facile synthesis of (n+3) and (n+4) ring enlarged lactones as well as of spiroketolactones from n-membered cycloalkanones. Synthesis 21: 3361–3370.

(88)   C.G. Francisco, A.J. Herrera, A. Martin, I. Perez-Martin and E. Suarez. 2007. Intramolecular 1,5-hydrogen atom transfer reaction promoted by phosphoramidyl and carbamoyl radicals: synthesis of 2-amino-C-glycosides. Tetrahedron Lett. 48: 6384–6388.

(89)   A. Martín, I. Pérez-Martín and E. Suárez. 2005. Intramolecular hydrogen abstraction promoted by amidyl radicals. Evidence for electronic factors in the nucleophilic cyclization of ambident amides to oxocarbenium ions. Org. Lett. 7: 2027–2030.

(90)   V.V. Zhdankin. 2009. Hypervalent iodine(III) reagents in organic synthesis. ARKIVOC (i): 1–62.

(91)   S.U. Tekale, S.S. Kauthale, S.A. Dake, S.R. Sarda and R.P. Pawar. 2012. Molecular iodine: an efficient and versatile reagent for organic synthesis. Curr. Org. Chem. 16: 1485–1501.

(92)   H. Veisi. 2011. Molecular iodine: recent application in heterocyclic synthesis. Curr. Org. Chem. 15: 2438–2468.

(93)   V.V. Zhdankin. 2011. Organoiodine(V) reagents in organic synthesis. J. Org. Chem. 76: 1185–1197.

(94)   H. Togo and S. Iida. 2006. Synthetic use of molecular iodine for organic synthesis. Synlett 14: 2159–2175.

(95)   S. Das, R. Borah, R. Devi and A. Thakur. 2008. Molecular iodine in protection and de-protection chemistry. Synlett 18: 2741–2762.

(96)   B. Alcaide, P. Almendros, G. Cabrero, R. Callejo, M.P. Ruiz, M. Arno and L.R. Domingo. 2010. Ring-expansion versus cyclization in 4-oxoazetidine-2-carbaldehydes catalyzed by molecular iodine: experimental and theoretical study in concert. Adv. Synth. Catal. 352: 1688–1700.

(97)   C.-C. Chen, S.-C. Yang and M.-J. Wu. 2011. Iodine-mediated cascade cyclization of enediynes to iodinated benzo[a] carbazoles. J. Org. Chem. 76: 10269–10274.

(98)   W.-C. Lee, H.-C. Shen, W.-P. Hu, W.-S. Lo, C. Murali, J.K. Vandavasi and J.-J. Wang. 2012. Iodine-catalyzed, stereo- and regioselective synthesis of 4-arylidene-4H-benzo[d][1,3]oxazines and their applications for the synthesis of quinazoline 3-oxides. Adv. Synth. Catal. 354: 2218–2228.

(99)   D. Bandyopadhyay and B.K. Banik. 2010. Microwave-induced stereocontrol of β-lactam formation with N-benzylidene-9,10-dihydrophenanthren-3-amine *via* Staudinger cycloaddition. Helv. Chim. Acta 93: 298–301.

(100)   D. Bandyopadhyay, M.A. Yanez and B.K. Banik. 2011. Microwave-induced stereoselectivity of β-lactam formation, effects of solvents. Heterocycl. Lett. 1: 65–67.

(101)   R. Andoh-Baidoo, R. Danso, S. Mukherjee, D. Bandyopadhyay and B.K. Banik. 2011. Microwave-induced N-bromosuccinimide-mediated novel synthesis of pyrroles *via* Paal-Knorr reaction. Heterocycl. Lett. 1: 107–109.

(102)   D. Bandyopadhyay, A. Banik, S. Bhatta and B.K. Banik. 2009. Microwave-assisted ruthenium trichloride-catalyzed synthesis of pyrrole fused with indole system in water. Heterocycl. Commun. 15: 121–122.

(103)   D. Abrego, D. Bandyopadhyay and B.K. Banik. 2011. Microwave-induced indium-catalyzed synthesis of pyrrole fused with indolinone in water. Heterocycl. Lett. 1: 94–95.

(104)   S. Rivera, D. Bandyopadhyay and B.K. Banik. 2009. Facile synthesis of N-substituted pyrroles *via* microwave-induced bismuth nitrate-catalyzed reaction. Tetrahedron Lett. 50: 5445–5448.

(105)   D. Bandyopadhyay, G. Rivera, I. Salinas, H. Aguilar and B.K. Banik. 2010. Remarkable iodine-catalyzed synthesis of novel pyrrole bearing N-polyaromatic β-lactams. Molecules 15: 1082–1088.

(106)   D. Bandyopadhyay, S. Mukherjee and B.K. Banik. 2010. An expeditious synthesis of N-substituted pyrroles *via* microwave-induced iodine-catalyzed reactions under solvent-less conditions. Molecules 15: 2520–2525.

(107)   F.F. Becker and B.K. Banik. 1998. Polycyclic aromatic compounds as anti-cancer agents: synthesis and biological evaluation of some chrysene derivatives. Bioorg. Med. Chem. Lett. 8: 2877–2880.

(108) F.F. Becker, C. Mukhopadhyay, L. Hackfeld, I. Banik and B.K. Banik. 2000. Polycyclic aromatic compounds as anti-cancer agents: synthesis and biological evaluation of dibenzofluorene derivatives. Bioorg. Med. Chem. 8: 2693–2699.

(109) B.K. Banik and F.F. Becker. 2001. Polycyclic aromatic compounds as anti-cancer agents: structure-activity relationships of chrysene and pyrene derivatives. Bioorg. Med. Chem. 9: 593–605.

(110) D. Bandyopadhyay, J.C. Granados, J.D. Short and B.K. Banik. 2012. Polycyclic aromatic compounds as anti-cancer agents: evaluation of synthesis and *in vitro* cytotoxicity. Oncol. Lett. 3: 45–49.

(111) B.K. Banik, F.F. Becker and I. Banik. 2004. Synthesis of anti-cancer β-lactams: mechanism of action. Bioorg. Med. Chem. 12: 2523–2528.

(112) I. Banik, F.F. Becker and B.K. Banik. 2003. Stereoselective synthesis of β-lactams with polyaromatic imines: entry to new and novel anti-cancer agents. J. Med. Chem. 46: 12–15.

(113) B.K. Banik, I. Banik and F.F. Becker. 2005. Stereocontrolled synthesis of anti-cancer β-lactams *via* the Staudinger reaction. Bioorg. Med. Chem. 13: 3611–3622.

(114) D. Bandyopadhyay, J. Cruz, R.N. Yadav and B.K. Banik. 2012. An expeditious iodine-catalyzed synthesis of 3-pyrrole-substituted 2-azetidinones. Molecules 17: 11570–11584.

(115) M. Ciufolini, N. Braun, S. Canesi, M. Ousmer, J. Chang and D. Chai. 2007. Oxidative amidation of phenols through the use of hypervalent iodine reagents: development and applications. Synthesis 24: 3759–3772.

(116) E. Miyazawa, T. Sakamoto and Y. Kikugawa. 2003. Synthesis of spirodienones by intramolecular *ipso*-cyclization of N-methoxy-(4-halogenophenyl)amides using [hydroxy(tosyloxy)iodo]benzene in trifluoroethanol. J. Org. Chem. 68: 5429–5432.

(117) Y. Misu and H. Togo. 2003. Novel preparation of 2,1-benzothiazine derivatives from sulfonamides with [hydroxy(tosyloxy)iodo]arenes. Org. Biomol. Chem. 1: 1342–1346.

(118) S.J. Walker and D.J. Hart. 2007. Synthesis of (–)-lapatin B. Tetrahedron Lett. 48: 6214–6216.

(119) K.D. Hesp and M. Stradiotto. 2009. Intramolecular hydroamination of un-activated alkenes with secondary alkyl- and arylamines employing [Ir(COD)Cl]$_2$ as a catalyst precursor. Org. Lett. 11: 1449–1452.

(120) I.K. Mangion, I.K. Nwamba, M. Shevlin and M.A. Huffman. 2009. Iridium-catalyzed X-H insertions of sulfoxonium ylides. Org. Lett. 11: 3566–3569.

(121) K.-I. Fujita, T. Fujii and R. Yamaguchi. 2004. Cp*Ir complex-catalyzed N-heterocyclization of primary amines with diols: a new catalytic system for environmentally benign synthesis of cyclic amines. Org. Lett. 6: 3525–3528.

(122) P. Eilbracht, L. Bärfacker, C. Buss, C. Hollmann, B.E. Kitsos-Rzychon, C.L. Kranemann, T. Rische, R. Roggenbuck and A. Schmidt. 1999. Tandem reaction sequences under hydroformylation conditions: new synthetic applications of transition metal catalysis. Chem. Rev. 99: 3329–3366.

(123) D.E. Fogg and E.N. dos Santos. 2004. Tandem catalysis: a taxonomy and illustrative review. Coord. Chem. Rev. 248: 2365–2379.

(124) P. Eilbracht and A.M. Schmidt. 2006. Synthetic applications of tandem reaction sequences involving hydroformylation. Top. Organomet. Chem. 18: 65–95.

(125) T.J.J. Muller. 2006. Sequentially palladium-catalyzed processes. Top. Organomet. Chem. 19: 149–205.

(126) A. Ajamian and J.L. Gleason. 2004. Two birds with one metallic stone: single-pot catalysis of fundamentally different transformations. Angew. Chem. Int. Ed. 43: 3754–3760.

(127) L.D. Field, B.A. Messerle and S.L. Wren. 2003. One-pot tandem hydroamination/hydrosilation catalyzed by cationic iridium(I) complexes. Organometallics 22: 4393–4395.

(128) A. Heutling, F. Pohlki, I. Bytschkov and S. Doye. 2005. Hydroamination/hydrosilylation sequence catalyzed by titanium complexes. Angew. Chem. Int. Ed. 44: 2951–2954.

(129) P.R. Burkholder. 1959. Anti-biotics: the exploitation of microbial antagonisms is having a challenging impact on medicine and society. Science 129: 1457–1465.

(130) R.B. Woodward. 1966. Recent advances in the chemistry of natural products. Science 153: 487–493.

(131) J.A. Kelly, P.C. Moews, J.R. Knox, J.-M. Frere and J.-M. Ghuysen. 1982. Penicillin target enzyme and the anti-biotic binding site. Science 218: 479–481.

(132) K. Dachs and E. Schwartz. 1962. Pyrrolidon, capryllactam und laurinlactam als neue grundstoffe für polyamidfasern. Angew. Chem. 74: 540–545.

(133) R. Greco, N. Lanzetta, G. Maglio, M. Malinconico, E. Martuscelli, R. Palumbo, G. Ragosta and G. Scarinzi. 1986. Rubber modification of polyamide 6 during caprolactam polymerization: influence of composition and functionalization degree of rubber. Polymer 27: 299–308.

(134) P.A. Magriotis. 2001. Recent progress in the enantioselective synthesis of β-lactams: development of the first catalytic approaches. Angew. Chem. Int. Ed. 40: 4377–4379.

(135) N.S. Chowdari, J.T. Suri and C.F. Barbas. 2004. Asymmetric synthesis of quaternary α- and β-amino acids and β-lactams *via* proline-catalyzed Mannich reactions with branched aldehyde donors. Org. Lett. 6: 2507–2510.

(136) J.D. White and Y. Choi. 2000. Catalyzed asymmetric Diels-Alder reaction of benzoquinone. Total synthesis of (–)-ibogamine. Org. Lett. 2: 2373–2376.

(137) G.R. Krow. 1981. Nitrogen insertion reactions of bridged bicyclic ketones. Regioselective lactam formation. Tetrahedron 37: 1283–1307.

(138) B.T. Smith, J.A. Wendt and J. Aubé. 2002. First asymmetric total synthesis of (+)-sparteine. Org. Lett. 4: 2577–2579.

(139) T. Gajda and A. Zwierzak. 1981. Phase-transfer catalyzed *N*-alkylation of carboxamides and sulfonamides. Synthesis 12: 1005–1008.

(140) K. Komeyama, T. Morimoto and K. Takaki. 2006. A simple and efficient iron-catalyzed intramolecular hydroamination of un-activated olefins. Angew. Chem. Int. Ed. 45: 2938–2941.

(141) T. Taniguchi, N. Goto, A. Nishibata and H. Ishibashi. 2010. Iron-catalyzed redox radical cyclizations of 1,6-dienes and enynes. Org. Lett. 12: 112–115.

(142) S. Singh, O.V. Singh and H. Han. 2007. Asymmetric synthesis of (+)-iso-6-cassine *via* stereoselective intramolecular amidomercuration. Tetrahedron Lett. 48: 8270–8273.

(143) N. Azizi, A. Khajeh-Amiri, H. Ghafuri, M. Bolourtchian and M. Saidi. 2009. Iron-catalyzed inexpensive and practical synthesis of *N*-substituted pyrroles in water. Synlett 14: 2245–2248.

(144) S. Maiti, S. Biswas and U. Jana. 2010. Iron(III)-catalyzed four-component coupling reaction of 1,3-dicarbonyl compounds, amines, aldehydes, and nitroalkanes: a simple and direct synthesis of functionalized pyrroles. J. Org. Chem. 75: 1674–1683.

(145) B. Das, G. Reddy, P. Balasubramanyam and B. Veeranjaneyulu. 2010. An efficient new method for the synthesis of poly-substituted pyrroles. Synthesis 10: 1625–1628.

(146) W. Imhof, E. Anders, A. Göbel and H. Görls. 2003. A theoretical study on the complete catalytic cycle of the hetero-Pauson-Khand-type [2+2+1]-cycloaddition reaction of ketimines, carbon monoxide and ethylene catalyzed by iron carbonyl complexes. Chem. Eur. J. 9: 1166–1181.

(147) C. Bolm, J. Legros, J.L. Paih and L. Zani. 2004. Iron-catalyzed reactions in organic synthesis. Chem. Rev. 104: 6217–6254.

(148) M.S. Sigman, C.E. Kerr and B.E. Eaton. 1993. Catalytic iron-mediated carbon-oxygen and carbon-carbon bond formation in [4+1] assembly of alkylidenebutenolides. J. Am. Chem. Soc. 115: 7545–7546.

(149) M.S. Sigman, B.E. Eaton, J.D. Heise and C.P. Kubiak. 1996. Low-temperature study of the iron-mediated [4+1]-cyclization of allenyl ketones with carbon monoxide. Organometallics 15: 2829–2832.

(150) M.S. Sigman and B.E. Eaton. 1994. The first iron-mediated catalytic carbon-nitrogen bond formation: [4+1]-cycloaddition of allenyl imines and carbon monoxide. J. Org. Chem. 59: 7488–7491.

(151) L. Li, M.-N. Zhao, Z.-H. Ren, J. Li and Z.-H. Guan. 2012. Synthesis of tetra-substituted NH pyrroles and poly-substituted furans *via* an addition and cyclization strategy. Synthesis 4: 532–540.

(152) D. Basavaiah and J.S. Rao. 2004. Applications of Baylis-Hillman acetates: one-pot, facile and convenient synthesis of substituted γ-lactams. Tetrahedron Lett. 45: 1621–1625.

(153) R. Ballini and M. Petrini. 2009. Nitroalkanes as key building blocks for the synthesis of heterocyclic derivatives. ARKIVOC (ix): 195–223.

(154) T. Taniguchi and H. Ishibashi. 2010. Iron-mediated radical nitro-cyclization reaction of 1,6-dienes. Org. Lett. 12: 124–126.

(155) R. Klvaňa, R. Pohl, J. Pawlas, J. Čejka. H. Dvořáková, R. Hrabal, S. Bohm, B. Kratochvil and J. Kuthan. 2000. Sterically crowded heterocycles. XII. Atropisomerism of (1-aryl-3,5-diphenyl-1*H*-pyrrol-2-yl)(phenyl)methanones. Collect. Czech. Chem. Comm. 65: 651–666.

(156) R.M. Carballo, M. Purino, M.A. Ramirez, V.S. Martin and J.I. Padron. 2010. Iron(III)-catalyzed consecutive aza-Cope-Mannich cyclization: synthesis of *trans*-3,5-dialkyl pyrrolidines and 3,5-dialkyl-2,5-dihydro-1*H*-pyrroles. Org. Lett. 12: 5334–5337.

(157) M.R. Gagne, L. Brard, V.P. Conticello, M.A. Giardello, C.L. Stern and T.J. Marks. 1992. Stereoselection effects in the catalytic hydroamination/cyclization of amino olefins at chiral organolanthanide centers. Organometallics 11: 2003–2005.

(158) G.A. Molander and E.D. Dowdy. 1998. Catalytic intramolecular hydroamination of hindered alkenes using organolanthanide complexes. J. Org. Chem. 63: 8983–8988.

(159) A.T. Gilbert, B.L. Davis, T.J. Emge and R.D. Broene. 1999. Synthesis, structure, and catalytic reactions of 1,2-bis(indenyl)ethane-derived lanthanocenes. Organometallics 18: 2125–2132.

(160) J.S. Ryu, T.J. Marks and F.E. McDonald. 2001. Organolanthanide-catalyzed intramolecular hydroamination/ cyclization of amines tethered to 1,2-disubstituted alkenes. Org. Lett. 3: 3091–3094.

(161) Y.K. Kim, T. Livinghouse and J.E. Bercaw. 2001. Intramolecular alkene hydroaminations catalyzed by simple amido derivatives of the group 3 metals. Tetrahedron Lett. 42: 2933–2935.

(162) G.A. Molander, E.D. Dowdy and S.K. Pack. 2001. A diastereoselective intramolecular hydroamination approach to the syntheses of (+)-, (±)-, and (–)-pinidinol. J. Org. Chem. 66: 4344–4347.

(163) S. Hong and T.J. Marks. 2002. Highly stereoselective intramolecular hydroamination/cyclization of conjugated aminodienes catalyzed by organolanthanides. J. Am. Chem. Soc. 124: 7886–7887.

(164) S. Hong and T.J. Marks. 2004. Organolanthanide-catalyzed hydroamination. Acc. Chem. Res. 37: 673–686.

(165) M.R. Gagne and T.J. Marks. 1989. Organolanthanide-catalyzed hydroamination. Facile, regiospecific cyclization of unprotected amino olefins. J. Am. Chem. Soc. 111: 4108–4109.

(166) M.R. Gagne, C.L. Stern and T.J. Marks. 1992. Organolanthanide-catalyzed hydroamination. A kinetic, mechanistic, and diastereoselectivity study of the cyclization of *N*-unprotected amino olefins. J. Am. Chem. Soc. 114: 275–294.

(167) M.R. Gagne, S.P. Nolan and T.J. Marks. 1990. Organolanthanide-centered hydroamination/cyclization of aminoolefins. Expedient oxidative access to catalytic cycles. Organometallics 9: 1716–1718.

(168) Y. Li, P.-F. Fu and T.J. Marks. 1994. Organolanthanide-catalyzed carbon-heteroatom bond formation. Observations on the facile, regiospecific cyclization of aminoalkynes. Organometallics 13: 439–440.

(169) Y. Li and T.J. Marks. 1996. Organolanthanide-catalyzed intramolecular hydroamination/cyclization of aminoalkynes. J. Am. Chem. Soc. 118: 9295–9306.

(170) V.M. Arredondo, F.E. McDonald and T.J. Marks. 1998. Organolanthanide-catalyzed intramolecular hydroamination/ cyclization of aminoallenes. J. Am. Chem. Soc. 120: 4871–4872.

(171) V.M. Arredondo, F.E. McDonald and T.J. Marks. 1999. Intramolecular hydroamination/cyclization of aminoallenes catalyzed by organolanthanide complexes. Scope and mechanistic aspects. Organometallics 18: 1949–1960.

(172) M.R. Bürgstein, H. Berberich and P.W. Roesky. 1998. (Aminotroponiminato)yttrium amides as catalysts in alkyne hydroamination. Organometallics 17: 1452–1454.

(173) Y. Fumoto, T. Eguchi, H. Uno and N. Ono. 1999. Synthesis of pyrrolostatin and its analogues. J. Org. Chem. 64: 6518–6521.

(174) A.V. Gulevich, A.G. Zhdanko, R.V.A. Orru and V.G. Nenajdenko. 2010. Isocyanoacetate derivatives: synthesis, reactivity, and application. Chem. Rev. 110: 5235–5331.

(175) (a) C. Agami, F. Couty and N. Rabasso. 2002. An efficient asymmetric synthesis of azetidine 2-phosphonic acids. Tetrahedron Lett. 43: 4633–4636. (b) C. Agami, F. Couty and G. Evano. 2002. A straightforward synthesis of enantiopure 2-cyano azetidines from β-amino alcohols. Tetrahedron: Asymmetry 13: 297–302.

(176) A. Noble and J.C. Anderson. 2013. Nitro-Mannich reaction. Chem. Rev. 113: 2887–2939.

(177) S.M.-C. Pelletier, P.C. Ray and D.J. Dixon. 2009. Nitro-Mannich/lactamization cascades for the direct stereoselective synthesis of pyrrolidin-2-ones. Org. Lett. 11: 4512–4515.

(178) S.M.-C. Pelletier, P.C. Ray and D.J. Dixon. 2011. Diastereoselective synthesis of 1,3,5-trisubstituted 4-nitropyrrolidin-2-ones via a nitro-Mannich/lactamization cascade. Org. Lett. 13: 6406–6409.

(179) T. Sakai, K.-I. Yamada and K. Tomioka. 2008. Base-induced sequential cyclization-rearrangement of enantioenriched 3-aminoalkanoates to five- and seven-membered lactams. Chem. Asian J. 3: 1486–1493.

(180) (a) R. Verma and S.K. Ghosh. 1999. Desymmetrization of 3-dimethyl(phenyl)silyl glutaric anhydride with Evans' oxazolidinone: an application to stereocontrolled synthesis of the anti-fungal agent (+)-preussin. J. Chem. Soc. Perkin Trans. 1: 265–270. (b) R. Chowdhury and S. Ghosh. 2011. Enantioselective route to β-silyl-δ-keto esters by organocatalyzed regioselective Michael addition of methyl ketones to a (silylmethylene)malonate and their use in natural product synthesis. Synthesis 12: 1936–1945.

(181) R. Verma and S.K. Ghosh. 1997. A silicon controlled total synthesis of the anti-fungal agent (+)-preussin. Chem. Commun. 17: 1601–1602.

(182) A. Ates and C. Quinet. 2003. Efficient intramolecular hydroamination of un-activated alkenes catalyzed by butyllithium. Eur. J. Org. Chem. 9: 1623–1626.

(183) M.E. Jung and G. Pizzi. 2005. gem-Di-substituent effect: theoretical basis and synthetic applications. Chem. Rev. 105: 1735–1766.

(184) C. Quinet, P. Jourdain, C. Hermans, A. Ates, I. Lucas and I.E. Marko. 2008. Highly efficient, base-catalyzed, intramolecular hydroamination of non-activated olefins. Tetrahedron 64: 1077–1087.

(185) H. Fujita, M. Tokuda, M. Nitta and H. Suginome. 1992. Stereoselective cyclization of δ-alkenylamines catalyzed with butyllithium. Synthesis of cis-N-methyl-2,5-disubstituted pyrrolidines. Tetrahedron Lett. 33: 6359–6362.

(186) F.A. Davis, R.E. Reddy, J.M. Swewczyk and P.S. Portonovo. 1993. Asymmetric synthesis of sulfinimines: chiral ammonia imine synthons. Tetrahedron Lett. 34: 6229–6232.

(187) T. Balasubramanian and A. Hassner. 1996. Synthesis of chiral non-racemic 2-arylpyrrolines by a [3+2]-cycloaddition route. Tetrahedron Lett. 37: 5755–5758.

(188) A. Hassner, E. Ghera, T. Yechezkel, V. Kleiman, T. Balasubramanian and D. Ostercamp. 2000. Stereoselective and enantioselective synthesis of five-membered rings via conjugate additions of allylsulfone carbanions. Pure Appl. Chem. 72: 1671–1683.

(189) P.D. Howes, D. Andrews, G. Bravi, L. Chambers, S. Fernandes, R. Guidetti, D. Haigh, D. Hartley, D. Jackson, V. Lovegrove, K. Medhurst, P. Shah, M.J. Slater and R. Stocker. 27th–31st August, 2006. Acyl pyrrolidines as hepatitis C NS5B polymerase inhibitors: lead optimization via N1 and C4 modifications. 1st European Chemistry Congress; Budapest, Hungary.

(190) J.E. Davoren, D.L. Gray, A.R. Harris, D.M. Nason and W. Xu. 2010. Remarkable [3+2]-annulations of electron rich olefins with un-stabilized azomethine ylides. Synlett 16: 2490–2492.

(191) V.F. Mironov, A.T. Gubaidullin, L.M. Burnaeva, I.A. Litvinov, G.A. Ivkova, S.V. Romanov, T.A. Zyablikova, A.I. Konovalov and I.V. Konovalova. 2004. Reactions of 2-alkoxy-4-oxo-5,6-benzo-1,3,2-dioxaphosphorinanes with imines. Synthesis and steric structure of 6,7-benzo-1,4,2-oxazaphosphepine derivatives. Russ. J. Gen. Chem. 74: 32–47.

(192) M.S.T. Morin, D.J. St-Cyr and B.A. Arndtsen. 2010. Horner-Wadsworth-Emmons reagents as azomethine ylide analogues: pyrrole synthesis via [3+2]-cycloaddition. Org. Lett. 12: 4916–4919.

(193) A.P. Krapcho. 1982. Synthetic applications of dealkoxycarbonylations of malonate esters, β-keto esters, α-cyano esters and related compounds in dipolar aprotic media - part I. Synthesis 10: 805–822.

(194) (a) I.R. Baxendale, C.D. Buckle, S.V. Ley and L. Tamborini. 2009. A base-catalyzed one-pot three-component coupling reaction leading to nitro substituted pyrroles. Synthesis 9: 1485–1493. (b) A.Z. Halimehjani, I.N.N. Namboothiri and S.E. Hooshmand. 2014. Part I: Nitroalkenes in the synthesis of heterocyclic compounds. RSC Adv. 4: 48022–48084.

(195) U. Jahn, F. Kafka, R. Pohl and P.G. Jones. 2009. N,3,4-Trisubstituted pyrrolidines by electron transfer-induced oxidative cyclizations of N-allylic β-amino ester enolates. Tetrahedron 65: 10917–10929.

(196) L. Lu, G. Chen and S. Ma. 2006. Ring-opening cyclization of alkylidenecyclopropyl ketones with amines. An efficient synthesis of 2,3,4-tri-substituted pyrroles. Org. Lett. 8: 835–838.

(197) L. Benati, G. Bencivenni, R. Leardini, M. Minozzi, D. Nanni, R. Scialpi, P. Spagnolo and G. Zanardi. 2006. Generation and cyclization of unsaturated carbamoyl radicals derived from S-4-pentynyl carbamothioates under tin-free conditions. J. Org. Chem. 71: 3192–3197.

(198) T. Taniguchi, T. Fujii, A. Idota and H. Ishibashi. 2009. Reductive addition of the benzenethiyl radical to alkynes by amine-mediated single electron transfer reaction to diphenyl disulfide. Org. Lett. 11: 3298–3301.

(199) H. Kim, D. Yoo, S.Y. Choi, Y.K. Chung and Y.G. Kim. 2008. Efficient and stereoselective synthesis of (2S,3S,4S)-3,4-dihydroxyglutamic acid via intramolecular epoxidation. Tetrahedron: Asymmetry 19: 1965–1969.

(200) V. Gandon, P. Bertus and J. Szymoniak. 2002. Conversion of imines into C,N-dimagnesiated compounds and trapping with electrophiles. One-pot access to 1-azaspirocyclic framework. Synthesis 8: 1115–1120.

(201) X. Fang, K. Liu and C. Li. 2010. Efficient regio- and stereoselective formation of azocan-2-ones via 8-endo-cyclization of α-carbamoyl radicals. J. Am. Chem. Soc. 132: 2274–2283.

(202) A.S. Azmi, P.A. Philip, F.W.J. Beck, Z. Wang, S. Banerjee, S. Wans, D. Yang, F.H. Sarkar and R.M. Mahammad. 2011. MI-219-zinc combination: a new paradigm in MDM2 inhibitor-based therapy. Oncogene 30: 117–126.

(203) P.B. Alper, C. Meyers, A. Lerchner, D.R. Siegel and E.M. Carreira. 1999. Facile, novel methodology for the synthesis of spiro[pyrrolidin-3,3'-oxindoles]: catalyzed ring-expansion reactions of cyclopropanes by aldimines. Angew. Chem. Int. Ed. 38: 3186–3189.

(204) G.S. Singh and Z.Y. Desta. 2012. Isatins as privileged molecules in design and synthesis of spiro-fused cyclic frameworks. Chem. Rev. 112: 6104–6155.

(205) E.P.J. Ng, Y.-F. Wang and S. Chiba. 2011. Manganese(III)-catalyzed formal [3+2]-annulation of vinyl azides and β-keto acids for the synthesis of pyrroles. Synlett 6: 783–786.

(206) Y.-F. Wang, K.K. Toh, S. Chiba and K. Narasaka. 2008. Mn(III)-catalyzed synthesis of pyrroles from vinyl azides and 1,3-dicarbonyl compounds. Org. Lett. 10: 5019–5022.

(207) K. Asahi and H. Nishino. 2009. Synthesis of bicyclo[3.1.0]hexan-2-ones by manganese(III) oxidation in ethanol. Synthesis 3: 409–423.

(208) M. Murakami, S. Kadowaki and T. Matsuda. 2005. Molybdenum-catalyzed ring-closing metathesis of allenynes. Org. Lett. 7: 3953–3956.

(209) (a) K.C. Majumdar, S. Muhuri, R.U. Islam and B. Chattopadhyay. 2009. Synthesis of five- and six-membered heterocyclic compounds by the application of the metathesis reactions. Heterocycles 78: 1109–1169. (b) M. Mori. 2010. Recent progress on enyne metathesis: its application to syntheses of natural products and related compounds. Materials 3: 2087–2140.

(210) N. Kaneta, K. Hikichi, S. Asaka, M. Uemura and M. Mori. 1995. Novel synthesis of di-substituted alkyne using molybdenum-catalyzed cross-alkyne metathesis. Chem. Lett. 24: 1055–1056.

(211) M. Nishida, H. Shiga and M. Mori. 1998. [2+2+2]-Co-cyclization using [Mo(CO)₆-p-ClPhOH]. J. Org. Chem. 63: 8606–8608.

(212) S. Kotha, E. Brahmachary and K. Lahiri. 2005. Transition metal-catalyzed [2+2+2]-cycloaddition and application in organic synthesis. Eur. J. Org. Chem. 22: 4741–4767.

(213) A. Kamal, M.S. Malik, A.A. Shaik and S. Azeeza. 2008. Synthesis of enantiomerically pure γ-azidoalcohols by lipase-catalyzed trans-esterification. Tetrahedron: Asymmetry 19: 1078–1083.

(214) V.B. Phapale, E. Bunuel, M. Garcia-Iglesias and D.J. Cardenas. 2007. Ni-catalyzed cascade formation of C(sp³)-C(sp³) bonds by cyclization and cross-coupling reactions of iodoalkanes with alkyl zinc halides. Angew. Chem. Int. Ed. 46: 8790–8795.

(215) P. Caramella and K.N. Houk. 1976. Geometries of nitrilium betaines. The clarification of apparently anomalous reactions of 1,3-dipoles. J. Am. Chem. Soc. 98: 6397–6399.

(216) R. Sustmann. 1974. Orbital energy control of cycloaddition reactivity. Pure Appl. Chem. 40: 569–593.

(217) D. Seebach and P. Konchel. 1984. 2-Nitro-2-propen-1-yl 2,2-dimethylpropanoate (NPP), a multiple coupling reagent. Helv. Chim. Acta 67: 261–283.

(218) M. Uemura, H. Yorimitsu and K. Oshima. 2006. Synthesis of Cp*CH₂PPh₂ and its use as a ligand for the nickel-catalyzed cross-coupling reaction of alkyl halides with aryl Grignard reagents. Chem. Commun. 45: 4726–4728.

(219) Y. Yoshitomi, K. Makino and Y. Hamada. 2007. Organocatalytic synthesis of (2S,3R)-3-hydroxy-3-methyl-proline (OHMePro), a component of polyoxypeptins, and relatives using OHMePro itself as a catalyst. Org. Lett. 9: 2457–2460.

(220) J.S. Reddy and B.V. Rao. 2007. A short, efficient, and stereoselective total synthesis of a pyrrolidine alkaloid: (–)-codonopsinine. J. Org. Chem. 72: 2224–2227.

(221) S. Vanlaer, A. Voet, C. Gielens, M. de Maeyer and F. Compernolle. 2009. Bridged 5,6,7,8-tetrahydro-1,6-naphthyridines, analogues of huperzine a: synthesis, modeling studies and evaluation as inhibitors of acetylcholinesterase. Eur. J. Org. Chem. 5: 643–654.

(222) P. Coutrot, S. Claudel, C. Didierjean and C. Grison. 2006. Stereoselective synthesis and glycosidase inhibitory activity of 3,4-dihydroxy-pyrrolidin-2-one, 3,4-dihydroxy-piperidin-2-one and 1,2-dihydroxy-pyrrolizidin-3-one. Bioorg. Med. Chem. Lett. 16: 417–420.

(223) C. Aubert, O. Buisine and M. Malacria. 2002. The behavior of 1,n-enynes in the presence of transition metals. Chem. Rev. 102: 813–834.

(224) Y. Sato, N. Imakuni, T. Hirose, H. Wakamatsu and M. Mori. 2003. Further studies on palladium-catalyzed bismetallative cyclization of enynes in the presence of Bu₃SnSiMe₃. J. Organomet. Chem. 687: 392–402.

(225) S. Diez-Gonzalez, N. Marion and S.P. Nolan. 2009. *N*-Heterocyclic carbenes in late transition metal catalysis. Chem. Rev. 109: 3612–3676.

(226) B.C. Bishop, I.F. Cottrell and D. Hands. 1997. Synthesis of 3-hydroxyalkylbenzo[*b*]furans *via* the palladium-catalyzed heteroannulation of silyl-protected alkynols with 2-iodophenol. Synthesis 11: 1315–1320.

(227) T. Konno, J. Chae, T. Ishihara and H. Yamanaka. 2004. A first regioselective synthesis of 3-fluoroalkylated benzofurans *via* palladium-catalyzed annulation of fluorine-containing internal alkynes with variously substituted 2-iodophenol. Tetrahedron 60: 11695–11700.

(228) M.L. Crawley, I. Goljer, D.J. Jenkins, J.F. Mehlmann, L. Nogle, R. Dooley and P.E. Mahaney. 2006. Regioselective synthesis of substituted pyrroles: efficient palladium-catalyzed cyclization of internal alkynes and 2-amino-3-iodoacrylate derivatives. Org. Lett. 8: 5837–5840.

(229) R.C. Larock, M.J. Doty and X. Han. 1998. Palladium-catalyzed heteroannulation of internal alkynes by vinylic halides. Tetrahedron Lett. 39: 5143–5146.

(230) G.R. Cook, P.S. Shanker and S.L. Peterson. 1999. Asymmetric synthesis of the balanol heterocycle *via* a palladium-mediated epimerization and olefin metathesis. Org. Lett. 1: 615–617.

(231) G.R. Cook and L. Sun. 2004. Nitrogen heterocycles *via* palladium-catalyzed carbocyclization. Formal synthesis of (+)-α-allokainic acid. Org. Lett. 6: 2481–2484.

(232) C. Palomo, A. Landa, A. Mielgo, M. Oiarbide, A. Puente and S. Vera. 2007. Water-compatible iminium activation: organocatalytic Michael reactions of carbon-centered nucleophiles with enals. Angew. Chem. Int. Ed. 46: 8431–8435.

(233) M. Kimura, T. Tamaki, M. Nakata, K. Tohyama and Y. Tamaru. 2008. Convenient synthesis of pyrrolidines by amphiphilic allylation of imines with 2-methylenepropane-1,3-diols. Angew. Chem. Int. Ed. 47: 5803–5805.

(234) S.E. Denmark, M.S. Dappen and C.J. Cramer. 1986. Intramolecular [4+2]-cycloadditions of nitroalkenes with olefins. J. Am. Chem. Soc. 108: 1306–1307.

(235) S.E. Denmark and M.E. Schnute. 1994. Nitroalkene [4+2]-cycloadditions with 2-(acyloxy)vinyl ethers. Stereoselective synthesis of 3-hydroxy-4-substituted pyrrolidines. J. Org. Chem. 59: 4576–4595.

(236) A. Furstner, H. Szillat, B. Gabor and R. Mynott. 1998. Platinum and acid-catalyzed enyne metathesis reactions: mechanistic studies and applications to the syntheses of streptorubin B and metacycloprodigiosin. J. Am. Chem. Soc. 120: 8305–8314.

(237) H. Kusama, Y. Yamashita and K. Narasaka. 1995. Synthesis of quinolines *via* intramolecular cyclization of benzylacetone oxime derivatives catalyzed with tetrabutylammonium perrhenate(VII) and trifluoromethanesulfonic acid. Chem. Lett. 24: 5–6.

(238) H. Kusama, Y. Yamashita and K. Narasaka. 1995. Beckmann rearrangement of oximes catalyzed with tetrabutylammonium perrhenate and trifluoromethanesulfonic acid. Bull. Chem. Soc. Jpn. 68: 373–377.

(239) H. Kusama, Y. Yamashita, K. Uchiyama and K. Narasaka. 1997. Transformation of oximes of phenethyl ketone derivatives to quinolines and azaspirotrienones catalyzed by tetrabutylammonium perrhenate and trifluoromethanesulfonic acid. Bull. Chem. Soc. Jpn. 70: 965–975.

(240) K. Narasaka. 2002. Metal-assisted amination with oxime derivatives. Pure Appl. Chem. 74: 143–149.

(241) S. Mori, K. Uchiyama, Y. Hayashi, K. Narasaka and E. Nakamura. 1998. S_N2 substitution on sp² nitrogen of protonated oxime. Chem. Lett. 27: 111–112.

(242) M. Yoshida, K. Uchiyama and K. Narasaka. 2000. Synthesis of dihydropyrroles and tetrahydropyridines by the cyclization of *O*-methylsulfonyloximes having an active methine group. Heterocycles 52: 681–691.

(243) V.M. Arredondo, S. Tian, F.E. McDonald and T.J. Marks. 1999. Organolanthanide-catalyzed hydroamination/cyclization. Efficient allene-based transformations for the syntheses of naturally occurring alkaloids. J. Am. Chem. Soc. 121: 3633–3639.

(244) S. Tian, V.M. Arredondo, C.L. Stern and T.J. Marks. 1999. Constrained geometry organolanthanide catalysts. Synthesis, structural characterization, and enhanced aminoalkene hydroamination/cyclization activity. Organometallics 18: 2568–2570.

(245) R. Pulz, A. Al-Harrasi and H.-U. Reissig. 2002. New polyhydroxylated pyrrolidines derived from enantiopure 3,6-dihydro-2*H*-1,2-oxazines. Org. Lett. 4: 2353–2355.

(246) Y.-B. Kang, Y. Tang and X.-L. Sun. 2006. Scandium triflate-catalyzed cycloaddition of imines with 1,1-cyclopropanediesters: efficient and diastereoselective synthesis of multi-substituted pyrrolidines. Org. Biomol. Chem. 4: 299–301.

(247) F. Aydogan, M. Basarir, C. Yolacan and A.S. Demir. 2007. New and clean synthesis of *N*-substituted pyrroles under microwave irradiation. Tetrahedron 63: 9746–9750.

(248) A. Majumder, R. Gupta and A. Jain. 2013. Microwave-assisted synthesis of nitrogen-containing heterocycles. Green Chem. Lett. Rev. 6: 151–182.

(249) P.W. Roesky, C.L. Stern and T.J. Marks. 1997. Ancillary ligand effects on organo-f-element reactivity and metallocenes with bridge-tethered donors. Organometallics 16: 4705–4711.

(250) F. Aydogan and A.S. Demir. 2005. Clean and efficient microwave-solvent-free conversion of homochiral amines, α-amino alcohols and α-amino acids to their chiral 2-substituted pyrrole derivatives. Tetrahedron 61: 3019–3023.

(251) D.P. Curran, N.A. Porter and B. Giese. 1996. Stereochemistry of radical reactions; VCH: Weinheim, 59.

(252) D.C. Spellmeyer and K.N. Houk. 1987. Force-field model for intramolecular radical additions. J. Org. Chem. 52: 959–974.

(253) A.L.J. Beckwith and C.H. Scheisser. 1985. A force-field study of alkenyl radical ring-closure. Tetrahedron Lett. 26: 373–376.

(254) H. Senboku, Y. Kajizuka, H. Hasegawa, H. Fujita, H. Suginome, K. Orito and M. Tokuda. 1999. Tandem cyclization of *N*-allylaminyl radicals: stereoselective synthesis of 1,2,5-trisubstituted pyrrolizidines. Tetrahedron 55: 6465–6474.

(255) W.R. Bowman, M.J. Broadhurst, D.R. Coghlan and K.A. Lewis. 1997. Cyclization of aminyl radicals derived from amino acids. Tetrahedron Lett. 38: 6301–6304.

(256) B.B. Snider and T. Liu. 1997. Synthesis of (±)-cylindricines A, D, and E. J. Org. Chem. 62: 5630–5633.

(257) G. Heuger, S. Kalsow and R. Gottlich. 2002. Copper(I) catalysts for the stereoselective addition of *N*-chloroamines to double bonds: a diastereoselective radical cyclization. Eur. J. Org. Chem. 11: 1848–1854.

(258) M. Newcomb, D.J. Marquardt and T.M. Deeb. 1990. *N*-Hydroxypyridine-2-thione carbamates. Syntheses of alkaloid skeletons by aminium cation radical cyclizations. Tetrahedron 46: 2329–2344.

(259) H. Senboku, H. Hasegawa, K. Orito and M. Tokuda. 1999. New stereoselective synthesis of (±)-*trans*-2-butyl-5-heptyl-1-methylpyrrolidine, ant venom alkaloid, by aminyl radical cyclization. Heterocycles 50: 333–340.

(260) I.G. Jung, J. Seo, S.I. Lee, S.Y. Choi and Y.K. Chung. 2006. Reductive cyclization of diynes and enynes catalyzed by allyl platinum *N*-heterocyclic carbene complexes. Organometallics 25: 4240–4242.

(261) I.G. Castro, A. Tillack, C.G. Hartung and M. Beller. 2003. From terminal alkynes directly to branched amines. Tetrahedron Lett. 44: 3217–3221.

(262) I. McCort, A. Duréault and J.-C. Depezay. 1996. Practical route to D-manno and D-gluco azasugars from C2 symmetric bis-aziridines. Tetrahedron Lett. 37: 7717–7720.

(263) S. Kobayashi, M. Sugiura, H. Kitagawa and W.W.-L. Lam. 2002. Rare-earth metal triflates in organic synthesis. Chem. Rev. 102: 2227–2302.

(264) S. Kobayashi, R. Akiyama, M. Kawamura and H. Ishitani. 1997. Lanthanide triflate-catalyzed three-component coupling reactions of aldehydes, hydroxylamines, and alkenes leading to isoxazolidine derivatives. Chem. Lett. 26: 1039–1040.

(265) S. Kobayashi and M. Kawamura. 1998. Catalytic enantioselective 1,3-dipolar cycloadditions between nitrones and alkenes using a novel heterochiral ytterbium(III) catalyst. J. Am. Chem. Soc. 120: 5840–5841.

(266) M. Kawamura and S. Kobayashi. 1999. A switch of enantiofacial selectivity in chiral ytterbium-catalyzed 1,3-dipolar cycloaddition reactions. Tetrahedron Lett. 40: 3213–3216.

(267) G.A. Molander and H. Hasegawa. 2004. Organolanthanide-catalyzed intramolecular 5-*endo-dig* hydroamination: an unusual *anti*-markovnikov cyclization. Heterocycles 64: 467–474.

(268) D. Tejedor, D. Gonzalez-Cruz, F. Garcia-Tellado, J. Juan Marrero-Tellado and M. Lopez Rodriguez. 2004. A diversity-oriented strategy for the construction of tetra-substituted pyrroles *via* coupled domino processes. J. Am. Chem. Soc.126: 8390–8391.

(269) A. Das, A. Kulkarni and B. Torok. 2012. Environmentally benign synthesis of heterocyclic compounds by combined microwave-assisted heterogeneous catalytic approaches. Green Chem. 14: 17–34.

(270) D. Lednicer. 2008. Indolones. In The Organic Chemistry of Drug Synthesis, Wiley-Interscience: Hoboken, New Jersey, 7: 148.

(271) M. Baumann, I.R. Baxendale, S.V. Ley and N. Nikbin. 2011. An overview of the key routes to the best selling 5-membered ring heterocyclic pharmaceuticals. Beilstein J. Org. Chem. 7: 442–495.

(272) M. Murakami, N. Hasegawa, I. Tomita, N. Inouye and Y. Ito. 1989. Fluoride-catalyzed Michael reaction of α-isocyanoesters with α,β-unsaturated carbonyl compounds. Tetrahedron Lett. 30: 1257–1260.

(273) M. Murakami, N. Hasegawa, M. Hayashi, N. Inouye and Y. Ito. 1991. Synthesis of (+)-α-allokainic acid *via* the zinc acetate-catalyzed cyclization of γ-isocyano silyl enolates. J. Org. Chem. 56: 7356–7360.

(274) H. Miyabe, R. Asada, A. Toyoda and Y. Takemoto. 2006. Enantioselective cascade radical addition-cyclization-trapping reactions. Angew. Chem. Int. Ed. 45: 5863–5866.

(275) R.I. Aminov, J.C. Chee-Sanford, N. Garrigues, B. Teferedegne, I.J. Krapac, B.A. White and R.I. Mackie. 2002. Development, validation and application of PCR primers for detection of tetracycline efflux genes of Gram negative bacteria. Appl. Environ. Microbiol. 68: 1786–1793.

(276) L. Feray and M.P. Bertrand. 2008. Dialkylzinc-mediated atom transfer sequential radical addition cyclization. Eur. J. Org. Chem. 18: 3164–3170.

(277) H. Ishibashi, M. Okano, H. Tamaki, K. Maruyama, T. Yakura and M. Ikeda. 1990. Total synthesis of (±)-cephalotaxine. J. Chem. Soc. Chem. Commun. 20: 1436–1437.

(278) H. Dong, M. Shen, J.E. Redford, B.J. Stokes, A.L. Pumphrey and T.G. Driver. 2007. Transition metal-catalyzed synthesis of pyrroles from dienyl azides. Org. Lett. 9: 5191–5194.

(279) T.G. Driver. 2010. Recent advances in transition metal-catalyzed *N*-atom transfer reactions of azides. Org. Biomol. Chem. 8: 3831–3846.

(280) F. Denes, A. Perez-Luna and F. Chemla. 2007. Diastereocontrolled synthesis of enantioenriched 3,4-disubstituted *β*-prolines. J. Org. Chem. 72: 398–406.

(281) R.W. Bates, J. Boonsombat, Y. Lu, J.A. Nemeth, K. Sa-Ei, P. Song, M.P. Cai, P.B. Cranwell and S. Winbush. 2008. *N,O*-Heterocycles as synthetic intermediates. Pure Appl. Chem. 80: 681–685.

(282) D.S. Black, G.L. Edwards, R.H. Evans, P.A. Keller and S.M. Laaman. 2000. Synthesis and reactivity of 1-pyrroline-5-carboxylate ester 1-oxides. Tetrahedron 56: 1889–1897.

(283) N. Sankuratri, E.G. Janzen, M.S. West and J.L. Poyer. 1997. Spin trapping with 5-methyl-5-phenylpyrroline *N*-oxide. A replacement for 5,5-dimethylpyrroline *N*-oxide. J. Org. Chem. 62: 1176–1178.

(284) C. Nsanzumuhire, J.-L. Clement, O. Ouari, H. Karoui, J.-P. Finet and P. Tordo. 2004. Synthesis of the *cis* diastereoisomer of 5-diethoxyphosphoryl-5-methyl-3-phenyl-1-pyrroline *N*-oxide (DEPMPPOc) and ESR study of its superoxide spin adduct. Tetrahedron Lett. 45: 6385–6389.

(285) R.A. Hunter, D.P.S. Macfarlane and R.J. Whitby. 2006. Organozirconium-mediated solution- and solid-phase synthesis of 3-benzyl pyrrolidines and other potentially neuroactive amines. Synlett 19: 3314–3318.

(286) T. Takahashi, D.Y. Kondakov and N. Suzuki. 1994. A novel type of zirconium-catalyzed or -promoted cyclization reaction. Organometallics 13: 3411–3412.

(287) K.S. Knight and R.M. Waymouth. 1994. Stereoselective cyclization *via* zirconocene-catalyzed intramolecular olefin allylation. Organometallics 13: 2575–2577.

(288) I. Ojima, M. Tzamarioudaki, Z. Li and R.J. Donovan. 1996. Transition metal-catalyzed carbocyclizations in organic synthesis. Chem. Rev. 96: 635–662.

(289) B.D. Stubbert and T.J. Marks. 2007. Mechanistic investigation of intramolecular aminoalkene and aminoalkyne hydroamination/cyclization catalyzed by highly electrophilic, tetravalent constrained geometry 4d and 5f complexes. Evidence for M-N σ-bonded insertive pathway. J. Am. Chem. Soc. 129: 6149–6167.

(290) S. Majumder and A.L. Odom. 2008. Group-4 dipyrrolylmethane complexes in intramolecular olefin hydroamination. Organometallics 27: 1174–1177.

Five-Membered *N*-Polyheterocycles

2.1 Introduction

Heterocyclic compounds are widely used in biological, chemical, and industrial settings. They form the core of several pharmaceutical agents and biologically active natural products and are also applied in corrosion inhibitors and herbicides. Various naturally occurring compounds possess benzo-fused heteroaromatic structures having a wide range of pharmaceutical applications (like anti-microbial and anti-biotic) [1a–f]. Many synthetic methods have been employed for the synthesis of 5-, 6-, 7-, and 8-membered benzo-fused heterocyclic compounds for their wide range of medicinal applications due to their ability to bind with multiple receptors with high affinity. These compounds are 'bicyclic privileged structures' and are defined as "a single molecular framework able to provide ligands for diverse receptors" [2–7].

Indole is one of the most commonly reported heterocyclic compounds. Indole moiety is present in numerous natural products which exhibit interesting biological activity (for example, reserpine, which is one of the first drugs to treat diseases of the central nervous system) [8–11]. Introduction of an indole moiety into pharmacologically active amino acid tryptophan is further evidence of its implicit relevance to life. The indole scaffold is successfully used in many pharmaceutical agents like new anti-migraine drugs and HIV-1 reverse transcriptase inhibitors (rescriptor), for instance. This indole moiety is also used in other fields of chemistry, such as polymer chemistry, for the fabrication of micro pH-sensors [12–16]. Therefore, researchers are interested in these structures and this chapter describes some relevant examples of synthesis of these molecules.

2.2 Metal- and non-metal-assisted synthesis of five-membered *N*-polyheterocycles

Aluminium-assisted synthesis

Ring-closing reactions to form an oxindole lactam ring occur at an un-functionalized position of an arene. These reactions have an advantage that the aromatic ring need not be rigged for cyclization, which facilitates the preparation of substrates. This reaction is exemplified by Friedel-Crafts cyclization of α-hydroxy or α-haloacetanilides **(Scheme 1)**. Also known as Stoll reaction, this method is used in the preparation of various oxindoles. Unfortunately, the need to utilize superstoichiometric or stoichiometric amounts of very strong Lewis acids severely limits the breadth of the functional groups that can be tolerated [17–19].

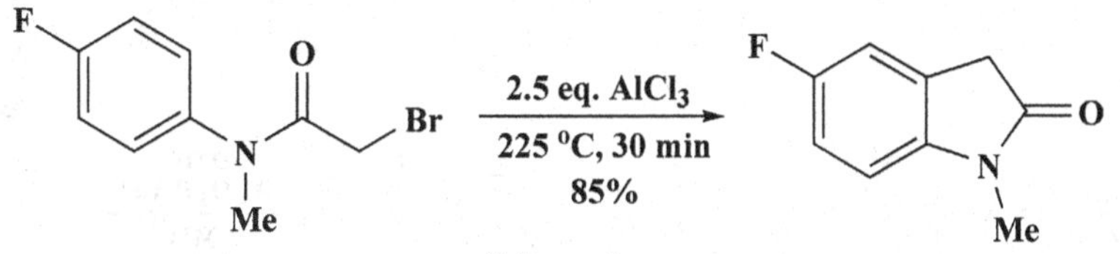

Scheme-1

Kabalka et al. [20] reported that the substituent present on the nitrogen atom affected the proportion of the open alkyne and the indole precursor. The reaction afforded indole if Y was an electron-withdrawing group, that is, sulfonamide and amide. The open form existed in various ratios when Y = H (**Scheme 2**) [21].

5-Membered nitrogen-containing heterocycles like indoles, pyrroles and carbazoles are important structural motifs in numerous biologically active compounds [22–24]. They are crucial for the pharmaceutical industry as they are present in the core structure of many drugs. A three-component reaction of vinyl and phenacyl bromides, acetylenes and pyridine has been reported for the preparation of indolizines, wherein basic alumina is used as the catalyst (**Scheme 3**) [25]. The condensation of pyridine and bromo-compound yielded *N*-alkylpyridinium salt *in situ*, which was then converted to a 1,3-dipole species under basic conditions. This dipolar intermediate, when reacted with acetylene under microwave-assisted solvent-free conditions by a dipolar cycloaddition reaction, formed indolizines in excellent yields (87–94%). For comparison purposes, the reactions were also performed in dry toluene under microwave-assisted conditions. The solvent-free reactions with all substrates were found to perform consistently better than those conducted in toluene which afforded 60–71% yields.

The *N*-(2-oxo-2-(*R*)-1-oxo-1-(*Z*)-2-(2-oxoindolin-3-ylidene)hydrazinyl)propan-2-ylamino)ethyl)-4-(5-oxo-4-(2-phenylhydrazono)-3-(trifluoromethyl)-4,5-dihydro-1*H*-pyrazol-yl)benzamide synthesized by the condensation of (*R*)-*N*-(2-(1-hydrazinyl-1-oxopropan-2-ylamino)-2-oxoethyl)-4-(5-oxo-4-(2-

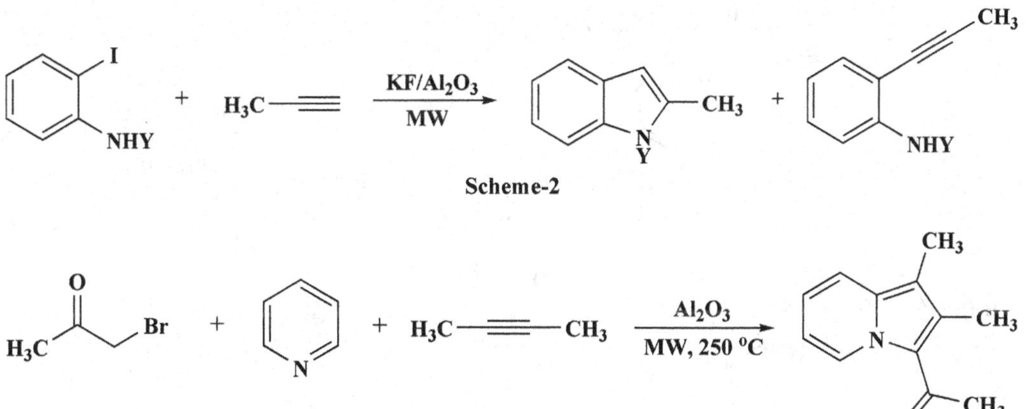

Scheme-2

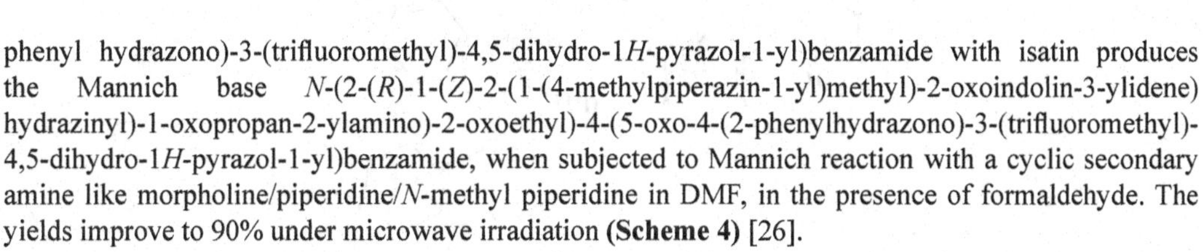

Scheme-3

phenyl hydrazono)-3-(trifluoromethyl)-4,5-dihydro-1*H*-pyrazol-1-yl)benzamide with isatin produces the Mannich base *N*-(2-(*R*)-1-(*Z*)-2-(1-(4-methylpiperazin-1-yl)methyl)-2-oxoindolin-3-ylidene) hydrazinyl)-1-oxopropan-2-ylamino)-2-oxoethyl)-4-(5-oxo-4-(2-phenylhydrazono)-3-(trifluoromethyl)-4,5-dihydro-1*H*-pyrazol-1-yl)benzamide, when subjected to Mannich reaction with a cyclic secondary amine like morpholine/piperidine/*N*-methyl piperidine in DMF, in the presence of formaldehyde. The yields improve to 90% under microwave irradiation (**Scheme 4**) [26].

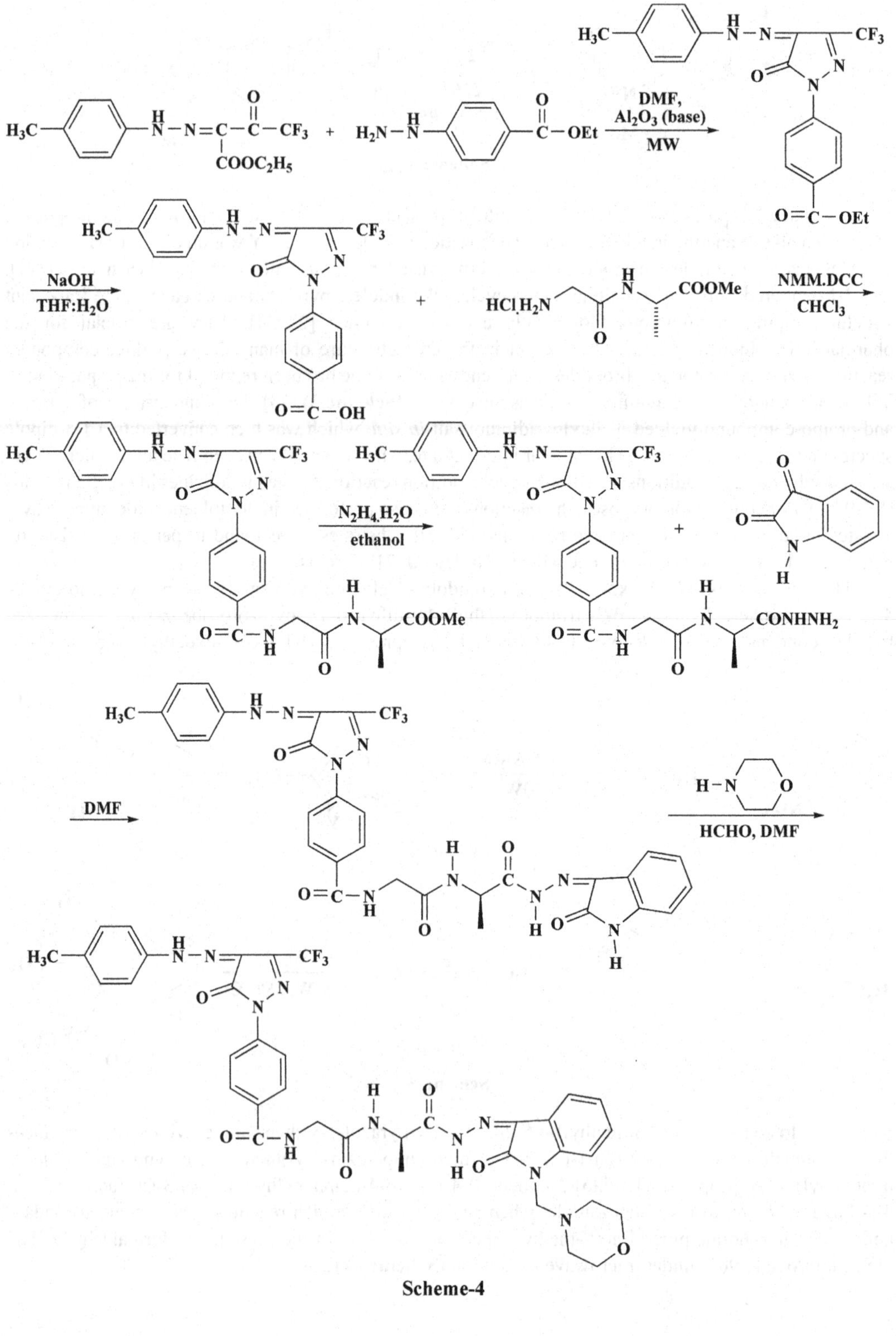

Scheme-4

Bismuth-assisted synthesis

Bi(OTf)$_3$·xH$_2$O and BiOClO$_4$·xH$_2$O catalyze electrophilic substitution reactions of indoles with many ketones and aldehydes at room temperature, under ultrasound irradiation, in acetonitrile to afford good to excellent yields of bis(indolyl)methanes. Even Bi(NO$_3$)$_3$·5H$_2$O or PANI-BC can be used to catalyze this reaction. Bis(indolyl)methanes are also synthesized efficiently from primary alcohols by a one-pot method under solvent-free conditions promoted in the presence of Bi(NO$_3$)$_3$·5H$_2$O [27]. The 3,3-di(3-indolyl)oxindoles can be prepared with high regioselectivity and excellent yields when indoles undergo a rapid condensation with isatin (1H-indole-2,3-dione) in the presence of Bi(OTf)$_3$·xH$_2$O catalyst **(Scheme 5)** [28].

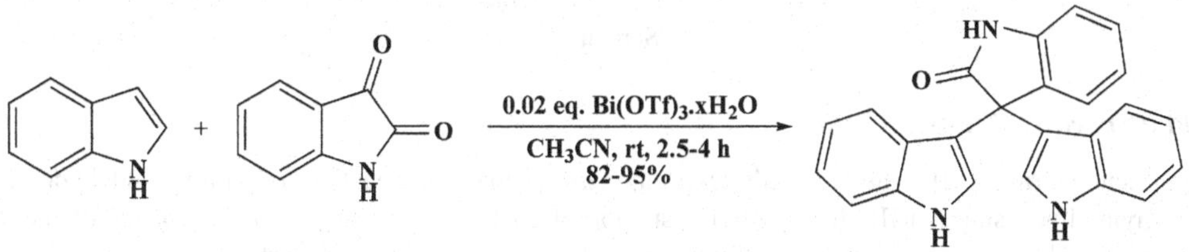

Scheme-5

Bis(indolyl)alkanes, which can be isolated from marine sources, serve as an important group of compounds [29]. Bis(indolyl)methanes are anti-fungals [30] and exhibit anti-proliferative activity in breast cancer [31]. They can be prepared by electrophilic substitution of indole with ketones or aldehydes **(Scheme 6)** [32–34]. This reaction is carried out efficiently in ionic liquids, with bismuth chloride acting as the catalyst [35]. Also, 3,3'-bis(indolyl)-4-chlorophenylmethane is formed in 88% yield when indole reacted with 4-chlorobenzaldehyde in the presence of the same catalyst (bismuth chloride), under microwave irradiation and solvent-free conditions [28, 36–37].

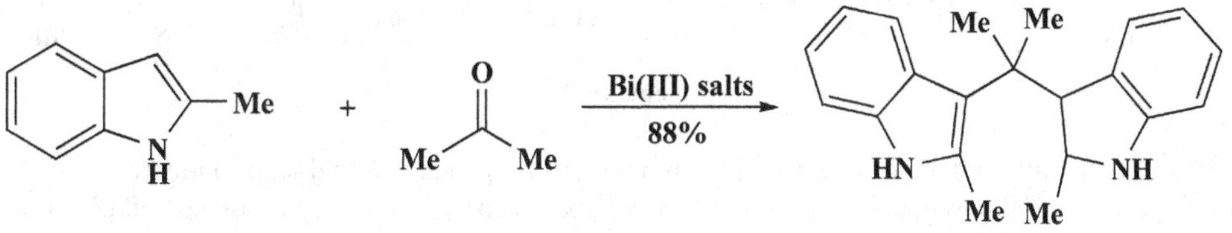

Scheme-6

Copper-assisted synthesis

Okano et al. [38] and Tokuyama et al. [39] synthesized aryne from an aryl bromide using magnesium. An arylmagnesium intermediate was produced when Boc-protected amine was cyclized successfully onto the aryne. Subsequently, an advanced indoline intermediate was generated for the synthesis of dictyodendrin A when cupric iodide and palladium catalyst were added along with iodoanisole, which, upon warming, cross-coupled with the organomagnesium intermediate **(Scheme 7)**.

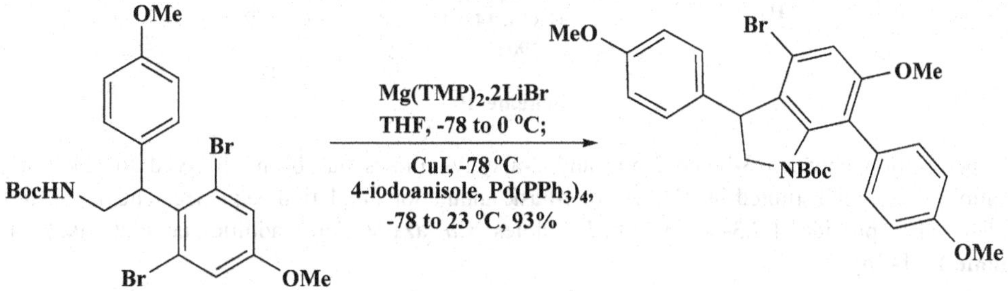

Scheme-7

Anilines can be condensed with 1,3-dicarbonyl compounds, followed by cyclization, in a one-pot sequence for synthesizing indoles. For example, the condensation of aniline with methyl acetoacetate performed at room temperature under solvent-free conditions in the presence of indium bromide yields the desired indole product when the resultant enamine carboxylate is cyclized under standard conditions **(Scheme 8)** [40–43].

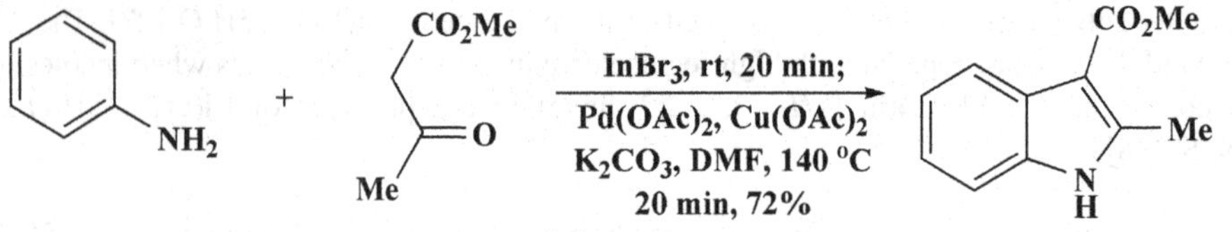

Scheme-8

Gold-assisted synthesis

Zhang and Corma [44] reported a three-component cyclization reaction of amine, aldehyde, and 2-ethynylaniline using Au(III) nanoparticles supported on CeO_2 or ZrO_2 for the synthesis of indoles **(Scheme 9)**. The reaction occurred by alkynylation of iminium ions through carbon-hydrogen activation of alkynes, followed by reductive cyclization. The overall reaction showed high turnover numbers and frequencies using only 0.35 mol% gold. Moreover, the air-stable catalysts (gold/CeO_2 or gold/ZrO_2) were easily recovered by filtration and reused at least three times without any significant loss in catalytic activity. The reaction mechanism was the same as the one involved with cationic gold(III) species under homogeneous conditions, the gold(III) species being stabilized by nanocrystalline ZrO_2 or CeO_2 [45a].

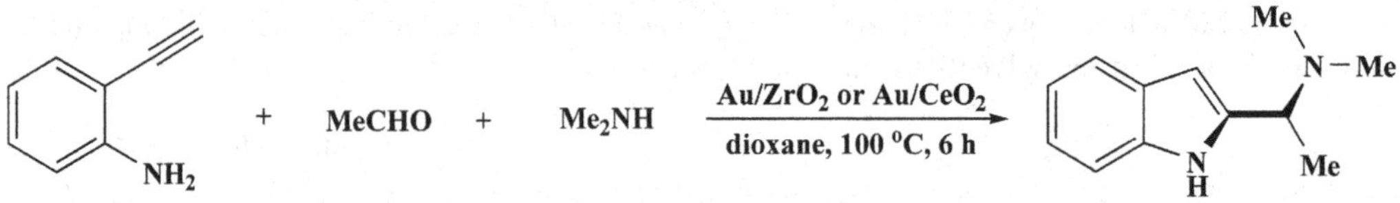

Scheme-9

This reaction allowed regioselectivity, high functional group tolerance and scope under relatively mild conditions. When 2-alkynylanilines react with 1,3-dicarbonyl compounds, they produce 3-alkenylindoles through Au-catalyzed sequential cyclization/alkenylation. The reaction temperature is increased for the sequential reaction **(Scheme 10)** [45b]. It has been observed that only n-Bu$_4$NaAuCl$_4$ catalyst can be recycled and reused for more than five runs without considerable loss in activity.

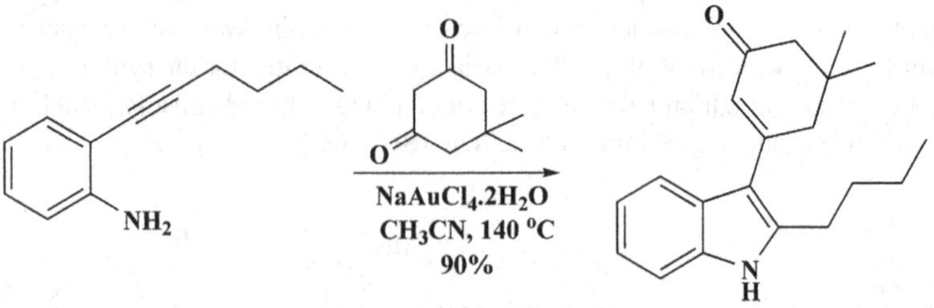

Scheme-10

The reaction between 3-buten-2-one and 2-alkynylanilines has been observed to result in the formation of 2,3-disubstituted indoles *via* one-flask annulation/alkylation sequence. The same starting materials also provide 1,2,3-trisubstituted indoles *via* aza-Michael addition/annulation/alkylation **(Scheme 11)** [45b].

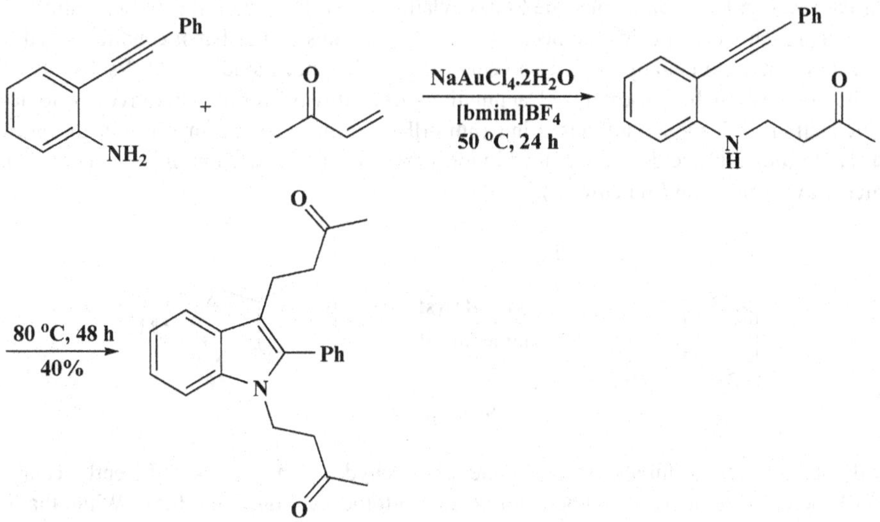

Scheme-11

Through another Au-catalyzed one-flask protocol, 2-alkynylanilines can be converted to 3-iodo and 3-bromoindoles. Since Fukuda and Utimoto [45c] and Fukuda et al. [45d] reported the first example of Au-catalyzed hydroamination, intramolecular hydroamination of alkynylamines has made significant progress. An alternative approach has also been developed with Au catalysts to afford indole scaffolds *via* Au-catalyzed cyclization of 2-alkynylaniles in ethanol or ethanol-water mixtures at room temperature [45e]. Good yields of indoles derivatives can be obtained by Au(III)-catalyzed annulation of 2-alkynylanilines at room temperature in ethanol or ethanol-water mixtures (**Scheme 12**) [45b].

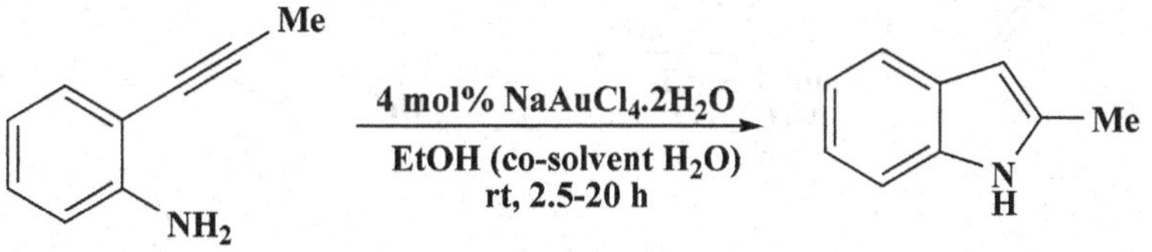

Scheme-12

Moderate to high yields of 3-alkylindoles can also be obtained by Au-catalyzed cyclization/conjugate addition of 2-alkynylanilines and α,β-enones (**Scheme 13**) [45b, 45f].

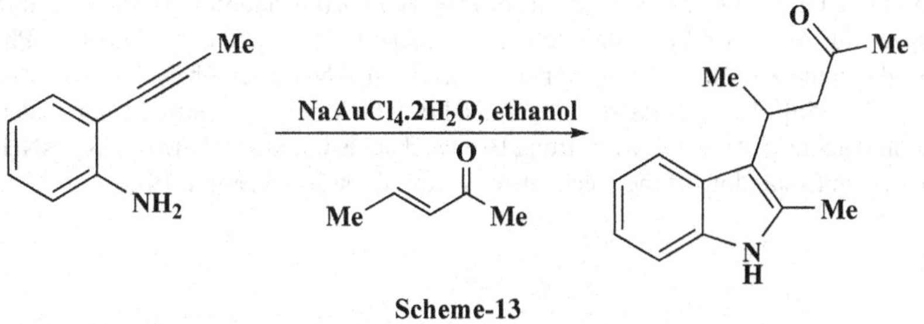

Scheme-13

When 2-alkynylaniline reacts with α,β-enone in an equimolecular ratio at 30 °C, satisfactory yields of indoles are produced. The reaction occurs smoothly even with less reactive α,β-enones like cyclic enones and di-substituted α,β-enones. Through Au-catalyzed sequential cyclization/alkylation, *N*-alkylation/ cyclization/alkylation or *N*-alkylation/cyclization reactions, different indoles can be synthesized by changing the reaction temperature and the 2-alkynylaniline to α,β-enone ratio. With an aim to develop a synthetic approach which avoids the need of protecting groups and/or harsh conditions and involves the use of more environmentally benign solvents [45g], it has been studied that 2-substituted indoles can also be synthesized by Au-catalyzed annulations of 2-ethynylaniline derivatives in ionic liquids [45h]. Cyclization of 2-alkynylanilines using [bmim]BF$_4$ as the reaction medium in the presence of NaAuCl$_4 \cdot$H$_2$O efficiently yields 2-substituted indoles. Moreover, the reaction times reduce significantly under microwave irradiation **(Scheme 14)** [45b].

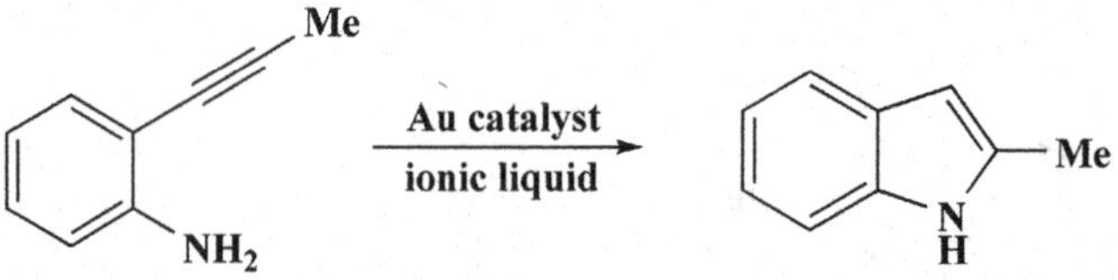

Scheme-14

Condensation of two furans with acetone or aldehydes can also be efficiently catalyzed by thallium(III) perchlorate, mercury(II) perchlorate, and *p*-toluenesulfonic acid [45i]. While thallium and mercury salts, due to their toxicity, are not an alternative to Brønsted acids with non-nucleophilic counter-ions, and ytterbium salts suffered from activity problems, the mild conditions of Au catalysis are an interesting synthetic alternative. Nair et al. [45j–k] synthesized tris-addition products in good to excellent yields when more reactive α,β-unsaturated aldehydes underwent an unusual hydroarylation with electron-rich arenes in the presence of 1 mol% Au(III) salt **(Scheme 15)** [45b].

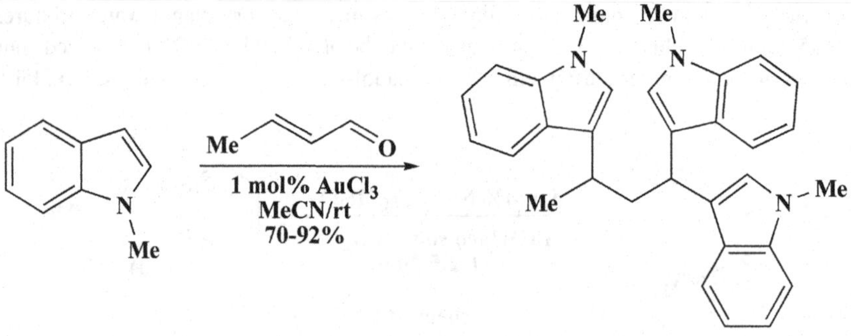

Scheme-15

Zhang et al. [45l] and Li et al. [45m] reported that hydroamination of alkenes was promoted by gold(I). Both intra- and intermolecular addition of tosylamide occurred in good to excellent yields at 85 °C with 5 mol% Ph$_3$PAuOTf. This mechanism involved a *trans*-addition. Hydroamination of olefins can be catalyzed by Au as well as catalytic amount of HOTf [45n]. The ^{31}P NMR of Ph$_3$PAuOTf at 85 °C suggested that the AuPPh$_3^+$ species did not interact with TsNH$_2$ but with cyclohexene or norbornene. Further, He et al. [45o] also synthesized Ph$_3$PAuNHTs which neither catalyzed nor underwent olefin hydroamination reaction but reacted with strong Brønsted acids immediately to release TsNH$_2$. However, a comprehensive understanding of the mechanism is still necessary **(Scheme 16)**.

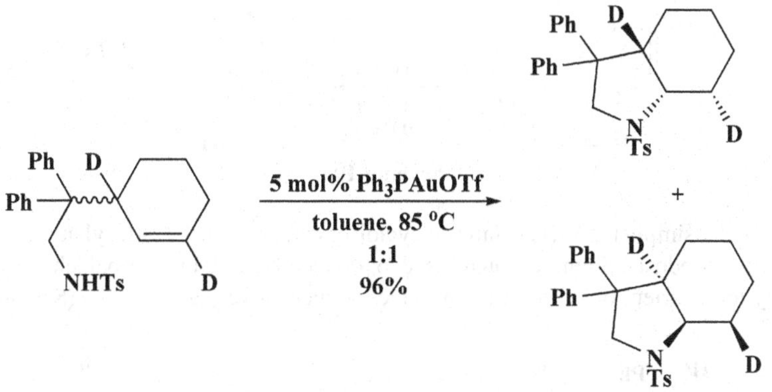

Scheme-16

Alfonsi et al. [45p] and Arcadi et al. [45q] developed a tandem reaction employing 2-alkynyl-phenylamines with α,β-enones, which afforded substituted indoles in the presence of catalytic amounts of NaAuCl₄ in C₂H₅OH. Similar results were obtained when an ionic liquid was used as a solvent **(Scheme 17)** [45o].

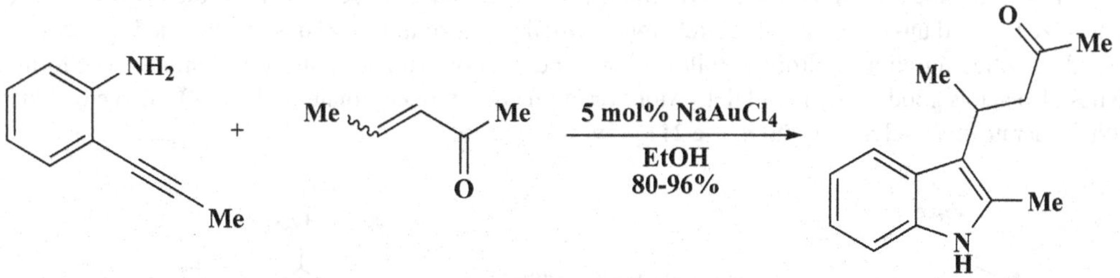

Scheme-17

Liu et al. [45r] reported Au-catalyzed annulations of enantioenriched allenes for an enantioselective total synthesis of (−)-rhazinilam with high *de* and yields. In another work by He et al., Au-catalyzed intramolecular nucleophilic substitution and hydration of enantioenriched alkynyl ether tethered with an amine functional group was found to afford enantioselective total synthesis of rachcinidine **(Scheme 18)** [45o].

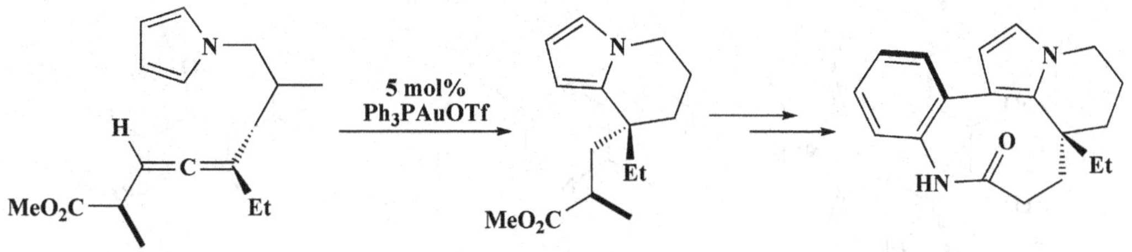

Scheme-18

Hashmi et al. [45s] reported that furylmethyl propargyl ether and amine underwent an Au-catalyzed [4+2]-cycloaddition reaction to give arenes in good to excellent yields. Another study, by Nakamura, found that intramolecular Diels-Alder reaction provided an intermediate which underwent gold(III)-promoted isomerization and finally synthesized arenes upon aromatization **(Scheme 19)** [45t].

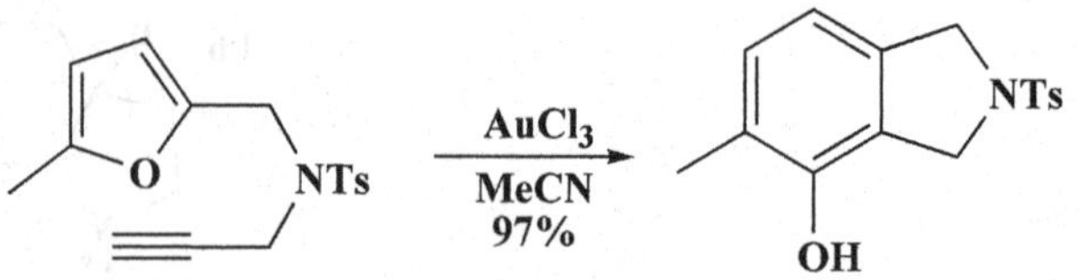

Scheme-19

An operationally simplistic Au(I)-catalyzed cycloisomerization of 1-(2-(tosylamino)phenyl)prop-2-yn-1-ols with NIS (*N*-iodosuccinimide) can be used to produce 1*H*-indole-2-carbaldehydes. The reactions occur efficiently for a variety of substrates to provide products in very good yields **(Scheme 20)** [45u].

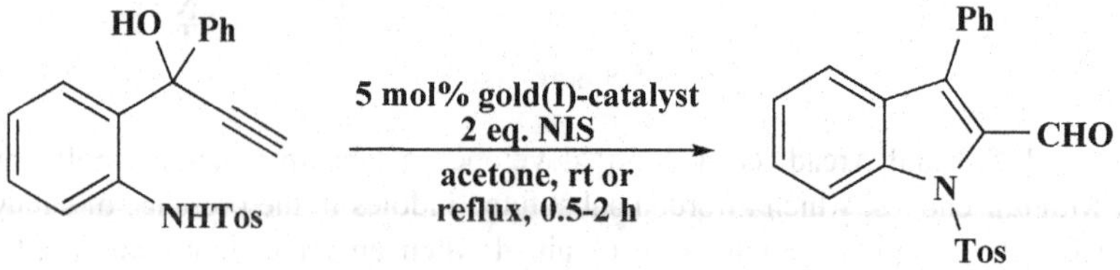

Scheme-20

Through double migratory cascade reaction, 1,4-bis-propargylic acetates afford dienes which, when hydrolyzed, afford un-symmetrical 1,2-diketones. Good yields of quinoxalines are obtained by a one-pot cascade double migration-hydrolysis followed by condensation with aromatic 1,2-diamines. The formed dienes also act as good partners in Diels-Alder reaction with *N*-phenylmaleimide to afford cycloadducts with moderate *endo*-selectivity **(Scheme 21)** [45v–w].

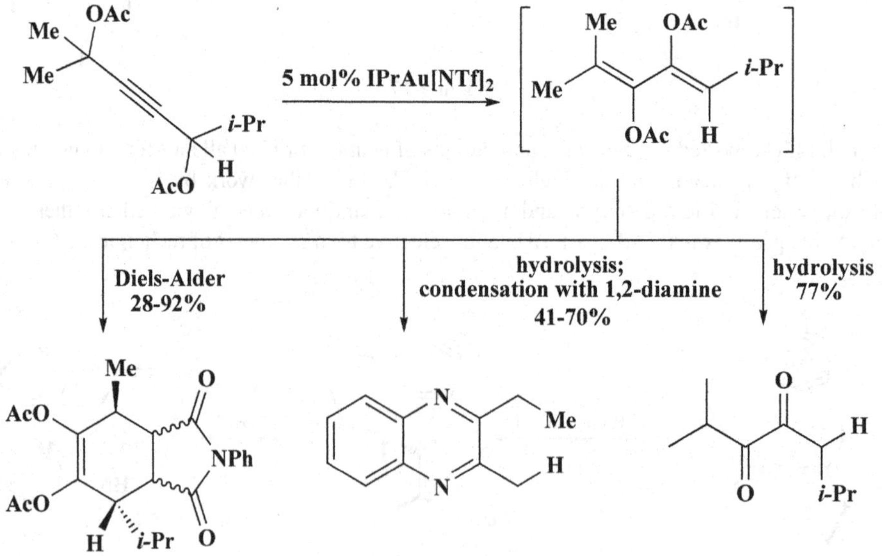

Scheme-21

Indium-assisted synthesis

Indium-catalyzed cyclization of 2-ethynylanilines having an aryl or alkyl group on the terminal alkyne result in good yields of various poly-functionalized indole derivatives **(Scheme 22)**. On the other hand, substrates without substituent on the triple bond or with a trimethylsilyl group produce good yields of poly-substituted quinoline derivatives *via* intermolecular dimerization [46–48].

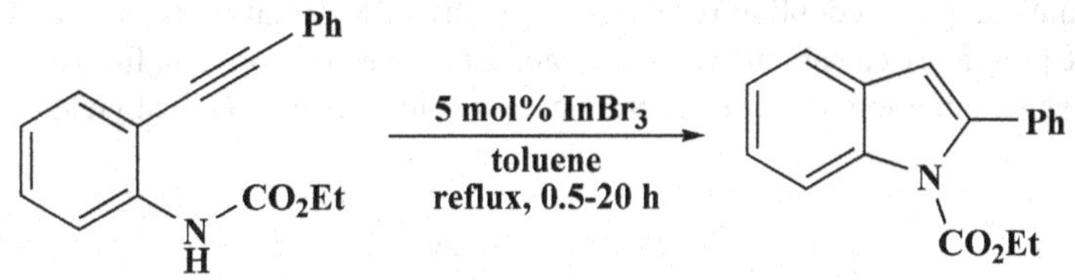

Scheme-22

Hirashita et al. [49] reported conversion of allylamine into indoline derivative *via* a Br-Li exchange, formation of an allylic indium compound, and subsequent intramolecular radical cyclization to provide a 5-*exo-trig* product **(Scheme 23)** [48].

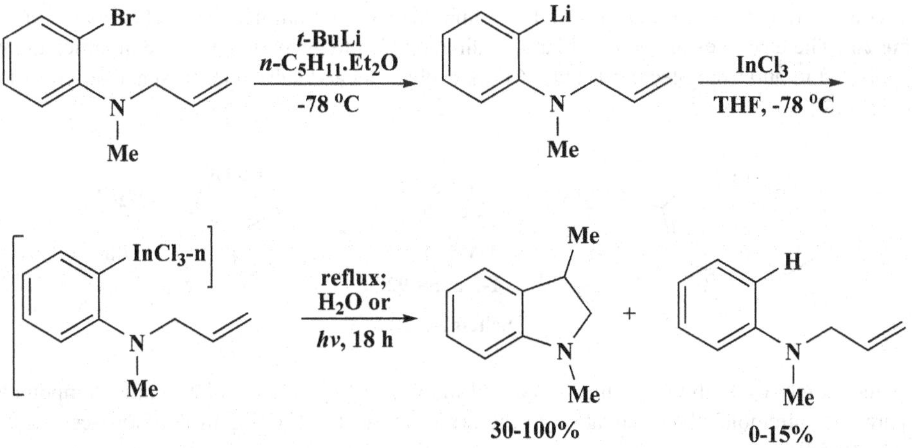

30-100%   0-15%

Scheme-23

Iodine-assisted synthesis

Iodine-mediated intramolecular cyclization of enamines under transition metal-free reaction conditions has been known to produce good to high yields of a wide range of 3*H*-indole derivatives containing multifunctional groups **(Scheme 24)** [50].

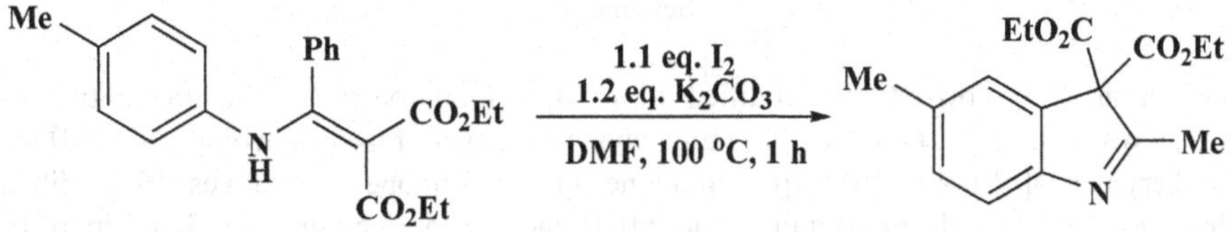

Scheme-24

Yue et al. [51] investigated iodine-assisted electrophilic cyclization for the synthesis of various 2,3-disubstituted indoles **(Scheme 25)**. An iodoirenium intermediate was produced when iodine coordinated with the carbon-carbon triple bond. Indole derivatives were formed by intramolecular cyclization of iodoirenium intermediate by the nucleophilic attack of amino group followed by loss of methyl iodide. Sakamoto et al. [52] studied a coupling of 3-iodoindoles with terminal acetylenes, wherein only indoles having a protected nitrogen atom with an electron-withdrawing 1-methanesulfonyl group provided satisfactory results. The C-I bond was electron-rich, with an electron-donating group on the nitrogen atom of the 3-iodoindole, and limited the further functionalization at the third position of the indole by palladium-catalyzed coupling reactions. Coupling of terminal acetylenes with *N,N*-dialkyl-*o*-iodoanilines in the presence of a palladium/copper catalyst has been reported to furnish *N,N*-dialkyl-*o*-(1-alkynyl)anilines which upon iodocyclization produce 3-iodoindoles in excellent yields [53–55].

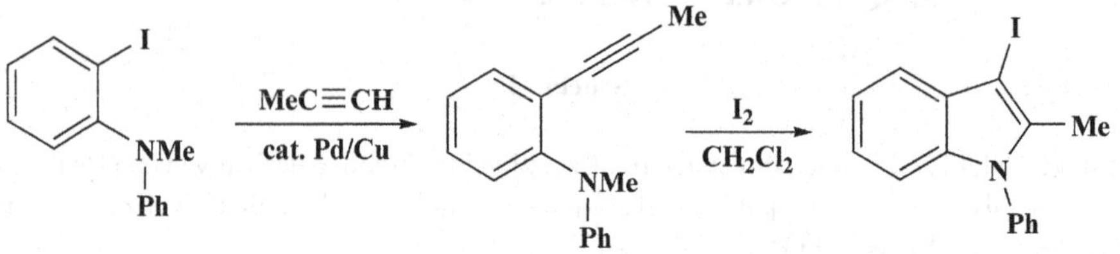

Scheme-25

He et al. [56] synthesized 3*H*-indoles by iodine-mediated intramolecular cyclization of enamines **(Scheme 26)**. The enamines undergo oxidative iodination [57] to generate an iodide intermediate which when subjected to an intramolecular Friedel-Crafts cyclization and subsequent aromatization yields the 3*H*-indoles [55].

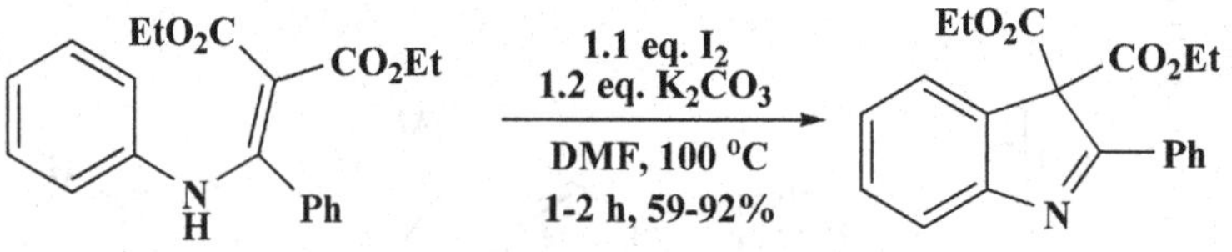

Scheme-26

A wide variety of *N*-alkylated and *N*-arylated indoles and pyrrole-fused aromatic compounds can be prepared by intramolecular cyclization in the presence of phenyliodine bis(trifluoroacetate) (PIFA) **(Scheme 27)** [58].

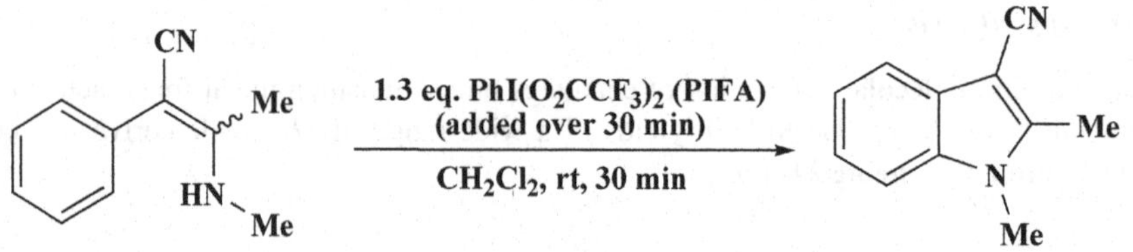

Scheme-27

Tellitu et al. [59] utilized the modified form of this approach for the preparation of many heterocyclic systems, e.g., naphtho-, benzo-, and heterocycle-fused pyrrolo[2,1-*c*][1,4]diazepines [60], 2,3-diarylbenzo[*b*]furans [61], pyrrolidinone or quinolinone derivatives [62], dibenzo[*a,c*] phenanthridines [63], thiazolo-fused quinolinones [64], indoline derivatives [65], 5-aroylpyrrolidinones [66], and indazolone derivatives [67]. Recent examples include the synthesis of indazol-3-ones from anthranilamides and indoline derivatives from anilides **(Scheme 28)** [68].

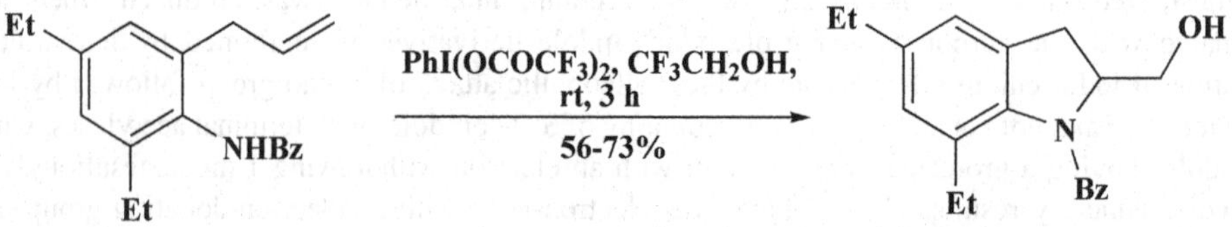

Scheme-28

Iridium-assisted synthesis

Intramolecular amination of *o*-homobenzyl-substituted aryl azides furnishes indolines under Ir catalysis. In this reaction, electron-poor azides have been found to be better suited, whereas little to no conversion has been reported with electron-rich azides (**Scheme 29**). Also, this reaction is limited to secondary benzylic carbon-hydrogen bonds, as both secondary alkyl and tertiary benzylic carbon-hydrogen bonds remain intact. The detection of indole by-products in varying amounts suggests that oxidation occurs as a major competing route. Indole becomes the major product with substituted aniline as the major by-product depending on the Ir catalyst used and the temperature. Intermolecular competition procedures suggest that the mechanism does not proceed *via* benzylic carbon-hydrogen bond activation nucleophilic addition sequence. Rather, an electrophilic Ir-nitrene species was suggested due to a more efficient reaction with electron-deficient substrates and the detection of anilines (a common nitrene decomposition product). The carbon-hydrogen bond amination was assumed to occur *via* hydrogen abstraction/radical rebound, or a concerted carbon-hydrogen insertion in clear contrast to the previously described Glorius system [69].

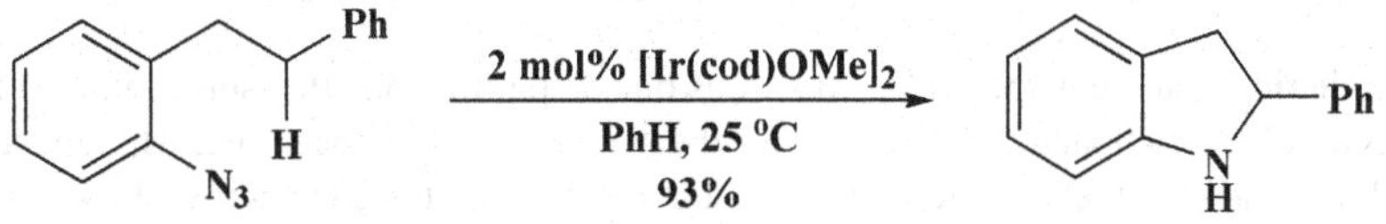

Scheme-29

Iron-assisted synthesis

Indole derivatives can be synthesized by iron-catalyzed intramolecular carbon-hydrogen amination reaction with commercially available iron(II) triflate catalyst (**Scheme 30**) [70].

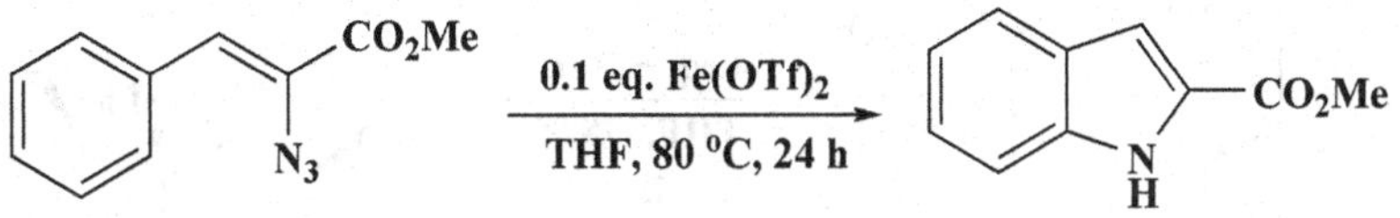

Scheme-30

Lithium-assisted synthesis

A variety of indole derivatives are prepared from 2-fluorophenyl imines. 2-Fluoroaniline-d_4 is used to investigate the mechanism of this indolization (**Scheme 31**) [71].

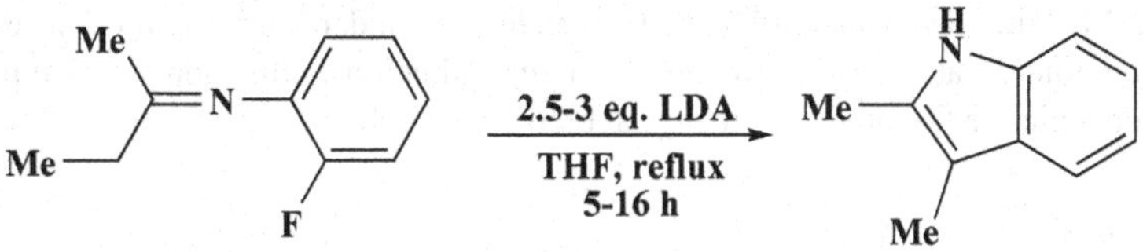

Scheme-31

Groth et al. [72] synthesized 3,3-disubstituted indulines from *N*-benzyl-*N*-allyl-2-bromoanilines by (−)-sparteine assisted intramolecular asymmetric carbolithiation in the presence of *t*-BuLi (**Scheme 32**). They studied the effect of the nature of the side chain on the yields and enantioselectivity of the product formed in detail [48].

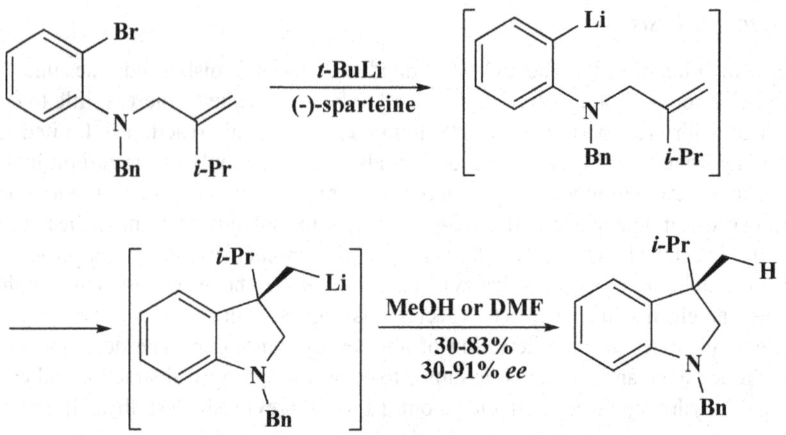

Scheme-32

Hydroindole rings can be prepared by numerous one-step protocols. Pearson et al. [73–74] reported an intramolecular [3+2]-cycloaddition reaction for the synthesis of (+)-coccinine, wherein stannyl imine was subjected to a tin-lithium exchange with *n*-BuLi at low temperatures **(Scheme 33)**. Hydroindole was formed in 45% yields when aza allyl anion underwent cycloaddition across the tethered olefinic bond. The product was presumably formed *via* a chair-like transition state as a single isomer.

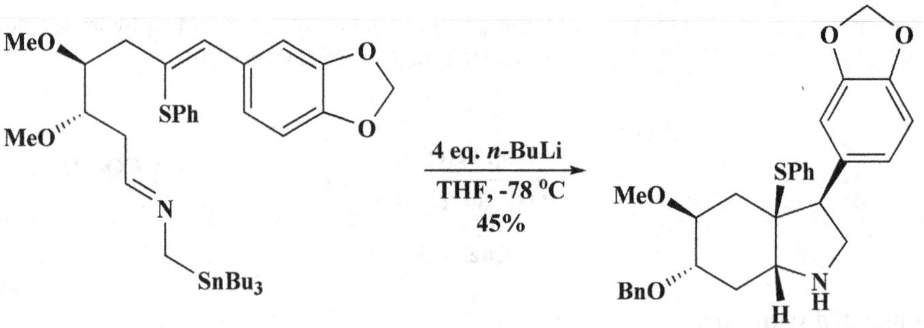

Scheme-33

Ogata et al. [75] reported a lithium-catalyzed asymmetric hydroamination/cyclization of amino-substituted stilbenes with a chiral bis-oxazoline ligand in enantioselectivities as high as 91% enantiomeric excess **(Scheme 34)**. The *exo*-cyclization product was formed under kinetic control when the reactions were carried out in toluene at –60 °C. However, the hydroamination/cyclization reaction in THF solution is reversible, producing the thermodynamically favored *endo* cyclization product [76].

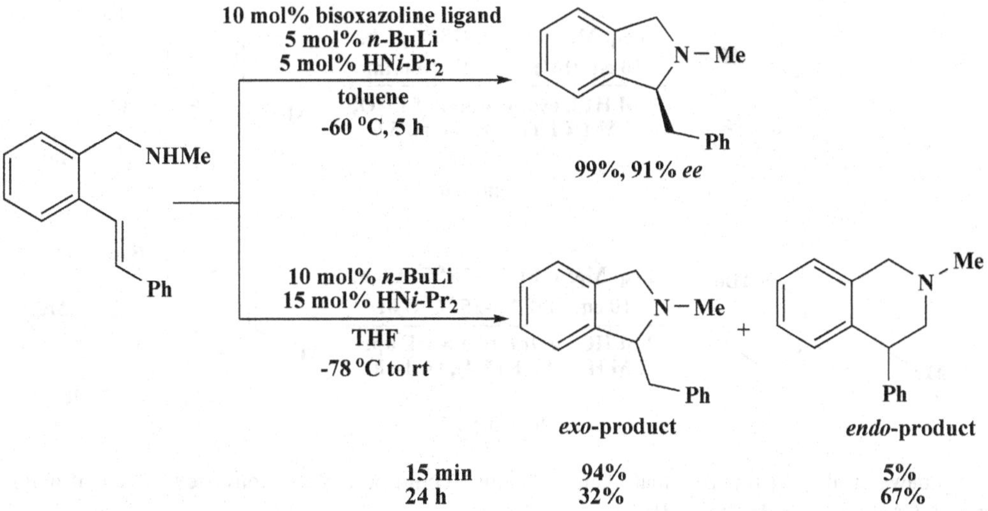

Scheme-34

Functionalized indoles can be synthesized *via* electrophilic activation of *N*-aryl amides and the addition of ethyl diazoacetate to these highly activated amides. This offers a great potential for the preparation of naturally occurring and pharmacologically active indole derivatives **(Scheme 35)** [77].

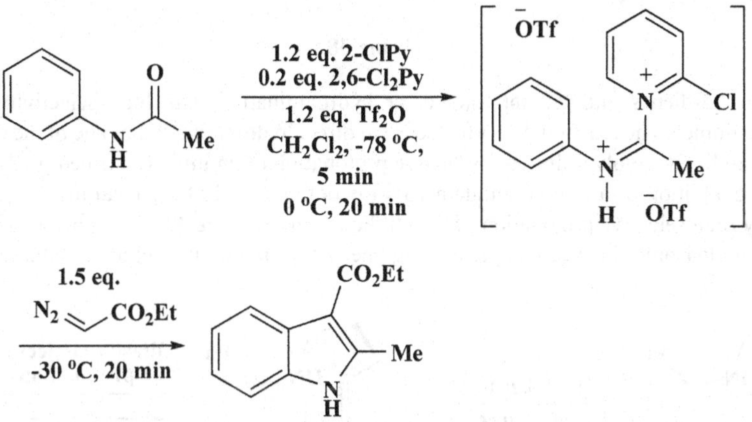

Scheme-35

A reaction between Boc-protected *o*-aminostyrenes and alkyllithiums, followed by the addition of specific electrophiles, sets up a cascade reaction between the reacted electrophile and the *o*-amino substituent, facilitating an *in situ* ring-closure and dehydration for the synthesis of an indole ring system **(Scheme 36–37)** [78].

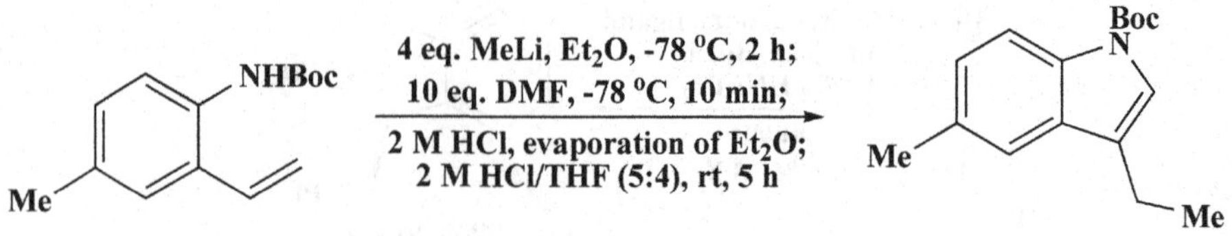

Scheme-36

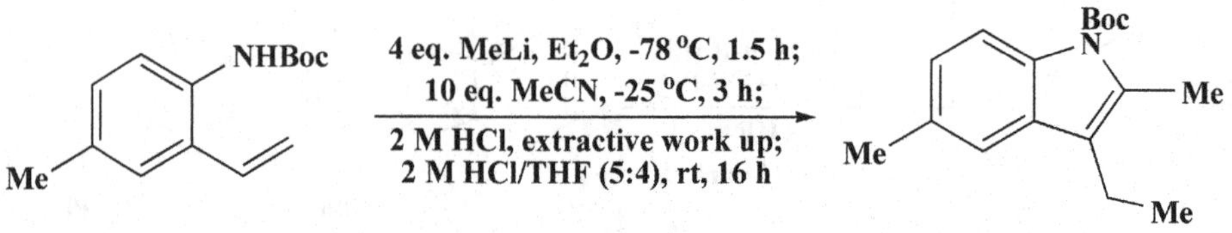

Scheme-37

Zabawa et al. [79] reported that cyclic sulfamide could be directly converted to vicinal diamine through LAH reduction (**Scheme 38**).

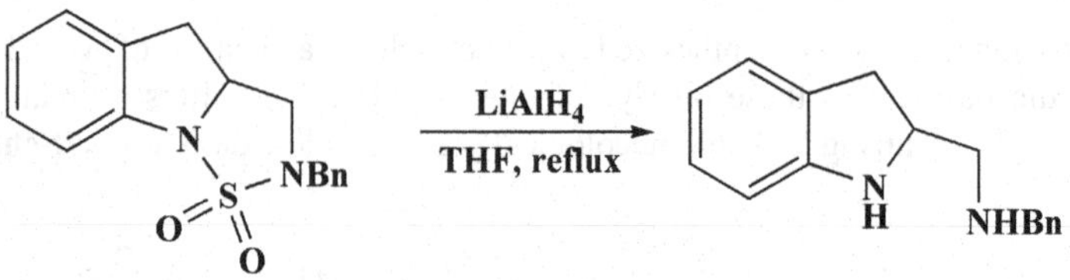

Scheme-38

Cyclohexa-2,5-dienes undergo intramolecular hydroamination with high selectivity to result in bicyclic allylic amines. The reaction does not occur *via* direct hydroamination of one of the diastereotopic olefins, but more likely involves diastereoselective protonation of an initially formed pentadienyl anion, followed by the addition of a lithium amide across the double bond of the resulting 1,3-diene, and the ultimate highly regioselective protonation of the allylic anion (**Scheme 39**) [80]. The cyclized product is obtained as a single isomer in excellent yield using phenylglycinol derived chiral auxiliaries [76].

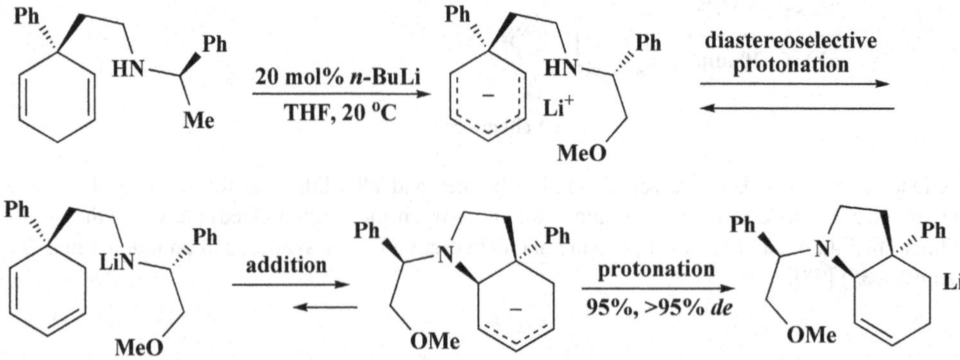

Scheme-39

Guthrie and Curran [81] reported radical and anionic cyclization of axially chiral atropisomers of substituted allylamine derivatives and observed high levels of chirality transfer from the N-Ar axis to the new stereocenter in the substituted dihydroindoles (**Scheme 40**) [48].

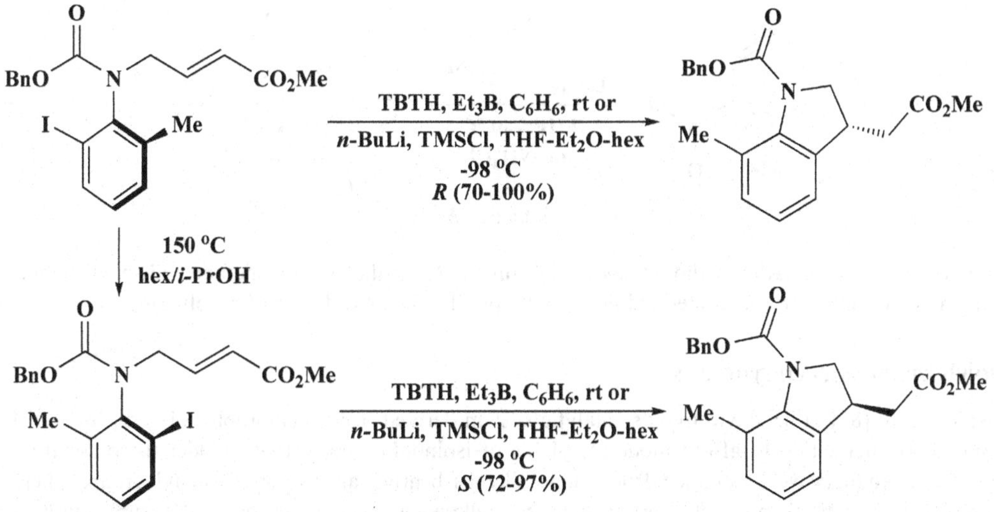

Scheme-40

Barluenga et al. [82] studied a direct protocol for intramolecular anionic carbon–carbon bond formation employing arynes and were able to develop a method for the synthesis of 3,4-disubstituted indoles by low-temperature lithiation. The *o*-fluoro anilines with pendant 2-bromo-allyl fragments produced aryne intermediates which were treated with *tert*-butyllithium for ring closure to form indole-3-methylidenes with a C(4) anion. Upon warming, these compounds were quenched with various electrophiles to synthesize the desired products. Notably, the reaction occurred selectively, forming aryne by elimination of aryl fluoride while simultaneously carrying out a lithium-halogen exchange to afford the vinyl anion (**Scheme 41**).

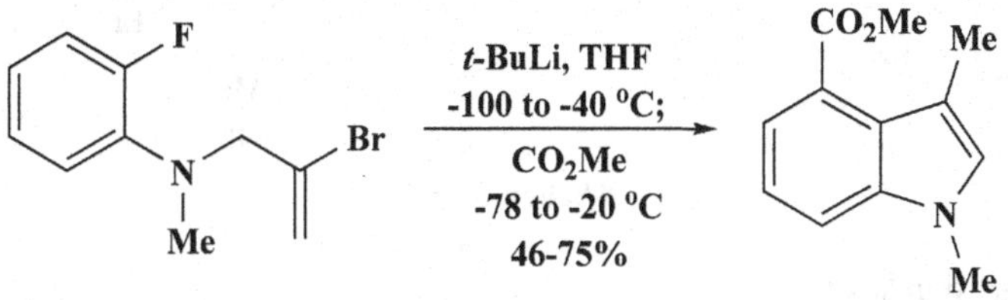

Scheme-41

Madelung synthesis [83] utilized for the synthesis of un-substituted or substituted indoles by intramolecular cyclization of *N*-phenylamides using a strong base at high temperature is confined to the synthesis of 2-alkenylindoles (not obtained easily *via* electrophile aromatic substitution) due to harsh reaction conditions (**Scheme 42**).

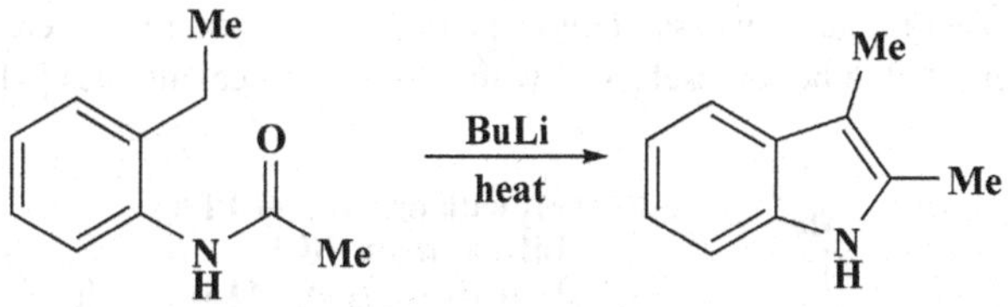

Scheme-42

Magnesium-assisted synthesis

Bartoli et al. [84] synthesized indole from nitroarenes using vinyl Grignard's reagent at low temperature. Three eq. of Grignard's reagent were required for a substantial yield of the products (**Scheme 43**).

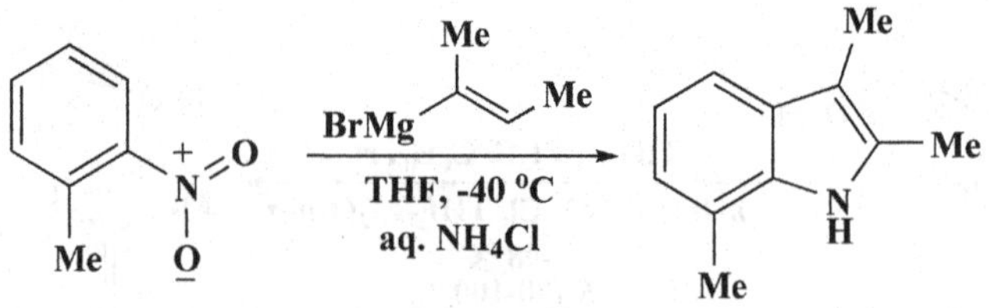

Scheme-43

However, the reaction failed without an *ortho* substituent. Nevertheless, this method had an advantage of being able to synthesize substituted indoles on both pyrole ring as well as carbocyclic ring.

Molybdenum-assisted synthesis

Nishida et al. [85] utilized Mortreux's catalytic system derived from *p*-chlorophenol and Mo(CO)$_6$ for alkyne co-trimerization to afford moderate yields of isoindoline derivatives. Under these conditions, trimerized products were obtained from mono-alkynes bearing an *o*-hydroxyphenyl group, whereas cross-alkyne metathesis products were formed from alkynes possessing *m*- or *p*-hydroxyphenyl group. The cyclized product along with by-products is obtained when the diyne undergoes co-trimerization with various mono-alkynes under similar reaction conditions **(Scheme 44)** [86]. This catalytic reaction occurs *via* molybdenacyclopentadiene complex as indicated by the isolation of the by-product. Using 15 eq. of monoalkyne improves the yields of the desired trimerized products. The same catalyst system can also be utilized for totally intramolecular [2+2+2]-cyclization reactions [87–89].

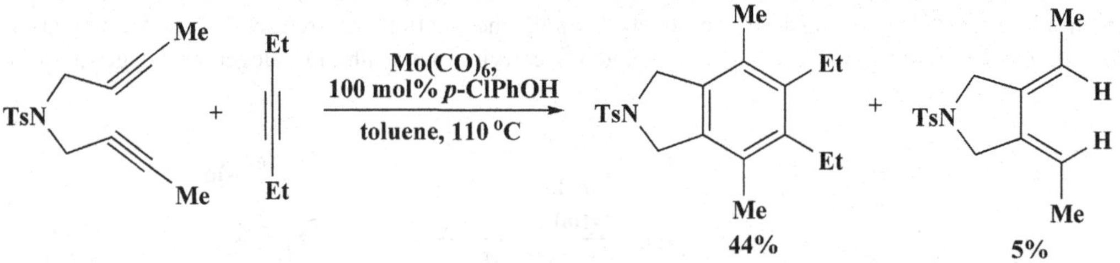

Scheme-44

Palladium-assisted synthesis

The best results are reported using one eq. of either *n*-Bu$_4$NCl or lithium chloride and an excess of the alkyne along with potassium or sodium carbonate or acetate base, occasionally adding 5 mol% triphenylphosphine [90–97]. Larock et al. [98–101] based their efficient and versatile palladium-catalyzed indole synthesis on this principle **(Scheme 45)**. Subsequently, more reproducible and higher yields were observed with lithium chloride compared to *n*-Bu$_4$NCl. However, the efficiency of this reaction was strongly linked to the utilization of soluble palladium catalysts associated with additives and phosphine ligands, as well as the use of a large excess of alkyne and base, resulting in severe practical drawbacks. Cyclization has been reported to be regioselective with un-symmetrical alkynes [41, 45].

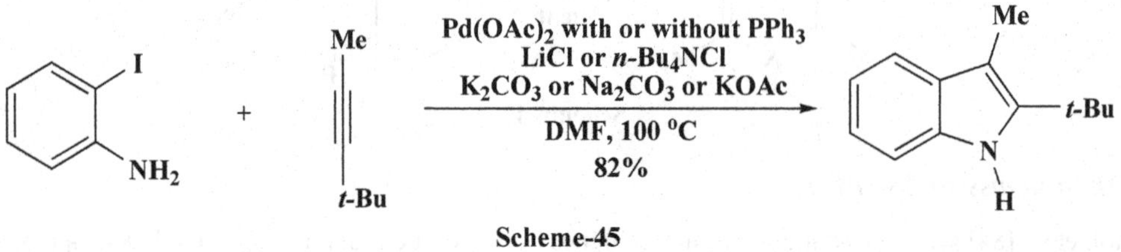

Scheme-45

α,β-Unsaturated carbonyl compounds and *o*-alkynylanilides undergo domino aminopalladation/ hydroarylation for the synthesis of 2-substituted 3-alkylindoles **(Scheme 46)** [102–103]. To prevent the synthesis of Heck products *via* β-hydride elimination, lithium bromide is needed. Excess amounts of lithium bromide inhibits the β-hydride elimination and allows the formation of the reduced product instead. This not only allows access to 3-alkyl substituted indoles, but also eliminates the need for stoichiometric oxidants [41].

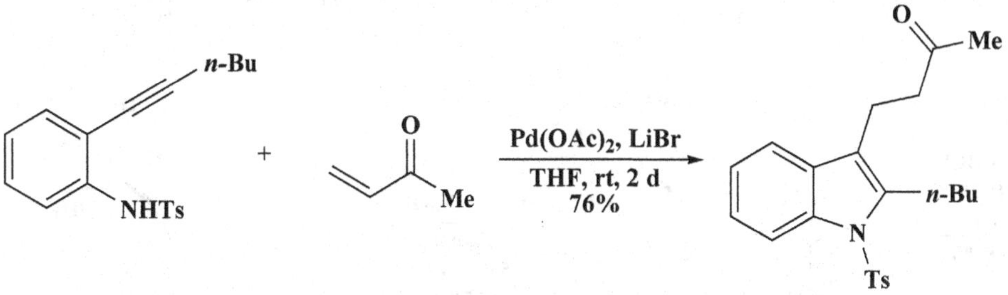

Scheme-46

Chen et al. [104] studied a palladium-catalyzed coupling for the preparation of indole core of rizatriptan. The iodoaniline derivative needed in this reaction was readily synthesized from triazolomethyl aniline through treatment with iodine monochloride in aqueous methanol. The bis-TES protected butynol served as the most efficient coupling partner and yielded the indole smoothly. Subsequently, the desired drug compound was generated by the removal of TES-groups and introduction of dimethylamino functionality **(Scheme 47)** [105].

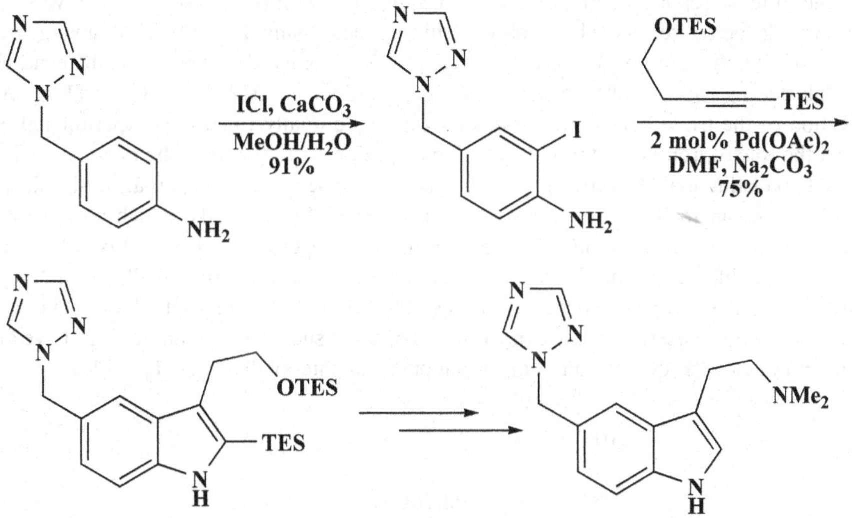

Scheme-47

Bz-substituted tryptophans [106]; optically active ring substituted tryptophans [107] **(Scheme 48)**; alkoxy-substituted indole bases 16-*epi*-Na-methylgardneral, 11-methoxymacroline and 11-methoxyaffinisine as well as indole alkaloids macralstonine and alstophylline [108]; 12-alkoxy-substituted indole alkaloids (+)-12-methoxyaffinisine, (+)-12-methoxy-Na-methylvellosimine, (–)-fuchsiaefoline [109]; (–)-5-methoxyindolmycin, and (–)-indolmycin [110]; 12-methoxy-substituted sarpagine indole alkaloids [111]; vincamajine-related indole alkaloids [112]; and the 5-HT1D receptor agonist MK-0462 are synthesized through the annulation desilylation sequence. This silylalkyne chemistry can also be used in solid-phase microwave-assisted synthesis of 5-carboxamido-*N*-acetyltryptamine derivatives [41].

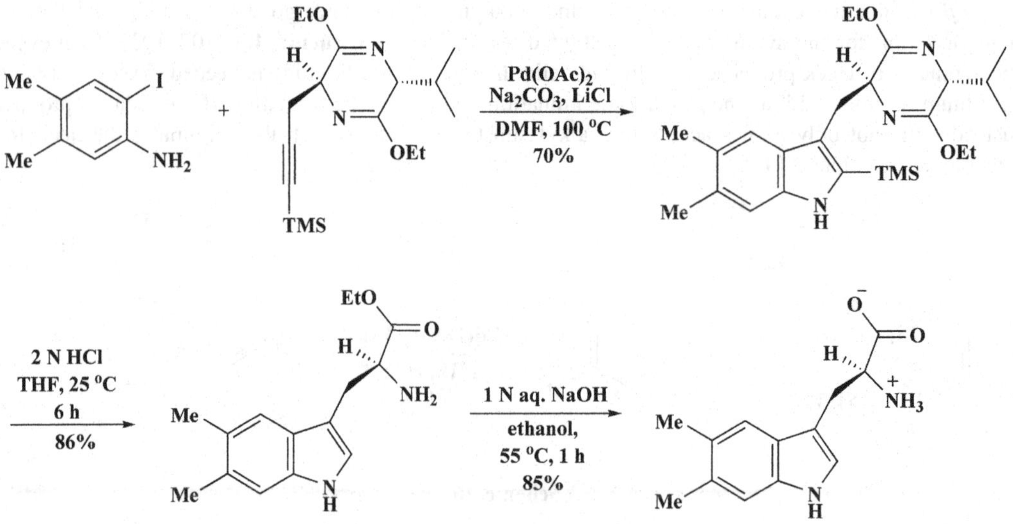

Scheme-48

In cases where *o*-(1-methylethenyl)indole is produced using *o*-[(3-hydroxy-3,3-dimethyl)prop-1-yl] trifluoroacetanilide [113], the olefinic double bond is formed by palladium-catalyzed cyclization followed by the hydrolysis of the amide bond and a dehydration process (**Scheme 49**). The *o*-alkynylanilines can be cyclized [114] in an acidic dichloromethane/hydrogen chloride two-phase system at room temperature in the presence of Bu$_4$NCl and palladium chloride. Usually, these conditions afford yields higher than or comparable to those reported with palladium chloride in acetonitrile at 60–80 °C. However, neutral conditions provide better results with *o*-alkynylanilines possessing electron-withdrawing groups on the alkyne functionality. Unprotected primary *o*-alkynylanilines can be cyclized with ferric chloride-palladium chloride (2 and 1 mol% respectively) in dichloroethane (DCE) at 80 °C [115]. Although the π-activation of the triple bond by iron species cannot be totally ruled out, experimental evidence suggests that the role of iron under these cyclization conditions is to facilitate the *in situ* re-oxidation of palladium(0) to palladium(II). Moreover, as these reactions are performed in open-air flasks, the presence of oxygen may account for the cascade re-oxidation of iron(II) to iron(III), which allows for the use of ferric chloride in catalytic amounts. Amatore et al. [116] reported the preparation of 2-substituted indoles from *o*-iodotrifluoroacetanilide and *o*-iodoaniline. This involved terminal alkynes in the presence of triphenylphosphinetrisulfonate sodium salt (TPPTS), Pd(OAc)$_2$ and triethylamine in water and acetonitrile without any copper promoter. In a number of other such studies, indole rings have also been produced by intramolecular cyclization of an organopalladate intermediate [41, 117–140].

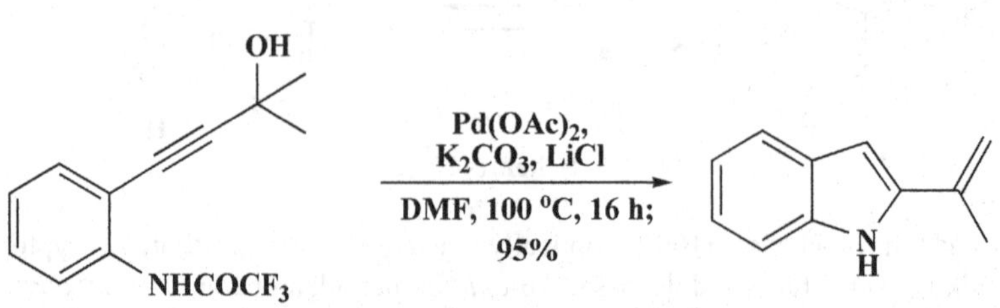

Scheme-49

Larock synthesis of indoles is based on the use of *N*-aroylbenzotriazoles and *o*-iodobenzoic acid as synthetic equivalents of *o*-iodoanilines [141–147]. The indoles are synthesized from *o*-iodobenzoic acid involving one-pot Curtius rearrangement or palladium-catalyzed indolization. Leogane and Lebel [148] observed that carboxylic acids, when coupled with alkynes and sodium azide, afforded indoles—an alternative to the use of aniline derivatives in this chemistry. In this case, a Curtius rearrangement occurred first to provide an *o*-iodo arylisocyanate *in situ*, which reacted with nucleophiles to form an NH unit required for palladium-catalyzed indole synthesis [149]. This reaction provided an intermediate for cyclization and coupling with amines to produce urea variants of the intermediate, thus yielding four separate units of *N*-substituted indoles in one-pot **(Scheme 50)** [41, 103].

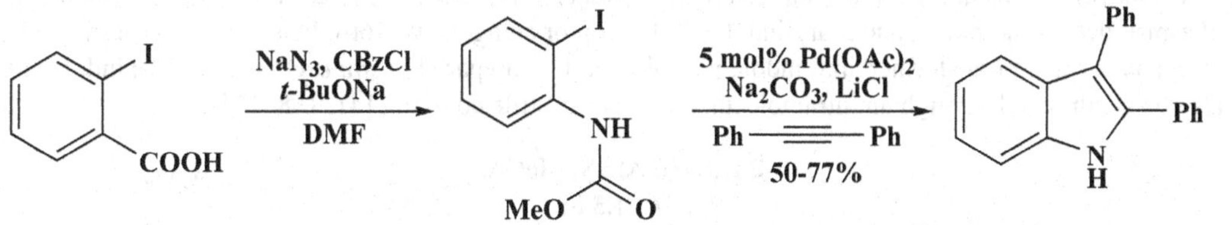

Scheme-50

Denmark and Baird [150] replaced the simple trimethylsilyl group used by Larock with a silyl ether and found that *N*-benzyl-2-iodoanilines reacted with an alkynyldimethylsilyl *tert*-butyl ether to provide indole-2-silanols after hydrolysis **(Scheme 51)**. The success of the reaction depended upon the use of *tert*-butoxysilyl ether. The silyl ether served two purposes: directing the heteroannulation, and allowing Si-based cross-coupling after unmasking the silanol. The sodium 2-indolylsilanolate salts have been found to be successfully engaged in cross-coupling reactions with aryl chlorides and bromides to provide *N*-benzyl-2,3-disubstituted indoles [41].

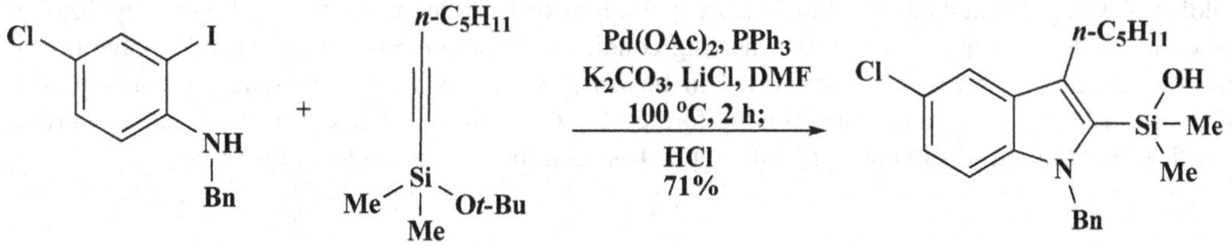

Scheme-51

Backvall and Andersson [151–152] reported a stereoselective palladium-catalyzed 1,4-addition of cyclic 1,3-dienes which yielded either lactam or pyrrolidine products. For example, when diene was treated with excess lithium chloride and catalytic palladium acetate, pyrrolidine was formed. Alternatively, treatment with excess of lithium acetate and catalytic palladium acetate diene afforded pyrrolidine. Both the reactions occurred *via anti*-aminopalladation to furnish an allylpalladium complex which was captured by an external nucleophile. The diene undergoes *syn*-addition to afford pyrrolidine by an outer-sphere attack of chloride ion on the allylpalladium complex, whereas *anti*-addition on diene produces pyrrolidine by inner-sphere attack of acetate **(Scheme 52)** [153–159].

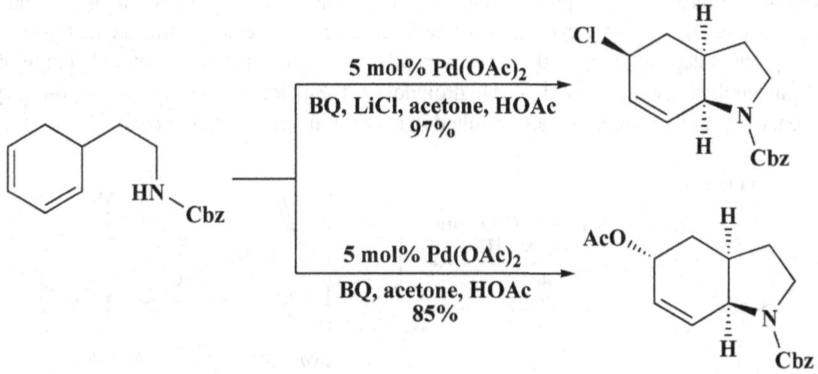

Scheme-52

Tin-assisted synthesis

Stille reaction is used in the key steps of synthesis of various biologically active compounds like borrerine [160] (a naturally occurring alkaloid), grossularines-1 [161–162], hyellazole and carazostatin [163] (poly-substituted carbazole alkaloids), tetracyclic oxazolocarbazoles [164] (formed as functionalized precursors of carbazoquinocins and antiostatins), (–)-tabersonine and (–)-vincadifformine [165] (members of aspirioderma alkaloids), and 2-arylindole NK1 receptor antagonists [166]. In a one-pot strategy [167], isocyanade can be transformed into indole *via* 2-iodoindole prepared *in situ* **(Scheme 53)**. An indole can also be synthesized through an intramolecular version of Stille coupling [41, 168–171].

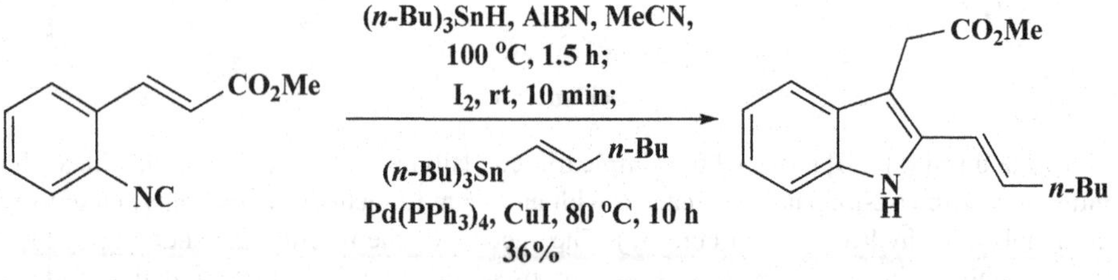

Scheme-53

Organotin hydrides serve as popular reagents in free radical reactions. However, removal of their derivatives from the reaction mixture is often difficult. The use of tin reagents in medicinal chemistry is limited due to tin residues in the final product. This issue is addressed using fluorous tin reagents. Olofsson et al. [172] studied the utility of a heavy fluorous tin hydride for halide reduction and cyclization reactions under continuous microwave heating conditions **(Scheme 54)**. To increase the solubility of the tin reagent, reactions were performed in benzotrifluoride (BTF). A three-phase extraction with dichloromethane, FC-84 and water was performed for the work-up of reaction mixtures. For further purification of products, circular chromatography has been found to be useful [173–174].

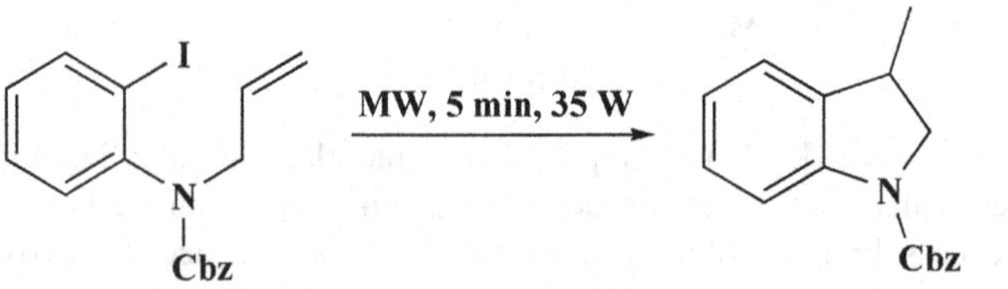

Scheme-54

Tungsten-assisted synthesis

Numerous cyclization reactions have been reported in the presence of transition metal catalysts. Both hydroindoline and hydroindole skeletons are generated by the reaction of a cyclic amine with an existing six-membered carbocycle through nucleophilic addition to a transition metal activated alkyne. Grandmare et al. [175] investigated W(CO)$_6$-catalyzed cyclization of acetylenic silyl enol ether to afford either 5-*exo* or 6-*endo*-*dig* cyclization product depending upon the reaction catalyst and solvent stoichiometry **(Scheme 55)**. The 10 mol% W(CO)$_6$ in tetrahydrofuran/water yielded hydroindole as the major product through 5-*exo*-*dig* cyclization. In contrast, 1 eq. of W(CO)$_6$ in toluene/water resulted in hydroquinoline through 6-*endo*-*dig* cyclization.

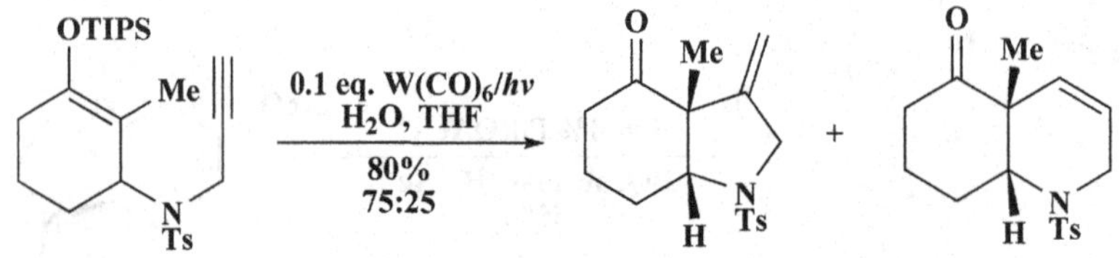

Scheme-55

Zinc-assisted synthesis

Fischer indole synthesis is the most important method for the preparation of substituted indole. Herein, phenylhydrazine reacts with pyruvic acid, followed by decarboxylation to synthesize an indole. This method follows a one-pot route under microwave irradiation **(Scheme 56)** [176].

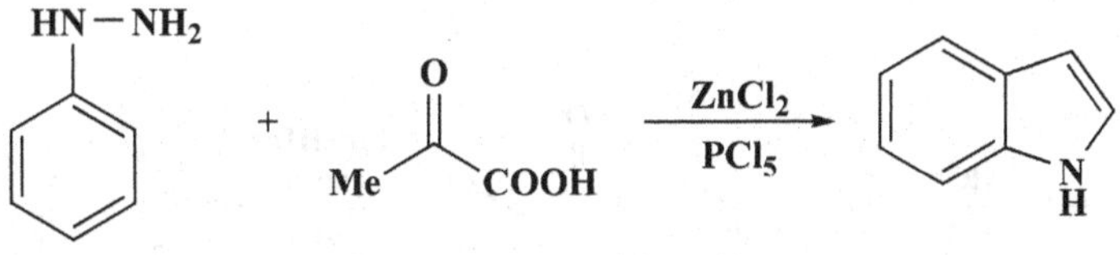

Scheme-56

Pummerer cyclization of amide yields dihydroindolone [177]. Subsequent to the optimization of known Pummerer promoters, it has been found that the highest yield of dihydroindolone (94%) is obtained by the reaction of amide with 1-(dimethyl-*tert*-butylsiloxy)-1-methoxyethylene in the presence of a catalytic amount of zinc iodide in dry CH_3CN **(Scheme 57)**. This reaction, wherein sulfoxides react with *O*-silylated ketene acetals, was initially reported by Kita et al. [178–179] for the synthesis of siloxy sulfides under mild conditions. The reaction occurs through the formation of a silyloxy sulfonium salt which undergoes subsequent elimination by a highly stereoselective de-protonation of the *anti*-periplanar α-methylene proton [180–181].

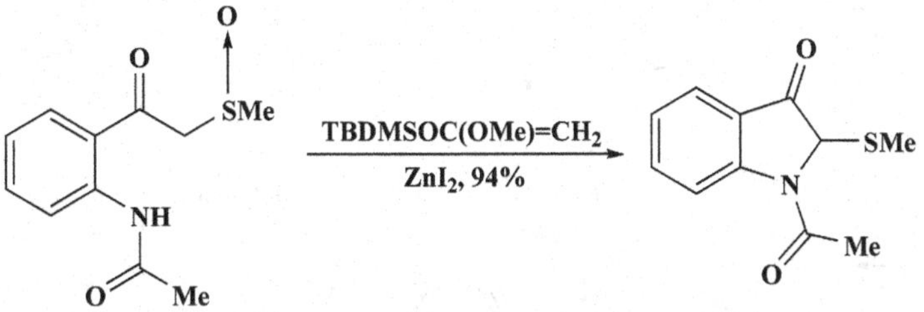

Scheme-57

N-Arylindole-3-carbonitriles can be synthesized from 2-aryl-3-arylamino-2-alkenenitriles through a one-pot method involving *N*-chlorosuccinimide- or *N*-bromosuccinimide-assisted halogenation followed by zinc acetate-catalyzed intramolecular cyclization. The arylnitrenium ion intermediates which undergo an electrophilic aromatic substitution affords cyclized *N*-arylindoles **(Scheme 58)** [182].

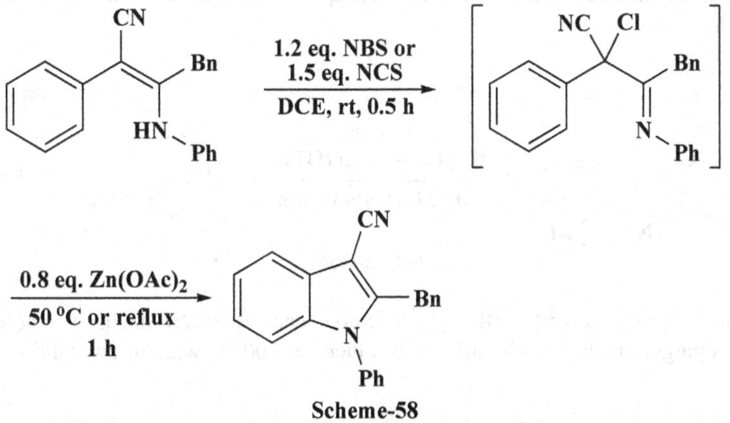

Scheme-58

Fluvastatin (Lescol), synthesized by a Fischer indole synthesis, acts as HMG-CoA reductase inhibitor **(Scheme 59)** [183]. However, it was reported in initial stages that a Bischler-Mohlau type reaction was used instead. The 3-substituted indole core required at an early stage in the synthesis can be prepared by zinc chloride-mediated Bischler-type indole synthesis [184] with stoichiometric amounts of aniline. The introduction of *syn*-diol pendant side-chainis was aided by a novel pathway to introduce a formyl substituent at the 2-position of the indole [105].

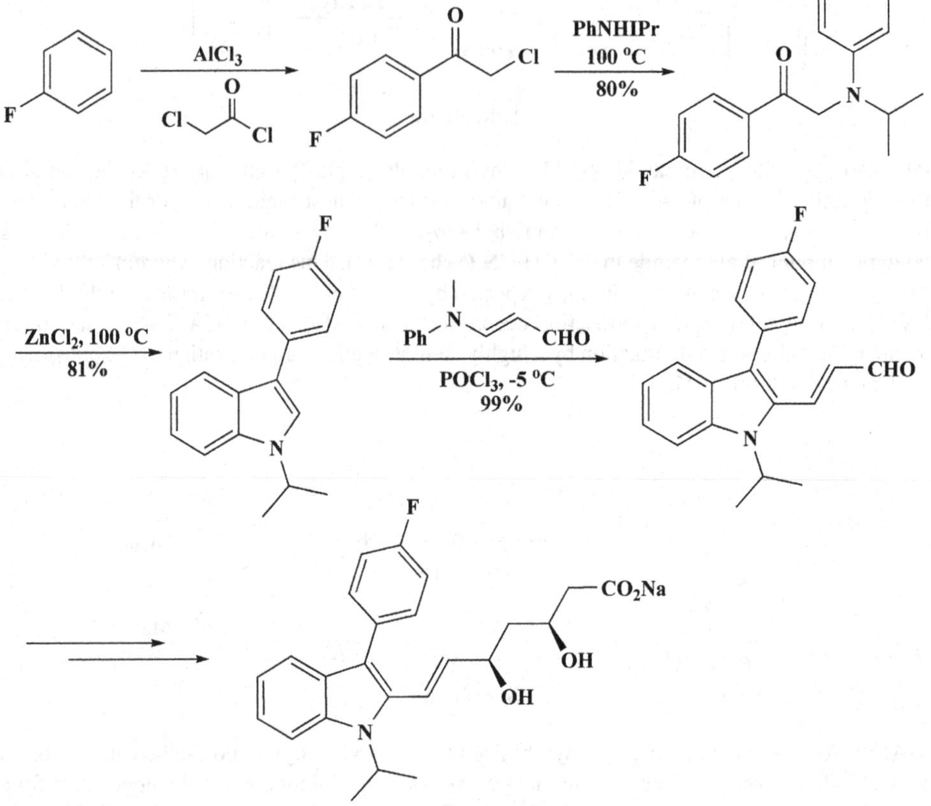

Scheme-59

Significant yields of 2,3-substituted indoles are obtained by Lewis acid-catalyzed cyclization of methyl phenyldiazoacetates with an *o*-imino group, synthesized from *o*-aminophenylacetic acid **(Scheme 60)** [185].

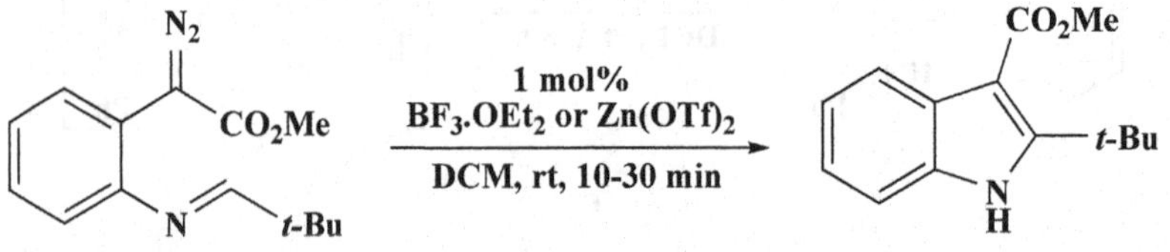

Scheme-60

Various indole products with different structures can be synthesized by $Zn(OTf)_2$-catalyzed cyclization of propargyl alcohols with anilines in toluene at 100 °C without any additive **(Scheme 61)** [186].

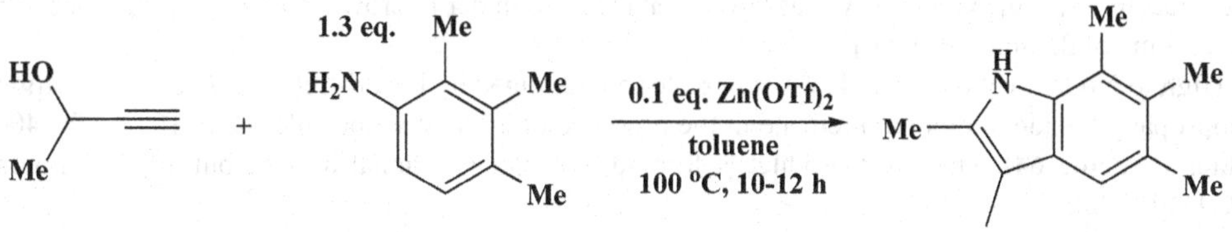

Scheme-61

Phenols or anilines have been studied to yield benzo derivatives—both those bearing a suitable substituent in the *ortho* position and those with an un-substituted *ortho* position can be cyclized to afford a variety of benzo heterocyclic compounds. In Reissert indole synthesis, an *o*-nitrotoluene is condensed with oxalic ester under the influence of a strong base. Subsequently, the nitroketoester undergoes simultaneous reduction and cyclization **(Scheme 62)** [105].

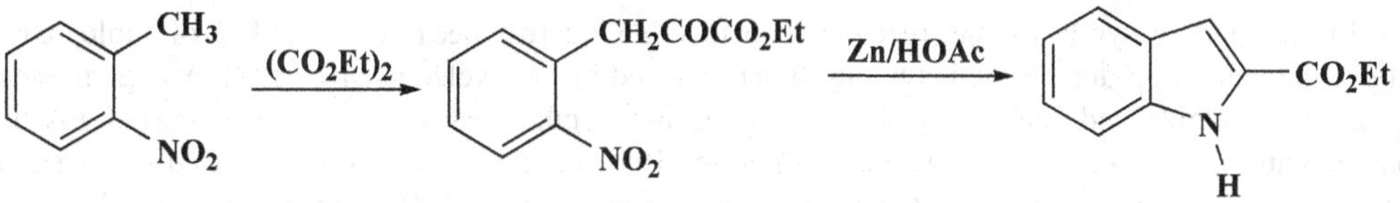

Scheme-62

Lipinska and Czarnocki [187–188] prepared 9-methoxyindolo[2,3-*a*]quinolizine from 2-acetylpyridines by microwave-induced heterogeneous catalytic Fischer indolization as the key step **(Scheme 63)**. Fisher indole synthesis (the protic or Lewis acid-catalyzed rearrangement of arylhydrazones into indoles) is among the most widely used methods for the synthesis of indoles. The synthesis of 2-(2-pyridyl)indoles from 2-acetylpyridines needs forced conditions and is regarded as a difficult example of Fisher indolization. Also, microwave heating is crucial for this synthesis. Many microwave-assisted protocols have been reported for the preparation of indoles. An initially reported solvent-free method used montmorillonite K10 clay modified with zinc chloride. However, higher yields were reported using 10 mol% zinc chloride (catalytic amounts) in triethylene glycol as a polar, high-boiling solvent. Thus, 52% yields of indole were obtained after 7 minutes upon dielectric heating of a mixture of phenylhydrazine and 2-acetylpyridine at 180 °C. The indole was formed in only 12% yield after 3 hours under otherwise identical conditions at 180 °C in an oil bath, which clearly showed the advantages of microwave flash heating. Furthermore, hydrazone of 2-acetylpyridine underwent Fischer indole reaction to produce 2-pyridoindole in the key step. The catalyst of choice was K-10 montmorillonite modified with zinc chloride (K-10/zinc chloride) as it was significantly greener than the previously reported catalysts for

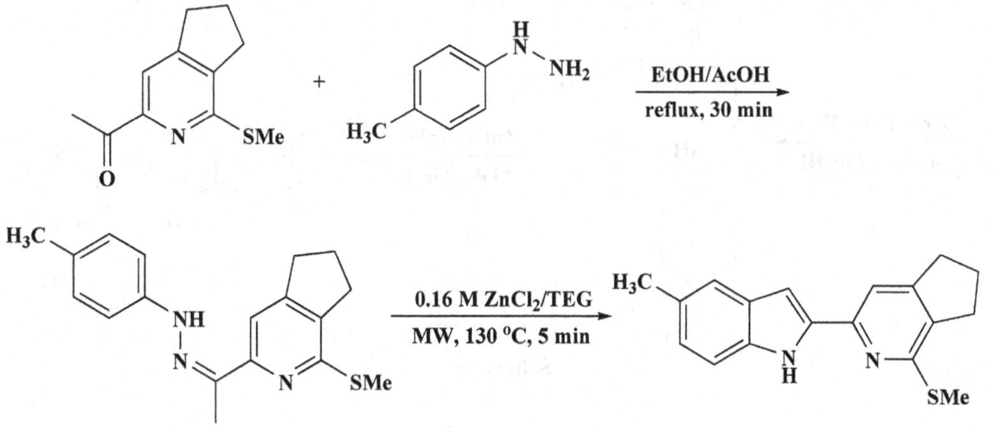

Scheme-63

similar reactions [189]. Moderate yields have been reported under microwave-assisted conditions after a reaction time of 2.5 minutes [190].

High to moderate yields (91–41%) of pyrroles are obtained by ligand-free 5-*endo-dig* cyclization of homopropargyl azide in dichloromethane in the presence of 20 mol% zinc chloride at 105 °C for 40–60 minutes **(Scheme 64)**. The same product is formed under conventional heating but after 16 hours at 75 °C [191–192].

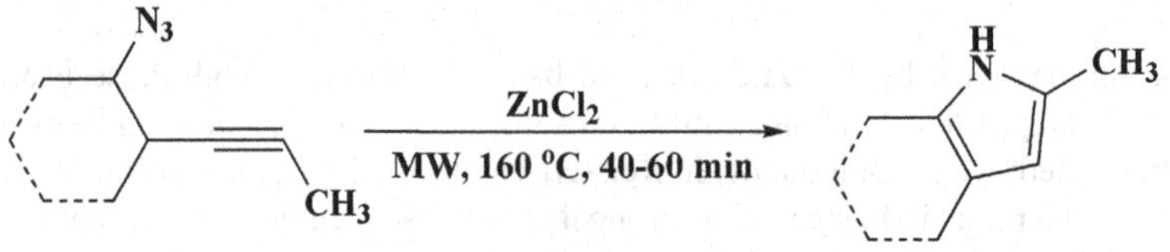

Scheme-64

Fischer indole synthesis has received considerable interest because a number of biologically interesting natural products contains highly functionalized indole skeletons [193–194]. A large number of indole alkaloids and their structural analogues are synthesized by microwave-assisted Fischer indole synthesis as a key step. Franco and Palermo [195] reported a microwave-assisted synthesis of iso-meridianins using $ZnCl_2$-promoted microwave-assisted Fischer indole synthesis as the key step **(Scheme 65)**. Iso-meridianins are close structural analogues of the naturally occurring indole alkaloids. Psammopemmins and Meridianins exhibit high anti-tumor activity as well [196–197]. Iso-meridianins contain a pyrimidine ring at the C-2 position of the indole, as against their parent compounds possessing a pyrimidine ring at the C-3 position. Phenylhydrazines and 2-aminopyrimidines are more easily available and considerably cheaper than functionalized indoles, therefore diversity-oriented and flexible synthesis can be achieved. The synthesis starts from commercially available isocytosine which is transformed into *N*-Boc-4-chloro analogue. Methyl ketone moiety is incorporated at the C-4 position through palladium-catalyzed cross-coupling of chloro derivative with tri-*n*-butyl(1-ethoxyvinyl)tin, followed by an acidic hydrolysis of the intermediate. The phenylhydrazones of compounds are then synthesized following standard methods [198a-b].

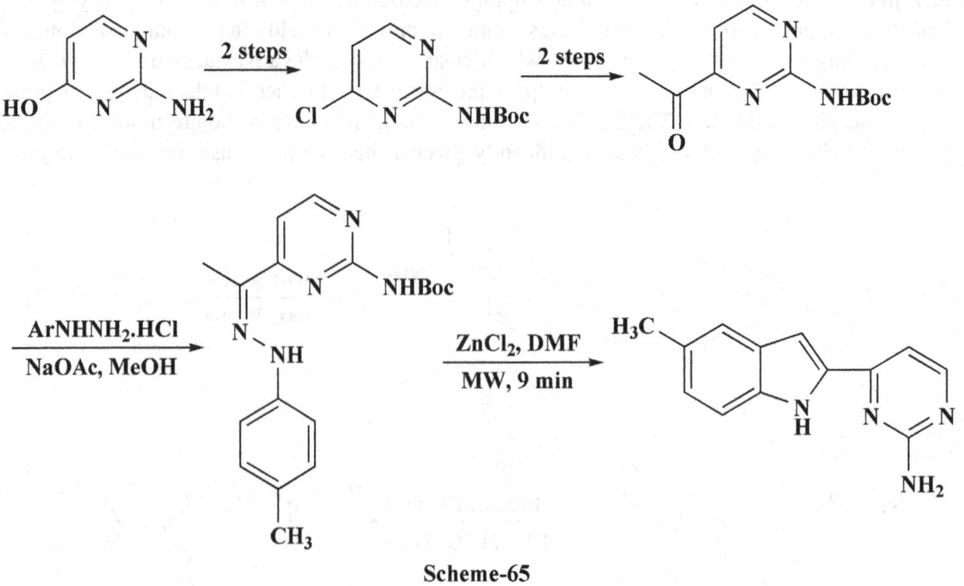

Scheme-65

Zirconium-assisted synthesis

Uesaka et al. [199] and Sato et al. [200] reported a zirconium-catalyzed cyclization reaction for the synthesis of bicyclic cores **(Scheme 66)**. The zirconacycle was obtained when diene was treated with Cp_2ZrCl_2 and underwent transmetallation with BuMgBr to afford magnesium complex. Various heterocycles were synthesized from this magnesium complex intermediate. For instance, hydroindoles were formed by a one-pot cascade reaction upon the exposure of the intermediate magnesium complex to several electrophiles. This protocol has also been expanded to the cyclization of enynes to afford hydroindoles [201–203].

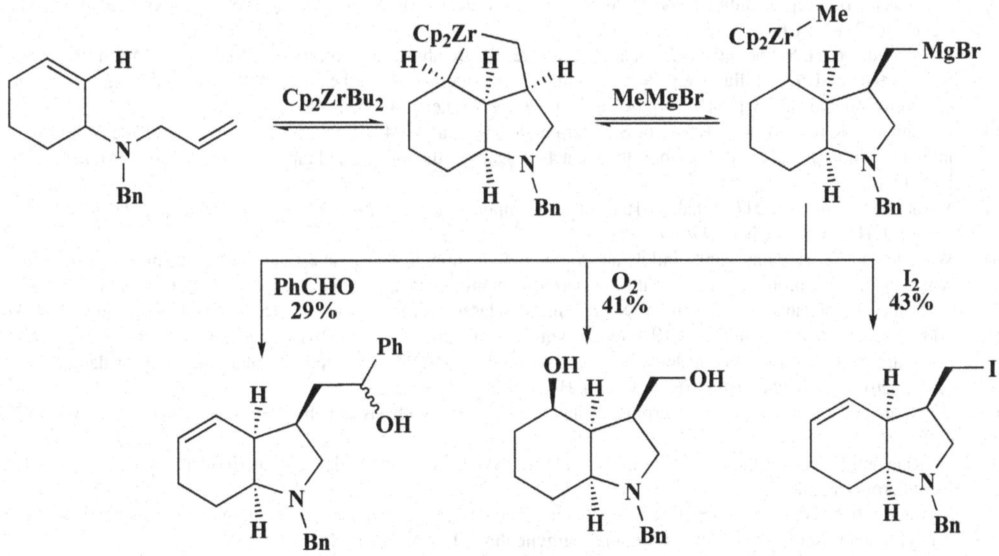

Scheme-66

Cao et al. [204], Ramanathan et al. [205] and Alex et al. [206] have published a few examples for the preparation of nitrogen-containing heterocycles using early transition metal catalysts. Synthesis of indoles from alkynes and substituted aromatic hydrazines through zirconium-catalyzed non-Fischer-type pathway has also been reported [207–208], wherein the authors extended the scope of zirconium-catalyzed indole synthesis for the preparation of tricyclic target compounds. Here, 1,7-annulated and *N*-phenyl-indoles were prepared using amidozirconium complex $[Zr(N_2XylNpy)(NMe_2)_2]$ as a catalyst **(Scheme 67)**. The successful extension of the zirconium-catalyzed indole synthesis to annulated tricyclic target compounds represented an unprecedented level of complexity for an early transition metal-catalyzed reaction.

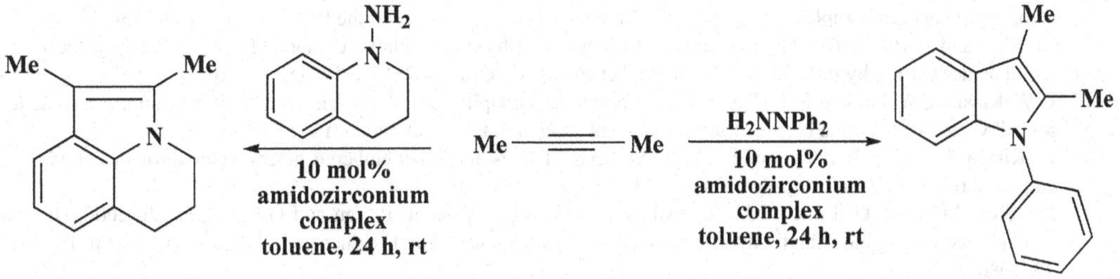

Scheme-67

References

(1)   (a) A. Chimirri, S. Grasso, A.M. Monforte, P. Monforte and M. Zappala. 1991. Anti-HIV agents II. Synthesis and *in vitro* anti-HIV activity of novel 1*H*,3*H*-thiazolo[3,4-*a*]benzimidazoles. Farmaco 46: 925–933. (b) N. Kaur. 2015. Insight into microwave-assisted synthesis of benzo derivatives of five-membered *N*,*N*-heterocycles. Synth. Commun. 45: 1269–1300. (c) N. Kaur. 2015. Synthesis of fused five-membered *N*,*N*-heterocycles using microwave irradiation. Synth. Commun. 45: 1379–1410. (d) N. Kaur. 2014. Microwave-assisted synthesis of seven-membered *S*-heterocycles. Synth. Commun. 44: 3201–3228. (e) N. Kaur. 2019. Copper-catalyzed synthesis of seven and higher-membered heterocycles. Synth. Commun. 49: 879–916. (f) N. Kaur. 2019. Copper-catalyzed synthesis of seven and higher-membered heterocycles. Synth. Commun. 49: 879–916.

(2)   K. Kubo, Y. Kohara, E. Imamiya, Y. Sugiura, Y. Inada, Y. Furukawa, K. Nishikawa and T. Naka. 1993. Non-peptide angiotensin II receptor antagonists. Synthesis and biological activity of benzimidazole carboxylic acids. J. Med. Chem. 36: 2182–2195.

(3)   J. Benavidesm, H. Schoemaker, C. Dana, Y. Claustre, M. Delahaye, M. Prouteau, P. Manoury, J. Allen, B. Scatton, S.Z. Langer and S. Arbilla. 1995. *In vivo* and *in vitro* interaction of the novel selective histamine H1 receptor antagonist mizolastine with H1 receptors in the rodent. F Arzneim. 45: 551–558.

(4)   K. Ishihara, T. Ichikawa, Y. Komuro, S. Ohara and K. Hotta. 1994. Effect on gastric mucus of the proton pump inhibitor leminoprazole and its cytoprotective action against ethanol-induced gastric injury in rats. F Arzneim. Drug Res. 44: 827–830.

(5)   V. Sharma, P. Kumar and D. Pathak. 2010. Biological importance of the indole nucleus in recent years: a comprehensive review. J. Heterocycl. Chem. 47: 491–502.

(6)   W. Gul and M.T. Hamann. 2005. Indole alkaloid marine natural products: an established source of cancer drug leads with considerable promise for the control of parasitic, neurological and other diseases. Life Sci. 78: 442–453.

(7)   M. Somei, F. Yamada, T. Kurauchi, Y. Nagahama, M. Hasegawa, K. Yamada, S. Teranishi, H. Sato and C. Kaneko. 2001. The chemistry of indoles. CIII. Simple syntheses of serotonin, *N*-methylserotonin, bufotenine, 5-methoxy-*N*-methyltryptamine, bufobutanoic acid, *N*-(indol-3-yl)methyl-5-methoxy-*N*-methyltryptamine, and lespedamine based on 1-hydroxyindole chemistry. Chem. Pharm. Bull. 49: 87–96.

(8)   F.E. Chen and J. Huang. 2005. Reserpine: a challenge for total synthesis of natural products. Chem. Rev. 105: 4671–4706.

(9)   M. Kale and K. Patwardhan. 2013. Synthesis of heterocyclic scaffolds with anti-hyperlipidemic potential: a review. Der Pharma Chemica 5: 213–222.

(10) S. Antoniotti, E. Genin, V. Michelet and J.-P. Genet. 2005. Highly efficient access to strained bicyclic ketals *via* gold-catalyzed cycloisomerization of bis-homopropargylic diols. J. Am. Chem. Soc. 127: 9976–9977.

(11) L.-P. Liu and G.B. Hammond. 2009. Highly efficient and tunable synthesis of dioxabicyclo[4.2.1]ketals and tetrahydropyrans *via* gold-catalyzed cycloisomerization of 2-alkynyl-1,5-diols. Org. Lett. 11: 5090–5092.

(12) B. Liu and J.K. de Brabander. 2006. Metal-catalyzed regioselective oxy-functionalization of internal alkynes: an entry into ketones, acetals, and spiroketals. Org. Lett. 8: 4907–4910.

(13) B. Alcaide, P. Almendros and J.M. Alonso. 2011. Gold-catalyzed cyclizations of alkynol-based compounds: synthesis of natural products and derivatives. Molecules 16: 7815–7843.

(14) L.-Z. Dai and M. Shi. 2009. Gold(I) catalysis: selective synthesis of six- or seven-membered heterocycles from epoxy alkynes. Eur. J. Org. Chem. 19: 3129–3133.

(15) Z. Shi and C. He. 2004. An Au-catalyzed cyclialkylation of electron rich arenes with epoxides to prepare 3-chromanols. J. Am. Chem. Soc. 126: 5965–5964.

(16) N. Sewald. 2003. Synthetic routes towards enantiomerically pure *β*-amino acids. Angew. Chem. Int. Ed. 42: 5794–5795.

(17) D. Ben-Ishai, N. Peled and I. Sataty. 1980. Inter vs intramolecular amidoalkylations of aromatics—a new synthesis of oxindoles, isoquinolones and benzazepinones. Tetrahedron Lett. 21: 569–572.

(18) M. Mori and Y. Ban. 1979. Reactions and syntheses with organometallic compounds. The intramolecular cyclization using arylpalladium complexes for generation of nitrogen-heterocycles. Tetrahedron Lett. 20: 1133–1136.

(19) M. Mori and Y. Ban. 1976. The reactions and syntheses with organometallic compounds IV. The new synthesis of oxindole derivatives by utilization of organonickel complex. Tetrahedron Lett. 17: 1807–1810.

(20) G.W. Kabalka, W. Lei and R.M. Pagni. 2001. Sonogashira coupling and cyclization reactions on alumina: a route to aryl alkynes, 2-substituted benzo[*b*]furans and 2-substituted indoles. Tetrahedron 57: 8017–8028.

(21) G. Kirsch, S. Hesse and A. Comel. 2004. Synthesis of five- and six-membered heterocycles through palladium-catalyzed reactions. Curr. Org. Synth. 1: 47–63.

(22) B. Torok, M. Abid, G. London, J. Esquibel, M. Torok, S.C. Mhadgut, P. Yan and G.K.S. Prakash. 2005. Highly enantioselective organocatalytic hydroxyalkylation of indoles with ethyl trifluoropyruvate. Angew. Chem. Int. Ed. 44: 3086–3089.

(23) V. Estevez, M. Villacampa and J.C. Menendez. 2010. Multi-component reactions for the synthesis of pyrroles. Chem. Soc. Rev. 39: 4402–4421.

(24)   K.D. Sheikh, P.P. Banerjee, S. Jagadeesh, S.C. Grindrod, L. Zhang, M. Paige and M.L. Brown. 2010. Fluorescent epigenetic small molecule induces expression of the tumor suppressor ras-association domain family 1A and inhibits human prostate xenograft. J. Med. Chem. 53: 2376–2382.

(25)   U. Bora, A. Saikia and R.C. Boruah. 2003. A novel microwave-mediated one-pot synthesis of indolizines *via* a three-component reaction. Org. Lett. 5: 435–438.

(26)   P.N. Reddy, L.K. Ravindranath, K.B. Chandrasekhar, P. Rameshbabu, G. Madhu and K.S.B. Aiswarya. 2012. Synthesis of novel Mannich bases containing pyrazolones and indole systems. Der Pharma Chemica 4: 1330–1338.

(27)   A.R. Khosropour, I. Mohammadpoor-Baltork, M.M. Khodaei and P. Ghanbary. 2007. Chemoselective one-pot conversion of primary alcohols to their bis(indolyl)methanes promoted by $Bi(NO_3)_3 \cdot 5H_2O$. Zeitschrift für Naturforschung B, 62b: 537–539.

(28)   J.A.R. Salvador, R.M.A. Pinto and S.M. Silvestre. 2009. Recent advances of bismuth(III) salts in organic chemistry: application to the synthesis of heterocycles of pharmaceutical interest. Curr. Org. Synth. 6: 426–470.

(29)   G. Bifulco, I. Bruno, R. Riccio, J. Lavayre and G. Bourdy. 1995. Further brominated bis- and tris-indole alkaloids from the deep-water new *Caledonian*marine sponge *Orina* sp. J. Nat. Prod. 58: 1254–1260.

(30)   G. Sivaprasad, P.T. Perumal, V.R. Prabavathy and N. Mathivanan. 2006. Synthesis and anti-microbial activity of pyrazolylbis-indoles—promising anti-fungal compounds. Bioorg. Med. Chem. Lett. 16: 6302–6305.

(31)   C. Hong, G.L. Firestone and L.F. Bjeldanes. 2002. Bcl-2 family mediated apoptotic effects of 3,3′-diindolylmethane (DIM) in human breast cancer cells. Biochem. Pharmacol. 63: 1085–1097.

(32)   I. Mohammadpoor-Baltork, H.R. Memarian, A.R. Khosropour and K. Nikoofar. 2006. $BiOClO_4 \cdot xH_2O$ and $Bi(OTf)_3$ as efficient and environmentally benign catalysts for synthesis of bis(indolyl)methanes in solution and under ultrasound irradiation. Lett. Org. Chem. 3: 768–772.

(33)   S. Palaniappan, C. Saravanan and V.J. Rao. 2005. Synthesis of polyaniline-bismoclite composite and its function as recoverable and reusable catalyst. J. Mol. Catal. A Chem. 229: 221–226.

(34)   D.K. Deodhar, R.P. Bhat and S.D. Samant. 2007. Bismuth(III) nitrate pentahydrate—an efficient catalyst for the synthesis of bis(indolyl)methanes under mild conditions. Indian J. Chem. 46B: 1455–1458.

(35)   S.-J. Ji, M.-F. Zhou, D.-G. Gu, S.-Y. Wang and T.-P. Loh. 2003. Efficient synthesis of bis(indolyl)methanes catalyzed by Lewis acids in ionic liquids. Synlett 13: 2077–2079.

(36)   M. Xia, S.-H. Wang and W.-B. Yuan. 2004. Lewis acid-catalyzed electrophilic substitution of indole with aldehydes and Schiff's bases under microwave solvent-free irradiation. Synth. Commun. 34: 3175–3182.

(37)   R.C. Larock, E.K. Yum, M.J. Doty and K.K.C. Sham. 1996. Synthesis of aromatic heterocycles *via* palladium-catalyzed annulation of internal alkynes. J. Org. Chem. 60: 3270–3271.

(38)   K. Okano, H. Fujiwara, T. Noji, T. Fukuyama and H. Tokuyama. 2010. Total synthesis of dictyodendrin A and B. Angew. Chem. Int. Ed. 49: 5925–5929.

(39)   H. Tokuyama, K. Okano, H. Fujiwara, T. Noji and T. Fukuyama. 2011. Total synthesis of dictyodendrins A-E. Chem. Asian J. 6: 560–572.

(40)   R. Zimmer, C.U. Dinesh, E. Nandanan and F.A. Khan. 2000. Palladium-catalyzed reactions of allenes. Chem. Rev. 100: 3067–3126.

(41)   S. Cacchi and G. Fabrizi. 2011. Palladium-catalyzed reactions. Chem. Rev. 111: 215–283.

(42)   W.-M. Dai, D.-S. Guo, L.-P. Sun and X.-H. Huang. 2003. Microwave-assisted solid-phase organic synthesis (MASPOS) as a key step for an indole library construction. Org. Lett. 5: 2919–2922.

(43)   C. Koradin, W. Dohle, A.L. Rodriguez, B. Schmid and P. Knochel. 2003. Synthesis of poly-functional indoles and related heterocycles mediated by cesium and potassium bases. Tetrahedron 59: 1571–1587.

(44)   X. Zhang and A. Corma. 2008. Supported gold(III) catalysts for highly efficient three-component coupling reactions. Angew. Chem. Int. Ed. 47: 4358–4361.

(45)   (a) L. Djakovitch, N. Batail and M. Genelot. 2011. Recent advances in the synthesis of *N*-containing heteroaromatics *via* heterogeneously transition metal-catalyzed cross-coupling reactions. Molecules 16: 5241–5267. (b) A. Arcadi. 2008. Alternative synthetic methods through new developments in catalysis by gold. Chem. Rev. 108: 3266–3325. (c) Y. Fukuda and K. Utimoto. 1991. Preparation of 2,3,4,5-tetrahydropyridines from 5-alkynylamines under the catalytic action of gold(III) salts. Synthesis 11: 975–978. (d) Y. Fukuda, K. Utimoto and H. Nozaki. 1987. Preparation of 2,3,4,5-tetrahydropyridines from 5-alkynylamines under the catalytic action of Au(III). Heterocycles 25: 297–300. (e) A. Arcadi, G. Bianchi and F. Marinelli. 2004. Gold(III)-catalyzed annulation of 2-alkynylanilines: a mild and efficient synthesis of indoles and 3-haloindoles. Synthesis 4: 610–618. (f) M. Alfonsi, A. Arcadi, M. Aschi, G. Bianchi and F. Marinelli. 2005. Gold-catalyzed reactions of 2-alkynyl-phenylamines with α,β-enones. J. Org. Chem. 70: 2273–2273. (g) K. Hiroya, S. Itoh and T. Sakamoto. 2005. Mild and efficient cyclization reaction of 2-ethynylaniline derivatives to indoles in aqueous medium. Tetrahedron 61: 10958–10964. (h) I. Ambrogio, A. Arcadi, S. Cacchi, G. Fabrizi and F. Marinelli. 2007. Gold-catalyzed synthesis of 2-substituted, 2,3-disubstituted and 1,2,3-trisubstituted indoles in [bmim]BF_4. Synlett 11: 1775–1779. (i) A.S.K. Hashmi, L. Schwarz, P. Rubenbauer and M.C. Blanco. 2006. The condensation of carbonyl compounds with electron rich arenes: mercury, thallium, gold or a proton? Adv. Synth. Catal. 348: 705–708. (j) V. Nair, N. Vidya and K.G. Abhilash. 2006. Efficient condensation reactions of electron rich arenes with aldehydes and enals promoted by gold(III) chloride: practical synthesis of triaryl- and triheteroarylmethanes and related compounds. Synthesis 21: 3647–3653. (k) V. Nair, N. Vidya and K.G. Abhilash. 2006. Gold(III) chloride-promoted addition of electron rich heteroaromatic compounds to the CC and

CO bonds of enals. Tetrahedron Lett. 47: 2871–2873. (l) J. Zhang, C.-G. Yang and C. He. 2006. Gold(I)-catalyzed intra- and intermolecular hydroamination of un-activated olefins. J. Am. Chem. Soc. 128: 1798–1799. (m) Z. Li, J. Zhang, C. Brouwer, C.-G. Yang, N.W. Reich and C. He. 2006. Brønsted acid-catalyzed addition of phenols, carboxylic acids, and tosylamides to simple olefins. Org. Lett. 8: 4175–4178. (n) D.C. Rosenfeld, S. Shekhar, A. Takemiya, M. Utsunomiya and J.F. Hartwig. 2006. Hydroamination and hydroalkoxylation catalyzed by triflic acid. Parallels to reactions initiated with metal triflates. Org. Lett. 8: 4179–4182. (o) Z. Li, C. Brouwer and C. He. 2008. Gold-catalyzed organic transformations. Chem. Rev. 108: 3239–3265. (p) M. Alfonsi, A. Arcadi, M. Aschi, G. Bianchi and F. Marinelli. 2005. Gold-catalyzed reactions of 2-alkynyl-phenylamines with α,β-enones. J. Org. Chem. 70: 2265–2273. (q) A. Arcadi, S. Di Giuseppe, F. Marinelli and E. Rossi. 2001. Gold-catalyzed sequential amination/ annulation reactions of 2-propynyl-1,3-dicarbonyl compounds. Adv. Synth. Catal. 343: 443–446. (r) Z. Liu, A.S. Wasmuth and S.G. Nelson. 2006. Au(I)-catalyzed annulation of enantioenriched allenes in the enantioselective total synthesis of (–)-rhazinilam. J. Am. Chem. Soc. 128: 10352–10353. (s) A.S.K. Hashmi, T.M. Frost and J.W. Bats. 2001. Gold catalysis: on the phenol synthesis. Org. Lett. 3: 3769–3771. (t) Y.Y. Nakamura. 2004. Transition metal-catalyzed reactions in heterocyclic synthesis. Chem. Rev. 104: 2127–2198. (u) P. Kothandaraman, S.R. Mothe, S.S.M. Toh and P.W.H. Chan. 2011. Gold-catalyzed cycloisomerizations of 1-(2-(tosylamino)phenyl)prop-2-yn-1- ols to 1H-indole-2-carbaldehydes and (E)-2-(iodomethylene)indolin-3-ols. J. Org. Chem. 76: 7633–7640. (v) T. de Haro, E. Gomez-Bengoa, R. Cribiu, X. Huang and C. Nevado. 2012. Gold-catalyzed 1,2-/1,2-bis-acetoxy migration of 1,4-bis-propargyl acetates: a mechanistic study. Chem. Eur. J. 18: 6811–6824. (w) R.K. Shiroodi and V. Gevorgyan. 2013. Metal-catalyzed double migratory cascade reactions of propargylic esters and phosphates. Chem. Soc. Rev. 42: 4991–5001.

(46)   N. Sakai, K. Annaka, A. Fujita, A. Sato and T. Konakahara. 2008. InBr$_3$-promoted divergent approach to poly- substituted indoles and quinolines from 2-ethynylanilines: switch from an intramolecular cyclization to an intermolecular dimerization by a type of terminal substituent group. J. Org. Chem. 73: 4160–4165.

(47)   D. Seomoon, K. Lee, H. Kim and P.H. Lee. 2007. Inter- and intramolecular palladium-catalyzed allyl cross-coupling reactions using allylindium generated *in situ* from allyl acetates, indium, and indium trichloride. Chem. Eur. J. 13: 5197–5206.

(48)   S. Nag and S. Batra. 2011. Applications of allylamines for the syntheses of aza-heterocycles. Tetrahedron 67: 8959–9061.

(49)   T. Hirashita, A. Hayashi, M. Tsuji, J. Tanaka and S. Araki. 2008. Radical reactions initiated by the photochemical cleavage of carbon-indium bonds of organoindium compounds. Tetrahedron 64: 2642–2650.

(50)   Z. He, H. Li and Z. Li. 2010. Iodine-mediated synthesis of 3H-indoles *via* intramolecular cyclization of enamines. J. Org. Chem. 75: 4296–4299.

(51)   D. Yue, T. Yao and R.C. Larock. 2006. Synthesis of 3-iodoindoles by the Pd/Cu-catalyzed coupling of N,N-dialkyl-2- iodoanilines and terminal acetylenes, followed by electrophilic cyclization. J. Org. Chem. 71: 62–69.

(52)   T. Sakamoto, T. Nagano, Y. Kondo and H. Yamanaka. 1988. Palladium-catalyzed coupling reaction of 3-iodoindoles and 3-iodobenzo[b]thiophene with terminal acetylenes. Chem. Pharm. Bull. 36: 2248–2252.

(53)   D. Yue and R.C. Larock. 2004. Synthesis of 3-iodoindoles by electrophilic cyclization of N,N-dialkyl-2-(1-alkynyl) anilines. Org. Lett. 6: 1037–1040.

(54)   S.A. Worlikar, B. Neuenswander, G.H. Lushington and R.C. Larock. 2009. Highly substituted indole library synthesis by palladium-catalyzed coupling reactions in solution and on a solid support. J. Comb. Chem. 11: 875–879.

(55)   P.T. Parvatkar, P.S. Parameswaran and S.G. Tilve. 2012. Recent developments in the synthesis of five- and six- membered heterocycles using molecular iodine. Chem. Eur. J. 18: 5460–5489.

(56)   Z. He, H. Li and Z. Li. 2010. Iodine-mediated synthesis of 3H-indoles *via* intramolecular cyclization of enamines. J. Org. Chem. 75: 4636–4639.

(57)   S. Stavber, M. Jereb and M. Zupan. 2008. Electrophilic iodination of organic compounds using elemental iodine or iodides. Synthesis 10: 1487–1513.

(58)   Y. Du, R. Liu, G. Linn and K. Zhao. 2006. Synthesis of N-substituted indole derivatives *via* PIFA-mediated intramolecular cyclization. Org. Lett. 8: 5919–5922.

(59)   I. Tellitu, A. Urrejola, S. Serna, I. Moreno, M.T. Herrero, E. Dominguez, R. SanMartin and A. Correa. 2007. On the phenyliodine(III)-bis(trifluoroacetate)-mediated olefin amidohydroxylation reaction. Eur. J. Org. Chem. 3: 437–444.

(60)   A. Correa, I. Tellitu, E. Dominguez, I. Moreno and R.S. Martin. 2005. An efficient, PIFA-mediated approach to benzo-, naphtho-, and heterocycle-fused pyrrolo[2,1-c][1,4]diazepines. An advantageous access to the anti-tumor anti-biotic DC-81. J. Org. Chem. 70: 2256–2264.

(61)   F. Churruca, R. SanMartin, I. Tellitu and E. Dominguez. 2005. A new, expeditious entry to the benzophenanthrofuran framework by a Pd-catalyzed C- and O-arylation/PIFA-mediated oxidative coupling sequence. Eur. J. Org. Chem. 12: 2481–2490.

(62)   S. Yan, S. Cao and J. Sun. 2017. Synthesis of seven-membered heterocycles *via* copper-catalyzed cross-coupling of terminal alkynes with diazo compounds and sequential Michael addition. Org. Biomol. Chem. 15: 5272–5274.

(63)   F. Churruca, R. SanMartin, M. Carril, M.K. Urtiaga, X. Solans, I. Tellitu and E. Dominguez. 2005. Direct, two-step synthetic pathway to novel dibenzo[a,c]phenanthridines. J. Org. Chem. 70: 3178–3187.

(64)   M.T. Herrero, I. Tellitu, S. Hernandez, E. Dominguez, I. Moreno and R. SanMartin. 2002. Novel applications of the hypervalent iodine chemistry. Synthesis of thiazolo-fused quinolinones. ARKIVOC (v): 31–37.

(65)   A. Correa, I. Tellitu, E. Dominguez and R. SanMartin. 2006. A metal-free approach to the synthesis of indoline derivatives by a phenyliodine(III) bis(trifluoroacetate)-mediated amidohydroxylation reaction. J. Org. Chem. 71: 8316–8319.

(66)   I. Tellitu, S. Serna, M.T. Herrero, I. Moreno, E. Dominguez and R. SanMartin. 2007. Intramolecular PIFA-mediated alkyne amidation and carboxylation reaction. J. Org. Chem. 72: 1526–1529.

(67)   A. Correa, I. Tellitu, E. Dominguez and R. SanMartin. 2006. An advantageous synthesis of new indazolone and pyrazolone derivatives. Tetrahedron 62: 11100–11105.

(68)   V.V. Zhdankin. 2009. Hypervalent iodine(III) reagents in organic synthesis. ARKIVOC (i): 1–62.

(69)   J.A. Jordan-Hore, C.C.C. Johansson, M. Gulias, E.M. Beck and M.J. Gaunt. 2008. Oxidative Pd(II)-catalyzed C-H bond amination to carbazole at ambient temperature. J. Am. Chem. Soc. 130: 16184–16186.

(70)   J. Bonnamour and C. Bolm. 2011. Iron(II) triflate as a catalyst for the synthesis of indoles by intramolecular C-H amination. Org. Lett. 13: 2012–2014.

(71)   L.V. Kudzma. 2003. Synthesis of substituted indoles and carbazoles from 2-fluorophenyl imines. Synthesis 11: 1661–1666.

(72)   U. Groth, P. Koettgen, P. Langenbach, A. Lindenmaier, T. Schuetz and M. Wiegand. 2008. Enantioselective synthesis of 3,3-disubstituted indolines *via* asymmetric intramolecular carbolithiation in the presence of (–)-sparteine. Synlett 9: 1301–1304.

(73)   W.H. Pearson and B.W. Lian. 1998. Application of the 2-azaallyl anion cycloaddition method to an enantioselective total synthesis of (+)-coccinine. Angew. Chem. Int. Ed. 37: 1724–1726.

(74)   W.H. Pearson and A. Aponick. 2006. Formal synthesis of *Aspidosperma* alkaloids *via* the intramolecular [3+2]-cycloaddition of 2-azapentdienyllithiums. Org. Lett. 8: 1661–1664.

(75)   T. Ogata, A. Ujihara, S. Tsuchida, T. Shimizu, A. Kaneshige and K. Tomioka. 2007. Catalytic asymmetric intramolecular hydroamination of aminoalkenes. Tetrahedron Lett. 48: 6648–6650.

(76)   T.E. Muller, K.C. Hultzsch, M. Yus, F. Foubelo and M. Tada. 2008. Hydroamination: direct addition of amines to alkenes and alkynes. Chem. Rev. 108: 3795–3892.

(77)   S.-L. Cui, J. Wang and Y.-G. Wang. 2008. Synthesis of indoles *via* domino reaction of N-aryl amides and ethyl diazoacetate. J. Am. Chem. Soc. 130: 13526–13527.

(78)   C.M. Coleman and D.F. O'Shea. 2003. New organolithium addition methodology to diversely functionalized indoles. J. Am. Chem. Soc. 125: 4054–4055.

(79)   T.P. Zabawa, D. Kasi and S.R. Chemler. 2005. Copper(II) acetate-promoted intramolecular diamination of un-activated olefins. J. Am. Chem. Soc. 127: 11250–11251.

(80)   R. Lebeuf, F. Robert, K. Schenk and Y. Landais. 2006. Desymmetrization of cyclohexa-2,5-dienes through a diastereoselective protonation-hydroamination cascade. Org. Lett. 8: 4755–4758.

(81)   D.B. Guthrie and D.P. Curran. 2009. Asymmetric radical and anionic cyclizations of axially chiral carbamates. Org. Lett. 11: 249–251.

(82)   J. Barluenga, F.J. Fananas, R. Sanz and Y. Fernandez. 2002. Synthesis of functionalized indole- and benzo-fused heterocyclic derivatives through anionic benzyne cyclization. Chem. Eur. J. 8: 2034–2046.

(83)   W. Madelung. 1912. Über eine neue darstellungsweise für substituierte indole. Berichte der deutschen chemischen Gesellschaft. 45: 1128–1134.

(84)   G. Bartoli, G. Palmieri, M. Bosco and R. Dalpozzo. 1989. The reaction of vinyl Grignard reagents with 2-substituted nitroarenes: a new approach to the synthesis of 7-substituted indoles. Tetrahedron Lett. 30: 2129–2132.

(85)   M. Nishida, H. Shiga and M. Mori. 1998. [2+2+2]-Co-cyclization using [Mo(CO)$_6$-p-ClPhOH]. J. Org. Chem. 63: 8606–8608.

(86)   B.R. Rosen, J.E. Ney and J.P. Wolfe. 2010. Use of aryl chlorides as electrophiles in Pd-catalyzed alkene difunctionalization reactions. J. Org. Chem. 75: 2756–2759.

(87)   S. Kotha, E. Brahmachary and K. Lahiri. 2005. Transition metal-catalyzed [2+2+2]-cycloaddition and application in organic synthesis. Eur. J. Org. Chem. 22: 4741–4767.

(88)   Y. Nakao, S. Ebata, A. Yada, T. Hiyama, M. Ikawa and S. Ogoshi. 2008. Intramolecular arylcyanation of alkenes catalyzed by nickel/AlMe$_2$Cl. J. Am. Chem. Soc. 130: 12874–12875.

(89)   Y. Nakao, T. Hiyama, J.-C. Hsieh and S. Ebata. 2010. Asymmetric synthesis of indolines bearing a benzylic quaternary stereocenter through intramolecular arylcyanation of alkenes. Synlett 11: 1709–1711.

(90)   A. Combs, S. Saubern, M. Rafalski and P.Y.S. Lam. 1999. Solid-supported arylheteroaryl C-N cross-coupling reactions. Tetrahedron Lett. 40: 1623–1626.

(91)   J.D. Ferguson. 2003. Focused™ microwave instrumentation from CEM Corporation. Mol. Divers. 7: 281–286.

(92)   L.-Z. Dai and M. Shi. 2008. Gold(I)-catalyzed reactions: substituents-dependent selective formation of bis-furans and 1,3-diketones from 1-alkynyl-2,3-epoxy alcohols. Tetrahedron Lett. 49: 6437–6439.

(93)   P. Walla and C.O. Kappe. 2004. Microwave-assisted Negishi and Kumada cross-coupling reactions of aryl chlorides. Chem. Commun. 35: 564–565.

(94)   J. Wannberg and M. Larhed. 2003. Increasing rates and scope of reactions: sluggish amines in microwave-heated aminocarbonylation reactions under air. J. Org. Chem. 68: 5750–5753.

(95)   N.-F.K. Kaiser, U. Bremberg, M. Larhed, C. Moberg and A. Hallberg. 2000. Fast, convenient, and efficient molybdenum-catalyzed asymmetric allylic alkylation under non-inert conditions: an example of microwave-promoted fast chemistry. Angew. Chem. 112: 3741–3744; Angew. Chem. Int. Ed. 39: 3596–3598.

(96)   A. Bengtson, A. Hallberg and M. Larhed. 2002. Fast synthesis of aryl triflates with controlled microwave heating. Org. Lett. 4: 1231–1233.

(97)   C.O. Kappe. 2004. Controlled microwave heating in modern organic synthesis. Angew. Chem. Int. Ed. 43: 6250–6284.

(98)   K.R. Roesch and R.C. Larock. 1999. Synthesis of isoindolo[2,1-*a*]indoles by the palladium-catalyzed annulation of internal alkynes. Org. Lett. 1: 1551–1553.

(99)   R.C. Larock, E.K. Yum and M.D. Refvik. 1998. Synthesis of 2,3-disubstituted indoles *via* palladium-catalyzed annulation of internal alkynes. J. Org. Chem. 63: 7652–7662.

(100)   R.C. Larock and E.K. Yum. 1991. Synthesis of indoles *via* palladium-catalyzed heteroannulation of internal alkynes. J. Am. Chem. Soc. 113: 6689–6690.

(101)   G. Zeni and R.C. Larock. 2006. Synthesis of heterocycles *via* palladium-catalyzed oxidative addition. Chem. Rev. 106: 4644–4680.

(102)   E.M. Beccalli, G. Broggini, M. Martinelli and G. Paladino. 2005. Pd-catalyzed intramolecular cyclization of pyrrolo-2-carboxamides: regiodivergent routes to pyrrolo-pyrazines and pyrrolo-pyridines. Tetrahedron 61: 1077–1082.

(103)   Z. Shen and X. Lu. 2006. Palladium(II)-catalyzed tandem intramolecular aminopalladation of alkynylanilines and conjugate addition for synthesis of 2,3-disubstituted indole derivatives. Tetrahedron 62: 10896–10899.

(104)   C.-Y. Chen, D.R. Lieberman, R.D. Larsen, R.A. Reamer, T.R. Verhoeven, P.J. Reider, I.F. Cottrell and P.G. Houghton. 1994. Synthesis of the 5-HT1D receptor agonist MK-0462 *via* a Pd-catalyzed coupling reaction. Tetrahedron Lett. 35: 6981–6984.

(105)   M. Baumann, I.R. Baxendale, S.V. Ley and N. Nikbin. 2011. An overview of the key routes to the best selling 5-membered ring heterocyclic pharmaceuticals. Beilstein J. Org. Chem. 7: 442–495.

(106)   T. Jeschke, D. Wensbo, U. Annby, S. Gronowitz and L.A. Cohen. 1993. A novel approach to Bz-substituted tryptophans *via* Pd-catalyzed coupling/annulation. Tetrahedron Lett. 34: 6471–6474.

(107)   C. Ma, X. Liu, S. Yu, S. Zhao and J.M. Cook. 1999. Concise synthesis of optically active ring α-substituted tryptophans. Tetrahedron Lett. 40: 657–660.

(108)   X. Liu, J.R. Deschamp and J.M. Cook. 2002. Regiospecific, enantiospecific total synthesis of the alkoxy-substituted indole bases, 16-*epi*-Na-methylgardneral, 11-methoxyaffinisine, and 11-methoxymacroline as well as the indole alkaloids alstophylline and macralstonine. Org. Lett. 4: 3339–3342.

(109)   H. Zhou, X. Liao and J.M. Cook. 2004. Regiospecific, enantiospecific total synthesis of the 12-alkoxy-substituted indole alkaloids, (+)-12-methoxy-Na-methylvellosimine, (+)-12-methoxyaffinisine, and (–)-fuchsiaefoline. Org. Lett. 6: 249–252.

(110)   N. Sutou, K. Kato and H. Akita. 2008. A concise synthesis of (–)-indolmycin and (–)-5-methoxyindolmycin. Tetrahedron: Asymmetry 19: 1833–1838.

(111)   H. Zhou, X. Liao, W. Yin, J. Ma and J.M. Cook. 2006. General approach for the synthesis of 12-methoxy-substituted sarpagine indole alkaloids including (–)-12-methoxy-Nb-methylvoachalotine, (+)-12-methoxy-Na-methylvellosimine, (+)-12-methoxyaffinisine, and (-)-fuchsiaefoline. J. Org. Chem. 71: 251–259.

(112)   J. Yu, X.Z. Wearing and J.M. Cook. 2005. A general strategy for the synthesis of vincamajine related indole alkaloids: stereocontrolled total synthesis of (+)-dehydrovoachalotine, (–)-vincamajinine, and (–)-11-methoxy-17-epivincamajine as well as the related quebrachidine diol, vincamajine diol, and vincarinol. J. Org. Chem. 70: 3963–3979.

(113)   J.S. Mahanty, M. De, P. Das and N.G. Kundu. 1997. Palladium-catalyzed heteroannulation with acetylenic carbinols as synthons- synthesis of quinolines and 2,3-dihydro-4(1*H*)-quinolones. Tetrahedron 53: 13397–13418.

(114)   O. Russo, S. Messaoudi, A. Hamze, N. Olivi, J.-F. Peyrat, J.-D. Brion, S.S. Sicsic, J. Berque-Bestel and M. Alami. 2007. Three-component one-pot process to propargylic amines and related amide and sulfonamide compounds: application to the construction of 2-(aminomethyl)benzofurans and indoles. Tetrahedron 63: 10671–10683.

(115)   S. Ye, Q. Ding, Z. Wang, H. Zhou and J. Wu. 2008. Tandem addition-cyclization reactions of 2-alkynylbenzenamines with isocyanates catalyzed by PdCl₂. Org. Biomol. Chem. 6: 4406–4412.

(116)   C. Amatore, E. Blart, J.P. Genet, A. Jutand, S. Lemaire-Audoire and M. Savignac. 1995. New synthetic applications of water-soluble acetate Pd/TPPTS catalyst generated *in situ*. Evidence for a true Pd(0) species intermediate. J. Org. Chem. 60: 682–6839.

(117)   Y. Kondo, R. Watanabe, T. Sakamoto and H. Yamanaka. 1989. Condensed heteroaromatic ring systmes. XVI. Synthesis of pyrrolo[2,3-*d*]pyrimidine derivatives. Chem. Pharm. Bull. 37: 2933–2936.

(118)   L. Xu, I.R. Lewis, S.K. Davidsen and J.B. Summers. 1998. Transition metal-catalyzed synthesis of 5-azaindoles. Tetrahedron Lett. 39: 5159–5162.

(119)   C.R. Hopkins and N. Collar. 2004. 6-Substituted 5*H*-pyrrolo[2,3-*b*]pyrazines *via* palladium-catalyzed heteroannulation from *N*-(3-chloropyrazin-2-yl)-methanesulfonamide and alkynes. Tetrahedron Lett. 45: 8087–8090.

(120)   J.C. Torres, R.A. Pilli, M.D. Vargas, F.A. Violante, S.J. Garden and A.C. Pinto. 2002. Synthesis of 1-ferrocenyl-2-aryl(heteroaryl)acetylenes and 2-ferrocenylindole derivatives *via* the Sonogashira-Heck-Cassar reaction. Tetrahedron 58: 4487–4492.

(121) P.J. Thomas, A.T. Axtell, J. Klosin, W. Peng, C.L. Rand, T.P. Clark, C.R. Landis and K.A. Abboud. 2007. Asymmetric hydroformylation of vinyl acetate: application in the synthesis of optically active isoxazolines and imidazoles. Org. Lett. 9: 2665–2668.

(122) T. Kalai, M. Balog, J. Jeko, W.L. Hubbell and K. Hideg. 2002. Palladium-catalyzed coupling reactions of paramagnetic vinyl halides. Synthesis 16: 2365–2372.

(123) B.C.J. van Esseveldt, F.L. van Delft, R. Gelder and F.P.J.T. Rutjes. 2003. Palladium-catalyzed synthesis of novel optically active tryptophan analogues. Org. Lett. 5: 1717–1720.

(124) S.S. Palimkar, P.H. Kumar, R.J. Lahoti and K.V. Srinivasan. 2006. Ligand-, copper-, and amine-free one-pot synthesis of 2-substituted indoles *via* Sonogashira coupling 5-*endo-dig* cyclization. Tetrahedron 62: 5109–5115.

(125) A.J. McCarroll, T.D. Bradshaw, A.D. Westwell, C.S. Matthews and M.F.G. Stevens. 2007. Quinols as novel therapeutic agents. 7.1 Synthesis of anti-tumor 4-[1-(arylsulfonyl-1*H*-indol-2-yl)]-4-hydroxycyclohexa-2,5-dien-1-ones by Sonogashira reactions. J. Med. Chem. 50: 1707–1710.

(126) J.H. Kim and S.-G. Lee. 2011. Palladium-catalyzed intramolecular trapping of the Blaise reaction intermediate for tandem one-pot synthesis of indole derivatives. Org. Lett. 13: 1350–1353.

(127) V.A. Rassadin, A.A. Tomashevskiy, V.V. Sokolov, A. Ringe, J. Magull and A. de Meijere. 2009. Facile access to bicyclic sultams with methyl 1-sulfonylcyclopropane-1-carboxylate moieties. Eur. J. Org. Chem. 16: 2635–2641.

(128) M.E. Krolski, A.F. Renaldo, D.E. Rudisill and J.K. Stille. 1988. Palladium-catalyzed coupling of 2-bromoanilines with vinylstannanes. A regiocontrolled synthesis of substituted indoles. J. Org. Chem. 53: 1170–1176.

(129) W.C. Frank, Y.C. Kim and R.F. Heck. 1978. Palladium-catalyzed vinylic substitution reactions with heterocyclic bromides. J. Org. Chem. 43: 2947–2949.

(130) P.J. Harrington, L.S. Hegedus and K.F. McDaniel. 1987. Palladium-catalyzed reactions in the synthesis of 3- and 4-substituted indoles. Total synthesis of the *N*-acetyl methyl ester of (+)-clavicipitic acids. J. Am. Chem. Soc. 109: 4335–4338.

(131) D.R. Adams, M.A.J. Duncton, J.R.A. Roffey and J. Spencer. 2002. Preparation of 6-chloro-5-fluoroindole *via* the use of palladium and copper-mediated heterocyclizations. Tetrahedron Lett. 43: 7581–7583.

(132) T. Sato, S. Ishida, H. Ishibashi and M. Ikeda. 1991. Regiochemistry of radical cyclizations (6-*exo*/7-*endo* and 7-*exo*/8-*endo*) of *N*-(*o*-alkenylphenyl)-2,2-dichloroacetamides. Chem. Soc. Perkin Trans. 1 22: 353–359.

(133) S. Hibino and E. Sugino. 1987. A facile and alternative synthesis of quinoline nucleus using thermal cyclization of 2-azahexatriene system generated from 2-alkenyl-acylaniline with POCl₃. Heterocycles 26: 1883–1889.

(134) M.K. Coooper and D.W. Yaniuk. 1981. Preparation and characterization of chelating monoolefin-aniline ligands and their platinum(II) complexes. J. Organomet. Chem. 221: 231–247.

(135) C. Subramanyam, M. Noguchi and S.M. Weinreb. 1989. An approach to amphimedine and related marine alkaloids utilizing an intramolecular Kondrat'eva pyridine synthesis. J. Org. Chem. 54: 5580–5585.

(136) J.E. Plevyak and R.F. Heck. 1978. Palladium-catalyzed arylation of ethylene. J. Org. Chem. 43: 2454–2456.

(137) M. Yamaguchi, M. Arisawa and M. Hirama. 1998. *ortho*-Vinylation reaction of anilines. Chem. Commun. 13: 1399–1400.

(138) L.S. Hegedus, G.F. Allen, J.J. Bozell and E.L. Waterman. 1978. Palladium-assisted intramolecular amination of olefins. Synthesis of nitrogen heterocycles. J. Am. Chem. Soc. 100: 5800–5807.

(139) L.S. Hegedus. 1983. Palladium-catalyzed synthesis of heterocycles. J. Mol. Catal. 19: 201–211.

(140) P.J. Harrington and L.S. Hegedus. 1984. Palladium-catalyzed reactions in the synthesis of 3- and 4-substituted indoles. Approaches to ergot alkaloids. J. Org. Chem. 49: 2657–2662.

(141) E.M. Beccalli, G. Broggini, M. Martinelli and S. Sottocornola. 2007. C-C, C-O, C-N Bond formation on sp² carbon by Pd(II)-catalyzed reactions involving oxidant agents. Chem. Rev. 107: 5318–5365.

(142) M. Gowan, A.S. Caille and C.K. Lau. 1997. Synthesis of 3-alkoxyindoles *via* palladium-catalyzed intramolecular cyclization of *N*-alkyl *ortho*-siloxyallylanilines. Synlett 11: 1312–1314.

(143) P.R. Weider, L.S. Hegedus, H. Asada and V. D'Andreq. 1985. Oxidative cyclization of unsaturated aminoquinones. Synthesis of quinolinoquinones. Palladium-catalyzed synthesis of pyrroloindoloquinones. J. Org. Chem. 50: 4276–4281.

(144) A. Kasahara, T. Izumi, S. Murakami, K. Miyamoto and T. Hino. 1989. A regiocontrolled synthesis of substituted indoles by palladium-catalyzed coupling of 2-bromonitrobenzenes and 2-bromoacetanilides. J. Heterocycl. Chem. 26: 1405–1413.

(145) M. Akazome, T. Kondo and Y. Watanabe. 1992. Novel synthesis of indoles *via* palladium-catalyzed reductive *N*-heterocyclization of *o*-nitrostyrene derivatives. Chem. Lett. 5: 769–772.

(146) M. Akazome, T. Kondo and Y. Watanabe. 1994. Palladium complex-catalyzed reductive *N*-heterocyclization of nitroarenes: novel synthesis of indole and 2*H*-indazole derivatives. J. Org. Chem. 59: 3375–3380.

(147) T. Fukuyama, A.A. Laird and L.M. Hotchkiss. 1985. *p*-Anisyl group: a versatile protecting group for primary alcohols. Tetrahedron Lett. 26: 6291–6292.

(148) O. Leogane and H. Lebel. 2008. One-pot multi-component synthesis of indoles from 2-iodobenzoic acid. Angew. Chem. Int. Ed. 47: 350–352.

(149) I. Nakamura, T. Nemoto, N. Shiraiwa and M. Terada. 2009. Palladium-catalyzed indolization of *N*-aroylbenzotriazoles with di-substituted alkynes. Org. Lett. 11: 1055–1058.

(150) S.E. Denmark and J.D. Baird. 2009. Preparation of 2,3-disubstituted indoles by sequential Larock heteroannulation and silicon-based cross-coupling reactions. Tetrahedron 65: 3120–3129.

(151) J.-E. Backvall and P.G. Andersson. 1990. Palladium-catalyzed stereocontrolled intramolecular 1,4-additions to cyclic 1,3-dienes involving amides as nucleophiles. J. Am. Chem. Soc. 112: 3683–3685.

(152) P.G. Andersson and J.-E. Backvall. 1992. Palladium-catalyzed tandem cyclization of 4,6- and 5,7-diene amides. A new route toward the pyrrolizidine and indolizidine alkaloids. J. Am. Chem. Soc. 114: 8696–8698.

(153) T. Fukuyama, X. Chen and G. Peng. 1994. A novel tin-mediated indole synthesis. J. Am. Chem. Soc. 116: 3127–3128.

(154) Y. Kobayashi and T. Fukuyama. 1998. Development of a novel indole synthesis and its application to natural products synthesis. J. Heterocycl. Chem. 35: 1043–1056.

(155) D.D. Hennnings, S. Iwasa and V.H. Rawal. 1997. Anion-accelerated palladium-mediated intramolecular cyclizations: synthesis of benzofurans, indoles, and a benzopyran. Tetrahedron Lett. 38: 6379–6382.

(156) N. Lachance, M. April and M.-A. Joly. 2005. Rapid and efficient microwave-assisted synthesis of 4-, 5-, 6- and 7-azaindoles. Synthesis 15: 2571–2577.

(157) C. Nevado, D.J. Cardenas and A.M. Echavarren. 2003. Reaction of enol ethers with alkynes catalyzed by transition metals: 5-*exo-dig* versus 6 *endo-dig* cyclizations *via* cyclopropyl platinum or gold carbene complexes. Chem. Eur. J. 9: 2627–2635.

(158) S.I. Lee, S.Y. Park, J.H. Park, I.G. Jung, S.Y. Choi, Y.K. Chung and B.Y. Lee. 2006. Rhodium *N*-heterocyclic carbene-catalyzed [4+2]- and [5+2]-cycloaddition reactions. J. Org. Chem. 71: 91–96.

(159) S. Diez-Gonzalez, N. Marion and S.P. Nolan. 2009. *N*-Heterocyclic carbenes in late transition metal catalysis. Chem. Rev. 109: 3612–3676.

(160) M. Somei, S. Sayama, K. Naka and F. Yamada. 1988. A convenient synthetic method of 2-substitute indoles and its application for the synthesis of natural alkaloid, borrerine. Heterocycles 27: 1585–1587.

(161) T. Choshi, S. Yamada, E. Sugino, T. Kuwada and S. Hibino. 1995. The total synthesis of the marine anti-tumor grossularine-2. Synlett 2: 147–148.

(162) T. Choshi, S. Yamada, E. Sugino, T. Kuwada and S. Hibino. 1995. Total synthesis of grossularines-1 and -2. J. Org. Chem. 60: 5899–5904.

(163) T. Choshi, T. Sada, H. Fujimoto, C. Nagayama, E. Sugino and S. Hibino. 1996. Total syntheses of carazostatin and hyellazole by allene-mediated electrocyclic reaction. Tetrahedron Lett. 37: 2593–2596.

(164) T. Choshi, H. Fujimoto, E. Sugino and S. Hibino. 1996. Synthesis of new tetracyclic oxazolocarbazoles as functionalized precursors to anti-oxidative agents, antiostatins and carbazoquinocins. Heterocycles 43: 1847–1854.

(165) T. Kobayashi, G. Peng and T. Fukuyama. 1999. Efficient total syntheses of (±)-vincadifformine and (−)-tabersonine. Tetrahedron Lett. 40: 1519–1522.

(166) K. Dinnel, G.G. Chicchi, M.J. Dhar, J.M. Elliot, G.J. Hollingworth, M.M. Kurtz, M.P. Ridgill, W. Rycroft, K.-L. Tsao, A.R. Williams and C.J. Swain. 2001. 2-Aryl indole NK$_1$ receptor antagonists: optimization of the 2-aryl ring and the indole nitrogen substituent. Bioorg. Med. Chem. Lett. 11: 1237–1240.

(167) H. Tokuyama, Y. Kaburagi, X. Chen and T. Fukuyama. 2000. Synthesis of 2,3-disubstituted indoles by palladium-mediated coupling of 2-iodoindoles. Synthesis 3: 429–434.

(168) Y. Yokoyama, M. Ikeda, M. Saito, T. Yoda, H. Suzuki and Y. Murakami. 1990. Palladium-catalyzed reaction of 3-bromoindole derivative with allyl esters in the presence of hexa-*n*-butyldistannane. Heterocycles 31: 1505–1511.

(169) T. Sakamoto, A. Yasuhara, Y. Kondo and H. Yamanaka. 1994. Synthesis of ethoxyethynylarenes by the palladium-catalyzed reaction of aryl iodides with ethoxy(trialkylstannyl)acetylenes. Chem. Pharm. Bull. 42: 2032–2035.

(170) R.L. Hudkins, J.L. Diebold and F.D. Marsh. 1995. Synthesis of 2-aryl- and 2-vinyl-1*H*-indoles *via* palladium-catalyzed cross-coupling of aryl and vinyl halides with 1-carboxy-2-(tributylstannyl)indole. J. Org. Chem. 60: 6218–6220.

(171) T. Choshi, S. Yamada, J. Nobuhiro, Y. Mihara, E. Sugino and S. Hibino. 1998. The anomalous Stille reaction of 5-stannylimidazole with 3-iodoindole. Heterocycles 48: 11–14.

(172) K. Olofsson, S.Y. Kim, M. Larhed, D.P. Curran and A. Hallberg. 1999. High-speed, highly fluorous organic reactions. J. Org. Chem. 64: 4539–4541.

(173) V. Khedkar, A. Tillack, M. Michalik and M. Beller. 2005. Convenient synthesis of tryptophols and tryptophol homologues by hydroamination of alkynes. Tetrahedron 61: 7622–7631.

(174) V. Khedkar, A. Tillack, M. Michalik and M. Beller. 2004. Efficient one-pot synthesis of tryptamines and tryptamine homologues by amination of chloroalkynes. Tetrahedron Lett. 45: 3123–3126.

(175) A. Grandmare, H. Kusama and N. Iwasawa. 2007. W(CO)$_5$(L)-catalyzed cyclization of ω-acetylenic silyl enol ethers for the preparation of nitrogen-containing cyclic compounds. Chem. Lett. 36: 66–67.

(176) G. Bratulescu. 2008. A new and efficient one-pot synthesis of indoles. Tetrahedron Lett. 49: 984–986.

(177) A. Padwa and A.G. Waterson. 2000. The thionium/*N*-acyliminium ion cyclization cascade as a strategy for the synthesis of azapolycyclic ring systems. Tetrahedron 56: 10159–10173.

(178) Y. Kita, H. Yasuda, O. Tamura, F. Itoh and Y. Tamura. 1984. The chemistry of *O*-silylated ketene acetal; Pummerer rearrangement of sulfoxides into α-siloxysulfides. Tetrahedron Lett. 25: 4681–4682.

(179) Y. Kita, O. Tamura, T. Miki and Y. Tamura. 1987. Chemistry of *O*-silylated ketene acetals: an efficient synthesis of α-thio-heterocycles from ω-amidosulfoxides by a novel intramolecular Pummerer-type rearrangement. Tetrahedron Lett. 28: 6479–6480.

(180) A. Lei, X. Lu and G. Liu. 2004. A novel highly regio- and diastereoselective haloamination of alkenes catalyzed by divalent palladium. Tetrahedron Lett. 45: 1785–1788.

(181) A. Padwa, D.M. Danca, J.D. Ginn and S.M. Lynch. 2001. Application of the tandem thionium/*N*-acyliminium ion cascade toward heterocyclic synthesis. J. Braz. Chem. Soc. 12: 571–585.

(182) Q. Yan, J. Luo, D. Zhang-Negrerie, H. Li, X. Qi and K. Zhao. 2011. Oxidative cyclization of 2-aryl-3-arylamino-2-alkenenitriles to *N*-arylindole-3-carbonitriles mediated by NXS/Zn(OAc)$_2$. J. Org. Chem. 76: 8690–8697.

(183) O. Repic, K. Prasad and G.T. Lee. 2001. The story of lescol: from research to production. Org. Process Res. Dev. 5: 519–527.

(184) R.E. Walkup and J. Linder. 1985. 2-Formylation of 3-arylindoles. Tetrahedron Lett. 26: 2155–2158.

(185) L. Zhou and M.P. Doyle. 2009. Lewis acid-catalyzed indole synthesis *via* intramolecular nucleophilic attack of phenyldiazoacetates to iminium ions. J. Org. Chem. 74: 9222–9224.

(186) M.P. Kumar and R.-S. Liu. 2006. Zn(OTf)$_2$-catalyzed cyclization of proparyl alcohols with anilines, phenols, and amides for synthesis of indoles, benzofurans, and oxazoles through different annulation mechanisms. J. Org. Chem. 71: 4951–4955.

(187) T.M. Lipinska. 2004. Microwave-induced solid-supported Fischer indolization, a key step in the total synthesis of the sempervirine type methoxy analogues. Tetrahedron Lett. 45: 8831–8834.

(188) T.M. Lipinska and S.J. Czarnocki. 2006. A new approach to difficult Fischer synthesis: the use of zinc chloride catalyst in triethylene glycol under controlled microwave irradiation. Org. Lett. 8: 367–370.

(189) R. Bently, T.S. Stevens and M. Thompson. 1970. The alkaloids of Gelsemium sempervirens. Part III. Sempervirine. J. Chem. Soc. 6: 791–795.

(190) A. Das, A. Kulkarni and B. Torok. 2012. Environmentally benign synthesis of heterocyclic compounds by combined microwave-assisted heterogeneous catalytic approaches. Green Chem. 14: 17–34.

(191) H.A. Saad, M.M. Youssef and M.A. Mosselhi. 2011. Microwave-assisted synthesis of some new fused 1,2,4-triazines bearing thiophene moieties with expected pharmacological activity. Molecules 16: 4937–4957.

(192) P. Wyrbek, A. Sniady, N. Bewick, Y. Li, A. Mikus, K.A. Wheeler and R. Dembinski. 2009. Microwave-assisted zinc chloride-catalyzed synthesis of substituted pyrroles from homopropargyl azides. Tetrahedron 65: 1268–1275.

(193) M. Somei and F. Yamada. 2005. Simple indole alkaloids and those with a non-rearranged monoterpenoid unit. Nat. Prod. Rep. 22: 73–103.

(194) M. Somei and F. Yamada. 2004. Simple indole alkaloids and those with a non-rearranged monoterpenoid unit. Nat. Prod. Rep. 21: 278–311.

(195) L.H. Franco and J.A. Palermo. 2003. Synthesis of 2-(pyrimidin-4-yl)indoles. Chem. Pharm. Bull. 51: 975–977.

(196) F.L. Hernandez, J.E. Bal de Kier, L. Puricelli, M. Tatian, A.M. Seldes and J.A. Palermo. 1998. Indole alkaloids from the tunicate aplidiummeridianum. J. Nat. Prod. 61: 1130–1132.

(197) M.S. Butler, R.J. Capon and C.C. Lu. 1992. Psammopemmins, novel brominated 4-hydroxyindole alkaloids from an antarctic sponge, *Psammopemma* sp. Aust. J. Chem. 45: 1871–1877.

(198) (a) N.B. Perry, L. Ettouati, M. Litaudon, J.W. Blunt, M.H.G. Munro, S. Parkin and H. Hope. 1994. Alkaloids from the antarctic sponge *Kirkpatrickia varialosa*. Tetrahedron 50: 3987–3992. (b) P. Appukkuttan and E. van der Eycken. 2006. Microwave-assisted natural product chemistry. Top. Curr. Chem. 266: 1–47.

(199) N. Uesaka, M. Mori, K. Okamura and T. Date. 1994. Novel synthesis of heterocycles using zirconium-catalyzed diene cyclization. J. Org. Chem. 59: 4542–4547.

(200) Y. Sato, T. Tamura and M. Mori. 2004. Arylnaphthalene lignans through Pd-catalyzed [2+2+2]-co-cyclization of arynes and diynes: total synthesis of Taiwanins C and E. Angew. Chem. Int. Ed. 43: 2436–2440.

(201) M. Mori, N. Uesaka, F. Saitoh and M. Shibasaki. 1994. Novel synthesis of nitrogen heterocycles using zirconium-promoted reductive cyclization. J. Org. Chem. 59: 5643–5649.

(202) M. Bandini and A. Eichholzer. 2009. Catalytic functionalization of indoles in a new dimension. Angew. Chem. Int. Ed. 48: 9608–9644.

(203) K. Kruger, A. Tillack and M. Beller. 2008. Catalytic synthesis of indoles from alkynes. Adv. Synth. Catal. 350: 2153–2167.

(204) C. Cao, Y. Shi and A.L. Odom. 2003. A titanium-catalyzed three-component coupling to generate α,β-unsaturated β-iminoamines. J. Am. Chem. Soc. 125: 2880–2881.

(205) B. Ramanathan, A. Keith, D. Armstrong and A.L. Odom. 2004. Pyrrole syntheses based on titanium-catalyzed hydroamination of diynes. Org. Lett. 6: 2957–2960.

(206) K. Alex, A. Tillack, N. Schwarz and M. Beller. 2008. Zinc-promoted hydrohydrazination of terminal alkynes: an efficient domino synthesis of indoles. Angew. Chem. Int. Ed. 47: 2304–2307.

(207) P.J. Walsh, M.J. Carney and R.G. Bergman. 1991. Generation, dative ligand trapping, and nitrogen-nitrogen bond cleavage reactions of the first monomeric η^1-hydrazido zirconocene complex, (Cp$_2$Zr:NNPh$_2$). A zirconium-mediated synthesis of indoles. J. Am. Chem. Soc. 113: 6343–6345.

(208) F. Basuli, H. Aneetha, J.C. Huffmann and D.J. Mindiola. 2005. A fluorobenzene adduct of Ti(IV), and catalytic carboamination to prepare α,β-unsaturated imines and triaryl-substituted quinolines. J. Am. Chem. Soc. 127: 17992–17993.

Five-Membered Fused *N*-Heterocycles

3.1 Introduction

For over a century, heterocyclic compounds have been the largest classical division of organic chemistry with industrial and biological importance [1a–c]. Numerous biologically active agrochemicals, pharmaceuticals, additives and modifiers applied in industrial applications such as information storage, cosmetics reprography and plastics are heterocycles. One striking structural feature of heterocyclic compounds lies in their ability to manifest substituents around a core scaffold in defined three dimensional representations [2a–i]. Heterocycles have contributed to the understanding of life processes as well as to the development of society from an industrial and biological point of view. Among over 20 million chemical compounds identified by the end of the second millennium, more than two-thirds are partially or fully aromatic and about half are heterocyclic [3a–e]. Presence of heterocyclic structures in organic compounds of interest in biology, electronics, pharmacology, optics and material sciences is widely known [4–8].

Although various methods have been developed for the synthesis of heterocyclic compounds, their functional group tolerance and substrate scope is often limited. The protocols developed in the recent years involved metal- and non-metal-assisted carbon-nitrogen bond forming reactions. These reactions tolerate a number of functional groups, occur under mild conditions and proceed with high stereoselectivity. The use of metals and non-metals allows highly convergent multi-component coupling approaches which form many bonds and/or stereocenters in one-pot protocol. This chapter describes the approaches for the preparation of fused five-membered nitrogen heterocycles through metal- and non-metal-assisted reactions [9–16].

3.2 Metal- and non-metal-assisted synthesis of five-membered *N*-heterocycles fused with other heterocycles

Aluminium-assisted synthesis

A collection of indolizines is obtained by three-component coupling of pyridine, α-acyl bromides, and internal alkynes under solvent-free conditions and microwave irradiation [17]. The 1,3-dipole is formed *in situ* from *N*-acyl pyridinium salt under basic alumina catalysis and the indolizine core is a result of subsequent [3+2]-cycloaddition reaction of the 1,3-dipole with alkyne. The reaction products can be easily separated from alumina by extraction with ethyl acetate or dichloromethane (**Scheme 1**).

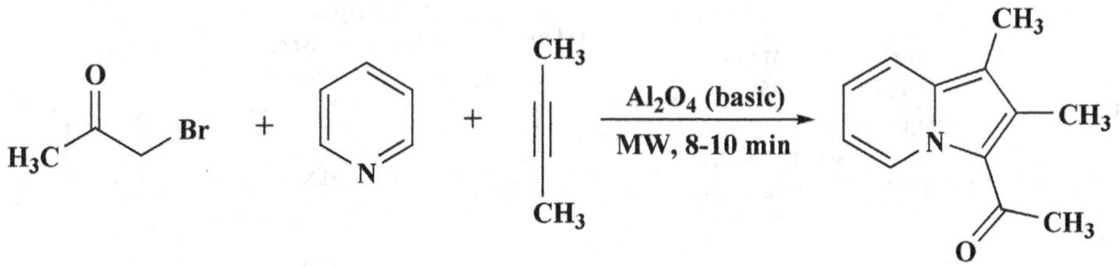

Scheme-1

Cesium-assisted synthesis

Ye et al. [18] generated a bicyclo [3.3.0] ring system with three continuous stereogenic centers from bis-allylamine by a metal-free catalytic intramolecular ylide annulation in a single manipulation (**Scheme 2**).

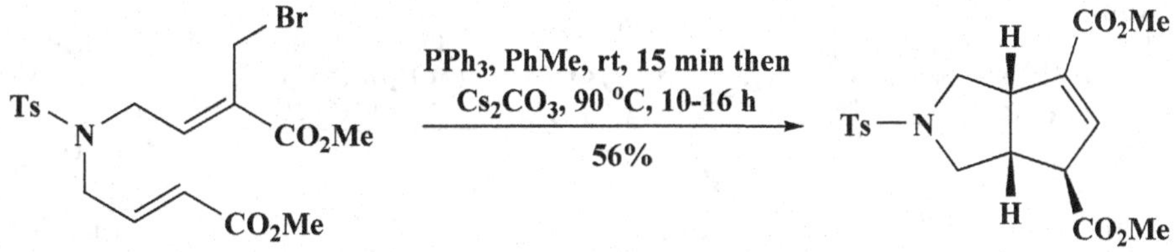

Scheme-2

Chromium-assisted synthesis

Korkowski et al. [19] studied cyclopropanes in carbocyclic systems. They observed that a stoichiometric amount (1.2 eq.) of Fischer carbene complex was required for cyclopropanation. Another study concluded that the metathesis product is formed in good yields when the alkene is substituted with a phenyl group (**Scheme 3**) [20].

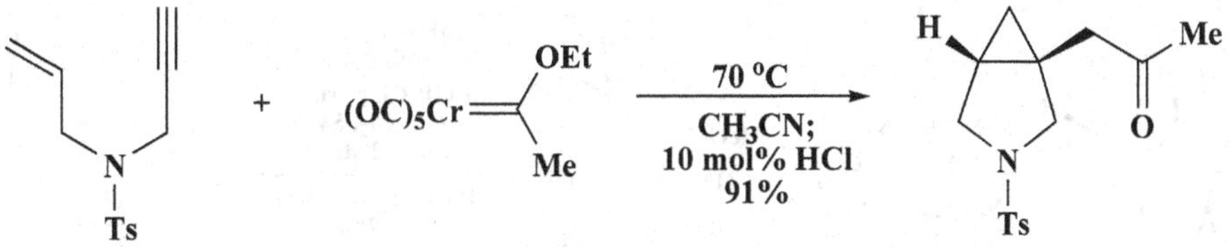

Scheme-3

The synthesis of salinosporamide A is illustrated in **Scheme 4**. It starts with the formation of amide, a substrate for the key In(OTf)$_3$-catalyzed cyclization, from chiral propargyl alcohol [21]. As per the method developed by Marshall [22], propargyl alcohol is transformed into mesylate which reacts with (*tert*-butyldimethylsilyloxy)acetaldehyde by forming allenylzinc species to afford alkyne as a 90:10 mixture of epimers. Upon removal of the PMB group, selective acetylation [23], and desilylation, the alkyne yields diol. The diol, on subsequent exposure to HIO$_4$ and CrO$_3$ [24] in aqueous acetone, produces carboxylic acid which can then be condensed with dimethyl-2-(4-methoxybenzylamino)malonate [25] through acid chloride. The amide partially undergoes cyclization to afford an inseparable 72:28 mixture during purification by column chromatography on silica gel. This mixture, when treated with catalytic amounts of In(OTf)$_3$ under reflux in toluene for the complete conversion of amide into acetate, provides quantitative yields. It is important to note that in this particular case, significant loss of enantiomeric purity of the substrate is not observed. As acetate is very labile under basic conditions, the acetoxy group is hydrolyzed under mild lipase-catalyzed reaction conditions to yield alcohol which is then oxidized

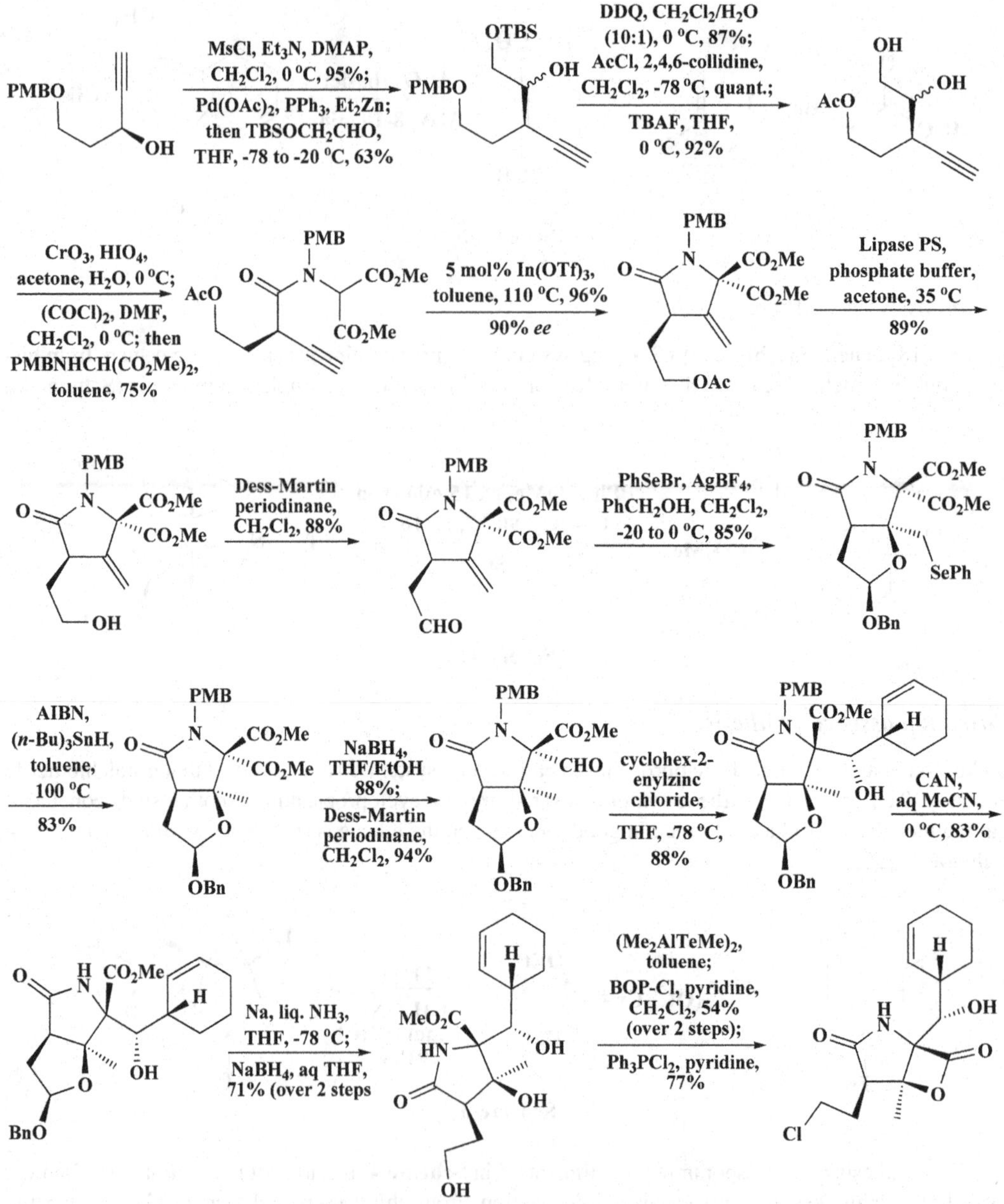

Scheme-4

into aldehyde. The aldehyde is then subjected to acetal-mediated cationic cyclization for the assembly of C3 quaternary center. Thus, aldehyde reacts with phenylselenenyl bromide and $AgBF_4$ in the presence of benzyl alcohol to provide a selenium derivative which undergoes radical deselenenylation to produce a deselenium derivative. The deselenium derivative is reduced by sodium borohydride, resulting in excellent differentiation of the geminal esters and aldehyde is formed as the sole product after oxidation with Dess-Martin periodinane. The aldehyde reacts with cyclohex-2-enylzinc chloride to yield a single stereoisomer. The removal of PMB group provides a product which is subjected to reductive ring-opening of the cyclic acetal to give an intermediate. Finally, this intermediate affords (−)-salinosporamide A upon dealkylative cleavage of methyl ester with $(Me_2AlTeMe)_2$, β-lactonization, and chlorination [26].

Copper-assisted synthesis

Yamada et al. [27] recognized an efficient method for synthesizing natural anti-tumor agents like yatakemycin and duocarmycins. The first key reaction in the synthesis of duocarmycin A is the intramolecular amination of aryldibromide, a challenging reaction due to the presence of an additional bromide the retention of which is required for ensuing transformations. Whereas palladium-catalyzed amination conditions fail to afford the desired cyclized product in decent yields, possibly due to the complications arising from unwanted oxidative addition to the remaining bromide by the palladium catalyst, the use of catalytic copper iodide with cesium acetate as base and without ligand cleanly provides the desired indoline which is then utilized for the synthesis of aryl bromide where, again, an intramolecular aryl amidation in stoichiometric amounts of copper iodide affords a tricyclic compound. This compound is finally converted into duocarmycin A. The synthesis of indolecarboxylic acid functionality again involves a successful implementation of copper-assisted aryl amination. When treated with cesium acetate and excess copper iodide at room temperature, the trimethoxy compound undergoes heterocyclization smoothly to afford the desired indole, a precursor of fragment trimethoxy indole, in nearly quantitative yields (**Scheme 5**) [28].

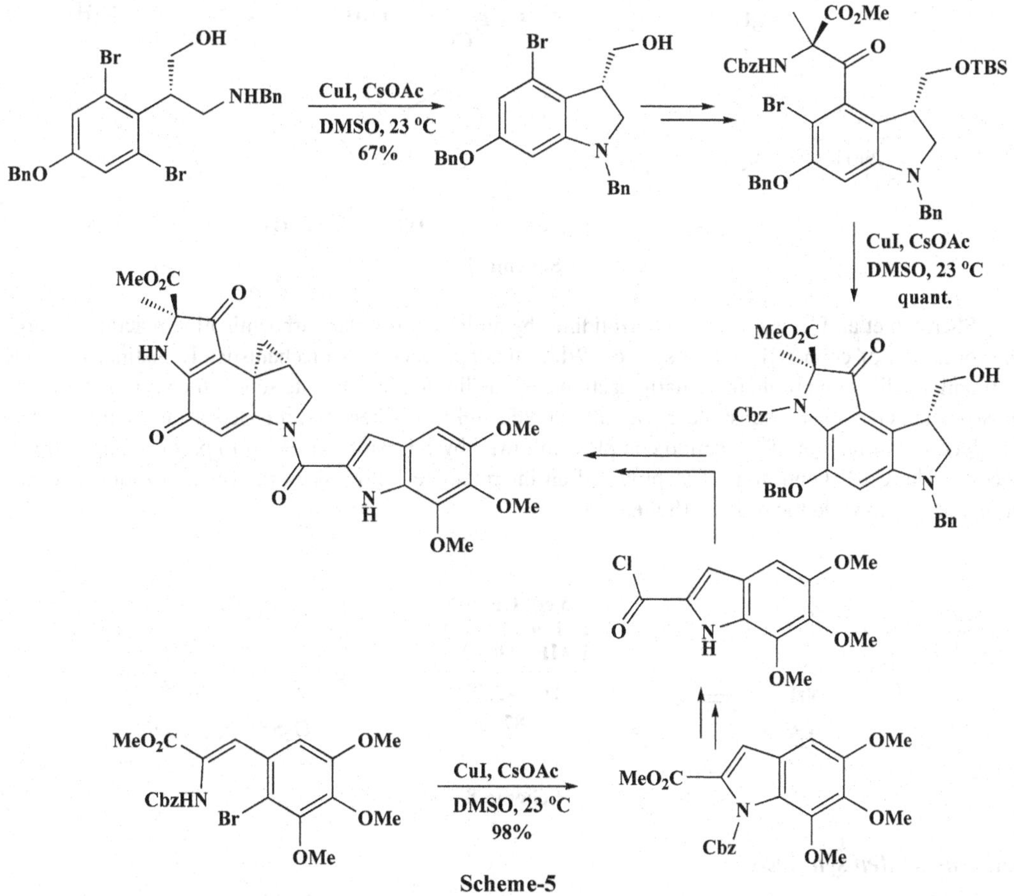

Scheme-5

Cai et al. [29] reported a tandem synthesis of fused heterocyclic scaffolds 4-oxo-indeno[1,2-b] pyrroles using cuprous iodide catalyst by reacting alkynyl decorated acetophenones and isocyanides. In this case, the target products were obtained from indoline-like intermediates through quick carbon-carbon coupling (**Scheme 6**) [30].

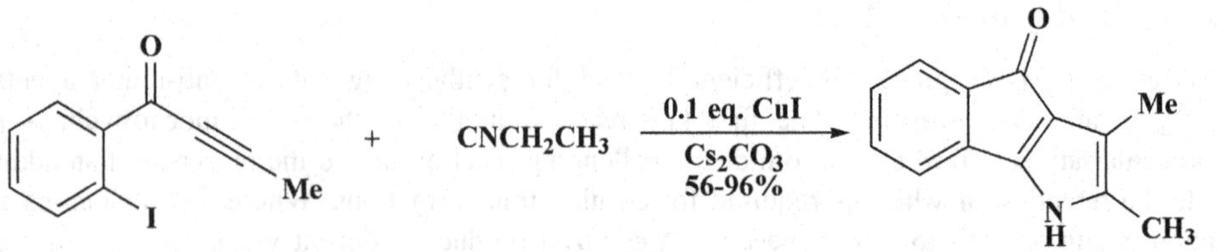

Scheme-6

The reduction of cycloadducts with tributyltin hydride affords an indolizidine alkaloid, gephyrotoxin (naturally found in the skin of poison-dart frogs) along with its diastereomers in a 4.8:1 mixture (**Scheme 7**) [31].

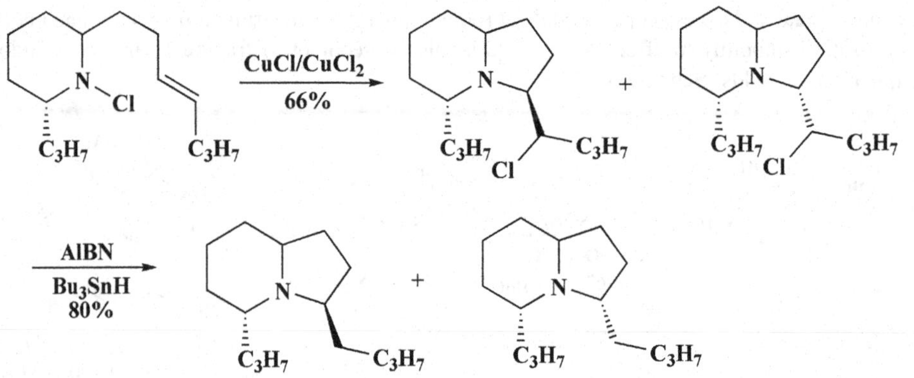

Scheme-7

Sherman et al. [32] synthesized pyrrolidines by an intramolecular carboamination reaction through the formation of copper(II)-amido species. While this approach was mechanistically distinct from the *anti*-aminopalladation route of Tamaru reactions, it was limited in substrate scope; the aryl functionality must be tethered to the substrate by a sulfonamide linkage. Chemler proposed a mechanism which involved the formation of Cu-amido complex, followed by migratory insertion to produce copper alkyl species. The Cu-C bond homolysis provided an intermediate followed by radical cyclization onto the aromatic ring to yield the product (**Scheme 8**).

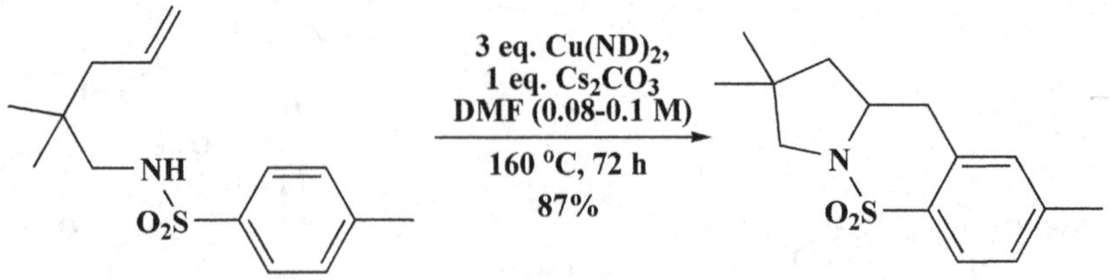

Scheme-8

Indium-assisted synthesis

The C3 quaternary center of Salinosporamide A is formed stereoselectively from a cyclized compound by intramolecular delivery of an oxygen atom from the C2 substituent to the *exo*-olefin, and another C4 quaternary center is a result of selective reduction of the ester located in the convex face of the rigid bicyclic structure (**Scheme 9**) [33–37]. Salinosporamide A is synthesized following the method developed by Reddy et al. [38] from an advanced intermediate by stereoselective cyclohexenylation and subsequent suitable functional group interconversions [39–45, 46a–b].

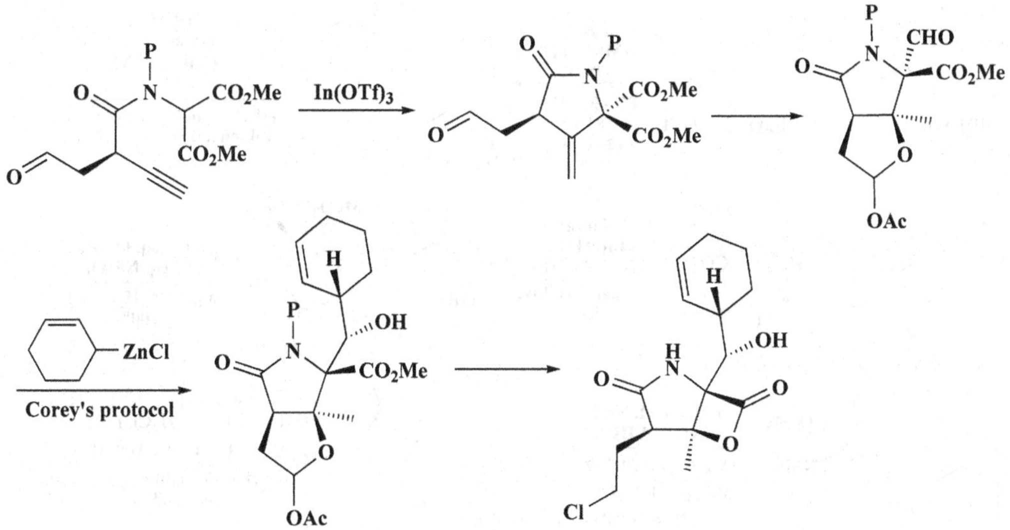

Scheme-9

Jayagobi et al. [47] and Chinnakali et al. [48–49] utilized *N*-prenylated aliphatic aldehydes for the synthesis of hexahydropyrrolo[3,4-*b*]quinolines in the presence of indium chloride **(Scheme 10)**.

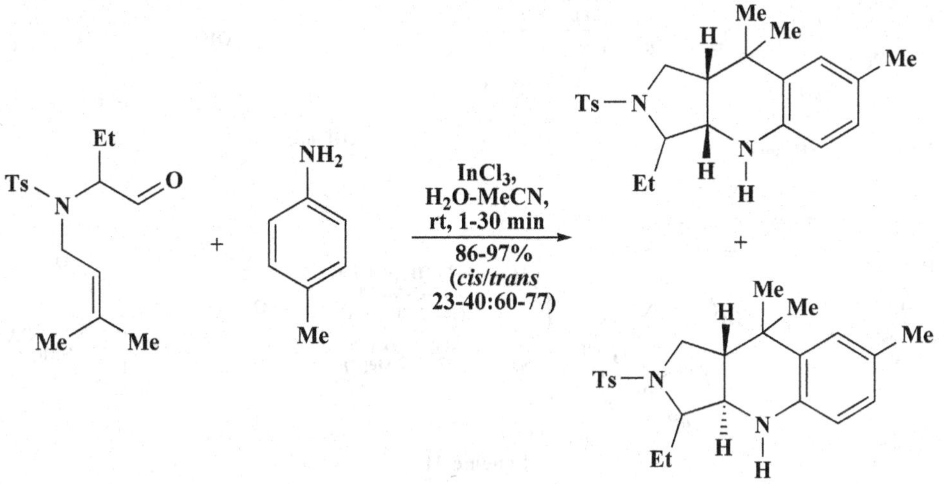

Scheme-10

Alkynol is first synthesized by a coupling of triflate and alkyne, both easily available from (*S*)-hydroxy-2-methylpropanoate, followed by desilylation **(Scheme 11)**. Jones oxidation of alkynol provides carboxylic acid which is condensed with dimethyl-2-(methylamino)malonate by acid chloride to afford amide. The stereo- and regioselective cyclization occurs efficiently to afford good yields of the product when the amide is treated with catalytic amounts of In(OTf)$_3$ in the presence of 1,8-diazabicyclo[5.4.0]undec-7-ene in boiling toluene. Particularly in this case, the reaction occurs without epimerization and with complete *E*-selectivity. In subsequent dihydroxylation of product, it has been found that OsO$_4$-*N*-methylmorpholine-*N*-oxide conditions promote a highly α-face selective dihydroxylation accompanied with concomitant lactonization to provide lactone as the sole product. On the basis of molecular mechanics calculations and nuclear Overhauser effect (NOE) experiments, the observed high diastereoselectivity can be explained by assuming as a preferred conformer, where the approach of OsO$_4$ is restricted to the α-face. The hydrolysis of lactone forms carboxylic acid which

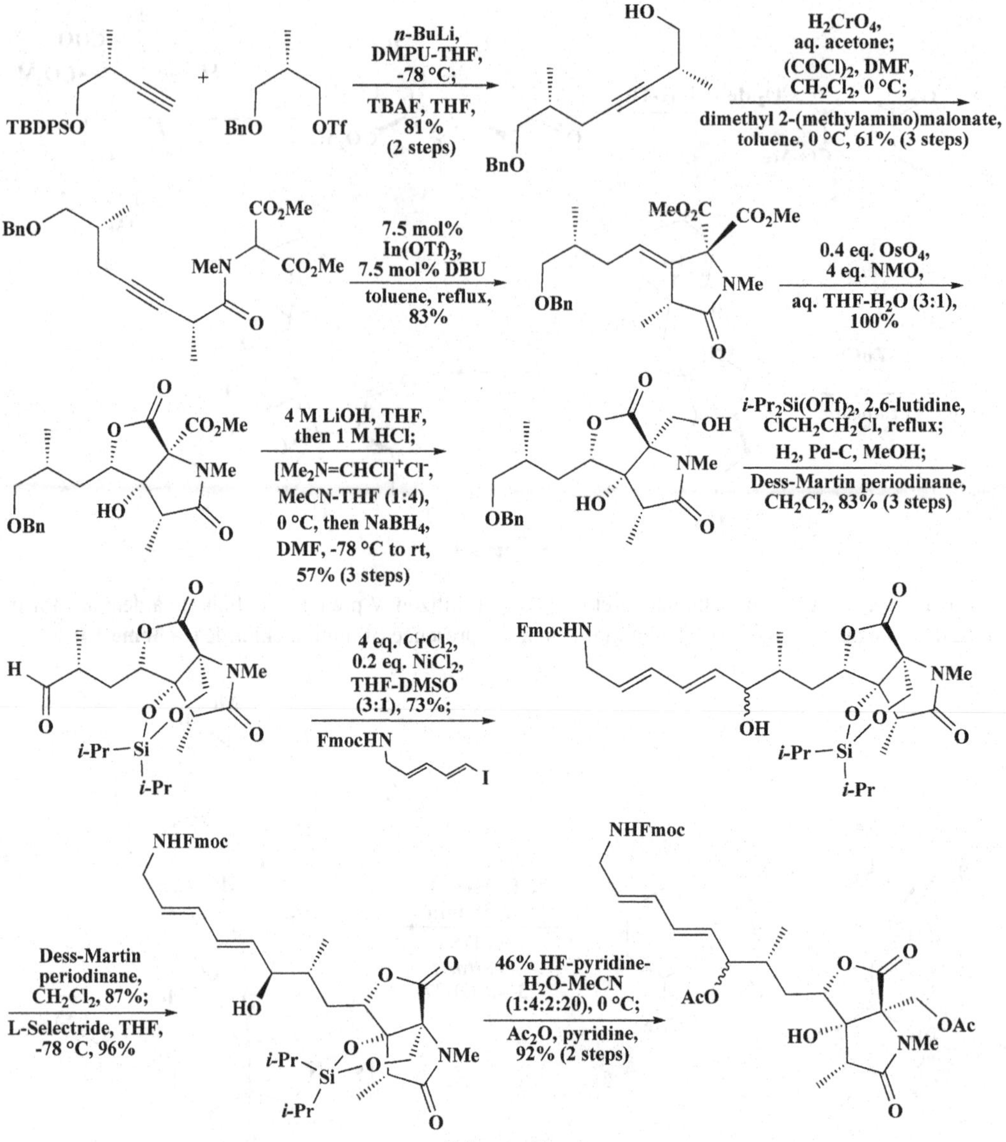

Scheme-11

has been chemoselectively transformed into *tert*-butyldimethylsilyl (TBS) ether by Fujisawa et al. successively [50]. Upon protection as its dioxasilinane, debenzylation and Dess-Martin oxidation, *tert*-butyldimethylsilyl (TBS) ether yields aldehyde. After considerable experimentation under several conditions, Nozaki-Hiyama-Kishi reaction of aldehyde turns out to be best achieved by conditions [51] using 0.2 eq. of nickel chloride and 4 eq. of $CrCl_2$ in tetrahydrofuran-dimethylsulfoxide at room temperature. Although no diastereoselectivity is observed in this coupling, Dess-Martin oxidation followed by L-selectride reduction allows a highly stereoselective production of the intermediate with the desired *R*-configuration. The intermediate, when exposed to hydrogen fluoride-pyridine followed by acetylation of the formed triol, produces the right-hand segment in almost quantitative yields [46a–b].

Iodine-assisted synthesis

A single diastereoisomer is obtained by iodocyclization of (3*R*)-configured β-amino esters **(Scheme 12)** [52].

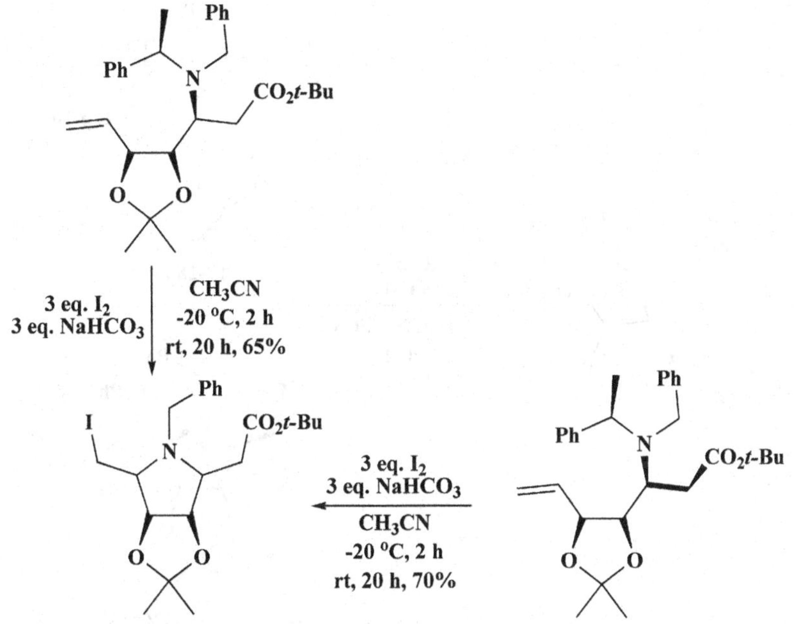

Scheme-12

Kim et al. [53] reported a 5-*endo-trig* cyclization reaction to synthesize numerous highly substituted indolizines from allyl acetates in a one-pot protocol (**Scheme 13**). An intermediate was formed when *in situ* generated iodoiranium intermediate underwent 5-*endo-trig* iodocyclization. Good yields of these products by subsequent isomerization and aromatization by loss of hydrogen iodide have also been reported [52].

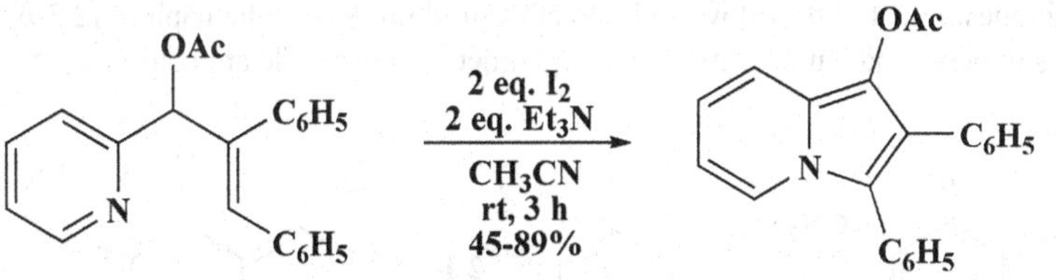

Scheme-13

Quiclet-Sire and Zard [54] synthesized fused 4,5-dihydro-3*H*-pyrazoles in good to excellent yields when hydrazones were reacted with iodine (**Scheme 14**). The diazo intermediate was produced by iodination of hydrazones followed by expulsion of hydrogen iodide. The dihydro-pyrazole derivatives were formed by intramolecular cycloaddition of the diazo intermediate [52].

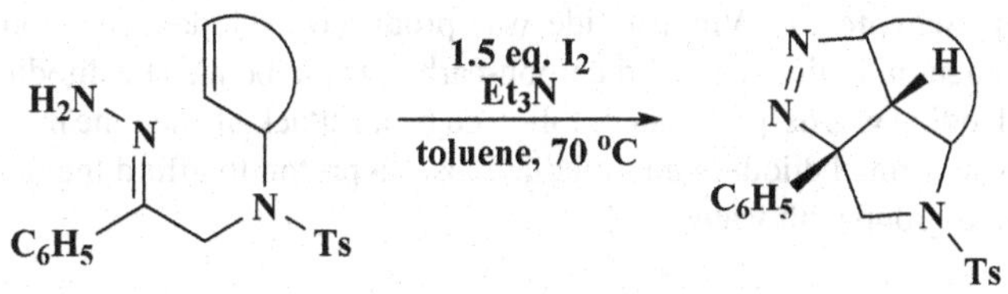

Scheme-14

Davies et al. [55] synthesized polyhydroxy pyrrolidines by ring-closing alkene iodoamination reaction (**Scheme 15**). The intermediate produced when *in situ* generated iodoiranium intermediate underwent intramolecular cyclization and lost *N*-α-methyl-benzyl protecting group *via* S$_N$1 for the formation of

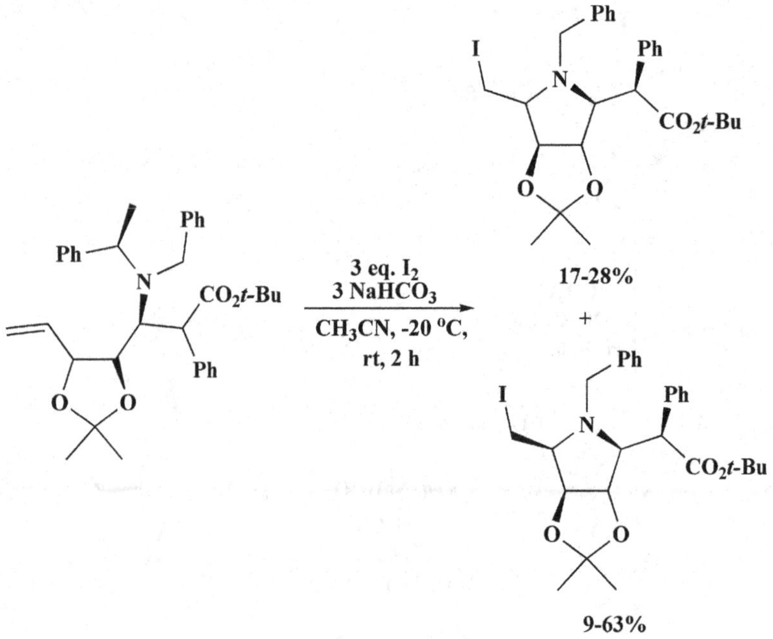

Scheme-15

iodomethyl pyrrolidines. Functional group manipulations with silver acetate occurred through the formation of aziridinium ion. Polyhydroxylated pyrrolidines can be obtained by further de-protection [52].

Naphtho[2,3-*b*]indolizine-6,11-dione derivatives can be synthesized by iodine-assisted oxidation of 2-alkyl-1,4-naphthoquinones in the presence of substituted pyridines. Other oxidant systems ($Fe(ClO_4)_3$/ iodine or manganese (IV) oxide/I$^-$) can also be utilized for the purpose of this reaction. However, 2-alkylbenzoquinones react to afford lower yields of desired products. The naphtho[2,3-*b*]indolizine-6,11-dione derivatives undergo iodination with iodine and dibenzoyl peroxide at position 12 (**Scheme 16**) [56].

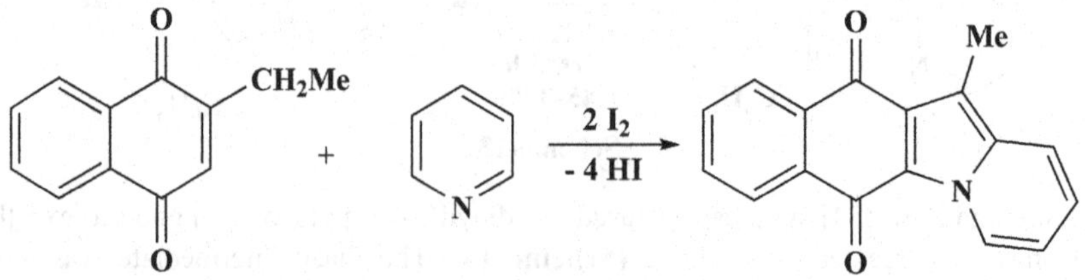

Scheme-16

Kim et al. [57] reported the synthesis of 3-acylated indolizines from alkynes through iodine-induced hydrative cyclization (**Scheme 17**). Vinyl iodide was produced by iodocyclization of alkynes by the attack of pyridyl nitrogen on iodine-activated carbon-carbon triple bond. The diiodide intermediate was generated when vinyl iodide was de-protonated, followed by an attack of enamine intermediate on another molecule of iodine. The formed diiodide intermediate loses a proton to afford the desired indolizines by the hydrolysis of diiodo group with water [52].

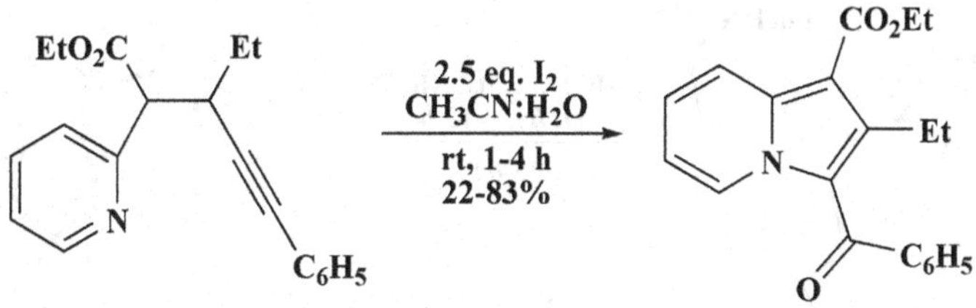

Scheme-17

Singh et al. [58] reported a stereoselective synthesis of aza-sterpurane and aza-triquinane from allylamide **(Scheme 18)**. This strategy involved *in situ* preparation of cyclohexa-2,4-dienones bearing an allylamine chain, followed by intramolecular [4+2]-cycloaddition which afforded a bicyclo[2.2.2]octenone-annulated pyrrolidine. The aza-sterpurane and aza-triquinane frameworks were constructed by further manipulation of the formed adduct, followed by photochemical sigmatropic shifts.

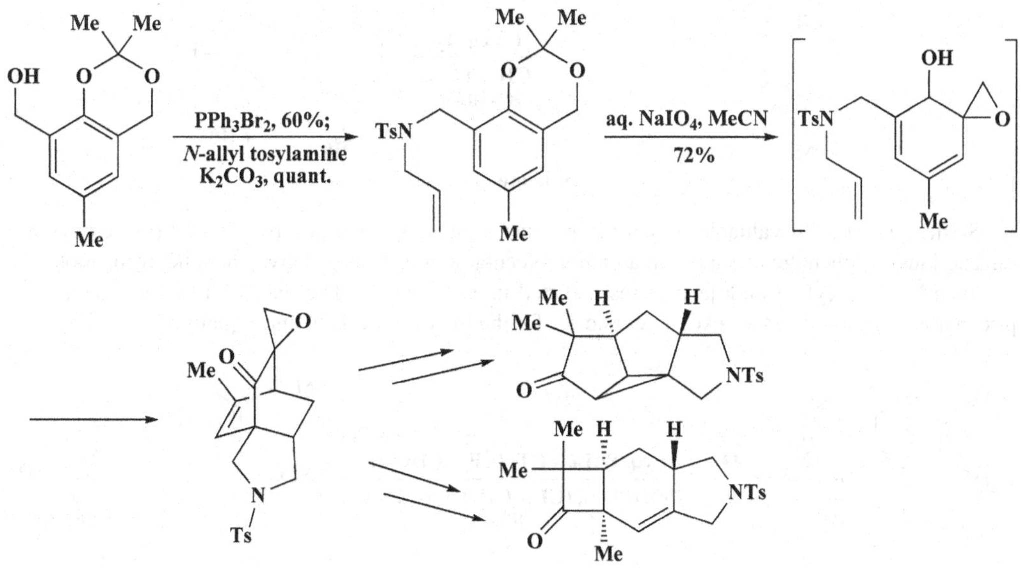

Scheme-18

Various researchers have studied cycloalkanones for the synthesis of (n+3) and (n+4) ring enlarged lactones as well as spiroketolactones by alkoxy radical fragmentation with (diacetoxyiodo)benzene in the presence of iodine [59]. Important synthesis protocols are based on the rearrangement or cyclization of the nitrogen-centered radicals produced in the reaction of specific amides with (diacetoxyiodo)benzene and iodine. Specific examples have been demonstrated by the preparation of bicyclic spirolactams from amides [60] **(Scheme 19)**, and the synthesis of oxa-azabicyclic systems by intramolecular hydrogen atom transfer promoted by phosphoramidyl and carbamoyl radicals produced from the appropriately substituted carbohydrates **(Scheme 20)** [61–62].

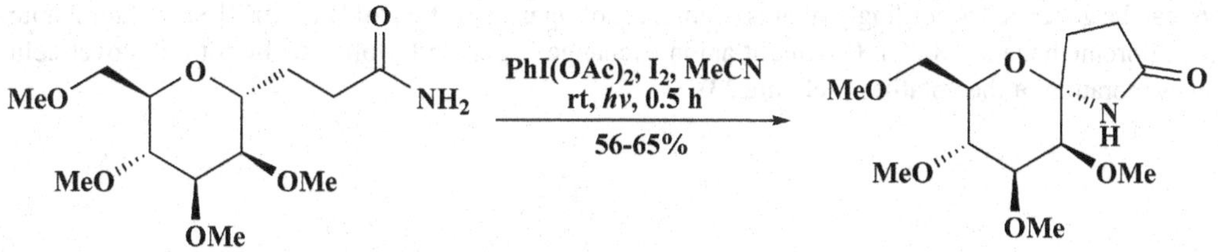

Scheme-19

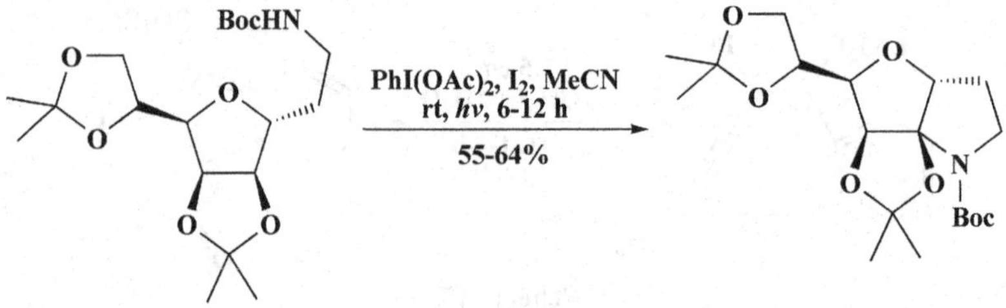

Scheme-20

Kim et al. [63] synthesized indolizines by 5-*endo-dig* cyclization of propargylic acetates **(Scheme 21)**. It has been established that initial activation of the carbon-carbon triple bond with iodine and subsequent intramolecular nucleophilic attack of pyridine nitrogen followed by dehydrogenation affords indolizines [52].

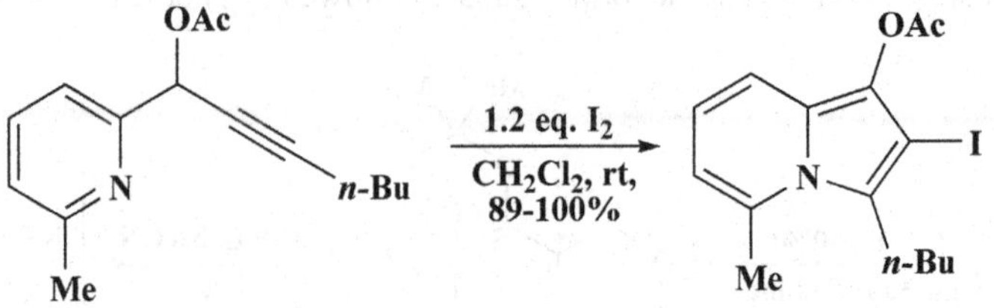

Scheme -21

Some synthetically valuable polycyclic products can be synthesized by BTI- or (diacetoxyiodo) benzene-induced phenolic oxidation in an intramolecular mode. The oxidative phenolic cyclizations are promoted by [bis(acyloxy)iodo]arenes (as depicted in **Scheme 22**). The oxidation of enamide affords spiroenamide which serves as a key intermediate for the total synthesis of annosqualine [62, 64].

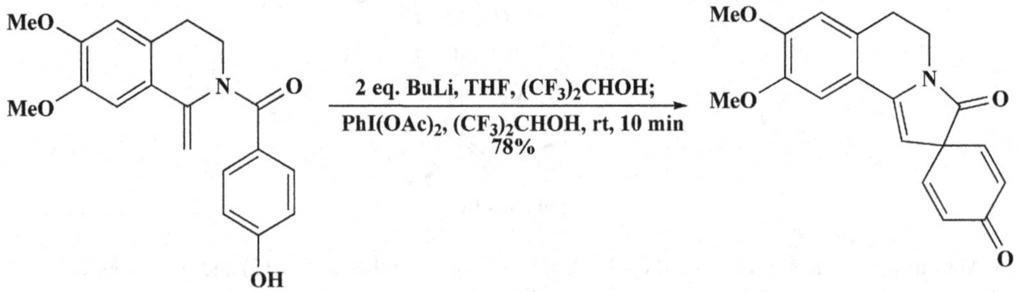

Scheme-22

Iridium-assisted synthesis

Hexafluorophosphate, trifluoromethane-sulfonate, and tetrafluoroborate are synthesized as per standard protocols from 2-[2-(diphenylphosphanyl)-phenyl]-4,5-dihydro-4-isopropyl-oxazole, [Ir(cod)Cl]$_2$, and the respective ammonium or silver salts [65]. Other iridium salts such as methylsulfonate, hexafluoroantimonate, trifluoroacetate, toluenesulfonate, and tetrakis(perfluoro-*tert*-butoxy)aluminate [66] can be generated accordingly (that is, from 2 eq. of ligand and 1 eq. of [Ir(cod)Cl]$_2$ stirred for 2 hours in dichloromethane at 48 °C). Overnight anion exchange is generally followed by filtration over celite and evaporation of the volatiles **(Scheme 23)**.

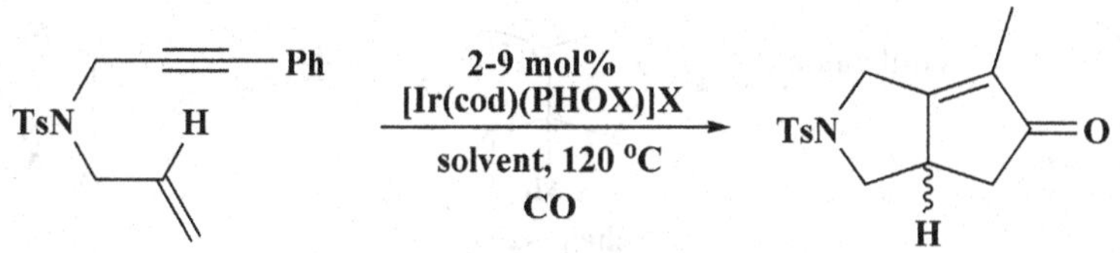

Scheme-23

An enantioselective intramolecular Pauson-Khand-type reaction catalyzed by chiral iridium diphosphine complex yields a variety of chiral bicyclic cyclopentenones. Excellent enantioselectivity is observed at low partial pressure of CO **(Scheme 24)** [67].

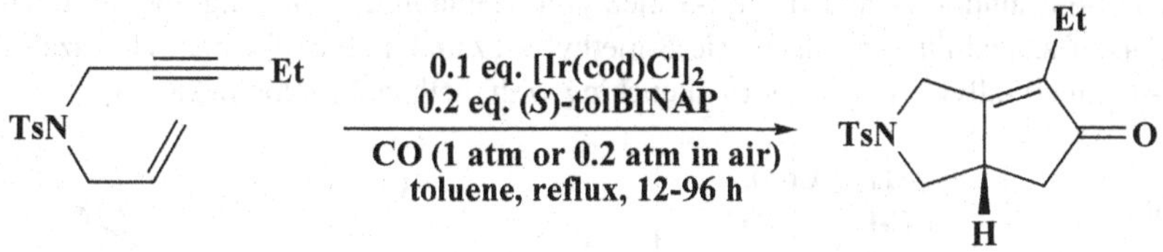

Scheme-24

Iron-assisted synthesis

Cyclobutane-fused bicyclic skeletons are produced by intramolecular [2+2]-cycloaddition of *N*-vinyl allylamine or *N,N*-bis-allylamine derivatives. Bouwkamp et al. [68] studied iron-catalyzed intramolecular [2+2]-cycloaddition reactions of bis-allylamines for the synthesis of 3-azabicyclo[0.2.3]heptane derivatives **(Scheme 25)**.

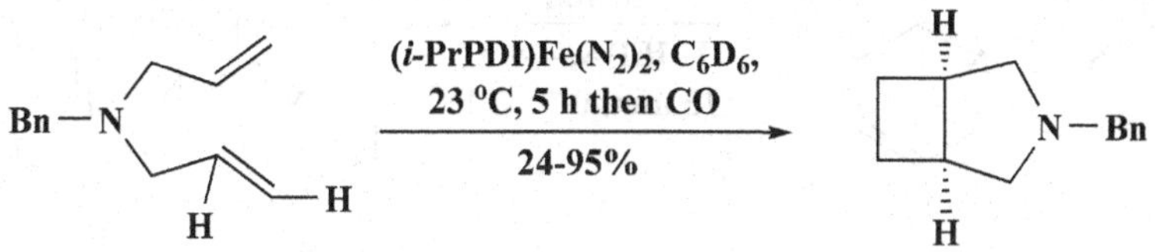

Scheme-25

Lanthanum-assisted synthesis

Pyrrolizidine derivatives are synthesized by organolanthanide-catalyzed bicyclization of aminoenynes, aminodiynes and aminodienes **(Scheme 26)** [69].

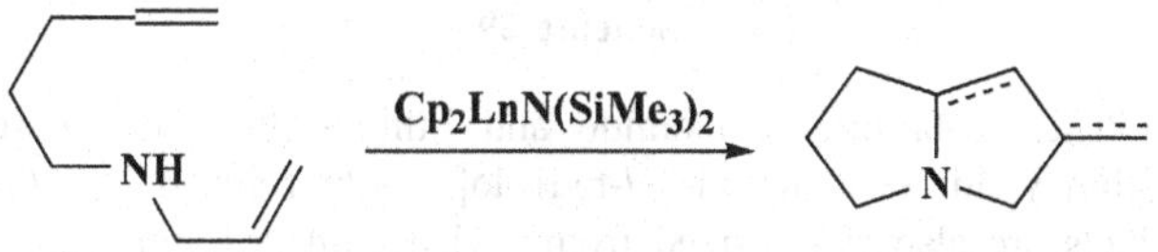

Scheme-26

The hetero-Pauson-Khand reaction, that is, the catalytic hetero-[2+2+1]-cycloaddition of a C-C multiple bond, a carbon-heteroatom bond and CO results in the synthesis of functionalized butyrolactams and butyrolactones **(Scheme 27)** [70–72].

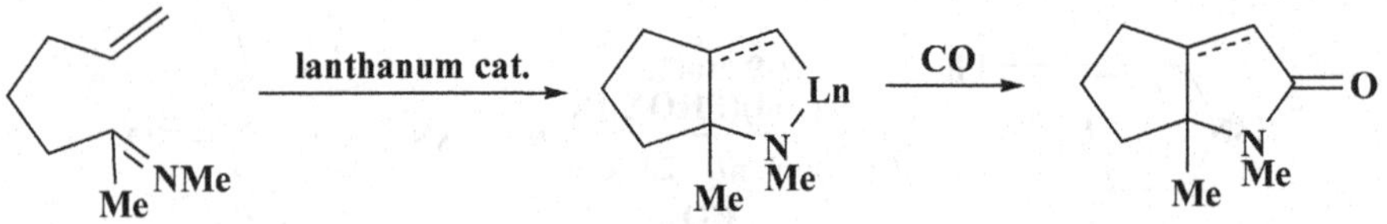

Scheme-27

Lithium-assisted synthesis

Bailey et al. [73] reported a method involving intramolecular carbolithiation of aryllithium (formed from (*N,N*-diallylamino)bromopyridine) for the synthesis of 3-substituted 4-, 5-, 6-, and 7-azaindolines (2,3-dihydro-1*H*-pyrrolopyridines). It was reported that although cyclization occurred to afford 1-allyl-3-methyl-6-azaindoline and 1-allyl-3-methyl-4-azaindoline following the protonation of the 3-CH$_2$Li group of azaindoline, the isomeric 3-methyl-7-azaindoline and 3-methyl-5-azaindoline were yielded as 3-methyl-*N*-allyl anions prior to quenching with methanol **(Scheme 28–29)**.

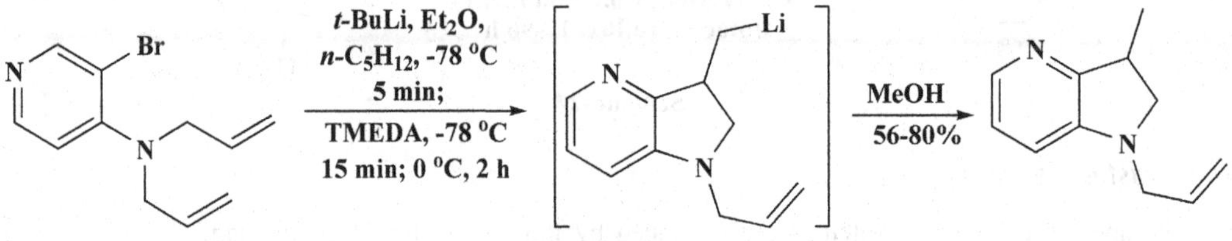

Scheme-28

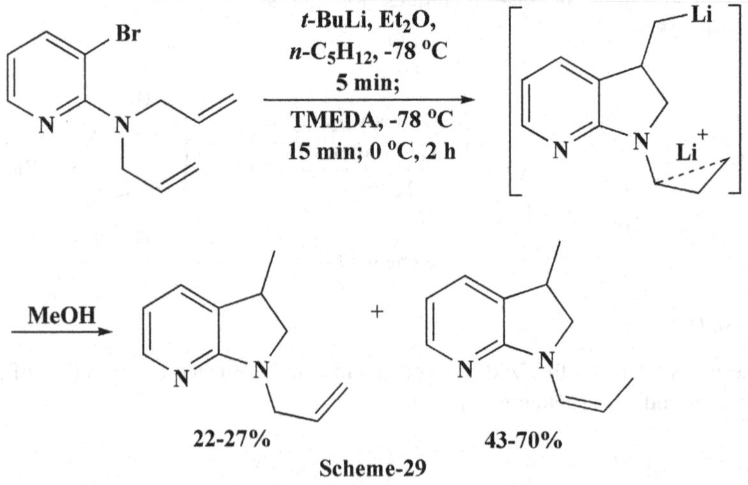

Scheme-29

When arylacetonitriles, 5-bromonicotinamide and lithium diisopropylamide (LDA) react *via* pyridyne mechanism, they yield 7-arylmethyl-1*H*-pyrrolo[3,4-*c*]pyridine-1,3-(2*H*)-diones **(Scheme 30)**. C-5 substitution products are also synthesized from arylacetonitriles and 5-chloro-3-pyridinol under similar conditions [74–75].

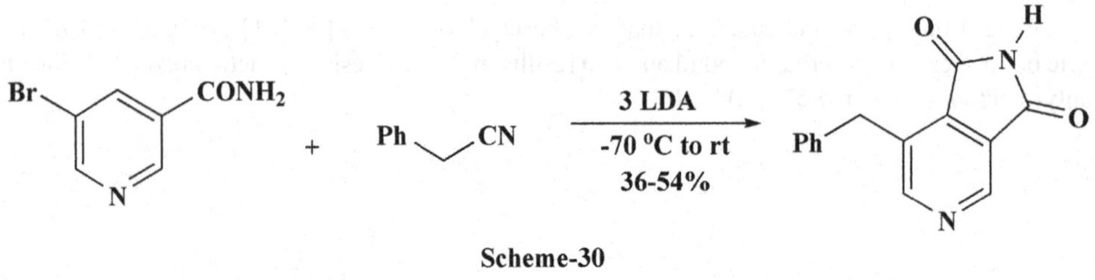

Scheme-30

Tsuchida et al. [76] investigated double cyclization of allylaminoalkenes through tandem aminolithiation-carbolithiation using lithium amide as a protonating agent as well as a lithiating agent to synthesize hexahydro-1*H*-pyrrolizine and bicyclic octahydro-indolizines in good diastereoselectivity and high yields (**Scheme 31**). The reaction was stopped using lithium amide in catalytic amounts, after the aminolithiation step, to afford the monocyclic product, whereas the bicyclic to monocyclic ratio increased when lithium amide was utilized in increased amounts. The yield of bicyclic product improved with increased diastereoselectivity when a bulkier amine (*tert*-butyltritylamine) was used.

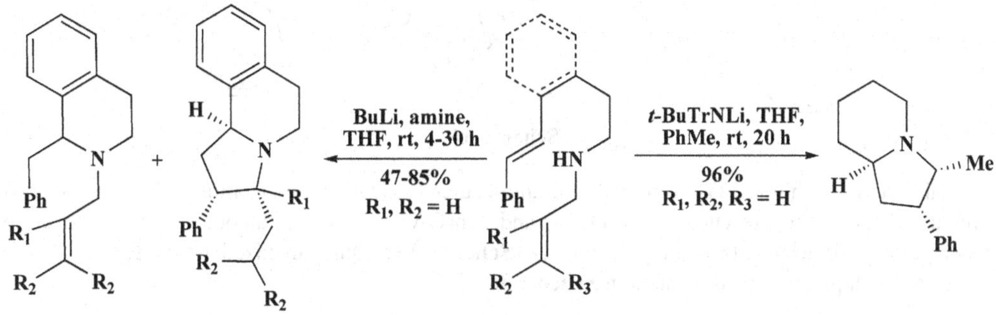

Scheme-31

Wang et al. [77] reported sequential I-Li exchange in vinyl iodide followed by intramolecular nucleophilic acyl substitution of β-amino-alkenyllithium ester to synthesize alkylidene aza-cycloketones with defined olefin geometry (**Scheme 32**). The total synthesis of allopumiliotoxin was performed by this key conversion.

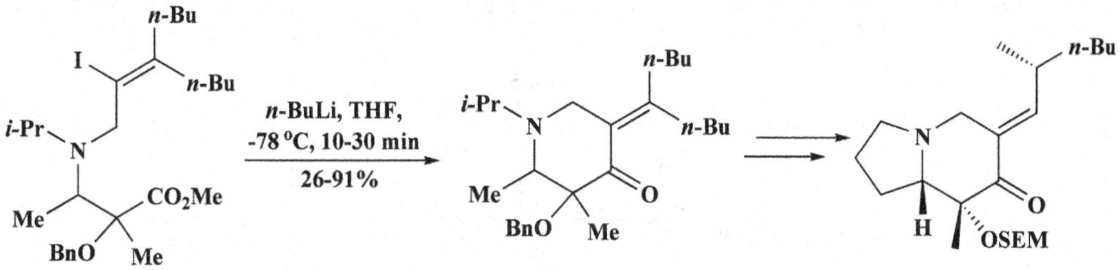

Scheme-32

Magnesium-assisted synthesis

The catalytic Diels-Alder reaction [78] of amidofuran substrate is used for the synthesis of the final product upon heating. When amidofuran is heated for 3 hours in toluene, no effect of the catalyst is reported with this substrate. This proposes that a carbonyl group inside the tether is needed for the activation of Lewis acid (**Scheme 33**).

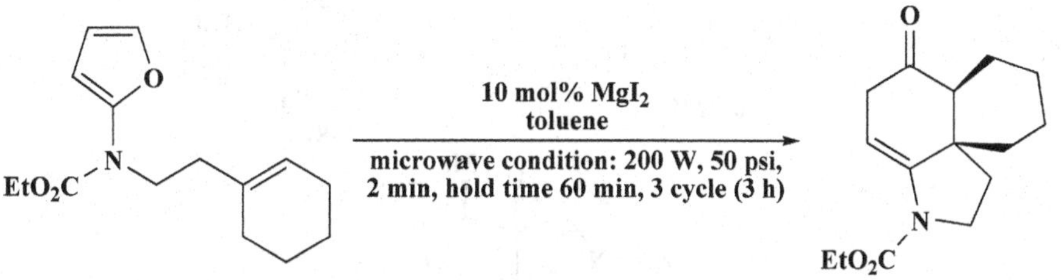

Scheme-33

Manganese-assisted synthesis

Oximes undergo palladium-catalyzed amino-Heck reaction and are successively treated with manganese(IV) oxide to provide good yields of 1-azaazulenes (**Scheme 34**) [79–80]. Cyclic amines other than isoquinolines and pyrroles, like azaazulene, aza-spiro compound, indoles and imidazoles can be prepared by an intramolecular amino-Heck reaction where many unsaturated fluorobenzoyloximes reacted in the presence of a palladium(0) catalyst [81–83].

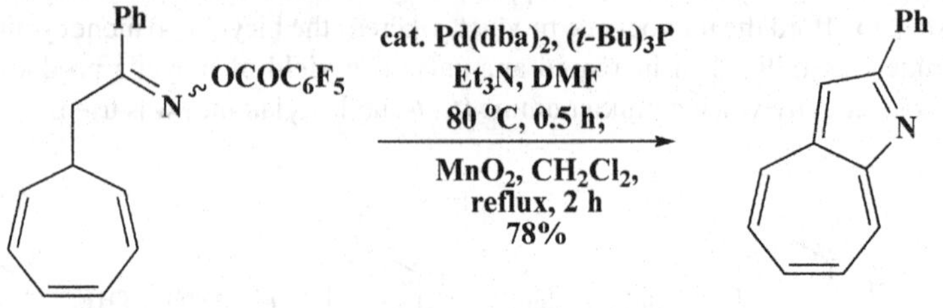

Scheme-34

Asahi and Nishino [84] reported intramolecular oxidative cyclization of *N*-propenyl-3-oxobutanamides in the presence of $Cu(OAc)_2$ and $Mn(OAc)_3$ radical initiators for the synthesis of 3-azabicyclo[3.1.0]hexan-2-ones in good yields (**Scheme 35**). Other cyclized products can also be obtained through pathway as illustrated in **Scheme 36**.

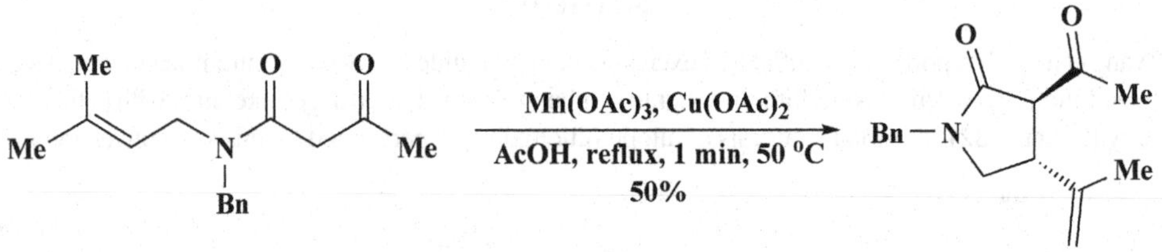

Scheme-35

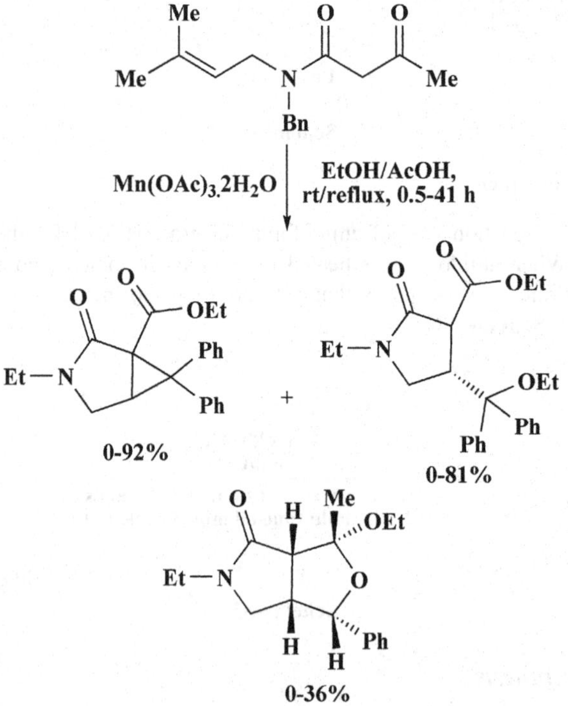

Scheme-36

Wipf and Maciejewski [85] demonstrated that the titanocene dichloride- and Mn metal-promoted radical annulation of epoxide tethered to substituted aminopyridine, generated from the allyl carbamate, formed the 3,3-disubstituted azaindoline (**Scheme 37**).

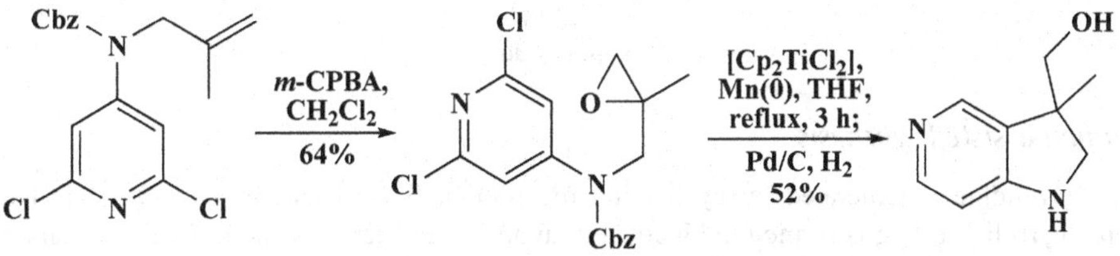

Scheme-37

Geny et al. [86] synthesized fused polycyclic cyclohexadienes by cyclization of enediyne precursors in the presence of *N*-heterocyclic carbenes. In comparison with established catalytic systems like CoI$_2$/manganese/triphenylphosphine and [(Cp)Co(CO)$_2$]/ferric chloride, the *N*-heterocyclic carbene ligand demonstrated superior activity in this reaction. IPr could be utilized catalytically with 1 eq. of manganese, while 2 eq. of triphenylphosphine along with 10 eq. of manganese were generally needed. The same research group also prepared cyclohexadiene from *N*-tethered enediyne through intramolecular [2+2+2]-cycloaddition in the presence of manganese, CoI$_2$ and *N*-heterocyclic carbene (C-23) (**Scheme 38**) [87].

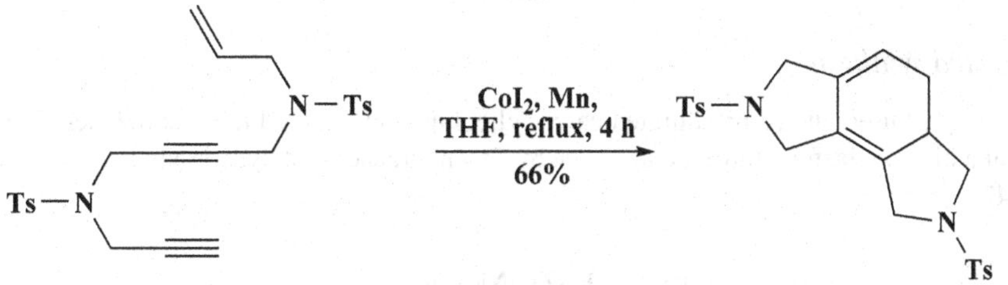

Scheme-38

Mercury-assisted synthesis

Karanjule et al. [88] were able to synthesize castanospermine analogues by mercury(II)-assisted cyclization of some unsaturated amines obtained from carbohydrates. In this reaction, as depicted in **Scheme 39**, the starting compound was converted into pyrrolidine by tandem mercuration-demercuration. Here, 5-*endo-trig* cyclization occurred exclusively. However, this regiochemistry was not common in intramolecular aminomercuration reactions.

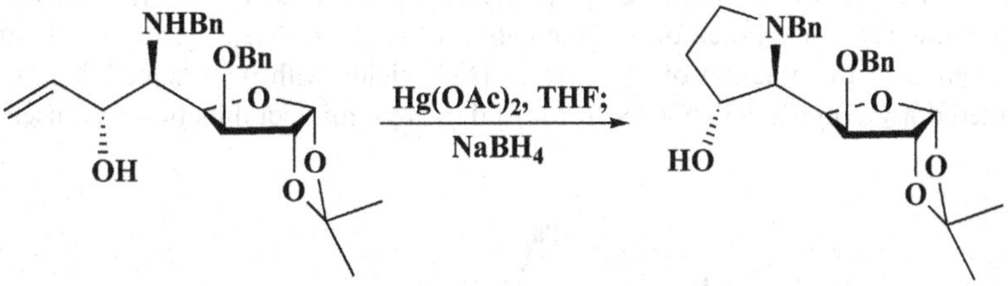

Scheme-39

Molybdenum-assisted synthesis

Gupta et al. [89] reported that allenamide underwent molybdenum-catalyzed [2+2+1]-cycloaddition to afford the final compound (**Scheme 40**).

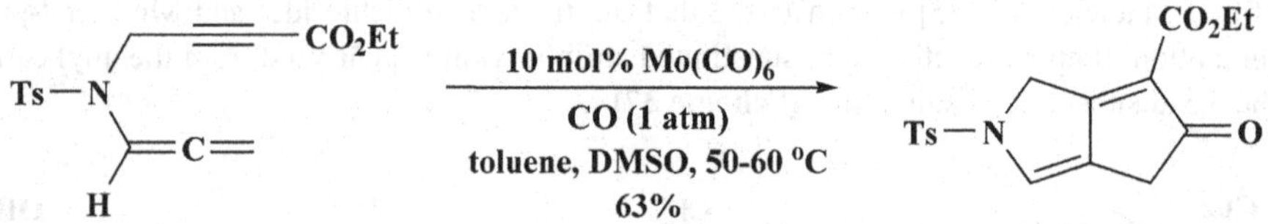

Scheme-40

Neodymium-assisted synthesis

Initially, pyrrolidine is generated by cyclization of aminodialkene (**Scheme 41**) [90–91]. The five-membered pyrrolidine ring is formed in kinetically favorable manner over the formation of an azocane, despite the fact that the sterically more encumbered 1,1-disubstituted olefin has to react first. Also, pyrrolizidine is formed by subsequent second hydroamination/cyclization through the formation of a highly organized chair-like transition state [92].

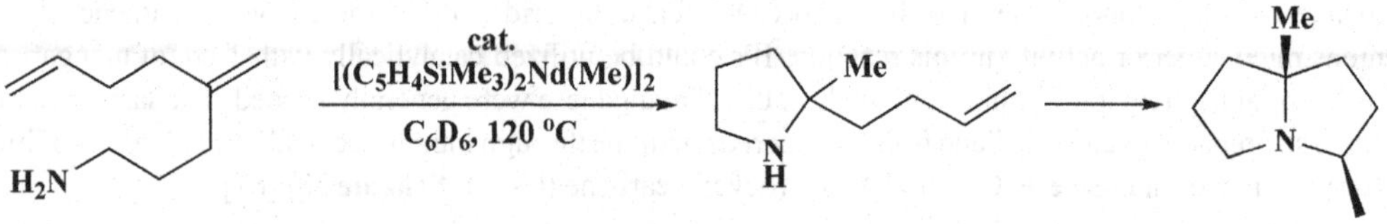

Scheme-41

Nickel-assisted synthesis

Joensuu et al. [93] used an α,β-unsaturated carbonyl compound tethered to a ketone electrophile for the synthesis of a bicyclic lactone through an amide by $Et_2ZnNi(acac)_2$-catalyzed reductive aldol cyclization (**Scheme 42**).

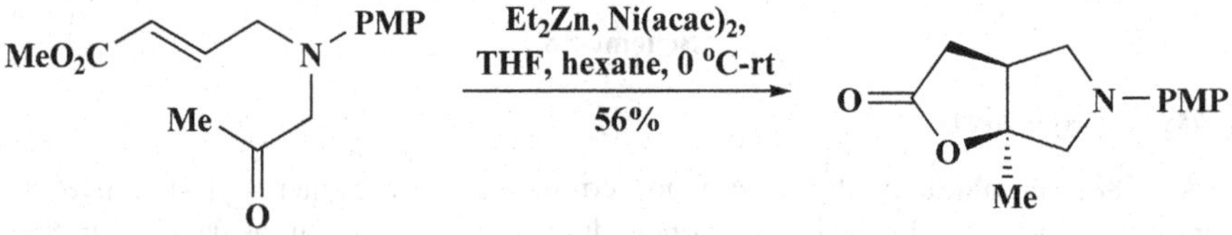

Scheme-42

The drawback of Ni-catalyzed cycloaddition with diynes is that the nickel catalyst (20–40 mol%) is required in large amounts. Chai et al. [94] reported that 1,6-enynes underwent cycloaddition to afford 1,5-cyclooctadienes along with some [(2+2)+2] products with lower loading of nickel/Me_2Zn (4 mol%) as catalyst (**Scheme 43**). The choice of a combination of different Me_2Zn and nickel complexes has a dramatic influence on the reaction of 1,6-enynes. High yields with poor selectivities were reported for carbon-tethered aryl enynes when a [Ni-(acac)$_2$]/dimethyl zinc combination was used. A complex

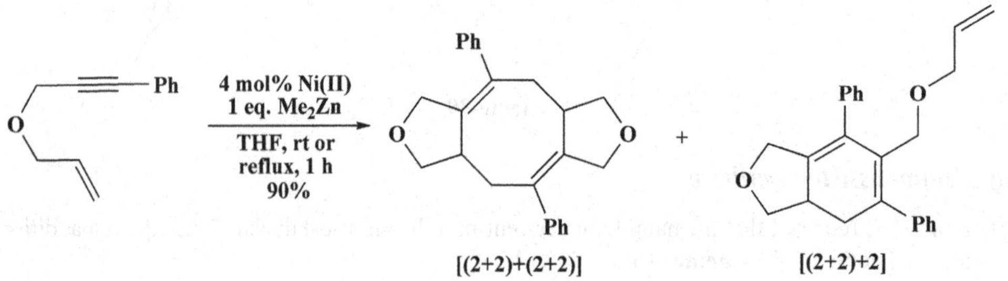

[(2+2)+(2+2)]   [(2+2)+2]

Scheme-43

reaction mixture or unidentifiable polymeric products were obtained using *N*-benzyl and *N*-tosyl enynes as substrates under these catalytic conditions. However, these substrates invariably favored the synthesis of [(2+2)+2] products in moderate yields with ratios of [(2+2)+2] cycloadducts ranging from 4:1 to 7:1 when [NiCl$_2$(PCy$_3$)$_2$]/dimethyl zinc combination was employed. On the other hand, when only 0.6 eq. of diethyl zinc or diethyl zinc instead of dimethyl zinc was used as the reducing reagent under the same reaction conditions, reductive eneyne cyclization products were formed exclusively [95].

Osmium-assisted synthesis

Takahashi et al. [96] used homochiral allylamine, obtained from L-xylose, as a precursor for the asymmetric preparation of naturally occurring polyhydroxylated pyrrolizidine alkaloids, (–)-7-*epi*-alexine and (+)-alexine, which are potent glycosidase inhibitors **(Scheme 44)**. In the very first stage, allylamines provide functionalized pyrrolidines which are manipulated to yield the desired alkaloids.

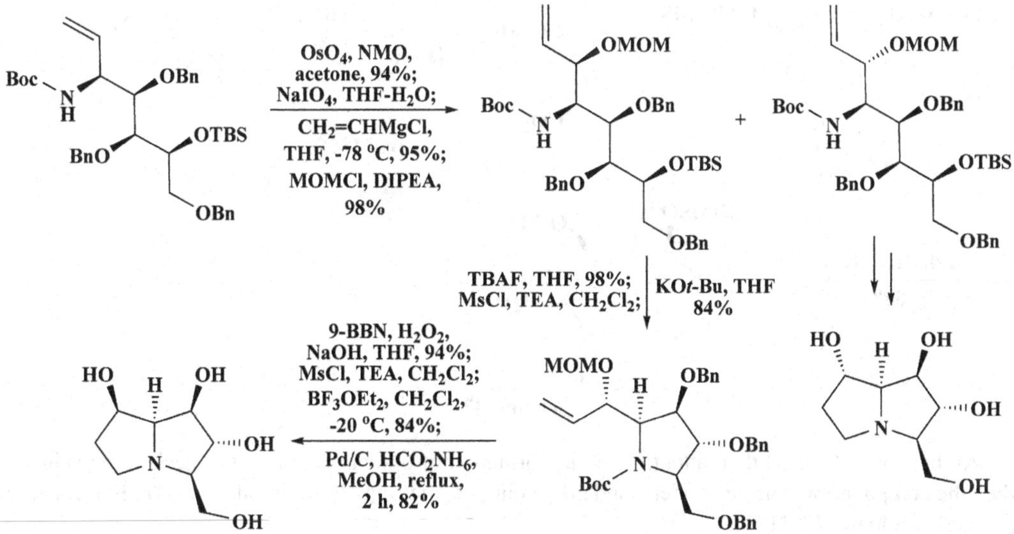

Scheme-44

β-Lactams can be oxidized with peracetic acid in an osmium trichloride-catalyzed reaction in acetic acid at room temperature to produce pyrrole **(Scheme 45)** [97].

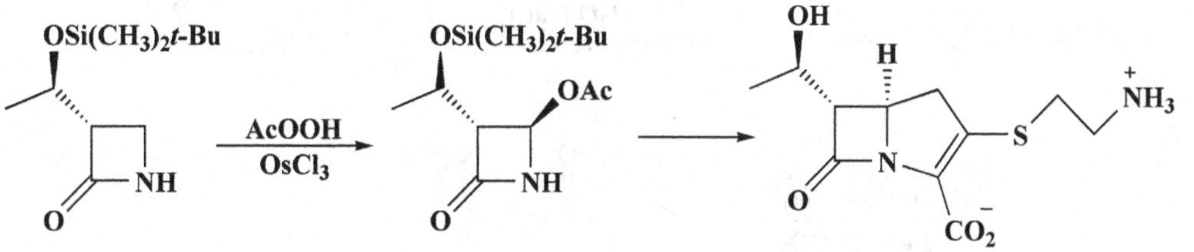

Scheme-45

Palladium-assisted synthesis

2-(1-Hydroxyallyl)-3-aminopyridines are used for the preparation of 1*H*-pyrrolo[3,2-*b*]pyridines **(Scheme 46)** [98]. The final products are obtained by the protection of the hydroxyl group with *tert*-butyldimethylsilyl group, followed by treatment with palladium(II)/benzoquinone [99].

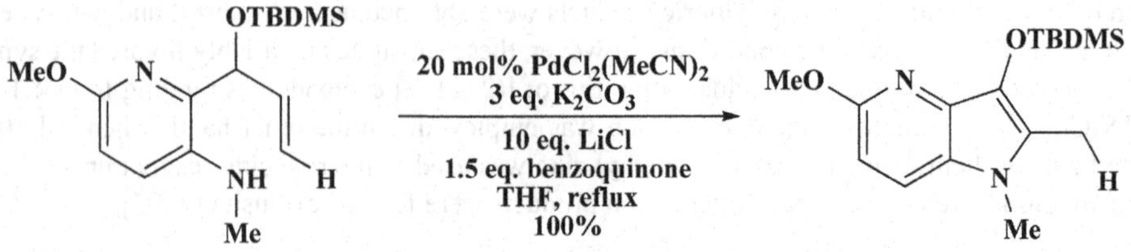

Scheme-46

The azabicyclic frameworks have been constructed by cycloaddition reactions which represent an attractive method as various stereocenters and bonds are generated in a single step during ring formation. For example, functionalized indolizidine is synthesized **(Scheme 47)** through intermolecular [3+2]-nitrone cycloaddition of a dienophile with either nitrone substrates [100].

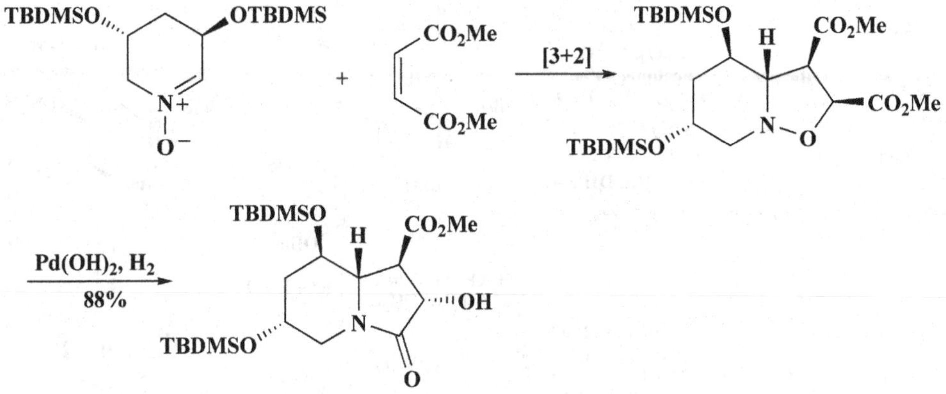

Scheme-47

Authors have focused their attention on the formation of 6-membered ring synthesis. Interestingly, when the compound was heated under standard conditions, the ring closure product smoothly formed in 75% yield **(Scheme 48)** [101].

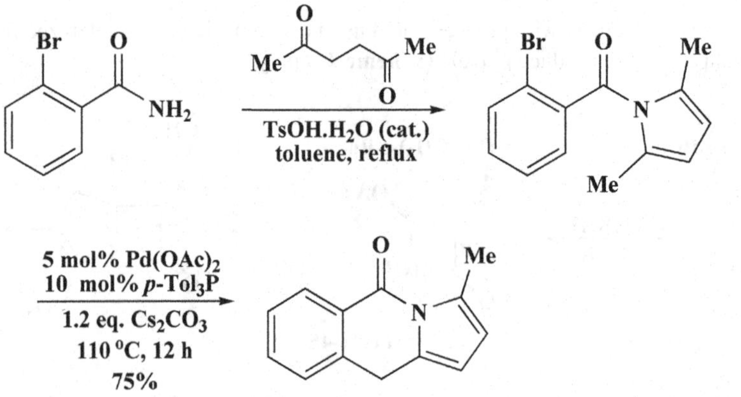

Scheme-48

Bis(allylamino)benzoquinone undergoes a tandem oxidative cyclization/olefin insertion reaction in the presence of $PdCl_2(MeCN)_2$/benzoquinone catalytic system to afford mitosenes (a large class of anti-neoplastic, anti-biotic pyrroloindoloquinones) **(Scheme 49)** [102]. The reaction occurs through the formation of a primary alkylpalladium intermediate by sequential elimination/addition of a hydridopalladium species [99].

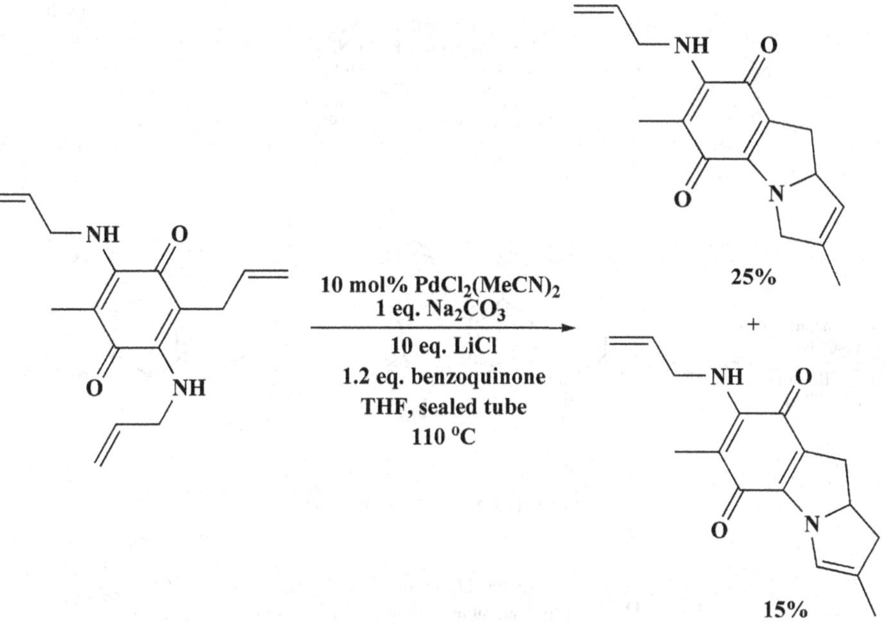

25%

+

15%

Scheme-49

Benzyl *N*-[2-(2,4-cyclohexadienyl)ethyl]carbamate undergoes intramolecular 1,4-chloroamidation for the total synthesis of lycorine, an alkaloid (**Scheme 50**) [99, 103].

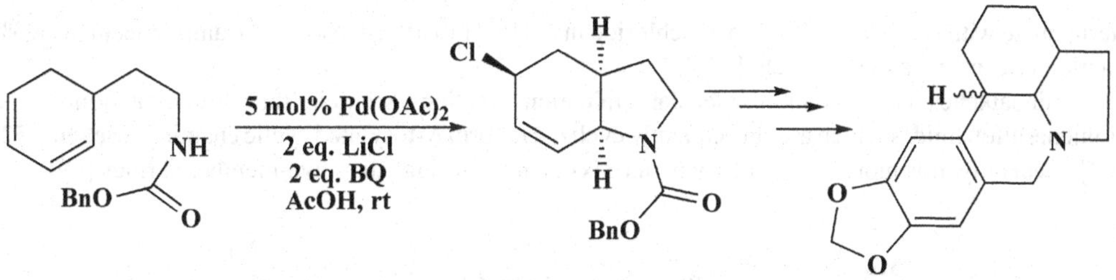

Scheme-50

Although 14-azacamptothecin was synthesized successfully, unfortunately no asymmetric induction was observed under Sharpless asymmetric dihydroxylation conditions, providing only the racemic product. The same result was reported using other chiral ligands like (DHQD)$_2$-PHAL. The two proximal nitrogen atoms (*N*-1 and *N*-14) in the rigid cyclic enol ether might disrupt the coordination of osmium with the chiral ligand during the AD reaction. As an alternative, the enantioselective dihydroxylation was performed early in the synthesis (**Scheme 51**) to afford the oxygenated stereogenic center in 94% enantiomeric excess. Due to the poor regioselectivity observed in *O*- and *N*-propargylation step, it was decided to synthesize the *O*- and *N*-allyl mixture and perform a Pd-catalyzed rearrangement of *O*-allyl to *N*-allyl derivative. The rearrangement was successful and formed the key *N*-allyl amide in a remarkable yield of 100%. These allyl amides were not previously used in the aza-Diels-Alder cascade reaction designed earlier with alkyne tethers. The allyl amide was treated with 3 eq. triphenylphosphine oxide and 1.5 eq. Tf$_2$O at 0 °C, followed by oxidation with freshly prepared manganese(IV) oxide to afford the pentacyclic intermediate in 89% yield. This conversion represented a novel application of the Hendrickson reagent where an allyl group underwent a highly efficient aza-Diels-Alder cascade reaction successfully. The enantioselective synthesis was completed upon deacetylation of pentacyclic

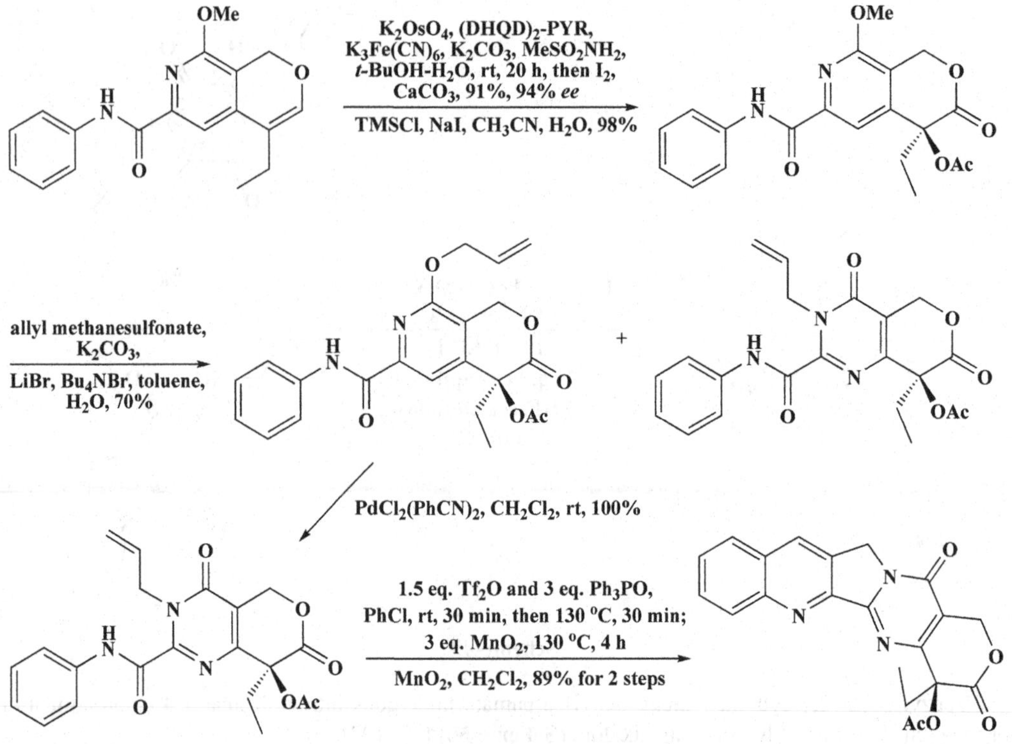

Scheme-51

intermediate with concentrated hydrogen chloride in C_2H_5OH to afford (*S*)-14-azacamptothecin in >99% enantiomeric excess and 95% yield [104].

Aminoalkenes undergo intramolecular amidation reactions wherein they are transformed into *p*-toluenesulfonamides which are subsequently cyclized to form *N*-tosylated cyclic enamines **(Scheme 52)** [105]. Generally, it is more difficult to generate six-membered rings than five-membered rings [99].

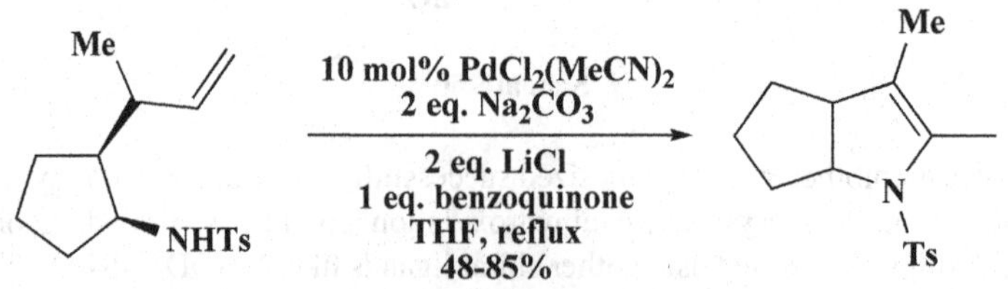

Scheme-52

In 1966, Wall et al. [106] discovered camptothecin which acts as a cytotoxic pentacyclic quinoline alkaloid for systematic screening of natural products for anti-cancer activity. This compound was naturally isolated from the stem and bark of *Camptotheca acuminata*, a tree native to China. CPT exhibits remarkable anti-cancer properties in preliminary clinical trials and targets and stabilizes the covalent binary complex formed between topoisomerase I (topo I) and DNA during DNA relaxation, leading to aptosis [107–109]. Due to promising bioactivity, medicinal and synthetic chemists have developed many methods for synthesizing camptothecin and several derivatives [110–111]. Although various impressive syntheses of CPT have been studied, involvement of long steps and difficult operations have rendered most of these impractical for large-scale preparation. The application and development of cascade reactions to form multiple carbon-carbon bonds in one operation is an attractive protocol in the preparation of complex molecular architectures. Fortunately, the short and efficient cascade pathway developed earlier by Zhou et al. [112] is generally applicable to this class of alkaloids and analogues. A mild Hendrickson reagent promoting the cascade reaction consisting of an imidate synthesis, an intramolecular aza-Diels-Alder reaction and an eliminative aromatization is likely to form the common tetracyclic A/B/C/D ring core of CPT family alkaloids in high yields. Thus, the total synthesis of camptothecin starts from the known chloropyridine (**Scheme 53**) which is carbonylated under carbon monoxide atmosphere, thus producing methyl ester in 97% yield. This methyl ester undergoes selective *O*-demethylation with iodotrimethylsilane, followed by *N*-propargylation of the formed pyridone with propargyl bromide, tetrabutylammonium bromide, potassium carbonate and lithium bromide in toluene to yield *N*-propargyl pyridone in 67% yield. The methyl ester can also be transformed into acyl chloride, followed by coupling with aniline, to provide the key amide precursor for the cascade reaction (overall 90%). With the newly developed cascade annulation protocol in hand, the advanced intermediate bearing A/B/C/D ring core of CPT is formed in 96% yields upon simple treatment of amide with the reagent (3 eq. triphenylphosphine oxide, 1.5 eq. Tf$_2$O and manganese(IV) oxide) at room temperature. For the completion of synthesis, a highly enantioselective Sharpless asymmetric dihydroxylation and an iodine/calcium carbonate-based hemiacetal oxidation provides a 83% yield of camptothecin with 95% enantiomeric excess [104].

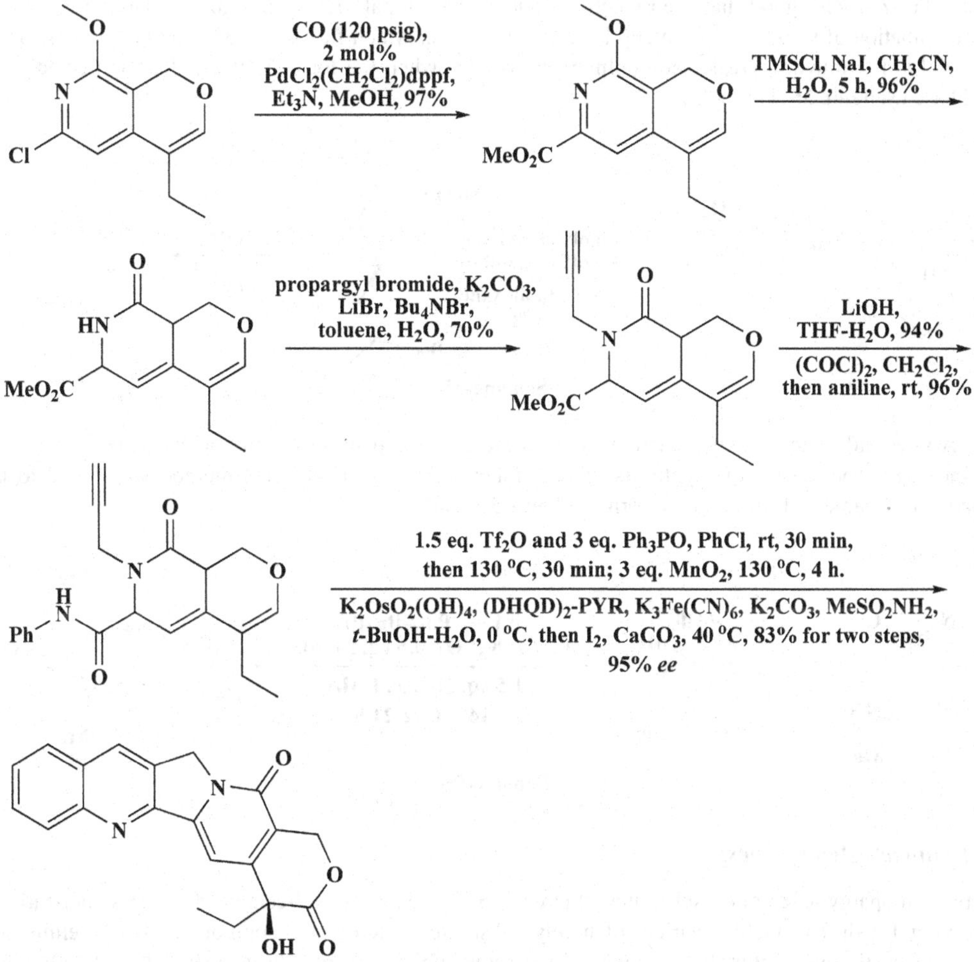

Scheme-53

Kozawa and Mori [113–115] used propargyl ester for the synthesis of heterocyclic compounds with different ring sizes by a ligand effect on palladium(0) (**Scheme 54**). Carbapenam was obtained

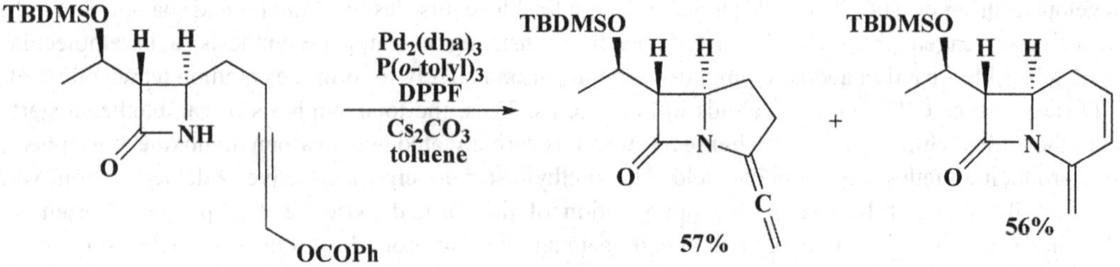

Scheme-54

in 57% yield when the reaction was performed using P(*o*-tolyl)$_3$ as a ligand, while carbacepham was formed in 56% yield when 1,1'-bis(diphenylphosphino)ferrocene was used. However, enamide was isolated as the sole elimination product from the palladium π-allyl complex when bidentate ligands like 1,1'-bis(diphenylphosphino)ferrocene were used.

Carboamination reactions of *N*-protected γ-aminoalkenes afford interesting compounds which are not obtained using γ-*N*-(arylamino)alkenes [116]. For example, tetrahydropyrroloisoquinolin-5-one can be synthesized with good diastereoselectivity and in substantial yields through a palladium-catalyzed carboamination of substrate with methyl-2-bromobenzoate, which furnishes 73% yield of intermediate. The yield of tetrahydropyrroloisoquinolin-5-one is 95% when the formed intermediate is treated with acid base **(Scheme 55)** [117a–b].

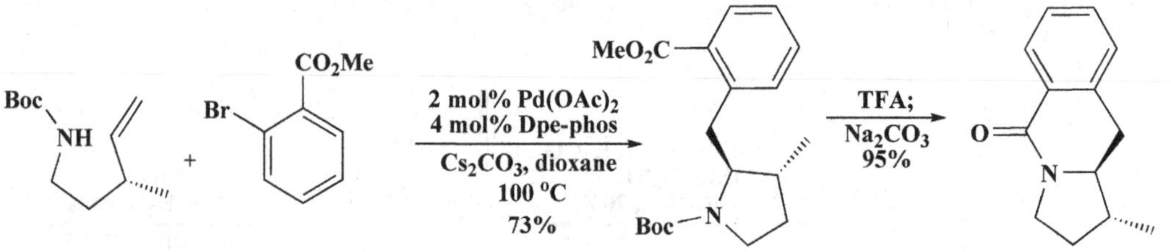

Scheme-55

A new, mild and efficient reaction of substituted 2-chloroanilines with acyclic as well as cyclic ketones was reported for the synthesis of poly-functionalized indoles. This method was found to be broadly applicable and simple to perform **(Scheme 56)** [118].

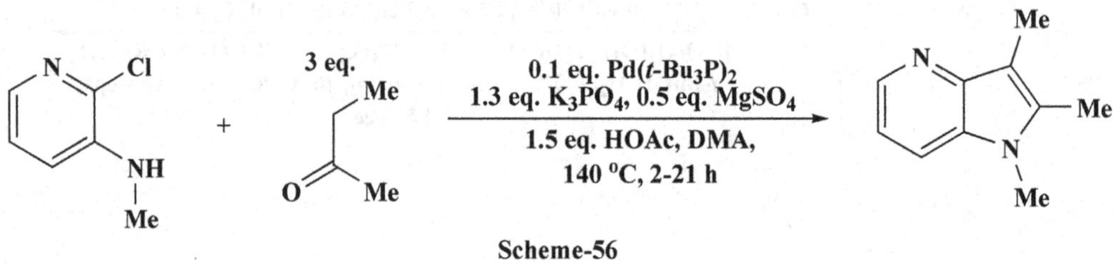

Scheme-56

Platinum-assisted synthesis

Tertiary propargylic alcohol substrates **(Scheme 57)** undergo metal-catalyzed cycloisomerizations involving 1,2-shift to yield a variety of highly substituted heterocyclic compounds. In a preliminary study, it was discovered that the treatment of hydrazone with platinum chloride (10 mol%) at 100 °C for 24 hours produces pyrrolone in 71% yields [119].

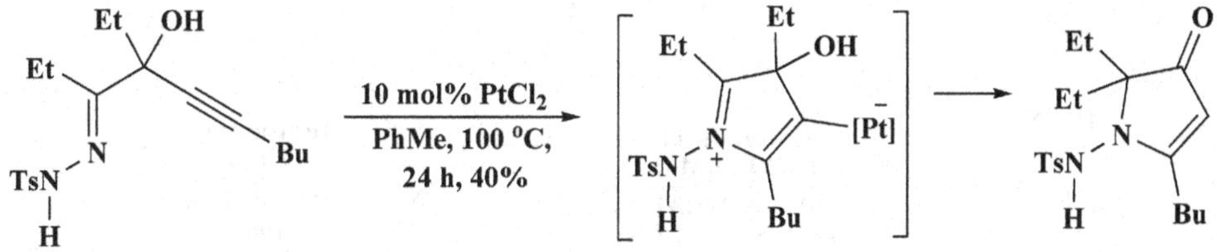

Scheme-57

Rhenium-assisted synthesis

Kusama et al. [120] synthesized bicyclic enol silyl ethers through stereoselective cyclization of aza-dienyne in the presence of gold(I) catalyst **(Scheme 58)**; however, aza-dienyne underwent a cascade cyclization in the presence of a Re catalyst to afford tricyclic compounds. The reaction occurred through *exo*-cyclization and a 1,2-hydrogen or 1,2-alkyl shift from the carbene complex intermediate bearing a bicyclo[3.3.0]octane framework.

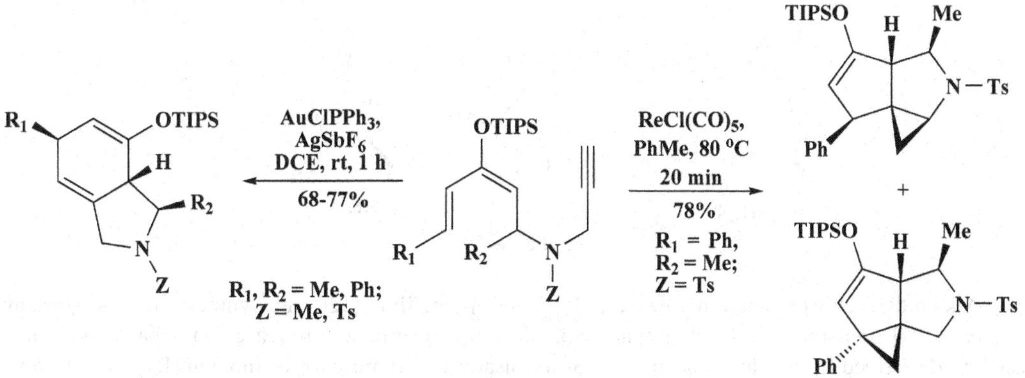

Scheme-58

Rhodium-assisted synthesis

Complex bioactive compounds can be synthesized by intramolecular alkylation. For example, the potent Jun N-terminal kinase inhibitor was originally synthesized in 6% overall yield and 14 linear steps [121]. The reaction illustrated in **Scheme 59** affords N-terminal kinase inhibitor in 13% overall yield and 11 linear steps by relying on carbon-hydrogen functionalization as the key step [122–123].

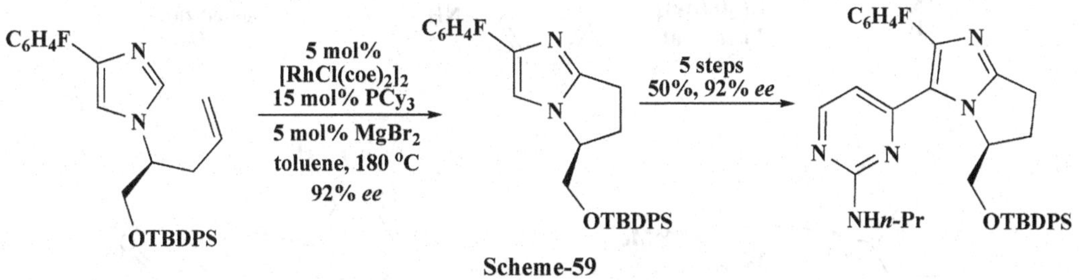

Scheme-59

Similarly, N-terminal lunase inhibitor can be prepared through **Scheme 60**. More highly substituted derivatives, which are very difficult to prepare by alternative methods, can be readily synthesized rapidly in 15 and 17% overall yields respectively and result in the identification of even more potent inhibitors [123–124].

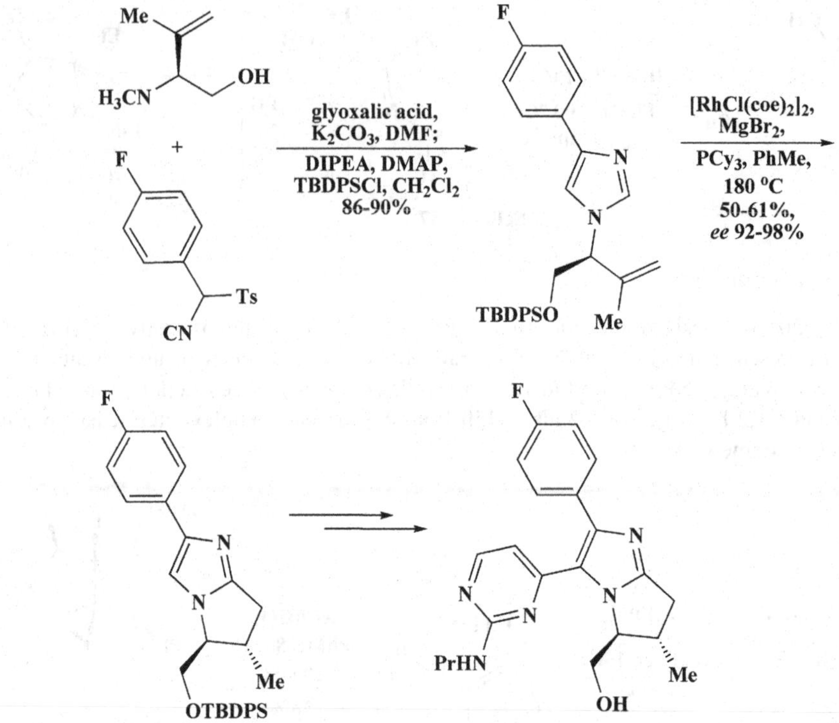

Scheme-60

Alkaloids such as (±)-strychnofoline [125] and (±)-horsfiline [126] are synthesized. Carreira and Meyers [127] prepared spiro-cyclopropaneoxindole from pyperilene through cyclopropanation, using carbenoid (derived by the decomposition of 3-diazoisatin in the presence of rhodium(II) catalyst), and performed ring-expansion with the help of imine to afford the spiro-pyrrolidine-oxindole which was used for preparing spirotryprostatin B **(Scheme 61)** [128].

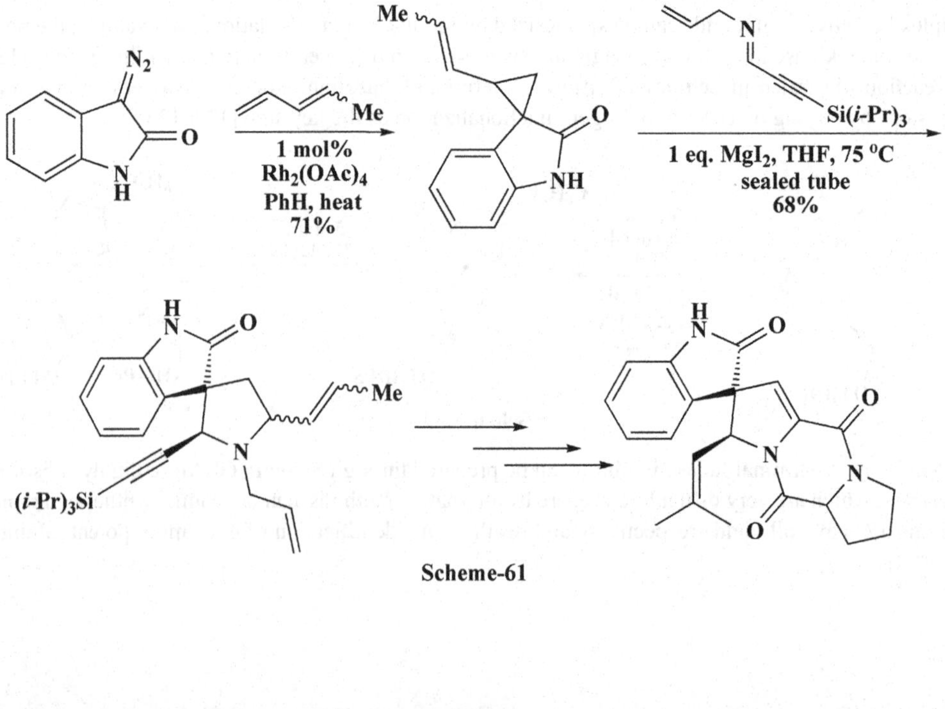

Scheme-61

Bates and Lim [129] reported a Wilkinson's catalyst-mediated double bond reduction of allylamine followed by intramolecular reductive amination for a highly diastereoselective synthesis of piperidine derivative **(Scheme 62)**. A bicyclic heterocyclic compound and the *Nuphar* alkaloid, nupharamine, were synthesized from the piperidine derivative. A saturated analogue was generated when the bicyclic heterocyclic compound was reduced with lithium aluminium hydride.

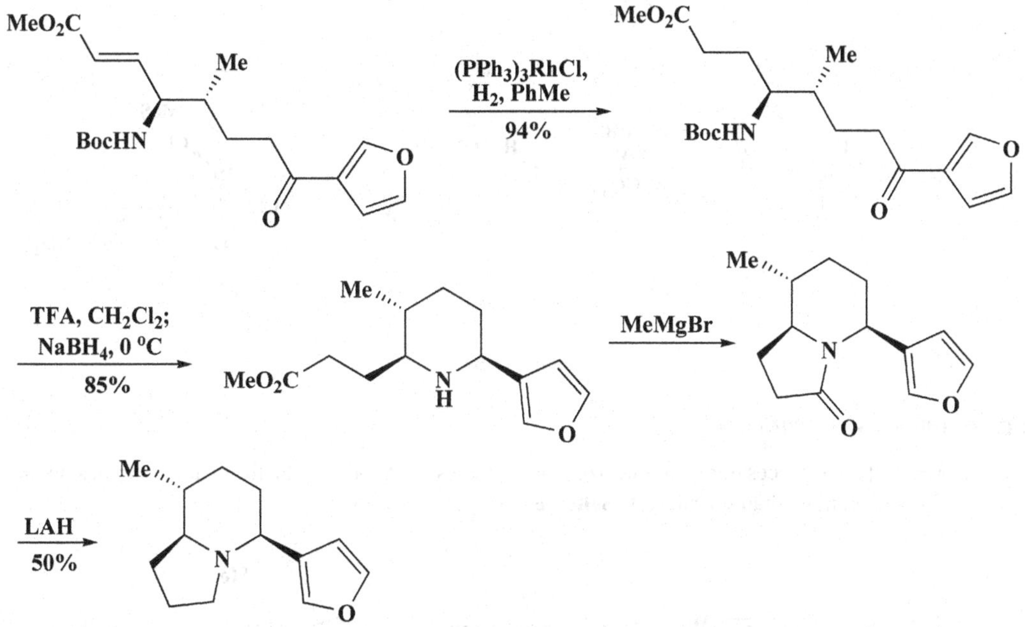

Scheme-62

Ruthenium-assisted synthesis

Five-membered cyclic carbonyl compounds are synthesized by a general synthetic method—catalytic [2+2+1]-cycloaddition reaction of two C-C multiple bonds with CO. In particular, Pauson-Khand reaction has been widely investigated and established as a powerful pathway to prepare cyclopentenones [130–147]. Many transition metals like titanium, cobalt, rhodium, ruthenium, and iridium have been used as catalysts in Pauson-Khand reaction. Bicyclic ketones containing a heterocyclic ring can be constructed by intramolecular Pauson-Khand reaction of allyl propargyl ether and amine **(Scheme 63)**. This reaction occurs through the formation of bicyclic metallacyclopentene intermediate which yields bicyclic ketones through carbon monoxide insertion.

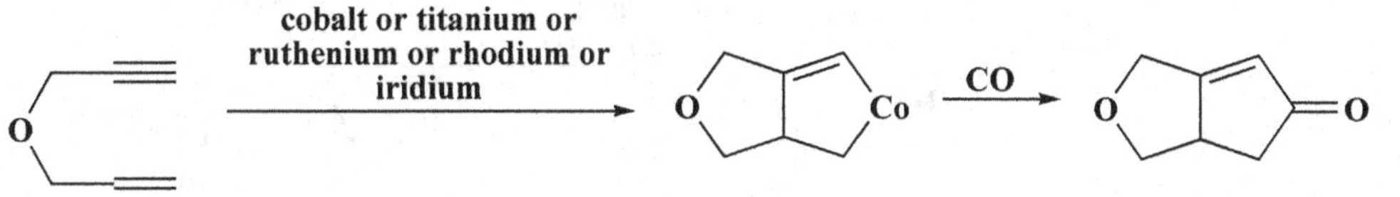

Scheme-63

Honda et al. [148] reported a successful diastereoselective synthesis of (–)-securinine in an optically pure form using ring-closing metathesis of dienyne as a key step. The dienyne containing di-substituted alkene and terminal alkene parts was synthesized from (+)-pipecolinic acid as the ruthenium-carbene complex first reacted with the terminal alkene to afford a furan ring. Good yields of the bicyclic compound were obtained by the metathesis of dienyne, using ruthenium catalyst [149]. This compound was oxidized with CrO_3 to afford lactone which was subsequently treated with *N*-bromosuccinimide and then trifluoracetic acid to synthesize (–)-securinine **(Scheme 64)**. Similarly, viroallosecurinine can also be synthesized [150a–b].

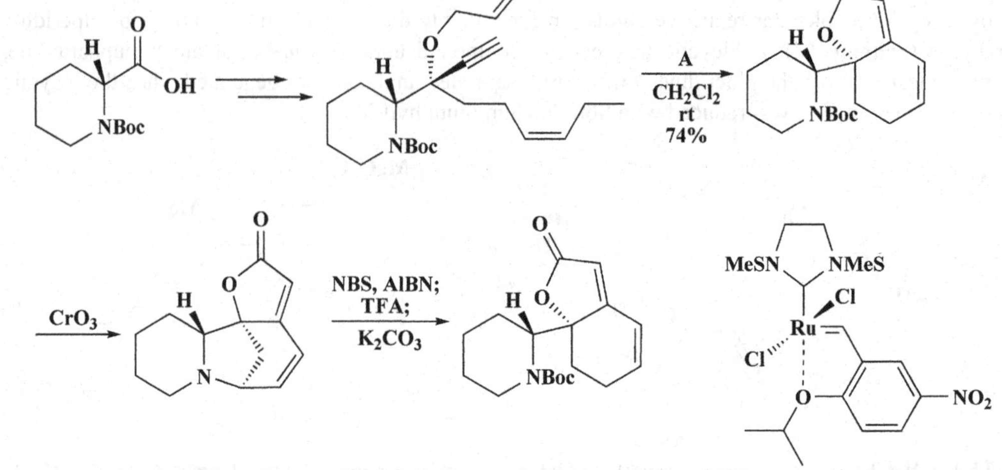

Scheme-64

Samarium-assisted synthesis

Li and Marks [151] successfully synthesized pyrrolizines in good yields from aminoenynes in the presence of an organolanthanide catalyst (**Scheme 65**).

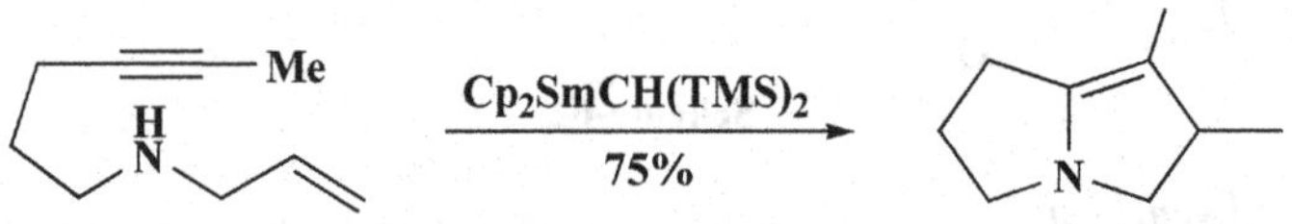

Scheme-65

Arredondo et al. [152] and Tian et al. [153] prepared bicyclic pyrrolizidine through tandem hydroamination/cyclization of allenylalkenylamine in the presence of samarium catalyst (**Scheme 66**). (+)-Xenovenine was formed by the hydrogenation of pyrrolizidine. On the other hand, more sterically hindered lanthanocene catalysts leave the alkene functionality untouched and react selectively with the allene functionality [95].

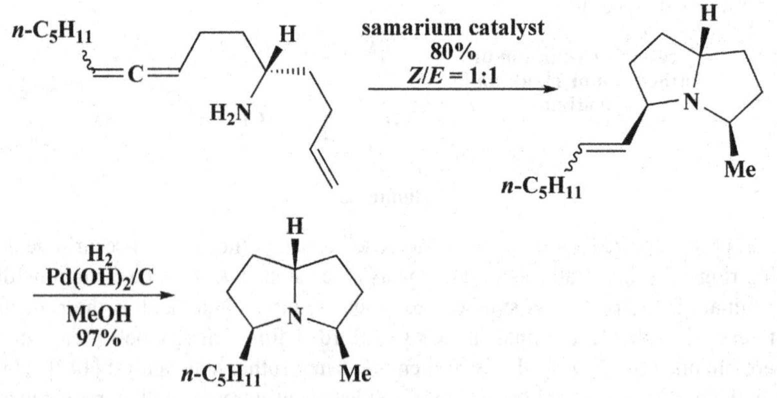

Scheme-66

Tin-assisted synthesis

Spiro[1,4-benzodiazepin-3,1-cyclohexane]-2-amine can be prepared [154] by reducing, with presumed subsequent ring-closure of 1-(2-nitrobenzylideneamino)-cyclohexanecarbonitrile. Moreover, 1-(2-nitrobenzylideneamino)-cyclohexanecarbonitrile can be synthesized from 1-aminocyclohexanecarbonitrile and 2-nitrobenzaldehyde (**Scheme 67**). However, the reaction affords 13-methoxy-7,11b-dihydro-13*H*-6,12[1:,2:]-benzeno-6*H*-quinazolino[3,4-*a*]quinazoline. The structure of this compound is determined by X-ray structural analysis. Aminonitrile, acting merely as a catalyst, does not become a component of the product formed [155].

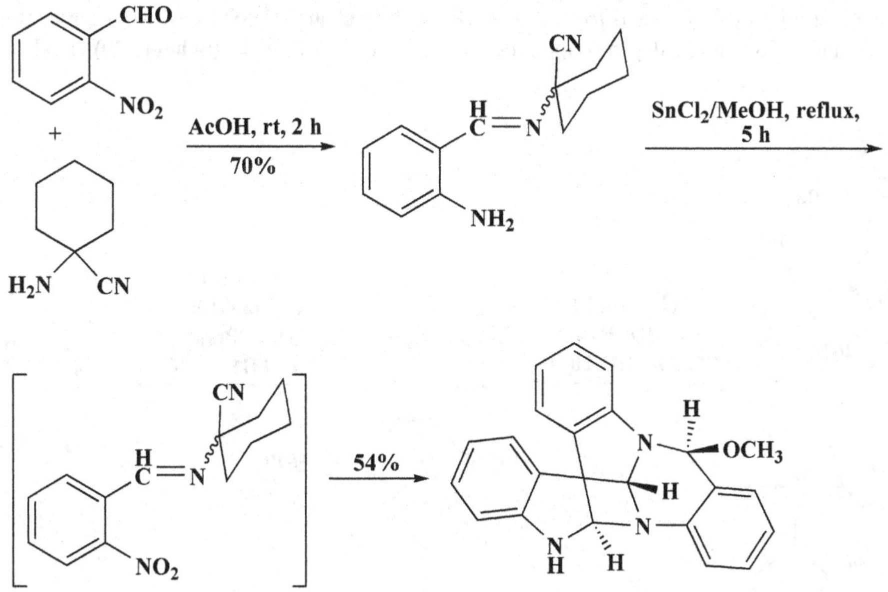

Scheme-67

Pyridine derivatives undergo several intramolecular aryl radical reactions [156–157]. Such reactions have earlier been reported for a fast synthesis of camptothecin, wherein it was produced when the tetracyclic intermediate was cyclized by a free radical reaction (**Scheme 68**) [75, 158].

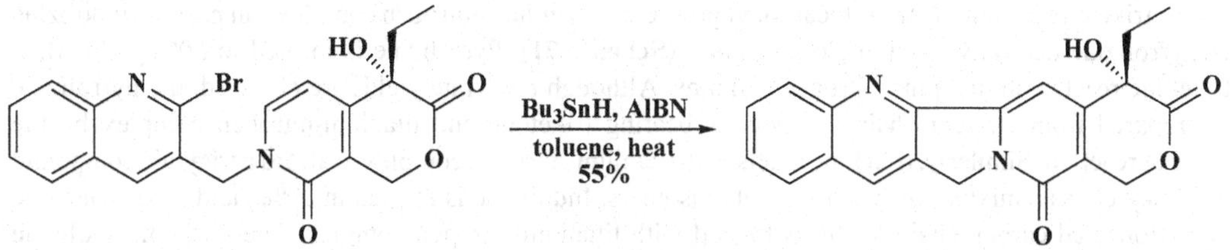

Scheme-68

Boivin et al. [159] and Zard [160] reported *N*-radical formation through the reactions of oxime derivatives with nickel powder and stannane and their addition to intramolecular alkenes (**Scheme 69**). However, both these reactions had some disadvantages in the operation procedure and the generality. In order to use alkylideneaminyl radicals more effectively as active species for organic synthesis, methods to synthesize cyclic imines from *γ,δ*-unsaturated *O*-2,4-dinitrophenyl oximes have been developed using a reductive radical generation protocol.

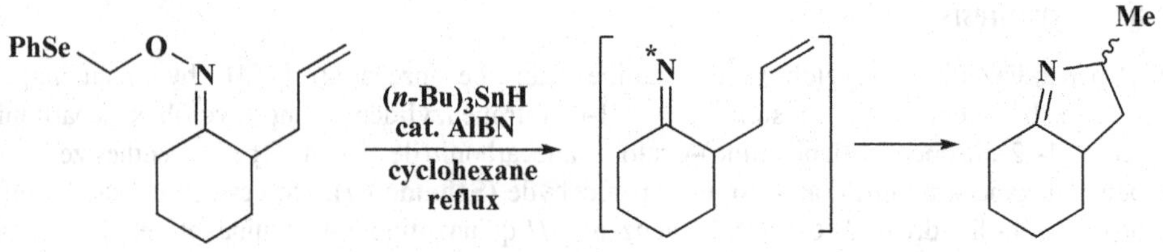

Scheme-69

Titanium-assisted synthesis

An inseparable mixture of *cis*- and *trans*-aza-bicyclo[3.1.0]hexanols is obtained from allylamine. The ratio of *cis*- and *trans* isomers depends upon the concentration of *i*-PrMgCl **(Scheme 70)** [161].

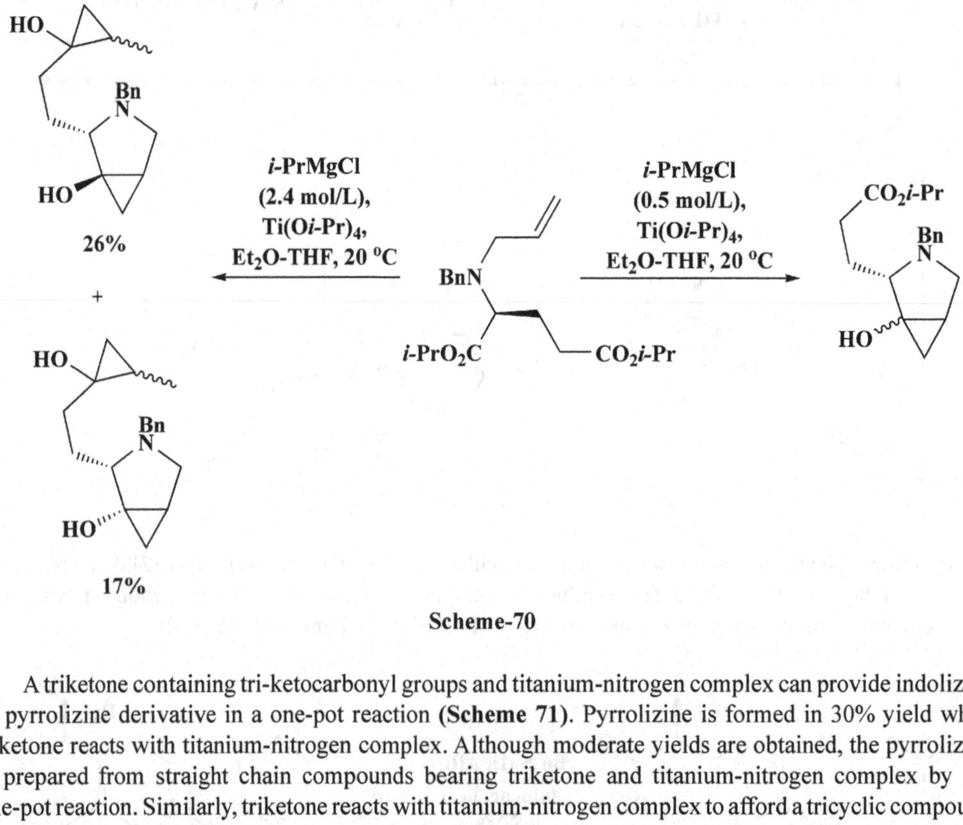

Scheme-70

A triketone containing tri-ketocarbonyl groups and titanium-nitrogen complex can provide indolizine or pyrrolizine derivative in a one-pot reaction **(Scheme 71)**. Pyrrolizine is formed in 30% yield when triketone reacts with titanium-nitrogen complex. Although moderate yields are obtained, the pyrrolizine is prepared from straight chain compounds bearing triketone and titanium-nitrogen complex by the one-pot reaction. Similarly, triketone reacts with titanium-nitrogen complex to afford a tricyclic compound in 31% yield as a mixture of two inseparable isomers. Indolizine is formed in 29% yield when triketone with elongated carbon chain length is treated with titanium-nitrogen complex. The yield of indolizine is increased to 41% when 2 eq. of titanium-nitrogen complex is used in this reaction [162]. Various indolizine and pyrrolizine derivatives can be synthesized easily from straight chain compounds having triketone and titanium-nitrogen complex, by such a one-pot reaction [163a–b].

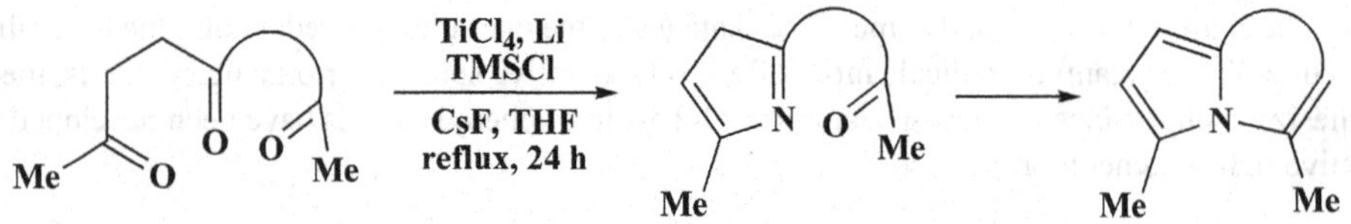

Scheme-71

1,4-Diketo-alkyne is reacted for the synthesis of indolizidines in 20% and 10% yields, along with pyrrole derivative in 11% yield. NOE experiments determine the structures of the formed compounds. The ester group of indolizine can be transformed into a nitrile group with the help of titanium nitrogen complex to produce indolizine **(Scheme 72)** [164–165].

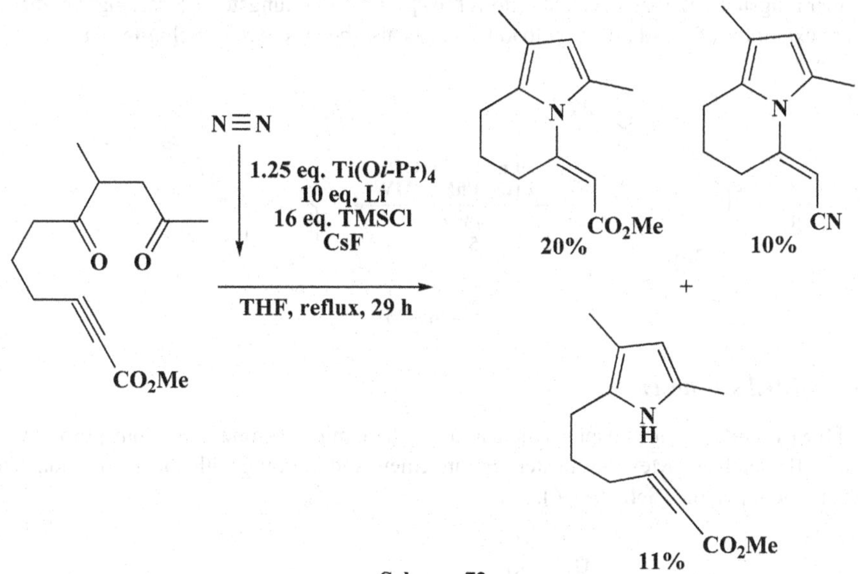

Scheme-72

Faler et al. [166] reported a titanium(II)-assisted coupling reaction of *N,N*-disubstituted carboxamide derivatives of amino acids and terminal olefin for producing a series of novel [3.1.0] bicyclic cyclopropylamines **(Schemes 73 and 74)** [161].

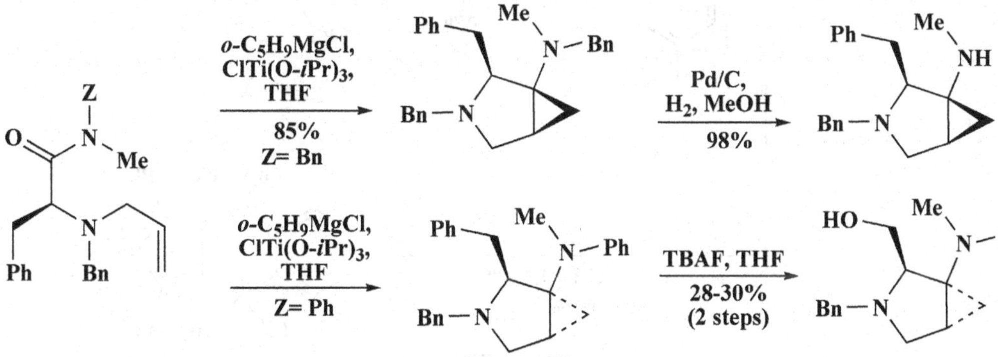

Scheme-73

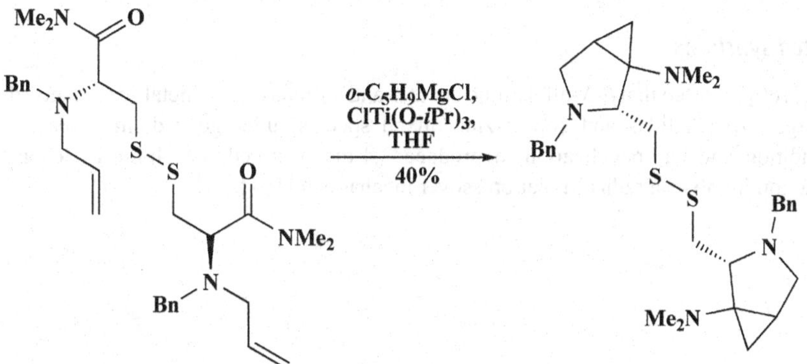

Scheme-74

Tungsten-assisted synthesis

Epoxide and carbonyl croups serve as internal nucleophiles to provide an oxonium cation which undergoes Prins cyclization to produce desired cycloadducts. Also, tricyclic indoles are formed by both intra- and intermolecular tungsten-catalyzed cycloadditions [167]. The use of tungsten-possessing vinylazomethine ylides in rare examples of 1,5-dipolar cycloaddition has also been reported (**Scheme 75**).

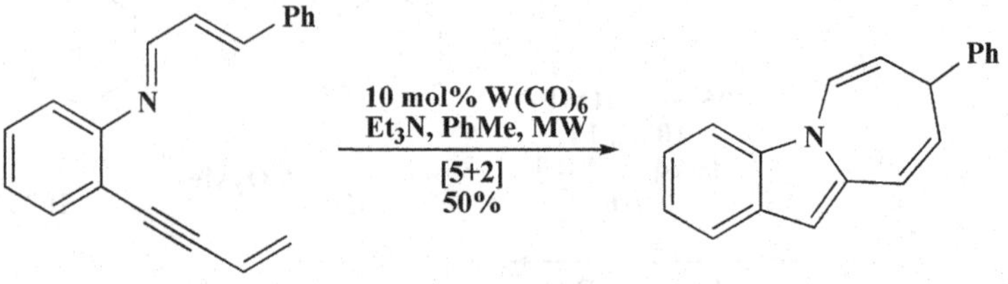

Scheme-75

Ytterbium-assisted synthesis

Yin et al. [168] reported that phosphoryl-substituted allenamide, synthesized from propargyl alcohol *via* ytterbium(III)-catalyzed Meyer-Schuster rearrangement and trapping with phosphoramidate, undergo cyclizations to yield pyrrole (**Scheme 76**).

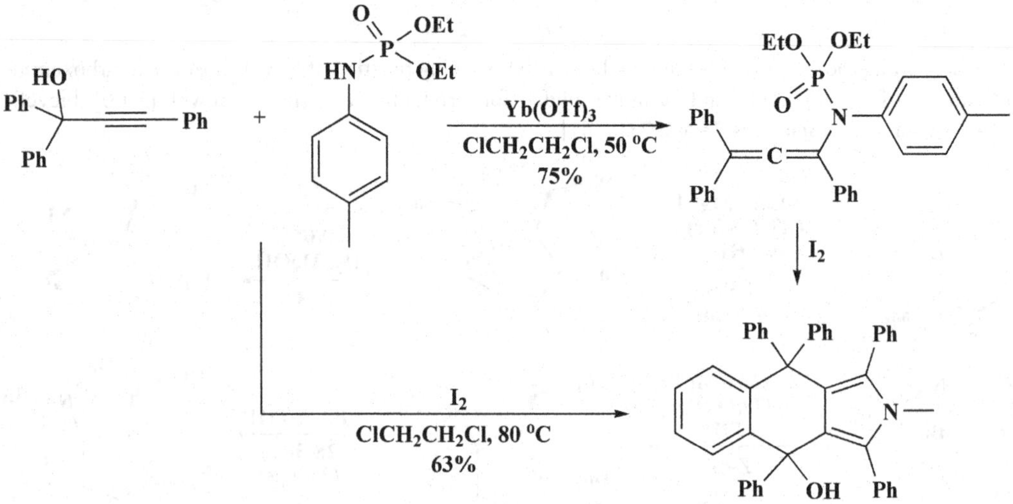

Scheme-76

Zinc-assisted synthesis

Denes et al. [169] reported that β-N-allylamino enoates and various organometallic reagents (diorganozinc reagents, organozinc halides and copper-zinc mixed species) undergo a domino reaction involving Michael addition and carbocyclization to produce 3,4-disubstituted pyrrolidines (**Scheme 77**). This domino reaction involves a radical-polar crossover mechanism [161].

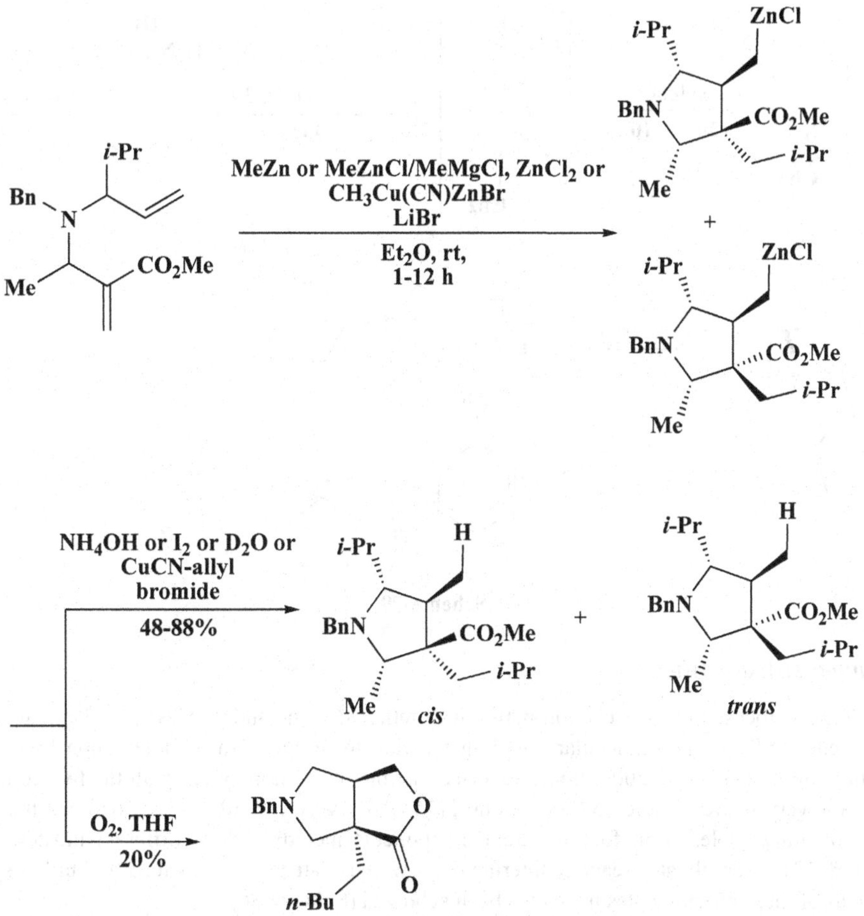

Scheme-77

There was a report that cyclization occurred on the nitrogen atom of oximes possessing alkenyl groups **(Scheme 78)**, as an example of C-N bond formation on the nitrogen atoms of oxime derivatives. However, none of these reports further investigated the reaction mechanisms or their general utility [170–172].

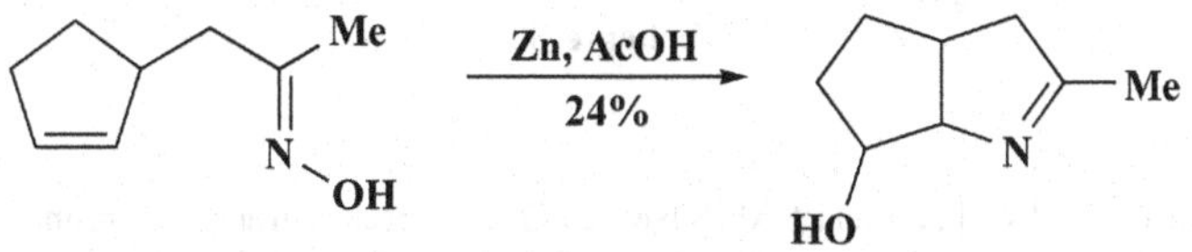

Scheme-78

Ji et al. [173] synthesized pyridine-substituted 3,6-diazabicyclo[3.2.0]heptanes which served as selective agonists for a nicotinic acetylcholine receptor **(Scheme 79)**. Allylamine afforded the bicyclic core of these compounds which underwent [1,3] dipolar cycloaddition to afford isoxazolidine. The cleavage of this isoxazolidine ring resulted in a product which furnished the required sub-unit after chiral resolution and cyclization [161].

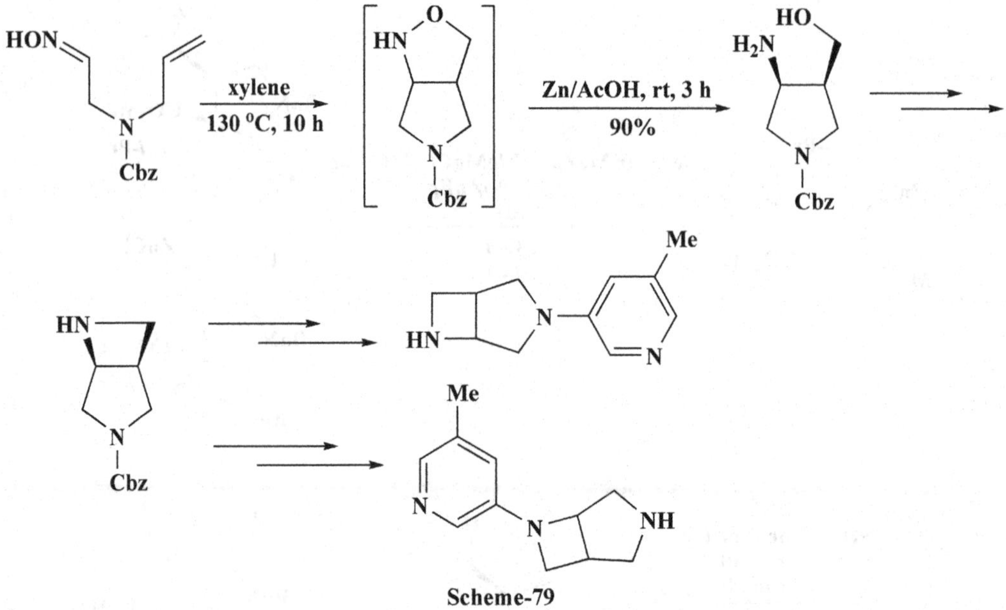

Scheme-79

Zirconium-assisted synthesis

A study in 2004 focused on the reaction between Si-tethered diyne and nitriles (in 1:3 molar ratio) by a zirconocene-mediated intermolecular coupling reaction to provide 5-azaindoles upon hydrolysis of the reaction mixture [174]. In 2009, it was reported that the reaction integrating all the five components together was very unpredictable and interesting [175–177]. Very recently, it was realized the one-pot synthesis of 5-azaindoles from four different components including three different nitriles and one alkyne [178–181]. Complicated reactive intermediates were isolated and characterized. Further synthetic applications of these intermediates have also been achieved **(Scheme 80)**.

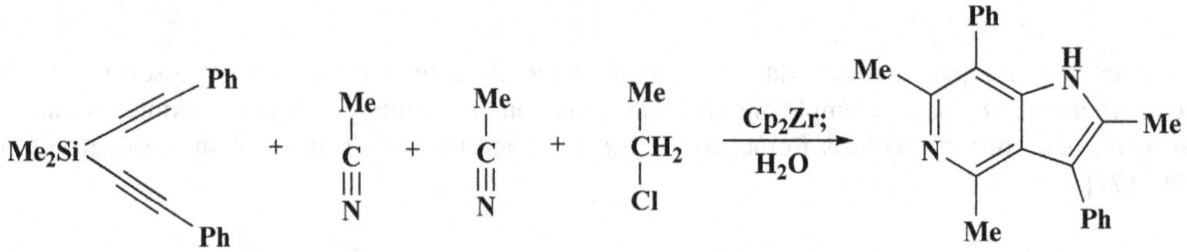

Scheme-80

References

(1) (a) J.A.R. Salvador, R.M.A. Pinto and S.M. Silvestreb. 2009. Recent advances of bismuth(III) salts in organic chemistry: application to the synthesis of heterocycles of pharmaceutical interest. Curr. Org. Synth. 6: 426–470. (b) N. Kaur. 2015. Six-membered heterocycles with three and four *N*-heteroatoms: microwave-assisted synthesis. Synth. Commun. 45: 151–172. (c) N. Kaur. 2015. Application of microwave-assisted synthesis in the synthesis of fused six-membered heterocycles with *N*-heteroatom. Synth. Commun. 45: 173–201.

(2) (a) D. Chernyak and V. Gevorgyan. 2010. Palladium-catalyzed intramolecular carbopalladation/cyclization cascade: access to polycyclic *N*-fused heterocycles. Org. Lett. 12: 5558–5560. (b) N. Kaur. 2015. Advances in microwave-assisted synthesis for five-membered *N*-heterocycles synthesis. Synth. Commun. 45: 432–457. (c) N. Kaur. 2014. Microwave-assisted synthesis of five-membered *S*-heterocycles. J. Iran. Chem. Soc. 11: 523–564. (d) N. Kaur. 2015. Review on the synthesis of six-membered *N,N*-heterocycles by microwave irradiation. Synth. Commun. 45: 1145–1182. (e) N. Kaur. 2015. Greener and expeditious synthesis of fused six-membered *N,N*-heterocycles using microwave irradiation. Synth. Commun. 45: 1493–1519. (f) N. Kaur. 2015. Applications of microwaves in the synthesis of polycyclic six-membered *N,N*-heterocycles. Synth. Commun. 45: 1599–1631. (g) N. Kaur. 2015.

Synthesis of five-membered *N,N,N*- and *N,N,N,N*-heterocyclic compounds: applications of microwaves. Synth. Commun. 45: 1711–1742. (h) N. Kaur. 2018. Ruthenium catalysis in six-membered *O*-heterocycles synthesis. Synth. Commun. 48: 1551–1587. (i) N. Kaur. 2018. Green synthesis of three to five-membered *O*-heterocycles using ionic liquids. Synth. Commun. 48: 1588–1613.

(3)   (a) N. Krause, V. Belting, C. Deutsch, J. Erdsack, H.-T. Fan, B. Gockel, A. Hoffmann-Röder, N. Morita and F. Volz. 2008. Golden opportunities in catalysis. Pure Appl. Chem. 80: 1063–1069. (b) N. Kaur. 2015. Recent impact of microwave-assisted synthesis on benzo derivatives of five-membered *N*-heterocycles. Synth. Commun. 45: 539–568. (c) N. Kaur and D. Kishore. 2014. Microwave-assisted synthesis of seven- and higher-membered *N*-heterocycles. Synth. Commun. 44: 2577–2614. (d) N. Kaur and D. Kishore. 2014. Microwave-assisted synthesis of six-membered *S*-heterocycles. Synth. Commun. 44: 2615–2644. (e) N. Kaur and D. Kishore. 2014. Microwave-assisted synthesis of seven- and higher-membered *O*-heterocycles. Synth. Commun. 44: 2739–2755.

(4)   A. Das, S.M.A. Sohel and R.-S. Liu. 2010. Carbo- and heterocyclization of oxygen- and nitrogen-containing electrophiles by platinum, gold, silver and copper species. Org. Biomol. Chem. 8: 960–979.

(5)   J.-M. Weibel, A. Blanc and P. Pale. 2008. Ag-mediated reactions: coupling and heterocyclization reactions. Chem. Rev. 108: 3149–3173

(6)   J.-J. Li, T.-S. Mei and J.-Q. Yu. 2008. Synthesis of indolines and tetrahydroisoquinolines from arylethylamines by PdII-catalyzed C-H activation reactions. Angew. Chem. 120: 6552–6555.

(7)   R.K. Friedman, K.M. Oberg, D.M. Dalton and T. Rovis. 2010. Phosphoramidite-rhodium complexes as catalysts for the asymmetric [2+2+2]-cycloaddition of alkenyl isocyanates and alkynes. Pure Appl. Chem. 82: 1353–1364.

(8)   N. Krause, Ö. Aksin-Artok, V. Breker, C. Deutsch, B. Gockel, M. Poonoth, Y. Sawama, Y. Sawama, T. Sun and C. Winter. 2010. Combined coinage metal catalysis for the synthesis of bioactive molecules. Pure Appl. Chem. 82: 1529–1536.

(9)   A. Padwa, D.M. Danca, J.D. Ginn and S.M. Lynch. 2001. Application of the tandem thionium/*N*-acyliminium ion cascade toward heterocyclic synthesis. J. Braz. Chem. Soc. 12: 571–585.

(10)   D.-H. Zhang, Z. Zhang and M. Shi. 2012. Transition metal-catalyzed carbocyclization of nitrogen and oxygen-tethered 1,n-enynes and diynes: synthesis of five- or six-membered heterocyclic compounds. Chem. Commun. 48: 10271–10279.

(11)   B. Montaignac, V. Östlund, M.R. Vitale, V.R. Vidal and V. Michelet. 2012. Copper(I)-amine metallo-organocatalyzed synthesis of carbo- and heterocyclic systems. Org. Biomol. Chem. 10: 2300–2306.

(12)   A. Furstner and G. Seidel. 1995. Palladium-catalyzed arylation of polar organometallics mediated by 9-methoxy-9-borabicyclo 3.3.1 nonane - Suzuki reactions of extended scope. Tetrahedron 51: 11165–11176.

(13)   A. Mori, T. Kondo, T. Kato and Y. Nishihara. 2001. Palladium-catalyzed cross-coupling poly-condensation of bis-alkynes with dihaloarenes activated by tetrabutylammonium hydroxide or silver(I) oxide. Chem. Lett. 286–287.

(14)   Y. Bai, J. Zeng, J. Ma, B.K. Gorityala and X.-W. Liu. 2010. Quick access to drug like heterocycles: facile silver-catalyzed one-pot multi-component synthesis of aminoindolizines. J. Comb. Chem. 12: 696–699

(15)   A. Minatti and K. Muniz. 2007. Intramolecular aminopalladation of alkenes as a key step to pyrrolidines and related heterocycles. Chem. Soc. Rev. 36: 1142–1152.

(16)   G. Balme, E. Bossharth and N. Monteiro. 2003. Pd-assisted multi-component synthesis of heterocycles. Eur. J. Org. Chem. 21: 4101–4111.

(17)   U. Bora, A. Saikia and R.C. Boruah. 2003. A novel microwave-mediated one-pot synthesis of indolizines *via* a three-component reaction. Org. Lett. 5: 435–438.

(18)   L.-W. Ye, X.-L. Sun, Q.-G. Wang and Y. Tang. 2007. Phosphine-catalyzed intramolecular formal [3+2]-cycloaddition for highly diastereoselective synthesis of bicyclo[n.3.0] compounds. Angew. Chem. Int. Ed. 46: 5951–5954.

(19)   P.F. Korkowski, T.R. Hoye and D.B. Rydberg. 1988. Fischer carbene-mediated conversions of enynes to bi- and tricyclic cyclopropane containing carbon skeletons. J. Am. Chem. Soc. 110: 2676–2678.

(20)   S.T. Diver and A.J. Giessert. 2004. Enyne metathesis (enyne bond reorganization). Chem. Rev. 104: 1317–1382.

(21)   Y. Kiyotsuka, J. Igarashi and Y. Kobayashi. 2002. A study toward a total synthesis of fostriecin. Tetrahedron Lett. 43: 2725–2729.

(22)   J.A. Marshall. 2007. Chiral allylic and allenic metal reagents for organic synthesis. J. Org. Chem. 72: 8153–8166.

(23)   K. Ishihara, H. Kurihara and H. Yamamoto. 1993. An extremely simple, convenient, and selective method for acetylating primary alcohols in the presence of secondary alcohols. J. Org. Chem. 58: 3791–3793.

(24)   M.A. Kinsella, V.J. Kalish and S.M. Weinreb. 1990. Approaches to the total synthesis of the anti-tumor anti-biotic echinosporin. J. Org. Chem. 55: 105–111.

(25)   S. Husinec, I. Juranic, A. Llobera and A.E.A. Porter. 1988. Bis(methoxycarbonyl)carbene insertion into N-H bonds: a facile route to *N*-substituted aminomalonic esters. Synthesis 9: 721–723.

(26)   K. Takahashi, M. Midori, K. Kawano, J. Ishihara and S. Hatakeyama. 2008. Entry to heterocycles based on indium-catalyzed Conia-ene reactions: asymmetric synthesis of (–)-salinosporamide A. Angew. Chem. Int. Ed. 47: 6244–6246.

(27)   K. Yamada, T. Kurokawa, H. Tokuyama and T. Fukuyama. 2003. Total synthesis of the duocarmycins. J. Am. Chem. Soc. 125: 6630–6631.

(28)   G. Evano, N. Blanchard and M. Toumi. 2008. Copper-mediated coupling reactions and their applications in natural products and designed bio-molecules synthesis. Chem. Rev. 108: 3054–3131.

(29)   Q. Cai, F.T. Zhou, T.F. Xu, L.B. Fu and K. Ding. 2011. Copper-catalyzed tandem reactions of 1-(2-iodoary)-2-yn-1-ones with isocyanides for the synthesis of 4-oxo-indeno[1,2-*b*]pyrroles. Org. Lett. 13: 340–343.

(30)   Y. Liu and J.-P. Wan. 2011. Tandem reactions initiated by copper-catalyzed cross-coupling: a new strategy towards heterocycle synthesis. Org. Biomol. Chem. 9: 6873–6894.

(31)   C.A. Broka and K.K. Eng. 1986. A short total synthesis of (+)-gephyrotoxin-223AB. J. Org. Chem. 51: 5043–5045.

(32)   E.S. Sherman, P.H. Fuller, D. Kasi and S.R. Chemler. 2007. Pyrrolidine and piperidine formation *via* copper(II) carboxylate-promoted intramolecular carboamination of un-activated olefins: diastereoselectivity and mechanism. J. Org. Chem. 72: 3896–3905.

(33)   R.H. Feling, G.O. Buchanan, T.J. Mincer, C.A. Kauffman, P.R. Jensen and W. Fenical. 2003. Salinosporamide A: a highly cytotoxic proteasome inhibitor from a novel microbial source, a marine bacterium of the new genus *Salinospora*. Angew. Chem. Int. Ed. 42: 355–357.

(34)   D. Chauhan, L. Catley, G. Li, K. Podar, T. Hideshima, M. Velankar, C. Mitsiades, N. Mitsiades, H. Yasui, A. Letai, H. Ovaa, C. Berkers, B. Nicholson, T.-H. Chao, S.T.C. Neuteboom, P. Richardson, M.A. Palladino and K.C. Anderson. 2005. A novel orally active proteasome inhibitor induces apoptosis in multiple myeloma cells with mechanisms distinct from bortezomib. Cancer Cell 8: 407–419.

(35)   V.R. Macherla, S.S. Mitchell, P.R. Manam, K.A. Reed, T.-H. Chao, B. Nicholson, G. Deyanat-Yazdi, B. Mai, P.R. Jensen, W. Fenical, S.T.C. Neuteboom, K.S. Lam, M.A. Palladino and B.C.M. Potts. 2005. Structure-activity relationship studies of salinosporamide A (NPI-0052), a novel marine derived proteasome inhibitor. J. Med. Chem. 48: 3684–3687.

(36)   M. Groll, R. Huber and B.C.M. Potts. 2006. Crystal structures of salinosporamide A (NPI-0052) and B (NPI-0047) in complex with the 20S proteasome reveal important consequences of *β*-lactone ring-opening and a mechanism for irreversible binding. J. Am. Chem. Soc. 128: 5136–5141.

(37)   L.R. Reddy, J.-F. Fournier, B.V.S. Reddy and E.J. Corey. 2005. New synthetic route for the enantioselective total synthesis of salinosporamide A and biologically active analogues. Org. Lett. 7: 2699–2701.

(38)   L.R. Reddy, P. Saravanan and E.J. Corey. 2004. A simple stereocontrolled synthesis of salinosporamide A. J. Am. Chem. Soc. 126: 6230–6231.

(39)   A. Endo and S.J. Danishefsky. 2005. Total synthesis of salinosporamide A. J. Am. Chem. Soc. 127: 8298–8299.

(40)   T. Ling, V.R. Macherla, R.R. Manam, K.A. McArthur and B.C.M. Potts. 2007. Enantioselective total synthesis of (–)-salinosporamide A (NPI-0052). Org. Lett. 9: 2289–2292.

(41)   T. Fukuda, K. Sugiyama, S. Arima, Y. Harigaya, T. Nagamitsu and S. Omura. 2008. Total synthesis of salinosporamide A. Org. Lett. 10: 4239–4242.

(42)   N.P. Mulholland, G. Pattenden and I.A.S. Walters. 2006. A concise total synthesis of salinosporamide A. Org. Biomol. Chem. 4: 2845–2846.

(43)   G. Ma, H. Nguyen and D. Romo. 2007. Concise total synthesis of (±)-salinosporamide A, (±)-cinnabaramide A, and derivatives *via* a bis-cyclization process: implications for a biosynthetic pathway? Org. Lett. 9: 2143–2146.

(44)   V. Caubert, J. Massé, P. Retailleau and N. Langlois. 2007. Stereoselective formal synthesis of the potent proteasome inhibitor: salinosporamide A. Tetrahedron Lett. 48: 381–384.

(45)   I.V. Margalef, L. Rupnicki and H.W. Lam. 2008. Formal synthesis of salinosporamide A using a nickel-catalyzed reductive aldol cyclization-lactonization as a key step. Tetrahedron 64: 7896–7901.

(46)   (a) S. Hatakeyama. 2009. Indium-catalyzed Conia-ene reaction for alkaloid synthesis. Pure Appl. Chem. 81: 217–226. (b) S. Hatakeyama. 2014. Total synthesis of biologically active natural products based on highly selective synthetic methodologies. Chem. Pharm. Bull. 62: 1045–1061.

(47)   M. Jayagobi, M. Poornachandran and R. Raghunathan. 2009. A novel heterotricyclic assembly through intramolecular imino-Diels-Alder reaction: synthesis of pyrrolo[3,4-*b*]quinolines. Tetrahedron Lett. 50: 648–650.

(48)   K. Chinnakali, D. Sudha, M. Jayagobi, R. Raghunathan and H.K. Fun. 2009. 3-Benzyl-7-methyl-9-phenyl-2-tosyl-2,3,3a,4,9,9a-hexahydro-1*H*-pyrrolo[3,4-*b*]quinoline. Acta Crystallogr. Sect. E Struct. Rep. 65: 2924–2925.

(49)   K. Chinnakali, D. Sudha, M. Jayagobi, R. Raghunathan and H.K. Fun. 2009. 3-Benzyl-7-methoxy-9-phenyl-2-tosyl-2,3,3a,4,9,9a-hexahydro-1*H*-pyrrolo[3,4-*b*]quinoline. Acta Crystallogr. Sect. E Struct. Rep. 65: 2956–2957.

(50)   T. Fujisawa, T. Mori and T. Sato. 1983. Chemoselective reduction of carboxylic acids into alcohols using *N,N*-dimethylchloromethyleniminium chloride and sodium borohydride. Chem. Lett. 12: 835–838.

(51)   J.S. Panek and P. Liu. 2000. Total synthesis of the actin-depolymerizing agent (–)-mycalolide A: application of chiral silane-based bond construction methodology. J. Am. Chem. Soc. 122: 11090–11097.

(52)   P.T. Parvatkar, P.S. Parameswaran and S.G. Tilve. 2012. Recent developments in the synthesis of five- and six-membered heterocycles using molecular iodine. Chem. Eur. J. 18: 5460–5489.

(53)   I. Kim, H.K. Won, J. Choi and G.H. Lee. 2007. A novel and efficient approach to highly substituted indolizines *via* 5-*endo-trig* iodocyclization. Tetrahedron 63: 12954–12960.

(54)   B. Quiclet-Sire and S.Z. Zard. 2006. Observations on the reaction of hydrazones with iodine: interception of the diazo intermediates. Chem. Commun. 17: 1831–1832.

(55)   S.G. Davies, R.L. Nicholson, P.D. Price, P.M. Roberts, A.J. Russell, E.D. Savory, A.D. Smith and J.E. Thomson. 2009. Iodine-mediated ring-closing iodoamination with concomitant *N*-debenzylation for the asymmetric synthesis of polyhydroxylated pyrrolidines. Tetrahedron: Asymmetry 20: 758–772.

(56) A. Citterio, M. Fochi, A. Marion, A. Mele, R. Sebastiano and M. Delcanale. 1998. Synthesis of naphtho[2,3-*b*] indolizine-6,11-dione derivatives by iodine oxidation of 2-alkyl-1,4-naphthoquinones in the presence of substituted pyridines. 48: 2003–2017.

(57) I. Kim, S.G. Kim, J.Y. Kim and G.H. Lee. 2007. A novel approach to 3-acylated indolizine structures *via* iodine-mediated hydrative cyclization. Tetrahedron Lett. 48: 8976–8981.

(58) V. Singh, B.C. Sahu and S.M. Mobin. 2008. Intramolecular cycloaddition in cyclohexa-2,4-dienone and photochemical reactions: an efficient route to azatriquinane and azasterpurane frameworks. Synlett 8: 1222–1224.

(59) T. Pradhan and A. Hassner. 2007. A facile synthesis of (n+3) and (n+4) ring enlarged lactones as well as of spiroketolactones from n-membered cycloalkanones. Synthesis 21: 3361–3370.

(60) A. Martín, I. Pérez-Martín and E. Suárez. 2005. Intramolecular hydrogen abstraction promoted by amidyl radicals. Evidence for electronic factors in the nucleophilic cyclization of ambident amides to oxocarbenium ions. Org. Lett. 7: 2027–2030.

(61) C.G. Francisco, A.J. Herrera, A. Martin, I. Perez-Martin and E. Suarez. 2007. Intramolecular 1,5-hydrogen atom transfer reaction promoted by phosphoramidyl and carbamoyl radicals: synthesis of 2-amino-C-glycosides. Tetrahedron Lett. 48: 6384–6388.

(62) V.V. Zhdankin. 2009. Hypervalent iodine(III) reagents in organic synthesis. ARKIVOC (i): 1–62.

(63) I. Kim, J. Choi, H.K. Won and G.H. Lee. 2007. Expeditious synthesis of indolizine derivatives *via* iodine-mediated 5-*endo-dig* cyclization. Tetrahedron Lett. 48: 6863–6867.

(64) H. Shigehisa, J. Takayama and T. Honda. 2006. The first total synthesis of (±)-annosqualine by means of oxidative enamide-phenol coupling: pronounced effect of phenoxide formation on the phenol oxidation mechanism. Tetrahedron Lett. 47: 7301–7306.

(65) D.G. Blackmond, A. Lightfoot, A. Pfaltz, T. Rosner, P. Schnider and N. Zimmermann. 2000. Enantioselective hydrogenation of olefins with phosphinooxazoline-iridium catalysts. Chirality 12: 442–449.

(66) P. Schnider, G. Koch, R. Prétôt, G. Wang, F.M. Bohnen, C. Kruger and A. Pfaltz. 1997. Enantioselective hydrogenation of imines with chiral (phosphanodihydrooxazole)iridium catalysts. Chem. Eur. J. 3: 887–892.

(67) T. Shibata, N. Toshida, M. Yamasaki, S. Maekawa and K. Takagi. 2005. Iridium-catalyzed enantioselective Pauson-Khand-type reaction of 1,6-enynes. Tetrahedron 61: 9974–9979.

(68) M.W. Bouwkamp, A.C. Bowman, E. Lobkovsky and P.J. Chirik. 2006. Iron-catalyzed [2π+2π]-cycloaddition of α,ω-dienes: the importance of redox-active supporting ligands. J. Am. Chem. Soc. 128: 13340–13341.

(69) Y. Li and T. Marks. 1998. Organolanthanide-catalyzed intra- and intermolecular tandem C-N and C-C bond-forming processes of aminodialkenes, aminodialkynes, aminoalkeneynes, and aminoalkynes. New regiospecific approaches to pyrrolizidine, indolizidine, pyrrole, and pyrazine skeletons. J. Am. Chem. Soc. 120: 1757–1771.

(70) M.M.-C. Lo and G.C. Fu. 2002. Cu(I)/Bis(azaferrocene)-catalyzed enantioselective synthesis of β-lactams *via* couplings of alkynes with nitrones. J. Am. Chem. Soc. 124: 4572–4573.

(71) R. Shintani and G.C. Fu. 2003. Catalytic enantioselective synthesis of β-lactams: intramolecular Kinugasa reactions and interception of an intermediate in the reaction cascade. Angew. Chem. Int. Ed. 42: 4082–4085.

(72) M.-C. Ye, J. Zhou, Z.-Z. Huang and Y. Tang. 2003. Chiral tris(oxazoline)/Cu(II)-catalyzed coupling of terminal alkynes and nitrones. Chem. Commun. 20: 2554–2555.

(73) W.F. Bailey, P.D. Salgaonkar, J.D. Brubaker and V. Sharma. 2008. Preparation of the isomeric azaindoline family by intramolecular carbolithiation. Org. Lett. 10: 1071–1074.

(74) A. Wang, S. Tandel, H. Zhang, Y. Huang, T.C. Holdeman and E.R. Biehl. 1998. Preparation of 7-arylmethyl-1*H*-pyrrolo[3,4-*c*]pyridine-1,3-(2*H*)-diones and α-aryl-3-hydroxy-5-pyridylacetonitriles using arynic methodology. Tetrahedron 54: 3391–3400.

(75) B. Chattopadhyay and V. Gevorgyan. 2012. Transition metal-catalyzed denitrogenative transannulation: converting triazoles into other heterocyclic systems. Angew. Chem. Int. Ed. 51: 862–872.

(76) S. Tsuchida, A. Kaneshige, T. Ogata, H. Baba, Y. Yamamoto and K. Tomioka. 2008. Consecutive cyclization of allylaminoalkene by intramolecular aminolithiation-carbolithiation. Org. Lett. 10: 3635–3638.

(77) B. Wang, Z. Zhong and G.-Q. Lin. 2009. Efficient construction of stereodefined α-alkylidene aza-cycloketones *via* β-amino-alkenyllithium: straightforward and protection free synthesis of allopumiliotoxin 267A. Org. Lett. 11: 2011–2014.

(78) A. Padwa, M.A. Brodney, S.M. Lynch, P. Rashatasakhon, Q. Wang and H. Zhang. 2004. A new strategy toward indole alkaloids involving an intramolecular cycloaddition/rearrangement cascade. J. Org. Chem. 69: 3735–3745.

(79) M. Kitamura, S. Chiba, O. Saku and K. Narasaka. 2002. Palladium-catalyzed synthesis of 1-azaazulenes from cycloheptatrienylmethyl ketone *O*-pentafluorobenzoyl oximes. Chem. Lett. 31: 606–607.

(80) S. Chiba, M. Kitamura, O. Saku and K. Narasaka. 2004. Synthesis of 1-azaazulenes from cycloheptatrienylmethyl ketone *O*-pentafluorobenzoyloximes by palladium-catalyzed cyclization. Bull. Chem. Soc. Jpn. 77: 785–796.

(81) M. Kitamura, S. Zaman and K. Narasaka. 2001. Synthesis of spiro imines from oximes by palladium-catalyzed cascade reaction. Synlett 974–976.

(82) S. Zaman, M. Kitamura and K. Narasaka. 2003. Synthesis of polycyclic imines by palladium-catalyzed domino cyclization of di- and trienyl ketone *O*-pentafluorobenzoyloximes. Bull. Chem. Soc. Jpn. 76: 1055–1062.

(83) A.A. Aly, A.B. Brown, T.I. El-Emary, A.M.M. Ewas and M. Ramadane. 2009. Hydrazinecarbothioamide group in the synthesis of heterocycles. ARKIVOC (i): 150–197.

(84) K. Asahi and H. Nishino. 2009. Synthesis of bicyclo[3.1.0]hexan-2-ones by manganese(III) oxidation in ethanol. Synthesis 3: 409–423.

(85) P. Wipf and J.P. Maciejewski. 2008. Titanocene(III)-catalyzed formation of indolines and azaindolines. Org. Lett. 10: 4383–4386.

(86) A. Geny, S. Gaudrel, F. Slowinski, M. Amatore, G. Chouraqui, M. Malacria, C. Aubert and V. Gandon. 2009. A straightforward procedure for the [2+2+2]-cycloaddition of enediynes. Adv. Synth. Catal. 351: 271–275.

(87) S. Diez-Gonzalez, N. Marion and S.P. Nolan. 2009. *N*-Heterocyclic carbenes in late transition metal catalysis. Chem. Rev. 109: 3612–3676.

(88) N.S. Karanjule, S.D. Markad, V.S. Shinde and D.D. Dhavale. 2006. Intramolecular 5-*endo-trig* aminomercuration of *β*-hydroxy-*γ*-alkenylamines: efficient route to a pyrrolidine ring and its application for the synthesis of (+)-castanospermine and analogues. J. Org. Chem. 71: 4667–4670.

(89) A.K. Gupta, D.I. Park and C.H. Oh. 2005. The exclusive formation of cyclopentenones from molybdenum hexacarbonyl-catalyzed Pauson-Khand reactions of 5-allenyl-1-ynes. Tetrahedron Lett. 46: 4171–4174.

(90) G.A. Molander and E.D. Dowdy. 1998. Catalytic intramolecular hydroamination of hindered alkenes using organolanthanide complexes. J. Org. Chem. 63: 8983–8988.

(91) G.A. Molander and E.D. Dowdy. 1999. Lanthanide-catalyzed hydroamination of hindered alkenes in synthesis: rapid access to 10,11-dihydro-5*H*-dibenzo-[*a*,*d*]cyclohepten-5,10-imines. J. Org. Chem. 64: 6515–6517.

(92) T.E. Muller, K.C. Hultzsch, M. Yus, F. Foubelo and M. Tada. 2008. Hydroamination: direct addition of amines to alkenes and alkynes. Chem. Rev. 108: 3795–3892.

(93) P.M. Joensuu, G.J. Murray, E.A.F. Fordyce, T. Luebbers and H.W. Lam. 2008. Diastereoselective nickel-catalyzed reductive aldol cyclizations using diethylzinc as the stoichiometric reductant: scope and mechanistic insight. J. Am. Chem. Soc. 130: 7328–7338.

(94) Z. Chai, H.F. Wang and G. Zhao. 2009. Ni-catalyzed carbocyclization of 1,6-enynes mediated by dialkylzinc reagents: Me$_2$Zn or Et$_2$Zn makes a difference. Synlett 11: 1785–1790.

(95) Z.-X. Yu, Y. Wang and Y. Wang. 2010. Transition metal-catalyzed cycloadditions for the synthesis of eight-membered carbocycles. Chem. Asian J. 5: 1072–1088.

(96) M. Takahashi, T. Maehara, T. Sengoku, N. Fujita, K. Takabe and H. Yoda. 2008. New asymmetric strategy for the total synthesis of naturally occurring (+)-alexine and (–)-7-*epi*-alexine. Tetrahedron 64: 5254–5261.

(97) S.I. Murahashi, T. Saito, T. Naota, H. Kumobayashi and S. Akutagawa. 1991. Osmium-catalyzed oxidation of *β*-lactams with peroxides. Tetrahedron Lett. 32: 2145–2148.

(98) P. Zakrzewski, M. Gowan, L.A. Trimble and C.K. Lau. 1999. *o*-Hydroxyalkylation of aminopyridines: a novel approach to heterocycles. Synthesis 11: 1893–1902.

(99) E.M. Beccalli, G. Broggini, M. Martinelli and S. Sottocornola. 2007. C-C, C-O, C-N Bond formation on sp^2 carbon by Pd(II)-catalyzed reactions involving oxidant agents. Chem. Rev. 107: 5318–5365.

(100) A. Brandi, F. Cardona, S. Cicchi, F.M. Cordero and A. Goti. 2009. Stereocontrolled cyclic nitrone cycloaddition strategy for the synthesis of pyrrolizidine and indolizidine alkaloids. Chem. Eur. J. 15: 7808–7821.

(101) M. Lautens, J.-F. Paquin, S. Piguel and M. Dahlmann. 2001. Palladium-catalyzed sequential alkylation-alkenylation reactions and their application to the synthesis of fused aromatic rings. J. Org. Chem. 66: 8127–8134.

(102) P.R. Weider, L.S. Hegedus, H. Asada and V. D'Andreq. 1985. Oxidative cyclization of unsaturated aminoquinones. Synthesis of quinolinoquinones. Palladium-catalyzed synthesis of pyrroloindoloquinones. J. Org. Chem. 50: 4276–4281.

(103) J.-E. Backvall, P.G. Andersson, G.B. Stone and A. Gogoll. 1991. Synthesis of (+)-*α*- *and (+)-γ*-lycorane *via* a stereocontrolled organopalladium route. J. Org. Chem. 56: 2988–2993.

(104) Z. Moussa. 2012. The Hendrickson 'POP' reagent and analogues there of: synthesis, structure, and application in organic synthesis. ARKIVOC (i): 432–490.

(105) L.S. Hegedus and J.M. McKearin. 1982. Palladium-catalyzed cyclization of *ω*-olefinic tosamides. Synthesis of non-aromatic nitrogen heterocycles. J. Am. Chem. Soc. 104: 2444–2451.

(106) M.E. Wall, M.C. Wani, C.E. Cook, K.H. Palmer, A.T. McPhail and G.A. Sim. 1966. Plant anti-tumor agents. I. The isolation and structure of camptothecin, a novel alkaloidal leukemia and tumor inhibitor from *Camptotheca acuminata*. J. Am. Chem. Soc. 88: 3888–3890.

(107) Y.H. Hsiang, R. Hertzberg, S.M. Hecht and L.F. Liu. 1985. Camptothecin induces protein-linked DNA breaks *via* mammalian DNA topoisomerase I. J. Biol. Chem. 260: 14873–14878.

(108) K.W. Kohn and Y. Pommier. 2000. Molecular and biological determinants of the cytotoxic actions of camptothecins. Perspective for the development of new topoisomerase I inhibitors. Ann. N.Y. Acad. Sci. 922: 11–26.

(109) B.L. Staker, K. Hjerrild, M.D. Feese, C.A. Behnke, A.B. Burgin and L. Stewart. 2002. The mechanism of topoisomerase I poisoning by a camptothecin analog. Proc. Natl. Acad. Sci. 99: 15387–15392.

(110) W. Du. 2003. Towards new anti-cancer drugs: a decade of advances in synthesis of camptothecins and related alkaloids. Tetrahedron 59: 8649–8687.

(111) C.R. Hutchinson. 1981. Camptothecin: chemistry, biogenesis, and medicinal chemistry. Tetrahedron 37: 1047–1065.

(112) H.-B. Zhou, G.-S. Liu and Z.-J. Yao. 2007. Highly efficient and mild cascade reactions triggered by bis(triphenyl) oxodiphosphonium trifluoromethanesulfonate and a concise total synthesis of camptothecin. Org. Lett. 9: 2003–2006.

(113) Y. Kozawa and M. Mori. 2001. Construction of a carbapenam skeleton using palladium-catalyzed cyclization. Tetrahedron Lett. 42: 4869–4873.

(114) Y. Kozawa and M. Mori. 2002. Synthesis of different ring size heterocycles from the same propargyl alcohol derivative by ligand effect on Pd(0). Tetrahedron Lett. 43: 1499–1502.

(115) Y. Kozawa and M. Mori. 2003. Novel synthesis of carbapenam by intramolecular attack of lactam nitrogen toward η^1-allenyl and η^3-propargylpalladium complex. J. Org. Chem. 68: 8068–8074.

(116) M.B. Bertrand, M.L. Leathen and J.P. Wolfe. 2007. Mild conditions for the synthesis of functionalized pyrrolidines *via* Pd-catalyzed carboamination reactions. Org. Lett. 9: 457–460.

(117) (a) J.P. Wolfe. 2006. Stereoselective synthesis of saturated heterocycles *via* Pd-catalyzed alkene carboetherification and carboamination reactions. Synlett 4: 571–582. (b) J.P. Wolfe. 2008. Stereoselective synthesis of saturated heterocycles *via* Pd-catalyzed alkene carboetherification and carboamination reactions. Synlett 19: 2913–2937.

(118) M. Nazare, C. Schneider, A. Lindenschmidt and D.W. Will. 2004. A flexible, palladium-catalyzed indole and azaindole synthesis by direct annulation of chloroanilines and chloroaminopyridines with ketones. Angew. Chem. Int. Ed. 43: 4526–4528.

(119) C.R. Smith, E.M. Bunnelle, A.J. Rhodes and R. Sarpong. 2007. Pt-catalyzed cyclization/1,2-migration for the synthesis of indolizines, pyrrolones, and indolizinones. Org. Lett. 9: 1169–1171.

(120) H. Kusama, Y. Karibe, Y. Onizawa and N. Iwasawa. 2010. Gold-catalyzed tandem cyclization of dienol silyl ethers for the preparation of bicyclo[4.3.0]nonane derivatives. Angew. Chem. Int. Ed. 49: 4269–4272.

(121) P.P. Graczyk, A. Khan, G.S. Bhatia, V. Palmer, D. Medland, H. Numata, H. Oinuma, J. Catchick, A. Dunne, M. Ellis, C. Smales, J. Whitfield, S.J. Neame, B. Shah, D. Wilton, L. Morgan, T. Patel, R. Chung, H. Desmond, J.M. Staddon, N. Sato and A. Inoue. 2005. The neuroprotective action of JNK3 inhibitors based on the 6,7-dihydro-5*H*-pyrrolo[1,2-*a*]imidazole scaffold. Bioorg. Med. Chem. Lett. 15: 4666–4670.

(122) J.C. Rech, M. Yato, D. Duckett, B. Ember, P.V. LoGrasso, R.G. Bergman and J.A. Ellman. 2007. Synthesis of potent bicyclic bis-arylimidazole c-Jun N-terminal kinase inhibitors by catalytic C-H bond activation. J. Am. Chem. Soc. 129: 490–491.

(123) J.C. Lewis, R.G. Bergman and J.A. Ellman. 2008. Direct functionalization of nitrogen heterocycles *via* Rh-catalyzed C-H bond activation. Acc. Chem. Res. 41: 1013–1025.

(124) D. Castagnolo, F. Manetti, M. Radi, B. Bechi, M. Pagano, A. de Logu, R. Meleddu, M. Saddi and M. Botta. 2009. Synthesis, biological evaluation, and SAR study of novel pyrazole analogues as inhibitors of *Mycobacterium tuberculosis*: synthesis of rigid pyrazolones. Bioorg. Med. Chem. 17: 5716–5721.

(125) A. Lerchner and E.M. Carreira. 2002. First total synthesis of (±)-strychnofoline *via* a highly selective ring-expansion reaction. J. Am. Chem. Soc. 124: 14826–14827.

(126) C. Fischer, C. Meyers and E.M. Carreira. 2000. Efficient synthesis of (±)-horsfiline through the MgI_2-catalyzed ring-expansion reaction of a spiro[cyclopropane-1,3'-indol]-2'-one. Helv. Chim. Acta 83: 1175–1181.

(127) E.M. Carreira and C. Meyers. 2003. Total synthesis of (–)-spirotryprostatin B. Angew. Chem. Int. Ed. 42: 694–696.

(128) G.S. Singh and Z.Y. Desta. 2012. Isatins as privileged molecules in design and synthesis of spiro-fused cyclic frameworks. Chem. Rev. 112: 6104–6155.

(129) R.W. Bates and C.J. Lim. 2010. Synthesis of two *Nuphar* alkaloids by allenic hydroxylamine cyclization. Synlett 6: 866–868.

(130) H.-W. Fruhauf. 1997. Metal-assisted cycloaddition reactions in organotransition metal chemistry. Chem. Rev. 97: 523–596.

(131) O. Geis and H.-G. Schmalz. 1998. New developments in the Pauson-Khand reaction. Angew. Chem. Int. Ed. Engl. 37: 911–914.

(132) Y.K. Chung. 1999. Transition metal alkyne complexes: the Pauson-Khand reaction. Coord. Chem. Rev. 188: 297–341.

(133) S.-W. Kim, S.U. Son, S.I. Lee, T. Hyeon and Y.K. Chung. 2000. Cobalt on mesoporous silica: the first heterogeneous Pauson-Khand catalyst. J. Am. Chem. Soc. 122: 1550–1551.

(134) A.C. Comely, S.E. Gibson and N.J. Hales. 2000. Polymer-supported cobalt carbonyl complexes as novel solid-phase catalysts of the Pauson-Khand reaction. Chem. Commun. 4: 305–306.

(135) M. Hayashi, Y. Hashimoto, Y. Yamamoto, J. Usuki and K. Saigo. 2000. Phosphane sulfide/octacarbonyldicobalt-catalyzed Pauson-Khand reaction under an atmospheric pressure of carbon monoxide. Angew. Chem. Int. Ed. 39: 631–633.

(136) B.L. Pagenkopf, D.B. Belanger, D.J.R. O'Mahony and T. Livinghouse. 2000. (Alkylthio)alkynes as addends in the Co(0)-catalyzed intramolecular Pauson-Khand reaction. Substituent driven enhancements of annulation efficiency and stereoselectivity. Synthesis 7: 1009–1019.

(137) M.E. Krafft and L.V.R. Bonaga. 2000. Dodecacarbonyltetracobalt catalysis in the thermal Pauson-Khand reaction. Angew. Chem. Int. Ed. 39: 3676–3680.

(138) S.U. Son, S.I. Lee and Y.K. Chung. 2000. Cobalt on charcoal: a convenient and inexpensive heterogeneous Pauson-Khand catalyst. Angew. Chem. Int. Ed. 39: 4158–4160.

(139) M.E. Krafft, L.V.R. Bonaga and C. Hirosawa. 2001. Practical cobalt carbonyl catalysis in the thermal Pauson-Khand reaction: efficiency enhancement using Lewis bases. J. Org. Chem. 66: 3004–3020.

(140) S.-W. Kim, S.U. Son, S.S. Lee, T. Hyeon and Y.K. Chung. 2001. Colloidal cobalt nanoparticles: a highly active and reusable Pauson-Khand catalyst. Chem. Commun. 21: 2212–2213.

(141) A.C. Comely, S.E. Gibson, A. Stevenazzi and N.J. Hales. 2001. New stable catalysts of the Pauson-Khand annelation. Tetrahedron Lett. 42: 1183–1185.

(142) T. Shibata, N. Toshida and K. Takagi. 2002. Catalytic Pauson-Khand type reaction using aldehydes as a CO source. Org. Lett. 4: 1619–1621.

(143) S.E. Gibson, C. Johnstone and A. Stevenazzi. 2002. A stable catalyst of the Pauson-Khand annelation. Tetrahedron 58: 4937–4942.

(144) T. Shibata, N. Toshida and K. Takagi. 2002. Rhodium complex-catalyzed Pauson-Khand-type reaction with aldehydes as a CO source. J. Org. Chem. 67: 7446–7450.

(145) M. Ishizaki, H. Satoh and O. Hoshino. 2002. Intramolecular Pauson-Khand reaction of various 2-aryl-1,6-enynes: synthesis of bicyclic compounds bearing quaternary carbon center. Chem. Lett. 31: 1040–1041.

(146) B. Jiang and M. Xu. 2002. Catalytic diastereoselective Pauson-Khand reaction: an efficient route to enantiopure cyclopenta[c]proline derivatives. Org. Lett. 4: 4077–4080.

(147) K.H. Park, S.U. Son and Y.K. Chung. 2003. Immobilized heterobimetallic Ru/Co nanoparticle-catalyzed Pauson-Khand-type reactions in the presence of pyridylmethyl formate. Chem. Commun. 15: 1898–1899.

(148) T. Honda, H. Namiki, K. Kaneda and H. Mizutani. 2004. First diastereoselective chiral synthesis of (–)-securinine. Org. Lett. 6: 87–89.

(149) A. Michrowska, R. Bujok, S. Harutyunyan, V. Sashuk, G. Dolgonos and K. Grela. 2004. Nitro-substituted Hoveyda-Grubbs ruthenium carbenes: enhancement of catalyst activity through electronic activation. J. Am. Chem. Soc. 126: 9318–9325.

(150) (a) T. Honda, H. Namiki, M. Watanabe and H. Mizutani. 2004. First total synthesis of (+)-viroallosecurinine. Tetrahedron Lett. 45: 5211–5213. (b) M. Mori. 2010. Recent progress on enyne metathesis: its application to syntheses of natural products and related compounds. Materials 3: 2087–2140.

(151) Y. Li and T.J. Marks. 1996. Coupled organolanthanide-catalyzed C-N/C-C bond formation processes. Efficient regiospecific assembly of pyrrolizidine and indolizidine skeletons in a single catalytic reaction. J. Am. Chem. Soc. 118: 707–708.

(152) V.M. Arredondo, S. Tian, F.E. McDonald and T.J. Marks. 1999. Organolanthanide-catalyzed hydroamination/cyclization. Efficient allene-based transformations for the syntheses of naturally occurring alkaloids. J. Am. Chem. Soc. 121: 3633–3639.

(153) S. Tian, V.M. Arredondo, C.L. Stern and T.J. Marks. 1999. Constrained geometry organolanthanide catalysts. Synthesis, structural characterization, and enhanced aminoalkene hydroamination/cyclization activity. Organometallics 18: 2568–2570.

(154) E.A. Younes, A.Q. Hussein, A.M. Mitchell and F.R. Fronczek. 2011. Synthesis of 2-amino-1,4-benzodiazepin-5-ones from 2-nitrobenzoic acid and α-aminonitriles. ARKIVOC (ii): 322–330.

(155) P. Drabina and M. Sedlak. 2012. 2-Amino-2-alkyl(aryl)propanenitriles as key building blocks for the synthesis of five-membered heterocycles. ARKIVOC (i): 152–172.

(156) W. Zhang. 2004. Recent advances in the synthesis of biologically interesting heterocycles by intramolecular aryl radical reactions. Curr. Org. Chem. 8: 757–780.

(157) W.R. Bowman, M.O. Cloonan and S.L. Krintel. 2001. Synthesis of heterocycles by radical cyclization. J. Chem. Soc. Perkin Trans. 1 22: 2885–2902.

(158) D.L. Comins, H. Hong and G. Jianhua. 1994. Asymmetric synthesis of camptothecin alkaloids: a nine-step synthesis of (S)-camptothecin. Tetrahedron Lett. 35: 5331–5334.

(159) J. Boivin, E. Fouquet, A.-M. Schiano and S.Z. Zard. 1994. Iminyl radicals: Part III. Further synthetically useful sources of iminyl radicals. Tetrahedron 50: 1769–1776.

(160) S.Z. Zard. 1996. Iminyl radicals: a fresh look at a forgotten species (and some of its relatives). Synlett 12: 1148–1154.

(161) S. Nag and S. Batra. 2011. Applications of allylamines for the syntheses of aza-heterocycles. Tetrahedron 67: 8959–9061.

(162) M. Hori and M. Mori. 1995. Synthesis of heterocycles utilizing N_2-TiCl$_4$-Li-TMSCl. J. Org. Chem. 60: 1480–1481.

(163) (a) M. Mori. 2009. Synthesis of nitrogen heterocycles utilizing molecular nitrogen as a nitrogen source and attempt to use air instead of nitrogen gas. Heterocycles 78: 281–318. (b) M. Mori, M. Akashi, M. Hori, K. Hori, M. Nishida and Y. Sato. 2004. Nitrogen fixation: synthesis of heterocycles using molecular nitrogen as a nitrogen source. Bull. Chem. Soc. Jpn. 77: 1655–1670.

(164) A. Arcadi and E. Rossi. 1997. Sequential addition/elimination/annulation reactions of 4-pentynones with benzylamine and ammonia. Synlett 6: 667–668.

(165) A. Arcadi and E. Rossi. 1998. Synthesis of functionalized furans and pyrroles through annulation reactions of 4-pentynones. Tetrahedron 54: 15253–15272.

(166) C.A. Faler, B. Cao and M.M. Joullie. 2006. Synthesis of bicyclic cyclopropylamines from amino acid derivatives. Heterocycles 67: 519–522.

(167) H. Kusama, Y. Suzuki, J. Takaya and N. Iwasawa. 2006. Intermolecular 1,5-dipolar cycloaddition reaction of tungsten containing vinylazomethine ylides leading to seven-membered heterocycles. Org. Lett. 8: 895–897.

(168) G. Yin, Y. Zhu, L. Zhang, P. Lu and Y. Wang. 2011. Preparation of allenephosphoramide and its utility in the preparation of 4,9-dihydro-2*H*-benzo[*f*]isoindoles. Org. Lett. 13: 940–943.

(169) F. Denes, S. Cutri, A. Perez-Luna and F. Chemla. 2006. Radical-polar crossover domino reactions involving organozinc and mixed organocopper/organozinc reagents. Chem. Eur. J. 12: 6506–6513.

(170) R. Griot and T. Wagner-Jauregg. 1958. Über eine neuartige umlagerung des Δ2-cyclopentenyl-acetonoxims. Helv. Chim. Acta 41: 867–881.

(171) R. Griot and T. Wagner-Jauregg. 1959. Über eine neuartige isomerisierung des Δ²-cyclopentenyl-acetonoxims (darstellung und eigenschaften bicyclischer pyrrolin-derivate). 2. Mitteilung. Helv. Chim. Acta 42: 121–128.

(172) D. Schinzer and Y. Bo. 1991. Synthesis of heterocycles by tandem reactions: Beckmann rearrangements/allylsilane cyclizations. Angew. Chem. Int. Ed. Engl. 30: 687–688.

(173) J. Ji, M.R. Schrimpf, K.B. Sippy, W.H. Bunnelle, T. Li, D.J. Anderson, C. Faltynek, C.S. Surowy, T. Dyhring, P.K. Ahring and M.D. Meyer. 2007. Synthesis and structure-activity relationship studies of 3,6-diazabicyclo[3.2.0] heptanes as novel α4β2 nicotinic acetylcholine receptor selective agonists. J. Med. Chem. 50: 5493–5508.

(174) X. Sun, C.Y. Wang, Z. Li, S. Zhang and Z. Xi. 2004. Zirconocene-mediated intermolecular coupling of one molecule of Si-tethered diyne with three molecules of organonitriles: one-pot formation of pyrrolo[3,2-*c*]pyridine derivatives *via* cleavage of C≡N triple bonds of organonitriles. J. Am. Chem. Soc. 126: 7172–7173.

(175) W.-X. Zhang, S. Zhang, X. Sun, M. Nishiura, Z. Hou and Z. Xi. 2009. Zirconium- and silicon-containing intermediates with three fused rings in a zirconocene-mediated intermolecular coupling reaction. Angew. Chem. Int. Ed. 48: 7227–7231.

(176) S. Zhang, X. Sun, W.-X. Zhang and Z. Xi. 2009. One-pot multi-component synthesis of azaindoles and pyrroles from one molecule of a silicon-tethered diyne and three or two molecules of organonitriles mediated by zirconocene. Chem. Eur. J. 15: 12608–12617.

(177) S. Zhang, W.-X. Zhang and Z. Xi. 2010. Efficient one-pot synthesis of *N*-containing heterocycles by multi-component coupling of silicon-tethered diynes, nitriles, and isocyanides through intramolecular cyclization of iminoacyl Zr intermediates. Chem. Eur. J. 16: 8419–8426.

(178) S. Zhang, W.-X. Zhang, J. Zhao and Z. Xi. 2010. Cleavage and reorganization of Zr-C/Si-C bonds leading to Zr/Si-N organometallic and heterocyclic compounds. J. Am. Chem. Soc. 132: 14042–14045.

(179) S. Zhang, J. Zhao, W.-X. Zhang and Z. Xi. 2011. One-pot synthesis of pyrrolo[3,2-*d*]pyridazines and pyrrole-2,3-diones *via* zirconocene-mediated four-component coupling of Si-tethered diyne, nitriles, and azide. Org. Lett. 13: 1626–1629.

(180) S. Zhang, W.-X. Zhang, J. Zhao and Z. Xi. 2011. One-pot selective syntheses of 5-azaindoles through zirconocene-mediated multi-component reactions with three different nitrile components and one alkyne component. Chem. Eur. J. 17: 2442–2449.

(181) W.-X. Zhang, S. Zhang and Z. Xi. 2011. Zirconocene and Si-tethered diynes: a happy match directed toward organometallic chemistry and organic synthesis. Acc. Chem. Res. 44: 541–551.

4

Five-Membered Fused Polyheterocycles

4.1 Introduction

Molecules containing heterocyclic substructures have always attracted the attention of researchers for their biological activities [1a–e]. Interest in heterocyclic scaffolds containing pyrazole and indazole nucleus has been stimulated because of their properties suited for industrial, agricultural and medicinal purposes [2a–c]. The practical applications of indoles and pyrroles are heavily centered in the pharmaceutical area. It is well known that the indole nucleus is associated with a large number of important pharmacological properties [3a–d]—anti-bacterial [4], anti-cancer [5–6], anti-biotic [7–8], central nervous system modulating [9], etc.

The indole scaffold arguably represents one of the major vital structural subunits for the discovery of new drugs. The expression that numerous alkaloids enclose the indole nucleus, the recognition of the significance of the essential amino acid tryptophan in human nutrition, and the discovery of plant hormones have led to extensive research on indole derivatives, resulting in a vast number of biologically active natural and synthetic products with a wide range of therapeutic applications as anti-inflammatories, phosphodiesterase inhibitors, 5-hydroxy tryptamine receptor agonist and antagonists, cannabinoid receptors agonists and HMG-CoA reductase inhibitors. Numerous drugs, such as indomethacin, ergotamine, frovatriptan, ondansetron and tadalafil contain the indole substructure; many of these drugs' target receptors belong to the class of GPCRS (integral membrane G-protein-coupled receptors) and own a preserved building pocket that is recognized by the indole scaffold in a 'common' complementary binding domain.

Till now, several natural products containing the indole ring have been identified, such as the anti-inflammatory disclosed bis-indole compounds, which include topsentins, nortopsentins [10], dragmacidins, and their analogs. A novel indole alkaloid, martefragin A [11] isolated from a red alga, *Martensia fragilis*, is a potent inhibitor for lipid peroxidation. Indololactum V [12], a protein kinase C activator, also has indole moiety. Indole is also present in drugs with remarkable range of activities (for example, indomethacin [13], which is a steroidal anti-inflammatory agent [14–16]).

4.2 Metal- and non-metal-assisted synthesis of five-membered fused N-polyheterocycles

Bismuth-assisted synthesis

The reaction of tetrahydroisoquinoline-3-carboxylic acid with isatin was reported. Compound dipolarophile, reacted with isatin and tetrahydroisoquinoline-3-carboxylic acid in the presence of TiO$_2$-silica catalyst **(Scheme 1–2)** [17].

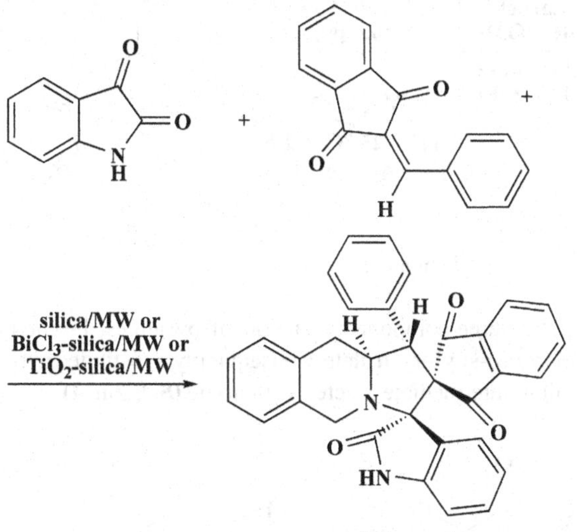

Scheme-1

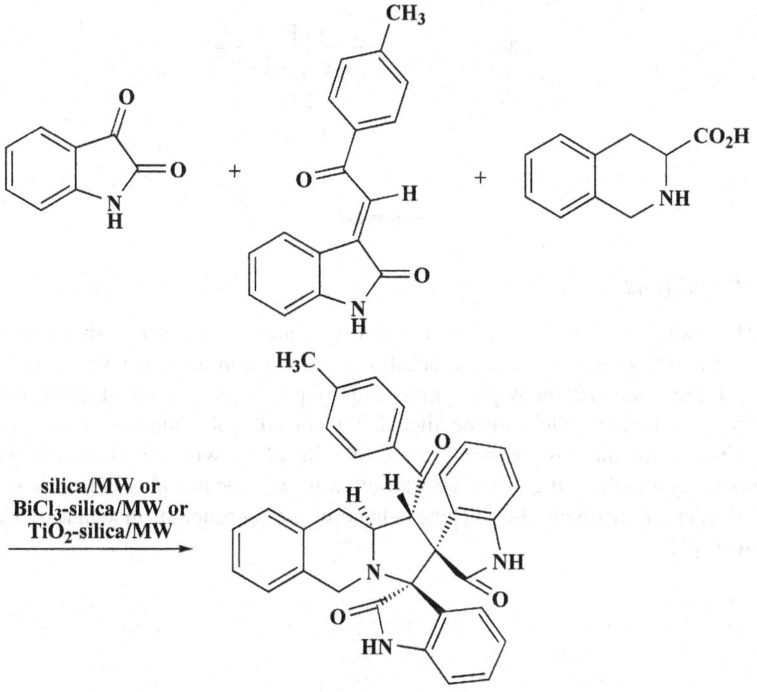

Scheme-2

Cesium-assisted synthesis

In 2009, Rogness and Larock [18] and Giacometti and Ramtohul [19] concomitantly reported reactions in which 1*H*-indole-2-carboxylates were reacted with arynes to afford indoloindolones (**Scheme 3**). A wide substrate scope was reported for this condensation reaction to occur effectively.

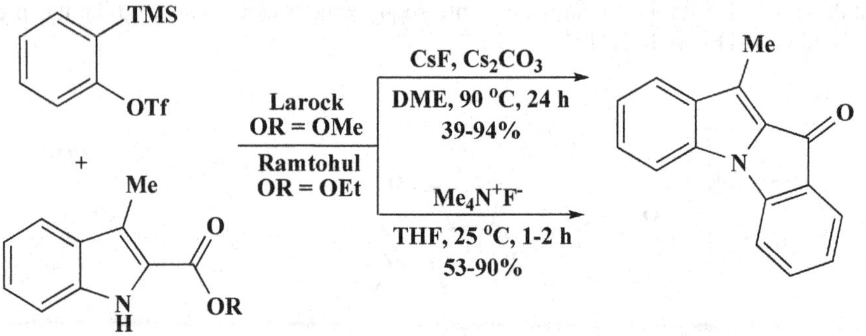

Scheme-3

Xie et al. [20], in 2008, reported a three-component reaction of pyridines, bromoacetophenone derivatives, and arynes produced *in situ* from *o*-silyl aryl triflate. Consequently, substituted pyridoisoindoles were produced when a dipolar pyridinium intermediate reacted with aryne (**Scheme 4**).

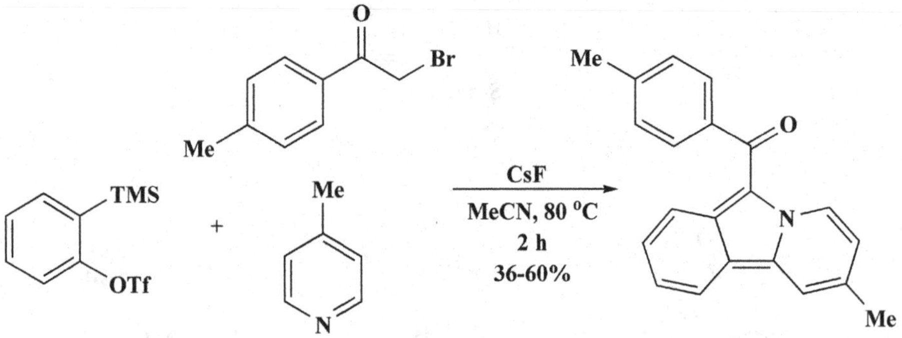

Scheme-4

Cobalt-assisted synthesis

Geny et al. [21] studied the effect of *N*-heterocyclic carbenes on the synthesis of fused polycyclic cyclohexadienes by cyclization of enediyne substrates. Direct comparison with established catalytic systems like CoI$_2$/manganese/triphenylphosphine and [(Cp)Co(CO)$_2$]/ferric chloride highlighted the superior activity of *N*-heterocyclic carbene ligand. Of considerable importance, IPr could be used catalytically with 1 eq. of manganese, while 2 eq. of PPh$_3$ along with 10 eq. of Mn were generally required. In another study, the same group also reported intramolecular [2+2+2]-cycloaddition for the conversion of *N*-tethered enediyne into cyclohexadiene in the presence of manganese, CoI$_2$ and NHC (C-23) (**Scheme 5**) [22].

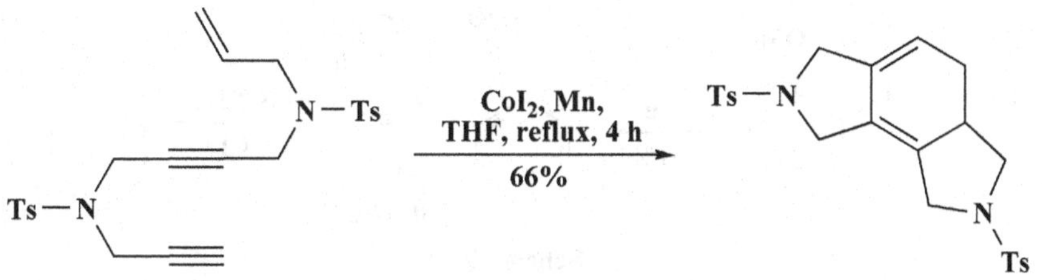

Scheme-5

Milbank et al. [23] utilized a iodo compound for the synthesis of 1*H*-pyrrolo[3,2-*f*]quinoline analogues and reported that they retained the enantiomeric characteristics and high selective cellular potencies of the broad class of CBI-toxins **(Scheme 6)**. The 1*H*-pyrrolo[3,2-*f*]quinolines formed stable cobalt complexes with various ancillary ligands. The cobalt-cyclen complexes were less cytotoxic compared to free effectors and displayed significant hypoxic cell-selective toxicity, representing their use as hypoxia-activated cytotoxins. These complexes also displayed efficient and almost quantitative release of their effectors on exposure to ionizing radiation, supporting the suitability of cobalt-cyclen 1*H*-pyrrolo[3,2-*f*]quinoline complexes for radiolytic release of cytotoxins [22].

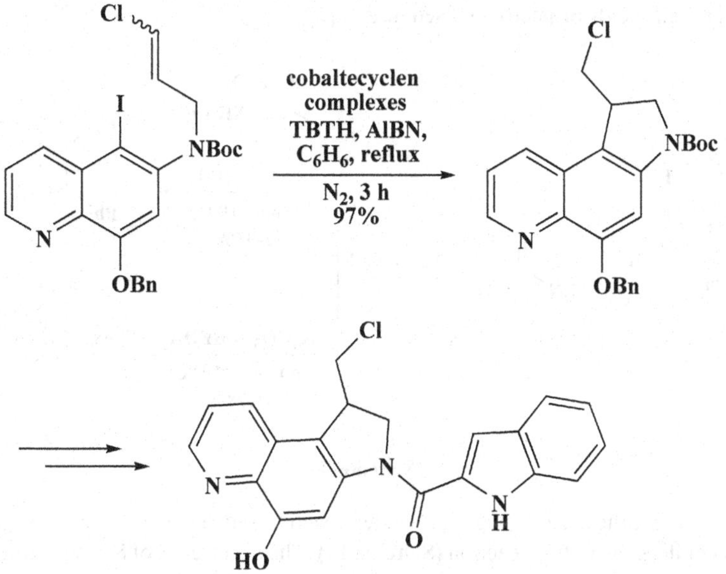

Scheme-6

Copper-assisted synthesis

By these methods, the key intermediate is generated by alternative pathways like Ullmann type reactions. However, these require more expensive starting materials and longer reaction sequences **(Scheme 7–8)** [24–26].

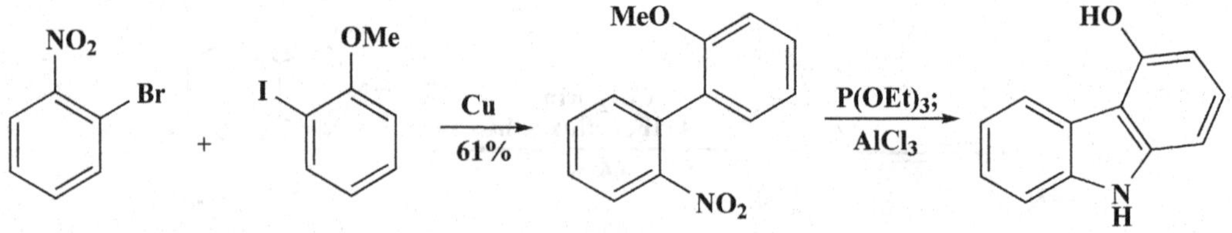

Scheme-7

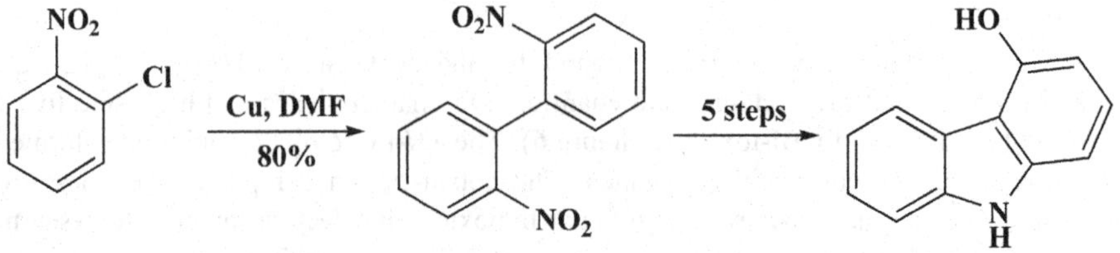

Scheme-8

Yuan et al. [27] and Li et al. [28] reported a cuprous iodide-catalyzed double carbon-nitrogen coupling of primary amides with 1,2-diiodobiphenyls or 1,4-diiodo-1,4-dialkenes to provide the carbazoles and pyrroles in good to fair yields respectively **(Scheme 9)** [29].

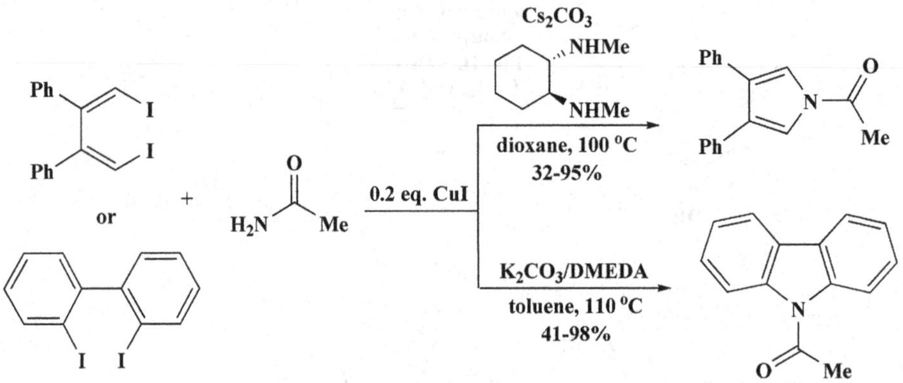

Scheme-9

Indolines can be synthesized through a two-step reaction sequence *via* a domino copper-catalyzed amidation/nucleophilic substitution reaction **(Scheme 10)**. The occurrence of S_N2 substitution is indicated strongly by the clean inversion observed at the mesylate center [30]. This reaction has been found to be compatible with a variety of nitrogen-protecting groups such as Cbz, Boc and acetyl, and effective for the synthesis of 6- and 7-membered rings as well [31].

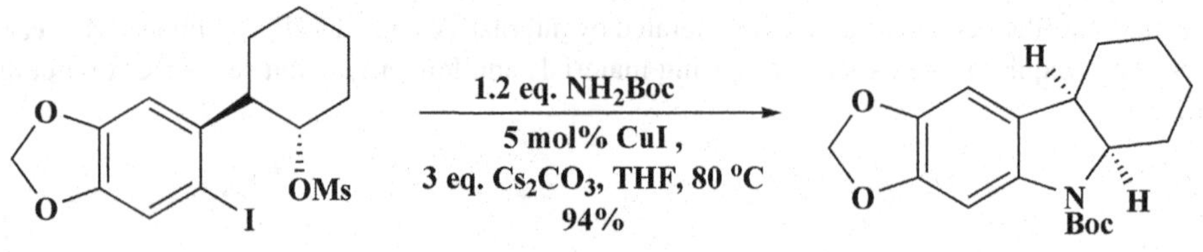

Scheme-10

Duocarmycin SA can be synthesized when the indoline intermediate is transformed into acyclic dehydroamino acid which undergoes cyclization to afford an intermediate employing the aforementioned arylation approach **(Scheme 11)** [32].

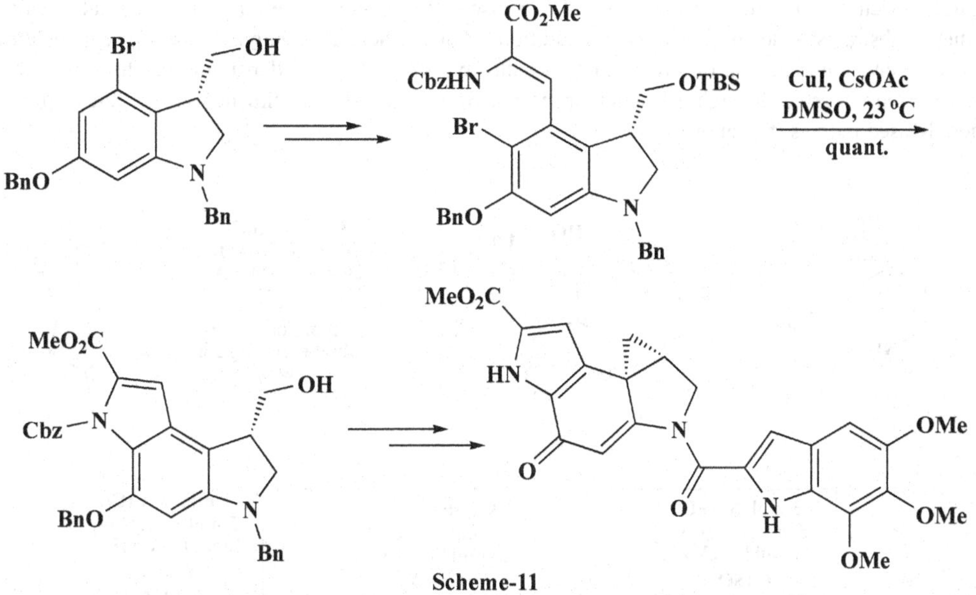

Scheme-11

Indolodioxane can be synthesized enantioselectively using hydrazines as reaction partners for carbon-nitrogen bond formation. This tricyclic compound acts as a potent anti-hypertensive agent among derivatives of non-natural hybrids of three 5-HT1A receptor binding molecules: spiroxatrine, 5-hydroxytryptamine and pindolol. The indole core is constructed from aryl iodide, employing a three-step sequence. An intermediate is formed in 74% yield through the copper-catalyzed formation of *N*-aryl *N*-Boc hydrazide **(Scheme 12)**; this substrate can then be utilized in Fischer's indole synthesis to afford the fully elaborated, conveniently substituted indole derivative [32–33].

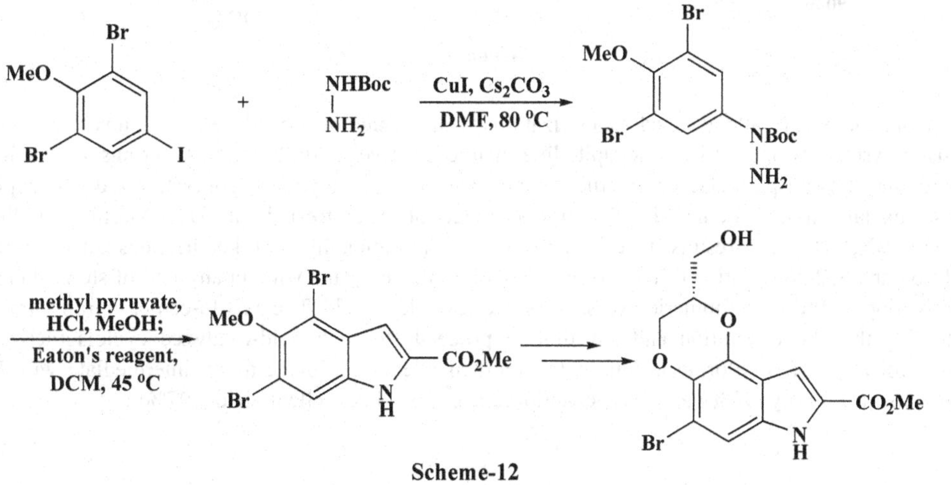

Scheme-12

Gold-assisted synthesis

Naturally occurring motifs like isoxazoles and oxazoles can be synthesized through gold-catalyzed cycloisomerization. The general protocol was established for the preparation of highly functionalized isoxazoles from alkynyl oxime ethers [34], or an intermolecular alkyne oxidation leads to 2,5-disubstituted oxazoles [35]. Nevertheless, while gold-based alkyne-oxygen cycloisomerization has become a hot topic in organic synthesis of late, only a few examples of alkyne-nitrogen coupling have been described [36–43]. Kothandaraman et al. [44] generated highly substituted indole skeletons [45] from easily available 2-tosylamino-phenylprop-1-yn-3-ols. This method showed a versatile approach for the synthesis of these naturally occurring motifs and made way for a fascinating study concerning the chemical reactivity of these substrates under gold-catalyzed conditions. Thus, whereas initially a 5-*exo-dig* cycloaddition afforded vinyl gold species, different reaction pathways were observed depending on the substituents. The reaction occurred through a Friedel-Crafts reaction when aryl substituent was present to provide indenyl-fused indoles (**Scheme 13**) [46].

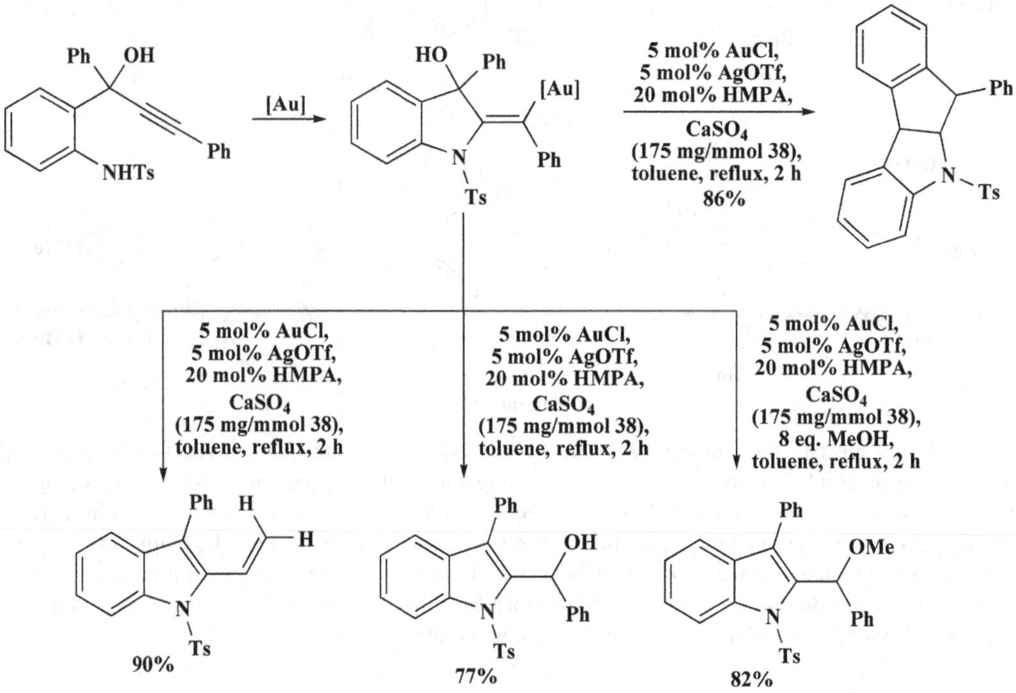

Scheme-13

Gold-catalyzed allene cycloisomerization with enantioenriched allene substrates yields (–)-isochrysotricine and (–)-isocyclocapitelline enantioselectively. Similarly, the A-D rings of azaspiracid containing trioxadispiroketal core structure are prepared employing a gold-catalyzed nucleophilic bis-spiroketalization (**Scheme 14**) [47a–d]. Epoxyalcohol, when treated with Dess-Martin periodinane (or with IBX, 88% yield), results in the formation of a configurationally stable ketone. Subsequent Grignard addition and S_N2-substitution of the tertiary alcohol to allene occurs without any loss of stereochemical information (> 98% enantiomeric excess) and with excellent yield. The axis-to-center chirality transfer following the above addition and substitution proceeds through gold-catalyzed cycloisomerization (0.05 mol% gold(III) chloride in tetrahydrofuran) of allene to provide a key intermediate with high stereochemical purity (96% *de*, > 98% enantiomeric excess) in excellent yields (97%).

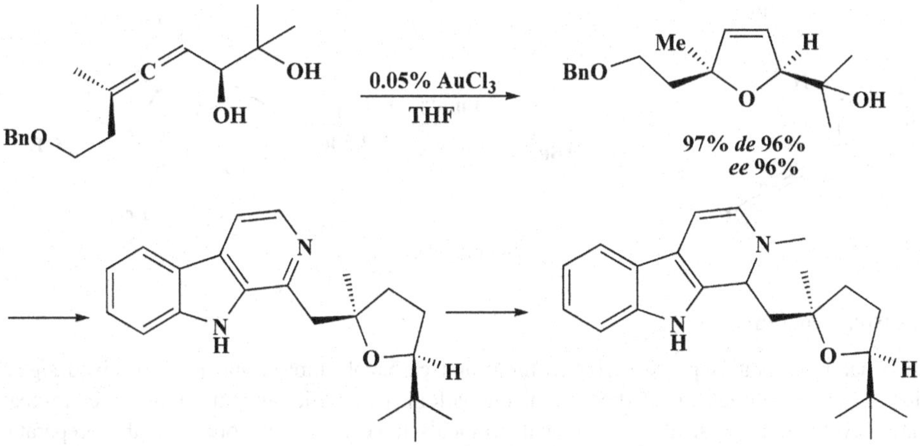

Scheme-14

Metal-catalyzed reaction of *N*-(2-alkynylphenyl)-*β*-lactams provides benzo-fused pyrrolizinones **(Scheme 15)**. Since gold salts have often been found to be less effective, platinum is a metal of choice [48a]. This cycloisomerization is viewed as a net intramolecular insertion of one end of the alkyne into the lactam amide bond with concurrent migration of the substituent at the alkyne terminus. The lactam undergoes an initial *5-endo-dig* cyclization of nitrogen to afford the metal-activated alkyne, followed by the fragmentation of the lactam amide bond and the formation of an acyl cation [48a–l].

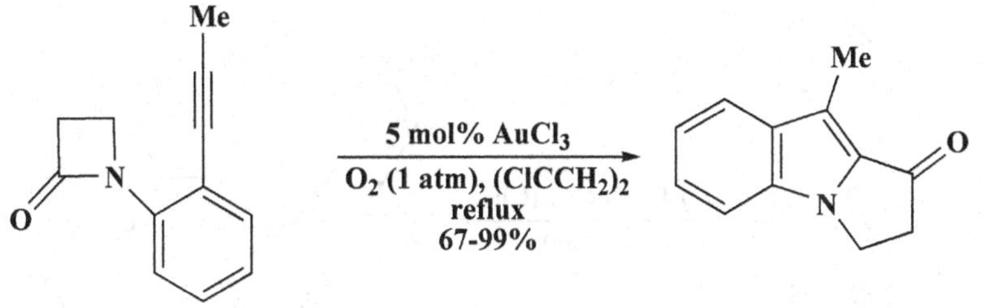

Scheme-15

Kusama et al. [48c] investigated the [3+2]-cycloaddition of *N*-(*o*-alkynylphenyl)imine through a metal possessing azomethine ylide *in situ* with olefins wherein gold(III) bromide and platinum chloride displayed better catalytic reactivity as compared to other metals. Schelwies et al. [48d] reported an intermolecular addition of ketones and aldehydes to 1,6-enynes to afford polycyclic products. The gold-carbene produced *in situ* attacked the carbonyl group to provide the final product. **Scheme 16** shows a useful [3+2]-cycloaddition of metal possessing azomethine ylides synthesized from *N*-(*o*-alkynylphenyl) imines with enol ethers. The formed cycloadducts undergo a 1,2-alkyl migration to afford the final product. A sequential gold(III) bromide-catalyzed cyclization of *N*-(*o*-ethynylphenyl) imines and [3+2]-cycloaddition reactions serve as a highly efficient protocol for the synthesis of tricyclic indole derivatives bearing a substituent at the 3-position of the indole nucleus. The previously reported method for the synthesis of 3-substituted tricyclic indoles involves a stoichiometric amount of W(CO)$_6$ [48e]. Several metal complexes exhibit much higher catalytic activity than the tungsten carbonyl complex [48f–k]; gold(III) bromide and platinum chloride have been found to be the best catalysts in terms of reaction time and product yield respectively [47c, 48l].

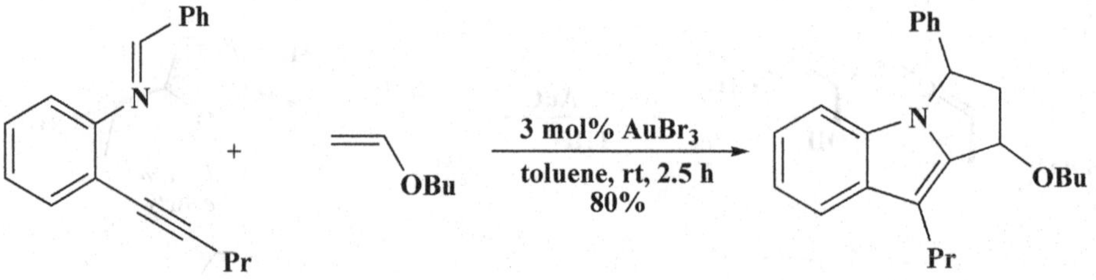

Scheme-16

Iron-assisted synthesis

Bioactive natural products possess dibenzofuran and carbazole motifs and have received significant attention as synthetic targets due to their promising biological activities and intriguing structural features. Till date, several non-classical and classical protocols have been developed for the preparation of dibenzofuran and carbazole frameworks [49–50]. The nucleophilic aromatic substitution of phenyl cation [51–54] is an efficient, straightforward and green pathway to synthesize dibenzofuran and carbazole cores. However, research in this area is still rare; the phenyl cation is unstable, hence limiting its application in organic synthesis. Herein, a $BF_3.OEt_2$-assisted intramolecular amination and oxylation strategy for the direct formation of carbazoles and dibenzofurans from biaryl triazenes under Lewis acid conditions has been described based on the synthesis of highly activated phenyl cation intermediates. Furthermore, this method has been successfully applied for the preparation of carbazole alkaloids like Clausine R, Clausine C, and Clauraila A (**Scheme 17**).

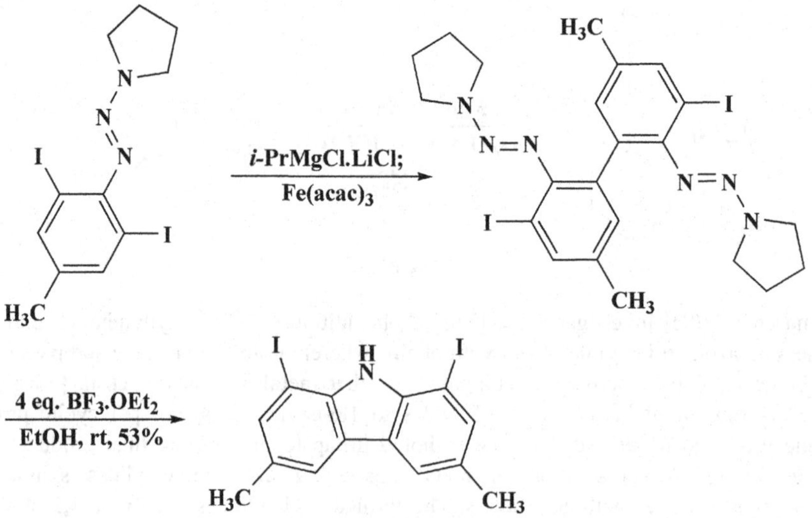

Scheme-17

Parvatkar et al. [55] developed a method for the direct synthesis of pyrrole compound through double reductive cyclization (**Scheme 18**). The condensation of *o*-nitrophenylacetic acid and *o*-nitrobenzaldehyde, followed by esterification provided stilbene. This stilbene was reduced with iron and acetic acid in hydrochloride acid to afford an intermediate in 74% yield. Further, the methylated compound was synthesized in 80% yield by regioselective methylation. For this synthesis, the overall yield was 42% in over 4 steps, the highest yield reported so far.

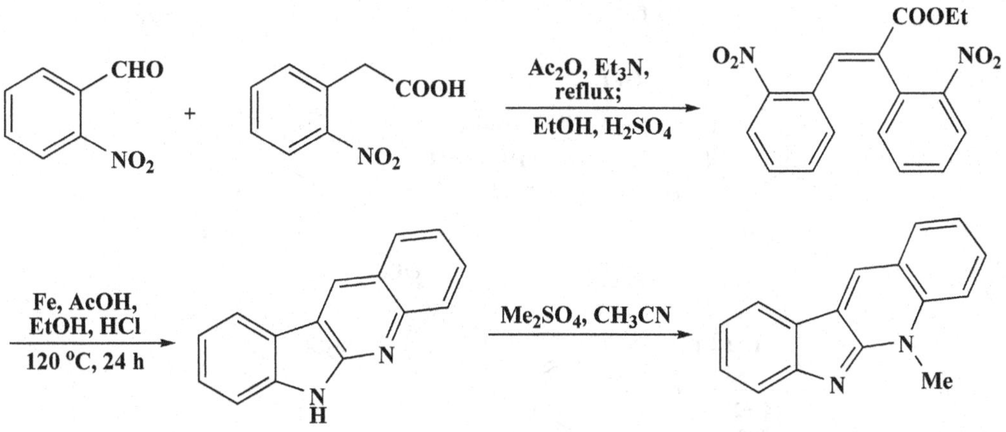

Scheme-18

N-(2-Nitro-3-methylphenyl)phthalimide is prepared when phthalic anhydride and 2-nitro-3-methylaniline are heated in *n*-amyl alcohol. The reaction between *N*-(2-nitro-3-methylphenyl)phthalimide and iron powder at 100 °C in 50 mol% aqueous CH_3COOH yields 6-methyl-11-oxoisoindolo[2,1-*a*] benzimidazole **(Scheme 19)** [56–57].

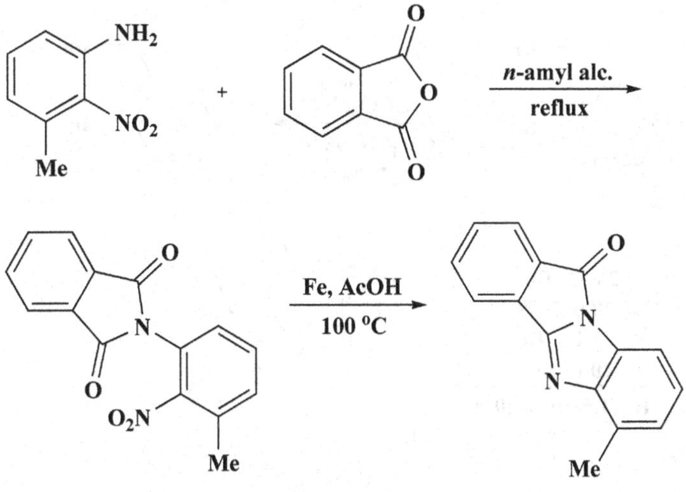

Scheme-19

Lithium-assisted synthesis

Le Strat and Maddaluno [58a] used allenamide for the preparation of 3-(2-ethoxy)vinyl indole **(Scheme 20)**. The major product obtained was *E*-isomer formed through an initial aryl lithium addition onto the central allenic carbon followed by elimination of an OEt group. The *E*-isomer acted as an excellent 1,3-diene undergoing high pressure or thermal [4+2]-cycloaddition with ethyl acrylate to provide tetrahydrocarbazoles [58b].

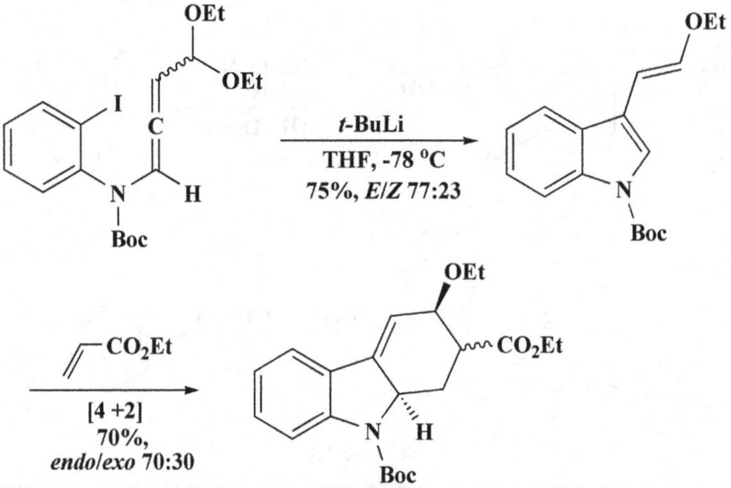

Scheme-20

4a,9a-Substituted 2,3,4,4a,9,9a-hexahydro-1*H*-carbazole serves as a promising amine functionality for chiral *ansa*-ammonium-borates. Literature procedures [59–61] for the synthesis of enantiopure (2*R*,4a*R*,9a*S*)-2-isopropyl-4a-methyl-9a-phenyl-2,3,4,4a,9,9a-hexahydro-1*H*-carbazole are depicted in **Scheme 21**. The standard protocol produces the final *ansa*-aminoborane with 34% yield.

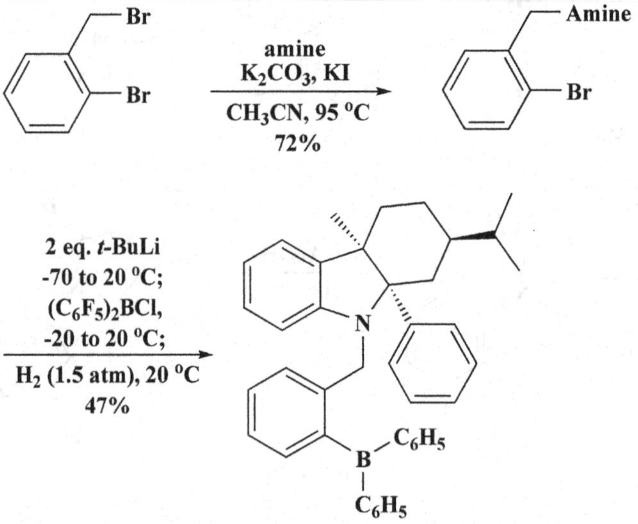

Scheme-21

In case of substrates with conflicting substitution patterns when the directing effect of a DMG overrules that of weaker one or if two groups with comparable directing abilities are present—it leads to decreased regioselectivity in the metalation process. This concept also has many applications in the synthesis of natural products for example, in the synthesis of eupolauramine. Biaryl is prepared by sequential cross-coupling and *ortho*-metalation, with subsequent acid-catalyzed cyclization to afford lactone **(Scheme 22)** [62].

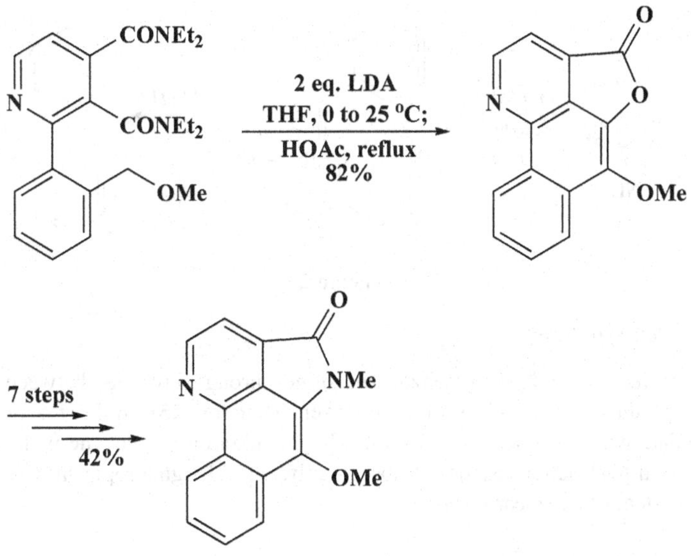

Scheme-22

The reactivity of arynes can be exploited in targeting more complex polycyclic structures with heterocycles at their core to build complexity through multiple C-C bond formations in a single step. Estevez et al. [63–64] reported a protocol for the synthesis of benzisoxindole by a [4+2]-cycloaddition reaction in their preparation of aristolactam (**Scheme 23–24**). The aryne reacted by a formal cycloaddition reaction through enamine-mediated dearomatization of the pendant aryl ring which formed the natural product. More than a decade later, en route to eupolauramine, Hoarau et al. [65] reported a related reaction to synthesize benzisoxindoles containing a fused pyridyl ring.

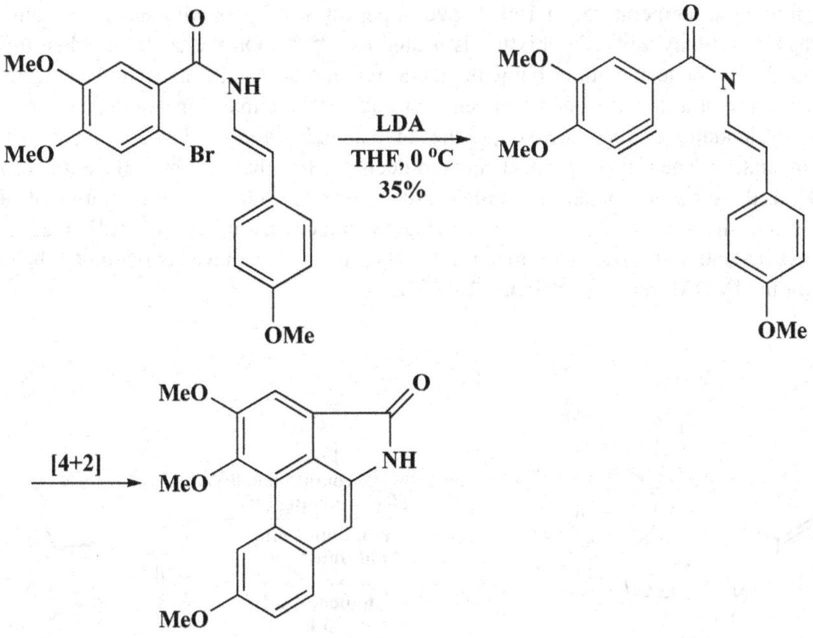

Scheme-23

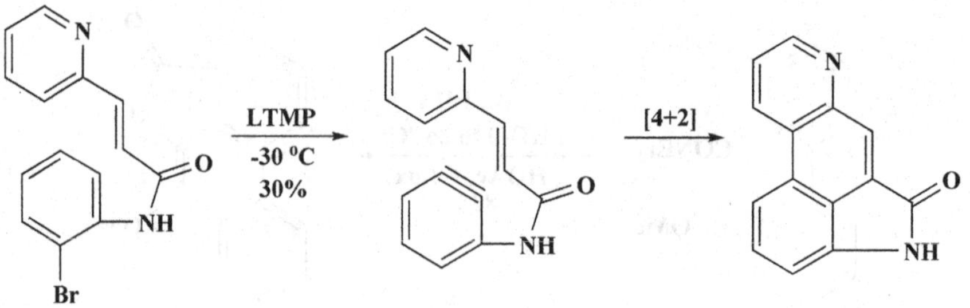

Scheme-24

Magnesium-assisted synthesis

Cant et al. [66] reported the synthesis of benzocarbazolines through a reaction between *N*-methyl pyrrole and aryne (2 eq.) produced from *o*-fluoro-bromobenzene **(Scheme 25)**. In this reaction, an ammonium tricyclic intermediate was generated by initial [4+2]-cycloaddition of benzyne with *N*-methyl pyrrole, followed by a second *N*-arylation reaction. Subsequently, a [3,3]-sigmatropic shift occurred to provide the indoline ring system of benzocarbazoline.

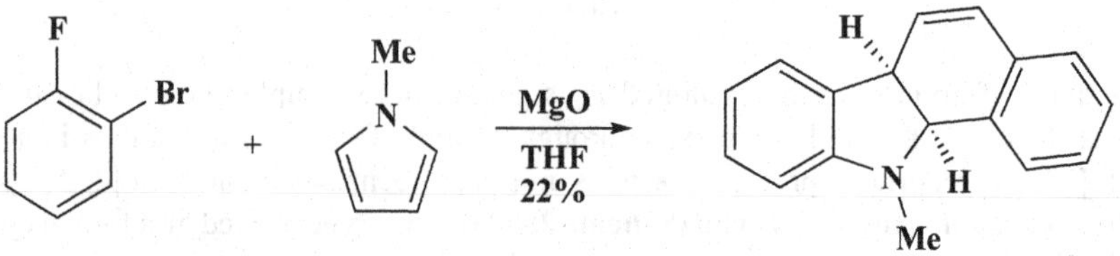

Scheme-25

When the IMDAF reaction with magnesium iodide as the additive is conducted in a microwave reactor for 3 hours at 200 W with varying amounts of magnesium iodide and reaction temperature, no product is formed at temperatures of 150 °C over a period of 3 hours without magnesium iodide. The product was formed only when the mixture is heated to 175 °C. On the contrary, when the reaction is performed under the same conditions but with magnesium iodide, as little as 10 mol% magnesium iodide can drastically promote the cycloaddition reaction and 100% conversion is observed at 150 °C. The Diels-Alder cycloadduct is precipitated out of the solvent readily as a solid by cooling the 0.1 M solution to room temperature. The ratio of product starts to decline when the amount of the additive is increased beyond 50% as the substrate contains multiple coordination sites. A higher concentration of catalyst leads to an altered coordination complex, thereby slowing down the reaction. The 10 mol% magnesium iodide with microwave conditions (200 W) in toluene at 150 °C for 3 hours have been found to be the optimum condition for the IMDAF reaction **(Scheme 26)** [67].

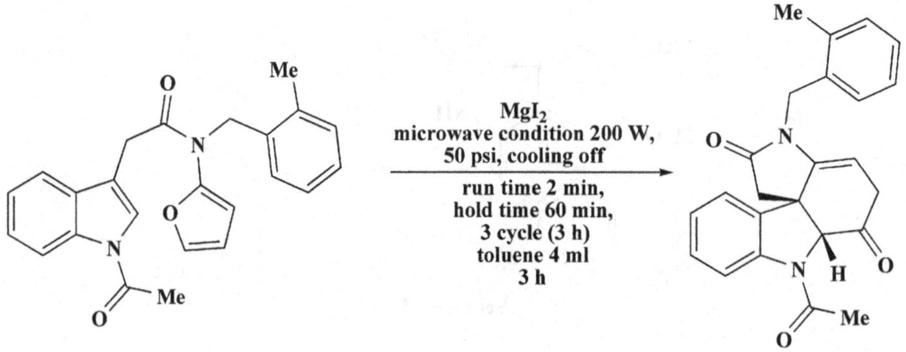

Scheme-26

Montagne et al. [68], Cariou et al. [69] and Mumford et al. [70] reported a reaction using 2-bromopropenylamines for the preparation of 2-methyleneaziridines by base-mediated nucleophilic displacement of the halide **(Scheme 27)**. Consequently, 2-methyleneaziridines have been used for the synthesis of various aza systems [22].

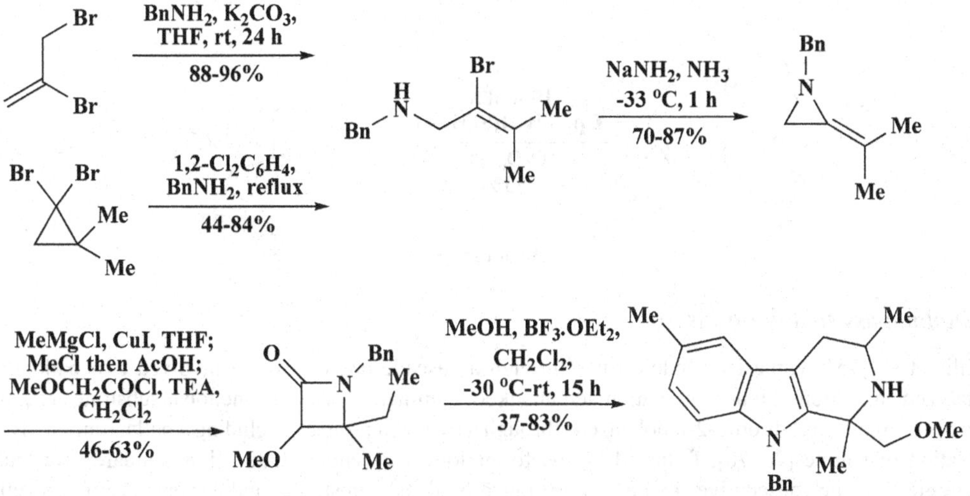

Scheme-27

Manganese-assisted synthesis

5-Stannylated *N*-Boc-protected 2,3-dihydro-1*H*-pyrrole can be prepared by direct lithiation-stannylation of *N*-Boc-pyrroline and is used in Stille cross-coupling with vinyl triflate to afford trienecarbarbamate. This compound is heated to affect an electrocyclic ring-closure and oxidized *in situ* with MnO$_2$ to afford the marine sponge metabolite (±)-*cis*-trikentrin A after Boc-de-protection and aromatization **(Scheme 28)**. (±)-*cis*-trikentrin B is synthesized from a related stannylated pyrroline as the starting compound [71–72].

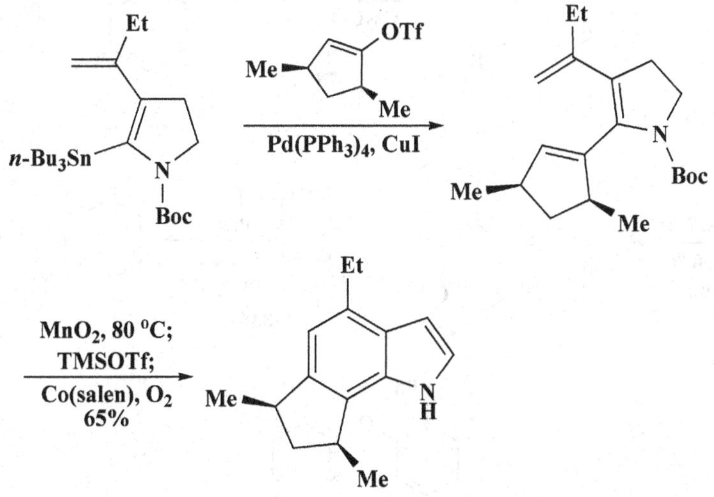

Scheme-28

Neodymium-assisted synthesis

Tricyclic and tetracyclic alkaloidal skeletons can be generated through this method [73]. High diastereoselectivities have been reported in the synthesis of benzo[*a*]quinolizine and pyrido[2,1-*a*] isoindolizine ring systems. Remarkably, selectivity and catalyst activity is not affected significantly by the electron-donating methoxy substitution of the aromatic ring **(Scheme 29)** [74].

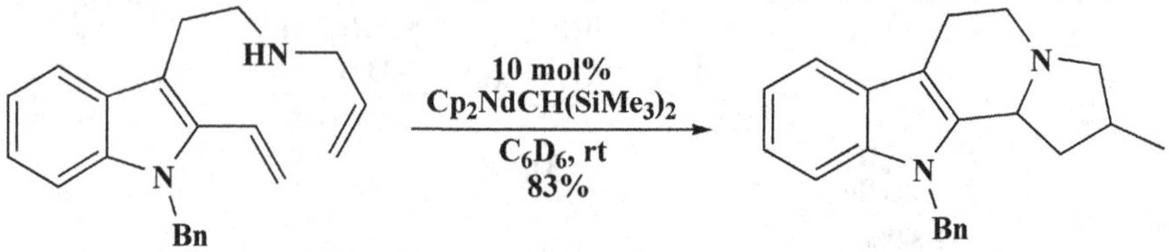

Scheme-29

Palladium-assisted synthesis

Willis et al. [75] synthesized indole rings through a cascade process which involved two palladium-catalyzed aminations, an aryl amination and an alkenyl amination. The bi-functional substrates required were prepared from 1-bromo-2-iodobenzene through a two-step protocol including a palladium-catalyzed arylation of ketone [76–78], followed by the formation of an enol triflate using standard procedures **(Scheme 30)**. The consecutive amination reactions provided *N*-substituted indoles under proper reaction conditions. This reaction is very general with respect to *N*-substituents: aromatic and aliphatic amines, carbamates, amides, hydrazones and sulfonamides can be successfully introduced. The required bis-activated carbon skeletons were prepared conveniently from *o*-dihaloarenes and ketones through a two-step synthesis. Subsequently, to address some limitations of this method such as difficult access to the substrates required for the preparation of 2- or 3-monosubstituted structures and the two-step synthesis required to prepare the alkenyl triflate substrates, the same group developed an improved approach to *N*-substituted indoles [79]. Because triflate to halide substitution provides more robust substrates, this

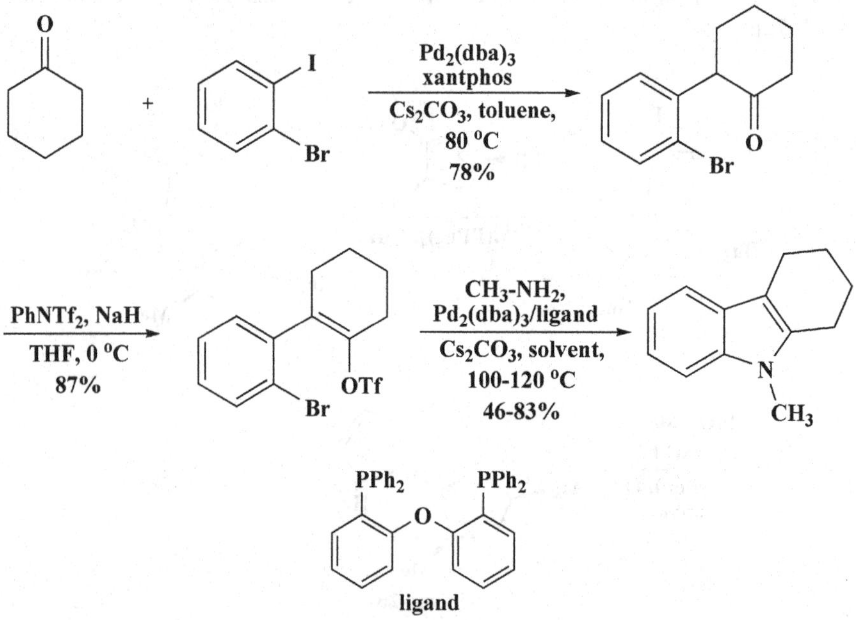

Scheme-30

was based on the reaction of *o*-(2-haloalkenyl)aryl halides conveniently prepared in a single step from the corresponding *o*-halobenzaldehydes with amines. All combinations of Cl and Br leaving groups were employed, and a variety of substituents on the alkene, arene, and amine were tolerated. This protocol has also been employed for the synthesis of indoles containing sterically demanding *N*-substituents [80], including the demethylasterriquinone A1. Both geometrical isomers of the substrates can be transformed into indole products as well [81–82].

Enaminones coupled with *o*-dibromobenzene in the presence of Cs_2CO_3, $Pd_2(dba)_3$ and a phosphine ligand yield *N*-aryl-enaminone (in the first step) which is cyclized by further addition of ligand and catalyst [83]. Chen et al. [84] also synthesized 1,2,3,4-tetrahydro-4-oxo-β-carbolines using bromoenaminones, whereas Yamazaki et al. [85] employed this reaction on immobilized enaminoesters for solid-phase synthesis. Since then, a number of structurally diverse *N*-(*o*-haloaryl)-enamines conjugated with ester and ketone groups (derived from 1,3-dicarbonyls, palladium-catalyzed oxidative amination of electron deficient olefins with *o*-haloanilines [86], Michael addition of *o*-haloanilines to ethynyl ketones and esters, and Buchwald/Hartwig amination) have been shown to function well in this application of the Heck reaction [87–90]. Such domino reactions in which *o*-haloenamines conjugated to carbonyl groups formed by Buchwald/Hartwig palladium-catalyzed carbon-nitrogen bond formation from vinylogous amides and vinylogous acyl chlorides or *o*-dibromobenzenes and *o*-haloanilines [91] underwent an *in situ* Heck cyclization have also been reported. The reaction can be performed as a domino process omitting the isolation of *N*-(*o*-bromophenyl)-α-phosphoryloxyenecarbamate intermediates (**Scheme 31**) [81, 92–93].

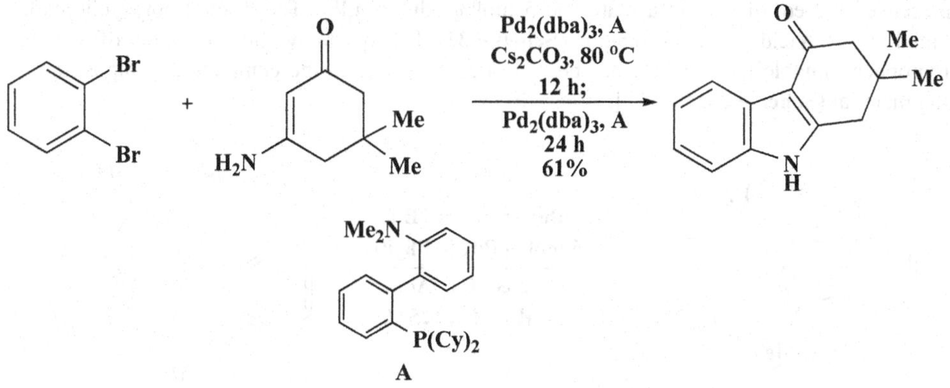

Scheme-31

The last reaction sequence has also been extended to the synthesis of pyrrolizidine, indoline, and indolizidine with the tolerance of aryl halides in titanium-catalyzed hydroamination step [94]. *O*-chloro-substituted 2-alkyl-1-arylalkynes are synthesized readily by Sonogashira coupling of 1-chloro-2-iodoarenes. Intermolecular hydroamination of *o*-chloro-substituted 2-alkyl-1-arylalkynes provides arylethylamines (after reduction with $NaBH_3CN$/zinc chloride) with high anti-Markovnikov regioselectivity in the case of sterically demanding amines but diminished regioselectivity for benzylamines and *n*-alkylamines. Indulines are formed in high yields by subsequent Buchwald-Hartwig coupling. This protocol, when varied for intramolecular hydroamination of *o*-halide-substituted aminoalkyl(phenyl)alkynes, provides cyclic arylethylamines (after reduction) which are readily cyclized by palladium-catalyzed cross-coupling to indolizidine and pyrrolizidine (**Scheme 32**). The intramolecular hydroamination proceeds readily with high regioselectivity without sterically demanding amines [74, 95].

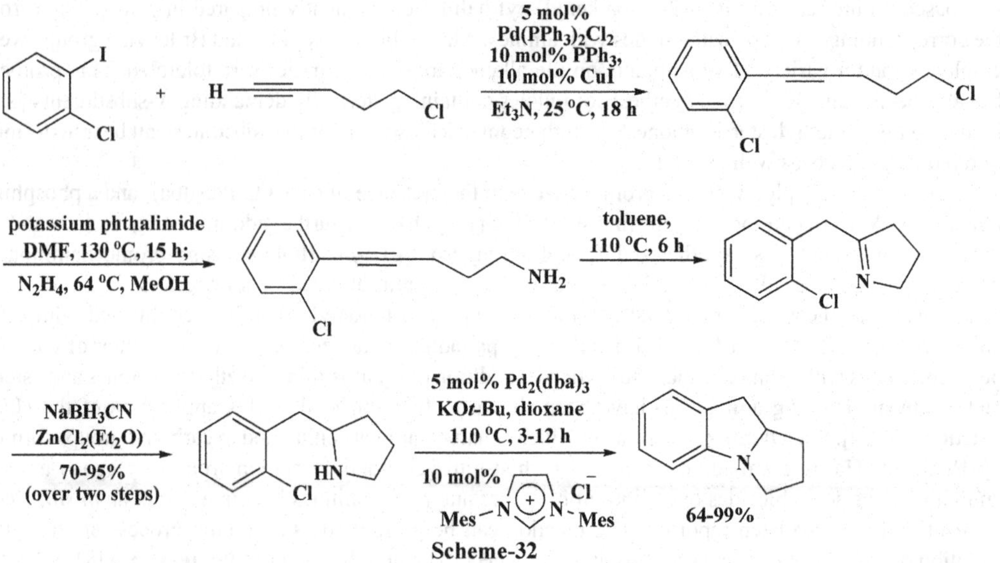

Scheme-32

N-Methyl-2-(*o*-bromophenyl)indole undergoes cascade annulations/cyclization with 5-decyne in the presence of 2 eq. of cesium acetate and 5 mol% PdCl₂(Ph₂P)₂. The desired polycyclic carbazole is formed in 90% yield in this cyclization **(Scheme 33)**. Unexpectedly, the cyclization of *N*-methyl-3-(*o*-bromophenyl)indole to tetracycle has been unsuccessful due to the complete decomposition of the starting material **(Scheme 34)** [96–97].

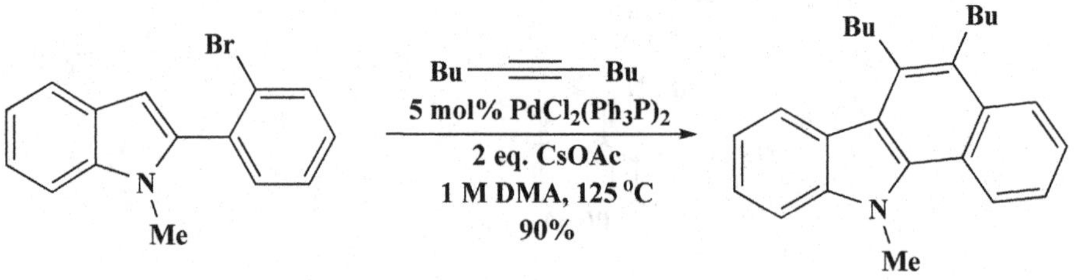

Scheme-33

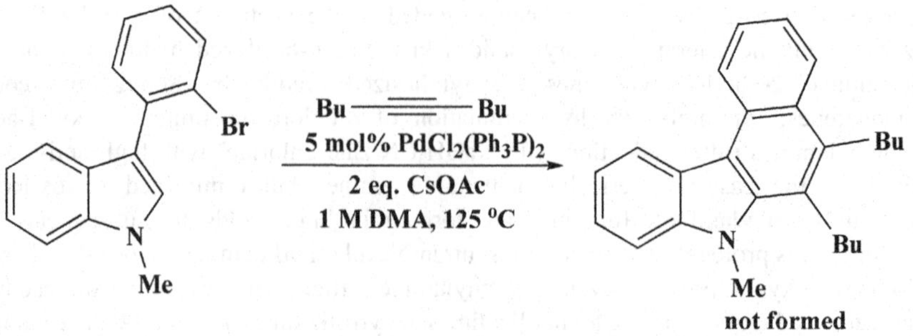

Scheme-34

3-Benzoyl derivatives undergoes arylation/cyclization with 5-decyne. However, in this case, instead of the targeted seven-membered heterocyclic compound, tetracyclic indenone product is formed in 80% yield by intramolecular arylation **(Scheme 35)** [97–98].

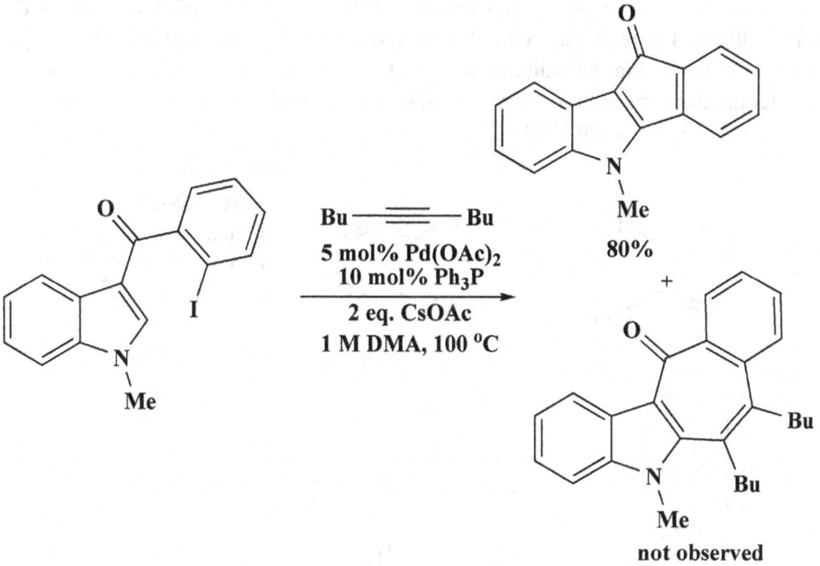

Scheme-35

Roesch et al. [99–100] studied the synthesis of isoindole[2,1-*a*]-indoles *via* annulation of internal alkynes by imines derived from *o*-iodoanilines (**Scheme 36**), as an extension of palladium-catalyzed annulation of *o*-iodoanilines and *o*-iodoanilides with internal alkynes to indoles. Indoles were formed by the cyclization of *o*-halo *N*-allylanilines when treated with tetrabutylammonium chloride, Pd(OAc)$_2$ and a base in dimethylformamide [101]. The same cyclization also occurred with *o*-halo-*N*-vinylanilines [102–103]. In another such study, Larock and Yum [104] reported the reaction of imines with 2 eq. of internal alkynes, expecting the formation of quinolines. However, quinolines were not formed, but isoindoloindoles could be isolated in good yields (50–80%) under the conditions used in the earlier annulation reaction. Internal alkynes possessing either a heterocyclic ring or a phenyl have been used in this annulation reaction [81, 93].

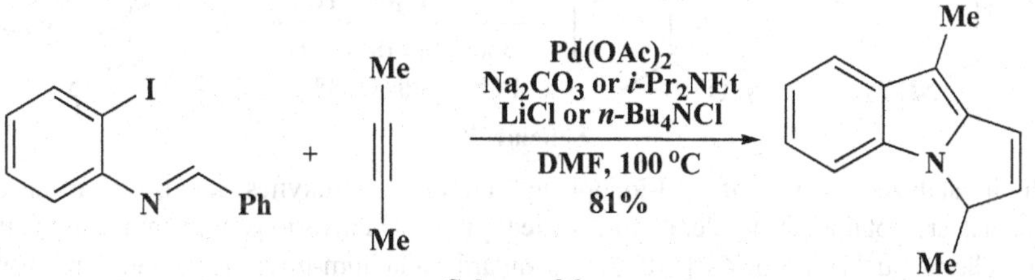

Scheme-36

Condensed nitrogen-containing heterocycles can be synthesized by sp^3 carbon-hydrogen bond activation in the presence of *p*-Tol$_3$P (10 mol%), palladium acetate (5 mol%) and cesium carbonate (1.2 eq.). The 2,5-unsymmetrically substituted monobromo derivatives undergo chemoselective sp^3 carbon-hydrogen bond activation (**Scheme 37**) [105].

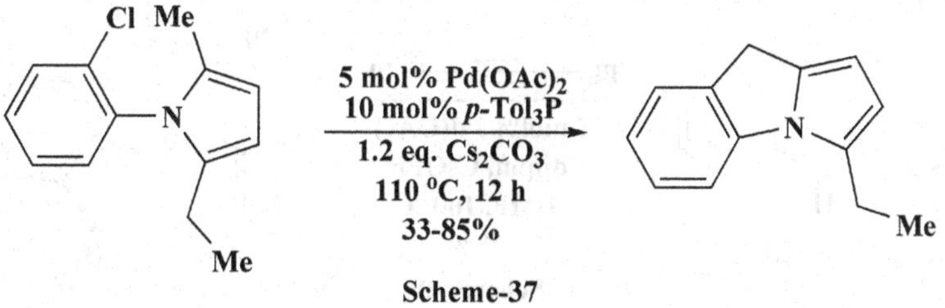

Scheme-37

Watanabe et al. [106–107] synthesized carbazoles through one-pot palladium-catalyzed reaction of anilines and aryl triflates. The reaction occurred *via* Buchwald-Hartwig amination to afford diphenylamine which underwent intramolecular palladium-catalyzed CH-CH coupling to form carbazole under aerobic atmosphere. Later on, the same authors employed this strategy for the preparation of Clausine L, a natural anti-platelet carbazole **(Scheme 38)** [108].

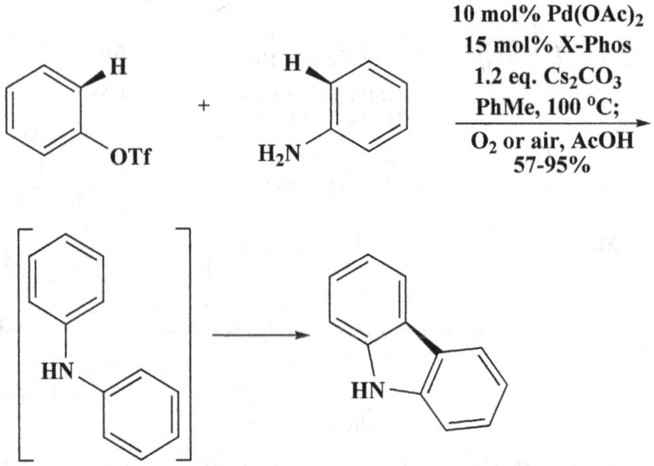

Scheme-38

Liu and Larock [109] synthesized carbazoles by a one-pot two-step reaction. The first step involved the addition of *o*-iodoaniline to a benzyne intermediate produced *in situ* from silylaryl triflates using cesium fluoride. The *N*-arylated *o*-iodoaniline thus formed underwent an intramolecular direct arylation in the presence of palladium catalyst to provide the desired carbazole. Various NH and *N*-substituted carbazoles were synthesized in good yields using this protocol **(Scheme 39)**. In addition, dibenzofurans and nitrogen-containing six-membered rings were also generated from phenol and benzylamines derivatives respectively.

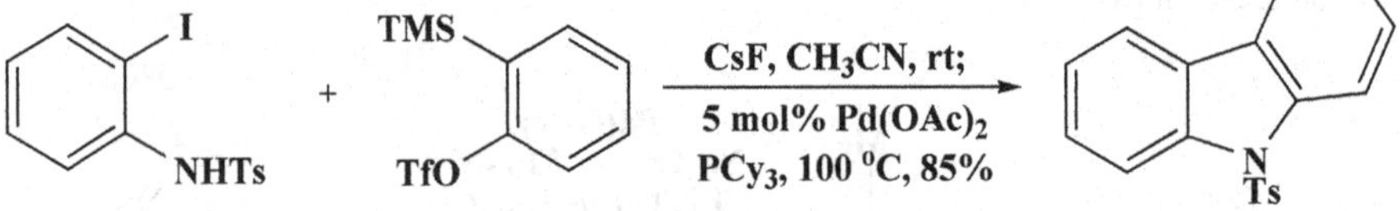

Scheme-39

Palladium-catalyzed reaction of *N*-(3-iodophenyl) anilines and alkynes can afford various carbazoles. These carbazoles are obtained in moderate to excellent yields when various *N*-phenyl-3-iodoanilines react with internal alkynes using Larock's [110–111] standard palladium-migration conditions **(Scheme 40)**. Moreover, various internal alkynes such as alkyl-, aryl-, dialkyl- and diaryl-substituted alkynes can be tolerated herein. The reaction tolerated both electron-donating and electron-withdrawing substituents on the aromatic ring. No substituent was crucial on the aniline nitrogen since the phenyl- and methyl-substituted amines formed none of the anticipated carbazole products. The reaction occurred by carbopalladation of alkyne, followed by nitrogen-directed vinyl to aryl palladium migration and direct arylation.

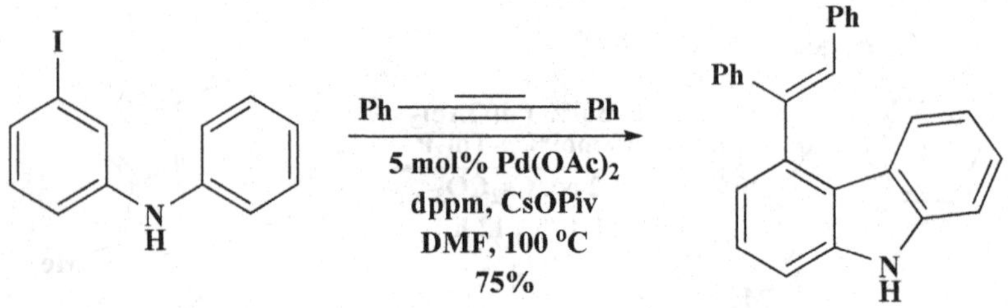

Scheme-40

Huang et al. [112] reported a reaction wherein 3-iodo-1-*p*-tosylindole was treated with norbornene in the presence of palladium acetate/1,2-bis(diphenylphosphino)ethane as catalyst **(Scheme 41)**. Followed by oxidative addition, the indol-3-ylpalladium intermediate was subjected to *syn*-addition on to norbornene to provide an alkylpalladium species. Since α-hydride elimination was not favored in this intermediate, the palladium was forced to undergo1,4-palladium migration to the 2-position of indole. Subsequently, the indol-2-ylpalladium species thus formed underwent intramolecular cyclization onto the tosyl group to afford the fused ring system.

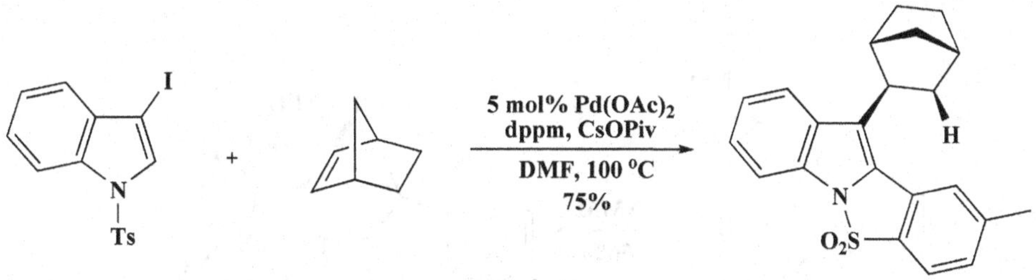

Scheme-41

Palladium-catalyzed tandem 1,4-migration/direct arylation of 3-arylindole was also investigated by Campo et al. [113] and Huang et al. [114]. The reaction occurred *via* initial oxidative addition of palladium(0) to the aryl iodide, followed by 1,4-migration of palladium to afford an indol-2-ylpalladium species. The fused tetracyclic product was formed when indol-2-ylpalladium species underwent direct arylation with *N*-benzyl group **(Scheme 42)**. Owing to relative ease of carbon-hydrogen activation for the electron-rich indole, shorter reaction times and higher yields compared to other arene systems were observed.

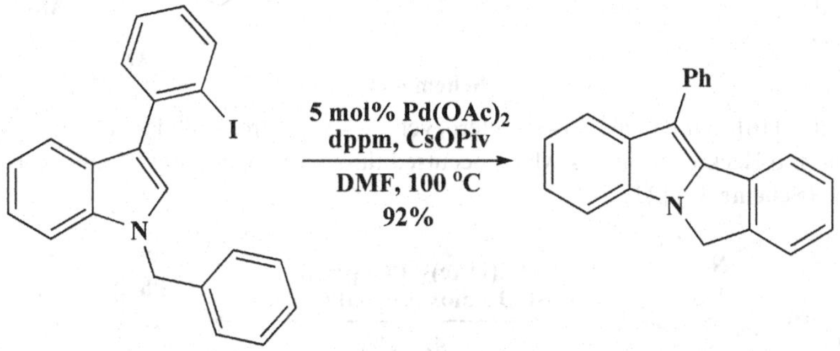

Scheme-42

Grigg et al. [115] developed a poly-component cascade involving various stereo- and regioselective 5- or 6-*exo-dig* cyclization, and direct arylation **(Scheme 43–44)**. The propargyl amides underwent a thallium carbonate-assisted isomerization to afford allenamides. An unstable enaminoindole was formed by subsequent oxidative addition and carbopalladation/cyclization onto the central allenic carbon. The enaminoindoles were trapped through Diels-Alder cycloadditions with *N*-methyl-maleimide (NMM) to provide *endo*-cycloadducts in refluxing acetonitrile. Three examples from propargyl amides to polycyclic products were reported without isolation of the diene intermediate.

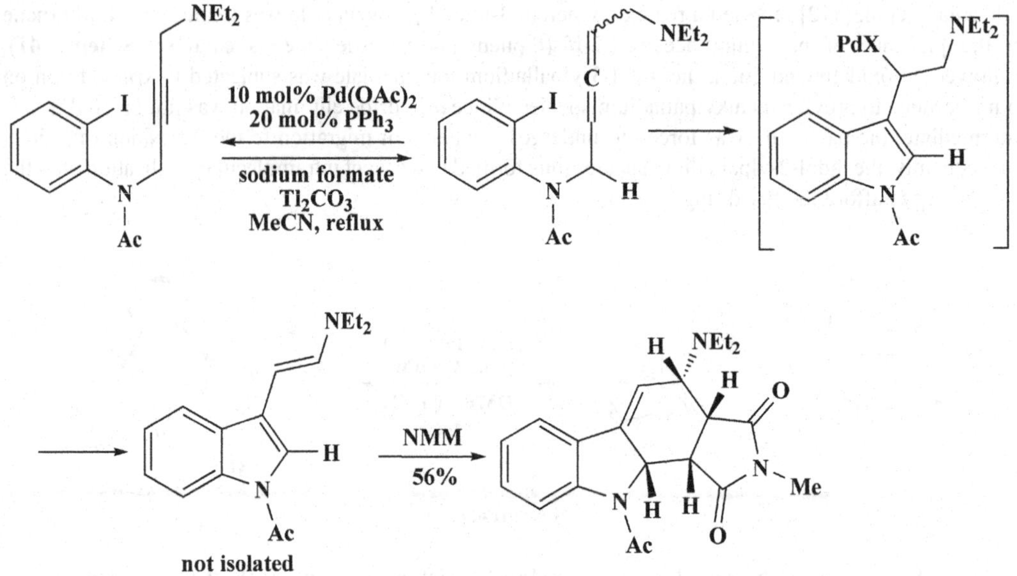

Scheme-43

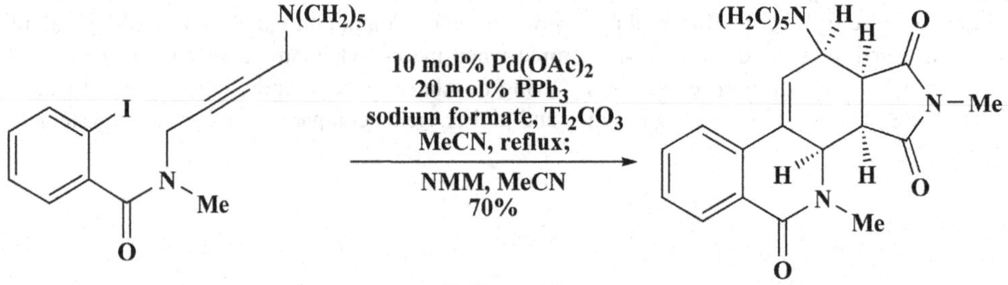

Scheme-44

Ohno et al. [116] synthesized fused heterocyclic systems from allylamines by two sequential palladium-catalyzed Heck cyclizations which occurred through 'zipper-mode' double carbon-hydrogen bond activation (**Scheme 45**) [22].

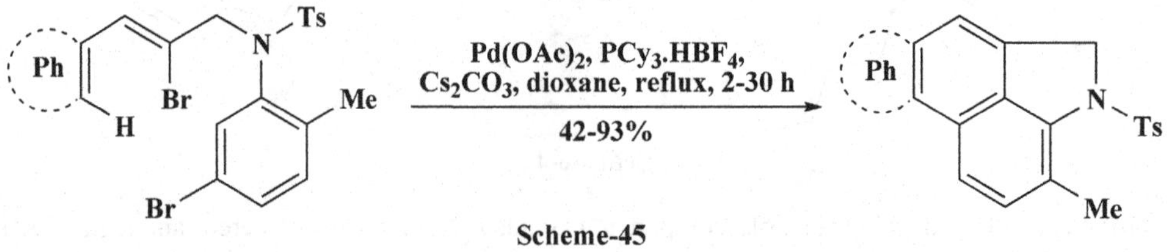

Scheme-45

Beccalli et al. [117] also prepared ethyl-3,4-dimethylpyrrolo[3,2-*b*]indole-1(4*H*)-carboxylate besides various heteropolycyclic systems under heating or microwave irradiation through intramolecular palladium-catalyzed coupling of vinyl bromide onto the 2-position of indole (**Scheme 46**) [22].

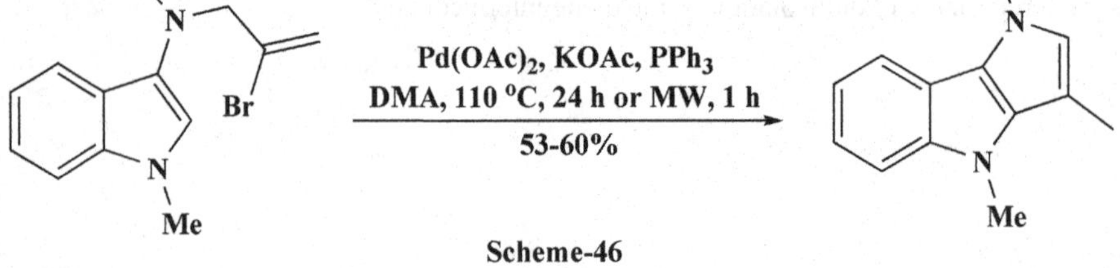

Scheme-46

Ohta et al. [118] studied the synthesis of polycyclic indole frameworks, wherein 2-(aminomethyl) indoles were produced *via* copper-catalyzed domino three-component coupling cyclization of 2-ethynylanilines with paraformaldehyde and *N*-butyl allylamine. A palladium-catalyzed carbon-hydrogen functionalization of the C-3 position of 2-(aminomethyl)indoles resulted in the formation of the required products **(Scheme 47)** [22].

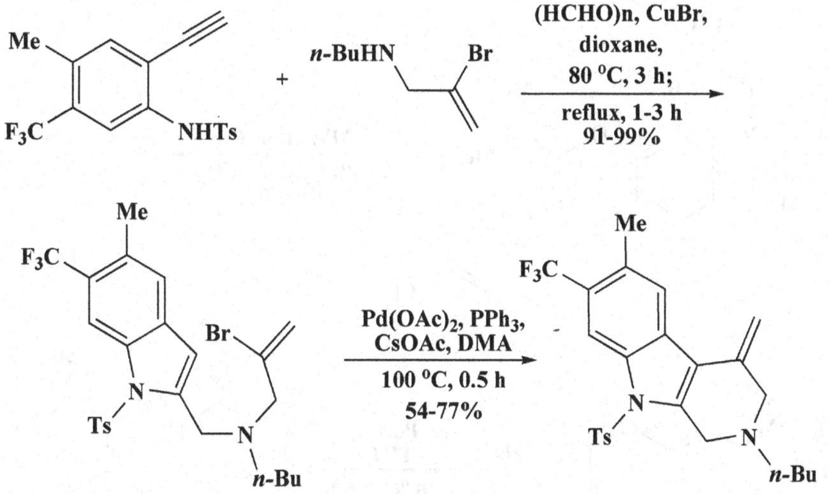

Scheme-47

The alkyl-substituted substrate under intramolecular Heck reaction conditions yields 1,2,3-triazole-fused isoindoline in 70% yield. Palladium acetate has been found to be the best catalyst as revealed by re-optimization. Furthermore, increased reaction yields of 87% have been observed when the reaction time was enhanced to 27 hours. Other iodotriazole substrates also provide similar yields **(Scheme 48)** [119–121].

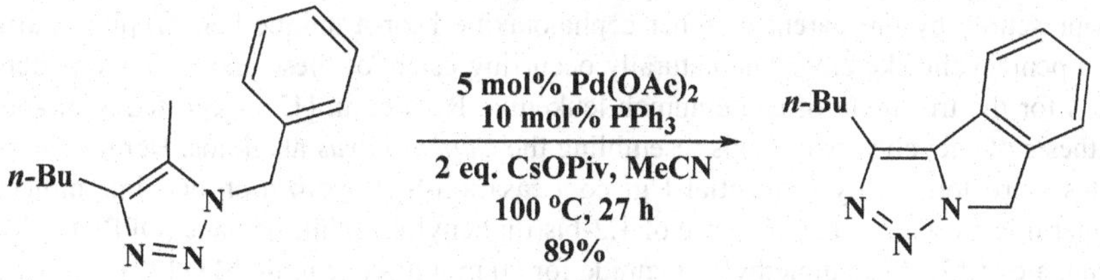

Scheme-48

(−)-Physostigmine, a powerful inhibitor of acetylcoline-esterase, can be prepared by asymmetric Heck reaction (95% enantiomeric excess) **(Scheme 49)** [122–123].

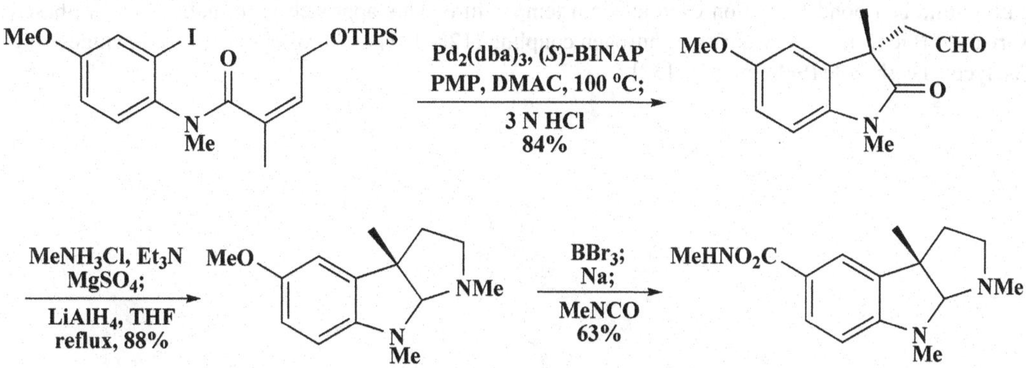

Scheme-49

Pyrano[2,3-*e*]indol-4(7*H*)-one is generated either by annelation of a *γ*-pyrone ring onto indole or by annelation of a pyrrole ring onto chromone. 7-Aminochromone affords the starting compound by the action of NBS followed by treatment with benzyl alcohol and phosgene and then allyl bromide. The chromone fragment is consequently annelated with pyrrole ring upon microwave treatment in the presence of tetrakis(triphenylphosphine)palladium *via* cross-coupling (**Scheme 50**) [124–125].

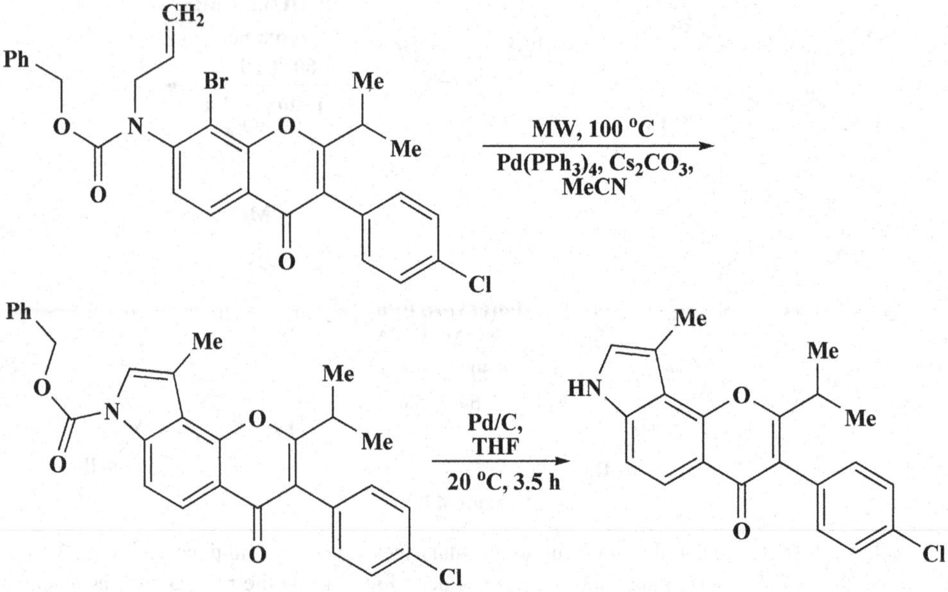

Scheme-50

Benzazepine is present in many naturally occurring compounds. For instance, the rare *Cephalotaxus* alkaloids represented by the parent member cephalotaxine bear a unique benzazepine scaffold which is fused to a pentacyclic skeleton. The naturally occurring esters of these alkaloids are under advanced clinical trials for the treatment of acute human leukemia. Riva et al. [126] reported a one-pot protocol for the synthesis of tricyclic compounds resembling the *Cephalotaxus* alkaloids. Herein, the pyrrolidine intermediates were formed by sequential Ugi condensation-S_N2 cyclization and intramolecular Heck reaction under microwaves in the presence of 1,2-bis(diphenylphosphino)ethane (DPPE), Pd(PPh$_3$)$_4$ and cesium carbonate at 120 °C in dimethylformamide for 60 minutes (**Scheme 51**) [127].

Various 2-aminobiphenyls undergo catalytic intramolecular aryl-nitrogen bond formation for the synthesis of carbazole. This protocol is again based on regioselective carbon-hydrogen bond activation by Pd(II), where aniline group is used as a tether to afford a trinuclear Pd(II) complex. The metal is oxidized selectively with PhI(OAc)$_2$ to yield a palladium(IV) intermediate readily, which induces intramolecular carbon-nitrogen bond formation even at room temperature. This approach is formally a Pd(IV)-based variant of Buchwald-Hartwig carbon-nitrogen coupling [128–131] and proves effective for synthesizing *N*-glycosyl carbazole (**Scheme 52**) [132].

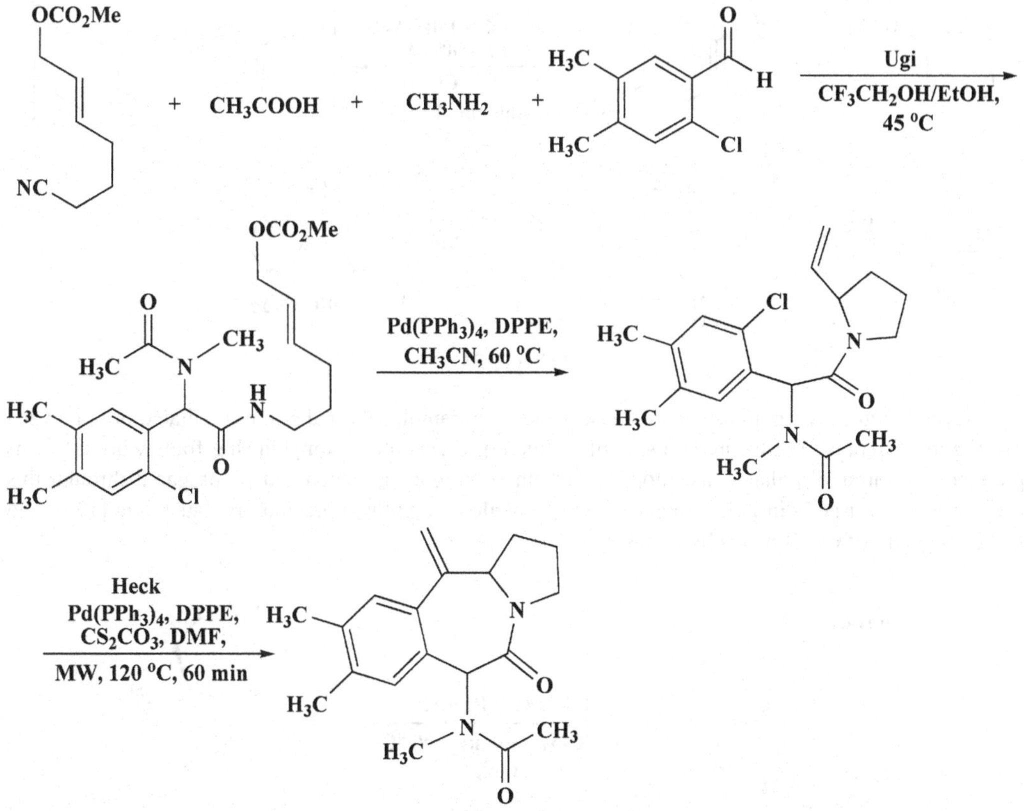

Scheme-51

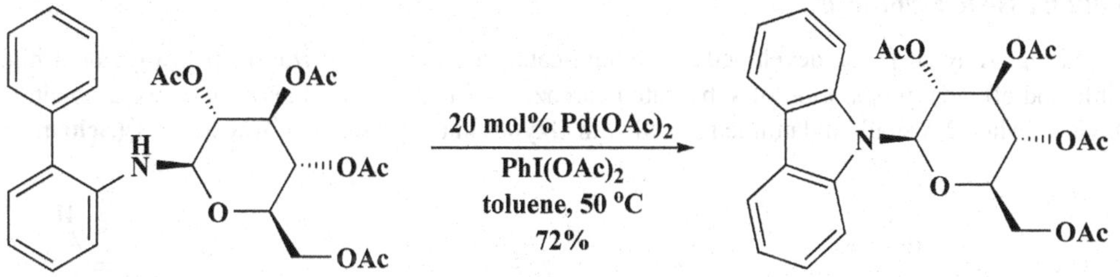

Scheme-52

Palladium-catalyzed carbon-heteroatom formation can be coupled with direct, oxidative coupling of two carbon-hydrogen bonds as well. Ohno et al. [106] reported this reaction for the synthesis of carbazoles from aniline and aryl-triflates (**Scheme 53**). This consecutive carbon-nitrogen and carbon–carbon bond formation reaction occur with a single palladium catalyst produced from palladium acetate where C-N coupling is carried out under an inert atmosphere, followed by the incorporation of acetic acid and oxygen to mediate the subsequent carbon-hydrogen activation/cyclization. This reaction can tolerate various substituents, with product selectivity favoring activation of the less sterically hindered carbon-hydrogen bond. Ryu et al. [133] reported a related approach for the preparation of benzothiophenes.

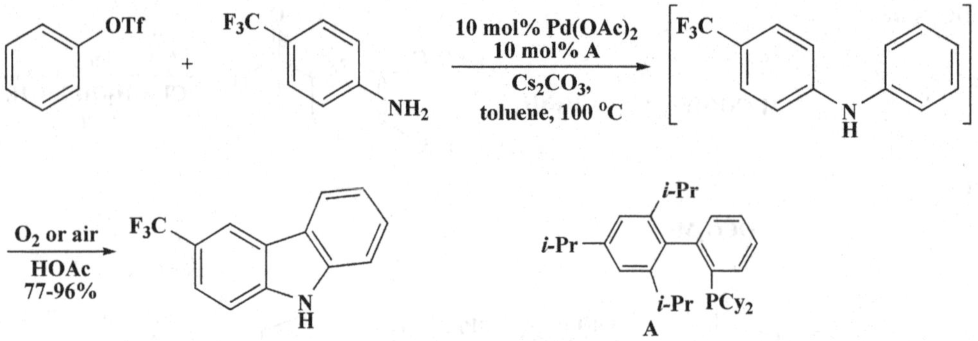

Scheme-53

Several natural compounds can be constructed by intramolecular palladium-mediated or -catalyzed *N*-arylation reactions. For example, pyrroloindoline represents the mitomycin ring framework which is produced by intramolecular *N*-arylation [134]. Other compounds which can be prepared through this reaction include asperlicin [135], the cryptocarya alkaloids cryptowoline and cryptaustoline [136], and the CPI subunit of CC-1065 (**Scheme 54**) [137].

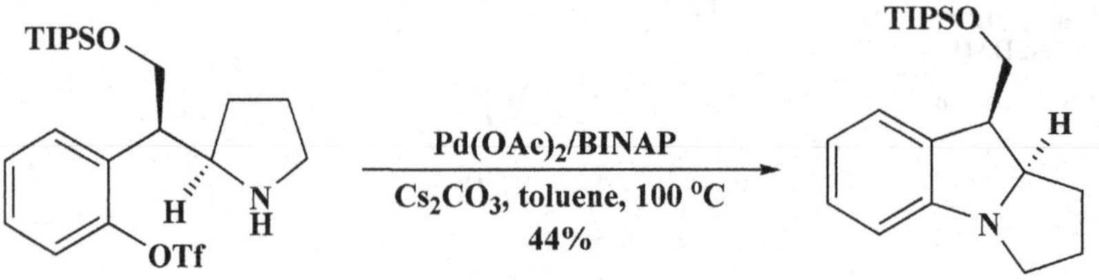

Scheme-54

Rhodium-assisted synthesis

Witulski and Alayrac [138] developed a rhodium-catalyzed cyclotrimerization pathway for a highly flexible and efficient preparation of substituted carbazoles. Functionalized carbazoles were obtained in high yields when 2,*N*-dialkynylaniline reacted with alkynes in the presence of $RhCl(PPh_3)_3$ (**Scheme 55**).

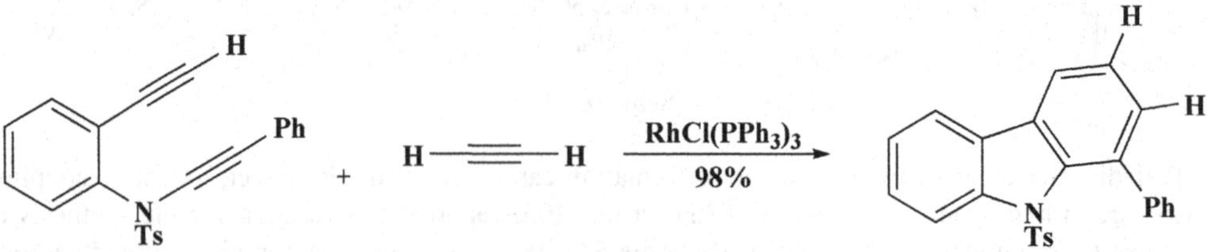

Scheme-55

The intramolecular tandem reaction is exemplified by rhodium(II)-catalyzed reaction of diazo keto ester. The push-pull dipole intermediate obtained from diazo keto ester is trapped either by an indole bond or by a tethered vinyl group, depending on the structure of diazo compound [139]. This result clearly demonstrates a critical role of the conformation of the cycloaddition transition state (**Scheme 56**).

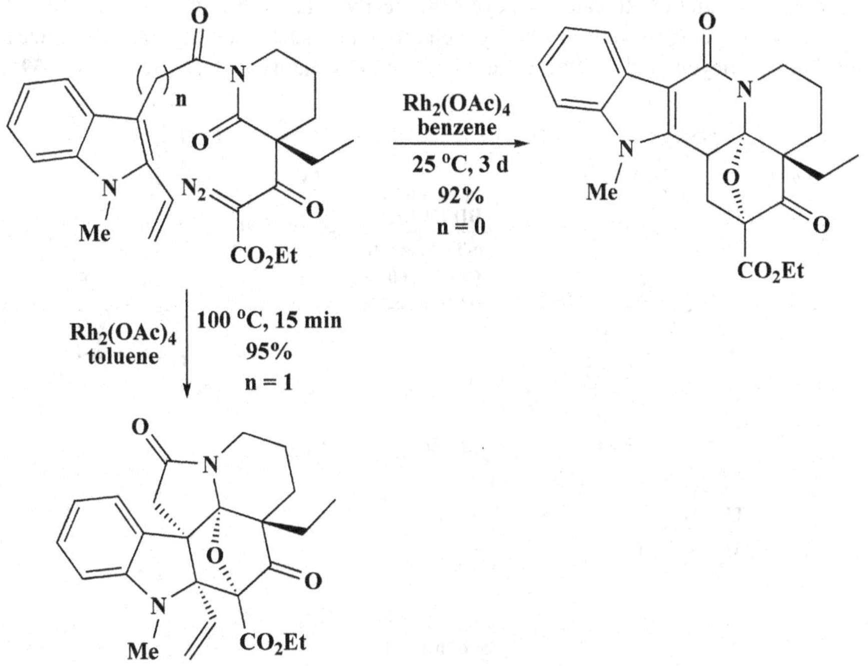

Scheme-56

A study on the reactivity of triaryl azides provided insights into the mechanism of the reaction [140]. Therein, the Hammett equation was used to analyze the ratio of products; the results obtained were anticipated to shed light on the nature of carbon-nitrogen bond formation. It was observed that the carbon-nitrogen bond formation proceeded through electrophilic aromatic substitution (EAS) and a linear correlation to σ *meta*-constants. The *meta*-position of methoxy group with respect to the reaction center made it an inductive electron-withdrawing group for EAS **(Scheme 57)** [141].

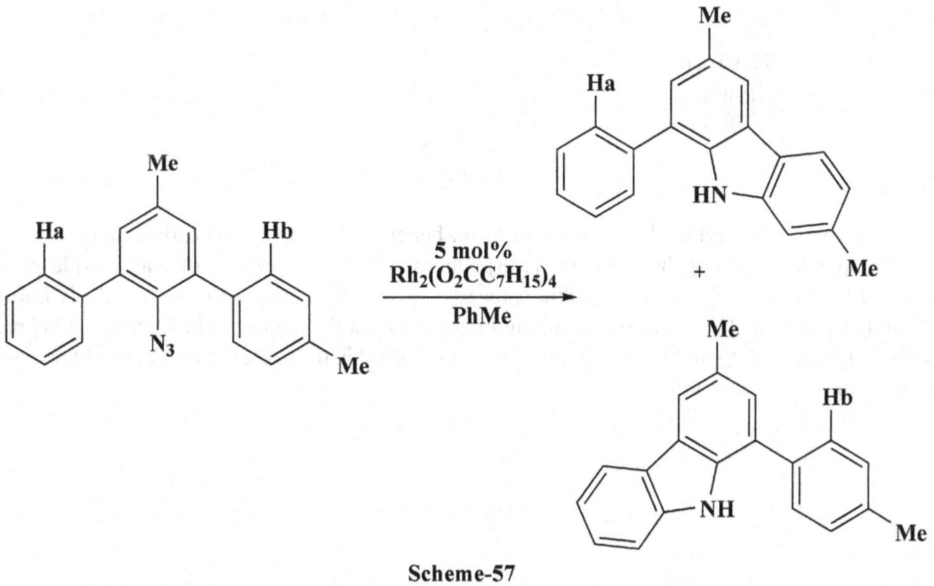

Scheme-57

Chiou et al. [142] studied Rh-catalyzed cyclohydrocarbonylation-bicyclization of *N*-allylic amides of arylacetic acids producing tricyclic aza-heterocyclic structures such as tetracyclic *β*-carboline alkaloid, harmicine as well as tricyclic indolizidine alkaloids, crispine A and its analogs **(Scheme 58–59)** [22].

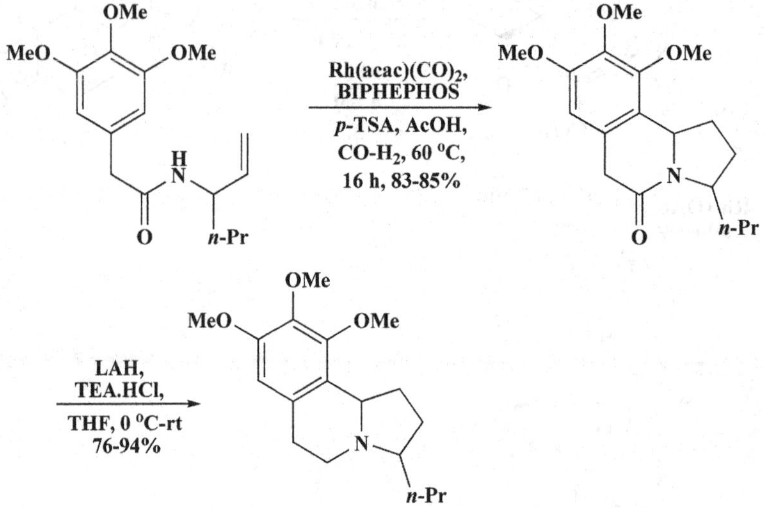

Scheme-58

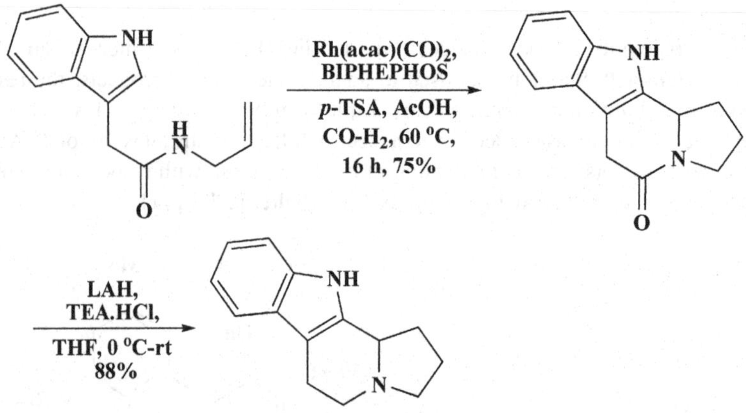

Scheme-59

The reactivity of biaryl azides shows significant benefits of using Rh(II) carboxylates as catalysts to afford nitrogen-containing heterocycles [143]. Photolysis or thermolysis of biaryl azide provides a nearly 50:50 mixture of the two possible carbazoles [144]. In contrast, the selectivity is improved significantly by rhodium(II) octanoate, resulting in carbazole as the major product. Driver [141] proved this selectivity, contending that the metal complex was involved in the carbon-nitrogen bond forming step of the mechanism **(Scheme 60)**.

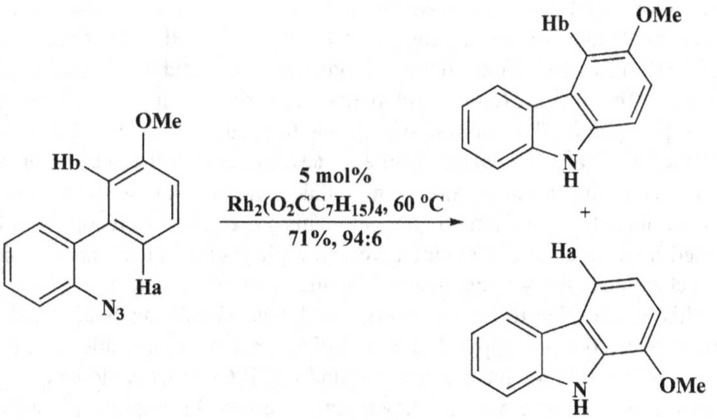

Scheme-60

Triazaindane derivative is afforded in 96% yield by cycloisomerization of a triaza substrate with 5 mol% RhCl(CO)(PPh$_3$)$_2$ under reflux in toluene **(Scheme 61)**. This protocol has also been extended for the preparation of macrocyclic *cis-* and *trans-*enediynes [145].

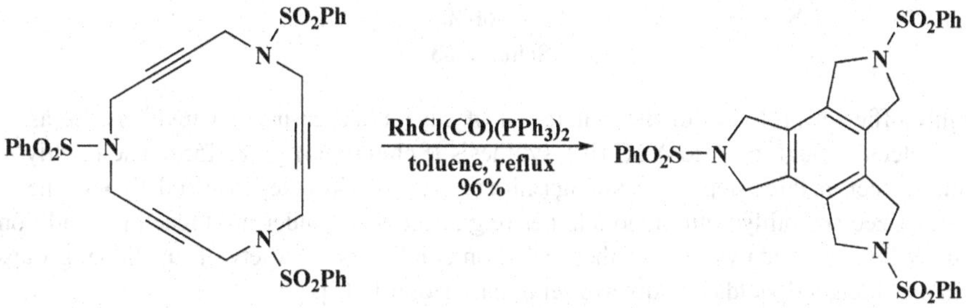

Scheme-61

The decomposition of α-diazoamides in the presence of rhodium(II) catalyst produces quinoline lactams **(Scheme 62)**. The lactam is formed as a sole product and is justified on the basis of stereoelectronic and steric effects as also exhibited by the less selective behavior of the ketone [146–151].

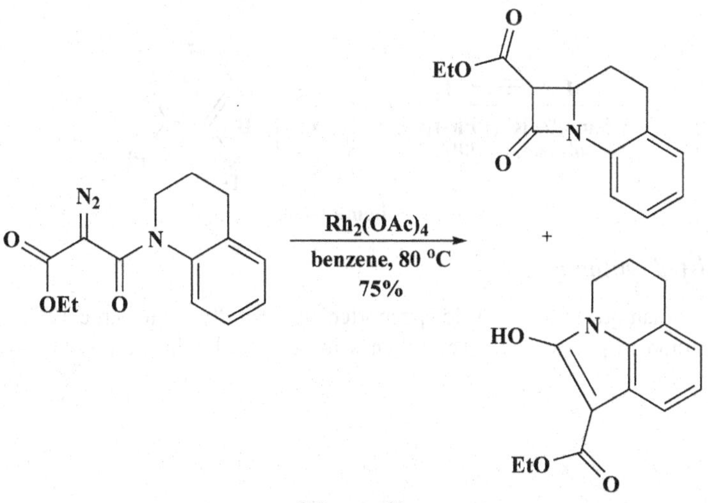

Scheme-62

Many late transition metal complexes have been screened for their ability to catalyze intramolecular alkylation of a benzimidazole bearing a pendant olefin, leading to the observation that Wilkinson's catalyst (RhCl(PPh$_3$)$_3$) affords a single cyclization product in 60% yield. Optimized reaction parameters led to the finding that [RhCl(coe)$_2$]$_2$ (coe = cyclooctene) is a highly effective rhodium pre-catalyst and PCy$_3$ is the optimal phosphine. This catalyst system has been utilized for the preparation of a variety of bi-, tri-, and tetracyclic 2-alkylimidazoles **(Scheme 63)**. In general, the cyclization affords products bearing a five-membered ring as the major isomer unless an overriding steric bias such as allylic α,α-dibranding or geminal alkene substitution prevails. This preference was reported for both homoallyl- and allyl-substituted imidazoles due to competitive and rapid olefin isomerization which generates an allyl-substituted cyclization precursor regardless of the initial olefin position. Extensive efforts to improve the efficiency of this reaction led to the discovery that Brønsted and Lewis acid additives, including magnesium bromide and 2,6-dimethylpyridinium chloride, provide remarkable increase in conversion and reaction rates [152–153]. Subsequently, it was found that [PCy$_3$H]Cl could be conveniently used to afford both phosphine and the additive. This modification rendered phosphine air stable for long-term storage and simplified the reaction setup [154].

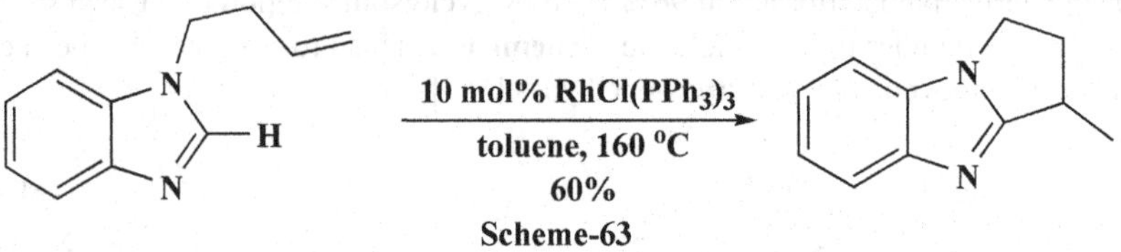

Scheme-63

A highly efficient protocol for the generation of substituted carbazoles involves the assembly of carbazole nucleus through an A to ABC ring synthesis **(Scheme 64)** [138, 155]. The readily available 2-iodoanilines reacts in three steps, with Sonogashira reaction as a key step to afford diynes. The dialkynes are cyclotrimerized smoothly with mono-alkynes (e.g., acetylene) under mild reaction conditions in the presence of Wilkinson's catalyst. Under these reaction conditions, a variety of functional groups can be tolerated to provide good yields of a diverse set of carbazoles [145].

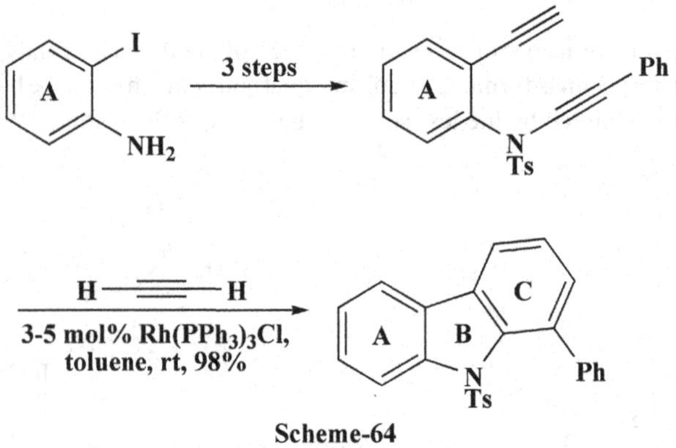

Scheme-64

Ruthenium-assisted synthesis

In their studies, Yamamoto et al. [156–157] reported that 1,6-diynes undergo [2+2+2]-cycloaddition with 2,5-dihydrofuran to provide polycyclic dienes in good yields in the presence of Cp*Ru(cod)Cl **(Scheme 65)**.

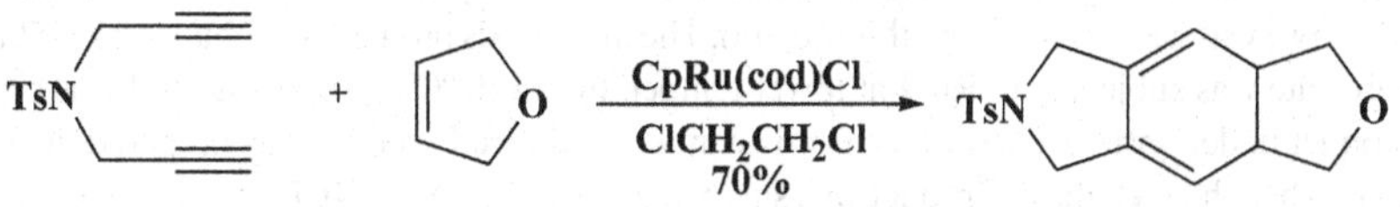

Scheme-65

It has also been observed that ring-closing metathesis (RCM) reactions are facilitated when two multi-component reactions are combined under microwave irradiation **(Scheme 66)** [158–159].

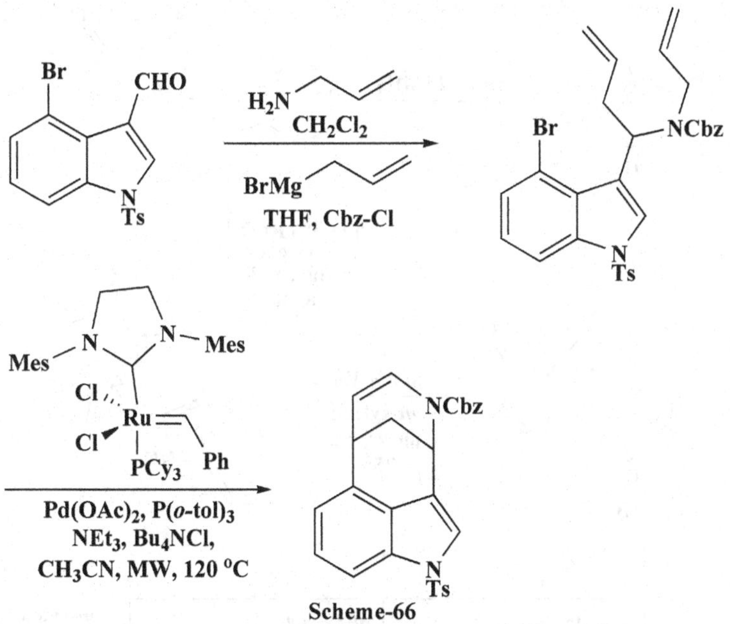

Scheme-66

Carbazoles are synthesized by reductive carbonylation of *o*-nitrobiphenyl, catalyzed by Fe(CO)$_5$, Ru$_3$(CO)$_{12}$ or palladium(II). Carbazole is prepared when 2-nitrobiphenyl reacts with Ru$_3$(CO)$_{12}$ and CO. The insertion into the aromatic carbon-hydrogen bond produces carbazole through intermediate formation of a singlet nitrene species—a ruthenium bound nitrene. The reaction occurs through the formation of Ru$_3$(μ_3-NC$_6$H$_4$-*o*-C$_6$H$_5$)$_2$(CO)$_9$ intermediate, a cluster where nitrene ligand is triply bridged with ruthenium atoms. The intermediate species is considerably stable and decomposes into the reaction products at 220 °C. Moreover, 2-aminobiphenyl is obtained as a by-product **(Scheme 67)** [160–161] and can be deoxygenated with carbon monoxide and iron pentacarbonyl to afford carbazole in 15% yield and 2-aminobiphenyl in 58% yield [162].

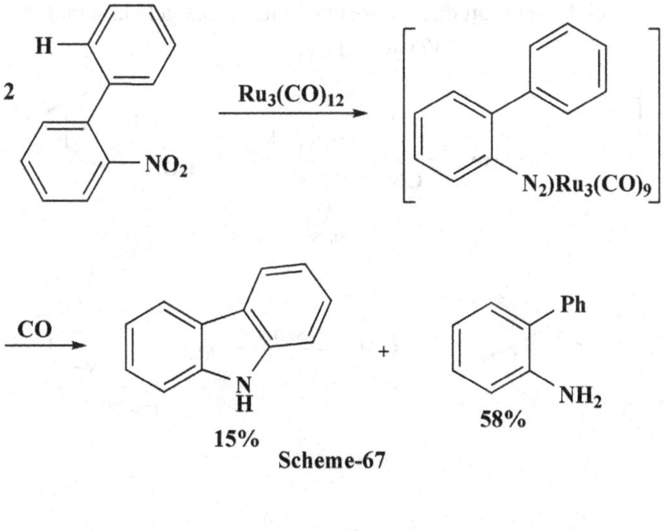

Scheme-67

In another pathway to the synthesis of harmicine, the reaction was completed within 17 hours to provide the desired product in 82% yield upon raising the catalyst loading of Ru-2 to 6 mol% **(Scheme 68)**. Two new carbon–carbon bonds were formed and a four-step formal synthesis of anti-parasitic natural product harmicine [163–166] was accomplished by this method. The metathesis product was obtained in 75% yield when the starting substrate was subjected to Ru-2 at lower temperatures (60 °C) in *m*-xylene and was converted into the desired product under various conditions. For instance, the desired product was prepared quantitatively in less than 5 hours when the metathesis product was subjected to Ru-2 catalyst [167]. A 1:1 mixture of the desired product and the metathesis product was obtained from same experiment without using a catalyst, showing the beneficial effect of adding catalyst to the non-metathetic part of the tandem sequence. The metathesis product was also converted into the desired product in less than 10 minutes upon addition of TFA (1 eq.) [168a–b].

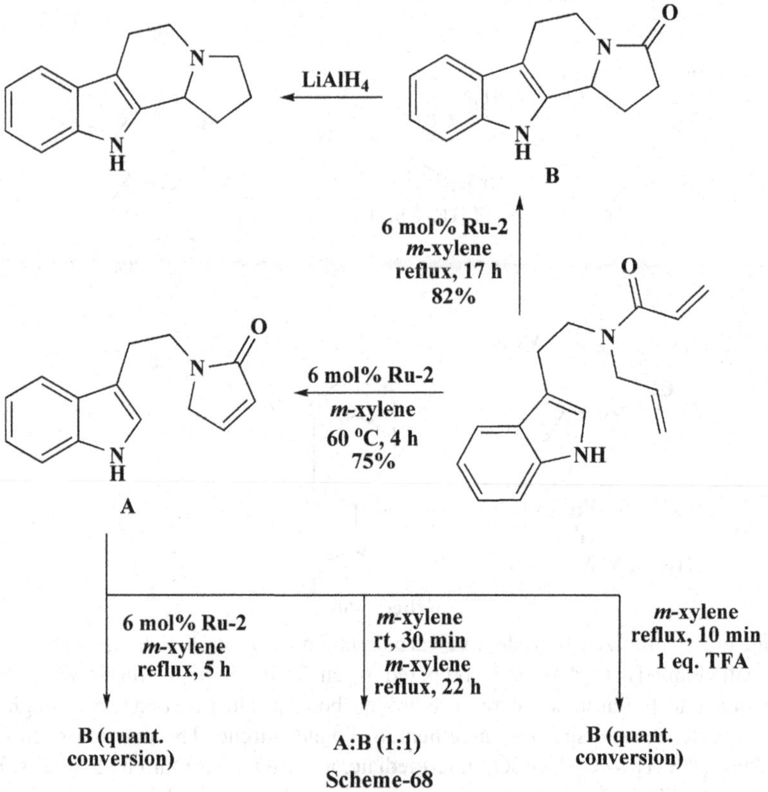

Scheme-68

The diastereoselective ring-closing metathesis of an enantiomerically pure substrate resulted in the total synthesis of aspidospermine **(Scheme 69)** [169]. Triene is formed through a multi-step route which includes a Ti-catalyzed enantioselective epoxidation. Catalytic cyclization of triene promoted by Ru carbene provides a diene in 87:13 d.r. after silyl ether hydrolysis; also, a diastereomerically pure sample (> 98:2 d.r.) of chiral cyclohexenol product is obtained after silica gel chromatography. Subsequent to

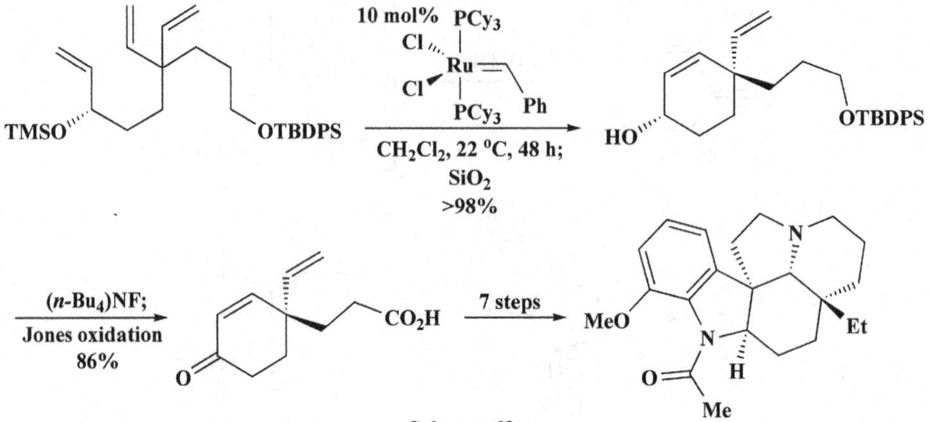

Scheme-69

the catalytic ring-closing metathesis, the original stereogenic center used to establish the all-carbon quaternary stereogenic center in diene is oxidatively removed. Enantiomerically pure γ,γ-disubstituted cyclohexenone is also obtained, which serves to complete the total synthesis of aspidospermine. An enantioselective desymmetrization process promoted by a chiral olefin metathesis catalyst, involving the enone corresponding to the triene, directly affords the α,β-unsaturated carbonyl, making the total synthesis scheme more concise [170].

Bressy et al. [171] synthesized polycyclic sultams and lactams from sulfonamides and amides by a one-pot procedure—a tandem RCM/isomerization followed by a sequential radical cyclization. This reaction has also been successfully applied on bis-allylsulfonamide to afford sultams (**Scheme 70**) [172].

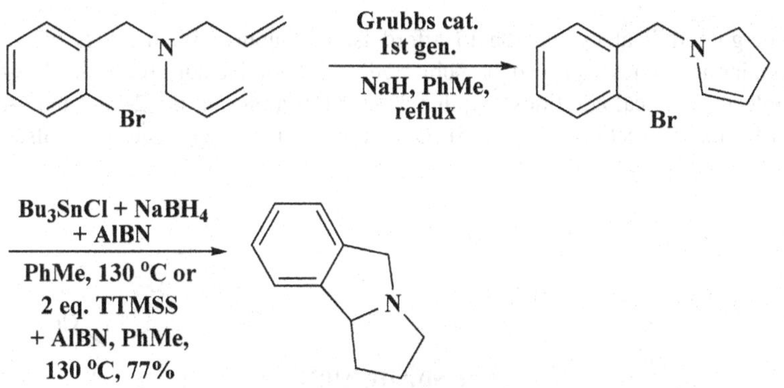

Scheme-70

Scandium-assisted synthesis

The cyclic compound is generated in 90% yield when 1,3-dimethylindole undergoes alkylative cyclization with N-Cbz-aziridine in the presence of Me$_3$SiCl and Sc(OTf)$_3$ (**Scheme 71**) [173–174].

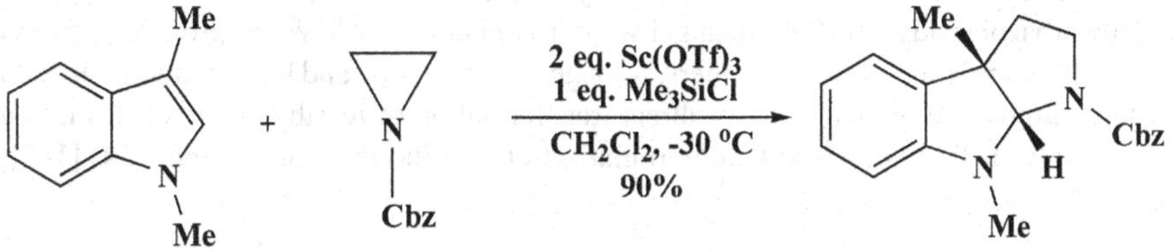

Scheme-71

Nüchter et al. [175–176], in their studies, investigated the parallel synthesis of a 36-member library of Biginelli dihydropyrimidines in a suitable multi-vessel rotor placed in a multimode microwave reactor. Given the fact that modern multimode MW reactors can operate with specifically designed 96-well plates under sealed-vessel conditions, the parallel approach offered a considerably higher throughput than the automated sequential technique, albeit at the cost of having less control over the reaction parameters for each individual vessel/well. One additional limitation of the parallel approach was that all reaction vessels during library synthesis were exposed to the same irradiation conditions in terms of MW power and reaction time, thus not allowing specific needs of individual building blocks to be addressed by varying temperature or time. A variety of other heterocyclic compounds have also been prepared by microwave-assisted cyclocondensation or cycloaddition protocols as shown in **Schemes 72.**

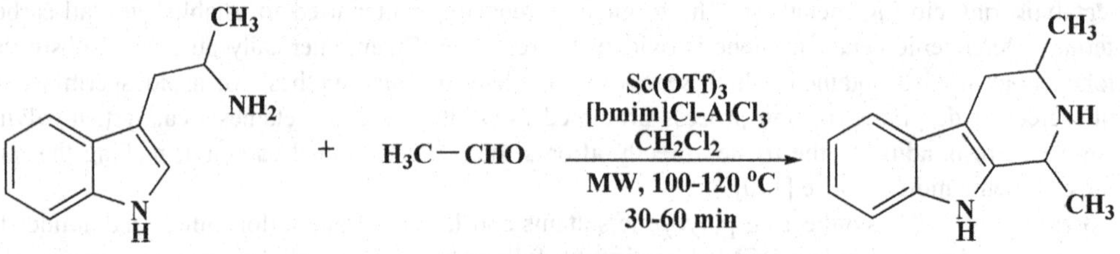

Scheme-72

Silicon-assisted synthesis

4-Pyridone undergoes radical cyclization to afford isoindolinone with the loss of a chlorine atom [177]. The isoquinolin-1-ones can also be subjected to intramolecular free radical arylations [178]. Several intramolecular radical additions to quinolines and pyridines at the 2-, 3-, and 4-positions have been shown to be facile reactions [179–185]. Occasionally, rearrangements have also been reported **(Scheme 73)**.

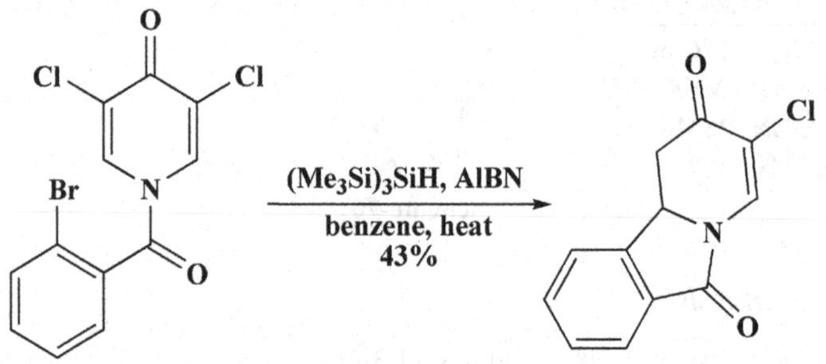

Scheme-73

Laronze-Cochard et al. [186] prepared 2,4-diaryl substituted carbazoles from tetrahydrocarbazole. The tetrahydrocarbazoles adsorbed on silica gel were irradiated under MWs to provide carboxylic acids in 71–86% yields as a mixture of diastereomers. A longer reaction time and higher irradiation temperature (200 °C) afforded moderate yield (41%) of direct transformation of tetrahydrocarbazoles into aromatic carbazoles by a solvent-free decarboxylation-aromatization domino reaction **(Scheme 74)** [187].

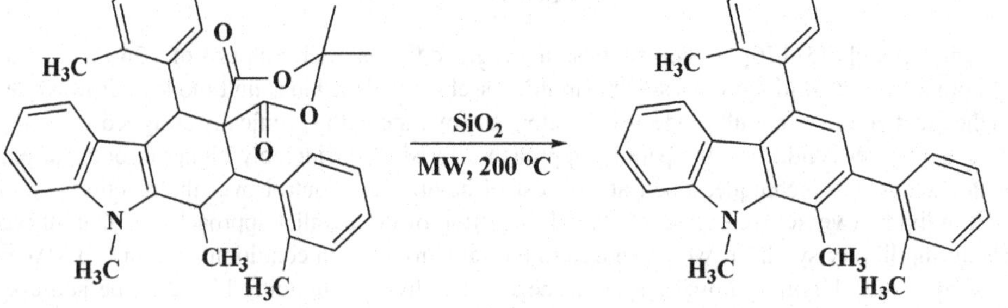

Scheme-74

Silver-assisted synthesis

Natural compounds possessing 2-aminomethylene indoline can be prepared by various protocols, including the preparation of vallesamidine by Dickman and Heathcock [188]. Herein, 2-aminomethylene indoline core was prepared *via* intramolecular bromoamination of enamide to result in excellent yields of aminal. The formation of vallesamidine was completed easily by two successive hydride reductions **(Scheme 75)**.

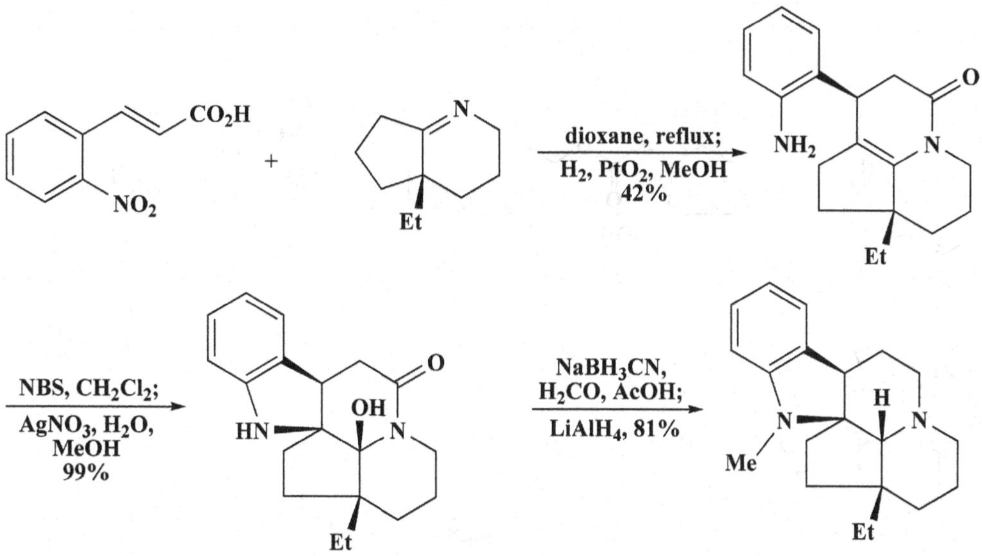

Scheme-75

Intramolecular reactions of *E*-alkenyl halides provide polycyclic indole derivatives. The tethered reaction components are incorporated through a judicious choice of *N*-substituents, which allows the tandem synthesis of second ring structures. Strategies involving heterocycle direct arylation can be utilized [189–190], such as the synthesis of polycyclic indole derivative in excellent yield **(Scheme 76)** [191–192].

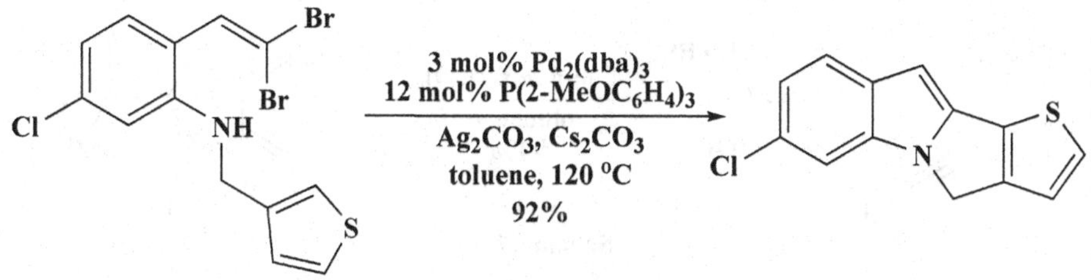

Scheme-76

Tin-assisted synthesis

Etodolac (Lodine) is a racemic tetrahydrocarbazole derivative which exhibits non-steroidal anti-inflammatory activities and is used to treat inflammation and impart relief from general pain. Two general pathways have been reported for the synthesis of this drug. The hydrazine is synthesized *via* reduction of diazonium salt with SnCl$_2$ and is subjected to a Fischer indole reaction with aldehyde. Etodolac is ultimately obtained when the indole formed is condensed with ethyl 3-oxopropanoate, followed by saponification of the ester **(Scheme 77)** [26, 193].

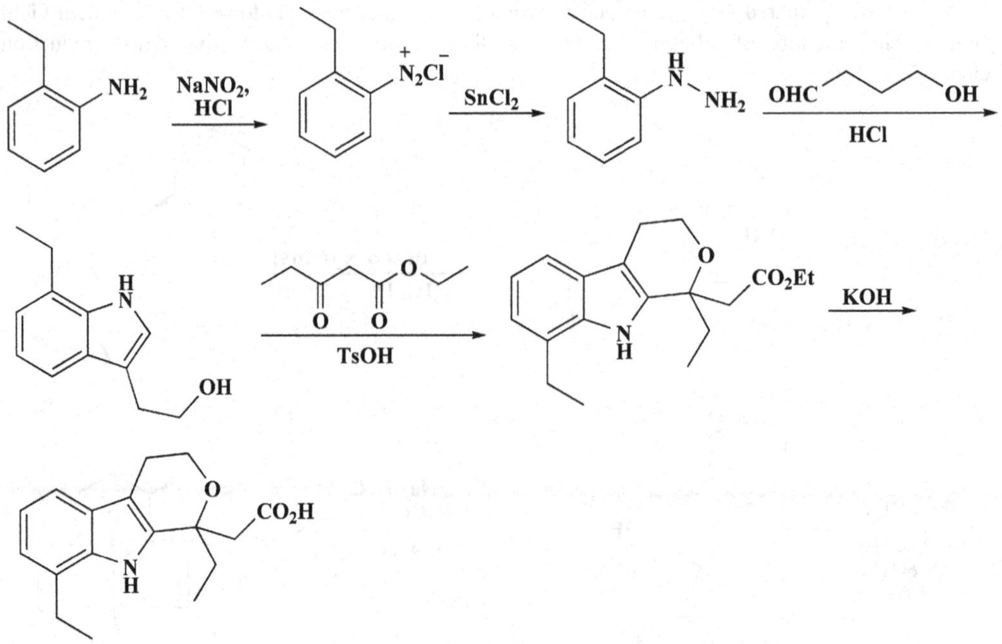

Scheme-77

Tungsten-assisted synthesis

Polycyclic indole derivatives are synthesized when *N*-(*o*-ethynylphenyl)imine reacts with electron-rich alkenes in the presence of tungsten [194]. The reaction occurs through the synthesis of tungsten-possessing azomethine ylide which undergoes [3+2]-cycloaddition with electron-rich alkenes (**Scheme 78**).

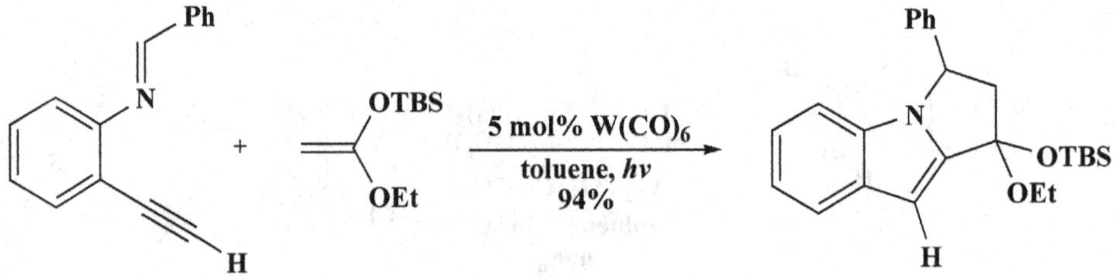

Scheme-78

When substrates containing an internal triple bond react with an alkene in stoichiometric amounts of tungsten complex, it results in the formation of [1,2]-alkyl-migrated products along with formal [4+2] cycloadducts in small amounts (**Scheme 79**) [123, 195–196].

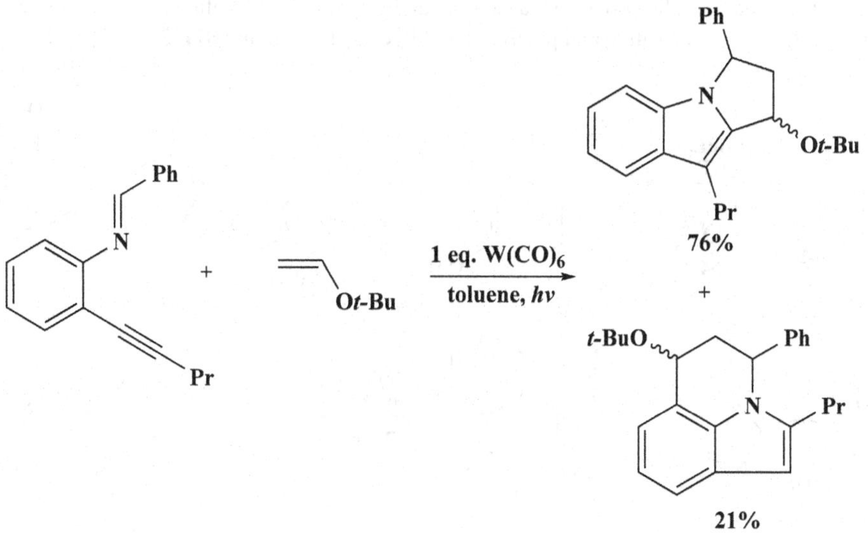

Scheme-79

Zinc-assisted synthesis

Carbazoles are similar to indoles. Therefore, it is feasible to use analogous protocols for their preparation. For carvedilol, the key intermediate is 4-hydroxy-9*H*-carbazole [24]. Like Fischer indole synthesis, condensation of phenylhydrazine with 1,3-cyclohexanedione yields cyclohexane-1,3-dione monophenyl hydrazone. The cyclohexane-1,3-dione monophenyl hydrazone then undergoes an acid-catalyzed Fischer indole synthesis to afford tetrahydro-4-oxocarbazole. This intermediate is dehydrogenated by various methods, including the use of sulfur, lithium chloride or cuprous chloride, bromine, lead dioxide, chloranil or Pd on charcoal. The desired 4-hydroxy compound is obtained efficiently and with high conversion rates in the presence of catalytic amounts Raney Ni in aqueous potassium hydroxide **(Scheme 80)** [26].

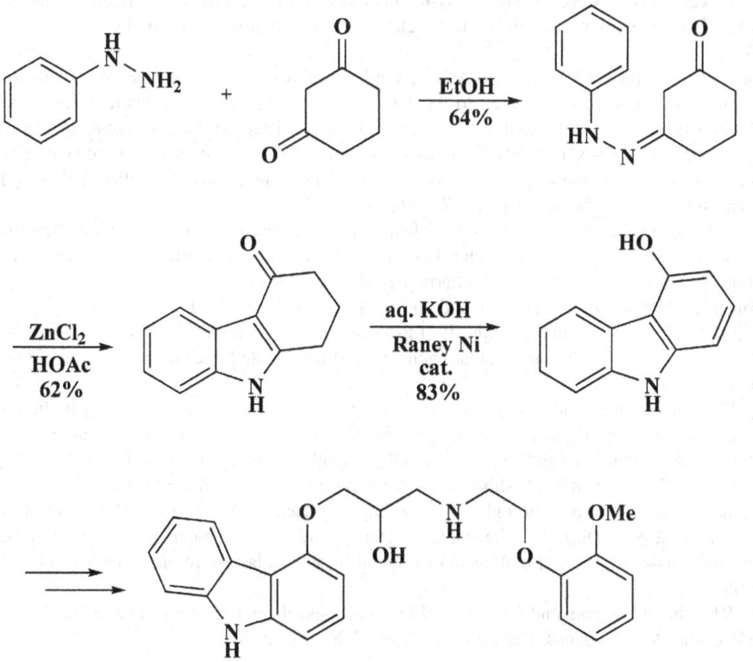

Scheme-80

Fully substituted tricyclic core of ondansetron can be synthesized by direct Fischer indole synthesis using a cyclic 1,3-dione derivative and phenylmethyl hydrazine **(Scheme 81)** [26, 197].

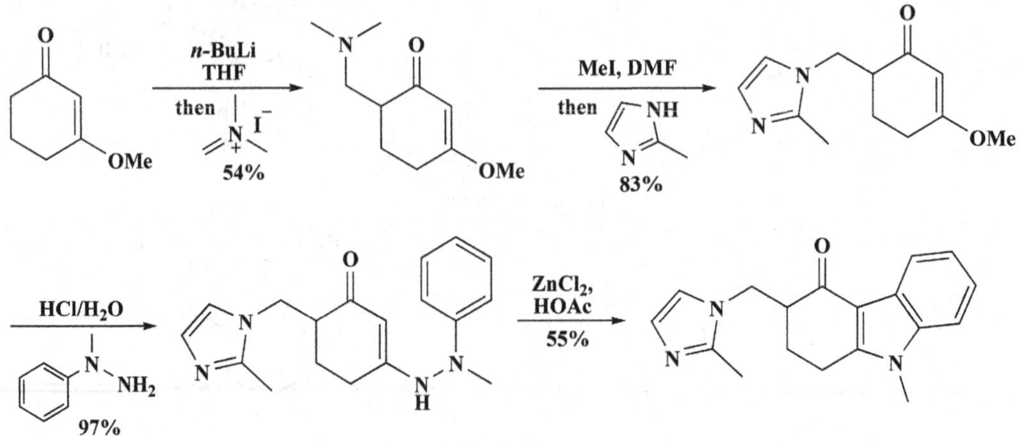

Scheme-81

References

(1)   (a) D.V. Lefemine, M. Dann, F. Barbatschi, W.K. Hausmann, V. Zbinovsky and P. Monnikendam. 1962. Isolation and characterization: mitiromycin and other anti-biotics. J. Am. Chem. Soc. 84: 3184–3185. (b) N. Kaur. 2018. Copper catalysts in the synthesis of five-membered *N*-polyheterocycles. Curr. Org. Synth. 15: 940–971. (c) N. Kaur. 2018. Recent developments in the synthesis of nitrogen-containing five-membered polyheterocycles using rhodium catalysts. Synth. Commun. 48: 2457–2474. (d) N. Kaur. 2018. Mercury-catalyzed synthesis of heterocycles. Synth. Commun. 48: 2715–2749. (e) N. Kaur. 2018. Photochemical irradiation: seven- and higher-membered *O*-heterocycles. Synth. Commun. 48: 2935–2964.

(2)   (a) N. Kaur, R. Tyagi and D. Kishore. 2016. Expedient protocols for the installation of 1,5-benzoazepino-based privileged templates on 2-position of 1,4-benzodiazepine through a phenoxyl spacer. J. Heterocycl. Chem. 53: 643–646. (b) N. Kaur. 2019. Nickel catalysis: six-membered heterocycle syntheses. Synth. Commun. 49: 1103–1133. (c) N. Kaur. 2019. Seven-membered *N*-heterocycles: metal and non-metal assisted synthesis. Synth. Commun. 49: 987–1030.

(3)   (a) N. Kaur, A. Aditi and D. Kishore. 2016. A facile synthesis of face 'D' quinolino annulated benzazepinone analogues with its quinoline framework appended to oxadiazole, triazole and pyrazole heterocycles. J. Heterocycl. Chem. 53: 457–460. (b) N. Kaur and D. Kishore. 2014. Microwave-assisted synthesis of six-membered *O,O*-heterocycles. Synth. Commun. 44: 3082–3111. (c) N. Kaur and D. Kishore. 2014. Microwave-assisted synthesis of six-membered *O*-heterocycles. Synth. Commun. 44: 3047–3081. (d) Y.Y. Nakamura. 2004. Transition metal-catalyzed reactions in heterocyclic synthesis. Chem. Rev. 104: 2127–2198.

(4)   N. Kaur. 2013. A new approach to anti-HIV chemotherapy devised by linking the vital fragments of active RT inhibitors such as etravirine to the molecular framework of anti-HIV prone privileged nucleus of 1,4-benzodiazepine as possible substitute to 'HAART'. Int. J. Pharm. Bio. Sci. 4: 309–317.

(5)   J.F. Collinus. 1965. Anti-biotics, proteins, and nucleic acids. Brit. Med. Bull. 21: 223–228.

(6)   N. Kaur. 2017. Gold catalysts in the synthesis of five-membered *N*-heterocycles. Curr. Organocatal. 4: 122–154.

(7)   G. Stork and J. Ficini. 1961. Intramolecular cyclization of unsaturated diazoketones. J. Am. Chem. Soc. 83: 4678–4678.

(8)   R. Tyagi, N. Kaur, B. Singh and D. Kishore. 2013. A noteworthy mechanistic precedence in the exclusive formation of one regioisomer in the Beckmann rearrangement of ketoximes of 4-piperidones annulated to pyrazolo-indole nucleus by organocatalyst derived from TCT and DMF. Synth. Commun. 43: 16–25.

(9)   R. Tyagi, N. Kaur, B. Singh and D. Kishore. 2014. A novel synthetic protocol for the heteroannulation of oxocarbazole and oxoazacarbazole derivatives through corresponding oxoketene dithioacetals. J. Heterocycl. Chem. 51: 18–23.

(10)   B. Meseguer, D. Alonso-Diaz, N. Griebenow, T. Herget and H. Waldmann. 1999. Natural product synthesis on polymeric supports-synthesis and biological evaluation of an indolactam library. Angew. Chem. Int. Ed. Engl. 38: 2902–2906.

(11)   A.K. Jha, P.K. Shukla, N. Soni and A. Verma. 2015. Synthesis, characterization and biological evaluation of thiazole incorporated triazole compounds. Der Pharmacia Lett. 7: 67–74.

(12)   P.K. Shukla, N. Soni, A. Verma and A. Kumar. 2014. Synthesis, characterization and *in vitro* biological evaluation of a series of 1,2,4-triazoles derivatives and triazole-based Schiff bases. Der Pharma Chemica 6: 153–160.
(13)   P.K. Shukla, P. Pathak, A. Thakur and A.K. Jha. 2012. Synthesis, characterization and anti-microbial activity of 2-(5-mercapto-3-subsituted-1,5-dihydro-[1,2,4]triazole. J. Pharm. Herb. Res. 2: 22–26.
(14)   N. Kaur. 2015. Palladium-catalyzed approach to the synthesis of *S*-heterocycles. Catal. Rev. 57: 478–564.
(15)   N. Kaur. 2018. Photochemical reactions as key steps in five-membered *N*-heterocycles synthesis. Synth. Commun. 48: 1259–1284.
(16)   N. Kaur. 2018. Synthesis of six- and seven-membered heterocycles under ultrasound irradiation. Synth. Commun. 48: 1235–1258.
(17)   A.R. Suresh Babu and R. Raghunathan. 2007. TiO$_2$-silica-mediated one-pot three-component 1,3-dipolar cycloaddition reaction: a facile and rapid synthesis of dispiro acenaphthenone/oxindole[indanedione/oxindole]pyrroloisoquinoline ring systems. Tetrahedron 63: 8010–8016.
(18)   D.C. Rogness and R.C. Larock. 2009. Rapid synthesis of the indole-indolone scaffold *via* [3+2]-annulation of arynes by methyl indole-2-carboxylates. Tetrahedron Lett. 50: 4003–4008.
(19)   R.D. Giacometti and Y.K. Ramtohul. 2009. Synthesis of polycyclic indolone and pyrroloindolone heterocycles *via* the annulation of indole- and pyrrole-2-carboxylate esters with arynes. Synlett 12: 2010–2016.
(20)   C. Xie, Y. Zhang and P. Xu. 2008. Novel synthesis of pyrido[2,1-*a*]isoindoles *via* a three-component assembly involving benzynes. Synlett 20: 3115–3120.
(21)   A. Geny, S. Gaudrel, F. Slowinski, M. Amatore, G. Chouraqui, M. Malacria, C. Aubert and V. Gandon. 2009. A straightforward procedure for the [2+2+2]-cycloaddition of enediynes. Adv. Synth. Catal. 351: 271–275.
(22)   S. Nag and S. Batra. 2011. Applications of allylamines for the syntheses of aza-heterocycles. Tetrahedron 67: 8959–9061.
(23)   J.B.J. Milbank, R.J. Stevenson, D.C. Ware, J.Y.C. Chang, M. Tercel, G.-O. Ahn, W.R. Wilson and W.A. Denny. 2009. Synthesis and evaluation of stable bidentate transition metal complexes of 1-(chloromethyl)-5-hydroxy-3-(5,6,7-trimethoxyindol-2-ylcarbonyl)-2,3-dihydro-1*H*-pyrrolo[3,2-*f*]quinoline (seco-6-azaCBI-TMI) as hypoxia selective cytotoxins. J. Med. Chem. 52: 6822–6834.
(24)   K. Lauer and E. Kiegel. 16 June 1981. Process for the preparation of 4-hydroxycarbazole. U.S. Patent 4,273,711.
(25)   K. Vellaisamy, G. Li, C.-N. Ko, H.-J. Zhong, S. Fatima, H.-Y. Kwan, C.-Y. Wong, W.-J. Kwong, W. Tan, C.-H. Leung and D.-L. Ma. 2018. Cell imaging of dopamine receptor using agonist labeling iridium(III) complex. Chem. Sci. 9: 1119–1125.
(26)   M. Baumann, I.R. Baxendale, S.V. Ley and N. Nikbin. 2011. An overview of the key routes to the best selling 5-membered ring heterocyclic pharmaceuticals. Beilstein J. Org. Chem. 7: 442–495.
(27)   X.Y. Yuan, X.B. Xu, X.B. Zhou, J.W. Yuan, L.G. Mai and Y.Z. Li. 2007. Copper-catalyzed double *N*-alkenylation of amides: an efficient synthesis of di- or tri-substituted *N*-acylpyrroles. J. Org. Chem. 72: 1510–1513.
(28)   E.D. Li, X.B. Xu, H.F. Li, H.M. Zhang, X.L. Xu, X.Y. Yuan and Y.Z. Li. 2009. Copper-catalyzed synthesis of five-membered heterocycles *via* double C-N bond formation: an efficient synthesis of pyrroles, dihydropyrroles, and carbazoles. Tetrahedron 65: 8961–8968.
(29)   Y. Liu and J.-P. Wan. 2011. Tandem reactions initiated by copper-catalyzed cross-coupling: a new strategy towards heterocycle synthesis. Org. Biomol. Chem. 9: 6873–6894.
(30)   A. Minatti and S.L. Buchwald. 2008. Synthesis of indolines *via* a domino Cu-catalyzed amidation/cyclization reaction. Org. Lett. 10: 2721–2724.
(31)   D.S. Surry and S.L. Buchwald. 2010. Diamine ligands in copper-catalyzed reactions. Chem. Sci. 1: 13–31.
(32)   G. Evano, N. Blanchard and M. Toumi. 2008. Copper-mediated coupling reactions and their applications in natural products and designed bio-molecules synthesis. Chem. Rev. 108: 3054–3131.
(33)   J. Chae and S.L. Buchwald. 2004. Palladium-catalyzed regioselective hydrodebromination of dibromoindoles: application to the enantioselective synthesis of indolodioxane U86192A. J. Org. Chem. 69: 3336–3339.
(34)   M. Ueda, A. Sato, Y. Ikeda, T. Miyoshi, T. Naito and O. Miyata. 2010. Direct synthesis of tri-substituted isoxazoles through gold-catalyzed domino reaction of alkynyl oxime ethers. Org. Lett. 11: 2594–2597.
(35)   W. He, C. Li and L. Zhang. 2011. An efficient [2+2+1] synthesis of 2,5-disubstututed oxazoles *via* gold-catalyzed intermolecular alkyne oxidation. J. Am. Chem. Soc. 133: 8482–8485.
(36)   A.S.K. Hashmi, A.M. Schuster, M. Zimmer and F. Rominger. 2011. Synthesis of 5-halo-4*H*-1,3-oxazine-6-amines by copper-mediated domino reaction. Chem. Eur. J. 17: 5511–5515.
(37)   X. Gao, Y.-M. Pan, M.L. Li, L. Chen and Z.-P. Zhan. 2010. Facile one-pot synthesis of three different substituted thiazoles from propargylic alcohols. Org. Biomol. Chem. 8: 3259–3266.
(38)   M. Yoshimatsu, M. Matsui, T. Yamamoto and A. Sawa. 2010. Convenient preparation of 4-arylmethyl and 4-hetarylmethyl thiazoles by regioselective cycloaddition reactions of 3-sulfanyl- and selenylpropargyl alcohols. Tetrahedron Lett. 66: 7975–7987.
(39)   Y. Asanuma, S.-I. Fujiwara, T. Shin-Ike and N. Kambe. 2004. Selenoimidoylation of alcohols with selenium and isocyanides and its application to the synthesis of selenium-containing heterocycles. J. Org. Chem. 69: 4845–4848.
(40)   K. Wilckens, M. Uhlemann and C. Czekelius. 2009. Gold-catalyzed *endo*-cyclizations of 1,4-diynes to seven-membered ring heterocycles. Chem. Eur. J. 15: 13323–13326.

(41)   E.M. Bunnelle, C.R. Smith, S.K. Lee, S.W. Singaram, A.J. Rhodes and R. Sarpong. 2008. Pt-catalyzed cyclization/ migration of propargylic alcohols for the synthesis of 3(2*H*)-furanones, pyrrolones, indolizines, and indolizinones. Tetrahedron Lett. 64: 7008–7014.

(42)   K.-G. Ji, H.-T. Zhu, F. Yang, X.-Z. Shu, S.-C. Zhao, X.-Y. Liu, A. Shaukat and Y.-M. Liang. 2010. A novel iodine-promoted tandem cyclization: an efficient synthesis of substituted 3,4-diiodoheterocyclic compounds. Chem. Eur. J. 16: 6151–6154.

(43)   X. Zang, W.T. Teo, S.W.H. Chan and P.W.H. Chan. 2010. Brønsted acid-catalyzed cyclization of propargylic alcohols with thioamides. Facile synthesis of di- and tri-substituted thiazoles. J. Org. Chem. 75: 6290–6293.

(44)   P. Kothandaraman, W. Rao, S.J. Foo and P.W.H. Chan. 2010. Gold-catalyzed cycloisomerization reaction of 2-tosylamino-phenylprop-1-yn-3-ols *via* a versatile approach for indole synthesis. Angew. Chem. Int. Ed. 49: 4619–4623.

(45)   F.-D. Boyer, X. Le Goff and I. Hanna. 2008. Gold(I)-catalyzed cycloisomerization of 1,7- and 1,8-enynes: application to the synthesis of a new allocolchicinoid. J. Org. Chem. 74: 5163–5166.

(46)   B. Alcaide, P. Almendros and J.M. Alonso. 2011. Gold-catalyzed cyclizations of alkynol-based compounds: synthesis of natural products and derivatives. Molecules 16: 7815–7843.

(47)   (a) D.B. Dess and J.C. Martin. 1983. Readily accessible 12-I-5 oxidant for the conversion of primary and secondary alcohols to aldehydes and ketones. J. Org. Chem. 48: 4155–4156. (b) R.K. Boeckman, P. Shao and J.J. Mullins. 2000. The Dess-Martin periodinane: 1,1,1-triacetoxy-1,1-dihydro-1,2-benziodoxol-3(1*H*)-one. Org. Synth. 77: 141–152. (c) Z. Li, C. Brouwer and C. He. 2008. Gold-catalyzed organic transformations. Chem. Rev. 108: 3239–3265. (d) F. Volz and N. Krause. 2007. Golden opportunities in natural product synthesis: first total synthesis of (–)-isocyclocapitelline and (–)-isochrysotricine by gold-catalyzed allene cycloisomerization. Org. Biomol. Chem. 5: 1519–1521.

(48)   (a) G. Li, X. Huang and L. Zhang. 2008. Platinum-catalyzed formation of cyclic-ketone-fused indoles from *N*-(2-alkynylphenyl)lactams. Angew. Chem. Int. Ed. 47: 346–349. (b) B. Alcaide and P. Almendros. 2011. Gold-catalyzed heterocyclizations in alkynyl- and allenyl-*β*-lactams. Beilstein J. Org. Chem. 7: 622–630. (c) H. Kusama, Y. Miyashita, J. Takaya and N. Iwasawa. 2006. Pt(II)- or Au(III)-catalyzed [3+2]-cycloaddition of metal containing azomethine ylides: highly efficient synthesis of the mitosene skeleton. Org. Lett. 8: 289–292. (d) M. Schelwies, A.L. Dempwolff, F. Rominger and G. Helmchen. 2007. Gold-catalyzed intermolecular addition of carbonyl compounds to 1,6-enynes. Angew. Chem. Int. Ed. 46: 5598–5601. (e) H. Kusama, H. Funami, M. Shido, Y. Hara, J. Takaya and N. Iwasawa. 2005. Generation and reaction of tungsten-containing carbonyl ylides: [3+2]-cycloaddition reaction with electron rich alkenes. J. Am. Chem. Soc. 127: 2709–2716. (f) H. Kusama, H. Yamabe, Y. Onizawa, T. Hoshino and N. Iwasawa. 2005. Rhenium(I)-catalyzed intramolecular geminal carbofunctionalization of alkynes: tandem cyclization of *ω*-acetylenic dienol silyl ethers. Angew. Chem. Int. Ed. 44: 468–470. (g) L.L. Ouh, T.E. Muller and Y.K. Yan. 2005. Intramolecular hydroamination of 6-aminohex-1-yne catalyzed by Lewis acidic rhenium(I) carbonyl complexes. J. Organomet. Chem. 690: 3774–3782. (h) Y. Kuninobu, A. Kawata and K. Takai. 2005. Rhenium-catalyzed insertion of terminal acetylenes into a C-H bond of active methylene compounds. Org. Lett. 7: 4823–4825. (i) J. Muzart. 2005. Palladium-catalyzed reactions of alcohols. Part C: formation of ether linkages. Tetrahedron 61: 5955–6008. (j) E. Genin, S. Antoniotti, V. Michelet and J.-P. Genet. 2005. An IrI-catalyzed *exo*-selective tandem cycloisomerization/hydroalkoxylation of bis-homopropargylic alcohols at room temperature. Angew. Chem. Int. Ed. 44: 4949–4953. (k) T. Shibata, Y. Kobayashi, S. Maekawa, N. Toshida and K. Takagi. 2005. Iridium-catalyzed enantioselective cycloisomerization of nitrogen-bridged 1,6-enynes to 3-azabicylo[4.1.0]heptanes. Tetrahedron 61: 9018–9024. (l) A. Das, S.M.A. Sohel and R.-S. Liu. 2010. Carbo- and heterocyclization of oxygen and nitrogen-containing electrophiles by platinum, gold, silver and copper species. Org. Biomol. Chem. 8: 960–979.

(49)   H.-J. Knolker and K.R. Reddy. 2008. In The Alkaloids: Chemistry and Biology, Ed.: G. A. Cordell, Elsevier, Evanston, Illinois, 65, 1.

(50)   H.-J. Knolker and K.R. Reddy. 2002. Isolation and synthesis of biologically active carbazole alkaloids. Chem. Rev. 102: 4303–4428.

(51)   M. Fagnoni and A. Albini. 2005. Arylation reactions: the photo-S_N1 path *via* phenyl cation as an alternative to metal catalysis. Acc. Chem. Res. 38: 713–721.

(52)   S. Protti, M. Fagnoni and A. Albini. 2006. Benzyl (phenyl) *γ*- and *δ*-lactones *via* photoinduced tandem Ar-C, C-O bond formation. J. Am. Chem. Soc. 128: 10670–10671.

(53)   V. Dichiarante and M. Fagnoni. 2008. Aryl cation chemistry as an emerging versatile tool for metal-free arylations. Synlett 6: 787–800.

(54)   A. Fraboni, M. Fagnoni and A. Albini. 2003. A novel *α*-arylation of ketones, aldehydes, and esters *via* a photoinduced S_N1 reaction through 4-aminophenyl cations. J. Org. Chem. 68: 4886–4893.

(55)   P.T. Parvatkar, P.S. Parameswaran and S.G. Tilve. 2007. Double reductive cyclization: a facile synthesis of the indoloquinoline alkaloid cryptotackieine. Tetrahedron Lett. 48: 7870–7872.

(56)   S.K. Meegalla, G.J. Stevens, C.A. McQueen, A.Y. Chen, C. Yu, L.F. Liu, L.R. Barrows and E.J. LaVoie. 1994. Synthesis and pharmacological evaluation of isoindolo[1,2-*b*]quinazolinone and isoindolo[2,1-*a*]benzimidazole derivatives related to the anti-tumor agent batracylin. J. Med. Chem. 37: 3434–3439.

(57)   K.M. Dawood and B.F. Abdel-Wahab. 2010. Synthetic routes to benzimidazole-based fused polyheterocycles. ARKIVOC (i): 333–389.

(58)   (a) F. Le Strat and J. Maddaluno. 2002. New carbanionic access to 3-vinylindoles and 3-vinylbenzofurans. Org. Lett. 4: 2791–2793. (b) T. Lu, Z. Lu, Z.-X. Ma, Y. Zhang and R.P. Hsung. 2013. Allenamides: a powerful and versatile building block in organic synthesis. Chem. Rev. 113: 4862–4904.

(59)   H. Fritz and H.G. Gerber. 1975. Indole und indolalkaloide, XVII1) beeinflussung des chiroptischen verhaltens von tetrahydro-4a*H*-carbazolen durch die konfiguration von alkylsubstituenten in den positionen 2 und 4a. Liebigs Ann. Chem. 7-8: 1422–1434.

(60)   A. Furstner and P. Hannen. 2006. Platinum- and gold-catalyzed rearrangement reactions of propargyl acetates: total syntheses of (–)-α-cubebene, (–)-cubebol, sesquicarene and related terpenes. Chem. Eur. J. 12: 3006–3019.

(61)   J.G. Rodriguez and A. Urrutia. 1998. Synthesis of sterically hindered 4a,9a-disubstituted 1,2,3,4,4a,9a-hexahydrocarbazoles from 4a-methyl-1,2,3,4-tetrahydro-4a*H*-carbazole with organolithium reagents. Tetrahedron 54: 15613–15618.

(62)   P.E. Eaton, C.-H. Lee and Y. Xiong. 1989. Magnesium amide bases and amido-Grignards. *Ortho* magnesiation. J. Am. Chem. Soc. 111: 8016–8018.

(63)   J.C. Estevez, R.J. Estevez, E. Guitian, M.C. Villaverde and L. Castedo. 1989. Intramolecular aryne cycloaddition approach to aristolactams. Tetrahedron Lett. 30: 5785–5786.

(64)   J.C. Estevez, R.J. Estevez and L. Castedo. 1995. The intramolecular aryne cycloaddition approach to *Aporphinoids*. A new total synthesis of aristolactams and phenanthrene alkaloids. Tetrahedron 51: 10801–10810.

(65)   C. Hoarau, A. Couture, H. Cornet, E. Deniau and P. Grandclaudon. 2001. A concise total synthesis of the azaphenanthrene alkaloid eupolauramine. J. Org. Chem. 66: 8064–8069.

(66)   A.A. Cant, G.H.V. Bertrand, J.L. Henderson, L. Roberts and M.F. Greaney. 2009. The benzyne aza-Claisen reaction. Angew. Chem. Int. Ed. 48: 5199–5202.

(67)   A. Padwa and H. Zhang. 2007. Synthesis of some members of the hydroxylated phenanthridone subclass of the *Amaryllidaceae* alkaloid family. J. Org. Chem. 72: 2570–2582.

(68)   C. Montagne, N. Prevost, J.J. Shiers, G. Prie, S. Rahman, J.F. Hayes and M. Shipman. 2006. Generation and electrophilic substitution reactions of 3-lithio-2-methyleneaziridines. Tetrahedron 62: 8447–8457.

(69)   C.C.A. Cariou, G.J. Clarkson and M. Shipman. 2008. Rapid synthesis of 1,3,4,4-tetrasubstituted β-lactams from methyleneaziridines using a four-component reaction. J. Org. Chem. 73: 9762–9764.

(70)   P.M. Mumford, J.J. Shiers, G.J. Tarver, J.F. Hayes and M. Shipman. 2008. Synthesis of 1,1-disubstituted tetrahydro-β-carbolines from 2-methyleneaziridines. Tetrahedron Lett. 49: 3489–3491.

(71)   R.J. Huntley and R.L. Funk. 2006. Total syntheses of (±)-*cis*-trikentrin A and (±)-*cis*-trikentrin B *via* electrocyclic ring closures of 2,3-divinylpyrrolines. Org. Lett. 8: 3403–3406.

(72)   R. Chinchilla, C. Najera and M. Yus. 2007. Metalated heterocycles in organic synthesis: recent applications (AK-2215GR). ARKIVOC (x): 152–231.

(73)   G.A. Molander and S.K. Pack. 2003. Determining the scope of the lanthanide-mediated, sequential hydroamination/C-C cyclization reaction: formation of tricyclic and tetracyclic aromatic nitrogen heterocycles. Tetrahedron 59: 10581–10591.

(74)   T.E. Muller, K.C. Hultzsch, M. Yus, F. Foubelo and M. Tada. 2008. Hydroamination: direct addition of amines to alkenes and alkynes. Chem. Rev. 108: 3795–3892.

(75)   M.C. Willis, G.N. Brace and I.P. Holmes. 2005. Palladium-catalyzed tandem alkenyl and aryl C-N bond formation: a cascade *N*-annulation route to 1-functionalized indoles. Angew. Chem. Int. Ed. 44: 403–406.

(76)   J.M. Fox, X. Huang, A. Chieffi and S.L. Buchwald. 2000. Highly active and selective catalysts for the formation of α-aryl ketones. J. Am. Chem. Soc. 122: 1360–1370.

(77)   D.A. Culkin and J.F. Hartwig. 2003. Palladium-catalyzed α-arylation of carbonyl compounds and nitriles. Acc. Chem. Res. 36: 234–245.

(78)   M.S. Viciu, R.F. Germaneau and S.P. Nolan. 2002. Well-defined, air-stable (NHC)Pd(Allyl)Cl (NHC = *N*-heterocyclic carbene) catalysts for the arylation of ketones. Org. Lett. 4: 4053–4056.

(79)   M.C. Willis, G.N. Brace, T.J.K. Findlay and I.P. Holmes. 2006. 2-(2-Haloalkenyl)-aryl halides as substrates for palladium-catalyzed tandem C-N bond formation: efficient synthesis of 1-substituted indoles. Adv. Synth. Catal. 348: 851–856.

(80)   A.J. Fletcher, M.N. Bax and M.C. Willis. 2007. Palladium-catalyzed *N*-annulation routes to indoles: the synthesis of indoles with sterically demanding *N*-substituents, including demethylasterriquinone A1. Chem. Commun. 45: 4764–4766.

(81)   S. Cacchi and G. Fabrizi. 2011. Palladium-catalyzed reactions. Chem. Rev. 111: 215–283.

(82)   J. Barluenga and C. Valdes. 2005. Palladium-catalyzed alkenyl amination: from enamines to heterocyclic synthesis. Chem. Commun. 39: 4891–4901.

(83)   S.D. Edmonson, A. Mastriacchio and E.R. Parmee. 2000. Palladium-catalyzed coupling of vinylogous amides with aryl halides: applications to the synthesis of heterocycles. Org. Lett. 2: 1109–1112.

(84)   L.-C. Chen, S.-C. Yang and H.-M. Wang. 1995. Palladium-catalyzed synthesis of 1,2,3,4-tetrahydro-4-oxo-β-carbolines. Synthesis 4: 385–386.

(85)   K. Yamazaki, Y. Nakamura and Y. Kondo. 2003. Solid-phase synthesis of indolecarboxylates using palladium-catalyzed reactions. J. Org. Chem. 68: 6011–6019.

(86)   J.J. Bozell and L.S. Hegedus. 1981. Palladium-assisted functionalization of olefins: a new amination of electron deficient olefins. J. Org. Chem. 46: 2561–2563.

(87)   J.P. Michael, S.-F. Chang and C. Wilson. 1993. Synthesis of pyrrolo[1,2-*a*]indoles by intramolecular Heck reaction of *N*-(2-bromoaryl)enaminones. Tetrahedron Lett. 34: 8365–8368.

(88)   Y. Blache, M.-E. Sinibaldi-Troin, A. Voldoire, O. Chavignon, J.-C. Gramain, J.-C. Teulade and J.-P. Chapat. 1997. Compared reactivity of heterocyclic enaminones: photochemical and palladium-catalyzed synthesis of 6,7,8,9-tetrahydro-5*H*-pyrido[3,2-*b*]indol-9-ones. J. Org. Chem. 62: 8553–8556.

(89)   J. Maruyama, H. Yamashita, T. Watanabe, S. Arai and A. Nishida. 2009. Novel synthesis of fused indoles and 2-substituted indoles by the palladium-catalyzed cyclization of *N*-cycloalkenyl-*o*-haloanilines. Tetrahedron 65: 1327–1335.

(90)   H.-J. Knolker, K.R. Reddy and A. Wagner. 1998. Indoloquinones, part 5. Palladium-catalyzed total synthesis of the potent lipid peroxidation inhibitor carbazoquinocin C. Tetrahedron Lett. 39: 8267–8270.

(91)   B. Dajka-Halasz, K. Monsieurs, O. Elias, L. Karolykazy, P. Tapolcsanyi, B.U.W. Maes, Z. Riedl, G. Hajos, R.A. Dommisse, G.L.F. Lemiere, J. Kosmrlj and P. Matyus. 2004. Synthesis of 5*H*-pyridazino[4,5-*b*]indoles and their benzofuran analogues utilizing an intramolecular Heck-Type reaction. Tetrahedron 60: 2283–2291.

(92)   H. Fuwa and M. Sasaki. 2007. Strategies for the synthesis of 2-substituted indoles and indolines starting from acyclic α-phosphoryloxy enecarbamates. Org. Lett. 9: 3347–3350.

(93)   G. Kirsch, S. Hesse and A. Comel. 2004. Synthesis of five- and six-membered heterocycles through palladium-catalyzed reactions. Curr. Org. Synth. 1: 47–63.

(94)   I. Bytschkov, H. Siebeneicher and S. Doye. 2003. A flexible synthesis of indoline, indolizidine, and pyrrolizidine derivatives. Eur. J. Org. Chem. 15: 2888–2902.

(95)   I. Bytschkov and S. Doye. 2002. The Cp$_2$TiMe$_2$-catalyzed intramolecular hydroamination/cyclization of aminoalkynes. Tetrahedron Lett. 43: 3715–3718.

(96)   A.N. Ryabov, D.V. Gribkov, V.V. Izmer and A.Z. Voskoboynikov. 2002. Zirconium complexes with cyclopentadienyl ligands involving fused a thiophene fragment. Organometallics 21: 2842–2855.

(97)   N. Chernyak, D. Tilly, Z. Li and V. Gevorgyan. 2011. Cascade carbopalladation-annulation approach toward polycylic derivatives of indole and indolizine. ARKIVOC (v): 76–91.

(98)   O.B. Ostby, B. Dalhus, L.-L. Gundersen, F. Rise, A. Bast and G.R.M.M. Haenen. 2000. Synthesis of 1-substituted 7-cyano-2,3-diphenylindolizines and evaluation of anti-oxidant properties. Eur. J. Org. Chem. 22: 3763–3770.

(99)   K.R. Roesch and R.C. Larock. 2001. Synthesis of isoindolo[2,1-*a*]indoles by the palladium-catalyzed annulation of internal acetylenes. J. Org. Chem. 66: 412–420.

(100)   K.R. Roesch and R.C. Larock. 1999. Synthesis of isoindolo[2,1-*a*]indoles by the palladium-catalyzed annulation of internal alkynes. Org. Lett. 1: 1551–1553.

(101)   R.C. Larock and S. Basu. 1987. Synthesis of nitrogen heterocycles *via* palladium-catalyzed intramolecular cyclization. Tetrahedron Lett. 28: 5291–5294.

(102)   A. Kasahara, T. Izumi, S. Murakami, H. Yanai and M. Takatori. 1986. Synthesis of 3-substituted indoles by a palladium-assisted reaction. Bull. Soc. Chim. Jpn. 59: 927–928.

(103)   T. Sakamoto, T. Nagano, Y. Kondo and H. Yamanaka. 1990. Palladium-catalyzed cyclization of β-(2-halophenyl) amino substituted α,β-unsaturated ketones and esters to 2,3-disubstituted indoles. Synthesis 3: 215–218.

(104)   R.C. Larock and E.K. Yum. 1991. Synthesis of indoles *via* palladium-catalyzed heteroannulation of internal alkynes. J. Am. Chem. Soc. 113: 6689–6690.

(105)   H.C. Kolb, M.G. Finn and K.B. Sharpless. 2001. Click chemistry: diverse chemical function from a few good reactions. Angew. Chem. Int. Ed. 40: 2004–2021.

(106)   T. Watanabe, S. Ueda, S. Inuki, S. Oishi, N. Fujii and H. Ohno. 2007. One-pot synthesis of carbazoles by palladium-catalyzed *N*-arylation and oxidative coupling. Chem. Commun. 43: 4516–4518.

(107)   T. Watanabe, S. Oishi, N. Fujii and H. Ohno. 2009. Palladium-catalyzed direct synthesis of carbazoles *via* one-pot *N*-arylation and oxidative biaryl coupling: synthesis and mechanistic study. J. Org. Chem. 74: 4720–4726.

(108)   A.V. Gulevich and V. Gevorgyan. 2012. Synthesis of fused heterocycles *via* Pd-catalyzed multiple aromatic C-H activation reactions. Chem. Heterocycl. Compd. 48: 17–20.

(109)   Z. Liu and R.C. Larock. 2004. Synthesis of carbazoles and dibenzofurans *via* cross-coupling of *o*-iodoanilines and *o*-iodophenols with silylaryl triflates. Org. Lett. 6: 3739–3741.

(110)   J. Zhao and R.C. Larock. 2005. Synthesis of substituted carbazoles by a vinylic to aryl palladium migration involving domino C-H activation processes. Org. Lett. 7: 701–704.

(111)   R.C. Larock, N.G. Berriospena and C.A. Fried. 1991. Regioselective, palladium-catalyzed heteroannulation and carboannulation of 1,2-dienes using functionally substituted aryl halides. J. Org. Chem. 56: 2615–2617.

(112)   Q.H. Huang, A. Fazio, G.X. Dai, M.A. Campo and R.C. Larock. 2004. Pd-catalyzed alkyl to aryl migration and cyclization: an efficient synthesis of fused polycycles *via* multiple C-H activation. J. Am. Chem. Soc. 126: 7460–7461.

(113)   M.A. Campo, Q. Huang, T. Yao, Q. Tian and R.C. Larock. 2003. 1,4-Palladium migration *via* C-H activation, followed by arylation: synthesis of fused polycycles. J. Am. Chem. Soc. 125: 11506–11507.

(114)   Q. Huang, M.A. Campo, T. Yao, Q. Tian and R.C. Larock. 2004. Synthesis of fused polycycles by 1,4-palladium migration chemistry. J. Org. Chem. 69: 8251–8257.

(115) R. Grigg, V. Loganathan, V. Sridharan, P. Stevenson, S. Sukirthalingam and T. Worakun. 1996. Palladium-catalyzed tandem cyclization - anion capture processes. Part 2. Cyclization onto alkynes or allenes with hydride capture. Tetrahedron 52: 11479–11502.

(116) H. Ohno, M. Iuchi, N. Fujii and T. Tanaka. 2007. Zipper-mode double C-H activation: palladium-catalyzed direct construction of highly fused heterocyclic systems. Org. Lett. 9: 4813–4815.

(117) E.M. Beccalli, G. Broggini, M. Martinelli and S. Sottocornola. 2008. Microwave-assisted intramolecular cyclization of electron rich heterocycle derivatives by a palladium-catalyzed coupling reaction. Synthesis 1: 136–140.

(118) Y. Ohta, H. Chiba, S. Oishi, N. Fujii and H. Ohno. 2009. Construction of nitrogen heterocycles bearing an aminomethyl group by copper-catalyzed domino three-component coupling-cyclization. J. Org. Chem. 74: 7052–7058.

(119) G. Cignarella, P. Sanna. E. Miele, V. Anania and M.S. Desole. 1981. Diuretic agents related to indapamide. 1. Synthesis and activity of 1-substituted 2-(4-chloro-3-sulfamoylbenzamido)isoindolines. J. Med. Chem. 24: 1003–1006.

(120) P. Sanna, G. Cignarella, V. Ananiaa, R.M.S. Siri and M.S. Desole. 1985. Diuretic agents related to indapamide. II. Synthesis and pharmacological activity of 1-alkyl, 1-carboxyalkyl- and 1,3-dialkyl-substituted 2-(4-chloro-3-sulfamoylbenzamido)isoindolines. Farmaco Ed. Sci. 40: 777–785.

(121) J.M. Schulman, A.A. Friedman, J. Panteleev and M. Lautens. 2012. Synthesis of 1,2,3-triazole-fused heterocycles *via* Pd-catalyzed cyclization of 5-iodotriazoles. Chem. Commun. 48: 55–57.

(122) A. Ashimori, T. Matsuura, L.E. Overman and D.J. Poon. 1993. Catalytic asymmetric synthesis of either enantiomer of physostigmine. Formation of quaternary carbon centers with high enantioselection by intramolecular Heck reactions of (Z)-2-butenanilides. J. Org. Chem. 58: 6949–6951.

(123) I. Ojima, M. Tzamarioudaki, Z. Li and R.J. Donovan. 1996. Transition metal-catalyzed carbocyclizations in organic synthesis. Chem. Rev. 96: 635–662.

(124) B. Wang, Y. Meng, Y. Zhou, L. Ren, J. Wu, W. Yu and J. Chang. 2017. Synthesis of 5-amino and 3,5-diamino-substituted 1,2,4-thiadiazoles by I_2-mediated oxidative N-S bond formation. J. Org. Chem. 82: 5898–5903.

(125) O.A. Lozinskii, T.V. Shokol and V.P. Khilya. 2011. Synthesis and biological activity of chromones annelated at the C(7)-C(8) bond with heterocycles (review). Chem. Heterocycl. Compd. 47: 1055–1077.

(126) R. Riva, L. Banfi, A. Basso, V. Cerulli, G. Guanti and M. Pani. 2010. A highly convergent synthesis of tricyclic N-heterocycles coupling an Ugi reaction with a tandem S(N)2'-Heck double cyclization. J. Org. Chem. 75: 5134–5143.

(127) A. Sharma, P. Appukkuttana and E.V. Eycken. 2012. Microwave-assisted synthesis of medium-sized heterocycles. Chem. Commun. 48: 1623–1637.

(128) J. Yin and S.L. Buchwald. 2002. Pd-catalyzed intermolecular amidation of aryl halides: the discovery that xantphos can be *trans*-chelating in a palladium complex. J. Am. Chem. Soc. 124: 6043–6048.

(129) J.F. Hartwig. 1998. Filtering algorithm for noise reduction in phase-map images with 2 pi phase jumps. Angew. Chem. Int. Ed. 37: 2046–2050.

(130) S.L. Buchwald, C. Mauger, G. Mignani and U. Scholz. 2006. Industrial scale palladium-catalyzed coupling of aryl halides and amines - a personal account. Adv. Synth. Catal. 348: 23–39.

(131) J.P. Wolfe, S. Wagaw, J.F. Marcoux and S.L. Buchwald. 1998. Rational development of practical catalysts for aromatic carbon-nitrogen bond formation. Acc. Chem. Res. 31: 805–818.

(132) K. Muniz. 2009. High-oxidation-state palladium catalysis: new reactivity for organic synthesis. Angew. Chem. Int. Ed. 48: 2–14.

(133) C.-K. Ryu, I.H. Choi and R.E. Park. 2006. Synthesis of benzo[b]benzo[2,3-d]thiophen-6,9-diones *via* palladium(II) acetate-mediated cyclization of 5-arylthiocyclohexa-2,5-diene-1,4-diones. Synth. Commun. 36: 3319–3328.

(134) R.S. Colemen and W. Chen. 2001. A convergent approach to the mitomycin ring system. Org. Lett. 3: 1141–1144.

(135) F. He, B.M. Foxman and B.B. Snider. 1998. Total syntheses of (–)-asperlicin and (–)-asperlicin C. J. Am. Chem. Soc. 120: 6417–6418.

(136) S.S. Cammerer, M.S. Viciu, E.D. Stevens and S.P. Nolan. 2003. Use of (NHC)Pd(η³-allyl)Cl (NHC = N-heterocyclic carbene) in a palladium-mediated approach to cryptocaryaalkaloids. Synlett 12: 1871–1873.

(137) M.D. Ganton and M.A. Kerr. 2007. Aryl amidation routes to dihydropyrrolo[3,2-e]indoles and pyrrolo[3,2-f] tetrahydroquinolines: total synthesis of the (±)-CC-1065 CPI subunit. J. Org. Chem. 72: 574–582.

(138) B. Witulski and C. Alayrac. 2002. A highly efficient and flexible synthesis of substituted carbazoles by rhodium-catalyzed inter- and intramolecular alkyne cyclotrimerizations. Angew. Chem. Int. Ed. 41: 3281–3284.

(139) J.M. Mejia-Oneto and A. Padwa. 2004. Intramolecular [3+2]-cycloaddition reaction of push-pull dipoles across heteroaromatic π-systems. Org. Lett. 6: 3241–3244.

(140) B.J. Stokes, K.J. Richert, T.G. Driver and T.G. Driver. 2009. Examination of the mechanism of Rh(II)-catalyzed carbazole formation using intramolecular competition experiments. J. Org. Chem. 74: 6442–6451.

(141) T.G. Driver. 2010. Recent advances in transition metal-catalyzed N-atom transfer reactions of azides. Org. Biomol. Chem. 8: 3831–3846.

(142) W.-H. Chiou, G.-H. Lin, C.-C. Hsu, S.J. Chaterpaul and I. Ojima. 2009. Efficient syntheses of crispine-A and harmicine by Rh-catalyzed cyclohydrocarbonylation. Org. Lett. 11: 2659–2662.

(143) B.J. Stokes, B. Jovanovic, H. Dong, K.J. Richert, R.D. Riell, T.G. Driver and G. Driver. 2009. Rh(II)-catalyzed synthesis of carbazoles from biaryl azides. J. Org. Chem. 74: 3225–3228.

(144) J.S. Swenton, T.J. Ikeler and B.H. Williams. 1970. Photochemistry of singlet and triplet azide excited states. J. Am. Chem. Soc. 92: 3103–3109.

(145) S. Kotha, E. Brahmachary and K. Lahiri. 2005. Transition metal-catalyzed [2+2+2]-cycloaddition and application in organic synthesis. Eur. J. Org. Chem. 22: 4741–4767.

(146) A.G.H. Wee and S.C. Duncan. 2002. The rhodium(II)-catalyzed reaction of *N*-bis(trimethylsilylmethyl)diazoamides: steric, electronic and conformational effects. Tetrahedron Lett. 43: 6173–6176.

(147) M.P. Doyle, W. Hu, A.G.H. Wee, Z. Wang and S.C. Duncan. 2003. Influences of catalyst configuration and catalyst loading on selectivities in reactions of diazoacetamides. Barrier to equilibrium between diastereomeric conformations. Org. Lett. 5: 407–410.

(148) A.G.H. Wee and S.C. Duncan. 2005. The bis(trimethylsilyl)methyl group as an effective *N*-protecting group and site-selective control element in rhodium(II)-catalyzed reaction of diazoamides. J. Org. Chem. 70: 8372–8380.

(149) A.G.H. Wee, S.C. Duncan and G.-J. Fan. 2006. Intramolecular asymmetric C-H insertion of *N*-arylalkyl, *N*-bis(trimethylsilyl)methyldiazoamides mediated by chiral rhodium(II) catalysts. Synthesis of (*R*)-β-benzyl-γ-aminobutyric acid. Tetrahedron: Asymmetry 17: 297–307.

(150) D. Muroni and A. Saba. 2005. Quinoline β-lactams by Rh(II)-catalyzed highly stereoselective carbon-hydrogen intramolecular insertion (05-1345EP). ARKIVOC (xiii): 1–7.

(151) A. Brandi, S. Cicchi and F.M. Cordero. 2008. Novel syntheses of azetidines and azetidinones. Chem. Rev. 108: 3988–4035.

(152) K.L. Tan, R.G. Bergman and J.A. Ellman. 2002. Intermolecular coupling of isomerizable alkenes to heterocycles *via* rhodium-catalyzed C-H bond activation. J. Am. Chem. Soc. 124: 13964–13965.

(153) K.L. Tan, S. Park, J.A. Ellman and R.G. Bergman. 2004. Intermolecular coupling of alkenes to heterocycles *via* C-H bond activation. J. Org. Chem. 69: 7329–7335.

(154) J.C. Lewis, R.G. Bergman and J.A. Ellman. 2008. Direct functionalization of nitrogen heterocycles *via* Rh-catalyzed C-H bond activation. Acc. Chem. Res. 41: 1013–1025.

(155) S. Gowrisankar, H.S. Lee, J.M. Kim and J.N. Kim. 2008. Pd-mediated synthesis of 2-arylquinolines and tetrahydropyridines from modified Baylis-Hillman adducts. Tetrahedron Lett. 49: 1670–1673.

(156) Y. Yamamoto, H. Kitahara, R. Ogawa and K. Itoh. 1998. Cp*Ru(cod)Cl-catalyzed [2+2+2]-cycloaddition of 1,6-heptadiynes with allylic ethers. A decisive role of coordination to the ether oxygen atom. J. Org. Chem. 63: 9610–9611.

(157) Y. Yamamoto, H. Kitahara, R. Ogawa, H. Kawaguchi, K. Tatsumi and K. Itoh. 2000. Ru(II)-catalyzed cycloadditions of 1,6-heptadiynes with alkenes: new synthetic potential of ruthenacyclopentatrienes as bis-carbenoids in tandem cyclopropanation of bicycloalkenes and heteroatom-assisted cyclocotrimerization of 1,6-heptadiynes with heterocyclic alkenes. J. Am. Chem. Soc. 122: 4310–4319.

(158) J.D. Sunderhaus, C. Dockendorff and S.F. Martin. 2007. Applications of multi-component reactions for the synthesis of diverse heterocyclic scaffolds. Org. Lett. 9: 4223–4226.

(159) H. Hügel. 2009. Microwave multi-component synthesis. Molecules 14: 4936–4972.

(160) C. Crotti, S. Cenini, A. Bossoli, B. Rindone and F. Demartin. 1991. Synthesis of carbazole by Ru$_3$(CO)$_{12}$-catalyzed reductive carbonylation of 2-nitrobiphenyl: the crystal and molecular structure of Ru$_3$(μ$_3$-NC$_6$H$_4$-*o*-C$_6$H$_5$)$_2$(CO)$_9$. J. Mol. Catal. 70: 175–187.

(161) M. Pizzotti, S. Cenini, S. Quici and S. Tollari. 1994. Role of alkali halides in the synthesis of nitrogen-containing heterocycles by reductive carbonylation of aromatic nitro-derivatives catalyzed by Ru$_3$(CO)$_{12}$. J. Chem. Soc. Perkin Trans. 2 4: 913–917.

(162) H. Alper and J.T. Edward. 1970. Reactions of iron pentacarbonyl with compounds containing the N-O linkage. Can. J. Chem. 48: 1543–1549.

(163) W.S. Knowles. 2002. Asymmetric hydrogenations (Nobel lecture). Angew. Chem. Int. Ed. 41: 1998–2007.

(164) R. Noyori. 2002. Asymmetric catalysis: science and opportunities (Nobel lecture). Angew. Chem. Int. Ed. 41: 2008–2022.

(165) B. Sharpless. 2002. Searching for new reactivity (Nobel lecture). Angew. Chem. Int. Ed. 41: 2024–2032.

(166) R.H. Grubbs. 2006. Olefin metathesis catalysts for the preparation of molecules and materials (Nobel lecture). Angew. Chem. Int. Ed. 45: 3760–3765.

(167) Y. Chauvin. 2006. Olefin metathesis: the early days (Nobel lecture). Angew. Chem. Int. Ed. 45: 3740–3747.

(168) (a) R.R. Schrock. 2006. Multiple metal-carbon bonds for catalytic metathesis reactions (Nobel lecture). Angew. Chem. Int. Ed. 45: 3748–3759. (b) E. Ascic, J.F. Jensen and T.E. Nielsen. 2011. Synthesis of heterocycles through a ruthenium-catalyzed tandem ring-closing metathesis/isomerization/*N*-acyliminium cyclization sequence. Angew. Chem. Int. Ed. 50: 5188–5191.

(169) Y. Fukuda, M. Shindo and K. Shishido. 2003. Total synthesis of (–)-aspidospermine *via* diastereoselective ring-closing olefin metathesis. Org. Lett. 5: 749–751.

(170) A.H. Hoveyda, S.J. Malcolmson, S.J. Meek and A.R. Zhugralin. 2010. Catalytic enantioselective olefin metathesis in natural product synthesis. Chiral metal-based complexes that deliver high enantioselectivity and more. Angew. Chem. Int. Ed. Engl. 49: 34–44.

(171) C. Bressy, C. Menant and O. Piva. 2005. Synthesis of polycyclic lactams and sultams by a cascade ring-closure metathesis/isomerization and subsequent radical cyclization. Synlett 4: 577–582.

(172) K.C. Majumdar, S. Muhuri, R.U. Islam and B. Chattopadhyay. 2009. Synthesis of five- and six-membered heterocyclic compounds by the application of the metathesis reactions. Heterocycles 78: 1109–1169.

(173) M. Nakagawa and M. Kawahara. 2000. A concise synthesis of physostigmine from skatole and activated aziridine *via* alkylative cyclization. Org. Lett. 2: 953–955.

(174) S. Kobayashi, M. Sugiura, H. Kitagawa and W.W.-L. Lam. 2002. Rare-earth metal triflates in organic synthesis. Chem. Rev. 102: 2227–2302.

(175) M. Nüchter and B. Ondruschka. 2003. Tools for microwave-assisted parallel syntheses and combinatorial chemistry. Mol. Divers. 7: 253–264.

(176) M. Nüchter, U. Müller, B. Ondruschka, A. Tied and W. Lautenschläger. 2003. Microwave-assisted chemical reactions. 26: 1207–1216.

(177) W. Zhang and G. Pugh. 2003. Free radical reactions for heterocycle synthesis. Part 6: 2-Bromobenzoic acids as building blocks in the construction of nitrogen heterocycles. Tetrahedron 59: 3009–3018.

(178) K. Orito, Y. Satoh, H. Nishizawa, R. Harada and M. Tokuda. 2000. Aryl radical cyclizations: one-pot syntheses of protoberberine and pavine alkaloids. Org. Lett. 2: 2535–2537.

(179) M. Lucilia, E.N. da Mata, W.B. Motherwell and F. Ujjainwalla. 1997. Steric and electronic effects in the synthesis of biaryls and their heterocyclic congeners using intramolecular free radical [1,5] *ipso* substitution reactions. Tetrahedron Lett. 38: 137–140.

(180) S. Mukhopadhyay, G. Rothenberg, D. Gitis, M. Baidossi, D.E. Ponde and Y. Sasson. 2000. Regiospecific cross-coupling of haloaryls and pyridine to 2-phenylpyridine using water, zinc, and catalytic palladium on carbon. J. Chem. Soc. Perkin Trans. 2 9: 1809–1812.

(181) D.C. Harrowven, B.J. Sutton and S. Coulton. 2001. Intramolecular radical cyclizations to pyridines. Tetrahedron Lett. 42: 9061–9064.

(182) D.C. Harrowven, B.J. Sutton and S. Coulton. 2001. Intramolecular radical additions to quinolines. Tetrahedron Lett. 42: 2907–2910.

(183) Y.-T. Park, C.-H. Jung, M.-S. Kim, K.-W. Kim, N.W. Song and D. Kim. 2001. Photoreaction of 2-halo-*n*-pyridinylbenzamide: intramolecular cyclization mechanism of phenyl radical assisted with *n*-complexation of chlorine radical. J. Org. Chem. 66: 2197–2206.

(184) E. Bacque, M.E. Qacemi and S.Z. Zard. 2004. Tin-free radical cyclizations for the synthesis of 7-azaoxindoles, 7-azaindolines, tetrahydro[1,8]naphthyridines, and tetrahydro-5*H*-pyrido[2,3-*b*]azepin-8-ones. Org. Lett. 6: 3671–3674.

(185) B. Chattopadhyay and V. Gevorgyan. 2012. Transition metal-catalyzed denitrogenative transannulation: converting triazoles into other heterocyclic systems. Angew. Chem. Int. Ed. 51: 862–872.

(186) M. Laronze-Cochard, F. Cochard, E. Daras, A. Lansiaux, B. Brassart, E. Vanquelef, E. Prost, J. Nuzillard, B. Baldeyrou, J. Goosens, O. Lozach, L. Meijer, J. Riou, E. Henon and J. Sapi. 2010. Synthesis and biological evaluation of new penta- and heptacyclic indolo- and quinolinocarbazole ring systems obtained *via* Pd0-catalyzed reductive *N*-heteroannulation. Org. Biomol. Chem. 8: 4625–4636.

(187) A. Das, A. Kulkarni and B. Torok. 2012. Environmentally benign synthesis of heterocyclic compounds by combined microwave-assisted heterogeneous catalytic approaches. Green Chem. 14: 17–34.

(188) D.A. Dickman and C.H. Heathcock. 1989. Total synthesis of (+)-vallesamidine. J. Am. Chem. Soc. 111: 1528–1530.

(189) H.F. He, S. Dong, Y. Chen, Y. Yang, Y. Le and W. Bao. 2012. Synthesis of 6*H*-isoindolo[2,1-*a*]indol-6-ones through a sequential copper-catalyzed C-N coupling and palladium-catalyzed C-H activation reaction. Tetrahedron 68: 3112–3116.

(190) Z.J. Wang, F. Yang, X. Lv and W. Bao. 2011. Synthesis of un-symmetrical 2,2'-biindolyl derivatives by a Cu-catalyzed *N*-arylation/Pd-catalyzed direct arylation sequential process. J. Org. Chem. 76: 967–970.

(191) S.C. Bryan and M. Lautens. 2008. Silver-promoted domino Pd-catalyzed amination/direct arylation: access to polycyclic heteroaromatics. Org. Lett. 10: 4633–4636.

(192) C.J. Ball and M.C. Willis. 2013. Cascade palladium- and copper-catalyzed aromatic heterocycle synthesis: the emergence of general precursors. Eur. J. Org. Chem. 3: 425–441.

(193) C.A. Demerson, L.G. Humber, T. Dobson and I.L. Jirkovsky. 1976. Certain pyrano[3,4-*b*]indoles and thiopyrano[3,4-*b*] indoles. U.S. Patent 3,393,178, Feb 17.

(194) H. Kusama, J. Takaya and N. Iwasawa. 2002. A facile method for the synthesis of polycyclic indole derivatives: the generation and reaction of tungsten-containing azomethine ylides. J. Am. Chem. Soc. 124: 11592–11593.

(195) M. Lautens, W. Klute and W. Tam. 1996. Transition metal-mediated cycloaddition reactions. Chem. Rev. 96: 49–92.

(196) H.-W. Fruhauf. 1997. Metal-assisted cycloaddition reactions in organotransition metal chemistry. Chem. Rev. 97: 523–596.

(197) A.W. Oxford, C.D. Eldred, I.H. Coates, J.A. Bell, D.C. Humber, G.B. Ewan. April 19, 1988. Process for preparing tetrahydrocarbazolones. U.S. Patent 4,739,072.

5

Five-Membered N,N-Heterocycles

5.1 Introduction

Pyrazolones are used in drugs and are of pharmacological importance, for they exhibit a number of medically useful properties—anti-fungal, anti-microbial, anti-bacterial, anti-mycobacterial, anti-tumor, anti-platelet, anti-inflammatory, anti-depressant, gastric secretion stimulatory as well as anti-filarial. They also work as substrates for pigments, dyes, chelating agents and pesticides [1–3]. Moreover, pyrazolones act as inhibitors against CDK2, with remarkable activity against a number of human tumor cell lines, cannabinoid type-1 (CB1) receptor antagonists, and against tissue-nonspecific alkaline phosphatase (TNAP). In pesticide chemistry, they have come to be used as fungicides, insecticides and herbicides. Pyrazole moiety is a core structure of a wide range of biologically active compounds such as blockbuster drugs like celecoxib and sildenafil (viagra). Whereas celecoxib exhibits anti-arthritic and analgesic activities and is a powerful COX-2 inhibitor, sildenafil (viagra) is a FDA approved drug used to treat erectile dysfunction [4–5].

The pharmacological activities of imidazole related drugs have encouraged medicinal chemists to produce a number of chemotherapeutic agents [6]. It has been found that the imidazole ring is present in many natural compounds like histamine, purine, nucleic acid, and histidine.

Benzimidazole is an important heterocycle known for its pharmacological activities and synthetic utility. It is used in a number of ways—as an anti-helminthic, anti-fungal, anti-histaminic, anti-HIV, anti-hypertensive, anti-ulcer, neuroleptic and cardiotonic. Extensive pharmacological and biochemical studies have confirmed that benzimidazoles are effective against many strains of microorganisms and hence have an important place among chemotherapeutic agents. The pharmacological importance of benzimidazoles is because of its close relationship with the structure of purines. The role of purines in biological system is well established and it has also been found that 5,6-dimethyl-1-(α-D-ribofuranosyl)benzimidazole is an integral constituent of Vitamin B12 [7–16].

5.2 Metal- and non-metal-assisted synthesis of five-membered heterocycles with two nitrogen atoms

Aluminium-assisted synthesis

Microwave-assisted solvent-free synthesis of 1,2,4,5-substituted and 2,4,5-substituted imidazoles has often been reported. Imidazoles can be obtained by condensation of a 1,2-dicarbonyl compound with an amine and an aldehyde using acidic alumina impregnated with ammonium acetate as the solid support. For instance, this synthesis protocol, through the condensation of 1,2-diarylethandienones with aldehydes and amines, provides 2,4,5-substituted **(Scheme 1)** or 1,2,4,5-substituted **(Scheme 2)** imidazoles respectively, employing ammonium acetate as the ammonia source. The solid-phase version of this reaction yields libraries of substituted imidazoles. This method involves impregnating the mixture of ammonium acetate (ammonia source) and the solid support with an ether solution of starting reagents, evaporating the solvent, and heating the solid residue in a microwave oven. A benzyl/benzaldehyde pair can be used for preliminary optimization of the reaction conditions. Among different supports tested, the desired product is formed on basic, acidic and neutral alumina, montmorillonite K10, bentonite, silica gel and montmorillonite KSF. It has been observed that the addition of a small amount (several drops) of acetic acid is needed for the reaction to occur on basic and neutral supports like neutral and basic alumina and silica gel; the reaction is successful without additional acid when acidic supports are used (montmorillonite K10, bentonite, acidic alumina, or montmorillonite KSF). Comparison of different supports has revealed that acidic alumina is the most suitable support. Optimum reaction conditions for the synthesis are found to be a reaction time of 20 minutes with MWI power of 130 W. However, no reaction occurs in the absence of heating arrangements or when the support is omitted from the reaction mixture. The optimum reaction conditions can then be applied for the synthesis of a series of 1,2,4,5-substituted imidazoles and 2,4,5-substituted imidazoles [17–18].

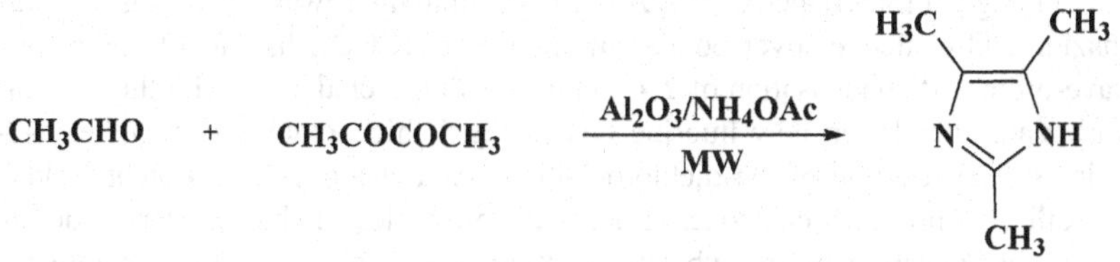

$$CH_3CHO \quad + \quad CH_3COCOCH_3 \quad \xrightarrow[\text{MW}]{Al_2O_3/NH_4OAc}$$

Scheme-1

$$CH_3CHO \quad + \quad CH_3COCOCH_3 \quad + \quad CH_3NH_2 \quad \xrightarrow[\text{MW}]{Al_2O_3/NH_4OAc}$$

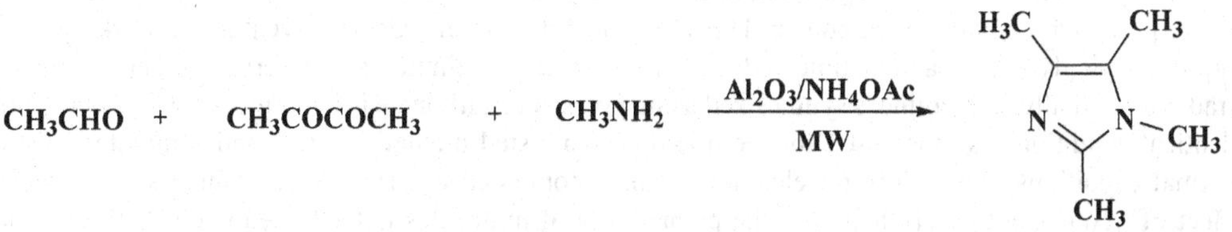

Scheme-2

The synthesis (microwave techniques) and anti-microbial activity of some 3-substituted phenyl-5-substituted phenyl-4,5-dihydro-pyrazole-1-carbothioamides and [3-(4-phenyl)-5-phenyl-4,5-dihydropyrazol-1-yl](pyridine-4-yl)methanones have also been performed by researchers. Substituted chalcones are synthesized when substituted acetophenones react with appropriately substituted benzaldehydes in ethanol. Substituted [3-(4-phenyl)-5-phenyl-4,5-dihydropyrazol-1-yl](pyridine-4-yl)methanones are obtained by the reaction of chalcones with isonicotinic acid hydrazide. Substituted 3,5-diphenyl-4,5-dihydro-1*H*-pyrazole-1-carbothioamides are formed when these chalcones react with thiosemicarbazide **(Scheme 3)** [19].

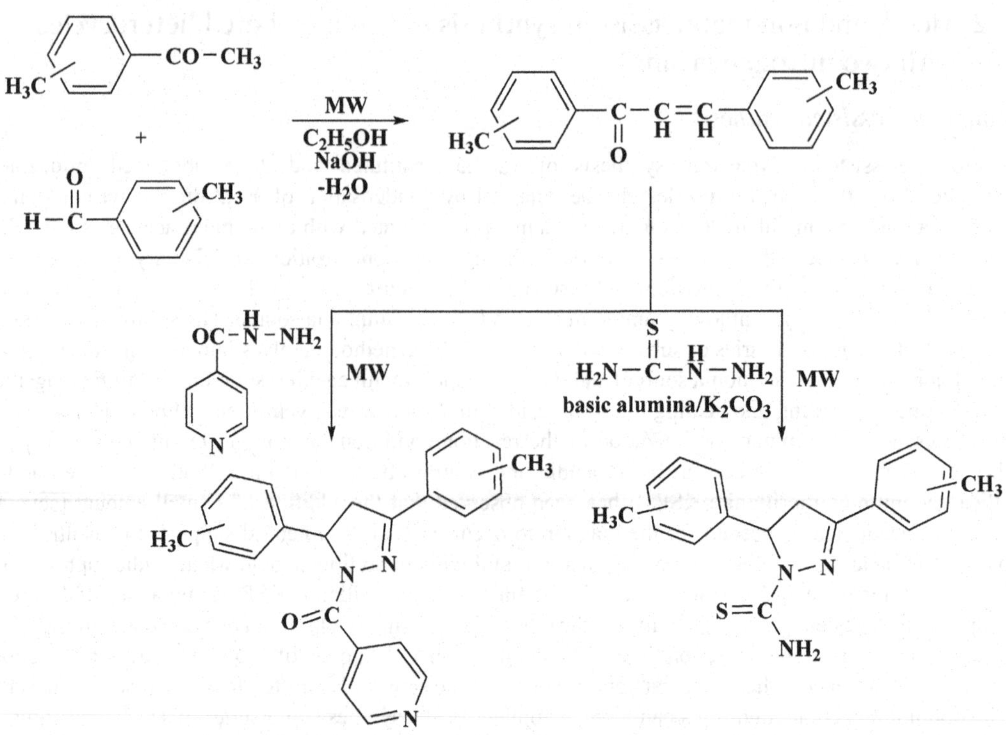

Scheme-3

Due to the biological importance of pyrazolines, a simple microwave synthesis of pyrazolines from thiosemicarbazide and chalcones over potassium carbonate (K_2CO_3) has also been reported. Solvent-free microwave-promoted condensation of 2,4-dichloro-5-fluoro chalcones with thiosemicarbazide over potassium carbonate affords ten new fluorine possessing 1-thiocarbamoyl-3,5-diphenyl-2-pyrazolines in 80–85% yields. The reaction of 2,4-dichloro-5-fluoroacetophenone with aromatic aldehydes in the presence of alkali also provides chalcones. Equimolar quantities of thiosemicarbazide and chalcones adsorbed over potassium carbonate and subjected to microwave irradiation yield pyrazolines **(Scheme 4)**. In a comparative study of potassium carbonate and basic alumina, it was found that the elution of product from a 1 g reaction batch with basic alumina needed about 25–30 ml of acetone, whereas only water was required with potassium carbonate. This eliminated the use of organic solvent in the work-up stage. Improved yields with reaction time reduced from hours to minutes are observed under microwave irradiation. Finally, pyrazoline is synthesized using potassium carbonate in a thermostatted oil-bath under identical conditions as those used for the microwave-assisted method (120 °C and 9 minutes). Under thermal conditions (27%), lower yields are obtained compared to MWI (82%), which shows that the effect of MWI is not purely thermal. The polarization of molecules is facilitated under MWI, causing rapid reaction to occur [20a–b].

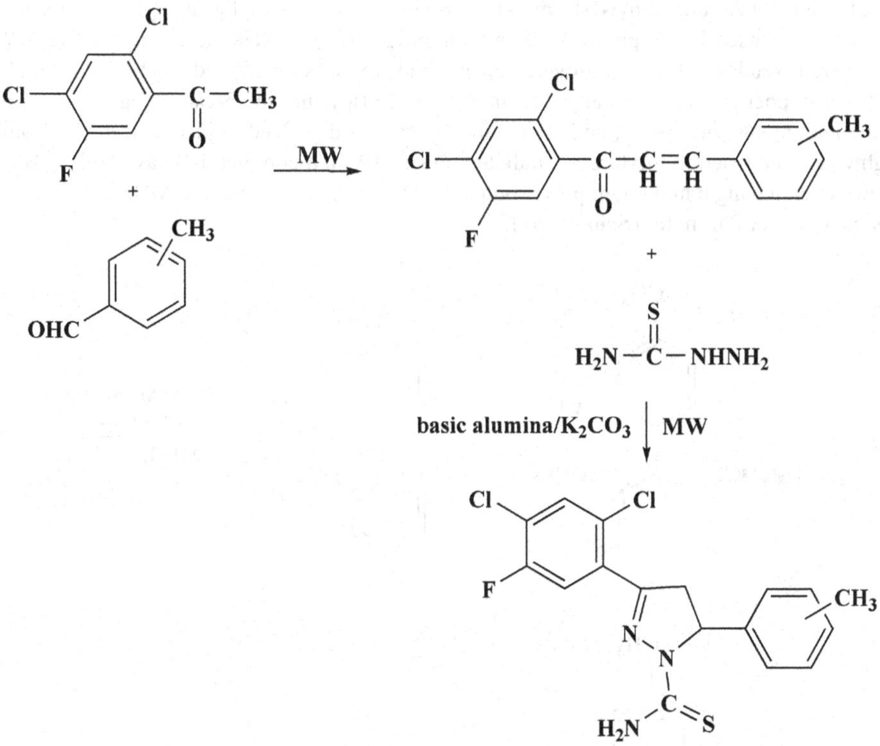

Scheme-4

Ondansetron is a prototypic 5-HT3 receptor antagonist. The imidazole moiety present in ondansetron is incorporated into the molecule upon substitution of trimethylammonium group of an advanced intermediate by 2-methylimidazole [21]. This substitution is carried out on cyclohexenone prior to the Fischer indole synthesis [22]. The imidazole required here is synthesized by various protocols. For example, the desired compound is obtained in excellent yields (95%) by the reaction of glyoxal, acetaldehyde and ammonium carbonate (**Scheme 5**). The 2-methylimidazole is formed in approximately 90% yield by condensation of ethylenediamine and acetic acid at high temperatures in the presence of γ-aluminium oxide as the catalyst [23–24].

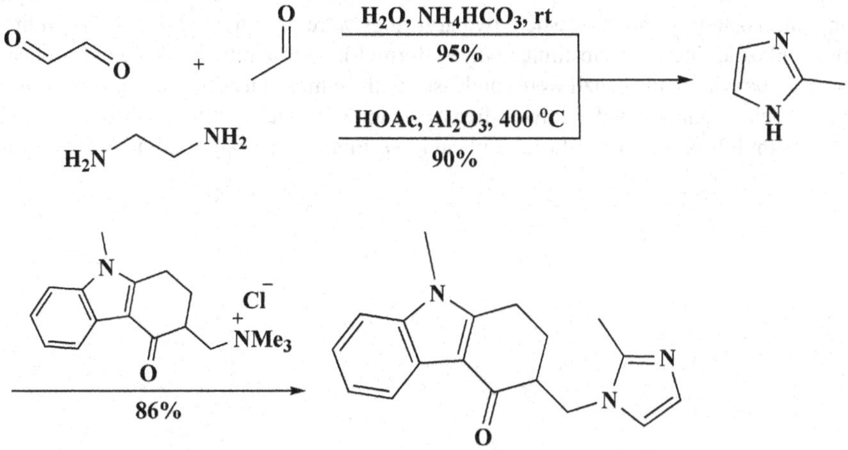

Scheme-5

Substituted benzil and ethyl-1-formyl-1,2,3,6-tetrahydro-4-methyl-6-phenyl-2-oxo-pyrimidine-5-carboxylates condensed in the presence of ammonium acetate in glacial acetic acid under MWI for 8 minutes and refluxed for 12 hours in conventional method produces substituted ethyl-1,2,3,6-tetrahydro-4-methyl-2-oxo-6-phenyl-1-(4,6-diphenyl-1*H*-imidazolyl-2-yl)pyrimidine-5-carboxylates. Subsequently, this compound, along with benzil, and ammonium acetate, is dissolved in glacial acetic acid and mixed thoroughly, and the reaction mixture is irradiated under MW in a commercially available IFB domestic microwave oven having a maximum power output of 110 W operating at 2450 MHz intermittently at 30 seconds intervals for 8 minutes **(Scheme 6)** [25].

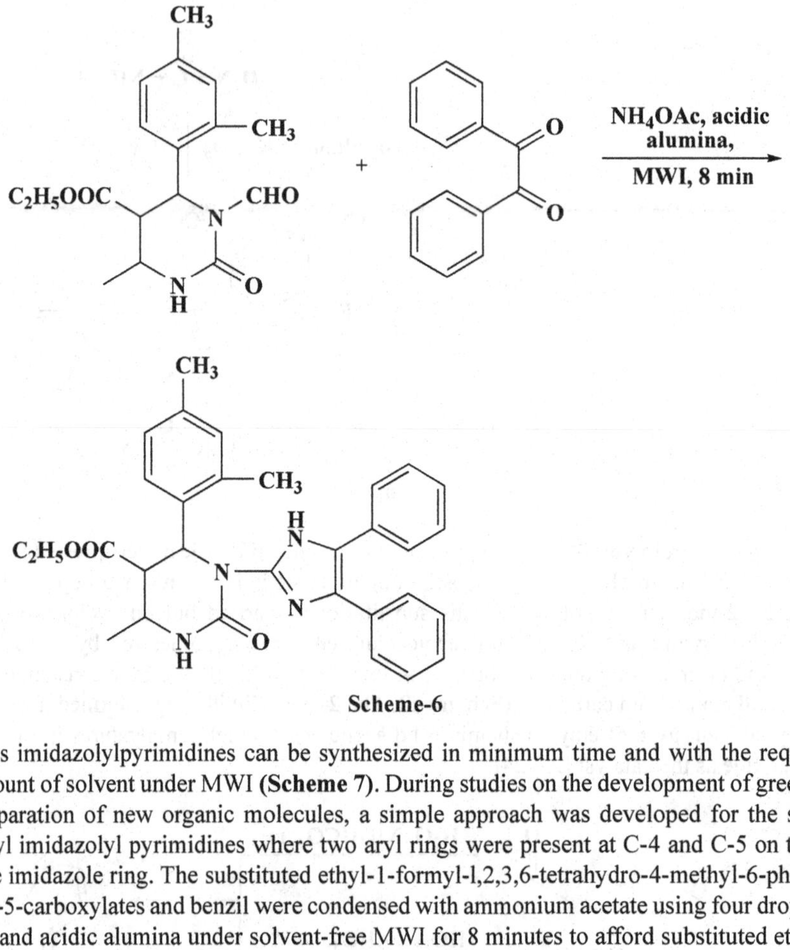

Scheme-6

Various imidazolylpyrimidines can be synthesized in minimum time and with the requirement of limited amount of solvent under MWI **(Scheme 7)**. During studies on the development of green protocols for the preparation of new organic molecules, a simple approach was developed for the synthesis of 4,5-diphenyl imidazolyl pyrimidines where two aryl rings were present at C-4 and C-5 on the opposite faces of the imidazole ring. The substituted ethyl-1-formyl-1,2,3,6-tetrahydro-4-methyl-6-phenyl-2-oxo-pyrimidine-5-carboxylates and benzil were condensed with ammonium acetate using four drops of glacial acetic acid and acidic alumina under solvent-free MWI for 8 minutes to afford substituted ethyl-1,2,3,6-tetrahydro-4-methyl-2-oxo-6-phenyl-1-(4,5-diphenyl-1*H*-imidazol-2-yl)pyrimidine-5-carboxylates [26].

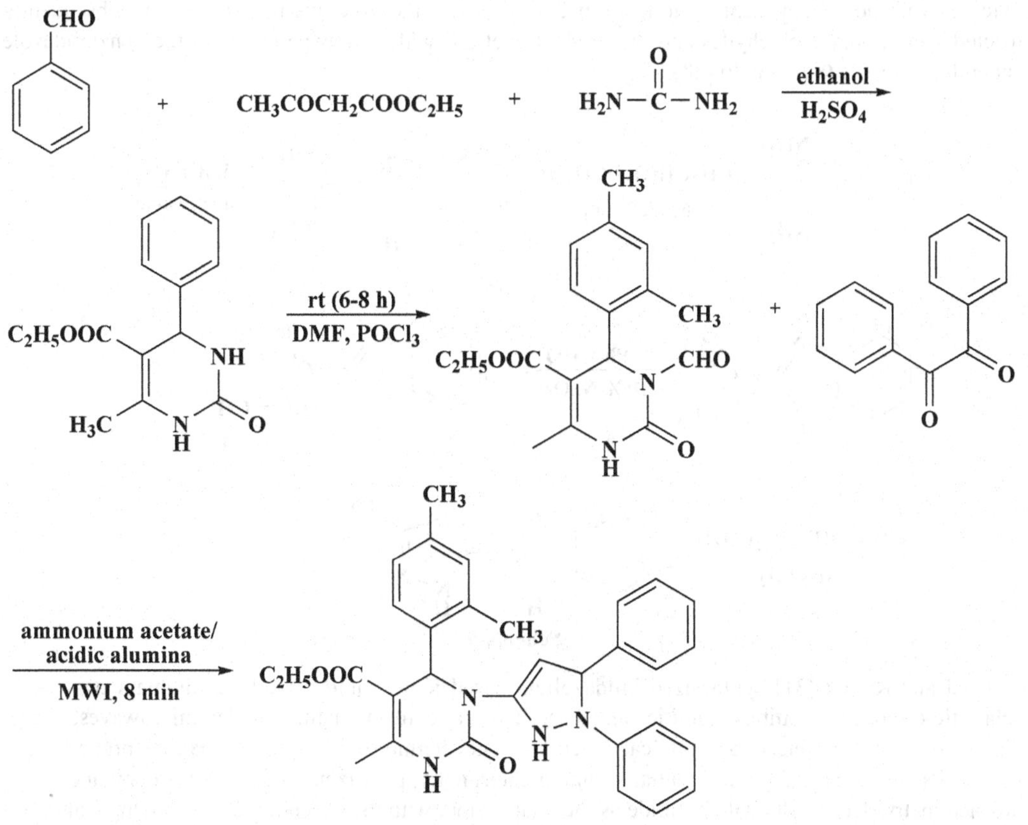

Scheme-7

Cesium-assisted synthesis

Internal olefins bearing *N*-allylureas can be utilized for the synthesis of various imidazolidin-2-ones **(Scheme 8)**. In all such cases, products are obtained in moderate to good yields and are reported as a single diastereomer. Moreover, both the electron-withdrawing 4-bromobenzonitrile and the electron-donating 2-bromo-6-methoxynaphthalene can be tolerated as coupling partners. Sodium *tert*-butoxide has also proved to be an acceptable base for providing carboamination products in some reactions [27–28].

A series of pyrazole derivatives of benzimidazole were prepared by Kalirajan et al. [29]. Substituted *o*-phenylenediamine was condensed with lactic acid under MWI and the obtained product was further

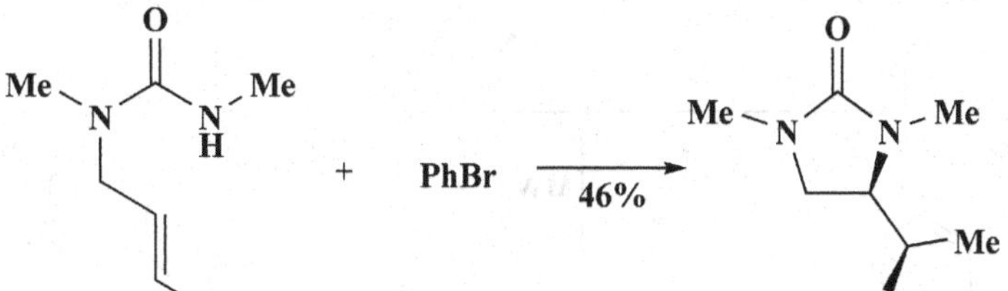

Scheme-8

oxidized with potassium dichromate to form 2-acetyl benzimidazole intermediate which subsequently reacted with aromatic aldehydes and the product cyclized with hydrazine to afford the benzimidazole appended pyrazole **(Scheme 9)** [30].

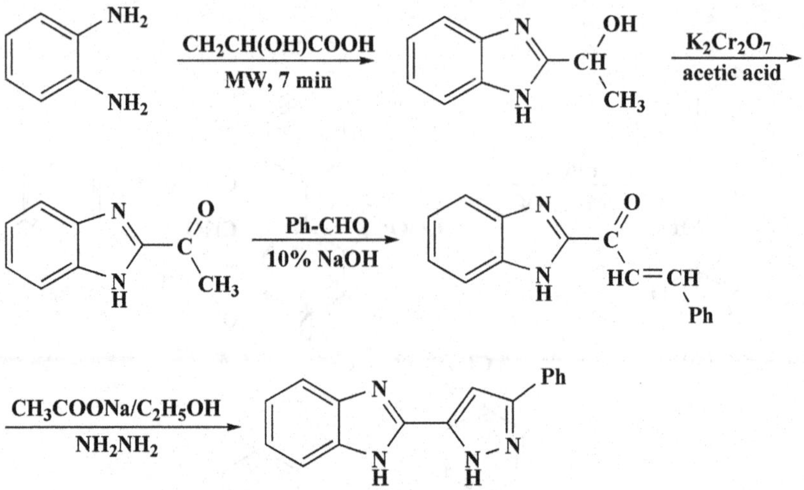

Scheme-9

Pal and Kumar [31] synthesized imidazolium-based ionic liquid crystalline dimmers possessing calamitic-discotic, calamitic-calamitic and discotic-discotic functionalities under microwaves. These dimers were not produced by classical reactions. The thermotropic liquid crystalline properties of these salts were reported by differential scanning calorimetry, polarizing optical microscopy and X-ray diffractometry. These salts with bromide as the counter ion, with the exception of one having calamitic-discotic units, were found to be mesom **(Scheme 10)**.

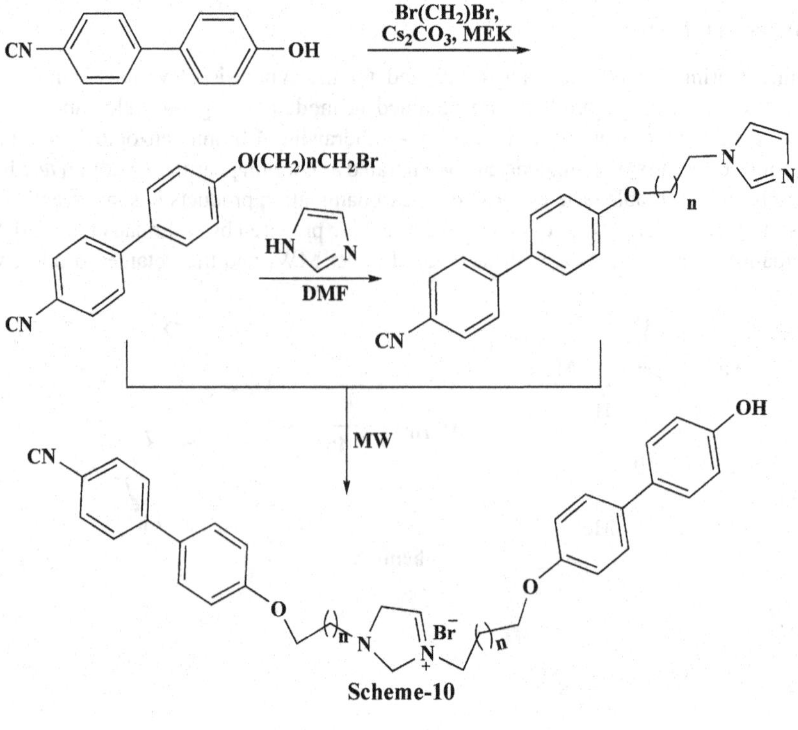

Scheme-10

Copper-assisted synthesis

Pyrazolo[3,4-*b*]pyridines [32] are an important class of heterocycles exhibiting a wide range of bioactivities and are structural analogs of purine bases. 5-Aminopyrazoles are useful intermediates for the preparation of pyrazolo[3,4-*b*]pyridines [32]. An efficient, one-pot synthesis of 5-sulfonamidopyrazoles can be performed from sulfonyl azides, terminal alkynes, and hydrazones. Hydrazones are easily synthesized from benzaldehydes and phenylhydrazines. They are used to trap the ketenimines produced *in situ* [33–34] and obtain cyclization products in a cascade process. This sequential process includes a three-component reaction with copper catalyst, an electrocyclic reaction with Lewis acid catalyst, and a dehydrogenation reaction (**Scheme 11**).

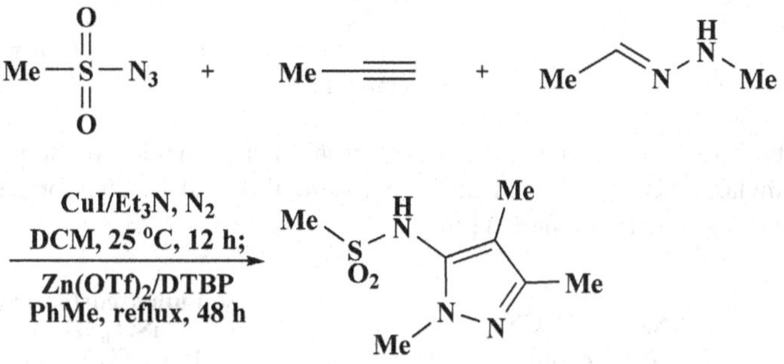

Scheme-11

The reaction of methyl isocyanoacetate with aromatic *N*-sulfonylimines using [(IPr)CuCl] as catalyst produces *N*-sulfonylimidazolines efficiently (**Scheme 12**). The reaction occurs through a copper-boryl complex which is easily synthesized under these conditions, followed by the insertion of the carbonyl function into a copper-boron bond to form a metal-carbon σ bond. Subsequently, a carbon-boron bond is formed by the reaction of the diboron reagent and an additional aldehyde [35–41].

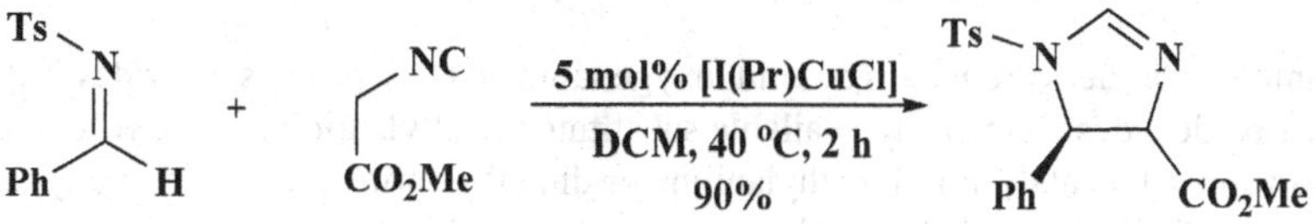

Scheme-12

Chiba et al. [42] synthesized azaspirocyclohexadienones by a copper-catalyzed reaction of α-azido-*N*-arylamides under an oxygen atmosphere. The use of $^{18}O_2$ showed that one of the oxygen atoms from the oxygen molecule formed the resulting carbonyl group of the azaspirodienones (**Scheme 13**) [43].

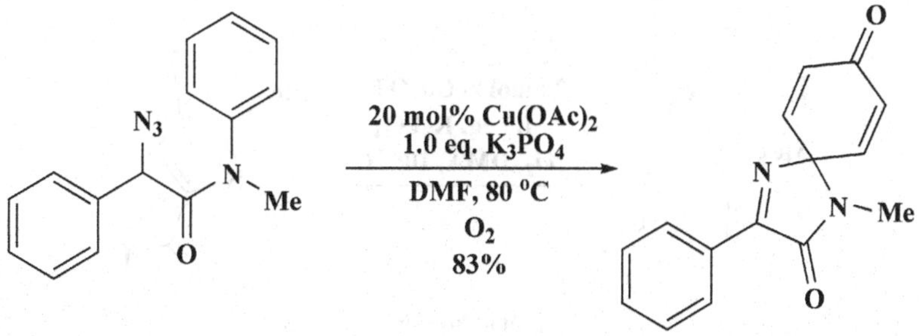

Scheme-13

Bonin et al. [44] reported a direct catalytic preparation of 1-arylimidazole-4-carboxylates, starting from *N*-formylglycine esters and *N*-arylformamides. Firstly, a mixture of two formamides was converted into a crude mixture of two isocyanides, followed by its transformation into the final imidazole in the presence of a catalyst. A new catalytic system containing cupric oxide (10 mol%)/proline (20 mol%) allowed the reaction to occur at room temperature with great efficiency **(Scheme 14)**. Generally, this reaction is efficient for the preparation of alkyl 1-arylimidazole-4-carboxylates [45].

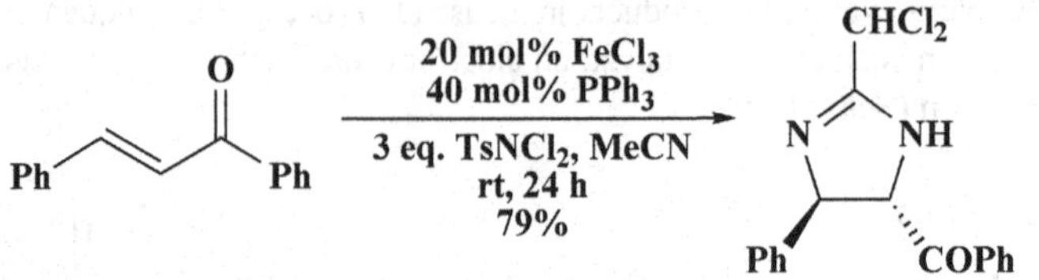

Scheme-14

The analog bearing triazole ring is prepared from methyl azide which is synthesized from methyliodide. Methylazide is not purified due to its instability and is, therefore, utilized directly in copper-assisted click coupling **(Scheme 15)** [46].

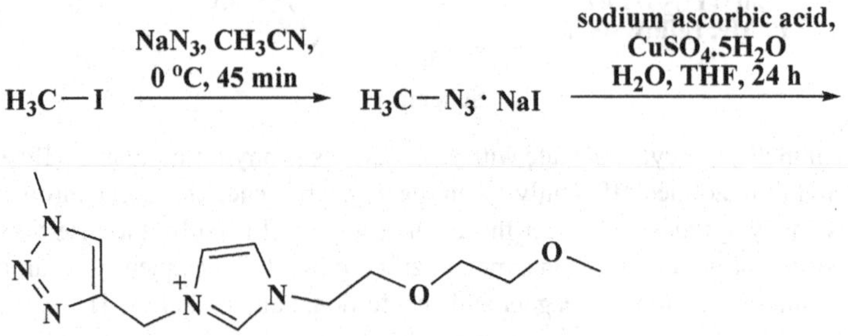

Scheme-15

An intramolecular dehydrogenative aminooxygenation (IDA) process provides 1,2-disubstituted imidazole-4-carbaldehydes from easily available substituted *N*-allylamidines. The reaction is performed with 20 mol% copper(II) catalyst in dimethylaniline or dimethylformamide under dioxygen atmosphere and is environmentally benign and efficient because no additional inorganic or organic oxidant is required. Substituted imidazole-4-carbaldehydes can be prepared in moderate to good yields by a reaction that possesses a good functionality tolerance and a broad substrate scope. These formed substituted imidazole-4-carbaldehydes serve as versatile synthetic intermediates. This method has opened a new route towards the direct formation of formyl-group substituted aromatic nitrogen heterocycles from acyclic substrates. Mechanistic studies have shown that the carbonyl oxygen in aldehyde is derived from dioxygen by a reaction which occurs *via* peroxycopper(III) intermediate **(Scheme 16)** [47].

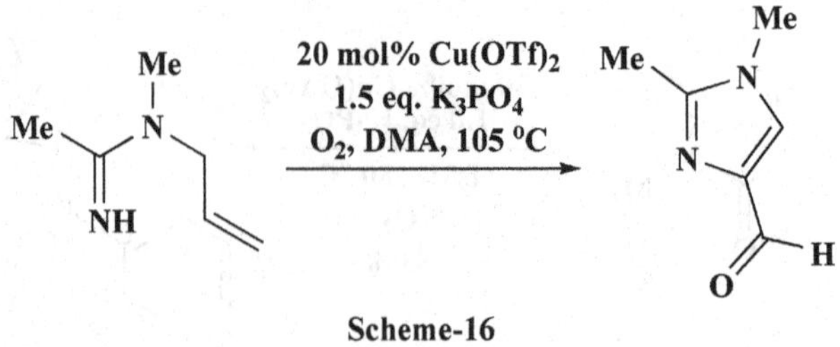

Scheme-16

The monomeric analog bears a monomethoxy-poly(ethylene glycol) chain adjacent to the triazole ring. This is done to see the effect of a polymeric backbone on the activity and stability of the resulting catalyst. Furthermore, the effect of hyper-branched backbone has been reported on the basis of the comparison of a linear analog with the hyper-branched one. For this purpose, PEG-azide prepared from PEG-Ts has been studied **(Scheme 17)** [48].

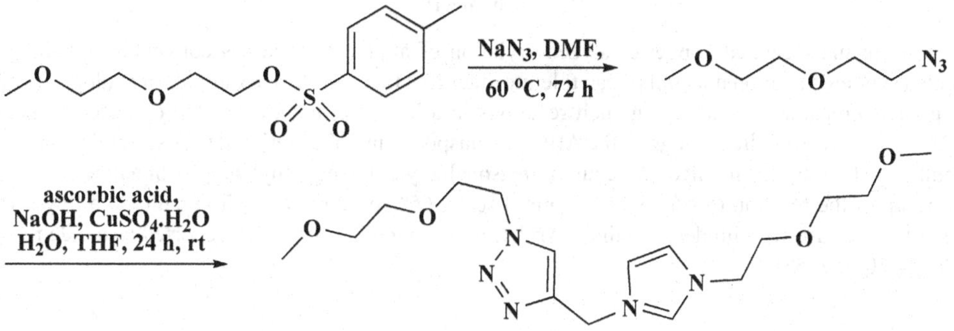

Scheme-17

Cu catalysts are attractive due to their low cost. However, highly selective catalysis was not observed in initial studies involving copper. Benito-Garagorri et al. [49], in 2006, reported that a *de* up to 99% was achieved by employing cuprous chloride/triphenylphosphine catalyst, although the substrate affected the diastereoselectivity of the products. Nevertheless, copper carbine complex is more suitable, demonstrating excellent *trans* selectivity in the reaction of methyl isocyanoacetate and *N*-tosylaldimines derived from benzaldehydes pivalaldehyde **(Scheme 18)** [45].

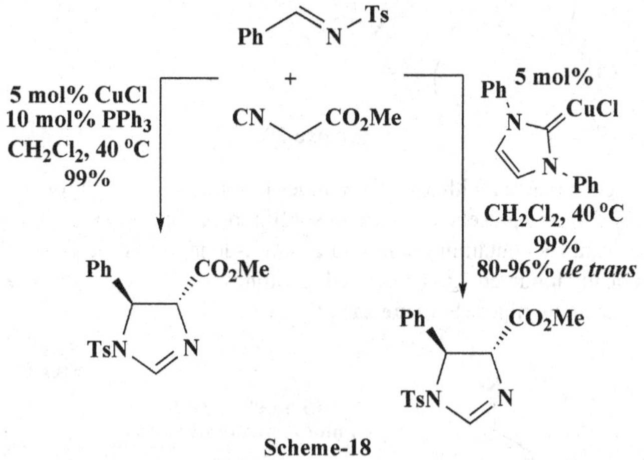

Scheme-18

Although amides, amines, azoles or hydrazines are used as nitrogen sources in most copper-catalyzed cascade reactions involving carbon-nitrogen coupling, some unconventional *N*-possessing moieties like amidines also act as nitrogen sources to afford several heterocyclic functionalities by copper-catalyzed transformation. Gong et al. [50] carried out a comprehensive and pioneering exploration of this subject, hence establishing a class of useful methods for preparing heterocycles **(Scheme 19)**. Furthermore, the reaction behavior of properly elaborated aliphatic reactants has also been explored. The synthesis of 2,4-disubstituted imidazolones from 2-bromo-3-alkylacrylic acids and amidines in the presence of cupric oxide and cesium carbonate at 80 °C has been found effective to secure expected results for most reactions [51].

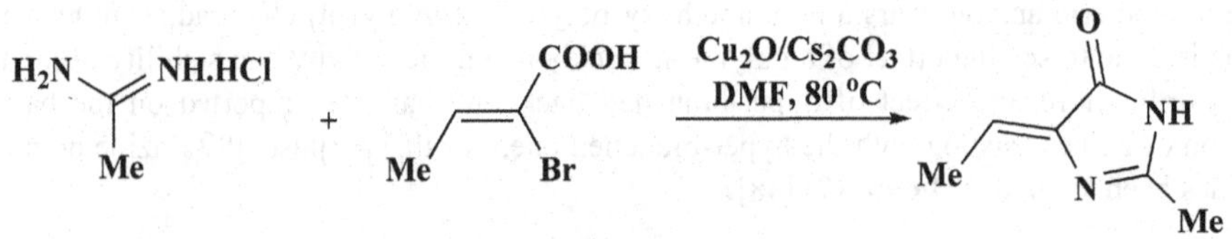

Scheme-19

Mild ligand-mediated copper-catalyzed arylation of aliphatic amines is convenient for diversity-oriented synthesis as aliphatic amines can tolerate other functional groups. Examples of analogs of natural compounds prepared by employing such reactions in a key step includes cyclopropanated analogs of iprodione and hydrophilic analogs of the ABCB1 transporter inhibitor tariquidar. A selective *N*-arylation of bulky cyclopropylamino alcohol occurs in reasonable yield using ethylene glycol as the ligand. The scaling up of the reaction (53% on 15–25 mmol scale, 65% on 2 mmol scale) results in a lower yield, as is often the case with hindered amines. Arylation of ligand has also been reported as a side reaction **(Scheme 20)** [52–53].

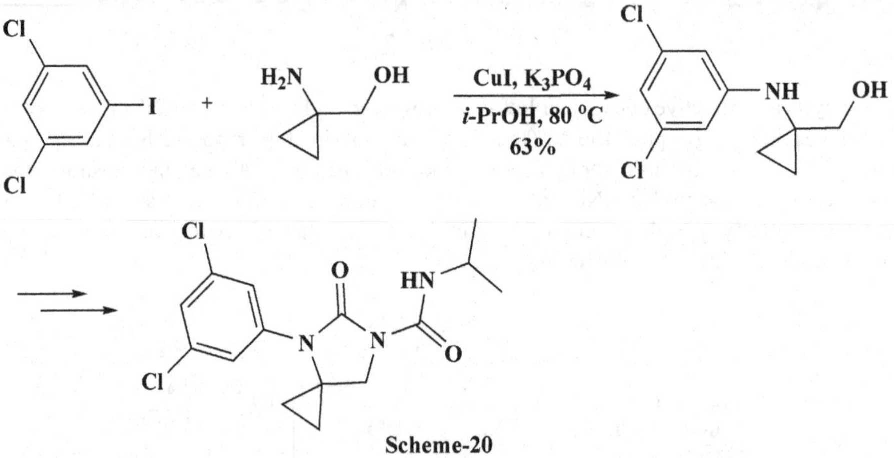

Scheme-20

Ethyl isocyanoacetate reacted with aryl isocyanides to form ethyl 1-arylimidazole-4-carboxylates in tetrahydrofuran at 80 °C in the presence of a catalyst (10 mol% cupric oxide/20 mol% phenanthroline). Several aromatic isocyanides containing either an electron-donating group or an electron-withdrawing group and even a sterically hindered 2,6-substituted substrates were acceptable and a series of imidazoles were synthesized in excellent yields **(Scheme 21)** [45, 54].

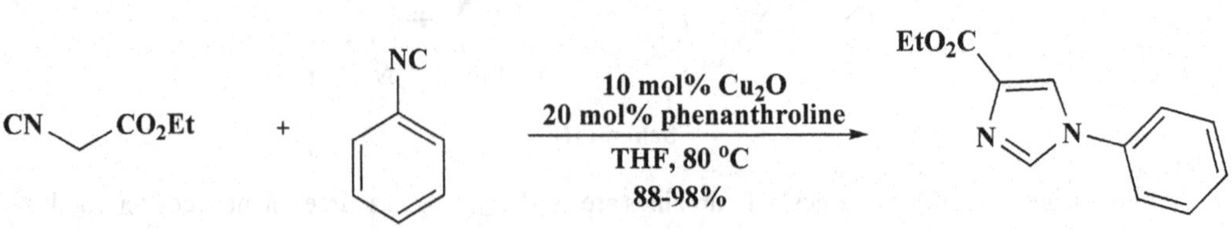

Scheme-21

Microwave assisted organic synthesis and organic reaction enhancement for the conversion of L-ascorbic acid and D-isoascorbic acid to various heterocyclic compounds *via* their monophenylhydrazones and bis(phenylhydrazones) have been achieved. Monophenylhydrazones are converted into olefinic compounds, and their oximes into 4-(L-*threo*-2,3-diacetoxy-(1-hydroxypropyl-2-phenyl-1,2,3-triazole-5-carboxylic acid lactone in the presence of acetic anhydride. Bis-phenylhydrazones treated with alcoholic cupric chloride forms dioxane which is acetylated to yield 5-*O*-acetyl-3,6-anhydro-3-phenylazo-L-xylo-2-hexulosono-1,4-lactone phenylhydrazone and degrades into 1-(2-phenylhydrazide)-

2-(2-phenylhydrazone)mesoxalic acid. The compounds formed are converted into 1-phenyl-3-(L-*threo* or D-*erythro*-glycerol-1-yl)-pyrazoline-4,5-dione phenylhydrazone which yields 3-carboxaldehyde-1-phenyl-pyrazoline-4,5-dione-4-phenylhydrazone upon degradation. All these reactions afford faster, cleaner, and higher or similar yields of products under microwave. Hydrazones can be transformed into oximes by 6–7 minutes of microwave irradiation. Oximes irradiated in the presence of acetic anhydride under microwaves for 11 minutes produce 4-[L-*threo*-2,3-diacetoxy(1-hydroxypropyl)]-2-phenyl-1,2,3-triazole-2-carboxylic acid lactone in 70% yield *via* tri-*O*-acetyl derivative. Bis-phenylhydrazone is oxidized with cupric chloride under microwave irradiation for 1 minute to yield 3,6-anhydro-3-phenylazo-L-xylo-2-hexulosono-1,4-lactone-2-phenylhydrazone (90% yield), which can be acetylated with acetic anhydride to afford 5-*O*-acetyl-3,6-anhydro-3-phenylazo-L-xylo-2-hexulosono-1,4-lactone phenylhydrazone. The degradation of dioxane with sodium hydroxide for 2 minutes under microwave irradiation provides 71% yield of 1-(2-phenylhydrazide)-2-(2-phenylhydrazone)mesoxalic acid. Ring-opening and cyclization of bis-phenylhydrazones under microwave irradiation and alkaline conditions for 1 minute, followed by acidification, affords 93% and 90% of the respective pyrazole derivatives. 3-(1,2-*O*-Isopropylidene-L-*threo*-glycerol-1-yl)-1-phenyl-4,5-pyrazolinedione-4-phenylhydrazone is formed upon the protection of pyrazole derivatives with acetone under microwave irradiation in the presence of sulfuric acid for 3 minutes. Alternatively, 3-(1,2-*O*-isopropylidene-3-*O*-p-tolylsulfonyl-L-*threo*-glycerol-1-yl)-1-phenyl-pyrazole-4,5-dione-4-phenylhydrazone can also be synthesized when dioxane is treated with *p*-toluenesulfonyl chloride in pyridine **(Scheme 22)**. The periodate oxidation of pyrazole derivatives provides 3-carboxaldehyde-1-phenylpyrazoline-4,5-dione-4-phenylhydrazone [55].

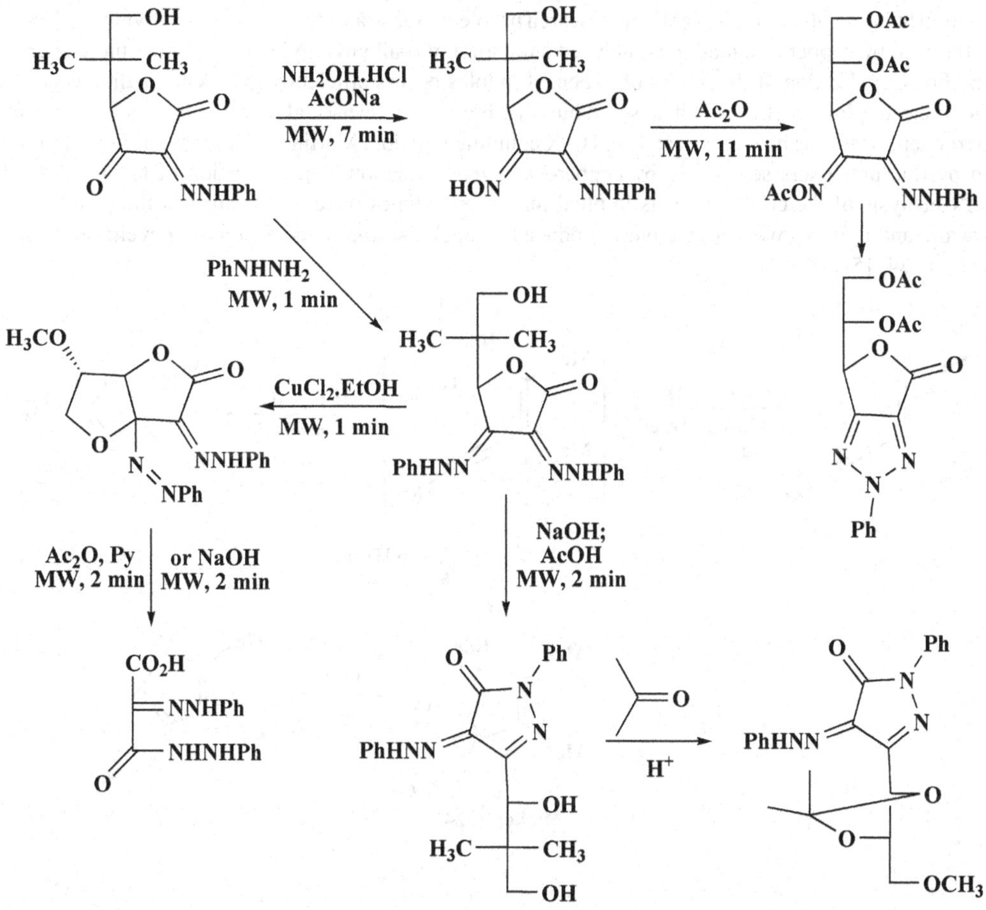

Scheme-22

Ackermann [56] found that indole synthesis involved a mechanism consisting of an intermolecular hydroamination, and a subsequent intramolecular metal-catalyzed carbon-nitrogen bond forming reaction was predominantly operative, although no intermediate was isolated [57]. Hence, two different mechanisms have been proposed for domino reactions: (1) initial carbon-nitrogen coupling followed by intramolecular hydroamidation of the alkyne portion, and (2) intermolecular hydroamidation with subsequent intramolecular amidation (**Scheme 23**).

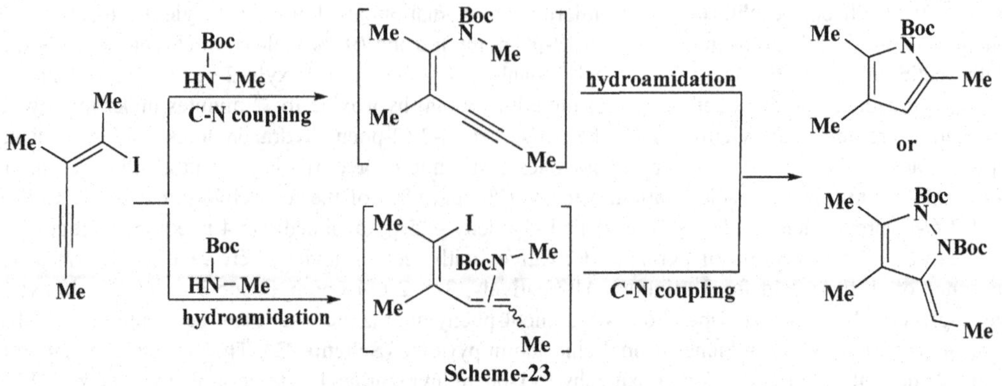

Scheme-23

The development of a protocol for the synthesis of multiple classes of nitrogen heterocycles from common building blocks is a sought after research problem. Substrates for Cacchi indole synthesis [58] are synthesized by copper or palladium-catalyzed amination of *o*-alkynyl aryl halides. Under these coupling conditions, cyclization of the *N*-aryl or *N*-benzyl indoles occurs efficiently [59]. Vinyl halide is another type of electrophilic partner and it acts as a building block in copper-catalyzed domino reaction to afford heterocyclic scaffolds by copper-catalyzed C-X coupling reaction. Martin et al. [57] synthesized pyrazole and pyrrole precursors selectively by copper-catalyzed amidation/hydroamidation of haloenynes. The NMR analysis of the crude products implied that intermediates were first formed in the reaction, and pyrazole and pyrroles were respectively produced through 5-*exo-dig* and 5-*endo-dig* cyclization routes (**Scheme 24**) [51, 60–64].

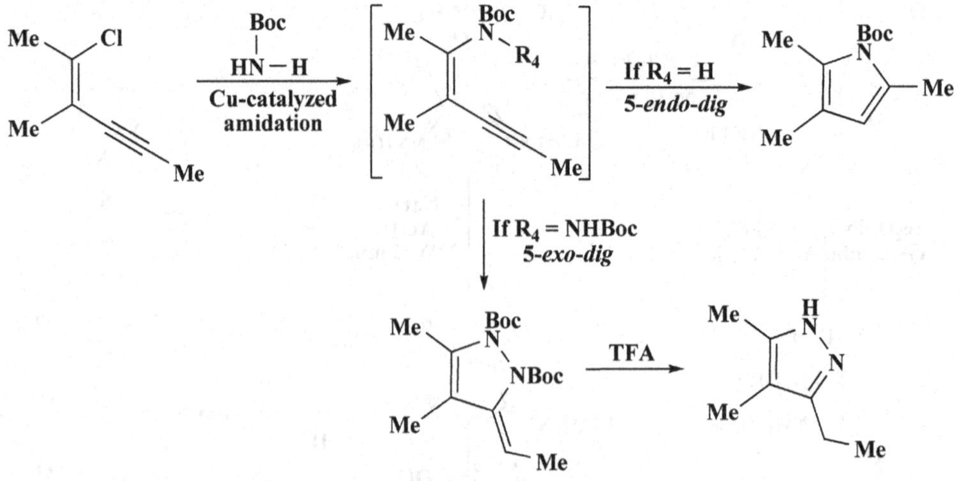

Scheme-24

Gallium-assisted synthesis

A one-pot protocol [65] for the synthesis of hydantoins, with 47–94% yields, has been realized as a multi-component reaction starting directly from ammonia, carbonyl compounds and carbon dioxide. The first step provides an imine, the second step produces the aminonitrile, and the last step yields the required hydantoin. The carbonyl compound provides imine in liquid ammonia at –78 °C in the presence of gallium(III) triflate catalyst. Subsequently, hydrogen cyanide dissolved in dichloromethane is added at –78 °C. Then the reaction mixture is heated to room temperature, carbon dioxide is introduced into the reaction solution, and Hunig's base is added. Without gallium(III) triflate, the reaction yield is as low as 9% **(Scheme 25)** [66–67].

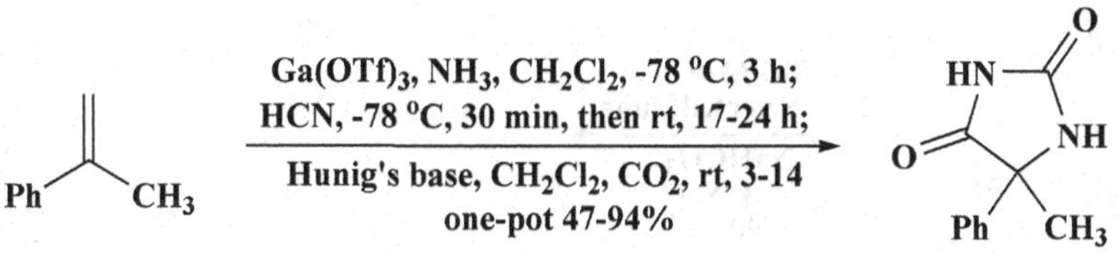
Ga(OTf)3, NH3, CH2Cl2, -78 °C, 3 h;
HCN, -78 °C, 30 min, then rt, 17-24 h;
Hunig's base, CH2Cl2, CO2, rt, 3-14
one-pot 47-94%
Ph
CH3
HN
O
NH
O
Ph
CH3

Scheme-25

Gold-assisted synthesis

Manzo et al. [68a] reported an intramolecular hydroamination of allenamides in the presence of gold(III) catalyst to provide a mixture of *cis*- and *trans*-2-vinylimidazolidines *via* 5-*exo-trig* transition states **(Scheme 26)**. It was observed that the *cis*- and *trans*-2-vinylimidazolidines were formed by cyclizations through gold(II)-complexes [68b].

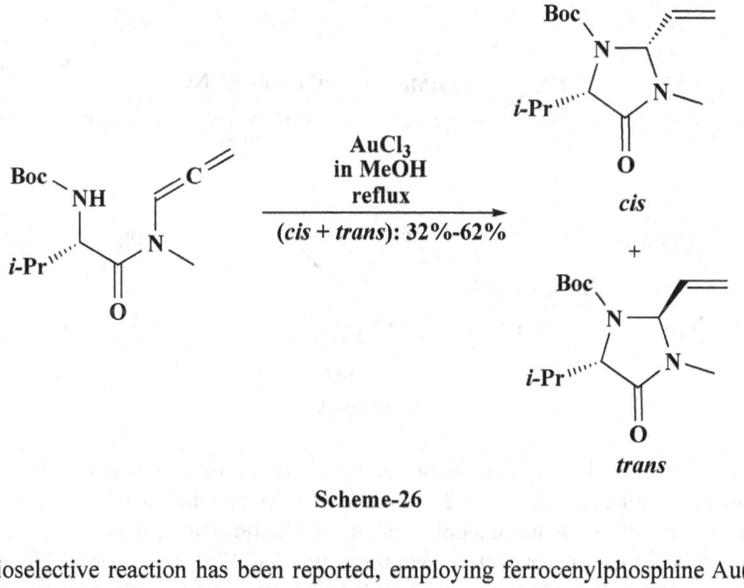
Boc
NH
i-Pr
O
C
N
AuCl3
in MeOH
reflux
(cis + trans): 32%-62%
Boc
N
N
i-Pr
O
cis
+
Boc
N
N
i-Pr
O
trans

Scheme-26

An enantioselective reaction has been reported, employing ferrocenylphosphine Au(I) complexes analogous to those utilized in asymmetric aldol reaction. However, whereas ionic complexes [Au]BF_4 formed in a $Au(CNCy)_2BF_4$ system display better enantio- and diastereoselectivity in oxazoline synthesis, the use of neutral complex gold(III) chloride formed from $AuCl(Me_2S)$ is more effective in the present reaction **(Scheme 27)** [45, 69–70].

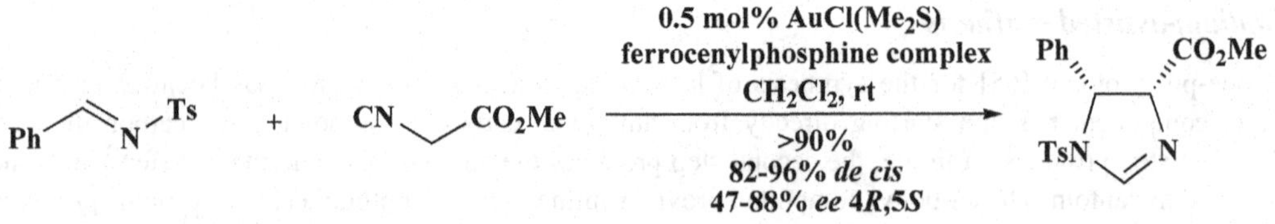

Scheme-27

Qian et al. [71] studied Au-catalyzed fluoroamination of alkyne **(Scheme 28)**. In this method, a hydroamination product pyrazole was synthesized simultaneously, which was further transformed into fluorinated pyrazole by electrophilic fluorination [72].

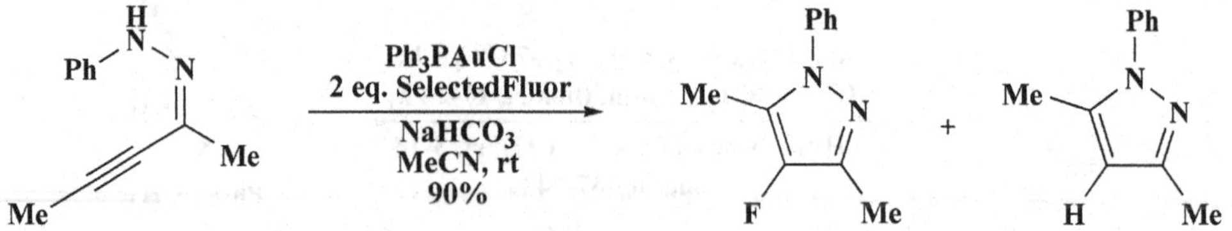

Scheme-28

Transition metal catalysts greatly increase the reactivity of substrate through metal coordination to the isocyano group. In 1996, Hayashi et al. [73] used various silver, copper, palladium, gold, nickel, and rhodium complexes in base-free reactions of several *N*-sulfonylaldimines with methyl isocyanoacetate. Less electrophilic imines obtained from anilines or $Ph_2P(O)NH_2$ and aromatic aldehydes were unreactive under these reaction conditions. $AuCl(CNCy)_2$ showed the best diastereoselectivity, providing *cis* isomers in high *de* despite the fact that *trans* isomers were thermodynamically more stable. The *trans* isomers were formed from *cis* isomers by epimerization with triethylamine in high *de* **(Scheme 29)** [45].

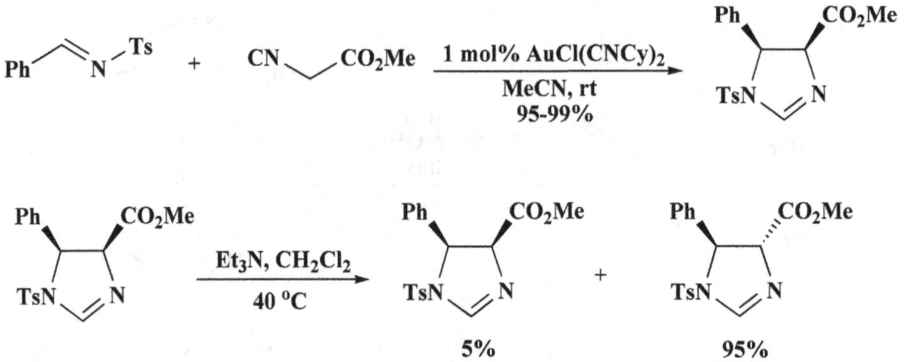

Scheme-29

The reaction occurs with *N*-sulfonylimines prepared from aromatic, heterocycle, α,β-unsaturated, and *t*-Bu-aldehydes. In all cases, the *trans*-2-imidazolines were obtained with high selectivity and yield [74]. Ruthenium(II) complex-catalyzed aldol reactions of *N*-sulfonylimines are highly efficient in terms of stereoselectivity and yield under neutral conditions. Ito et al. [75] developed an elegant asymmetric synthesis procedure using cationic bis(cyclohexyl isocyanide)gold(I) tetrafluoroborate-catalyzed aldol reaction and chiral ferrocenylphosphine ligands. The reaction of aldehydes and methyl isocyanoacetate in the presence of cationic chiral Au complexes displayed high enantioselectivity with enantiomeric excess as high as 97–99%. To probe the ability of the above catalyst with chiral ferrocenylphosphine ligand in order to induce enantioselectivity in similar aldol reactions with imines, such reactions were examined under identical conditions. When methyl isocynaoactate was treated with *N*-tosyl-*p*-chlorobenzaldimine in the presence of cationic chiral Au(I) complex, the enantiomeric excess values of *trans* and *cis*-2-

imidazolines were low (14%). Interestingly, when the neutral Au(I) complex Me₂SAuCl with chiral ferrocenylphosphine was used as the catalyst, the asymmetric induction of this reaction could be brought to good levels **(Scheme 30)**.

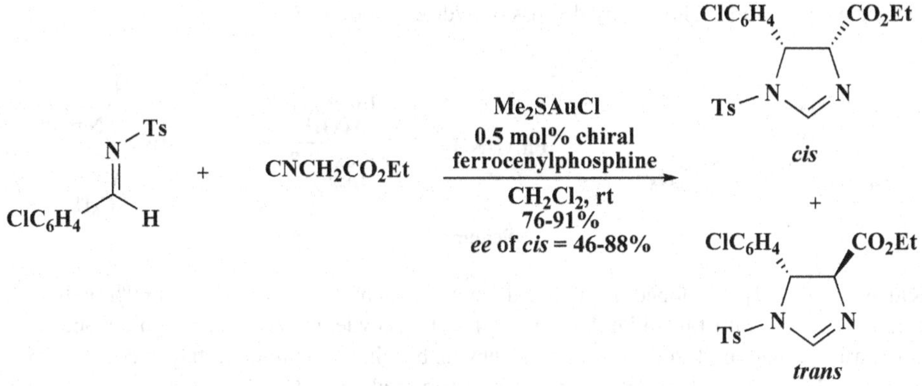

Scheme-30

Iodine-assisted synthesis

For large-scale production of active pharmaceutical ingredients, various industrial synthesis methods use enzyme-catalyzed reactions in organic media. For instance, Schering-Plough synthesized an azole anti-fungal agent in quantities of the order of hundred kilograms wherein the pivotal synthesis step was an enzymatic desymmetrization of a symmetrical diol by Novozym with acetonitrile as solvent and vinyl acetate as acyl donor to afford the monoester in high enantiopurity. Acetonitrile was chosen as the solvent because the subsequent iodocyclization to yield the final compound was performed in acetonitrile and simple filtration of the enzyme beads facilitated the reaction sequence to be telescoped into a single step [76]. Bristol-Myers Squib pharmaceutical research group reported a number of plant-scale chemoenzymatic syntheses in organic media as depicted in **Scheme 31** [77–79].

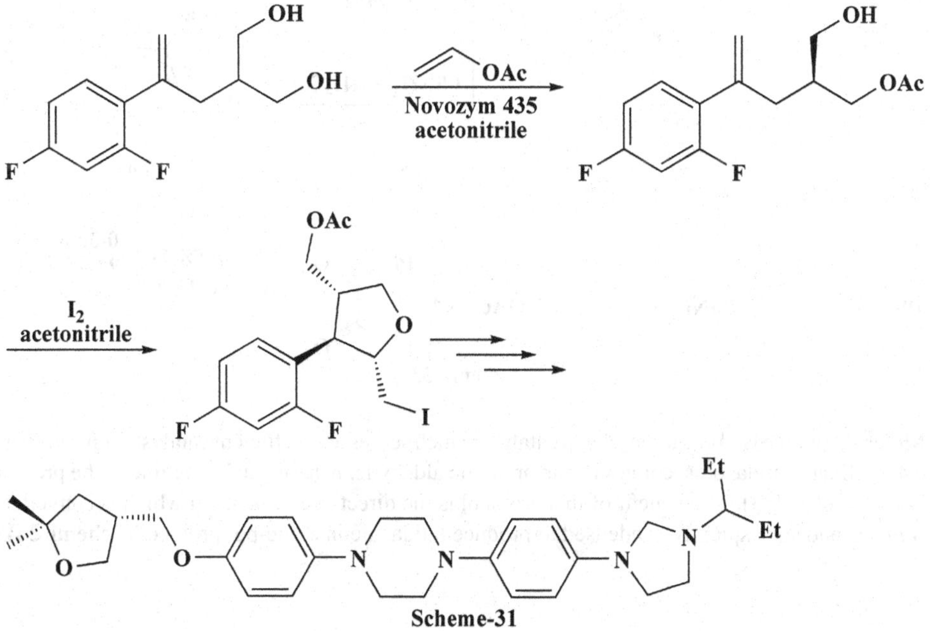

Scheme-31

Ponnala and Sahu [80] investigated the synthesis of pyrazoles from chalcones and phenyl hydrazine by heating them at reflux temperature in acetic acid, using stoichiometric amounts of iodine (**Scheme 32**). Chalcones were treated with phenyl hydrazine to produce hydrazone *in situ* in refluxing acetic acid, which reacted with iodine to synthesize an iodoiranium ion intermediate. The products were obtained by intramolecular cyclization followed by the loss of hydrogen iodide [81].

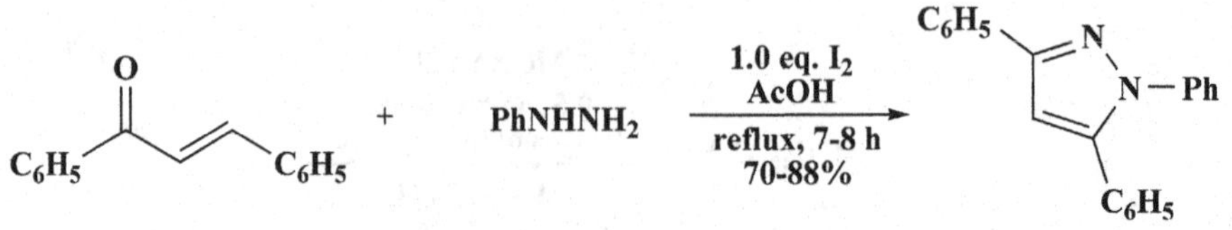

Scheme-32

Kidwai et al. [82] developed a one-pot three-component reaction, using 5 mol% iodine, for the preparation of 2,4,5-trisubstituted imidazoles from aryl aldehyde, benzil and ammonium acetate, and of 1,2,4,5-tetrasubstituted imidazoles from aryl aldehyde, benzil, ammonium acetate and aniline. A second synthesis of 1,2,3,5-tetraaryl imidazoles was also performed, including *in situ* oxidation of benzoin to benzil (**Scheme 33**). Iodine-catalyzed the reaction by bonding with the carbonyl oxygen which facilitated the synthesis of diamine intermediate and then condensed with iodine-activated carbonyl carbon of 1,2-diketone to afford the cyclized intermediate. Imidazoles were formed by the dehydration of cyclized intermediate followed by [1,5]-sigmatropic shift of the intermediate. Ammonium acetate, benzil, aniline and aryl aldehyde provided 1,2,4,5-tetrasubstituted imidazloes in a similar manner [81].

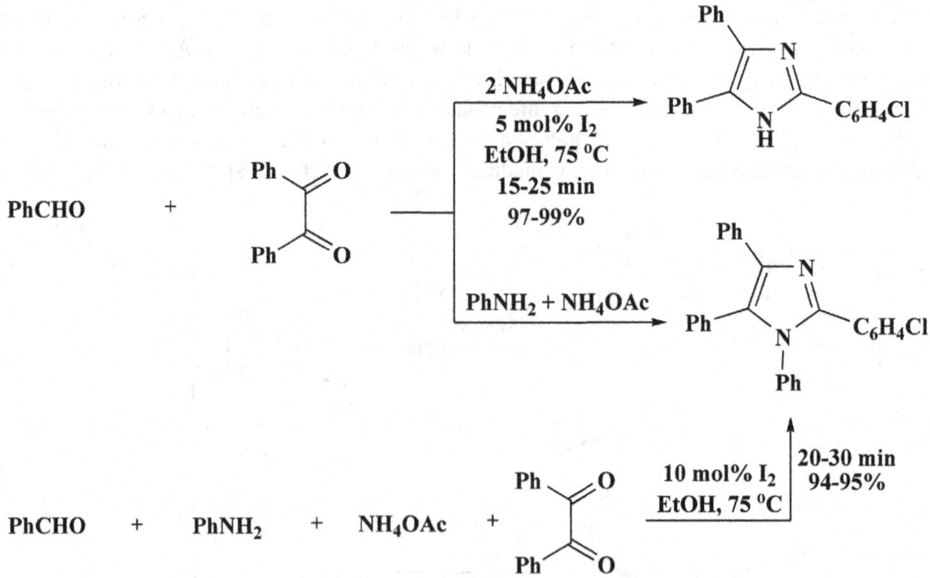

Scheme-33

Molecular iodine is a non-toxic, cheap catalyst, which serves as an efficient catalyst for the preparation of 1,2,4,5-tetraarylimidazoles, employing an aromatic aldehyde, benzoin and an amine in the presence of ammonium acetate [83]. The benefit of this protocol is the direct use of benzoin which is converted into benzil *in situ* and subsequently condensed to produce imidazole in a one-pot protocol (**Scheme 34**) [84].

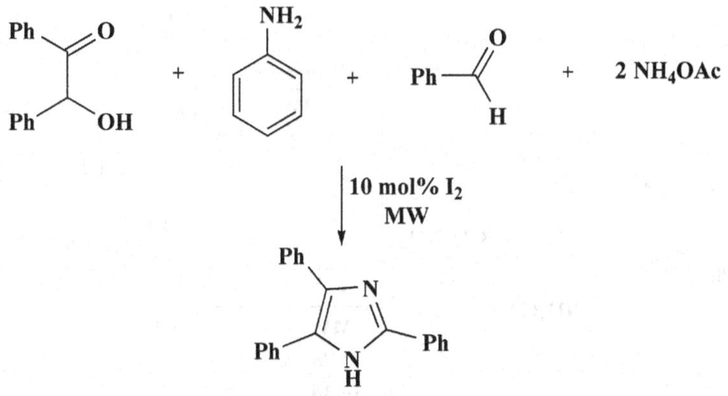

Scheme-34

Iridium-assisted synthesis

Pyrazoles can be synthesized from hydrazines and 1,3-diols **(Scheme 35)**. In order to drive the reaction to completion, an open system is used to allow the hydrogen gas formed to escape from the reaction mixture. Initially, pyrazole is generated and the formation of an intermediate is observed by GC-MS. However, the conversion stops after 2–3 hours. When the reaction is repeated with 3-pentanone as a hydrogen scavenger, the ketone reacts as N-alkylating reagent instead. The saturated pyrazolidine is formed using water as the solvent, whereas a better conversion into the desired pyrazole is observed with mesitylene (at reflux temperature). Ultimately, 1-phenyl pyrazole containing partially reduced 4,5-dihydro-1-phenyl-$1H$-pyrazole as an impurity is collected as a red oil after column chromatography (yield ~ 35%). A similar result has been reported for this reaction carried out at 140 °C. The reaction solvents, temperatures and additives (acids or bases) when varied, yields between 6 and 17% were obtained. The best result was obtained when the reaction was performed with catalytic amounts of trifluoroacetic acid at 140 °C. However, in all cases, the synthesis of aniline was faster than the desired reaction and suppression of this side reaction was apparently not possible [85].

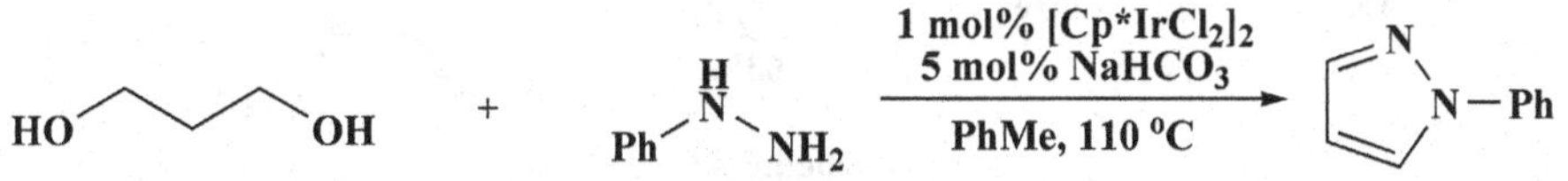

Scheme-35

Iron-assisted synthesis

A four-component reaction of ammonium acetate, benzil, primary amines and benzaldehydes yields 1,2,4,5-tetrasubstituted imidazoles. This microwave-assisted reaction is catalyzed by silica gel or HY-zeolite **(Scheme 36)** [86]. The solvent-free method provides the target compounds in 6 minutes in yields higher than those reported by conventional heating. The four-component cyclocondensation into tetra-substituted imidazoles can also be performed employing a Fe^{3+}/K-10 catalyst as reported by Raghuvanshi and Singh [87]. In this study, K-10 montmorillonite catalyst impregnated with ferric chloride was utilized. The products were obtained in 3–4 minutes in good yields under MWI and solvent-free conditions [88].

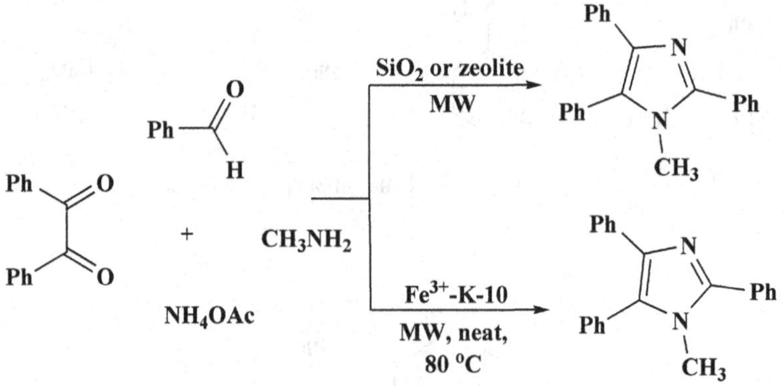

Scheme-36

The electrophilic diamination reaction of ketones and α,β-unsaturated carboxylic esters is promoted in the presence of ferric chloride-triphenylphosphine complex **(Scheme 37–38)**. The reaction uses easily available CH_3CN and $TsNCl_2$ as nitrogen sources and forms imidazolidines with high stereo- and regioselectivity (*trans*) in good yields [89–91].

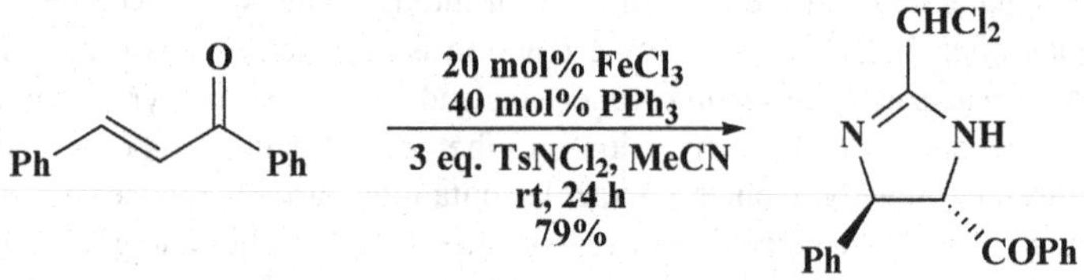

Scheme-37

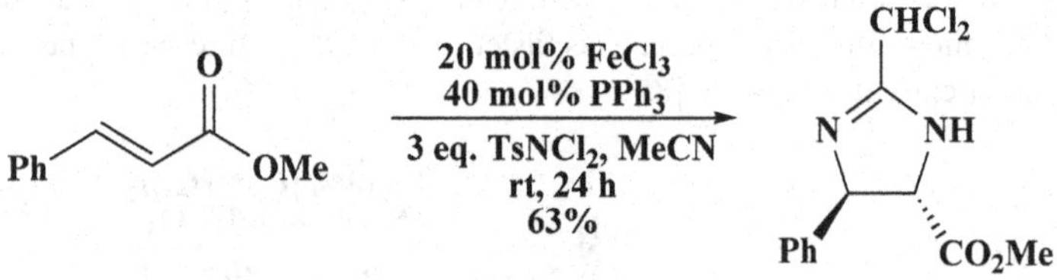

Scheme-38

Lanthanum-assisted synthesis

One-pot synthesis of poly-substituted imidazoles in excellent yields can be carried out using lanthanum chloride catalyst under solvent-free conditions. This protocol uses urea/thiourea as a source of ammonia (an environmentally benign source) for the preparation of poly-substituted imidazoles. The products do not require further purification and the synthesis is facile. The use of lanthanum chloride as a catalyst and the use of urea/thiourea as an efficient source of ammonia under MWI has been reported in several reactions. A one-pot, four-component reaction of aldehydes, 1,2-diketones, amines and urea/thiourea under MWI and solvent-free conditions affords poly-substituted imidazoles in 82–94% yields **(Scheme 39)**. Initially, for the optimization of the catalyst, the reaction of benzil, benzaldehyde, aniline and urea/thiourea is screened in the presence of different Lewis acids (35 mol%) under microwave irradiation and solvent-free conditions. Thus, a mixture of an aldehyde, benzil, amine and urea/thiourea reacts in the presence of lanthanum chloride under MWI and solvent-free conditions to provide poly-substituted imidazoles [92].

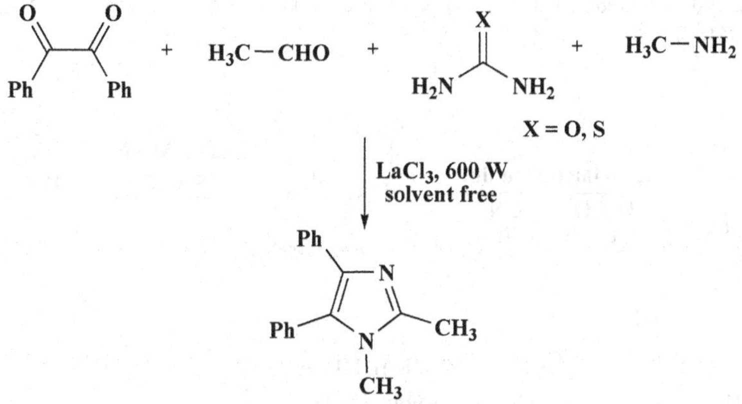

Scheme-39

Lithium-assisted synthesis

When a tetrazole ring is introduced at the beginning of protocol, the heterocyclic ring passing through all subsequent steps often needs protection. One common protecting group is trityl group as utilized in the preparation of losartan (**Scheme 40**) [24, 93–94].

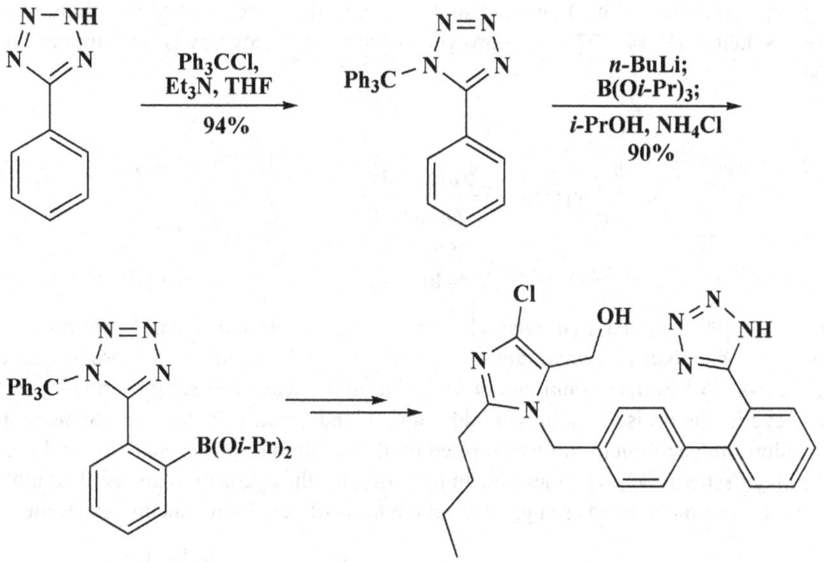

Scheme-40

Although base-catalyzed reaction has not been developed systematically, either imidazolones or 5-aminooxazoles are obtained as products. For example, without optimization, Solomon et al. [95] synthesized oxazole in 16% yield through the reaction of ethyl isocyanoacetate with benzyl isocyanate (**Scheme 41**) [45].

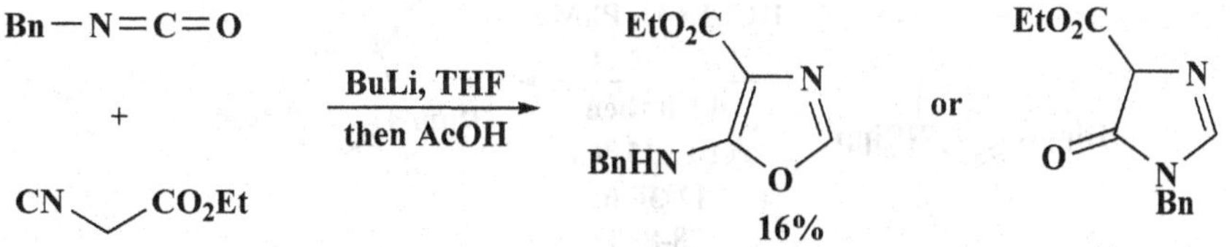

Scheme-41

The ligand hydrogenated in 4 and 5 positions of the imidazole ring is prepared *via* multistep synthesis (**Scheme 42**) [96].

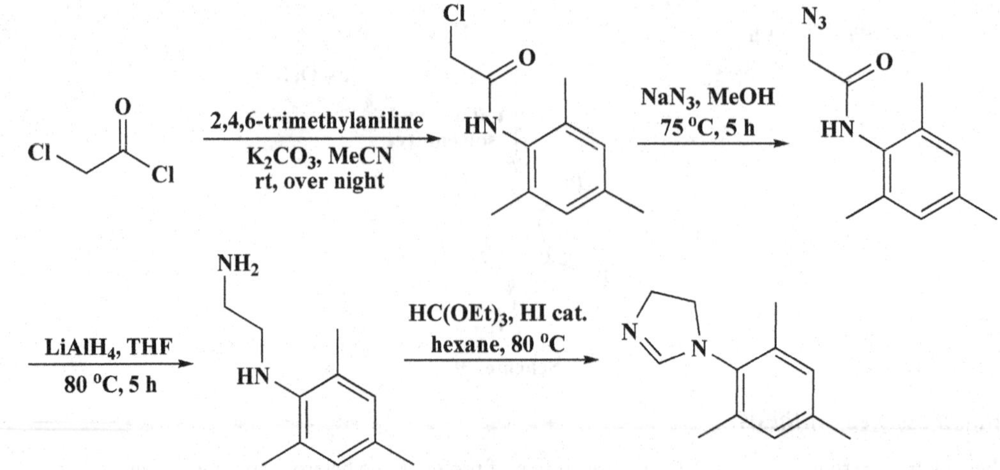

Scheme-42

N-Substituted isocyanoacetamides are cyclized under certain conditions by either the oxygen or the nitrogen atom to provide imidazolinones. *N*-Substituted 1-isocyano-1-cyclohexanecarboxamides (α,α-disubstituted isocyanoacetates) undergo smooth cyclization when treated with butyllithium at low temperatures (**Scheme 43**) [45, 97–98]. However, behavior of secondary *N*-aryl amides has not been investigated.

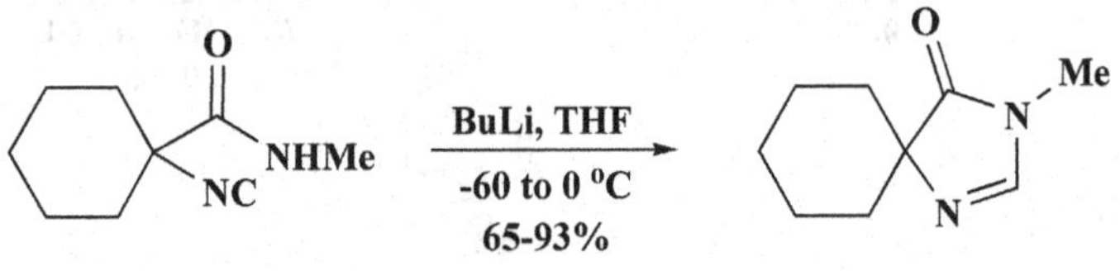

Scheme-43

Jazzar et al. [99] studied hydrogen chloride-mediated intramolecular hydroiminiumation and 3-amidiniumation of alkenyl-formamidines, -aldimines, and -amidines to produce the alkenyl-formamidinium, -aldiminium, and -amidinium salts which underwent regioselective ring-closure reactions to synthesize cyclic dihydroisoquinolinium, aldiminium, and imidazolinium salts (**Scheme 44**). On the other hand, chloro-imidazolinium salt was formed by the addition of phosgene to alkenyl urea followed by gentle heating (**Scheme 45**). The mechanism proposed for the cyclization involved an intramolecular proton transfer to the double bond as suggested on the basis of deuterium labeling experiments [100].

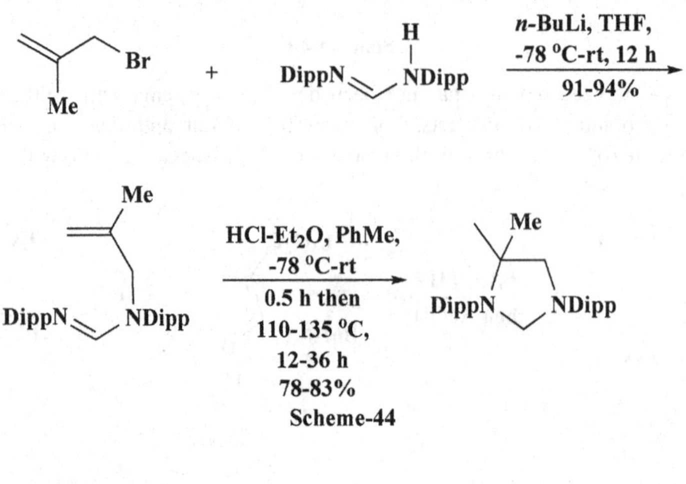

Scheme-44

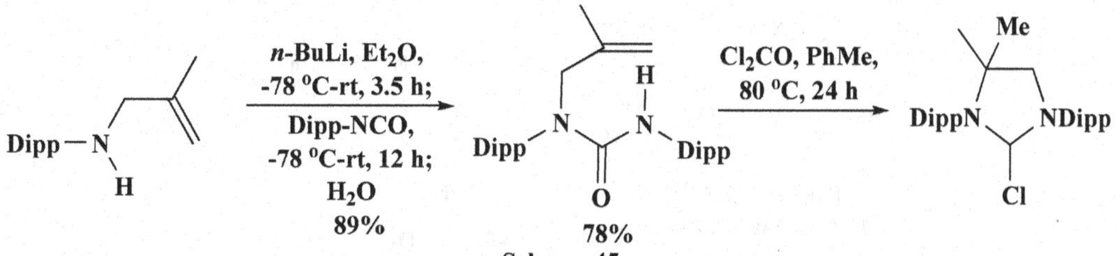

Scheme-45

Molybdenum-assisted synthesis

An improved and rapid synthesis of 2,4,5-trisubstituted imidazoles in good yields can be performed by a three-component one-pot condensation of aryl aldehydes, benzil and ammonium acetate, using $(NH_4)_6$. $Mo._7O_2.4H_2O$ as a catalyst, under solvent-free and MWI conditions **(Scheme 46)** [101].

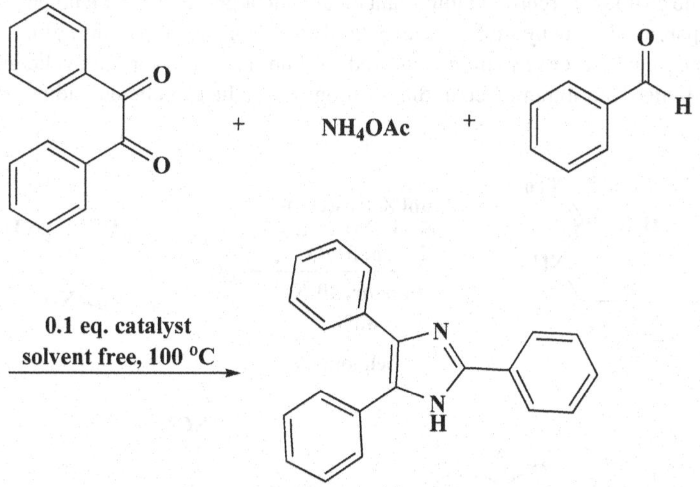

Scheme-46

Pyrazines and isomeric dihydropyrazines can be synthesized from simple 2-arylazirines in the presence of group VI metal hexacarbonyls. The azirine containing a substituent (Schiff base, aldehyde, α,β-unsaturated ester) at the saturated carbon of the three-membered ring participates in intramolecular cycloaddition to result in high yields of five-membered ring heterocycles. The stereochemistry about the double bond proves to be an important factor affecting the reaction course in the case of ketovinylazirines **(Scheme 47)** [102–103].

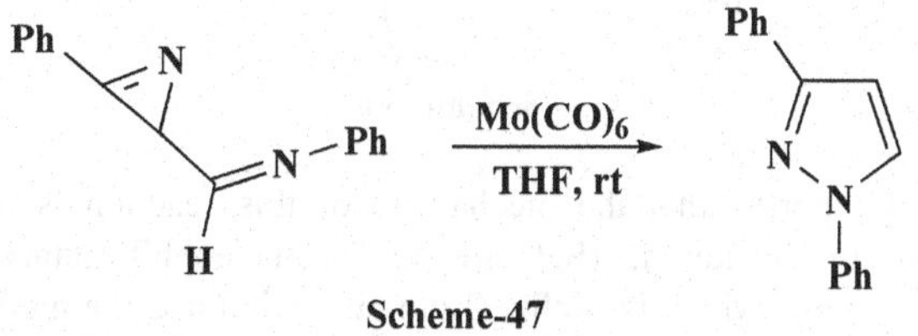

Scheme-47

Rhodium-assisted synthesis

The transannulation of triazole with benzonitrile can be performed using two different procedures—conventional and microwave. It has been reported that both protocols are equally efficient in furnishing imidazole as the transannulated product in high yields **(Scheme 48)** [104–106].

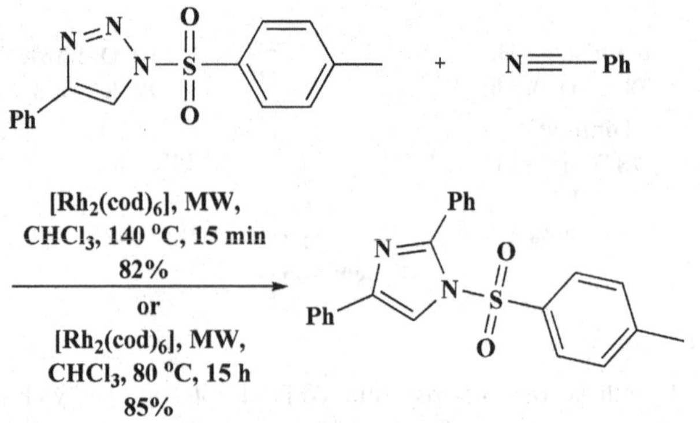

Scheme-48

Various cyclic guanidines represent important medicinal targets and exhibit potent biological activity. They can be prepared using many of the same methods used for the preparation of imidazolidin-2-ones. An acyclic guanidine undergoes rhodium-catalyzed C-H amination to form a cyclic guanidine. Zhao et al. [107] reported inter- and intramolecular diaminations to produce cyclic guanidines **(Scheme 49–50)**.

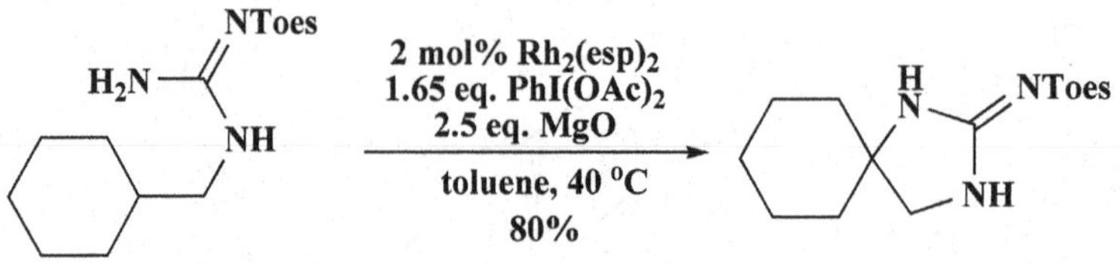

Scheme-49

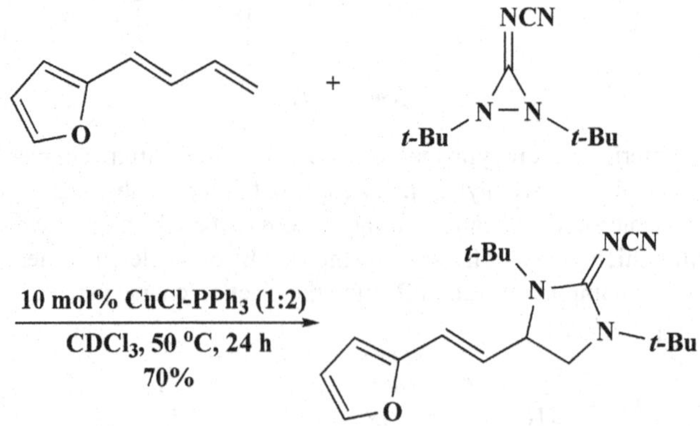

Scheme-50

Connell et al. [108] reported that the mechanism of this reaction is related to the analogous transannulation of nitiles with diazoketones **(Scheme 51)**. The nucleophilic attack of nitrile on the rhodium carbenoid [109] affords an ylide which is cyclized into zwitterion and the metal is subsequently lost to afford imidazole. In other words, the imidazole is produced by the cyclization of rhodium carbenoid and then reductive elimination [106, 110].

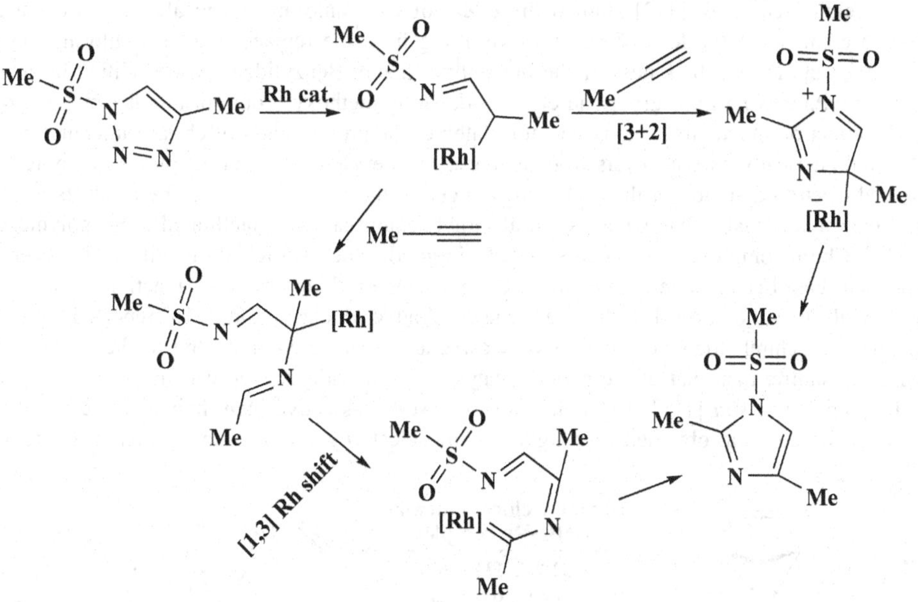

Scheme-51

Rh(II)-catalyzed reaction of readily available and stable 1-sulfonyl triazoles with nitriles can produce imidazoles in good to excellent yields. Herein, rhodium iminocarbenoids are formed as intermediates **(Scheme 52)** [111]. The protocols can be applied to transannulation of differently C4-substituted *N*-sulfonyl-1,2,3-triazoles with a number of nitriles. These reactions are very general with respect to the nitrile and the triazole components. Both the conventional and the microwave heating methods provide high to excellent yields of the diversely substituted transannulation products [106]. In contrast to pyridotriazoles, triazoles do not undergo transannulation with terminal alkynes into pyrroles under these reaction conditions.

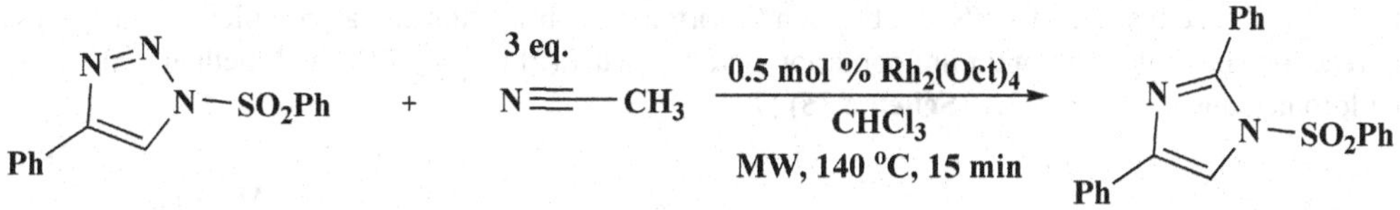

Scheme-52

Ruthenium-assisted synthesis

Metal-catalyzed reaction of isocyanoacetates with imines has an interesting feature in that the configuration of the major isomer (*cis* or *trans*) is dependent on the choice of catalyst. Lin et al. [74] synthesized *trans*-imidazoline through a highly efficient pathway using RuH$_2$(PPh$_3$)$_4$ as the catalyst. FeH$_2$(dppe)$_2$, IrH$_5$(*i*-Pr$_3$P)$_2$ and some simple palladium complexes also displayed *trans* selectivity in catalysis, but with lower efficiencies than ruthenium **(Scheme 53)** [45].

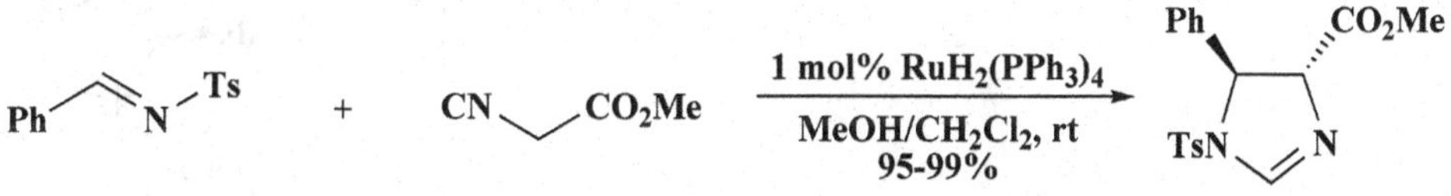

Scheme-53

Van Veldhuizen et al. [112] studied the effect of electronic and steric alteration on the parent catalyst to compensate the loss of reactivity resulting from the replacement of a chlorine ion by an aryl oxide group and the bulkiness of the binaphthyl ligand. Benzylidene was modified by installing an electron-withdrawing nitro group, an electron-donating methoxy group, or a bulky phenyl group to see if the effect on the achiral analog was translated to the present class of chiral ruthenium catalysts. Also, enantiomerically pure catalysts were synthesized to evaluate their effect on the catalytic activity of reduced electron donation of the aryl oxide oxygen to the ruthenium core. The catalysts resulted in conversions of less than 10%, whereas chiral molybdenum catalysts facilitated rapid polymerization [113–115]. Chiral complexes also yielded ARCM in good *ee* and sufficient quantities. However, chiral molybdenum-based catalysts are generally the complexes of choice for such reactions. The synthesis of these catalysts is, however, lengthy. Nevertheless, Van Veldhuizen et al. [116] reported a new, more readily available chiral bidentate *N*-heterocyclic carbene-ruthenium complex, the synthesis of which was considerably shorter than that of the parent catalyst. The chloride version was not stable on silica but could be prepared *in situ* [117]. The iodine analogs were less active than their chlorine counterparts; however, the product was obtained with higher enantioselectivity in case of the former **(Scheme 54)**.

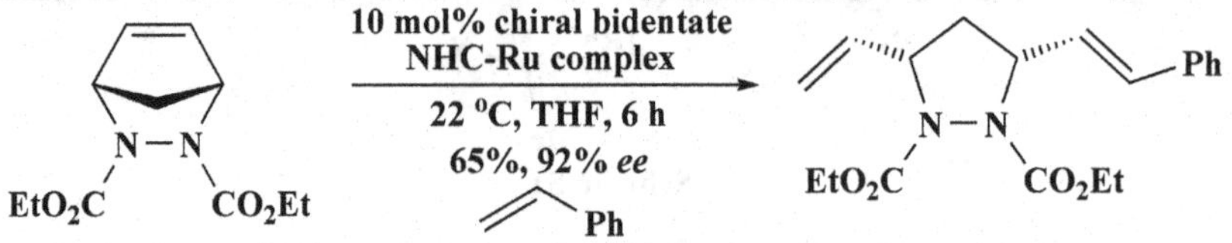

Scheme-54

The synthesis of five-membered *N*-heterocycles like imidazoline and oxazoline has drawn the attention of researchers due to their wide application in the preparation of pharmacologically active compounds [118–119]. Aldehydes and isocyanoacetate undergo Au complex-catalyzed aldol reactions to afford 4,5-disubstituted 2-oxazolines [120–121]. The reaction is catalyzed successfully, employing a dihydridic ruthenium complex [122]. When *N*-aryl or *N*-alkyl-substituted imines are used instead of aldehydes in this reaction, no products are obtained. Due to the low reactivity of imines in nucleophilic addition, a strong electron-withdrawing group, sulfonyl, was integrated with the nitrogen atom of imines for the activation of C=N double bond. Fortunately, transition metal complexes also catalyze the reaction smoothly. The optimum condition involves $RuH_2(PPh_3)_4$ at 25 °C and methanol along with dichloromethane in the ratio 3:1 **(Scheme 55)** [74–75].

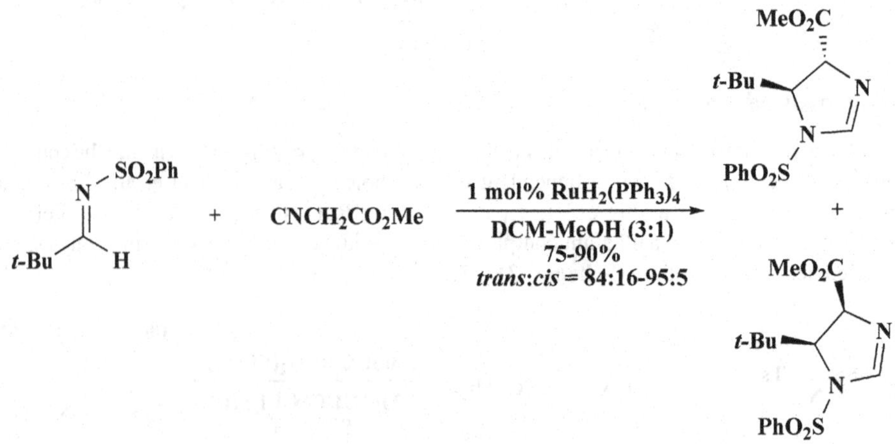

Scheme-55

Samarium-assisted synthesis

In 1946, Senkus [123] and Johnson [124–125] studied the reduction of β-nitroamines using nitro-Mannich reaction. Therein, nitroamines were reduced to polyamines through high pressure (34 bar) catalytic hydrogenation over Raney nickel. Much milder reducing conditions reported recently have shown tolerance of a variety of functional groups. Adams et al. [126] reported the first acyclic diastereoselective nitro-Mannich reactions for the reduction of nitro-Mannich products using a single electron transfer reduction in the presence of SmI$_2$. Originally, this protocol was reported by Sturgess and Yarberry [127] for the reduction of β-nitroamines (derived from aza-Michael additions) into nitroalkenes, and has also been employed for the synthesis of cyclic ureas (**Scheme 56**) [128–129]. Yamada et al. [130] also reported successful execution of this SmI$_2$ reduction method.

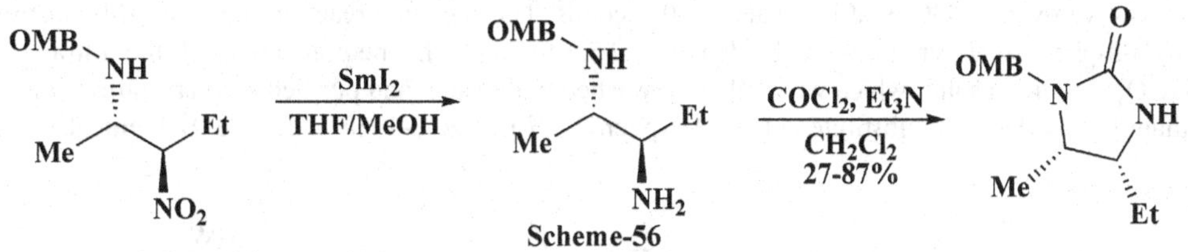

Scheme-56

Scandium-assisted synthesis

Imidazo[1,2-*a*]pyridine can be synthesized by three-component condensation of perfluoroalkanesulfonyl-protected hydroxybenzaldehydes, isonitriles and 2-aminopyridines (**Scheme 57**) [131]. The condensed products are subsequently used in palladium-catalyzed cross-coupling reactions with boronic acids to produce imidazo[1,2-*a*]pyridine.

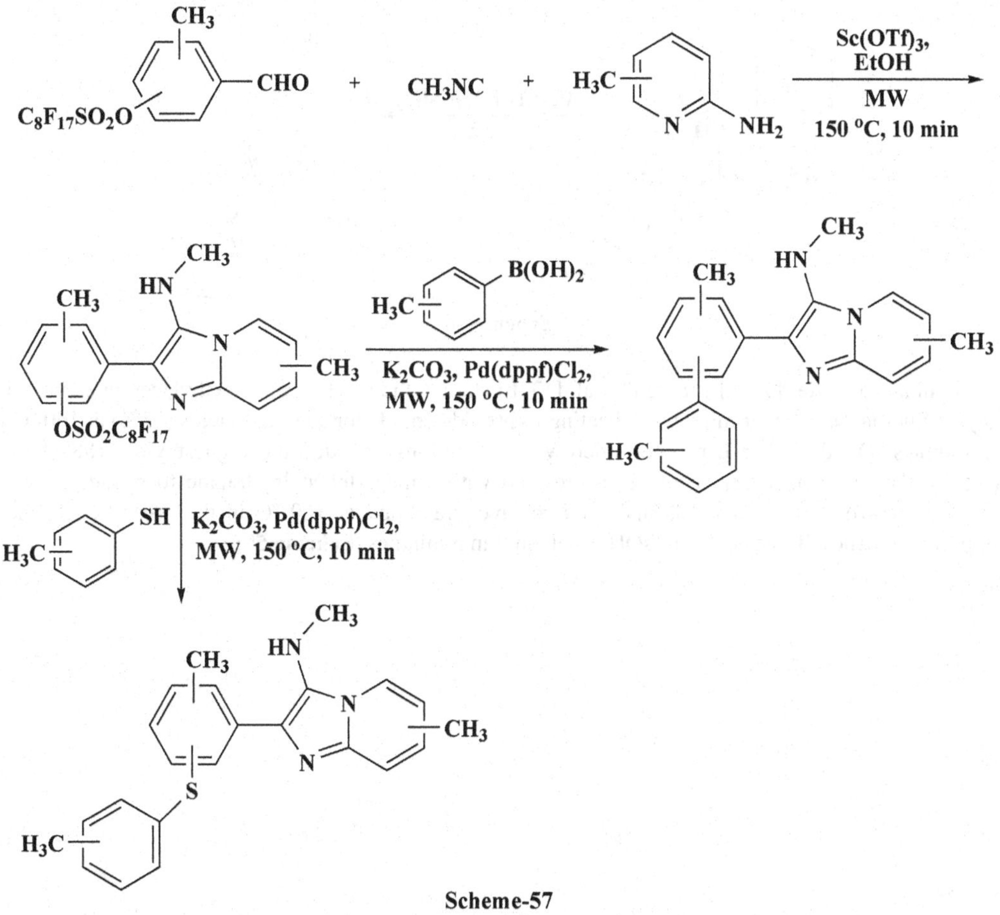

Scheme-57

Silicon-assisted synthesis

1-(4-Substituted phenyl)-3-phenyl-1*H*-pyrazole-4-carbaldehydes have been prepared and tested for their analgesic and anti-inflammatory properties. Vilsmeier-Haack reagent is used for the synthesis of pyrazoles. Substituted phenyl hydrazine, acetophenone and dimethylformamide are irradiated under microwaves at 200 W intermittently at 10-second intervals; the specified reaction time for hydrazone intermediate is 0.30 minutes. The reaction mixture is then cooled with cold water, and the obtained precipitate is filtered, washed with water, and purified by recrystallization in ethanol to yield hydrazone intermediate. The hydrazone intermediate is added portion-wise into Vilsmeier-Haack reagent. After the complete addition, the reaction flask is kept at room temperature for 5 minutes and silica gel is added and properly mixed till free flowing power is obtained. This powder is then intermittently irradiated in a microwave oven at 400 W at intervals of 30 seconds. The specified reaction time for 1-(4-substituted phenyl)-3-phenyl-1*H*-pyrazole-4-carbaldehyde is 5 minutes. The neutralization of the filtrate with NaHCO$_3$ provides a solid, which when filtered, washed with water, and purified by recrystallization from methanol, provides 1-(4-substituted phenyl)-3-phenyl-1*H*-pyrazole-4-carbaldehyde (**Scheme 58**) [132].

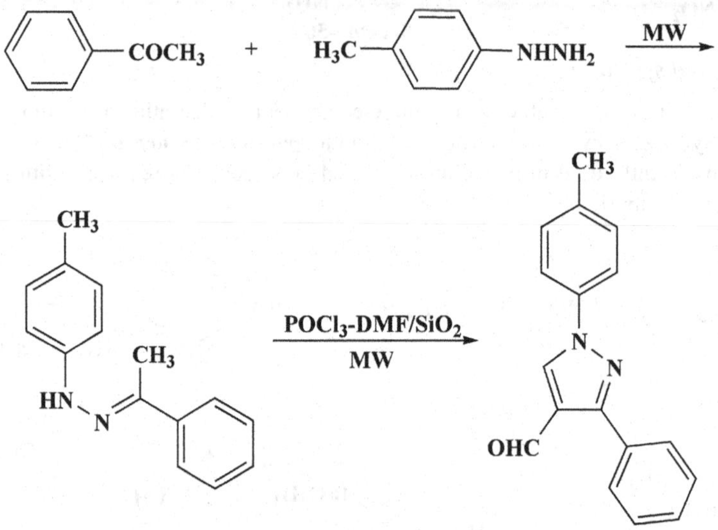

Scheme-58

Paul and Jennifer [133a] and Paul et al. [133b] studied the reaction of 4-methylacetophenone with ethyl trifluoroacetate under microwave heating to provide enol ketone in high yields (95%) at 160 °C in 10 minutes. On the other hand, non-microwave conditions afforded the highest yield (88%) in 5 days. In the subsequent step, enol ketone reacted with 4-methylphenylhydrazine to produce several 1,5-diarylpyrazoles in 95% yields under microwave irradiation at 160 °C in the presence of silica-supported toluenesulfonic acid (Si-TsOH) in ethanol in 5 minutes (**Scheme 59**).

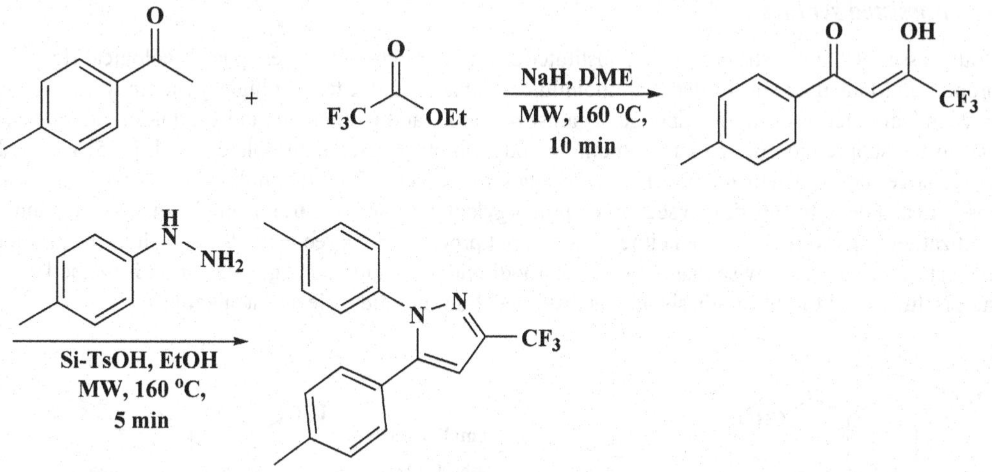

Scheme-59

The purification issue of present reaction can be solved by a solid-phase 'catch and release' protocol wherein a two-component two-step synthesis of 1-alkyl-4-imidazolecarboxylates [134a–b] is performed. A collection of isonitriles is immobilized onto a solid support in the first step through reaction with *N*-methyl aminomethylated polystyrene (commercially available). Subsequently, various amines are used for simultaneous derivatization and the release of desired imidazoles back into solution; only the derivatized material is released from the resin, hence ensuring high purity of the desired product. Microwave dielectric heating accelerates both the steps of the reaction, reducing the reaction time from 60 hours to 70 minutes (**Scheme 60**).

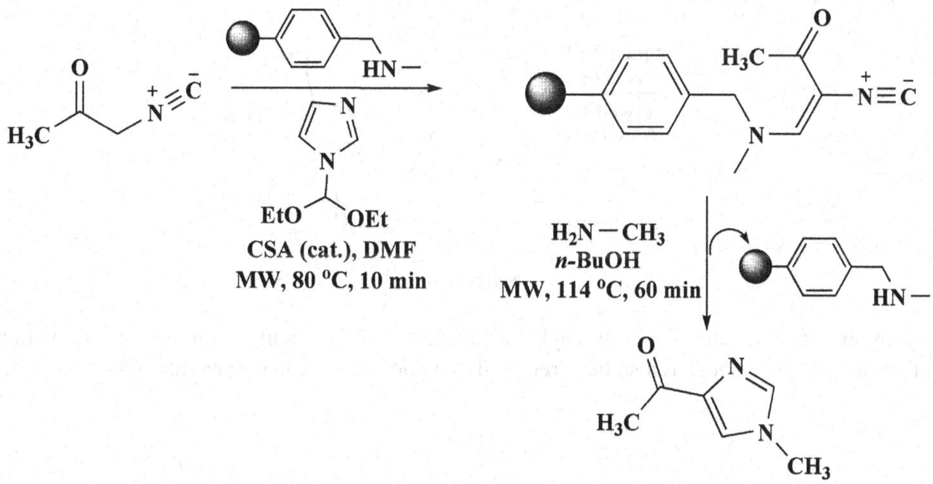

Scheme-60

Silver-assisted synthesis

Both classes of 1-unsubstituted and 1-substituted 2-aminoimidazoles possess pharmacological significant properties. Many poly-substituted 2-aminoimidazole alkaloids, due to their high cytotoxicity are involved in chemical defense of marine sponges against predators and pathogenic and fouling microorganisms, and in the suppression of spatial competition and epibiotic bacteria. Ermolatev et al. [135] reported a short, novel and efficient method for the synthesis of diverse 2-aminoimidazoles from thioureas and easily available poly-substituted secondary propargylamines (**Scheme 61**). Both the guanylation and the cyclization steps were performed either in a one-pot process with a recoverable AgI salt as the promoter and catalyst or in a step-wise manner with a carbodiimide activator and AgI catalyst. This method can be successfully used for the synthesis of tri-substituted 2-aminoimidazole amine alkaloids.

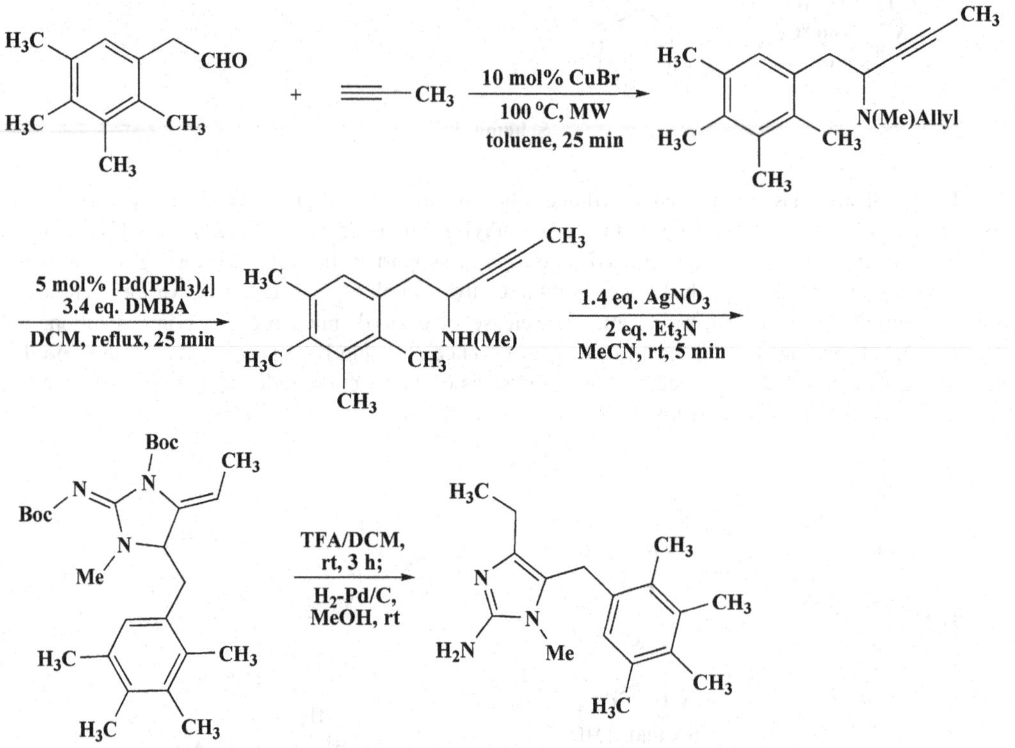

Scheme-61

Synthesis of pyrazole from styrenyl hydrazone in 73% yields, without the formation of 5,6-dihydropyridazin-4-one, has also been reported by various researchers (**Scheme 62**) [136–137].

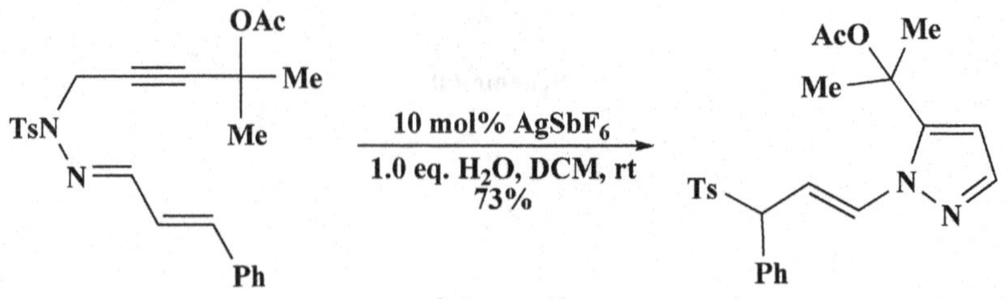

Scheme-62

Ethyl isocyanoacetate undergoes self-condensation to provide quantitative yields of ethyl 1-ethoxycarbonylimidazole-4-carboxylate in the presence of silver acetate as the catalyst **(Scheme 63)** [138]. This reaction also occurs with other isocyanoacetates competitively when silver acetate is used as a catalyst and becomes predominant if other components are not sufficiently reactive [45].

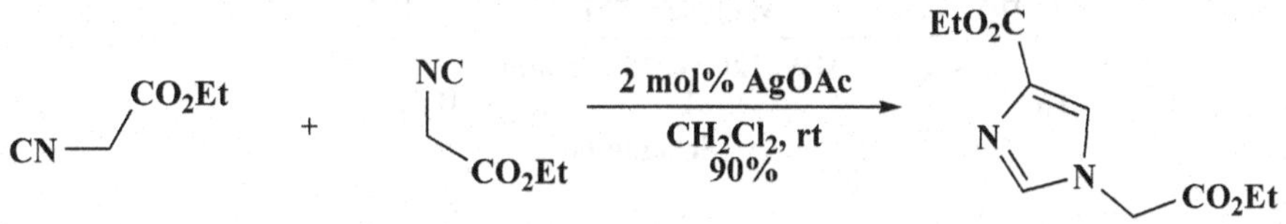

Scheme-63

Methyl isocyanoacetate, through self-condensation in the presence of 0.5 eq. of a strong base, results in the synthesis of isocyanomethyl oxazole, without any acylating agent. Oxazole is also used in reactions with acylating agents to provide bi- and quarter-oxazoles **(Scheme 64)** [45, 139].

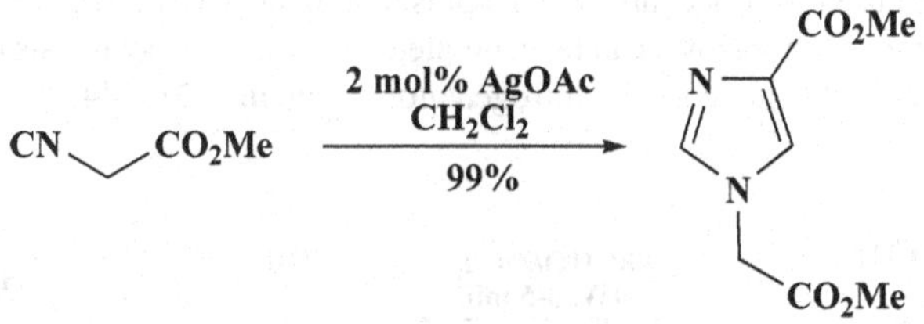

Scheme-64

Elders et al. [140] reported the synthesis of highly substituted 2-imidazolines through the reaction of aldehydes, amines and isocyanides bearing an acidic α-proton, in the presence of AgOAc as the catalyst. The reactions of ketones as oxo compounds instead of aldehydes, which are usually difficult, can also be promoted by AgOAc. The scope of the reaction has also been extended towards less α-acidic isocyanoacetates [141]. The reaction between benzylamine, acetone and methyl isocyanoacetates occurs in several solvents, methanol being the most effective and practical. This multi-component reaction occurs without any side process, is well-suited for the effective synthesis of imidazoline, and can be accelerated with AgOAc due to the coordination of the isocyanide carbon with silver, hence enhancing the NC electrophilicity and α-acidity **(Scheme 65)** [45, 142]. Nevertheless, reaction times are higher (up to 2 weeks) with less reactive substrates.

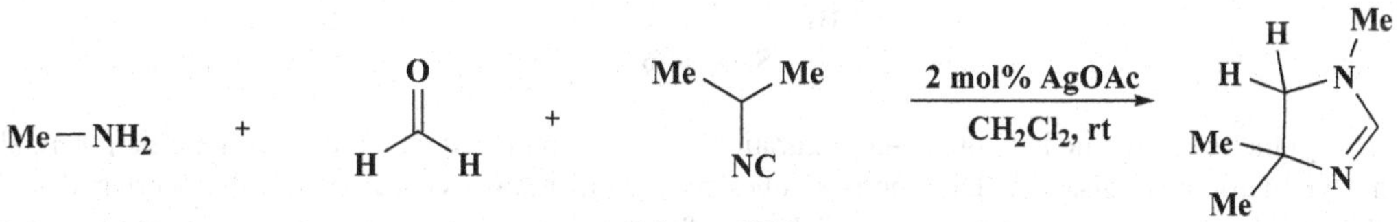

Scheme-65

Sulfur-assisted synthesis

Hoz et al. [143] studied the synthesis of 2-imidazolines. Microwave-assisted cyclization of nitriles with ethylenediamine, which aromatize in MagtrieveTM (oxidant) and toluene, furnished imidazoles in 75–105 minutes **(Scheme 66)**. The use of manganese(IV) oxide under conventional heating needed longer reaction times (24–48 hours) for high yields (76–93%).

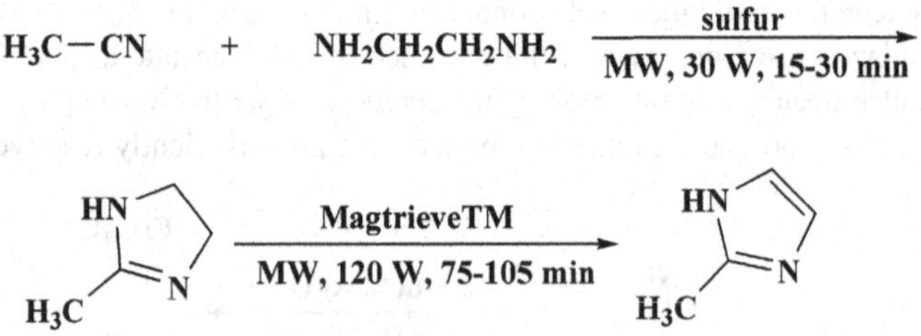

$$H_3C-CN \quad + \quad NH_2CH_2CH_2NH_2 \xrightarrow[\text{MW, 30 W, 15-30 min}]{\text{sulfur}}$$

Scheme-66

Zinc-assisted synthesis

Pyrazole can be prepared by elimination/aromatization of a cycloadduct intermediate. Mild and simple microwave-mediated solvent-free protocols have been used for the preparation of pyrazoles from α,β-unsaturated carbonyl compounds. Thieno[2,3-*b*]thiophene coupled with aldehyde provides α,β-unsaturated carbonyl compounds under microwave irradiation in $ZnCl_2$. Pyrazole derivatives were prepared following the classical procedure (ketone plus hydrazine derivatives in ethanol at reflux in very good yield). Furthermore, bis-pyrazoles can be generated by a step-wise synthesis of hydrazone followed by Michael 1,4-addition of the nucleophilic nitrogen atom (**Scheme 67**) [144].

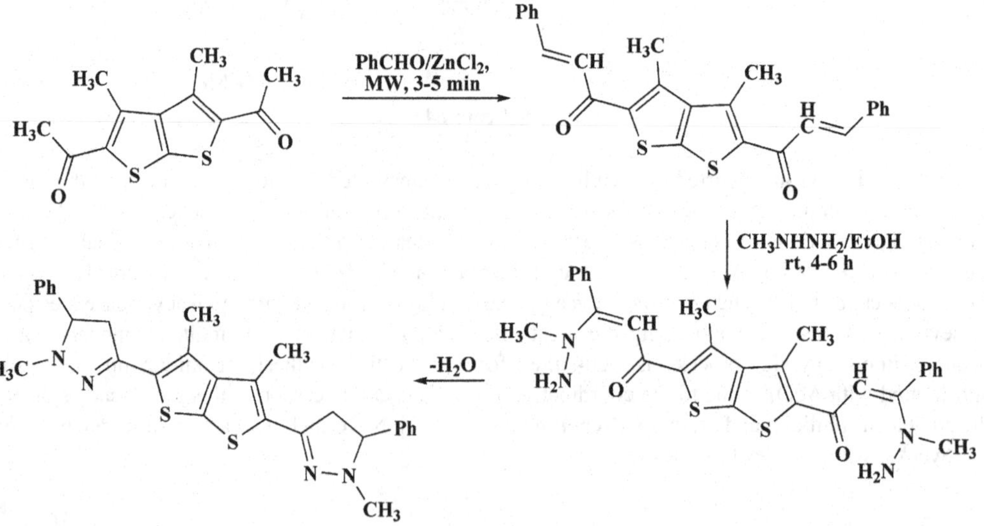

Scheme-67

Synthesis of diimine in (+)-(*S*,*S*)-configuration has been optimized with meso- as it has no potential in asymmetric catalysis. (*S*,*S*)-Diimine is synthesized in quantitative yields when enantioenriched (+)-(*S*,*S*)- aniline is condensed with glyoxal (**Scheme 68**). However, extensive optimization is required for the cyclization and purification of this product. Berthon-Gelloz et al. [145] modified the cyclization protocol using zinc chloride to coordinate the diimine in the reactive *s-cis* conformation which provided pure (*S*,*S*,*S*,*S*)-product in 87% yields after recrystallization [146].

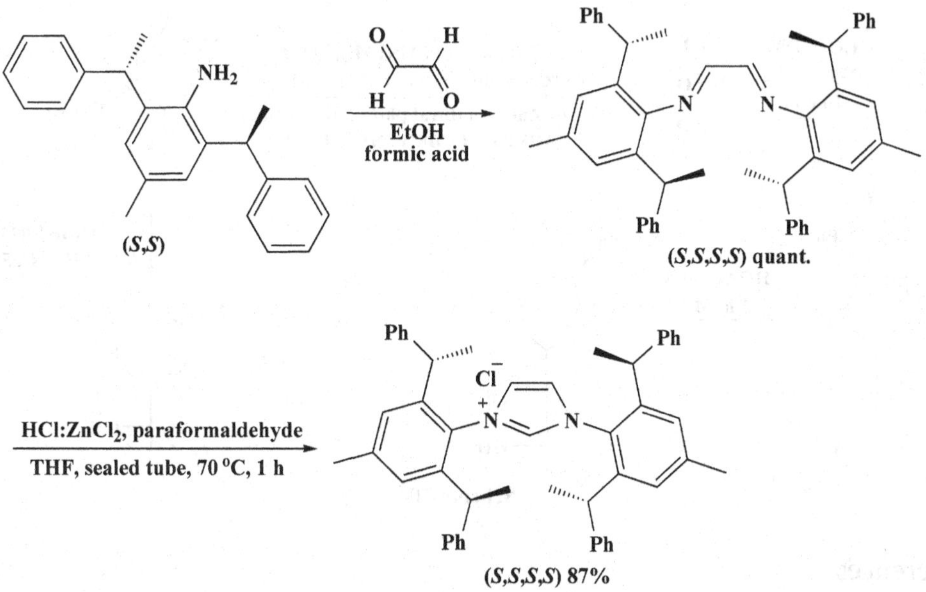

Scheme-68

Researchers have developed interest in cyclic ureas like imidazolidine-2-one for their capability to serve as intermediates for biologically active molecules, including DMP 450 and DMP 323, HIV protease inhibitors, pharmaceuticals, fine chemicals, pesticides and cosmetics [147]. A microwave-assisted method has been reported for the direct synthesis of cyclic ureas in the presence of zinc oxide **(Scheme 69)**, wherein the reaction is accelerated, thus reducing the reaction time, as well as eliminating the formation of by-products, compared to conventional heating procedures [148–149].

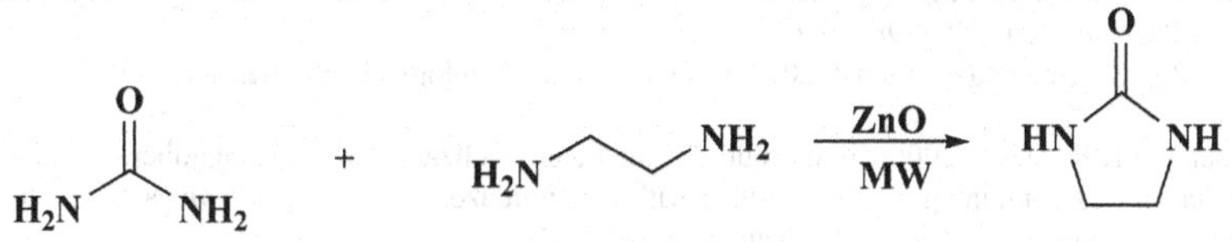

Scheme-69

The synthesis of dihydropyrimidinones is of importance to researchers due to their diverse biological properties (anti-bacterial, anti-viral and anti-hypertensive) as well as their efficacy as α-antagonists and calcium channel blockers [150–151]. The following method [152–153] of synthesizing dihydropyrimidinones, based on the Biginelli reaction, was developed as an efficient and convenient one-pot preparation of acyclic nucleosides having imidazole and dihydropyrimidinone bases. Herein, 2,2-dimethyl-[1,3]dioxolan-4-carbaldehyde was treated with benzil in ammonium hydroxide for 10 minutes at 100 °C to afford dioxolan imidazole derivative in 87% yields after purification [154–157]. Subsequently, free *C*-nucleosides were formed in 76% yields upon acid hydrolysis of the dioxolan imidazole derivative with 80% acetic acid at 80 °C **(Scheme 70)** [158].

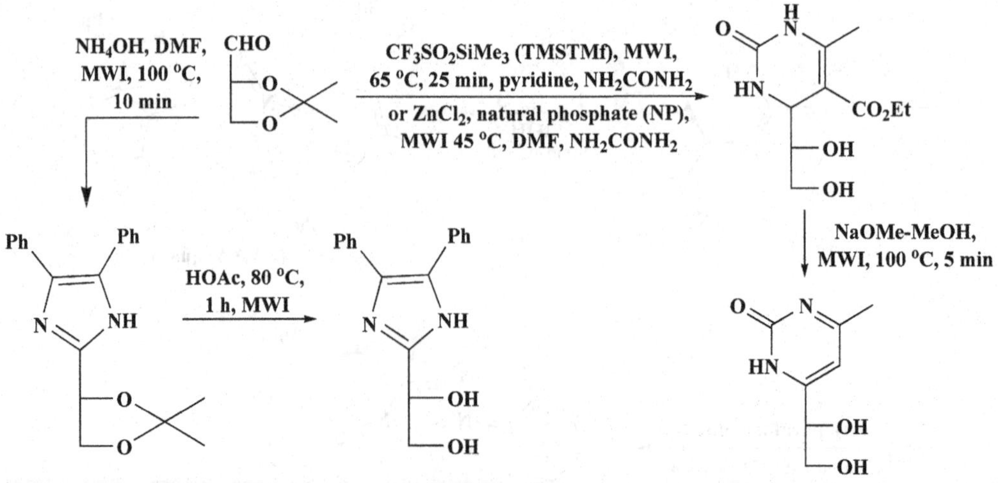

Scheme-70

References

(1) D. Castagnolo, F. Manetti, M. Radi, B. Bechi, M. Pagano, A. de Logu, R. Meleddu, M. Saddi and M. Botta. 2009. Synthesis, biological evaluation, and SAR study of novel pyrazole analogues as inhibitors of *Mycobacterium tuberculosis*: synthesis of rigid pyrazolones. Bioorg. Med. Chem. 17: 5716–5721.

(2) F.A. Pasha, M. Muddassar, M.M. Neaz and S.J. Cho. 2009. Pharmacophore and docking based combined *in silico* study of KDR inhibitors. J. Mol. Graph. Model. 28: 54–61.

(3) H.H. Zoorob, M.S. Elsherbini and W.S. Hamama. 2012. Utility of cyclododecanone as synthon to synthesize fused heterocycles. J. Org. Chem. 2: 63–68.

(4) K. Rehse, J. Kotthaus and L. Kadembashi. 2009. New 1*H*-pyrazole-4-carboxamides with anti-platelet activity. Arch. Pharm. Chem. Life Sci. 342: 27–33.

(5) G.P. Lahm, T.P. Selby, J.H. Freudenberger, T.M. Stevenson, B.J. Myres, G. Seburyamo, B.K. Smith, L. Flexner, C.E. Clark and D. Cordova. 2005. Insecticidal anthranilic diamides: a new class of potent ryanodine receptor activators. Bioorg. Med. Chem. Lett. 15: 4898–4906.

(6) K. Shalini, P.K. Sharma and N. Kumar. 2010. Imidazole and its biological activities: a review. Der Chemica Sinica 1: 36–47.

(7) K.G. Desai and K.R. Desai. 2006. Green route for the heterocyclization of 2-mercaptobenzimidazole into β-lactam segment derivatives containing -CONH- bridge with benzimidazole: screening *in vitro* anti-microbial activity with various microorganisms. Bioorg. Med. Chem. 14: 8271–8279.

(8) Z. Kazimierczuk, J.A. Upcroft, P. Upcroft, A. Gorska, B. Starosciak and A. Laudy. 2002. Synthesis, anti-protozoal and anti-bacterial activity of nitro- and halogeno-substituted benzimidazole derivatives. Acta Biochim. Pol. 49: 185–195.

(9) Z.M. Nofal, H.H. Fahmy and H.S. Mohamed. 2002. Synthesis, anti-microbial and molluscicidal activities of new benzimidazole derivatives. Arch. Pharm. Res. 25: 28–38.

(10) M. Pedini, B.G. Alunni, A. Ricci, L. Bastianini and E. Lepri. 1994. New heterocyclic derivatives of benzimidazole with germicidal activity-XII - synthesis of *N*1-glycosyl-2-furyl-benzimidazoles. Farmaco 49: 823–827.

(11) K.F. Ansari and C. Lal. 2009. Synthesis and biological activity of some heterocyclic compounds containing benzimidazole and β-lactam moiety. J. Chem. Sci. 121: 1017–1025.

(12) L. Garuti, M. Roberti and G. Gentilomi. 2000. Synthesis and anti-viral assays of some 2-substituted benzimidazoles *N*-carbamates. Farmaco 55: 35–39.

(13) D.P.A. Thakur, S.G. Wadodkar and C.T. Chopade. 2012. Synthesis and anti-inflammatory activity of some benzimidazole-2-carboxylic acids. Int. J. Drug Dev. Res. 4: 303–309.

(14) O.O. Guven, T. Erdogan, H. Goeker and S. Yildiz. 2007. Synthesis and anti-microbial activity of some novel phenyl and benzimidazole substituted benzyl ethers. Bioorg. Med. Chem. Lett. 17: 2233–2236.

(15) G.A. Kilcigil and N. Altanlar. 2006. Synthesis and anti-fungal properties of some benzimidazole derivatives. Turk J. Chem. 30: 223–228.

(16) M. Boiani and M. Gonzalez. 2005. Imidazole and benzimidazole derivatives as chemotherapeutic agents. Mini Rev. Med. Chem. 5: 409–424.

(17) A.Y. Usyatinsky and Y.L. Khmelnitsky. 2000. Microwave-assisted synthesis of substituted imidazoles on a solid support under solvent-free conditions. Tetrahedron Lett. 41: 5031–5034.

(18)   P.N. Reddy, L.K. Ravindranath, K.B. Chandrasekhar, P. Rameshbabu, G. Madhu and K.S.B. Aiswarya. 2012. Synthesis of novel Mannich bases containing pyrazolones and indole systems. Der Pharma Chemica 4: 1330–1338.

(19)   R. Chawla, U. Sahoo, A. Arora, P.C. Sharma and V. Radhakrishnan. 2010. Microwave-assisted synthesis of some novel 2-pyrazoline derivatives as possible anti-microbial agents. Acta Poloniae Pharm. Drug Res. 67: 55–61.

(20)   (a) V.M. Patel and K.R. Desai. 2004. Eco-friendly synthesis of pyrazoline derivatives over potassium carbonate. ARKIVOC (i): 123–129. (b) S.P. Sakthinathan, G. Vanangamudi and G. Thirunarayanan. 2012. Synthesis, spectral studies and anti-microbial activities of some 2-naphthyl pyrazoline derivatives. Spectrochimica Acta Part A: Mol. Biomol. Spectr. 95: 693–700.

(21)   M.B. Tyers, I.H. Coates, D.C. Humber, G.B. Ewan, J.A. Bell. June 28, 1988. Method for Treating Nausea and Vomiting. U.S. Patent 4,753,789.

(22)   A.C.-M. Daugan. Jan 12, 1999. Tetracyclic derivatives; process of preparation and use. U.S. Patent 5,859,006.

(23)   K.M. Gitis, N.I. Raevskaya and G.V. Isagulyants. 1992. The staged synthesis of 2-methylimidazole from ethylenediamine and acetic acid in the presence of a bi-functional aluminoplatinum catalyst. Russ. Chem. Bull. 41: 1551–1554.

(24)   M. Baumann, I.R. Baxendale, S.V. Ley and N. Nikbin. 2011. An overview of the key routes to the best selling 5-membered ring heterocyclic pharmaceuticals. Beilstein J. Org. Chem. 7: 442–495.

(25)   A.K. Rathod and G.B. Kulkarni. 2011. Synthesis and characterizations of diphenyl imidazolylpyrimidines-5-carboxylates (DPIPC) derivatives and their anti-fungal and anti-bacterial activity under conventional and microwave irradiation method. Int. J. Pharm. Tech. Res. 3: 435–441.

(26)   T.A. Katte, T.A. Reekie, W.T. Jorgensen and M. Kassiou. 2016. The formation of seven-membered heterocycles under mild Pictet-Spengler conditions: a route to pyrazolo[3,4]benzodiazepines. J. Org. Chem. 81: 4883–4889.

(27)   I.P. Beletskaya and A.V. Cheprakov. 2000. The Heck reaction as a sharpening stone of palladium catalysis. Chem. Rev. 100: 3009–3066.

(28)   D.L. Thorn and R. Hoffman. 1978. The olefin insertion reaction. J. Am. Chem. Soc. 100: 2079–2090.

(29)   R. Kalirajan, L. Rathore, S. Jubie, B. Gowramma, S. Gomathy, S. Sankar and K. Elango. 2010. Microwave-assisted synthesis and biological evaluation of pyrazole derivatives of benzimidazoles. Indian J. Pharm. 44: 358–390.

(30)   A. Chawla, G. Kaur and A.K. Sharma. 2012. Green chemistry as a versatile technique for the synthesis of benzimidazole derivatives: review. Int. J. Pharm. Phytopharmacol. Res. 2: 148–159.

(31)   S.K. Pal and S. Kumar. 2006. Microwave-assisted synthesis of novel imidazolium-based ionic liquid crystalline dimers. Tetrahedron Lett. 47: 8993–8997.

(32)   O. Bruno, C. Brullo, F. Bondavalli, S. Schenone, A. Ranise, N. Arduino, M.B. Bertolotto, F. Montecucco, L. Ottonello, F. Dallegri, M. Tognolini, V. Ballabeni, S. Bertoni and E. Barocelli. 2007. Synthesis and biological evaluation of N-pyrazolyl-N'-alkyl/benzyl/phenylureas: a new class of potent inhibitors of interleukin 8-induced neutrophil chemotaxis. J. Med. Chem. 50: 3618–3626.

(33)   P. Lu and Y.G. Wang. 2010. Strategies for heterocyclic synthesis *via* cascade reactions based on ketenimines. Synlett 2: 165–173.

(34)   V.V. Rostovtsev, L.G. Green, V.V. Fokin and K.B. Sharpless. 2002. A stepwise Huisgen cycloaddition process: copper(I)-catalyzed regioselective "ligation" of azides and terminal alkynes. Angew. Chem. Int. Ed. 41: 2596–2599.

(35)   J.A. Bull, M.G. Hutchings, C. Luján and P. Quayle. 2008. New reactivity patterns of copper(I) and other transition metal NHC complexes: application to ATRC and related reactions. Tetrahedron Lett. 49: 1352–1356.

(36)   J.A. Bull, M.G. Hutchings and P. Quayle. 2007. A remarkably simple and efficient benzannulation reaction. Angew. Chem. Int. Ed. 46: 1869–1872.

(37)   S. Zheng, F. Li, J. Liu and C. Xia. 2007. A novel and efficient (NHC)CuI (NHC = N-heterocyclic carbene) catalyst for the oxidative carbonylation of amino compounds. Tetrahedron Lett. 48: 5883–5886.

(38)   J. Haider, K. Kunz and U. Scholz. 2004. Highly selective copper-catalyzed monoarylation of aniline. Adv. Synth. Catal. 346: 717–722.

(39)   D.S. Laitar, E.Y. Tsui and J.P. Sadighi. 2006. Catalytic diboration of aldehydes *via* insertion into the copper-boron bond. J. Am. Chem. Soc. 128: 11036–11037.

(40)   D.S. Laitar, E.Y. Tsui and J.P. Sadighi. 2006. Copper(I) β-boroalkyls from alkene insertion: isolation and rearrangement. Organometallics 25: 2405–2408.

(41)   V. Lillo, M.R. Fructos, J. Ramirez, A.A.C. Braga, F. Maseras, M.M. Diaz-Requejo, P.J. Perez and E. Fernandez. 2007. A valuable, inexpensive CuI/N-heterocyclic carbene catalyst for the selective diboration of styrene. Chem. Eur. J. 13: 2614–2621.

(42)   S. Chiba, L. Zhang and J.-Y. Lee. 2010. Copper-catalyzed synthesis of azaspirocyclohexadienones from α-azido-N-arylamides under an oxygen atmosphere. J. Am. Chem. Soc. 132: 7266–7267.

(43)   Z. Shi, C. Zhang, C. Tanga and N. Jiao. 2012. Recent advances in transition metal-catalyzed reactions using molecular oxygen as the oxidant. Chem. Soc. Rev. 41: 3381–3430.

(44)   M.-A. Bonin, D. Giguere and R. Roy. 2007. N-Arylimidazole synthesis by cross-cycloaddition of isocyanides using a novel catalytic system. Tetrahedron 63: 4912–4917.

(45)   A.V. Gulevich, A.G. Zhdanko, R.V.A. Orru and V.G. Nenajdenko. 2010. Isocyanoacetate derivatives: synthesis, reactivity, and application. Chem. Rev. 110: 5235–5331.

(46) S. Mori, T. Mori and Y. Mukoyama. 1993. Elution behavior of polyethylene glycols on a hydrophilic polymer gel column used for size exclusion chromatography. J. Liquid Chromatography 16: 2269–2279.

(47) H. Wang, Y. Wang, C. Peng and J. Zhang. 2010. A direct intramolecular C-H amination reaction co-catalyzed by copper(II) and iron(III) as part of an efficient route for the synthesis of pyrido[1,2-*a*]benzimidazoles from *N*-aryl-2-aminopyridines. J. Am. Chem. Soc. 132: 13217–13219.

(48) C.L. Cioffi, W.T. Spencer, J.J. Richards and R.J. Herr. 2004. Generation of 3-pyridyl biaryl systems *via* palladium-catalyzed Suzuki cross-couplings of aryl halides with 3-pyridylboroxin. J. Org. Chem. 69: 2210–2212.

(49) D. Benito-Garagorri, V. Bocokie and K. Kirchner. 2006. Copper(I)-catalyzed diastereoselective formation of oxazolines and *N*-sulfonyl–2-imidazolines. Tetrahedron Lett. 47: 8641–8644.

(50) X.Y. Gong, H.J. Yang, H.X. Liu, Y.Y. Jiang, Y.F. Zhao and H. Fu. 2010. Simple and efficient copper-catalyzed approach to 2,4-disubstituted imidazolones. Org. Lett. 12: 3128–3131.

(51) Y. Liu and J.-P. Wan. 2011. Tandem reactions initiated by copper-catalyzed cross-coupling: a new strategy towards heterocycle synthesis. Org. Biomol. Chem. 9: 6873–6894.

(52) F. Brackman, M. Es-Sayed and A. de Meijere. 2005. Synthesis of spirocyclopropanated analogues of iprodione. Eur. J. Org. Chem. 11: 2250–2258.

(53) G. Evano, N. Blanchard and M. Toumi. 2008. Copper-mediated coupling reactions and their applications in natural products and designed bio-molecules synthesis. Chem. Rev. 108: 3054–3131.

(54) C. Kanazawa, S. Kamijo and Y. Yamamoto. 2006. Synthesis of imidazoles through the copper-catalyzed cross-cycloaddition between two different isocyanides. J. Am. Chem. Soc. 128: 10662–10663.

(55) A.P. Green and N.J. Turner. 2016. Bio-catalytic retrosynthesis: redesigning synthetic routes to high-value chemicals. Perspect. Sci. 9: 42–48.

(56) L. Ackermann. 2005. General and efficient indole syntheses based on catalytic amination reactions. Org. Lett. 7: 439–442.

(57) R. Martin, R.M. Rivero and S.L. Buchwald. 2006. Domino Cu-catalyzed C-N coupling/hydroamidation: a highly efficient synthesis of nitrogen heterocycles. Angew. Chem. Int. Ed. 45: 7079–7082.

(58) A. Arcadi, S. Cacchi and F. Marinelli. 1989. Palladium-catalyzed coupling of aryl and vinyl triflates or halides with 2-ethynylaniline: an efficient route to functionalized 2-substituted indoles. Tetrahedron Lett. 30: 2581–2584.

(59) Z.-Y. Tang and Q.-S. Hu. 2006. Efficient synthesis of 2-substituted indoles based on palladium(II) acetate/tri-*tert*-butylphosphine-catalyzed alkynylation/amination of 1,2-dihalobenzenes. Adv. Synth. Catal. 348: 846–850.

(60) Y. Yu, G.A. Stephenson and D. Mitchell. 2006. A regioselective synthesis of 3-benzazepinones *via* intramolecular hydroamidation of acetylenes. Tetrahedron Lett. 47: 3811–3814.

(61) K. Hiroya, S. Itoh and M. Ozawa. 2004. Development of an efficient procedure for indole ring synthesis from 2-ethynylaniline derivatives catalyzed by Cu(II) salts and its application to natural product synthesis. J. Org. Chem. 69: 1126–1136.

(62) L. Xu, I.R. Lewis, S.K. Davidsen and J.B. Summers. 1998. Transition metal-catalyzed synthesis of 5-azaindoles. Tetrahedron Lett. 39: 5159–5162.

(63) A. de Meijere, P. von Zezschwitz and S. Bräse. 2005. The virtue of palladium-catalyzed domino reactions—diverse oligocyclizations of acyclic 2-bromoenynes and 2-bromoenediynes. Acc. Chem. Res. 38: 413–422.

(64) L.F. Tietze. 1996. Domino reactions in organic synthesis. Chem. Rev. 96: 115–136.

(65) R.G. Murray, D.M. Whitehead, F. LeStart and S.J. Conway. 2008. Facile one-pot synthesis of 5-substituted hydantoins. Org. Biomol. Chem. 6: 988–991.

(66) G. Meng and M. Szostak. 2017. Site-selective C-H/C-N activation by cooperative catalysis: primary amides as arylating reagents in directed C-H arylation. ACS Catal. 7: 7251–7256.

(67) P. Drabina and M. Sedlak. 2012. 2-Amino–2-alkyl(aryl)propanenitriles as key building blocks for the synthesis of five-membered heterocycles. ARKIVOC (i): 152–172.

(68) (a) A.M. Manzo, A.D. Perboni, G. Broggini and M. Rigamonti. 2009. Gold-catalyzed intramolecular hydroamination of α-amino allenamides as a route to enantiopure 2-vinylimidazolidinones. Tetrahedron Lett. 50: 4696–4699. (b) T. Lu, Z. Lu, Z.-X. Ma, Y. Zhang and R.P. Hsung. 2013. Allenamides: a powerful and versatile building block in organic synthesis. Chem. Rev. 113: 4862–4904.

(69) X. Zhou, Y. Lin, L. Dai, J. Sun, L. Xia and M. Tang. 1999. A catalytic enantioselective access to optically active 2-imidazoline from *N*-sulfonylimines and isocyanoacetates. J. Org. Chem. 64: 1331–1334.

(70) X.-T. Zhou, Y.-R. Lin and L.-X. Dai. 1999. A simple and practical access to enantiopure 2,3-diamino acid derivatives. Tetrahedron: Asymmetry 10: 855–862.

(71) J. Qian, Y. Liu, J. Zhu, B. Jiang and Z. Xu. 2011. A novel synthesis of fluorinated pyrazoles *via* gold(I)-catalyzed tandem aminofluorination of alkynes in the presence of selectfluor. Org. Lett. 13: 4220–4223.

(72) G. Liu. 2012. Transition metal-catalyzed fluorination of multi carbon-carbon bonds: new strategies for fluorinated heterocycles. Org. Biomol. Chem. 10: 6243–6248.

(73) T. Hayashi, E. Kishi, V.A. Soloshonok and Y. Uozumi. 1996. Erythro-selective aldol-type reaction of *N*-sulfonylaldimines with methyl isocyanoacetate catalyzed by gold(I). Tetrahedron Lett. 37: 4969–4972.

(74) Y.-R. Lin, X.-T. Zhou, L.-X. Dai and J. Sun. 1997. Ruthenium complex-catalyzed reaction of isocyanoacetate and *N*-sulfonylimines: stereoselective synthesis of *N*-sulfonyl-2-imidazolines. J. Org. Chem. 62: 1799–1803.

(75) Y. Ito, M. Sawamura and T. Hayashi. 1986. Catalytic asymmetric aldol reaction: reaction of aldehydes with isocyanoacetate catalyzed by a chiral ferrocenylphosphine-gold(I) complex. J. Am. Chem. Soc. 108: 6405–6406.

(76) B. Morgan, D.R. Dodds, A. Zaks, D.R. Andrews and R. Klesse. 1997. Enzymatic desymmetrization of prochiral 2-substituted 1,3-propanediols: a practical chemoenzymatic synthesis of a key precursor of SCH51048, a broad-spectrum orally active anti-fungal agent. J. Org. Chem. 62: 7736–7743.

(77) R.N. Patel, A. Banerjee and L.J. Szarka. 1997. Stereoselective acetylation of racemic 7-[N,N'-bis(benzyloxycarbonyl)-N-(guanidinoheptanoyl)]-α-hydroxyglycine. Tetrahedron: Asymmetry 8: 1767–1771.

(78) R.N. Patel, C.M. Mcnamee and L.J. Szarka. 1992. Enantioselective enzymatic acetylation of racemic [4-[4α,6β (E)]]-6-[4,4-bis(4-fluorophenyl)-3-(1-methyl-1H-tetrazol-5-yl)-1,3-butadienyl]-tetrahydro-4-hydroxy-2H-pyran-2-one. Appl. Microbiol. Biotechnol. 38: 56–60.

(79) S.E. Milnera and A.R. Maguire. 2012. Recent trends in whole cell and isolated enzymes in enantioselective synthesis. ARKIVOC (i): 321–382.

(80) S. Ponnala and D.P. Sahu. 2006. Iodine-mediated synthesis of 2-arylbenzoxazoles, 2-arylbenzimidazoles, and 1,3,5-trisubstituted pyrazoles. Synth. Commun. 36: 2189–2194.

(81) P.T. Parvatkar, P.S. Parameswaran and S.G. Tilve. 2012. Recent developments in the synthesis of five- and six-membered heterocycles using molecular iodine. Chem. Eur. J. 18: 5460–5489.

(82) M. Kidwai, P. Mothsra, V. Bansal, R.K. Somvanshi, A.S. Ethayathulla, S. Dey and T.P. Singh. 2007. One-pot synthesis of highly substituted imidazoles using molecular iodine: a versatile catalyst. J. Mol. Catal. A: Chem. 265: 177–182.

(83) M. Kidwai and P. Mothsra. 2006. A one-pot synthesis of 1,2,4,5-tetraarylimidazoles using molecular iodine as an efficient catalyst. Tetrahedron Lett. 47: 5029–5031.

(84) G. Meng and M. Szostak. 2016. Rhodium-catalyzed C-H bond functionalization with amides by double C-H/C-N bond activation. Org. Lett. 18: 796–799.

(85) A. Bessmertnykh, F. Henin and J. Muzart. 2004. Palladium-catalyzed oxidation of benzylated aldose hemiacetals to lactones. Carbohydr. Res. 339: 1377–1380.

(86) S. Balalaie and A. Arabanian. 2000. One-pot synthesis of tetra-substituted imidazoles catalyzed by zeolite HY and silica gel under microwave irradiation. Green Chem. 2: 274–276.

(87) D.S. Raghuvanshi and K.N. Singh. 2010. A facile multi-component synthesis of tetra-substituted imidazoles using Fe^{3+}-K10 catalyst under solvent-free microwave conditions. Indian J. Chem. 49B: 1394–1397.

(88) A. Das, A. Kulkarni and B. Torok. 2012. Environmentally benign synthesis of heterocyclic compounds by combined microwave-assisted heterogeneous catalytic approaches. Green Chem. 14: 17–34.

(89) M. Johannsen and K.A. Jørgensen. 1998. Allylic amination. Chem. Rev. 98: 1689–1708.

(90) Y. Tamaru and M. Kimura. 1997. C-N bond formation reactions *via* transition metal catalysis. Synlett 7: 749–757.

(91) C. Bolm, J. Legros, J.L. Paih and L. Zani. 2004. Iron-catalyzed reactions in organic synthesis. Chem. Rev. 104: 6217–6254

(92) M.R. Manafi, P. Manafi and M.R. Kalaee. 2012. Versatile microwave-assisted and lanthanum chloride-catalyzed synthesis of poly-substituted imidazoles using urea/thiourea as benign source of ammonia. E-J. Chem. 9: 1773–1777.

(93) G.J. Griffiths, M.B. Hauck, R. Imwinkelried, J. Kohr, C.A. Roten, G.C. Stucky and J. Gosteli. 1999. Novel syntheses of 2-butyl-5-chloro-3H-imidazole-4-carbaldehyde: a key intermediate for the synthesis of the angiotensin II antagonist losartan. J. Org. Chem. 64: 8084–8089.

(94) Y. Wang, Y. Li, Y. Li, G. Zheng and Y. Li. Aug 10, 2006. Method for the production of losartan. WO Patent 2006/081807.

(95) D.M. Solomon, R.K. Rizvi and J.J. Kaminski. 1987. Observations on the reactions of isocyanoacetane esters with isothiocyanates and isocyanates. Heterocycles 26: 651–674.

(96) T. Hischer, D. Gocke, M. Fernández, P. Hoyos, A.R. Alcántara, J.V. Sinisterra, W. Hartmeier and M.B. Ansorge-Schumacher. 2005. Stereoselective synthesis of novel benzoins catalyzed by benzaldehyde lyase in a gel-stabilized two-phase system. Tetrahedron 61: 7378–7383.

(97) J.P. Chupp and K.L. Leschinsky. 1980. Heterocycles from substituted amides. VII Oxazoles from 2-isocyanoacetamides. J. Heterocycl. Chem. 17: 705–709.

(98) R. Bossio, S. Marcaccini, S. Paoli, R. Papaleo, R. Pepino and C. Polo. 1991. Studies on isocyanides and related compounds. Synthesis and cyclization of N-substituted 1-isocyano-1-cycloalkanecarboxamides. Liebigs Ann. Chem. 9: 843–849.

(99) R. Jazzar, J.-B. Bourg, R.D. Dewhurst, B. Donnadieu and G. Bertrand. 2007. Intramolecular "hydroiminiumation and -amidiniumation" of alkenes: a convenient, flexible, and scalable route to cyclic iminium and imidazolinium salts. J. Org. Chem. 72: 3492–3499.

(100) S. Nag and S. Batra. 2011. Applications of allylamines for the syntheses of aza-heterocycles. Tetrahedron 67: 8959–9061.

(101) J. Safari, S.D. Khalili and S.H. Banitaba. 2010. A novel and an efficient catalyst for one-pot synthesis of 2,4,5-trisubstituted imidazoles by using microwave irradiation under solvent-free conditions. J. Chem. Sci. 122: 437–441.

(102) H. Alper, J.E. Prickett and S. Wollowitz. 1977. Intermolecular and intramolecular cycloaddition reactions of azirines by group 6 metal carbonyls and by titanium tetrachloride. J. Am. Chem. Soc. 99: 4330–4333.

(103) F.D. Bellamy. 1978. Unexpected rearrangements of a (Z)-ketovinyiazirine. C-C versus C-N bond cleavage. Tetrahedron Lett. 19: 4577–4580.

(104) H.M.L. Davies and R.E.J. Beckwith. 2003. Catalytic enantioselective C-H activation by means of metal-carbenoid induced C-H insertion. Chem. Rev. 103: 2861–2904.

(105) M.P. Doyle and D.C. Forbes. 1998. Recent advances in asymmetric catalytic metal carbene transformations. Chem. Rev. 98: 911–936.

(106) B. Chattopadhyay and V. Gevorgyan. 2012. Transition metal-catalyzed denitrogenative transannulation: converting triazoles into other heterocyclic systems. Angew. Chem. Int. Ed. 51: 862–872.

(107) B. Zhao, H. Du and Y. Shi. 2008. Cu(I)-catalyzed cycloguanidination of olefins. Org. Lett. 10: 1087–1090.

(108) R. Connell, F. Scavo, P. Helquist and B. Åkermark. 1986. Functionalized oxazoles from rhodium-catalyzed reaction of dimethyl diazomalonate with nitriles. Tetrahedron Lett. 27: 5559–5562.

(109) K.J. Doyle and C.J. Moody. 1994. The rhodium carbenoid route to oxazoles. Synthesis of 4-functionalized oxazoles; three-step preparation of a bis-oxazole. Tetrahedron 50: 3761–3772.

(110) A. Padwa, J.M. Kassir and S.L. Xu. 1997. Cyclization reactions of rhodium carbene complexes. Effect of composition and oxidation state of the metal. J. Org. Chem. 62: 1642–1652.

(111) T. Horneff, S. Chuprakov, N. Chernyak, V. Gevorgyan and V. Fokin. 2008. Rhodium-catalyzed transannulation of 1,2,3-triazoles with nitriles. J. Am. Chem. Soc. 130: 14972–14974.

(112) J.J. van Veldhuizen, D.G. Gillingham, S.B. Garber, O. Kataoka and A.H. Hoveyda. 2003. Chiral Ru-based complexes for asymmetric olefin metathesis: enhancement of catalyst activity through steric and electronic modifications. J. Am. Chem. Soc. 125: 12502–12508.

(113) G.S. Weatherhead, J.G. Ford, E.J. Alexanian, R.R. Schrock and A.H. Hoveyda. 2000. Tandem catalytic asymmetric ring-opening metathesis/ring-closing metathesis. J. Am. Chem. Soc. 122: 1828–1829.

(114) D.S. La, E.S. Sattely, J.G. Ford, R.R. Schrock and A.H. Hoveyda. 2001. Catalytic asymmetric ring-opening metathesis/ cross metathesis (AROM/CM) reactions. Mechanism and application to enantioselective synthesis of functionalized cyclopentanes. J. Am. Chem. Soc. 123: 7767–7778.

(115) W.C.P. Tsang, J.A. Jernelius, G.A. Cortez, G.S. Weatherhead, R.R. Schrock and A.H. Hoveyda. 2003. An enantiomerically pure adamantylimido molybdenum alkylidene complex. An effective new catalyst for enantioselective olefin metathesis. J. Am. Chem. Soc. 125: 2591–2596.

(116) J.J. van Veldhuizen, J.E. Campbell, R.E. Giudici and A.H. Hoveyda. 2005. A readily available chiral Ag-based N-heterocyclic carbene complex for use in efficient and highly enantioselective Ru-catalyzed olefin metathesis and Cu-catalyzed allylic alkylation reactions. J. Am. Chem. Soc. 127: 6877–6882.

(117) D.G. Gillingham, O. Kataoka, S.B. Garber and A.H. Hoveyda. 2004. Efficient enantioselective synthesis of functionalized tetrahydropyrans by Ru-catalyzed asymmetric ring-opening metathesis/cross metathesis (AROM/ CM). J. Am. Chem. Soc. 126: 12288–12290.

(118) J.H. Kim, S. Grebies, M. Boultadakis-Arapinis, C. Daniliuc and F. Glorius. 2016. Rh(I)/NHC*-catalyzed site- and enantioselective functionalization of C(sp³)-H bonds toward chiral triarylmethanes. ACS Catal. 6: 7652–7656.

(119) X. Dong, X. Ma, H. Xu and Q. Ge. 2016. Comparative study of silica-supported copper catalysts prepared by different methods: formation and transition of copper phyllosilicate. Catal. Sci. Technol. 6: 4151–4158.

(120) F. Li, L. Wang, X. Han, P. He, Y. Cao and H. Li. 2016. Influence of support on the performance of copper catalysts for the effective hydrogenation of ethylene carbonate to synthesize ethylene glycol and methanol. RSC Adv. 6: 45894–45906.

(121) L.-X. Dai, Y.-R. Lin, X.-L. Hou and Y.-G. Zhou. 1999. Stereoselective reactions with imines. Pure Appl. Chem. 71: 1033–1040.

(122) H. Peng and G. Liu. 2011. Palladium-catalyzed tandem fluorination and cyclization of enynes. Org. Lett. 13: 772–775.

(123) M. Senkus. 1946. Reaction of primary aliphatic amines with formaldehyde and nitroparaffins. J. Am. Chem. Soc. 68: 10–12.

(124) H.G. Johnson. 1946. Reaction of aliphatic amines with formaldehyde and nitroparaffins. II. Secondary amines. J. Am. Chem. Soc. 68: 12–14.

(125) H.G. Johnson. 1946. The preparation and reduction of nitro amines obtained from aromatic amines, formaldehyde, and nitroparaffins. J. Am. Chem. Soc. 68: 14–18.

(126) H. Adams, J.C. Anderson, S. Peace and A.M.K. Pennel. 1998. The nitro-Mannich reaction and its application to the stereoselective synthesis of 1,2-diamines. J. Org. Chem. 63: 9932–9934.

(127) M.A. Sturgess and D.J. Yarberry. 1993. Rapid stereoselective reduction of thermally labile 2-aminonitroalkanes. Tetrahedron Lett. 34: 4743–4746.

(128) J.C. Anderson, A.J. Blake, G.P. Howell and C. Wilson. 2005. Scope and limitations of the nitro-Mannich reaction for the stereoselective synthesis of 1,2-diamines. J. Org. Chem. 70: 549–555.

(129) J.C. Anderson, S. Peace and S. Pih. 2000. The Lewis acid-catalyzed addition of 1-trimethylsilyl nitropropanate to imines. Synlett 6: 850–852.

(130) K.-I. Yamada, S.J. Harwood, H. Groger and M. Shibasaki. 1999. The first catalytic asymmetric nitro-Mannich-type reaction promoted by a new heterobimetallic complex. Angew. Chem. Int. Ed. 38: 3504–3506.

(131) Y. Lu and W. Zhang. 2004. Microwave-assisted synthesis of a 3-aminoimidazo[1,2-a]-pyridine/pyrazine library by fluorous multi-component reactions and subsequent cross-coupling reactions. QSAR Comb. Sci. 23: 827–835.

(132) T.P. Selvam, G. Saravanan, C.R. Prakash and P. Dinesh Kumar. 2011. Microwave-assisted synthesis, characterization and biological activity of novel pyrazole derivatives. Asian J. Pharm. Res. 1: 126–129.

(133) (a) S.H. Paul and M.F. Jennifer. 2006. Microwave-assisted synthesis utilizing supported reagents: a rapid and versatile synthesis of 1,5-diarylpyrazoles. Tetrahedron Lett. 47: 2443–2446. (b) S. Paul, M. Gupta and R. Gupta. 2000. Vilsmeier reagent for formylation in solvent-free conditions using microwaves. Synlett 8: 1115–1118.

(134) (a) L. de Luca, G. Giacomelli and A. Porcheddu. 2005. Synthesis of 1-alkyl-4-imidazolecarboxylates: a catch and release strategy. J. Comb. Chem. 7: 905–908. (b) E. Suna and I. Mutule. 2006. Microwave-assisted heterocyclic chemistry. Top. Curr. Chem. 266: 49–101.

(135) D.S. Ermolatev, J.B. Bariwal, H.P.L. Steenackers, S.C.J. de Keersmaecker and E.V. van der Eycken. 2010. Concise and diversity-oriented route toward poly-substituted 2-aminoimidazole alkaloids and their analogues. Angew. Chem. Int. Ed. 49: 9465–9468.

(136) Y.T. Lee and Y.K. Chung. 2008. Silver(I)-catalyzed facile synthesis of pyrazoles from propargyl *N*-sulfonylhydrazones. J. Org. Chem. 73: 4698–4701.

(137) D.-H. Zhang, Z. Zhang and M. Shi. 2012. Transition metal-catalyzed carbocyclization of nitrogen and oxygen-tethered 1,n-enynes and diynes: synthesis of five- or six-membered heterocyclic compounds. Chem. Commun. 48: 10271–10279.

(138) R. Grigg, M.I. Lansdell and M. Thornton-Pett. 1999. Silver acetate-catalyzed cycloadditions of isocyanoacetates. Tetrahedron 55: 2025–2044.

(139) K. Henneke, U. Schollkopf and T. Neudecker. 1979. Synthesen mit α-metallierten isocyaniden, XLII. Bi-, ter- und quateroxazole aus α-anionisierten isocyaniden und acylierungsmitteln; α-aminoketone und α,α-diaminoketone. Liebigs Ann. Chem. 9: 1370–1387.

(140) N. Elders, R.F. Schmitz, F.J.J. de Kanter, E. Ruijter, M.B. Groen and R.V.A. Orru. 2007. A resource efficient and highly flexible procedure for a three-component synthesis of 2-imidazolines. J. Org. Chem. 72: 6135–6142.

(141) R.S. Bon, B. van Vliet, N.E. Sprenkels, R.F. Schmitz, F.J.J. de Kanter, C.V. Stevens, M. Swart, F.M. Bickelhaupt, M.B. Groen and R.V.A. Orru. 2005. Multi-component synthesis of 2-imidazolines. J. Org. Chem. 70: 3542–3553.

(142) M. Syamala. 2009. Recent progress in three-component reactions: an update. Org. Prep. Proced. Int. 41: 1–68.

(143) A. de la Hoz, A. Diaz-Ortiz, M. del Carmen Mateo, M. Moral, A. Moreno, J. Elguero, C. Foces-Foces, M.L. Rodriguez and A. Sanchez-Migallon. 2006. Microwave-assisted synthesis and crystal structures of 2-imidazolines and imidazoles. Tetrahedron 62: 5868–5874.

(144) Y.N. Mabkhot, A. Barakat, A.M. Al-Majid, Z.A. Al-Othman and A.S. Alamary. 2011. A facile and convenient synthesis of some novel hydrazones, Schiff's base and pyrazoles incorporating thieno[2,3-*b*]thiophenes. Int. J. Mol. Sci. 12: 7824–7834.

(145) G. Berthon-Gelloz, M.A. Siegler, A.L. Spek, B. Tinant, J.N.H. Reek and I.E. Marko. 2010. IPr* an easily accessible highly hindered *N*-heterocyclic carbene. Dalton Trans. 39: 1444–1446.

(146) A. Albright, D. Eddings, R. Black, C. Welch, N.N. Gerasimchuk and R.E. Gawley. 2011. Design and synthesis of C2-symmetric *N*-heterocyclic carbene precursors and metal carbenoids. J. Org. Chem. 76: 7341–7351.

(147) C.N. Hodge, P.Y.S. Lam, C.J. Eyermann, P.K. Jadhav, Y. Ru, C.H. Fernandez, G.V. de Lucca, C.H. Chang, R.J.C. Kaltenbach and P.E. Aldrich. 1998. Calculated and experimental low-energy conformations of cyclic urea HIV protease inhibitors. J. Am. Chem. Soc. 120: 4570–4581.

(148) Y.J. Kim and R.S. Varma. 2004. Microwave-assisted preparation of cyclic ureas from diamines in the presence of ZnO. Tetrahedron Lett. 45: 7205–7208.

(149) V. Polshettiwar and R.S. Varma. 2008. Greener and expeditious synthesis of bioactive heterocycles using microwave irradiation. Pure Appl. Chem. 80: 777–790.

(150) C.O. Kappe. 2000. Recent advances in the Biginelli dihydropyrimidine synthesis. New tricks from an old dog. Acc. Chem. Res. 33: 879–888.

(151) C.O. Kappe. 1993. 100 years of the Biginelli dihydropyrimidine synthesis. Tetrahedron 49: 6937–6963.

(152) G.C. Rovnyak, S.D. Kimball, B. Beyer, G. Cucinotta, J.D. DiMarco, J. Gougoutas, A. Hedberg, M. Malley, J.P. McCarthy, R. Zhang and S. Moreland. 1995. Calcium entry blockers and activators: conformational and structural determinants of dihydropyrimidine calcium channel modulators. J. Med. Chem. 38: 119–129.

(153) S.K. De and R.A. Gibbs. 2005. Scandium(III) triflate as an efficient and reusable catalyst for synthesis of 3,4-dihydropyrimidin-2(1*H*)-ones. Synth. Commun. 35: 2645–2651.

(154) A.S. Paraskar and A. Sudalai. 2006. A novel Cu(OTf)$_2$-mediated three-component high yield synthesis of α-aminophosphonates. ARKIVOC (x): 183–189.

(155) F. Bossert and W. Vater. 1989. 1,4-Dihydropyridines—a basis for developing new drugs. Med. Res. Rev. 9: 291–324.

(156) I.C. Cotterill, A.Y. Usyatinsky, J.M. Arnold, D.S. Clark, J.S. Dordick, P.C. Michels and Y.L. Khmelnitsky. 1998. Microwave-assisted combinatorial chemistry synthesis of substituted pyridines. Tetrahedron Lett. 39: 1117–1120.

(157) J.J. Vanden Eynde, F. Delfosse, A. Mayence and Y. van Haverbeke. 1995. Old reagents, new results: aromatization of Hantzsch 1,4-dihydropyridines with manganese dioxide and 2,3-dichloro-5,6-dicyano-1,4-benzoquinone. Tetrahedron 51: 6511–6516.

(158) N.A. Al-Masoudi, B.A. Saeed, A.H. Essa and Y. Al-Soud. 2009. Microwave-assisted synthesis of acyclic C-nucleosides from 1,2- and 1,3-diketones. Nucleosides, Nucleotides and Nucleic Acids 28: 175–183.

Five-Membered *N,N*-Polyheterocycles

6.1 Introduction

Heterocyclic compounds, by virtue of their specific activities, can be employed in the treatment of infectious diseases. Review of literature indicates that nitrogen-containing heterocycles find a significant place in the development of pharmacologically important molecules. Benzimidazole has been reported to be pharmacologically and physiologically active, finding applications in the treatment of several diseases and conditions like diabetes, epilepsy and anti-fertility. These compounds are also known to possess other medically useful properties—anti-microbial, anti-cancer, etc. Anti-histamine astemizole [1a–f] is one of the notable clinical examples having a benzimidazole ring and is used to treat allergies, hives (urticaria) and other allergic inflammatory conditions, while also being useful in case of sneezing, runny nose, itching and watering of the eyes, and other allergic symptoms. The proton pump inhibitor, omeprazole [2], decreases the amount of acid produced in the stomach and is used to treat symptoms of gastroesophageal reflux disease (GERD) and other conditions caused by excess stomach acid. Lansoprazole [3] and pantoprazole [4] are used to treat and prevent stomach and intestinal ulcers, erosive esophagitis and other conditions caused by excessive stomach acid, such as Zollinger-Ellison syndrome. Triclabendazole [5] is a medication used to treat liver flukes, specifically fascioliasis and paragonimiasis, and aalbendazole [6] is used for the treatment of a variety of parasitic worm infestations and for giardiasis, trichuriasis, filariasis, neurocysticercosis, hydatid disease, pinworm disease, etc. Mebendazole (MBZ) [7] is another drug containing benzimidazole ring and helps in the treatment of a number of parasitic worm infestations, including ascariasis, hookworm infections, guinea worm infections, hydatid disease, and giardia, among others. Benzimidazole has attracted much attention in diverse fields of chemistry. This nucleus is present in anti-ulcerous drugs like omeprazole and in compounds with anti-parasitic, anti-tumoral and anti-viral properties.

Benzimidazoquinazolines [8–12], benzimidazoisoquinolines [13], and benzimidazo[2,1-*a*] isoindolones [14] have been reported as potent anti-tumor agents. Benzimidazo[2,1-*b*]quinazolines are powerful immuno-suppressors [15], and benzimidazo[2,1-*b*]benzo[*f*]isoquinoline ring system [16] is present in pharmacologically active compounds.

Over the past years, there has been a growing interest in the synthesis of benzimidazole-based heterocycles. There are a number of practically important routes to benzimidazole-based polyheterocycles – reactions between (i) benzaldehyde derivatives and benzimidazoles containing an activated methylene group at position 2, (ii) coumarins and *o*-phenylenediamines, (iii) 2-azidoanilines and substituted cinnamaldehydes, (iv) 2-(2-aminoaryl(hetaryl))benzimidazoles and haloketones, (v) *o*-phenylenediamines and phthalic anhydrides (vi) 2-(hydroxymethylene)-3-keto steroids and functionalized benzimidazoles; and (vii) metal-catalyzed cyclization of alkynylaniline derivatives. Such reactions provide convenient strategies for the synthesis of annulated benzimidazole polyheterocycles.

6.2 Metal- and non-metal-assisted synthesis of five-membered polyheterocycles with two nitrogen atoms

Aluminium-assisted synthesis

Several protocols have been reported for the synthesis of benzimidazoles [17–19]. These molecules possess interesting industrial applications and biological properties [20–21]. In connection with ongoing work on microwave and alum [22], the synthesis of some benzimidazoles by condensation of 1,2-phenylenediamine with *o*-esters in the presence of alum under microwave irradiation has been studied **(Scheme 1)**. This approach affords various advantages such as short reaction times, high yield of products and cleaner reaction. The catalyst is recovered and recycled without any reduction in activity in the subsequent reaction [23].

3-Chloro-4-flouro-*o*-phenylenediamine reacts with *p*-aminobenzoic acid followed by thioglycolic acid and an aromatic aldehyde in the presence of aluminium chloride to produce flouro-chlorobenzimidazolo-substituted thiazolidinone derivatives. 2-Substituted 3-(4-(4-chloro-5-fluoro-1*H*-benzo[*d*]imidazol-2-yl)phenyl)thiazolidin-4-one is prepared in three steps: synthesis of 4-(4-chloro-5-

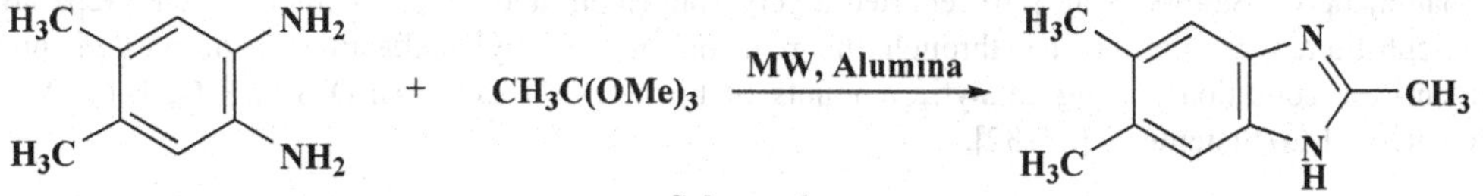

Scheme-1

fluoro-1*H*-benzo[*d*]imidazol-2-yl)benzenamine, synthesis of 2-substituted Schiff base, and synthesis of 2-substituted 3-(4-(4-chloro-5-fluoro-1*H*-benzo[*d*]imidazol-2-yl)phenyl)thiazolidin-4-one. Different aromatic aldehydes are substituted with 4-(4-chloro-5-fluoro-1*H*-benzo[*d*]imidazol-2-yl)benzenamine and thioglycolic acid to provide an intermediate compound using microwave and conventional protocol **(Scheme 2)** [24].

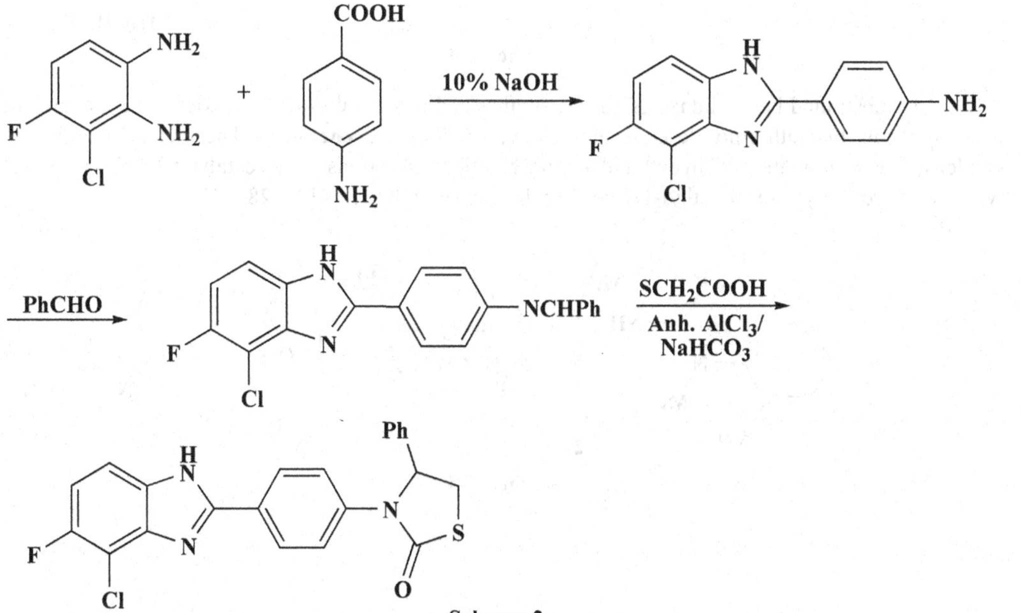

Scheme-2

Mathew et al. [25] synthesized (2*E*)-1-(1*H*-benzimidazol-2-yl)-3-phenylprop-2-en-1-ones. *o*-Phenylenediamine reacted with lactic acid to afford 2(α-hydroxyethyl)benzimidazole which was oxidized in the presence of potassium dichromate to yield 2-acetyl-benzimidazole. The chalcones were produced by Claisen-Schmidt condensation of 2-acetyl benzimidazole with appropriate aldehydes in the presence of a base (**Scheme 3**).

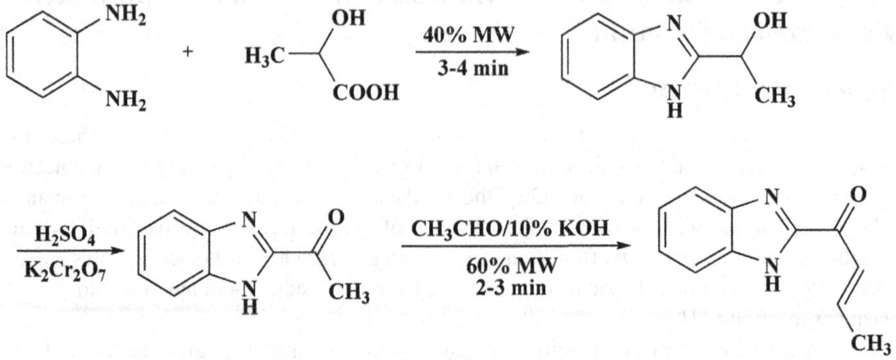

Scheme-3

Bismuth-assisted synthesis

Mohammadpoor-Baltork et al. [26] reported a very convenient and simple method for the preparation of 2-substituted benzimidazoles through the reaction of *o*-phenylenediamines with *o*-esters under solvent-free conditions, using catalytic amounts of bismuth(III) salts—Bi(OTf)$_3$·xH$_2$O, Bi(TFA)$_3$ or BiOClO$_4$·xH$_2$O (**Scheme 4**) [27–32].

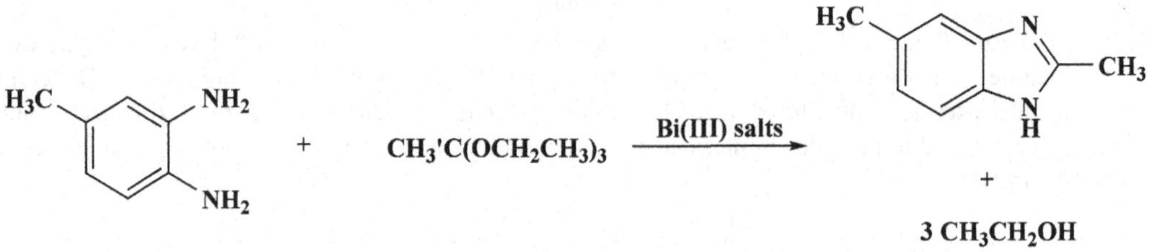

Scheme-4

1,2-Disubstituted benzimidazoles can be synthesized using a multistep protocol where cyclization is catalyzed by bismuth chloride under microwave irradiation (**Scheme 5**). The aryl-1,2-diamines are condensed with aromatic aldehydes under mild reaction conditions using catalytic Bi(OTf)$_3$·xH$_2$O in water to afford 2-aryl-1-arylmethyl-1*H*-benzimidazoles in high yields [27–28, 33].

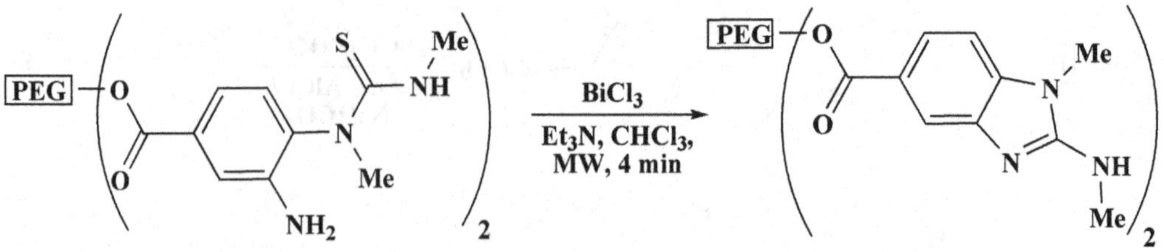

Scheme-5

Cerium-assisted synthesis

Kumar and Joshi [34] generated some benzimidazoles from various aldehydes and 1,2-diamines using ceric ammonium nitrate. The simplicity of the reaction conditions, no column chromatography, and shorter reaction times to get the pure products in high yields made this protocol more attractive (**Scheme 6–7**).

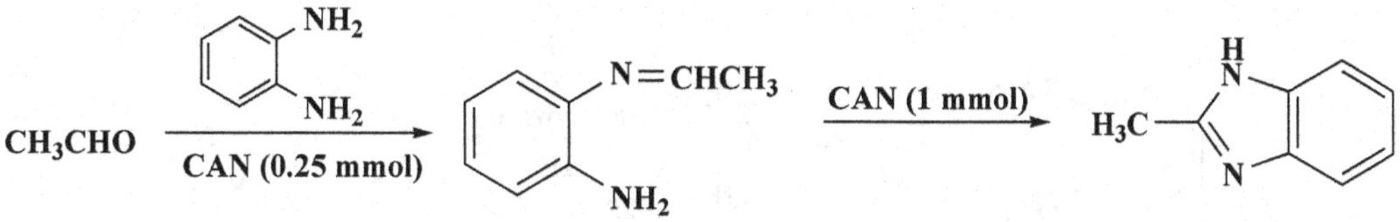

Scheme-6

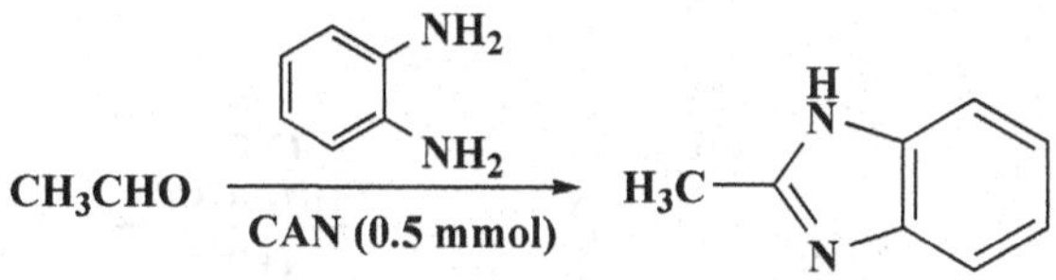

Scheme-7

Kidwai et al. [35] reported the synthesis of benzimidazoles from aldehydes and *o*-phenylenediamine in the presence of ceric ammonium nitrate in poly(ethylene glycol). This protocol provided a novel pathway with little catalyst loading for the preparation of benzimidazoles in good yields. Moreover, the solvent system was recovered and reutilized successfully (**Scheme 8**).

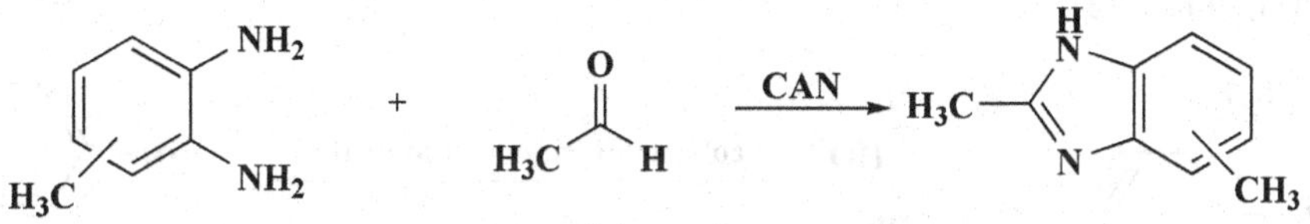

Scheme-8

Cesium-assisted synthesis

A reaction sequence has been used to produce aminobenzimidazoles and benzimidazoles by palladium-catalyzed intramolecular cyclization of arylbromide substituted guanidines and amidines respectively (**Scheme 9**) [36]. With arylbromides, Pd_2dba_3/triphenylphosphine or simple $Pd(PPh_3)_4$ catalysts are sufficient to mediate the cyclization. This method is equally applicable for the preparation of polycyclic benzimidazoles and indazoles [37].

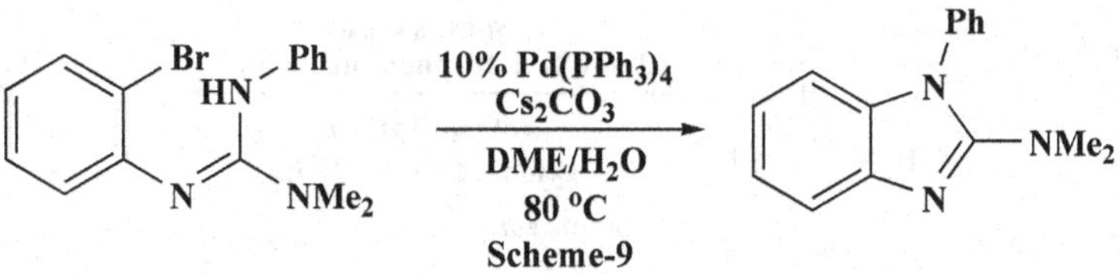

Scheme-9

Indole derivatives can be synthesized by a number of methods **(Scheme 10)**. In 2007, Jin and Yamamoto [38] studied 1,3-dipolar cycloaddition of diazo compounds with arynes to provide indazoles. The reaction was highly dependent upon the aryne stoichiometry; additional equivalents produced *N*-arylated indazole. In a study by Rogness and Larock [39], synthesis of isatins from the condensation of arynes with *N*-aryl methyl oxamide was observed. Zhao et al. [40] coupled tosyl hydrazones with arynes upon heating for the efficient synthesis of the desired diaza product **(Scheme 11)**.

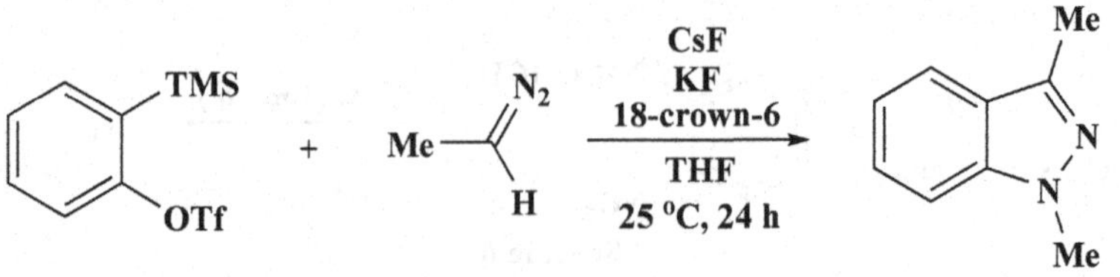

Scheme-10

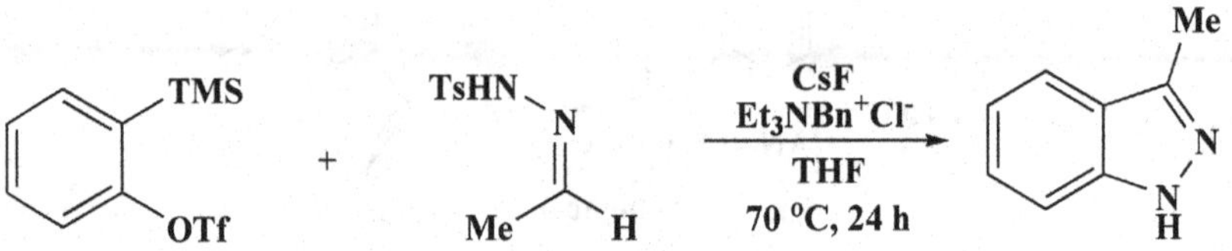

Scheme-11

Cobalt-assisted synthesis

A facile method for the preparation of 2-arylbenzimidazole through the reaction of *o*-phenylenediamine with various aromatic aldehydes in the presence of $CoCl_2$ hexahydrate catalyst was studied by Khan and Parvin [41] **(Scheme 12)**.

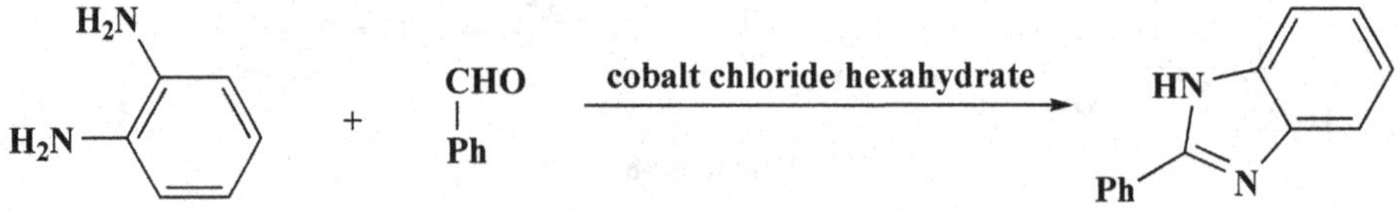

Scheme-12

Copper-assisted synthesis

Primary amides undergo domino cross-coupling with 1,2-dibromoarenes. 1,2-Dibromobenzene and a benzamidine react to form a bond by intermolecular copper-catalyzed cross-coupling to afford the N3-C3a bond of the benzimidazole ring [42–44]. The annulation with benzamidine produces benzimidazole in low yields **(Scheme 13)**.

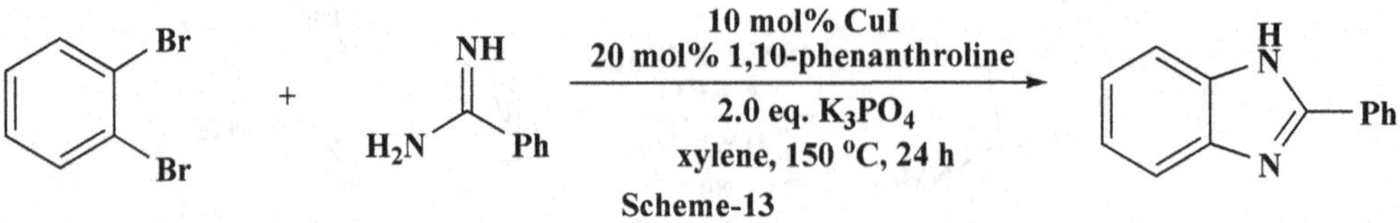

Scheme-13

In 2007, Zheng and Buchwald [45] reported the synthesis of *N*-alkylbenzimidazoles from *o*-haloanilides by copper-catalyzed amidation, followed by base- or acid-mediated cyclization. Among the ligands examined, diamine was found to be the most effective **(Scheme 14)** [46].

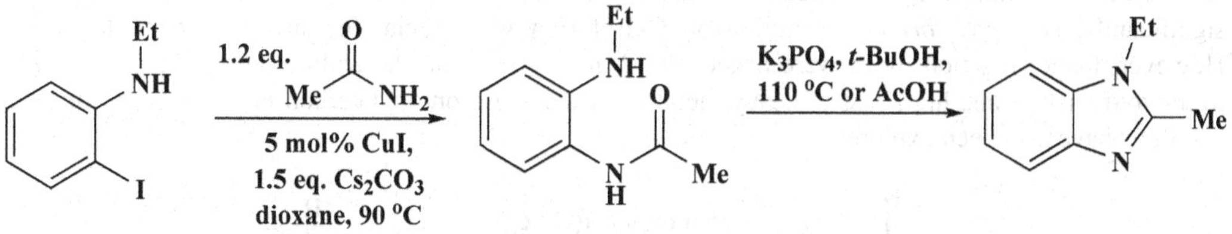

Scheme-14

O-halide (Br or I)-functionalized *N*-acyl aromatic amines in the presence of L-proline, cuprous iodide, and base undergo various *N*-amination-based tandem reactions to produce different heterocyclic compounds. Zhou et al. [47] reported a one-pot synthetic protocol for the synthesis of benzimidazoles while substitution was varied in both 1- and 2-positions of the benzimidazoles. The *ortho*-substituent effect of NHCOMe group was observed to be an important factor for the transformation **(Scheme 15)** [48].

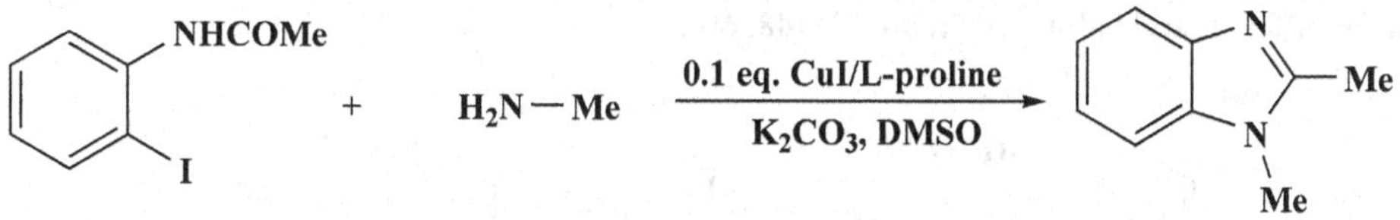

Scheme-15

A protocol for the synthesis of indazoles from benzophenione tosylhydrazones, using catalytic amounts of palladium acetate, copper acetate and AgOCOCF$_3$ in dimethylsulfoxide, was reported by Inamoto et al. [49] **(Scheme 16)**.

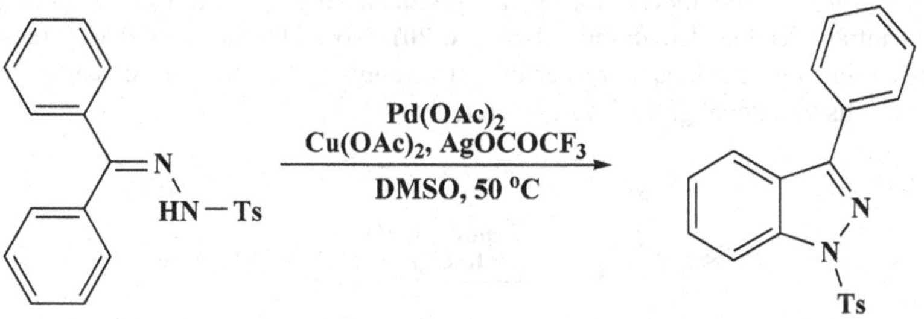

Scheme-16

Shen et al. [50] studied *N*-allenylation **(Scheme 17)**. Optically enriched allenyl halides underwent copper(I)-catalyzed stereospecific amidation to afford various chiral allenamides [51a–b]. The chirality information of the optically enriched allenyl halides was reliably transferred under these reaction conditions.

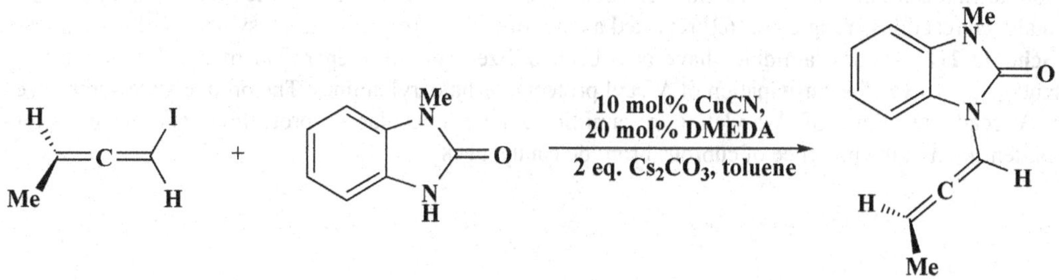

Scheme-17

Brasche and Buchwald [52] reported an aryl amination of amidines in the presence of Cu catalyst. The 2-alkyl- and 2-arylbenzimidazoles were formed in good yields when different amidines were treated with $Cu(OAc)_2$ and acetic acid under an oxygen atmosphere (**Scheme 18**). A variety of substituents (both electron-withdrawing and electron-donating) on either aryl ring did not affect the transformation significantly. However, *ortho*-substitution on C-aryl ring was crucial for any conversion to occur. However, longer reaction times were needed for *N*-methylation on the aniline nitrogen [53–59]. Due to the toxic nature and high cost of heavy metal catalysts, Cu-promoted carbon-nitrogen bond forming reactions have also been explored.

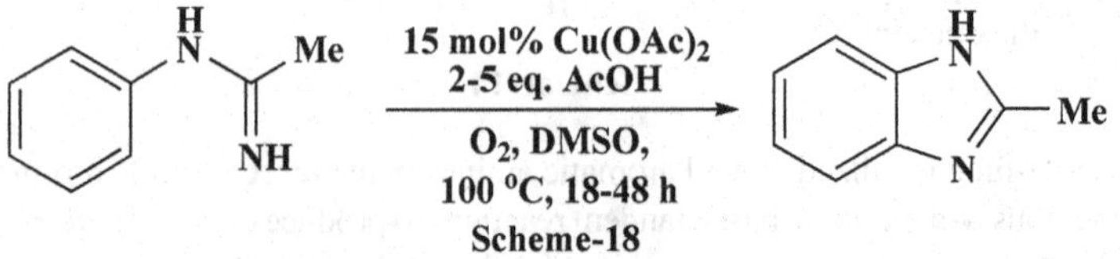

Scheme-18

Heterocyclic compounds are also synthesized by amination and subsequent intramolecular amidation of the carbamate fragment. Both iodide- and bromide-substituted substrates are good reactants, while slight changes in the reaction conditions like base and ligand are necessary to guarantee satisfactory yields for different aryl halides (**Scheme 19**) [48, 60].

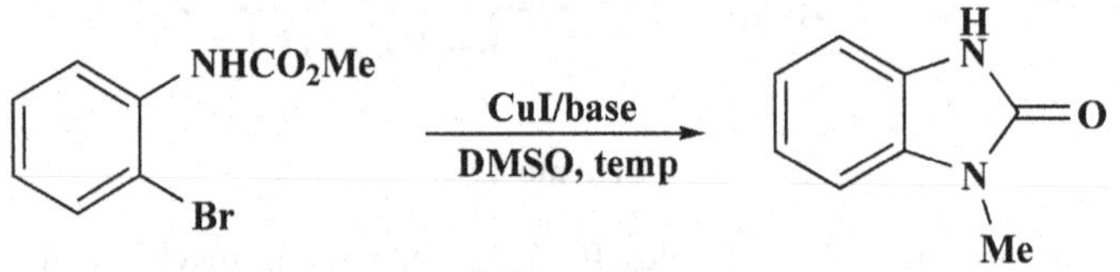

Scheme-19

Viña et al. [61] reported a CuO-catalyzed one-pot reaction for the preparation of indazoles from hydrazines and *o*-haloalkanoylphenones, with subsequent transformation through carbon-nitrogen amination and intramolecular dehydration (**Scheme 20**). No additional ligand was required in these reactions. The amination reactions were performed through carbon-fluoride or carbon-chloride bond cleavage and various functional groups were tolerated well [48].

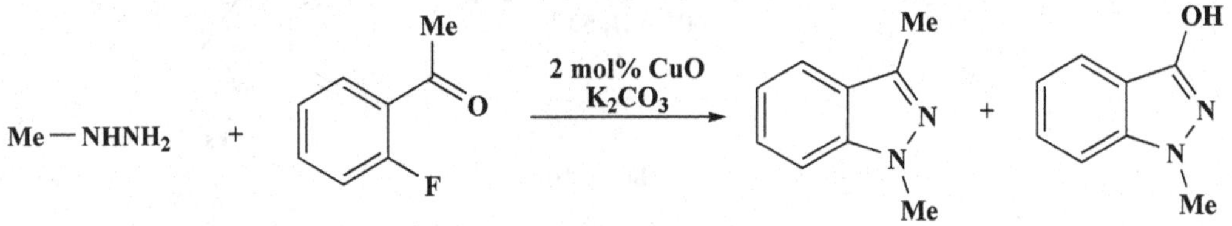

Scheme-20

Although amides, amines, azoles and hydrazines are used as nitrogen sources in most copper-catalyzed cascade reactions involving carbon-nitrogen coupling, some unconventional *N*-containing moieties like amidines also act as nitrogen sources to afford several heterocyclic motifs through copper-catalyzed reactions. Yang et al. [62] reported a class of useful protocols for the synthesis of heterocycles (**Scheme 21**). Besides, amidines have also been utilized for the preparation of benzimidazoles; the synthesis involves the participation of *N*-acyl protected *o*-haloaryl amines. The products are synthesized by cascade reactions of *N*-arylation of amidines, intramolecular de-protection, and deamination-condensation in the presence of cuprous bromide catalyst [48].

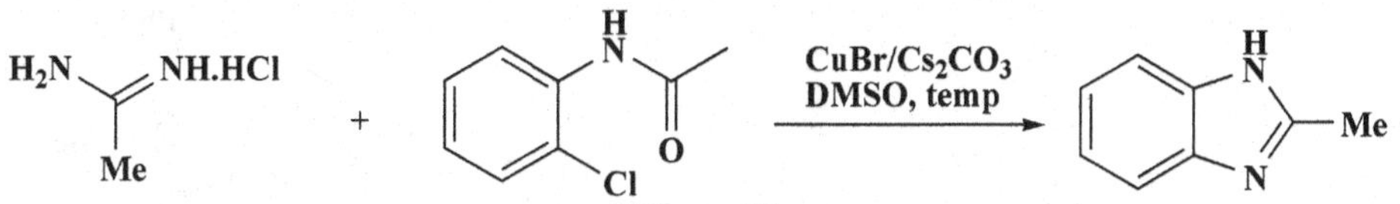

Scheme-21

Lv and Bao [63], while exploring new copper-catalyzed domino reactions, reported that diimine moiety was a versatile building block in designing Cu-catalyzed cascade reactions. 1,2-Disubstiuted benzimidazoles were synthesized by a nucleophilic attack on the electron-deficient carbon of diimide, with subsequent intramolecular carbon-nitrogen coupling of the imidamide intermediate. A significant benefit of this reaction was that imidazoles, amines and phenols could all be used as nucleophilic partners and enabled great structural diversity in the products (**Scheme 22**) [48].

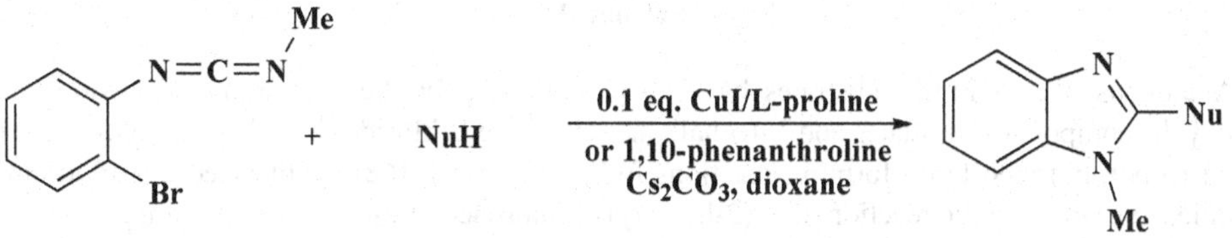

Scheme-22

The positions of the nucleophile and the diimine species can be exchanged for feasible cascade synthesis of analogous heterocyclic compounds. *O*-Haloaniline and *N,N*-disubstituted dimines react through a similar process to yield 2,3-dihydro-1*H*-benzo[*d*]imidazoles and benzimidazoles (**Scheme 23**) [48, 64].

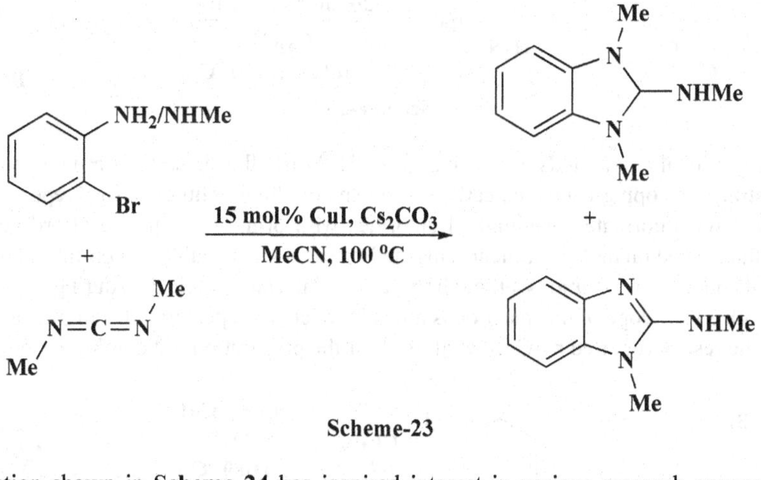

Scheme-23

The reaction shown in **Scheme 24** has inspired interest in various research groups for devising similar tandem reactions using amidines or guanidines as coupling partners of *o*-dihaloarenes to produce benzimidazole derivatives. Deng et al. [65–66] studied copper-catalyzed tandem reactions of amidines/ guanidines and dihaloarenes. The results of these protocols provided a novel pathway for the preparation of benzimidazoles. Moreover, interesting regioselectivity was observed in benzimidazole derivatives when substituted *o*-dihaloarenes were used. Regioselectivity in the synthesis of benzimidazoles was determined by the relative steric hindrance on the nucleophilic *N*-atoms in amidines or guanidines, as well as the relative reactivity of haloatoms [48].

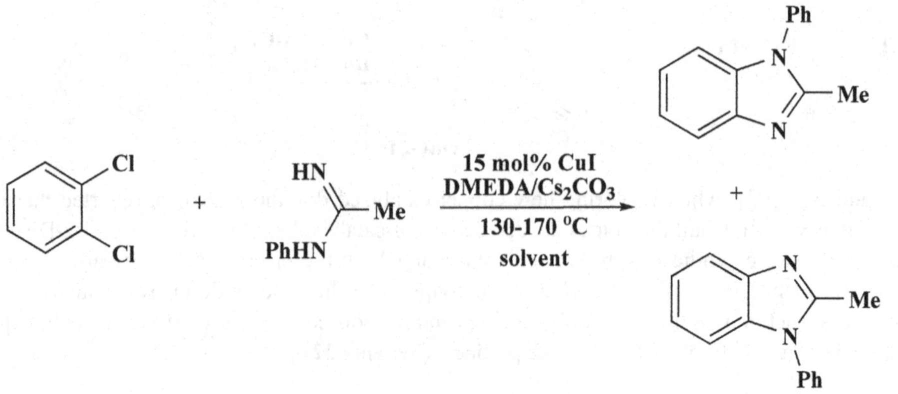

Scheme-24

Analogous *N*-containing substrates have also been used for the generation of a wide range of heterocyclic compounds. Imidates and *N*-(*o*-halophenyl)imidoyl chlorides have been employed in similar tandem carbon-nitrogen bond forming reactions [67]. Chen et al. [68] synthesized 2-(trifluoromethyl) benzimidazoles through the reaction of *N*-(2-haloaryl)trifluoroacetimidoyl chlorides with primary amines or anilines in the presence of copper catalyst **(Scheme 25)**. Substrates possessing aryl bromides, iodides and chlorides were used, each resulted in the synthesis of benzimidazoles in good yields. A similar protocol has been reported for the synthesis of 2-fluoroalkylbenzimidazoles in the presence of copper catalyst, although only the use of aryl iodide substrates has been described [69].

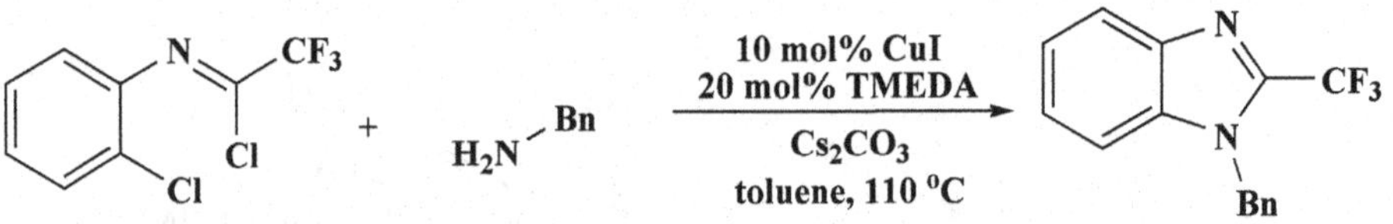

Scheme-25

As unconventional compounds possessing sp^2 C-Cl, *N*-aryl fluorinated acetimidoyl halides serve as effective substrates in copper-catalyzed carbon-nitrogen couplings. Zhu et al. [69] reported the reaction of *N*-(2-iodophenyl) fluorinated acetimidoyl chlorides with primary amines to afford benzimidazoles containing 2-fluorinated methyl substituent. This method involved a double carbon-nitrogen coupling and was completed under mild conditions although sp^2 carbon-chloride was cleaved during the transformation **(Scheme 26)** [48]. The scope of this reaction is much broader when performed at higher temperatures, as evident from the results reported by Chen et al. [68] for the preparation of homologous products.

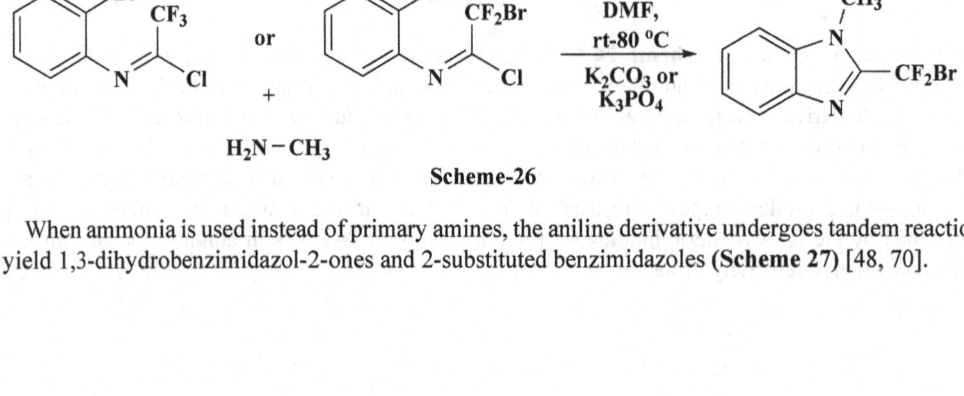

Scheme-26

When ammonia is used instead of primary amines, the aniline derivative undergoes tandem reactions to yield 1,3-dihydrobenzimidazol-2-ones and 2-substituted benzimidazoles **(Scheme 27)** [48, 70].

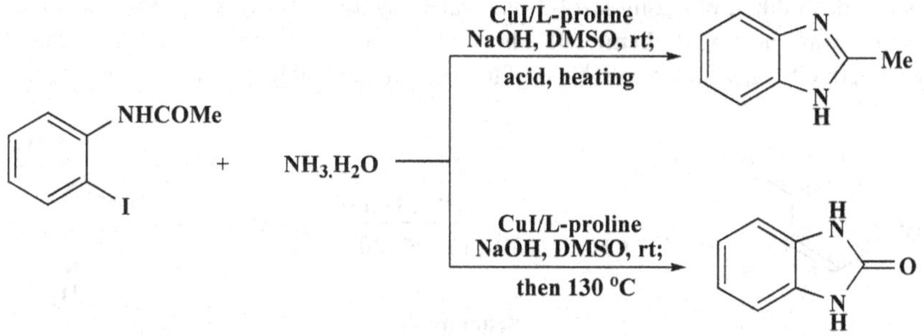

Scheme-27

The nucleophilic displacement of aryl fluorides with amines, followed by reduction and condensation with imidates produces benzimidazoles **(Scheme 28)** [71].

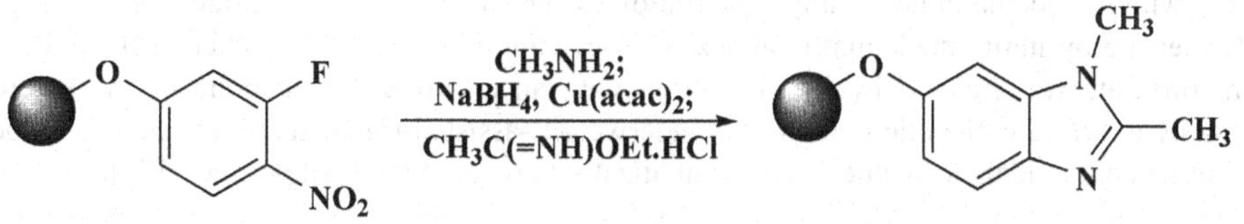

Scheme-28

Copper-catalyzed intramolecular *N*-arylation-cyclization of arylhydrazones produced *in situ* is a one-pot microwave-assisted two-step method for the synthesis of 1-arylindazoles. Traditionally, *N*-arylation reactions in the presence of cuprous iodide-diamine-complex catalyst occur at elevated temperatures (110 °C) and need long reaction times (24 hours) [72a–b]. The time of *N*-arylation can be reduced significantly to merely 10 minutes by a combination of microwave dielectric heating at 160 °C and a polar high boiling solvent *N*-methyl-2-pyrrolidone. Aryl bromides, iodides as well as chlorides are reactive in intramolecular *N*-arylation **(Scheme 29)**. Aryl hydrazines can be assembled with 2-haloacetophenones or 2-halobenzaldehydes under copper(I) catalysis to afford high yields of target molecules after two short microwave irradiation periods. For longer reaction times, regular heating has been reported. Aryl hydrazine reacts with 2-halobenzaldehyde under microwave irradiation to produce an aryl hydrazone as the intermediate which is not isolated. Subsequently, diamine ligand and cuprous iodide are added to the reaction mixture and irradiated continuously to provide indazole. The product is formed only in traces in the absence of diamine ligand [46, 73].

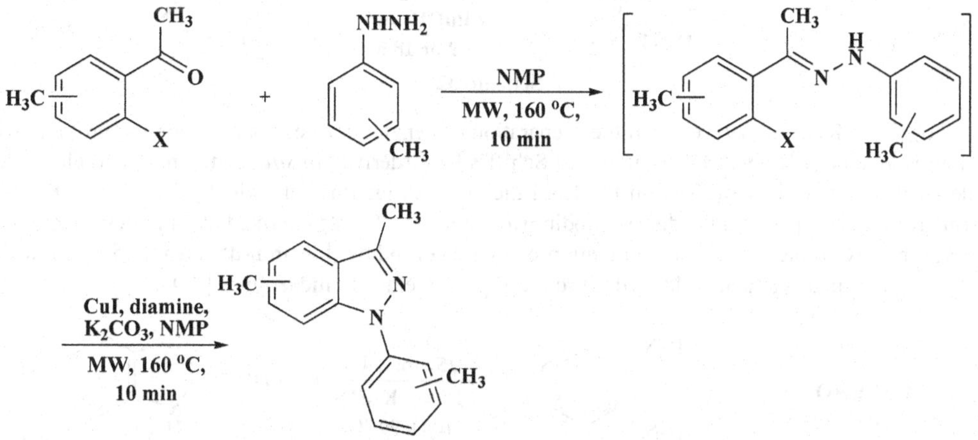

Scheme-29

N-Substituted 1,3-dihydrobenzimidazol-2-ones can be synthesized within minutes, in good to excellent yields, with microwave heating **(Scheme 30)**. The use of various functional groups has made this reaction particularly attractive for an efficient synthesis of medicinally and biologically active molecules [74].

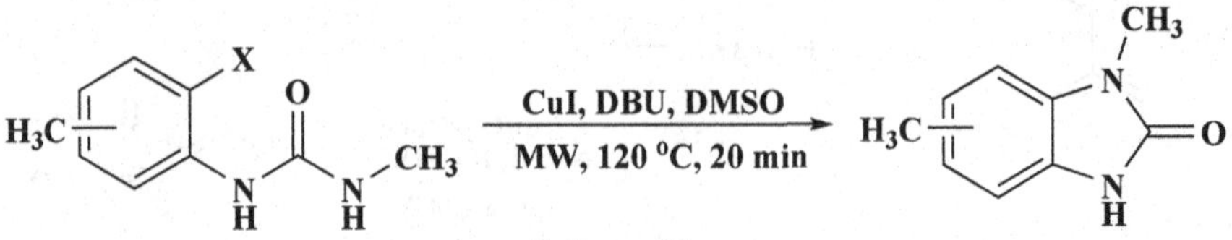

Scheme-30

Iodine-assisted synthesis

Masdeu et al. [75] developed an interesting method for the preparation of benzimidazolium salts from isocyanides and dihydropyridines **(Scheme 31)**. The α-haloiminium ion was produced when I$_2$ interacted with the double bond of dihydropyrimidines. The addition of isocyanides forms the nitrilium ion intermediate by initial nucleophilic attack of isocyanides followed by another nucleophilic attack. This intermediate was trapped by its interaction with an enamine double bond of the heterocyclic functionality to afford a bicyclic system that underwent I$_2$-assisted fragmentation to form a delocalized anion. Subsequent imidazole synthesis and aromatization provided the final products [76].

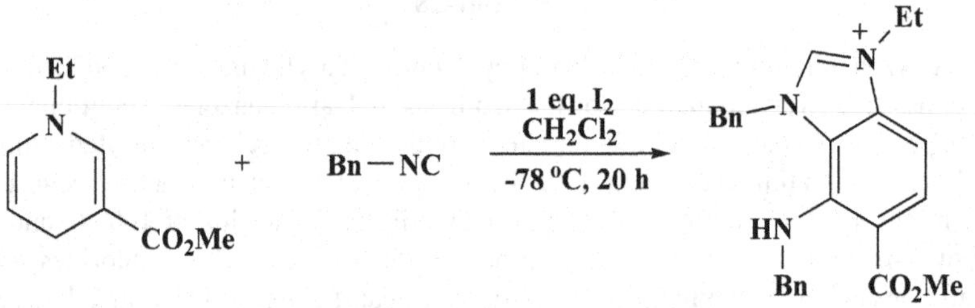

Scheme-31

Primary alcohols undergo direct one-pot oxidative conversion to form benzimidazoles **(Scheme 32)** [76–78].

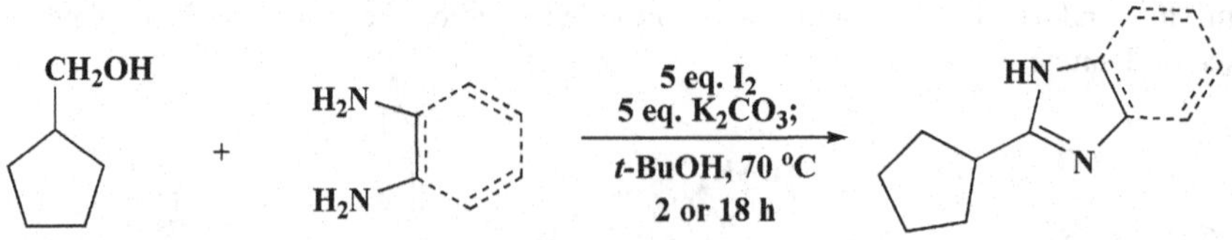

Scheme-32

Gogoi and Konwar [79] reported the preparation of benzimidazoles *via* an oxidation protocol with KI and I$_2$ in water **(Scheme 33)**. I$_2$-activated Schiff's base (derived *in situ* from amine and aldehydes) underwent intramolecular cyclization to afford the cyclized intermediate which reacted with KI$_3$ [80] (formed *in situ* from potassium iodide and iodine) to produce an *N*-iodo intermediate. The desired products were ultimately formed upon the elimination of hydrogen iodide. Electron-deficient and electron-rich heterocyclic, aliphatic and aryl aldehydes produced good yields of imidazolines [76].

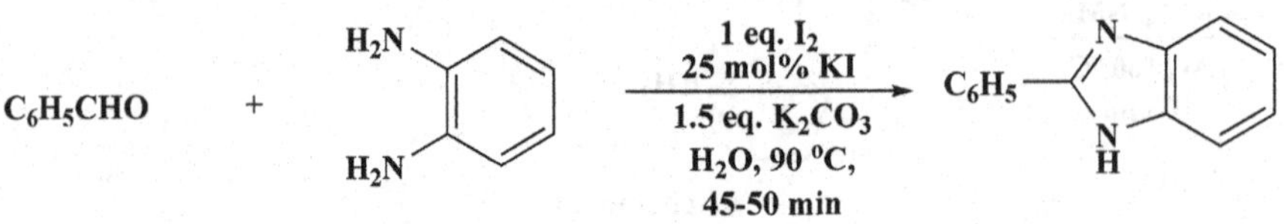

Scheme-33

Zhang et al. [81] reported an efficient pathway for the preparation of 2-substituted benzimidazoles (**Scheme 34**) through the condensation of 1,2-phenylene diamines with *o*-esters in the presence of 10 mol% I$_2$ [76].

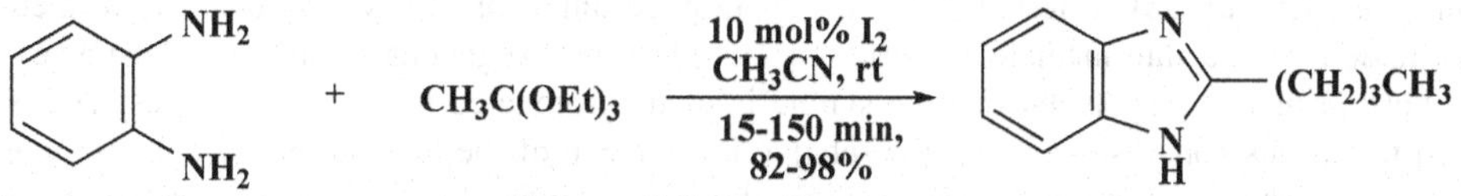

Scheme-34

Bis-benzimidazole can be synthesized in excellent yields when 3,3'-diaminobenzidine reacts with 2 eq. of triethyl *o*-valerate (**Scheme 35**) [76, 82].

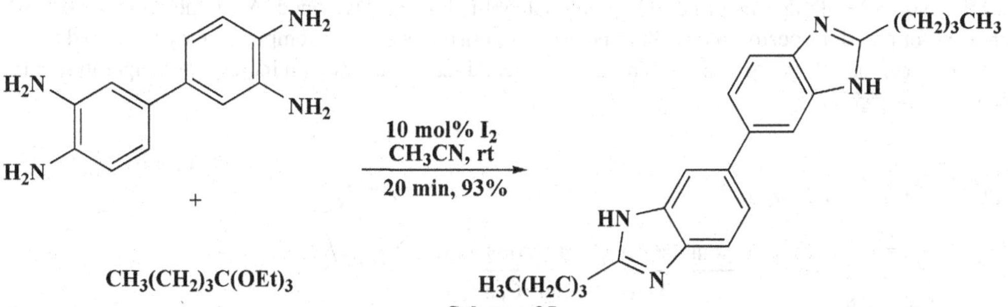

Scheme-35

Aryliodine(III) organosulfonates and [bis(acyloxy)iodo]arenes are commonly used as reagents in several fragmentations and cationic rearrangements. BTI, (diacetoxyiodo)benzene and [hydroxyl(tosyloxy)iodo]benzene act as excellent oxidants in Hofmann-type degradation of aromatic or aliphatic carboxamides into respective amines. The oxidative rearrangement of salicylamides or anthranilamides yields respective heterocyclic compounds [83]. Also alkyl carbamates of 1-protected indole-3-methylamines can be synthesized from acetamides (**Scheme 36**) [84–85].

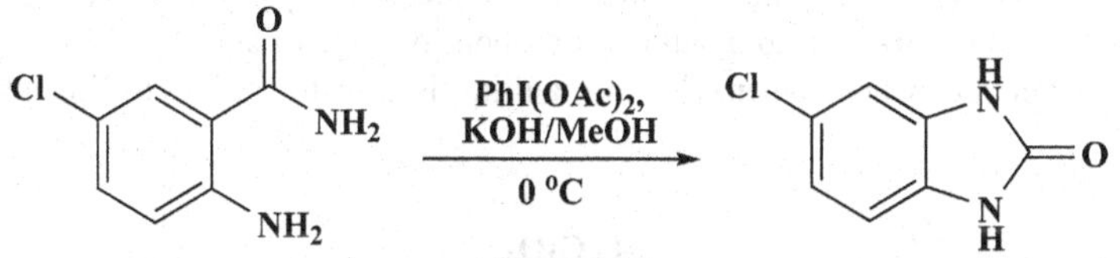

Scheme-36

Many researchers [86–92] have modified this protocol for the preparation of various heterocyclic systems such as naphtho-, benzo- and heterocycle-fused indazolone derivatives [93]. Specific examples include the synthesis of indolines from anilides [94] and of indazol-3-one derivatives from anthranilamides (**Scheme 37**) [85].

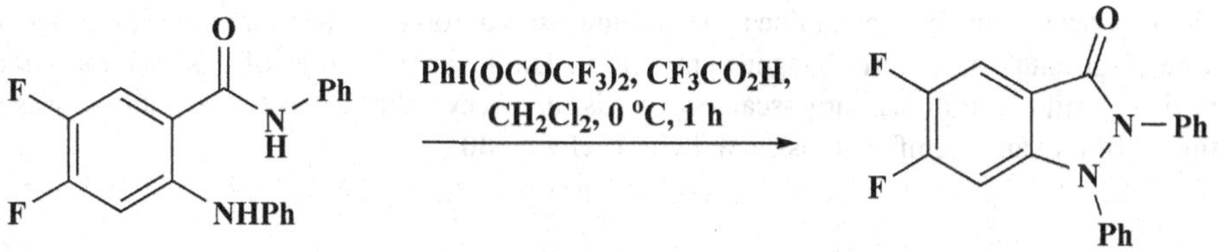

Scheme-37

Iridium-assisted synthesis

The reaction of benzimidazole and methyl cinnamyl carbonate in the presence of 4 mol% [Ir(COD)$_2$(ethylene)] catalyst without potassium phosphate has been studied at 50 °C and at room temperature (**Scheme 38**) to test whether the methoxide anion or methyl carbonate is formed upon the formation of the intermediate allyliridium complex, or exogenous potassium phosphate serves as the base primarily responsible for the synthesis of de-protonated heterocycle species. Two sets of data led to various conclusions about the relative importance of the counterions of allyl complex and potassium phosphate as base. First, the reaction of methyl cinnamyl carbonate and benzimidazole was 68% complete in the presence of parent catalyst [Ir(COD)$_2$(ethylene)] (4 mol%) after 5 hours at room temperature, while the same reaction was only 52% complete after 5 hours when performed at 50 °C. Because these reactions occurred without potassium phosphate, it was concluded that the methoxide or methyl carbonate produced *in situ* acted as a base for the synthesis of the benzimidazolate nucleophile, and the base-free allylations of heterocyclic nucleophiles were developed. As higher conversions were reported for reactions performed without potassium phosphate at room temperature as compared to those at 50 °C, it can be inferred that increasing catalyst deactivation occurs with increasing temperature during the reactions [95–96].

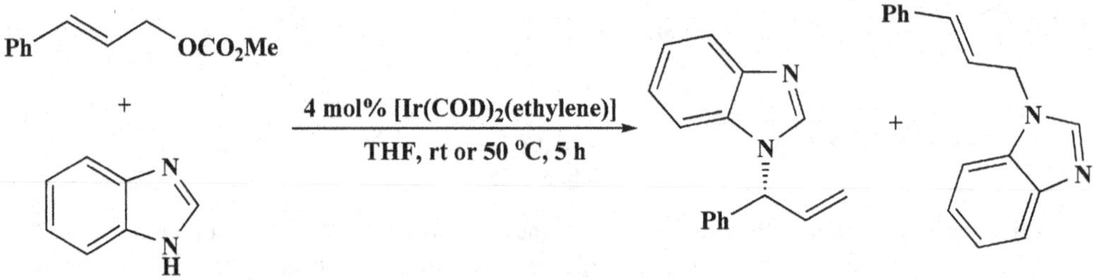

Scheme-38

Iron-assisted synthesis

o-Nitroanilines with at least one methine or methylene group present at the basic nitrogen also lead to reductive cyclization [97]. This type of compounds underwent pyrolysis to provide partly reduced heteroaromatic systems. Cyclization to a saturated carbon-hydrogen bond occurs at high temperatures in the presence of FeC$_2$O$_4$, when *N*-benzyl-2-nitroaniline is transformed into 2-phenylbenzimidazole (**Scheme 39**).

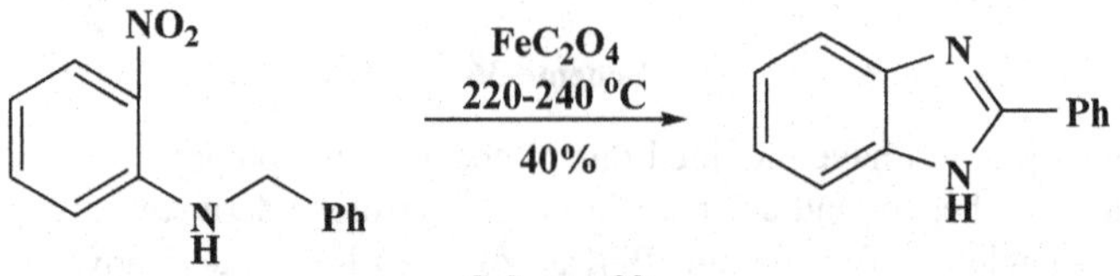

Scheme-39

Borhade et al. [98] reported a straightforward and convenient protocol for the preparation of benzimidazole from *o*-phenylenediamine and aryl aldehydes at room temperature in the presence of iron-containing magnesium oxide (iron/magnesium oxide). The notable features of this reaction were short reaction times, mild conditions, large-scale synthesis, and recyclable catalyst; the catalyst was reused many times without any significant loss in activity (**Scheme 40**).

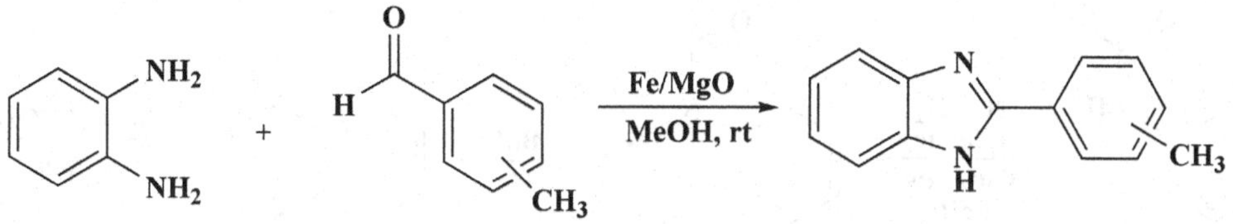

Scheme-40

Lead-assisted synthesis

Yan and Gstach [99] studied the synthesis of indazole on solid supports. The synthesis started with the coupling of hydrazone with unmodified Merrifield resin, following which the immobilized hydrazone was oxidized into an ester. A Lewis acid-catalyzed cyclization provided indazole which was ultimately cleaved from the solid support in a final purity of over 95%, in 79% yield **(Scheme 41)**.

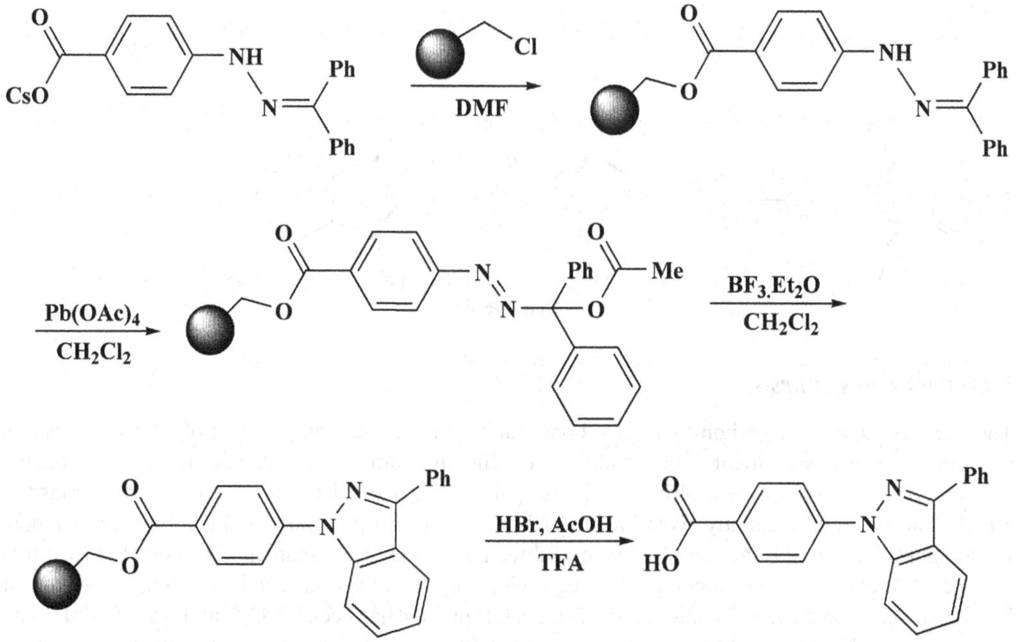

Scheme-41

Mercury-assisted synthesis

Studies have reported the synthesis of N-ferrocenyl substituted NHC precursors and the HgII-NHC complex [100]. One of the protocols for the synthesis of N-ferrocenyl-substituted benzimidazolium salts involves the coupling of the ferrocenyl group with a non-cyclic o-phenylenediamine precursor, ring-closure, and oxidation. Mercury(II)-N-heterocyclic carbene complex is formed from the salt by conventional methods using Hg(II) acetate as both metal and base source **(Scheme 42)**. These complexes were the first azolium compounds with directly attached N-ferrocenyl substituents. However, benzimidazol-2-ylidenes show a more significant electronic communication between the ferrocenyl substituent and the carbene moiety, compared to N-ferrocenyl derivatives. The directly linked ferrocenyl substituents mainly exhibit an electronic inductive donor effect. Along with mercury-N-heterocyclic carbene complexes, palladium and tungsten complexes of the same N-heterocyclic carbene-ligand systems have also been explored [101–102].

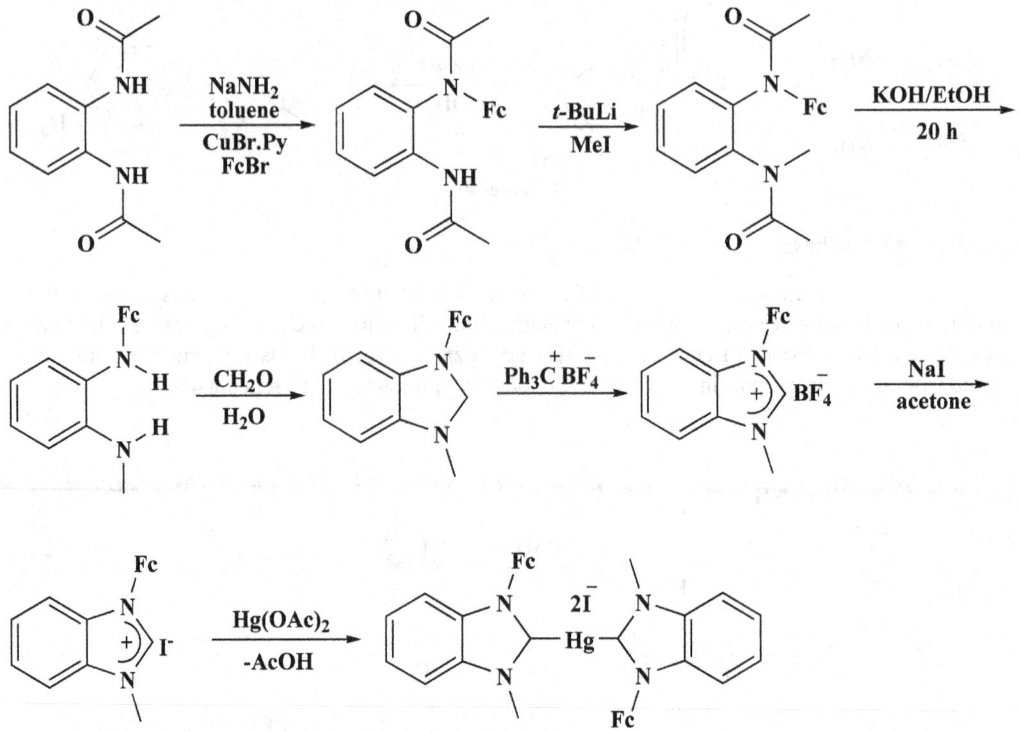

Scheme-42

Nickel-assisted synthesis

The regioselectivity of carbon-hydrogen bond functionalizations can be controlled with chelation assistance. Since this protocol is based on the coordinating ability of a tethered organic functionality, common directing groups typically bear a lone pair of electrons that can coordinate with transition metals. This method has greatly diversified the substrate scope for intermolecular carbon-hydrogen bond functionalization through the use of a range of directing groups. Kleinman and Dubeck [103] explored the use of directing groups to control the regioselectivity of transition metal insertion into a carbon-hydrogen bond. Since the publication of their study, this method has been widely employed. Coordinating groups help in directing such insertion so that either the thermodynamically or the kinetically favored five- or six-membered metallacycle is formed. One disadvantage of this method is that both mono- or di-functionalized products in *ortho* position with respect to the directing group are obtained in symmetrical substrates. However, sterics become the controlling factor for unsymmetrical substrates and the resultant functionalization occurs predominantly at the less hindered *ortho*-position (**Scheme 43**).

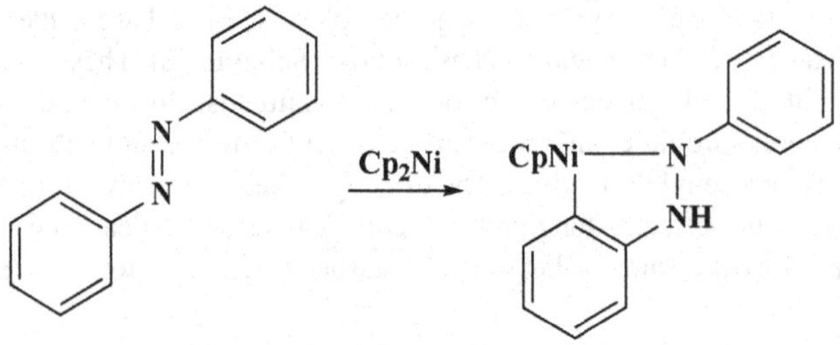

Scheme-43

Ruthenium-assisted synthesis

$Ru_3(CO)_{12}$ has been utilized as a catalyst in the deoxygenation of *o*-nitrobiphenyl to yield carbazole, as well as in reductive carbonylation of *o*-nitroazo derivatives to produce benzotriazoles in solvents like *o*-xylene or *o*-dichlorobenzene (at 70–80 atm, 180–200 °C). The addition of a base such as triethylamine serving as a deoxygenating agent for the nitro group increased the selectivity and conversion considerably [104]. $Ru_3(CO)_{12}$ acts as an efficient catalyst for the deoxygenation of 2-nitro-*N*-(phenylmethylene)-benzeneamine to yield 2-phenylbenzimidazole. The detection of carbon dioxide in the solution at the end of the reaction confirms the reducing action of carbon monoxide [105]. Moreover, amine is formed as the main by-product of this reaction due to hydrogen abstraction by the intermediate nitrene from traces of water or the solvent present in the medium **(Scheme 44)**.

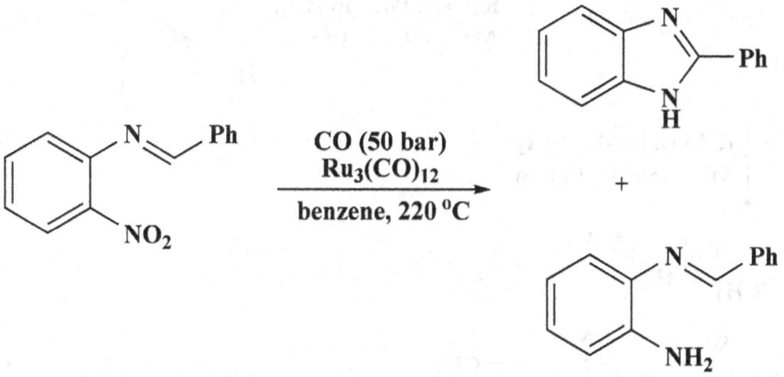

Scheme-44

Scandium-assisted synthesis

The Ugi four-component condensation, in which a ketone or aldehyde, an amine, an isocyanide and a carboxylic acid combine to produce α-acylaminoamide, is interesting because of its versatility—a wide range of products can be obtained by varying the substrates [106]. Heterocyclic amidines react with isocyanides and aldehydes in an Ugi-type three-component condensation in the presence of 5 mol% $Sc(OTf)_3$ as a catalyst **(Scheme 45)**, with reaction times up to 72 hours at room temperature, for the synthesis of fused 3-aminoimidazoles [107]. Ireland et al. [108] realized significantly higher reaction rates under sealed-vessel microwave conditions. A reaction time of merely 10 minutes at 160 °C in methanol (ethanol, in some cases) under microwave irradiation resulted in similar yields when the same reaction was performed at room temperature for 72 hours.

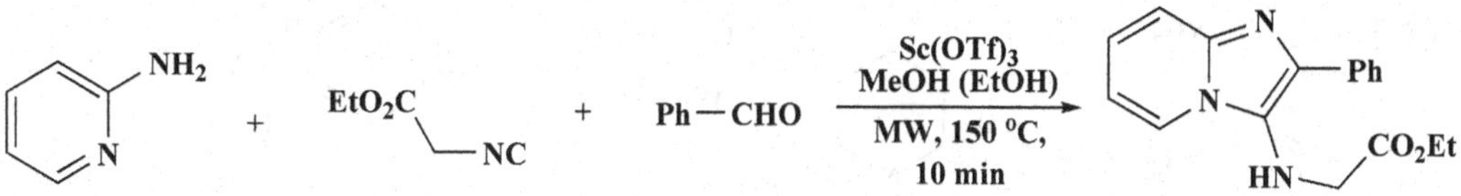

Scheme-45

Perfluoroalkanesulfonyl-protected hydroxybenzaldehydes, isonitriles and 2-aminopyridines subjected to three-component condensation afford imidazo[1,2-*a*]pyridine ring systems **(Scheme 46)** [109]. These condensed products undergo Pd-catalyzed cross-coupling with boronic acids to form imidazo[1,2-*a*]pyridine.

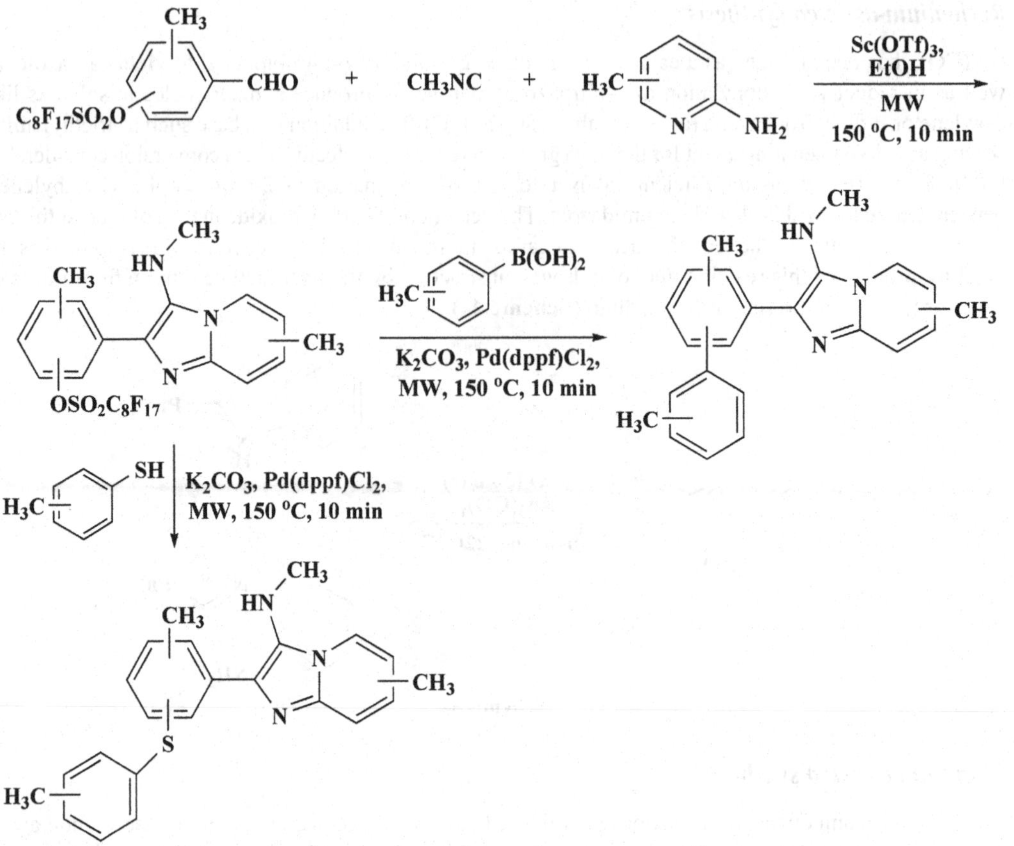

Scheme-46

Silicon-assisted synthesis

Guruswamy and Arul [110] investigated the preparation of several benzimidazoles from simple and substituted *o*-phenylenediamines and isonicotinic acid under microwave irradiation, using SiO_2/H_2SO_4 as the catalyst **(Scheme 47)**.

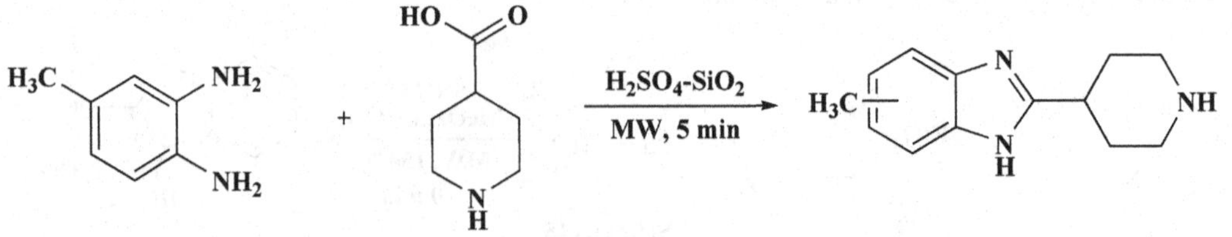

Scheme-47

Silver-assisted synthesis

2-Tosyl-2*H*-indazole can be synthesized through the reaction of tropone tosylhydrazone sodium salt with silver chromate **(Scheme 48)**. This reaction occurs through the cyclization of the hydrazyl radical intermediate which is stabilized by the canonical formula [111–112].

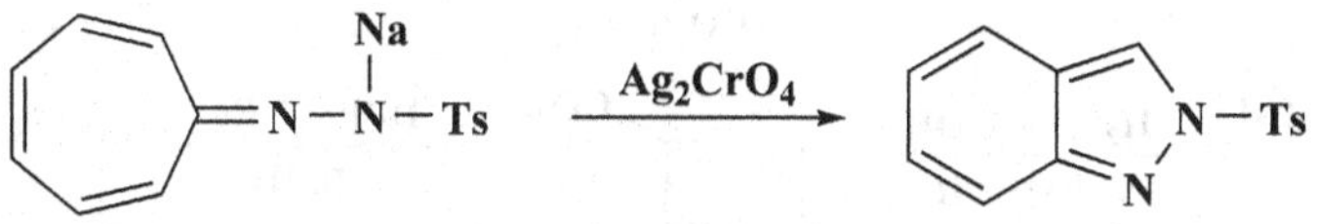

Scheme-48

Tin-assisted synthesis

The carbonylation and reductive *N*-hetero-cyclization of 2-nitrostyrenes with carbon monoxide in the presence of elemental selenium catalyst produces moderate to good yields of indoles [113]. The natural product arcyriaflavin A can be synthesized through nitrene insertion [114]. Moreover, many heterocycles are synthesized from *o*-nitroaromatics by transition metal-catalyzed reductive *N*-heteroannulation. For example, 1-phenyl-2*H*-indazole can be synthesized from *N*-(2-nitrobenzylidene)amine **(Scheme 49)**.

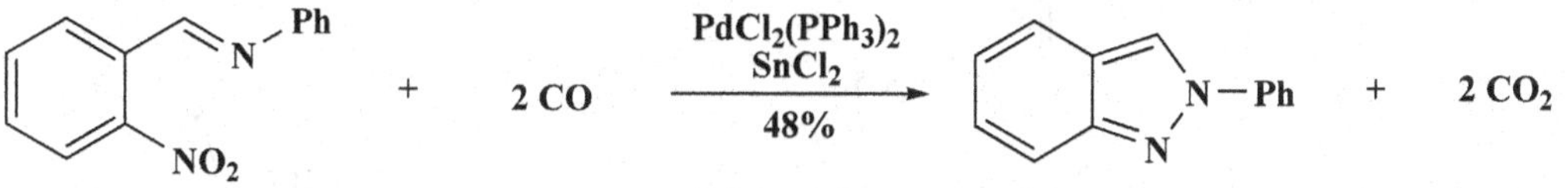

Scheme-49

Grigg and Sansano [115] reported a hydrostannylation-carbopalladation/cyclization cascade in the presence of palladium to provide small and large nitrogen-containing heterocycles where cyclization occurred at the *α*-allenic carbon **(Scheme 50)**. The allylstannanes were obtained as a mixture of E/Z isomers when a series of allenamides were subjected to highly regioselective palladium [palladium(II) or palladium(0)]-catalyzed hydrostannylations. Both E/Z isomeric allylstannanes underwent a sequence of oxidative addition, carbopalladation/cyclization and elimination assisted by palladium(0). The elimination of *n*-Bu$_3$SnPdX was faster than *β*-hydride (HPdX) elimination, thereby providing moderate to good yields of the final products [51b].

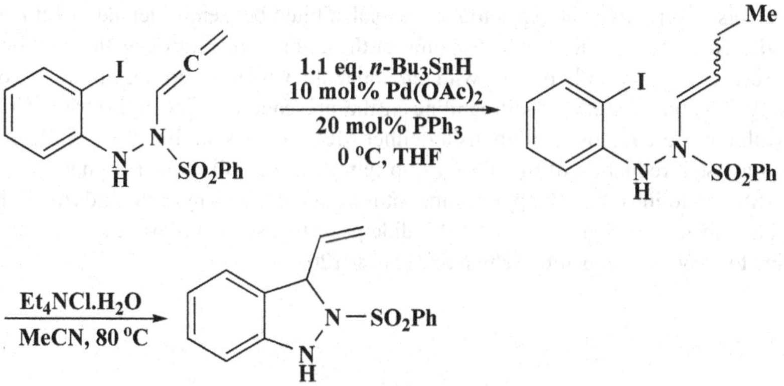

Scheme-50

The benzimidazole present in angiotensin II antagonist candesartan (Atacand) can be synthesized from highly substituted diaminobenzene derivatives, which is in turn formed by a tin-mediated nitro-reduction **(Scheme 51)**. This synthesis is concluded by the installation of an acetal side-chain and tetrazole ring, the former cleaved under physiological conditions to make candesartan (a pro-drug) in the same way

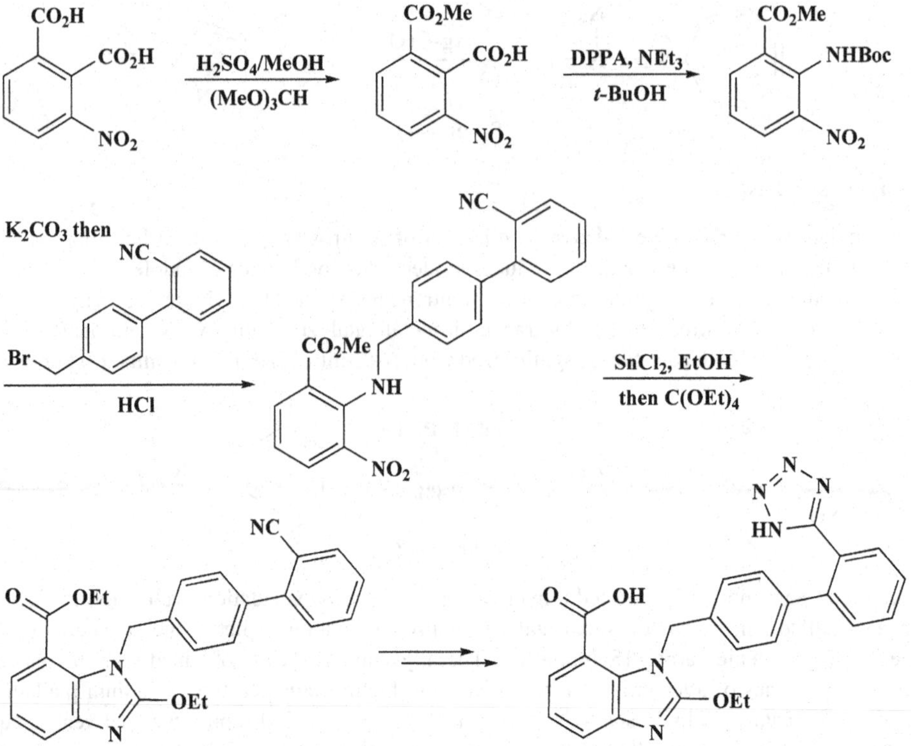

Scheme-51

as olmesartan [116]. The benzimidazole precursor is synthesized from 1,2,3-trisubstituted nitrophthalic acid which is monoesterified selectively and subjected to a Curtius rearrangement, a cumbersome method at scaled up levels. Alternative pathways to such tri-substituted benzenes include direct *ortho*-lithiation of 1,3-disubstituted benzenes followed by trapping of the anion with a nitrogen electrophile [117–118].

The synthesis of pazopanib begins with an indazole which is produced by diazotization and spontaneous cyclization of 2-ethyl-5-nitroaniline in the presence of *tert*-butyl nitrite. This indazole is methylated with complete regioselectivity using either Meerwein's salt, dimethyl sulfate or trimethyl-*o*-formate. A tin-assisted reduction of the nitro group unmasks the aniline which undergoes nucleophilic aromatic substitution to introduce the pyrimidine into the synthesis of pyrimidoindazole. The secondary amine function is then methylated with methyl iodide prior to a second S_NAr reaction with sulfonamide derived aniline to provide pazopanib (**Scheme 52**) [118–120].

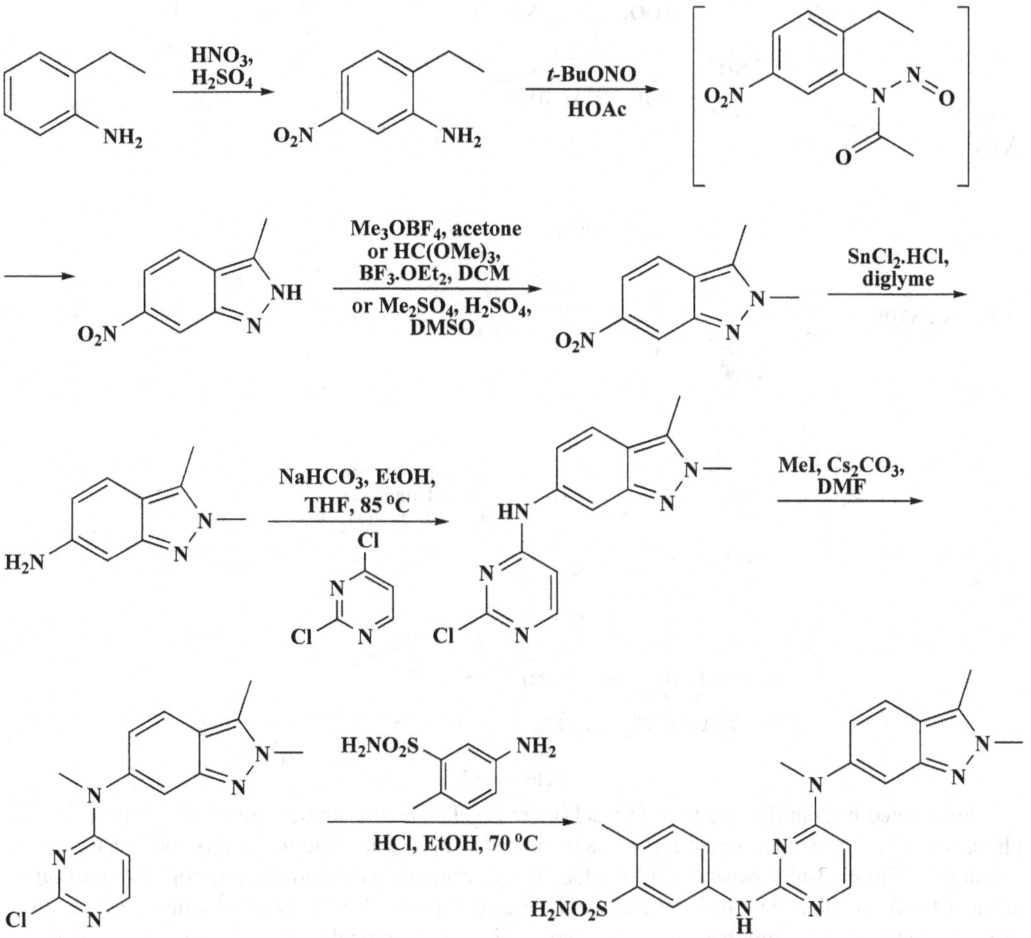

Scheme-52

Another protocol for the synthesis of benzimidazole uses resins as the substrate. The secondary amine bound to a resin is acylated with 4-fluoro-3-nitro-benzoic acid. The fluorine is displaced with a primary amine to produce a common intermediate *o*-nitroaniline. To obtain the remaining five cores, the nitro group in the adduct is reduced with buffered tin chloride in DMF to afford phenylenediamine which is subsequently cyclized efficiently using triphosgene to provide a product which forms benzimidazol-2-one upon cleavage. The benzimidazole is thus formed when the suspension in trimethyl-*o*-formate is heated, followed by resin cleavage (**Scheme 53**) [121].

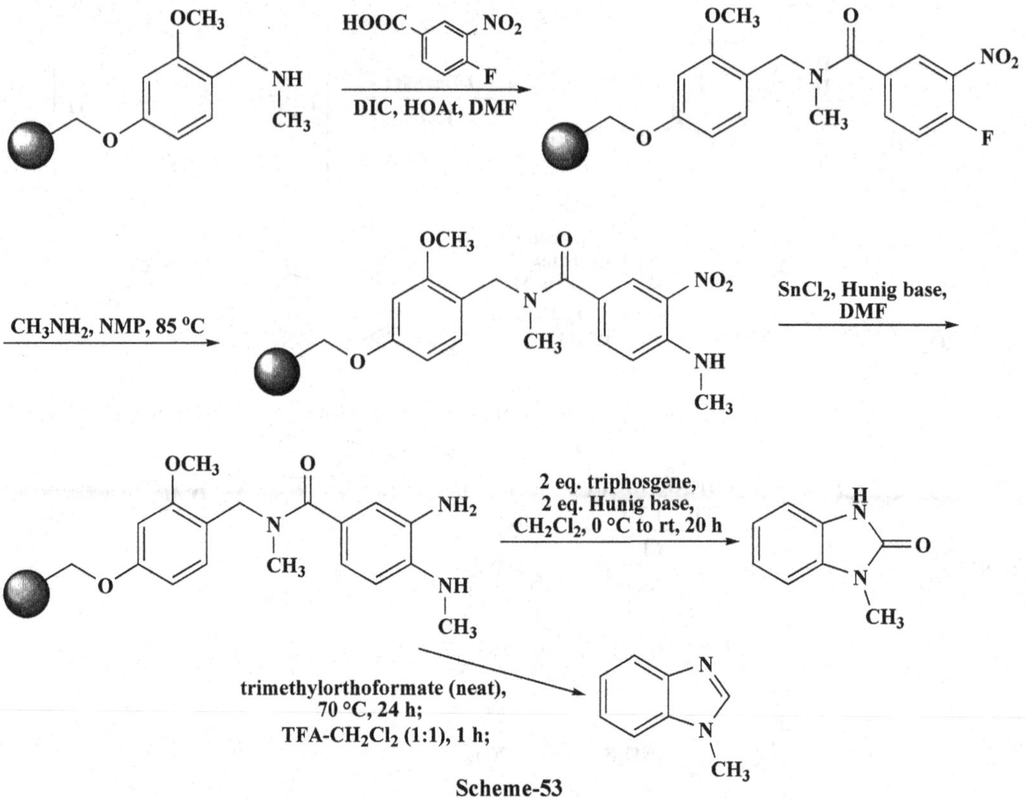

Scheme-53

Substituted benzimidazoles used in Boehringer Ingelheim contribution are synthesized by a solid-phase protocol. The Sieber amide resin functionalized with alanine is used for this solid-phase library synthesis. 4-Fluoro-3-nitrobenzoic acid is added to the reagent *via* an aromatic nucleophilic substitution at the 4-fluoro position. The resin-bound aromatic acid is coupled with specified amines and the nitro group is reduced to a 2-aminobenzimidazole intermediate *via* treatment with cyanogens bromide. Finally, the relevant class of compounds is realized for analysis upon addition of various aromatic acid chlorides and their release from the resin (**Scheme 54**) [122].

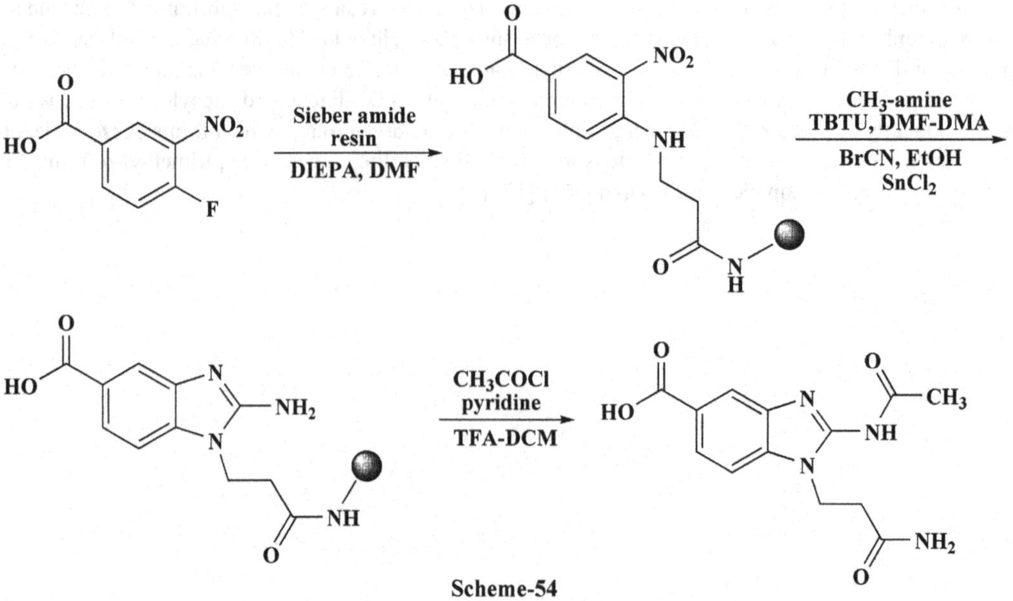

Scheme-54

Huang and Scarborough [123] reported an example of solid-supported preparation of benzimidazoles, wherein a five-membered ring compound was synthesized on the site of the solid-phase. The benzimidazoles were ultimately prepared upon treatment of amino compound with trimethyl-*o*-formate (**Scheme 55**).

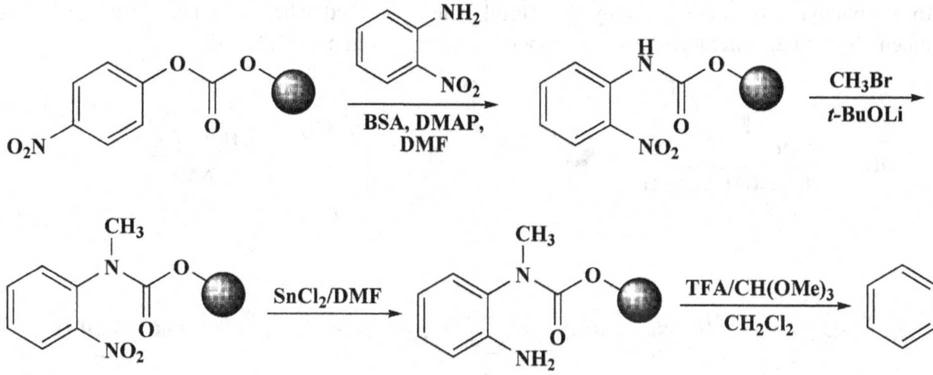

Scheme-55

Krchňák et al. [124] synthesized benzimidazoles on 4-(4-formyl-3-methoxyphenoxy)butyryl resin. In this method, after RA, the BAL amine was treated with *o*-fluoronitrobenzene and the *o*-nitroaniline formed was reduced with tin(II) chloride. Subsequently, cleavage from the linker and cyclization into benzimidazole occurred upon acylation with a carboxylic acid chloride and treatment with acetic acid at 80 °C (**Scheme 56**).

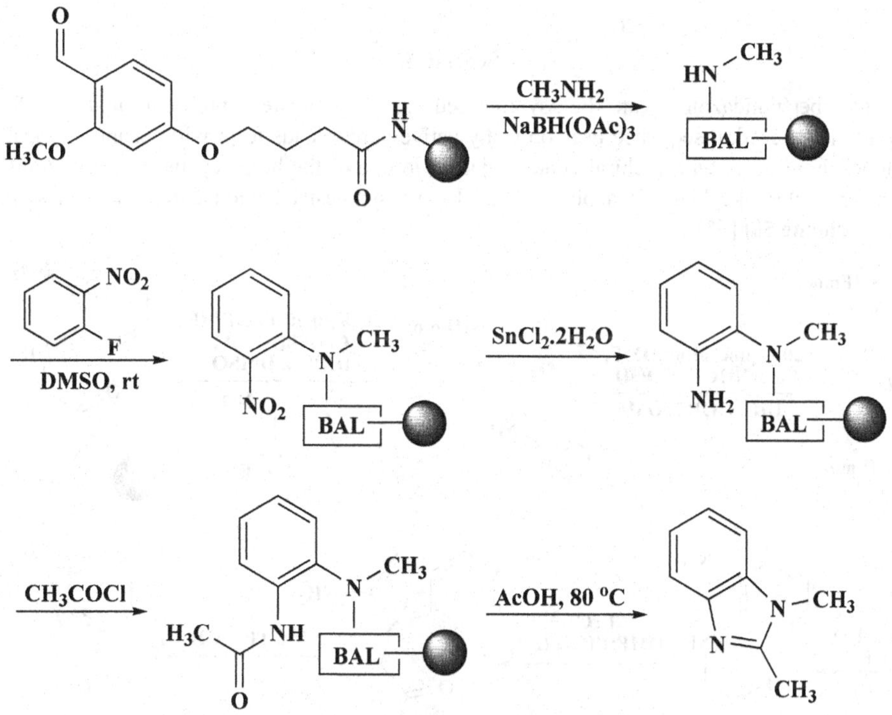

Scheme-56

The Wang resin (*p*-benzyloxybenzyl alcohol resin) can be esterified with 4-fluoro-3-nitrobenzoic acid under standard coupling conditions (4-dimethylaminopyridine, *N,N*'-dicyclohexylcarbodiimide), and the resin-bound intermediate produced in quantitative yields. Nitroanilines are obtained by nucleophilic aromatic displacement of the activated fluoride at the *ipso*-position with commercially available amines. The nitro moiety of nitroanilines is then reduced using tin(II) chloride in NMP (1-methyl-2-pyrrolidinone) which forms *o*-phenylenediamines. Finally, benzimidazole is formed when anilines are treated with aromatic aldehydes and 2,3-dichloro-5,6-dicyanobenzoquinone (**Scheme 57**) [125].

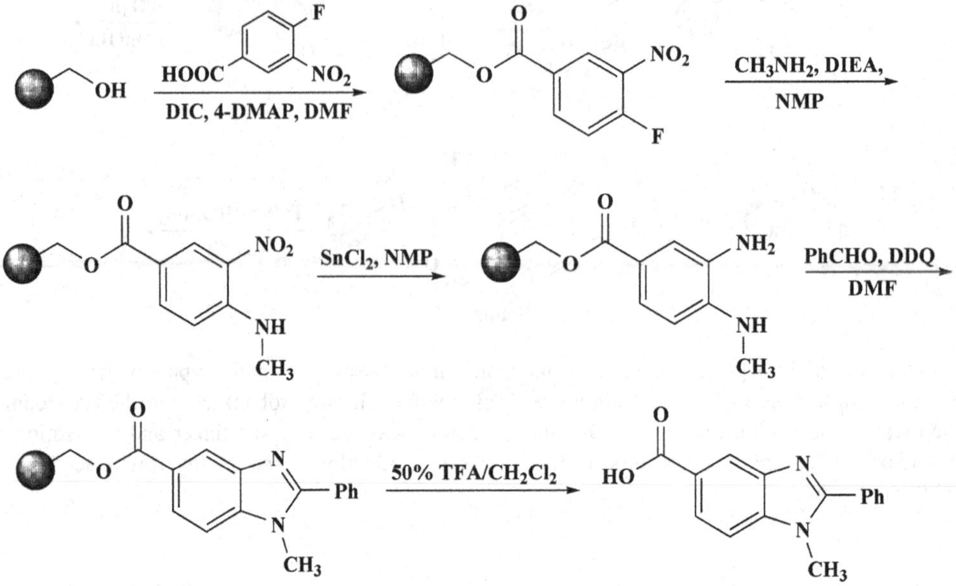

Scheme-57

2-Aminobenzimidazoles can be synthesized in a multistep protocol under solid-phase conditions. This reaction sequence uses optically active amino acids to provide enantiomerically pure 2-aminobenzimidazoles with a chiral center adjacent to one of the heterocyclic nitrogen atoms. Other chiral heterocycles like benzimidazolones can also be synthesized upon extending this solid-phase protocol (**Scheme 58**) [95].

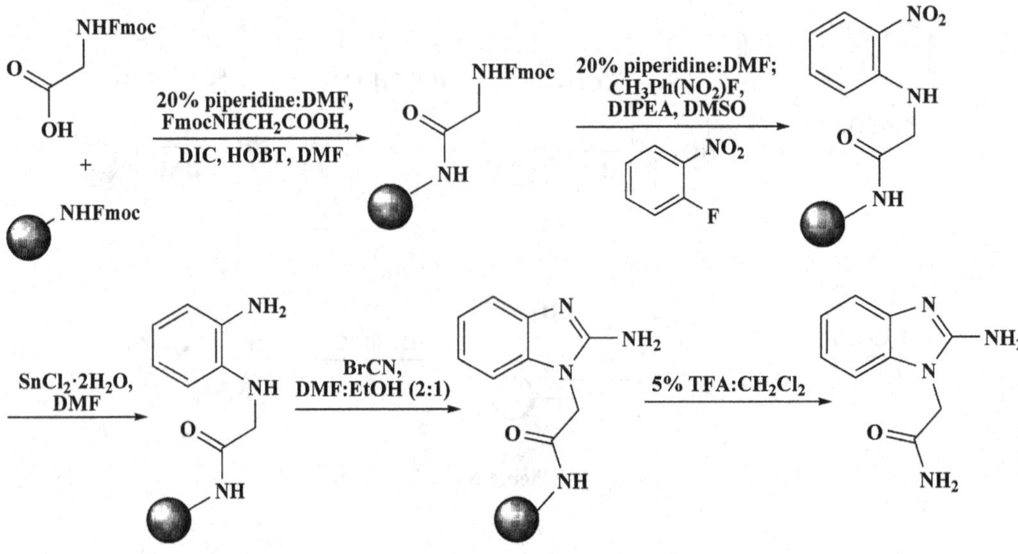

Scheme-58

VanVliet et al. [126] synthesized 2-substituted benzimidazoles directly from 2-nitroanilines by *in situ* reduction and cyclization under microwave irradiation (**Scheme 59**).

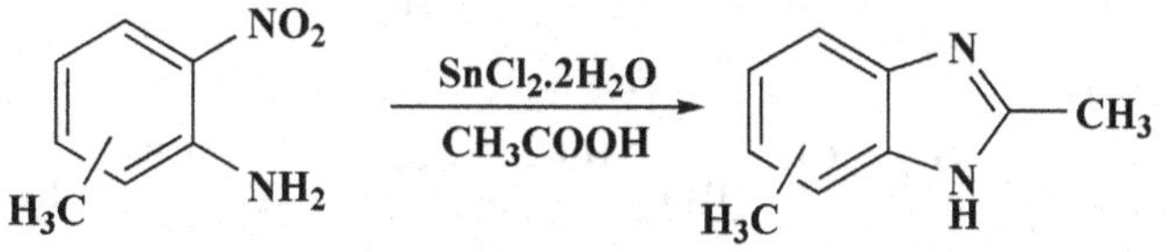

Scheme-59

Tumelty et al. [127] were successful in substituting unprotected ArgoGelTM Rink Fmoc resin with fluoronitrobenzoic acid. Diamines formed by nucleophilic aromatic substitution followed by reduction reacted with bromoacetic acid anhydride, and benzimidazoles were ultimately obtained after the displacement of bromide with a nucleophilic monomer (**Scheme 60**).

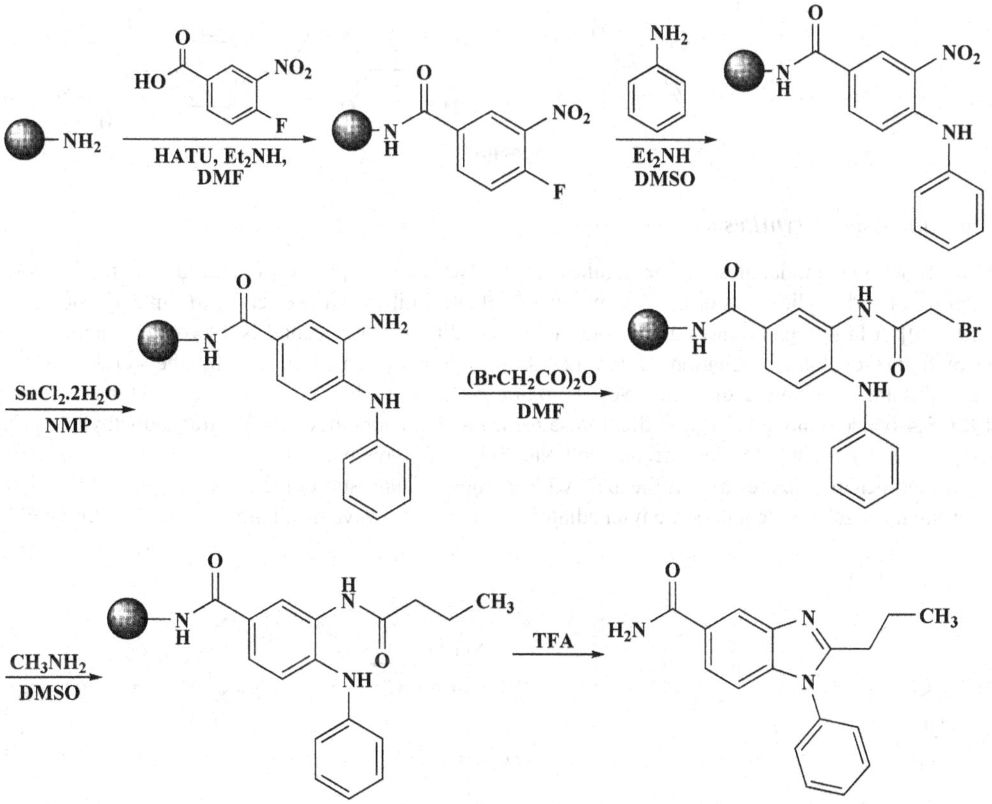

Scheme-60

Quinoxalines are generally less investigated compounds. However, growing efforts are being made to characterize and synthesize these compounds of late. 4-Amino-3-nitrophenol is reduced with stannous chloride in hydrochloric acid to provide 3,4-diaminophenol. The substrate 3-methylquinoxaline-2(1*H*)-one is prepared when *o*-phenylenediamine reacts with pyruvic acid under microwave heating in the presence of 6N hydrochloric acid. 4-{[-3-Methyl-2-oxoquinoxaline–1(2*H*)methyl]amino}benzoic acid is synthesized by Mannich reaction of 3-methylquinoxaline-2(1*H*)-one, formaldehyde and 4-aminobenzoic acid. Consequently, 1-({[4-(1*H*-benzimidazol-2-yl)phenyl]amino}-methyl)-3-methylquinoxaline-2(1*H*)-one is formed when 4-{[-3-methyl-2-oxoquinoxaline-1(2*H*)methyl]amino}benzoic acid reacts with *o*-phenylenediamine (**Scheme 61**) [128].

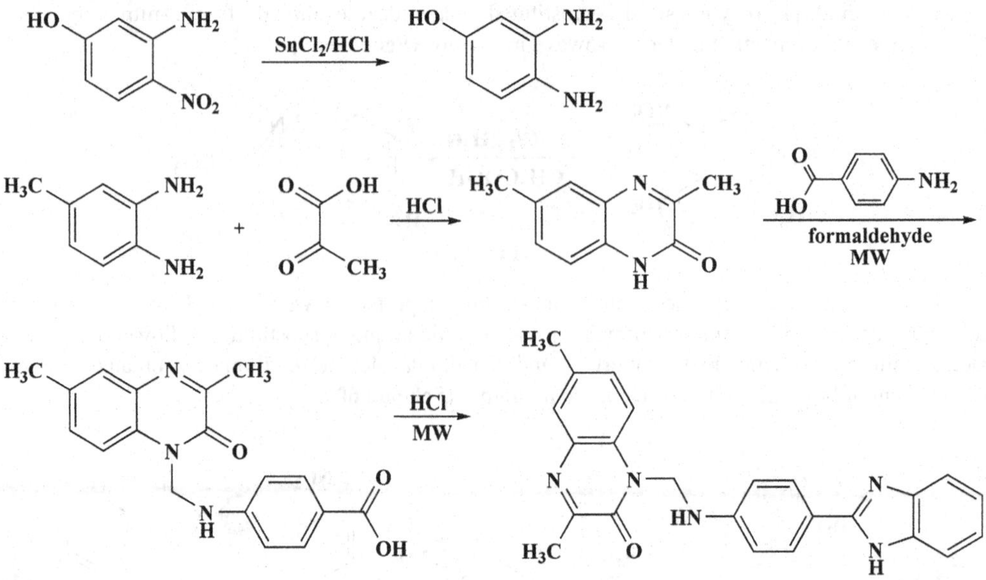

Scheme-61

Titanium-assisted synthesis

2-Substitued benzimidazoles can be synthesized by two general protocols. The first method involves coupling of carboxylic acids or their derivatives (imidates, nitriles or *o*-esters) and *o*-phenylenediamine [129–130] at high temperatures and strong acidic conditions. The second is a two-step method based on oxidative cyclohydrogenation of aniline Schiff's bases (formed *in situ* by the condensation of aldehydes and *o*-phenylenediamine). Several oxidative reagents like nitrobenzene [131], benzofuroxan [132], 1,4-benzoquinone [133], 2,3-dichloro-5,6-dicyanobenzoquinone [134], tetracyanoethylene [135], manganese(IV) oxide [136], lead acetate and $NaHSO_3$ [137] have been utilized for this reaction. Herein, substituted benzimidazoles are synthesized with the help of clay-supported titanium catalyst by stirring substituted benzaldehyde and *o*-phenylenediamine under microwave irradiation at 50 °C (**Scheme 62**).

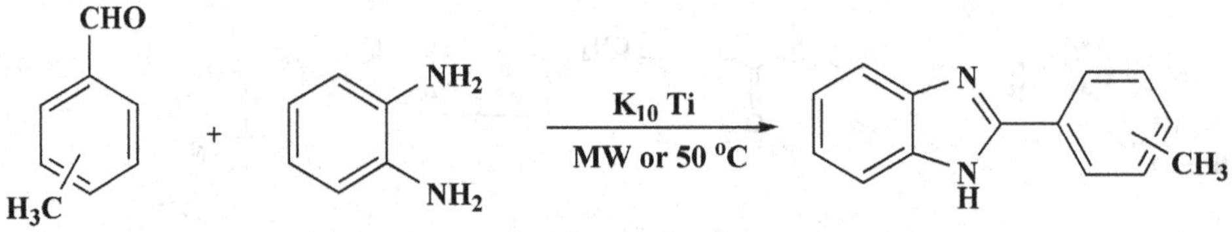

Scheme-62

Nagawade and Shinde [138] prepared differently substituted benzimidazoles from aldehydes and *o*-phenylenediamine under solvent-free conditions in the presence of titanium(IV) chloride catalyst. This protocol was applicable to unsaturated, aromatic as well as aliphatic aldehydes (**Scheme 63**).

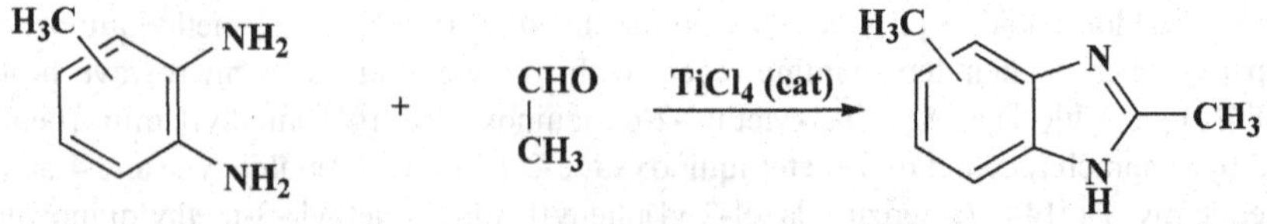

Scheme-63

Zinc-assisted synthesis

2-Arylaminobenzimidazoles can be synthesized on SynPhaseTM crowns through the formation of *o*-nitroaniline intermediate. It was also possible to prepare thiobenzimidazoles by following an analogous route **(Scheme 64)** [139–141].

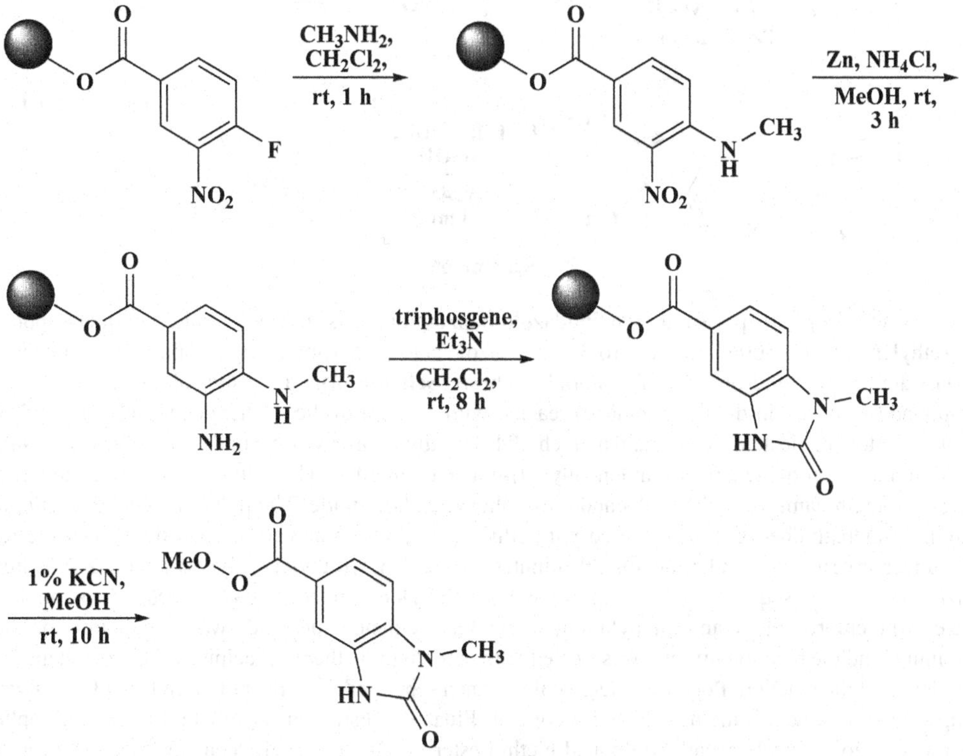

Scheme-64

2-(Arylamino)benzimidazoles are synthesized by cyclocondensation of poly(ethylene glycol)-supported *o*-phenylenediamines with isothiocyanates under microwaves, followed by separation of the product from the polymer support [142]. For considerable cyclization, either reflux in methanol for 4 hours or microwave heating for 10 minutes are required **(Scheme 65)**. The isolation of products can be simplified substantially through a soluble poly(ethylene glycol) matrix as poly(ethylene glycol)-bound products can be precipitated selectively from a suitable combination of solvents. Similarly, MeO-poly(ethylene glycol)-OH can be removed from the homogeneous solution by precipitation and filtration after microwave-assisted cleavage of the substituted benzimidazoles from the polymer support. All polymer-supported intermediates and the polymer support itself remain stable under microwaves. It is easy to monitor soluble polymer-supported reactions by conventional analytical methods, unlike the solid-phase synthesis. Later on, Lin and Sun [143] reported that the conversion of diamine into benzimidazoles (through the formation of *N,N*-disubstituted thiourea intermediate) under microwaves is facilitated by bismuth chloride and mercury(II) chloride [144] as catalysts. Microwave-assisted combinatorial synthesis of libraries of hydantoins [145–146] and thiohydantoins [147] in soluble poly(ethylene glycol)-matrix has also been studied by researchers. Another method involves a parallel preparation of 1,5-disubstituted thiohydantoins and hydantoins [148] in the presence of polyphosphoric ester under solvent-free conditions. Furthermore, an entire class of 3,5,5-trisubstituted hydantoins can be produced by the condensation of substituted benzils with ureas under microwaves, followed by *N*3-alkylation [72b, 149].

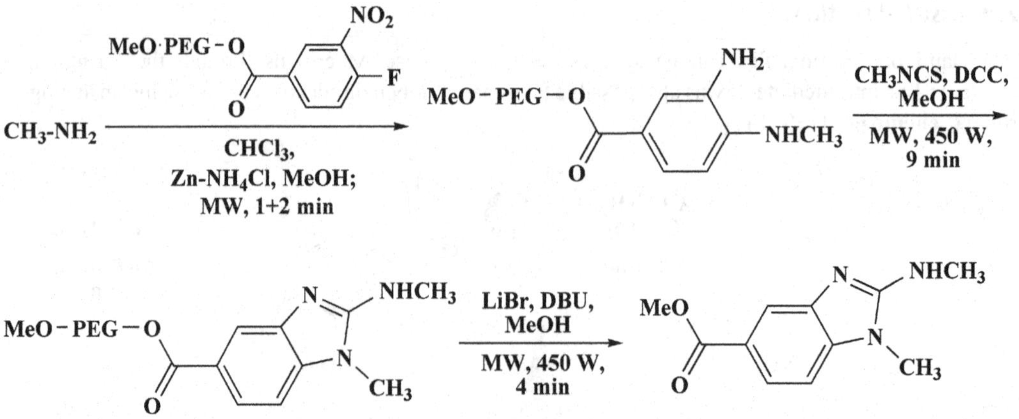

Scheme-65

Kini et al. [150] prepared benzimidazolo-benzothiophenes using soluble polymer-supported poly(ethylene glycol) 5000 and 4-fluoro-3-nitrobenzoic acid as starting materials along with substituted primary amines by liquid-phase combinatorial synthesis (Scheme 66). Firstly, the polymer bound diamino compound (dissolved in dichloromethane) reacted with 1.2 mol of dicyclohexyl carbodiimide, 1.2 mol of 4-mercaptobenzoic acid (MBA) and a pinch of 4-dimethylaminopyridine under microwave irradiation for 20 minutes to provide a poly(ethylene glycol)-bound compound. The solution was filtered to remove excess 4-dimethylaminopyridine salts and dicyclohexyl carbodiimide. The poly(ethylene glycol)-bound 3-amino-4-mercaptobenzene was treated with ethylene dichloride and trifluoroacetic acid in the ratio 10:1 under microwave irradiation for 20 minutes to yield poly(ethylene glycol)-bound 2-substituted benzimidazole. Subsequently, the solution of poly(ethylene glycol)bound mercaptobenzimidazole reacted with chloroacetone and triethylamine in dichloromethane under microwave irradiation for about 10 minutes and the reaction mixture was treated with cold diethyl ether to precipitate the product after the completion of the reaction. Polyethylene glycol was then separated from the poly(ethylene glycol)-bound compound using sodium methoxide and methanol. Finally, substituted 2-(3-methyl-benzo[*b*]thiophen-6-yl)-3*H*-benzoimidazole-5-carboxylic acid methyl ester was synthesized when substituted 2-(4-aceto-methyl-thio-phenyl)-1*H*-benzoimidazole-5-carboxylic acid methyl ester was treated with polyphosphoric acid (PPA) and heated in a water bath for 4 hours.

Aqueous solution ion exchange and microwave solid-state methods have often been investigated for the preparation of transition metal/Y zeolites. The activity of these zeolites has been reported for the transformation of acids into benzimidazoles through a reaction of 3-nitrobenzoic acid and 4-methyl-1,2-phenylenediamine. The yield of these reactions increases as CuY < FeY < NiY < CoY < NaY < CrY < MnY < ZnY for both the methods. Higher activity is observed in zeolites prepared by solid state ion-exchange compared to aqueous solution exchange. The products are obtained in 69–83% yields in the presence of ZnY zeolite. For this reaction, the Lewis sites have been found to be better than Brønsted sites. Differently substituted benzimidazoles can be prepared under solvent-free conditions from aldehydes and *o*-phenylenediamine in the presence of $BF_3 \cdot OEt_2$ catalyst in very good yields [151–152]. This protocol is applicable to substituted *o*-phenylenediamines as well as unsaturated, aromatic and aliphatic aldehydes without significant differences (Scheme 67).

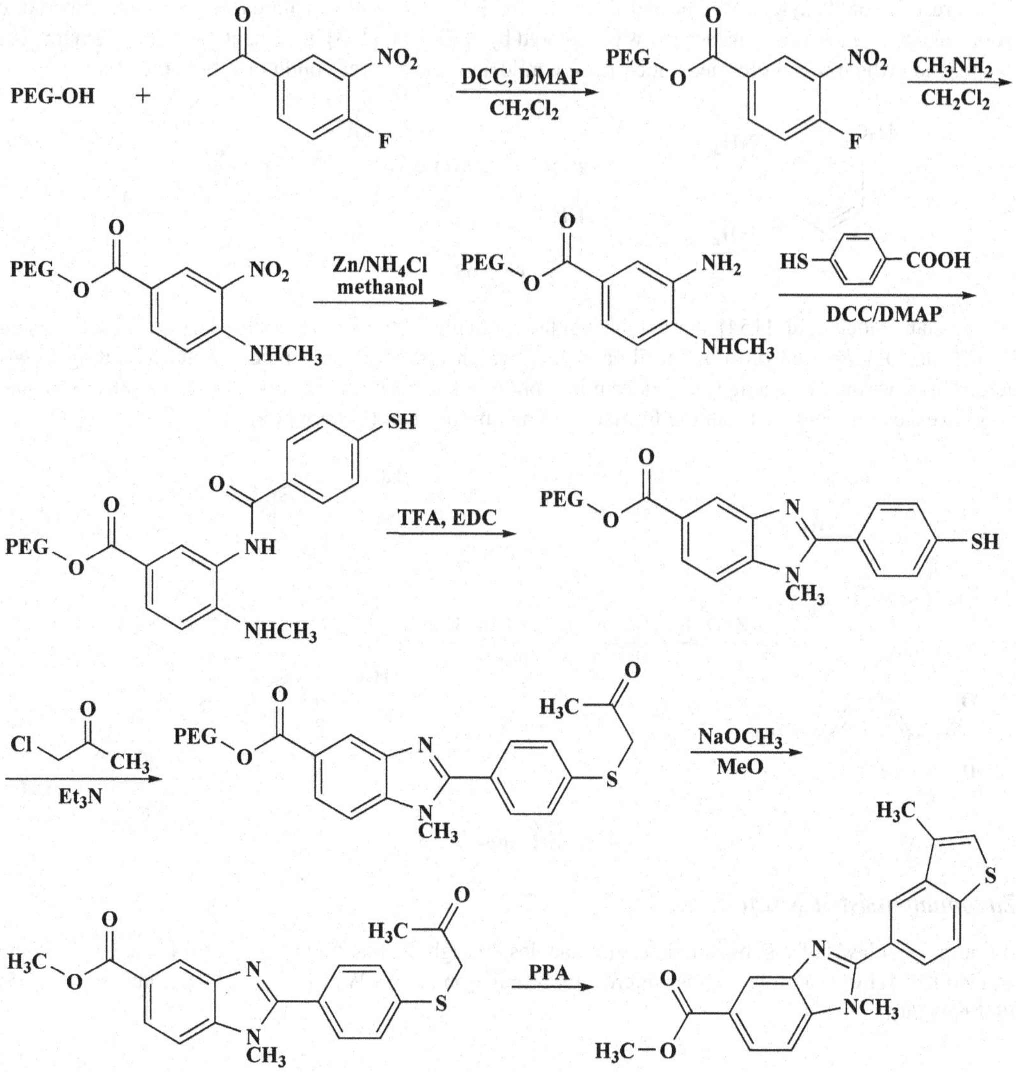

Scheme-66

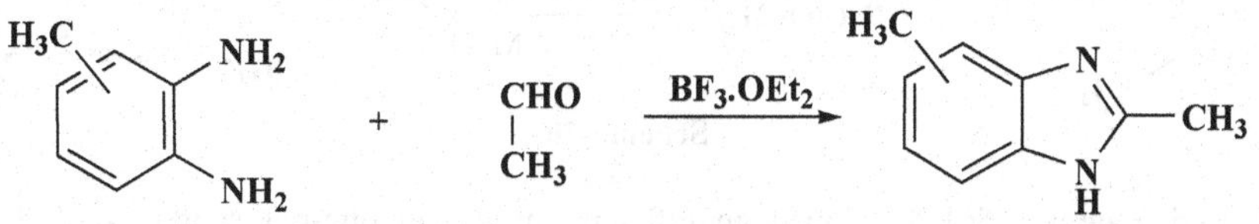

Scheme-67

A reaction of phenylenediamine and aldehydes for the synthesis of benzimidazoles at room temperature using catalytic amounts of zinc acetate was explored by Patil et al. [153]. Excellent yields were obtained in this reaction which was performed under neutral, mild and solvent-free conditions (**Scheme 68**).

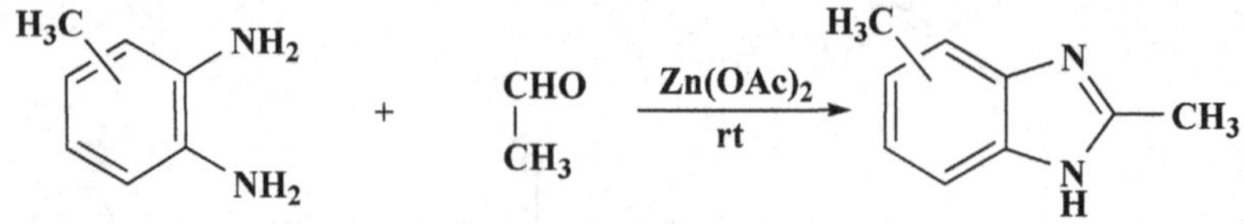

Scheme-68

Rostamizadeh et al. [154] studied the synthesis of substituted 2-arylbenzimidazoles in water, using $ZrOCl_2 \cdot nH_2O$ supported on montmorillonite K10 which served as an efficient water-tolerating Lewis acid. The reaction was carried out under mild conditions to result in remarkable chemoselectivity and good to excellent yields without the formation of any by-product (**Scheme 69**).

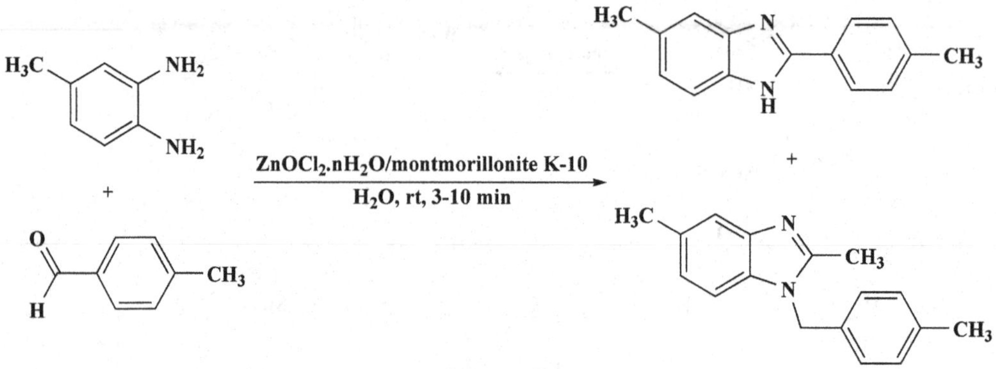

Scheme-69

Zirconium-assisted synthesis

A one-pot synthesis of 2-substituted benzimidazoles through the reaction of o-ester, o-phenylenediamine and $ZrOCl_2 \cdot 8H_2O$ (10 mol%) under microwave irradiation at 160 W was realized by Zhang et al. [155] (**Scheme 70**).

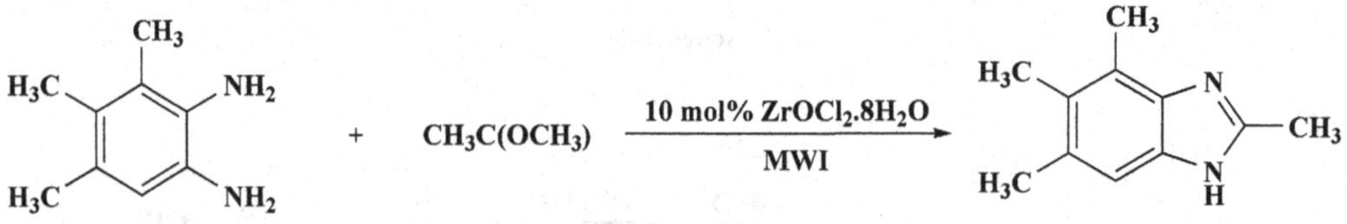

Scheme-70

Reddy and Nagraj [156] developed an efficient, mild and one-pot synthesis of 2-substituted benzimidazoles from o-phenylenediamine and o-esters like o-acetate, o-formate and o-valerate using $ZrOCl_2 \cdot 8H_2O$ under microwave irradiation at room temperature (**Scheme 71**) [157].

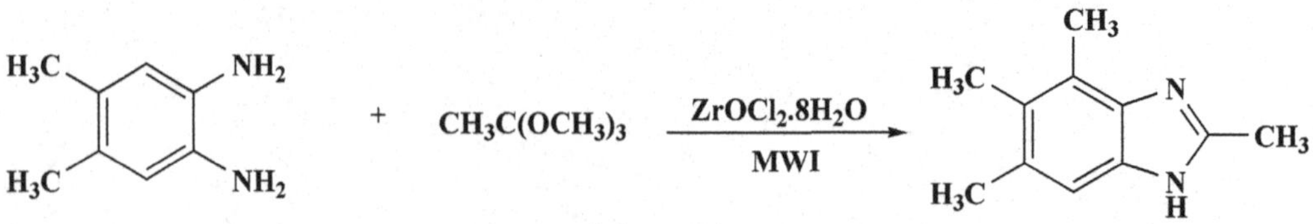

Scheme-71

References

(1)   (a) N. Kaur. 2018. Synthesis of seven- and higher-membered nitrogen-containing heterocycles using photochemical irradiation. Synth. Commun. 48: 2815–2849. (b) N. Kaur. 2018. Ruthenium-catalyzed synthesis of five-membered *O*-heterocycles. Inorg. Chem. Commun. 99: 82–107. (c) R. Dua and S.K. Srivastava. 2010. Synthesis, characterization and anti-microbial activity. Int. J. Pharm. Bio. Sci. 1: 35–39. (d) N. Kaur, P. Bhardwaj, M. Devi, Y. Verma and P. Grewal. 2019. Synthesis of five-membered *O,N*-heterocycles using metal and non-metal. Synth. Commun. 49: 1345–1384. (e) N. Kaur. 2019. Synthetic routes to seven and higher membered *S*-heterocycles by use of metal- and nonmetal-catalyzed reactions. Phosphorus, Sulfur, and Silicon and the Related Elements 194: 186–209. (f) N. Kaur. 2019. Synthesis of six-membered *N*-heterocycles using ruthenium catalysts. Catal. Lett. 14: 1513–1539.

(2)   C.J. Chen, B.A. Song and S. Yang. 2007. Synthesis and anti-fungal activities of 5-(3,4,5-trimethoxyphenyl)-2-sulfonyl-1,3,4-thiadiazole and 5-(3,4,5-trimethoxyphenyl)-2-sulfonyl-1,3,4-oxadiazole derivatives. Bioorg. Med. Chem. 15: 3981–3989.

(3)   S.R. Pattan, P. Kekare and N.S. Dighe. 2009. Synthesis and biological evaluation of some 1,3,4-thiadiazoles. J. Chem. Pharm. Res. 1: 191–198.

(4)   M. Moise, V. Sunel, L. Profire, M. Popa, J. Desbrieres and C. Peptu. 2009. Synthesis and biological activity of some new 1,3,4-thiadiazole and 1,2,4-triazole compounds containing a phenylalanine moiety. Molecules 14: 2621–2631.

(5)   M. Behrouzi-Fardmoghadam, F. Poorrajab, S.K. Ardestani, S. Emami, A. Shafiee and A. Foroumadi. 2008. Synthesis and *in vitro* anti-leishmanial activity of 1-[5-(5-nitrofuran-2-yl)-1,3,4-thiadiazol-2-yl]- and 1-[5-(5-nitrothiophen-2-yl)-1,3,4-thiadiazol-2-yl]-4-aroylpiperazines. Bioorg. Med. Chem. 16: 4509–4515.

(6)   E.E. Oruc, S. Rollas, F. Kandermirli, N. Shvets and A.S. Dimoglo. 2004. 1,3,4-Thiadiazole derivatives. Synthesis, structure elucidation, and structure-anti-tuberculosis activity relationship investigation. J. Med. Chem. 47: 6760–6676.

(7)   S. Karakus, U. Coruh and B. Barlas-Durgun. 2010. Synthesis and cytotoxic activity of some 1,2,4-triazoline-3-thione and 2,5-disubstituted-1,3,4-thiadiazole derivatives. Marm. Pharm. J. 14: 84–90.

(8)   P. Helissey, S. Cros and S. Giorgi-Renault. 1994. Synthesis, anti-tumor evaluation and SAR of new 1*H*-pyrrolo[3,2-*c*]quinoline-6,9-diones and 11*H*-indolo[3,2-*c*]quinoline-1,4-diones. Anticancer Drug Des. 9: 51–67.

(9)   M.F. Brana, J.M. Castellano, G. Keilhauer, A. Machuca, Y. Martin, C. Redondo, E. Schlick and N. Walker. 1994. Benzimidazo[1,2-*c*]quinazolines: a new class of anti-tumor compounds. Anticancer Drug Des. 9: 527–538.

(10) J.F. Riou, P. Helissey, L. Grondard and S. Giorgi-Renault. 1991. Inhibition of eukaryotic DNA topoisomerase I and II activities by indoloquinolinedione derivatives. Mol. Pharmacol. 40: 699–706.

(11) E. Ibrahim, A.M. Montgomerie, A.H. Sneddon, G.R. Proctor and B. Green. 1988. Synthesis of indolo[3,2-*c*]quinolines and indolo[3,2-*d*]benzazepines and their interaction with DNA. Eur. J. Med. Chem. 23: 183–188.

(12) P. Vivas-Mejía, O. Cox and F.A. Gonzalez. 1998. Inhibition of human topoisomerase II by anti-neoplastic benzazolo[3,2-*a*]quinolinium chlorides. Mol. Cell. Biochem. 178: 203–212.

(13) L.W. Deady, T. Rodemann, G.J. Finalay, B.C. Baguley and W.A. Denny. 2001. Synthesis and cytotoxic activity of carboxamide derivatives of benzimidazo[2,1-*a*]isoquinoline and pyrido[3',2':4,5]imidazo[2,1-*a*]isoquinoline. Anticancer Drug Des. 15: 339–346.

(14) W.R. Wand, S.D. Harrison, K.S. Gilbert, W.R. Laster and D.P. Griswold. 1991. Antitumor drug cross-resistance *in vivo* in a cisplatin-resistant murine P388 leukemia. Cancer Chemother. Pharmacol. 27: 456–463.

(15) R.D. Carpenter, K.S. Lam and M.J. Kurth. 2007. Microwave-mediated heterocyclization to benzimidazo[2,1-*b*]quinazolin-12(5*H*)-ones. J. Org. Chem. 72: 284–287.

(16) K. Panda, J.R. Suresh, H. Ila and H. Junjappa. 2003. Heteroaromatic annulation of 2-methyl/2-cyanomethylbenzimidazole dianions with α-oxoketene dithioacetals: a highly regioselective synthetic protocol for 1,2- and 2,3-substituted/annulated pyrido[1,2-*a*]benzimidazoles. J. Org. Chem. 68: 3498–3506.

(17) T.B. Christopher and A.B. Shirley. 2002. An intramolecular palladium-catalyzed aryl amination reaction to produce benzimidazoles. Tetrahedron Lett. 43: 1893–1895.

(18) R. Devalla and K. Ethirajulu. 1995. Synthesis of 2-substituted benzoxazoles and benzimidazoles based on mass spectral *ortho* interactions. J. Chem. Soc. Perkin Trans. 2 7: 1497–1501.

(19) M.R. Deluca and S.M. Kewin. 1997. The *para*-toluenesulfonic acid-promoted synthesis of 2-substituted benzoxazoles and benzimidazoles from diacylated precursors. Tetrahedron 53: 457–464.

(20) D. Bouyssi, N. Monteiro and G. Balme. 2011. Amines as key building blocks in Pd-assisted multi-component processes. Beilstein J. Org. Chem. 7: 1387–1406.

(21) M. Benchidmi, A. El Kihel, E.M. Essassi, N. Knouzi, L. Toupet, R. Danion-Bougot and R. Carrié. 2010. Nitration de benzimidazoles substitues. Bull. Soc. Chim. Belg. 104: 605–611.

(22) J. Azizian, A.A. Mohammadi and A.R. Karimi. 2003. Synthesis of some novel *γ*-spiroiminolactones from reaction of cyclohexyl isocyanide and dialkyl acetylene dicarboxylates with 1-benzylisatin and tryptanthrine. Synth. Commun. 33: 387–391.

(23) R. Kalirajan, L. Rathore, S. Jubie, B. Gowramma, S. Gomathy, S. Sankar and K. Elango. 2010. Microwave-assisted synthesis and biological evaluation of pyrazole derivatives of benzimidazoles. Indian J. Pharm. 44: 358–390.

(24) Z. Lu and S.S. Stahl. 2012. Intramolecular Pd(II)-catalyzed aerobic oxidative amination of alkenes: synthesis of six-membered *N*-heterocycles. Org. Lett. 14: 1234–1237.

(25)   B. Mathew, G. Unnikirishnan, V.P. Shafeer, C. Mohammed Musthafa and P. Femina. 2011. Microwave-assisted synthesis, physicochemical properties and anti-microbial activity of benzimidazole chalcones. Der Pharma Chemica 3: 627–631.

(26)   I. Mohammadpoor-Baltork, A.R. Khosropour and S.F. Hojati. 2007. Mild and efficient synthesis of benzoxazoles, benzothiazoles, benzimidazoles, and oxazolo[4,5-*b*]pyridines catalyzed by Bi(III) salts under solvent-free conditions. Montash. Chem. 138: 663–667.

(27)   J.A.R. Salvador, R.M.A. Pinto and S.M. Silvestre. 2009. Recent advances of bismuth(III) salts in organic chemistry: application to the synthesis of heterocycles of pharmaceutical interest. Curr. Org. Synth. 6: 426–470.

(28)   R.C. Larock, E.K. Yum, M.J. Doty and K.K.C. Sham. 1996. Synthesis of aromatic heterocycles *via* palladium-catalyzed annulation of internal alkynes. J. Org. Chem. 60: 3270–3271.

(29)   M. Boiani and M. Gonzalez. 2005. Imidazole and benzimidazole derivatives as chemotherapeutic agents. Mini Rev. Med. Chem. 5: 409–424.

(30)   N.R. Thimmegowda, S.N. Swamy, C.S.A. Kumar, Y.C.S. Kumar, S. Chandrappa, G.W. Yip and K.S. Rangappa. 2008. Synthesis, characterization and evaluation of benzimidazole derivative and its precursors as inhibitors of MDA-MB-231 human breast cancer cell proliferation. Bioorg. Med. Chem. Lett. 18: 432–435.

(31)   G. Navarrete-Vázquez, R. Cedillo, A. Hernandez-Campos, L. Yepez, F. Hernandez-Luis, J. Valdez, R. Morales, R. Cortes, M. Hernandez and R. Castillo. 2001. Synthesis and anti-parasitic activity of 2-(trifluoromethyl)benzimidazole derivatives. Bioorg. Med. Chem. Lett. 11: 187–190.

(32)   A. Kamal, K.L. Reddy, V. Devaiah, N. Shankaraiah and M.V.R. Rao. 2006. Recent advances in the solid-phase combinatorial synthetic strategies for the quinoxaline-, quinazoline- and benzimidazole-based privileged structures. Mini Rev. Med. Chem. 6: 71–89.

(33)   J.S. Yadav, B.V.S. Reddy, K. Prematatha and K.S. Shankar. 2008. Bismuth(III)-catalyzed rapid and highly efficient synthesis of 2-aryl-1-arylmethyl-1*H*-benzimidazoles in water. Can. J. Chem. 86: 124–128.

(34)   R. Kumar and Y.C. Joshi. 2007. Mild and efficient one-pot synthesis of imidazolines and benzimidazoles from aldehydes. E-J. Chem. 4: 606–610.

(35)   M. Kidwai, A. Jahan and D. Bhatnagar. 2010. Polyethylene glycol: a recyclable solvent system for the synthesis of benzimidazole derivatives using CAN as catalyst. J. Chem. Sci. 122: 607–612.

(36)   G. Evindar and R.A. Batey. 2003. Copper and palladium-catalyzed intramolecular aryl guanidinylation: an efficient method for the synthesis of 2-aminobenzimidazoles. Org. Lett. 5: 133–136.

(37)   C. Venkatesh, S.G.M. Sundaram, H. Ila and H. Junjappa. 2006. Palladium-catalyzed intramolecular *N*-arylation of heteroarenes: a novel and efficient route to benzimidazo[1,2-*a*]quinolines. J. Org. Chem. 71: 1280–1283.

(38)   T. Jin and Y. Yamamoto. 2007. An efficient, facile, and general synthesis of 1*H*-indazoles by 1,3-dipolar cycloaddition of arynes with diazomethane derivatives. Angew. Chem. Int. Ed. 46: 3323–3325.

(39)   D.C. Rogness and R.C. Larock. 2011. Synthesis of *N*-arylisatins by the reaction of arynes with methyl 2-oxo-2-(arylamino)acetates. J. Org. Chem. 76: 4980–4986.

(40)   P. Li, J. Zhao, C. Wu, R.C. Larock and F. Shi. 2011. Synthesis of 3-substituted indazoles from arynes and *N*-tosylhydrazones. Org. Lett. 13: 3340–3343.

(41)   A.T. Khan and T. Parvin. 2009. A simple and convenient one-pot synthesis of benzimidorole derivatives using coball(II) chearide hexahydrate as catalyst. Cheminform 40: 2339–2346.

(42)   G. Altenhoff and F. Glorius. 2004. A domino copper-catalyzed C-N and C-O cross-coupling for the conversion of primary amides into benzoxazoles. Adv. Synth. Catal. 346: 1661–1664.

(43)   A. Klapars, X. Huang and S.L. Buchwald. 2002. A general and efficient copper catalyst for the amidation of aryl halides. J. Am. Chem. Soc. 124: 7421–7428.

(44)   R.D. Viirre, G. Evindar and R.A. Batey. 2008. Copper-catalyzed domino annulation approaches to the synthesis of benzoxazoles under microwave-accelerated and conventional thermal conditions. J. Org. Chem. 73: 3452–3459

(45)   N. Zheng and S.L. Buchwald. 2007. Copper-catalyzed regiospecific synthesis of *N*-alkylbenzimidazoles. Org. Lett. 9: 4749–4751.

(46)   D.S. Surry and S.L. Buchwald. 2010. Diamine ligands in copper-catalyzed reactions. Chem. Sci. 1: 13–31.

(47)   B.L. Zhou, Q.L. Yuan and D.W. Ma. 2007. Synthesis of 1,2-disubstituted benzimidazoles by a Cu-catalyzed cascade aryl amination/condensation process. Angew. Chem. Int. Ed. 46: 2598–2601.

(48)   Y. Liu and J.-P. Wan. 2011. Tandem reactions initiated by copper-catalyzed cross-coupling: a new strategy towards heterocycle synthesis. Org. Biomol. Chem. 9: 6873–6894.

(49)   K. Inamoto, T. Saito, M. Katsuno, T. Sakamoto and K. Hiroya. 2007. Palladium-catalyzed C-H activation/intramolecular amination reaction: a new route to 3-aryl/alkylindazoles. Org. Lett. 9: 2931–2934.

(50)   L. Shen, R.P. Hsung, Y. Zhang, J.E. Antoline and X. Zhang. 2005. Copper-catalyzed stereospecific *N*-allenylations of amides. Syntheses of optically enriched chiral allenamides. Org. Lett. 7: 3081–3084.

(51)   (a) Y. Tang, L. Shen, B.J. Dellaria and R.P. Hsung. 2008. Saucy-Marbet rearrangements of alkynyl halides in the synthesis of highly enantiomerically enriched allenyl halides. Tetrahedron Lett. 49: 6404–6409. (b) T. Lu, Z. Lu, Z.-X. Ma, Y. Zhang and R.P. Hsung. 2013. Allenamides: a powerful and versatile building block in organic synthesis. Chem. Rev. 113: 4862–4904.

(52)   G. Brasche and S.L. Buchwald. 2008. C-H Functionalization/C-N bond formation: copper-catalyzed synthesis of benzimidazoles from amidines. Angew. Chem. Int. Ed. 47: 1932–1934.

(53) P. Thansandote and M. Lautens. 2009. Construction of nitrogen-containing heterocycles by C-H bond functionalization. Chem. Eur. J. 15: 5874–5883.

(54) Q. Xiao, W.-H. Wang, G. Liu, F.-K. Meng, J.-H. Chen, Z. Yang and Z.-J. Shi. 2009. Direct imidation to construct 1*H*-benzo[*d*]imidazole through PdII-catalyzed C-H activation promoted by thiourea. Chem. Eur. J. 15: 7292–7296.

(55) Y. Tan and J.F. Hartwig. 2010. Palladium-catalyzed amination of aromatic C-H bonds with oxime esters. J. Am. Chem. Soc. 132: 3676–3677.

(56) K. Inamoto, C. Hasegawa, K. Hiroya and T. Doi. 2008. Palladium-catalyzed synthesis of 2-substituted benzothiazoles *via* a C-H functionalization/intramolecular C-S bond formation process. Org. Lett. 10: 5147–5150.

(57) L.L. Joyce and R.A. Batey. 2009. Heterocycle formation *via* palladium-catalyzed intramolecular oxidative C-H bond functionalization: an efficient strategy for the synthesis of 2-aminobenzothiazoles. Org. Lett. 11: 2792–2795.

(58) K. Inamoto, T. Saito, K. Hiroya and T. Doi. 2010. Palladium-catalyzed intramolecular amidation of C(sp^2)-H bonds: synthesis of 4-aryl-2-quinolinones. J. Org. Chem. 75: 3900–3903.

(59) (a) R.K. Kumar, M.A. Ali and T. Punniyamurthy. 2011. Pd-catalyzed C-H activation/C-N bond formation: a new route to 1-aryl-1*H*-benzotriazoles. Org. Lett. 13: 2102–2105. (b) Z. Shi, C. Zhang, C. Tanga and N. Jiao. 2012. Recent advances in transition metal-catalyzed reactions using molecular oxygen as the oxidant. Chem. Soc. Rev. 41: 3381–3430.

(60) B.L. Zhou, Q.L. Yuan and D.W. Ma. 2007. Cascade coupling/cyclization process to *N*-substituted 1,3-dihydrobenzimidazol-2-ones. Org. Lett. 9: 4291–4294.

(61) D. Viña, E. Olmo, J.L. López-Pérez and A.S. Feliciano. 2007. Regioselective synthesis of 1-alkyl- or 1-aryl-1*H*-indazoles *via* copper-catalyzed cyclizations of 2-haloarylcarbonylic compounds. Org. Lett. 9: 525–528.

(62) D.S. Yang, H. Fu, L.M. Hu, Y.Y. Jiang and Y.F. Zhao. 2008. Copper-catalyzed synthesis of benzimidazoles *via* cascade reactions of *o*-haloacetanilide derivatives with amidine hydrochlorides. J. Org. Chem. 73: 7841–7844.

(63) X. Lv and W.L. Bao. 2009. Copper-catalyzed cascade addition/cyclization: an efficient and versatile synthesis of *N*-substituted 2-heterobenzimidazoles. J. Org. Chem. 74: 5618–5621.

(64) G.D. Shen and W.L. Bao. 2010. Synthesis of benzoxazole and benzimidazole derivatives *via* ligand-free copper(I)-catalyzed cross-coupling reaction of *o*-halophenols or *o*-haloanilines with carbodiimides. Adv. Synth. Catal. 352: 981–986.

(65) X.H. Deng, H. McAllister and N.S. Mani. 2009. CuI-catalyzed amination of arylhalides with guanidines or amidines: a facile synthesis of 1*H*-2-substituted benzimidazoles. J. Org. Chem. 74: 5742–5745.

(66) X.H. Deng and N.S. Mani. 2010. Reactivity-controlled regioselectivity: a regiospecific synthesis of 1,2-disubstituted benzimidazoles. Eur. J. Org. Chem. 4: 680–686.

(67) C.J. Ball and M.C. Willis. 2013. Cascade palladium- and copper-catalyzed aromatic heterocycle synthesis: the emergence of general precursors. Eur. J. Org. Chem. 3: 425–441.

(68) M.W. Chen, X.G. Zhang, P. Zhong and M.L. Hu. 2009. Copper-catalyzed tandem C-N bond formation reaction: selective synthesis of 2-(trifluoromethyl)benzimidazoles. Synthesis 9: 1431–1436.

(69) J. Zhu, H. Xie, Z. Chen, S. Li and Y. Wu. 2009. Synthesis of 2-fluoroalkylbenzimidazoles *via* copper(I)-catalyzed tandem reactions. Chem. Commun. 17: 2338–2340.

(70) X.Q. Diao, Y.J. Wang, Y.W. Jiang and D.W. Ma. 2009. Assembly of substituted 1*H*-benzimidazoles and 1,3-dihydrobenzimidazol-2-ones *via* CuI/l-proline-catalyzed coupling of aqueous ammonia with 2-iodoacetanilides and 2-iodophenylcarbamates. J. Org. Chem. 74: 7974–7977.

(71) A. Nefzi, M.A. Giulianotti and R.A. Houghten. 2000. Solid-phase synthesis of branched thiohydantoin benzimidazolinethiones and branched thiohydantoin tetrahydroquinoxalinediones. Tetrahedron Lett. 41: 2283–2287.

(72) (a) C. Pabba, H.-J. Wang, S.R. Mulligan, Z.-J. Chen, T.M. Stark and B.T. Gregg. 2005. Microwave-assisted synthesis of 1-aryl-1*H*-indazoles *via* one-pot two-step Cu-catalyzed intramolecular-*N*-arylation of arylhydrazones. Tetrahedron. Lett. 46: 7553–7557. (b) E. Suna and I. Mutule. 2006. Microwave-assisted heterocyclic chemistry. Top. Curr. Chem. 266: 49–101.

(73) P. Nilsson, K. Olofsson and M. Larhed. 2006. Microwave-assisted and metal-catalyzed coupling reactions. Top. Curr. Chem. 266: 103–144.

(74) Z. Li, H. Sun, H. Jiang and H. Liu. 2008. Copper-catalyzed intramolecular cyclization to *N*-substituted 1,3-dihydrobenzimidazol-2-ones. Org. Lett. 10: 3263–3266.

(75) C. Masdeu, E. Gomez, N.A.O. Williams and R. Lavilla. 2007. Double insertion of isocyanides into dihydropyridines: direct access to substituted benzimidazolium salts. Angew. Chem. Int. Ed. 46: 3043–3046.

(76) P.T. Parvatkar, P.S. Parameswaran and S.G. Tilve. 2012. Recent developments in the synthesis of five- and six-membered heterocycles using molecular iodine. Chem. Eur. J. 18: 5460–5489.

(77) M. Ishihara and H. Togo. 2007. Direct oxidative conversion of aldehydes and alcohols to 2-imidazolines and 2-oxazolines using molecular iodine. Tetrahedron 63: 1474–1480.

(78) Y.-M. Ren and C. Cai. 2008. Iodine as an efficient catalyst for the synthesis of benzimidazoles and imidazolines from primary alcohols and diamines. Org. Prep. Proced. Int. 40: 101–105.

(79) P. Gogoi and D. Konwar. 2006. An efficient and one-pot synthesis of imidazolines and benzimidazoles *via* anaerobic oxidation of carbon-nitrogen bonds in water. Tetrahedron Lett. 47: 79–82.

(80) P.H. Svensson and L. Kloo. 2003. Synthesis, structure, and bonding in polyiodide and metal iodide-iodine systems. Chem. Rev. 103: 1649–1684.

(81)   Z.-H. Zhang, J.-J. Li, Y.-Z. Gao and Y.-H. Liu. 2007. Synthesis of 2-substituted benzimidazoles by iodine-mediated condensation of *o*-esters with 1,2-phenylenediamines. J. Heterocycl. Chem. 44: 1509–1512.

(82)   H.-S. Wang and J.E. Zeng. 2008. Iodine-catalyzed one-pot synthesis of 3,4-dihydroquinazolin-4-ones from anthranilic acids, *o*-esters and amines under solvent-free conditions. Chin. J. Chem. 26: 175–178.

(83)   O. Prakash, H. Batra, H. Kaur, P.K. Sharma, V. Sharma, S.P. Singh and R.M. Moriarty. 2001. Hypervalent iodine oxidative rearrangement of anthranilamides, salicylamides and some *β*-substituted amides: a new and convenient synthesis of 2-benzimidazolones, 2-benzoxazolones and related compounds. Synthesis 4: 541–543.

(84)   H. Song, W. Chen, Y. Wang and Y. Qin. 2005. Preparation of alkyl carbamate of 1-protected indole-3-methylamine as a precursor of indole-3-methylamine. Synth. Commun. 35: 2735–2748.

(85)   V.V. Zhdankin. 2009. Hypervalent iodine(III) reagents in organic synthesis. ARKIVOC (i): 1–62.

(86)   I. Tellitu, A. Urrejola, S. Serna, I. Moreno, M.T. Herrero, E. Dominguez, R. SanMartin and A. Correa. 2007. On the phenyliodine(III)-bis(trifluoroacetate)-mediated olefin amidohydroxylation reaction. Eur. J. Org. Chem. 3: 437–444.

(87)   A. Correa, I. Tellitu, E. Dominguez, I. Moreno and R.S. Martin. 2005. An efficient, PIFA-mediated approach to benzo-, naphtho-, and heterocycle-fused pyrrolo[2,1-*c*][1,4]diazepines. An advantageous access to the anti-tumor anti-biotic DC–81. J. Org. Chem. 70: 2256–2264.

(88)   F. Churruca, R. SanMartin, I. Tellitu and E. Dominguez. 2005. A new, expeditious entry to the benzophenanthrofuran framework by a Pd-catalyzed *C*- and *O*-arylation/PIFA-mediated oxidative coupling sequence. Eur. J. Org. Chem. 12: 2481–2490.

(89)   S. Hernandez, R. SanMartin, I. Tellitu and E. Dominguez. 2003. Toward safer methodologies for the synthesis of polyheterocyclic systems: intramolecular arylation of arenes under Mizoroki-Heck reaction conditions. Org. Lett. 5: 1095–1098.

(90)   F. Churruca, R. SanMartin, M. Carril, M.K. Urtiaga, X. Solans, I. Tellitu and E. Dominguez. 2005. Direct, two-step synthetic pathway to novel dibenzo[*a,c*]phenanthridines. J. Org. Chem. 70: 3178–3187.

(91)   M.T. Herrero, I. Tellitu, S. Hernandez, E. Dominguez, I. Moreno and R. SanMartin. 2002. Novel applications of the hypervalent iodine chemistry. Synthesis of thiazolo-fused quinolinones. ARKIVOC (v): 31–37.

(92)   I. Tellitu, S. Serna, M.T. Herrero, I. Moreno, E. Dominguez and R. SanMartin. 2007. Intramolecular PIFA-mediated alkyne amidation and carboxylation reaction. J. Org. Chem. 72: 1526–1529.

(93)   A. Correa, I. Tellitu, E. Dominguez and R. SanMartin. 2006. An advantageous synthesis of new indazolone and pyrazolone derivatives. Tetrahedron 62: 11100–11105.

(94)   A. Correa, I. Tellitu, E. Dominguez and R. SanMartin. 2006. A metal-free approach to the synthesis of indoline derivatives by a phenyliodine(III) bis(trifluoroacetate)-mediated amidohydroxylation reaction. J. Org. Chem. 71: 8316–8319.

(95)   J. Lee, A. Doucette, N.S. Wilson and J. Lord. 2001. Solid-phase synthesis of chiral 2-amino-benzimidazoles. Tetrahedron Lett. 42: 2635–2638.

(96)   L.M. Stanley and J.F. Hartwig. 2009. Regio- and enantioselective *N*-allylations of imidazole, benzimidazole, and purine heterocycles catalyzed by single-component metallacyclic iridium complexes. J. Am. Chem. Soc. 131: 8971–8983.

(97)   R.H. Smith and H. Suschitzky. 1961. Syntheses of heterocyclic compounds - I. Tetrahedron 16: 80–84.

(98)   A.V. Borhade, D.R. Tope and D.R. Patil. 2012. An efficient synthesis of benzimidazole by cyclization-oxidation processes using Fe/MgO as a heterogeneous recyclable catalyst. J. Chem. Pharm. Res. 4: 2501–2506.

(99)   B. Yan and H. Gstach. 1996. An indazole synthesis on solid support monitored by single bead FTIR microspectroscopy. Tetrahedron Lett. 37: 8325–8328.

(100)   B. Bildstein, M. Malaun, H. Kopacka, K.-H. Ongania and K. Wurst. 1998. Imidazoline-2-ylidene metal complexes with pendant ferrocenyl substituents. J. Organomet. Chem. 552: 45–61.

(101)   B. Bildstein, M. Malaun, H. Kopacka, K.-H. Ongania and K. Wurst. 1999. *N*-Heterocyclic carbenes with *N*-ferrocenyl-*N'*-methyl-substitution: synthesis, reactivity, structure and electrochemistry. J. Organomet. Chem. 572: 177–187.

(102)   S. Budagumpi and S. Endud. 2013. Group XII metal-*N*-heterocyclic carbene complexes: synthesis, structural diversity, intramolecular interactions, and applications. Organometallics 32: 1537–1562.

(103)   J.P. Kleinman and M. Dubeck. 1963. The preparation of cyclopentadienyl[*o*-(phenylazo)phenyl]nickel. J. Am. Chem. Soc. 85: 1544–1545.

(104)   M. Pizzotti, S. Cenini, R. Psaro and S. Costanzi. 1990. Useful synthesis of benzotriazole photostabilizers by reductive carbonylation of *ortho*-nitrophenylazo compounds, catalyzed by $Ru_3(CO)_{12}$ in the presence of NEt_3. J. Mol. Catal. 63: 299–304.

(105)   C. Crotti, S. Cenini, F. Ragaini, F. Porta and S. Tollari. 1992. Ruthenium carbonyl-catalyzed deoxygenation by carbon monoidde of *o*-substituted nitrobenzenes. Synthesis of benzimidazoles. J. Mol. Catal. 72: 283–298.

(106)   I. Ugi. 2001. Recent progress in the chemistry of multi-component reactions. Pure Appl. Chem. 73: 187–191.

(107)   C. Blackburn, B. Guan, P. Fleming, K. Shiosaki and S. Tsai. 1998. Parallel synthesis of 3-aminoimidazo[1,2-*a*] pyridines and pyrazines by a new three-component condensation. Tetrahedron Lett. 39: 3635–3638.

(108)   S.M. Ireland, H. Tye and M. Whittaker. 2003. Microwave-assisted multi-component synthesis of fused 3-aminoimidazoles. Tetrahedron Lett. 44: 4369–4371.

(109)   Y. Lu and W. Zhang. 2004. Microwave-assisted synthesis of a 3-aminoimidazo[1,2-*a*]-pyridine/pyrazine library by fluorous multi-component reactions and subsequent cross-coupling reactions. QSAR Comb. Sci. 23: 827–835.

(110) B. Guruswamy and R. Arul. 2011. A microwave-assisted synthesis of benzimidazole derivatives using solid support B. Der Pharma Chemica 3: 483–486.

(111) K. Saito, T. Toda and T. Mukai. 1984. Reactions of sodium salts of tosylhydrazone compounds with silver chromate. Formation of indazole, pyrazole, and benzonitrile derivatives. Bull. Chem. Soc. Jpn. 57: 1567–1569.

(112) M. Alvarez-Corral, M. Munoz-Dorado and I. Rodriguez-Garcia. 2008. Silver-mediated synthesis of heterocycles. Chem. Rev. 108: 3174–3198.

(113) Y. Nishiyama, R. Maema, K. Ohno, M. Hirose and N. Sonoda. 1999. Synthesis of indoles: selenium-catalyzed reductive *N*-heterocyclization of 2-nitrostyrenes with carbon monoxide. Tetrahedron Lett. 40: 5717–5720.

(114) M. Adeva, F. Buono, E. Caballeno, M. Medarde and F. Tome. 2000. New synthetic approach to arcyriaflavin-A and un-symmetrical analogs. Synlett 6: 832–834.

(115) R. Grigg and J.M. Sansano. 1996. Sequential hydrostannylation-cyclization of δ- and ω-allenyl aryl halides. Cyclization at the proximal carbon. Tetrahedron 52: 13441–13454.

(116) V. Snieckus. 1990. Directed *ortho* metalation. Tertiary amide and *O*-carbamate directors in synthetic strategies for poly-substituted aromatics. Chem. Rev. 90: 879–933.

(117) M. Porcs-Makkay, T. Mezei and G. Simig. 2007. New practical synthesis of the key intermediate of candesartan. Org. Process Res. Dev. 11: 490–493.

(118) M. Baumann, I.R. Baxendale, S.V. Ley and N. Nikbin. 2011. An overview of the key routes to the best selling 5-membered ring heterocyclic pharmaceuticals. Beilstein J. Org. Chem. 7: 442–495.

(119) B. Gabriele, R. Mancuso, L. Veltri, V. Maltese and G. Salerno. 2012. Synthesis of substituted thiophenes by palladium-catalyzed heterocyclodehydration of 1-mercapto-3-yn-2-ols in conventional and non-conventional solvents. J. Org. Chem. 77: 9905–9909.

(120) P.A. Harris, A. Boloor, M. Cheung, R. Kumar, R.M. Crosby, R.G. Davis-Ward, A.H. Epperly, K.W. Hinkle, R.N. Hunter, J.H. Johnson, V.B. Knick, C.P. Laudeman, D.K. Luttrell, R.A. Mook, R.T. Nolte, S.K. Rudolph, J.R. Szewczyk, A.T. Truesdale, J.M. Veal, L. Wang and J.A. Stafford. 2008. Discovery of 5-[[4-[(2,3-dimethyl-2*H*-indazol-6-yl)methylamino]-2-pyrimidinyl]amino]-2-methyl-benzenesulfonamide (Pazopanib), a novel and potent vascular endothelial growth factor receptor inhibitor. J. Med. Chem. 51: 4632–4640.

(121) V. Purandare, A. Gao and M.A. Poss. 2002. Solid-phase synthesis of 'diverse' heterocycles. Tetrahedron Lett. 43: 3903–3906.

(122) R.J. Snow, A. Abeywardane, S. Campbell, J. Lord, M.A. Kashem, H.H. Khine, J. King, J.A. Kowalski, S.S. Pullen, T. Roma, G.P. Roth, C.R. Sarko, N.S. Wilson, M.P. Winters, J.P. Wolak and C.L. Cywin. 2007. Hit-to-lead studies on benzimidazole inhibitors of ITK: Discovery of a novel class of kinase inhibitors. Bioorg. Med. Chem. Lett. 17: 3660–3665.

(123) W.L. Huang and R.M. Scarborough. 1999. A new "traceless" solid-phase synthesis strategy: synthesis of a benzimidazole library. Tetrahedron Lett. 40: 2665–2668.

(124) V. Krchňák, J. Smith and J. Vágner. 2001. A solid-phase traceless synthesis of 2-arylaminobenzimidazoles. Tetrahedron Lett. 42: 1627–1630.

(125) D. Vourloumis, M. Takahashi, K.B. Simonsen, B.K. Ayida, S. Barluenga, G.C. Wintersa and T. Hermannb. 2003. Solid-phase synthesis of benzimidazole libraries biased for RNA targets. Tetrahedron Lett. 44: 2807–2811.

(126) D.S. VanVliet, P. Gillespie and J.J. Scicinski. 2005. Rapid one-pot preparation of 2-substituted benzimidazoles from 2-nitroanilines using microwave conditions. Tetrahedron Lett. 46: 6741–6743.

(127) D. Tumelty, M.K. Schwarz and M.C. Needels. 1998. Solid-phase synthesis of substituted 1-phenyl-2-aminomethyl-benzimidazoles and 1-phenyl–2-thiomethyl-benzimidazoles. Tetrahedron Lett. 39: 7467–7470.

(128) M.A. Amin and M.M. Youssef. 2012. Use of modern technique for synthesis of quinoxaline derivatives as potential anti-virus compounds. Der Pharma Chemica 4: 1323–1329.

(129) J.B. Wright. 1951. The chemistry of the benzimidazoles. Chem. Rev. 48: 397–541.

(130) T.A. Fairley, R.R. Tidwell, I. Donkor, N.A. Naiman, K.A. Ohemeng, R.J. Lombardy, J.A. Bentley and M. Cory. 1993. Structure, DNA minor groove binding, and base pair specificity of alkyl- and aryl-linked bis(amidinobenzimidazoles) and bis(amidinoindoles). J. Med. Chem. 36: 1746–1753.

(131) R.S. Harapanhalli, L.W. McLaughlin, R.W. Howell, D.V. Rao, S.J. Adelstein and A.I. Kassis. 1996. [^{125}I/^{127}I] Iodohoechst 33342: synthesis, DNA binding, and biodistribution. J. Med. Chem. 39: 4804–4809.

(132) F. Patzold, F. Zeuner, T.H. Heyer and H.-J. Niclas. 1992. Dehydrogenations using benzofuroxan as oxidant. Synth. Commun. 22: 281–288.

(133) E. Verner, B.A. Katz, J.R. Spencer, D. Allen, J. Hataye, W. Hruzewicz, H.C. Hui, A. Kolesnikove, Y. Li, C. Luong, A. Martelli, K. Radika, R. Rai, M. She, W. Shrader, P.A. Sprengeler, S. Trapp, J. Wang, W.B. Young and R.I. Mackman. 2001. Development of serine protease inhibitors displaying a multi-centered short (<2.3 Å) hydrogen bond binding mode: inhibitors of urokinase-type plasminogen activator and factor Xa. J. Med. Chem. 44: 2753–2771.

(134) J.J. Vanden Eynde, F. Delfosse, P. Lor and Y. van Haverbeke. 1995. 2,3-Dichloro–5,6-dicyano–1,4-benzoquinone, a mild catalyst for the formation of carbon-nitrogen bonds. Tetrahedron 51: 5813–5818.

(135) H. Chikashita, S. Nishida, M. Miyazaki, Y. Morita and K. Itoh. 1987. *In situ* generation and synthetic application of 2-phenylbenzimidazoline to the selective reduction of carbon-carbon double bonds of electron deficient olefins. Bull. Chem. Soc. Jpn. 60: 737–746.

(136) I. Bhatnagar and M.V. George. 1968. Oxidation with metal oxides-II. Tetrahedron 24: 1293–1298.

(137) F.F. Stephens and J.D. Bower. 1949. The preparation of benziminazoles and benzoxazoles from Schiff's bases. Part I. J. Chem. Soc. 2971–2972.

(138) R.R. Nagawade and D.B. Shinde. 2006. TiCl$_4$-promoted synthesis of benzimidazole derivatives. Indian J. Chem. 46B: 349–351.

(139) J.P. Kilburn, J. Lau and R.C.F. Jones. 2000. Solid-phase synthesis of substituted 2-aminomethylbenzimidazoles. Tetrahedron Lett. 41: 5419–5421.

(140) A. Mazurov. 2000. Traceless synthesis of benzimidazoles on solid support. Bioorg. Med. Chem. Lett. 10: 67–70.

(141) Y.C. Chi and C.M. Sun. 2000. Soluble polymer-supported synthesis of a benzimidazole library. Synlett 5: 591–594.

(142) P.M. Bendale and C.-M. Sun. 2002. Rapid microwave-assisted liquid-phase combinatorial synthesis of 2-(arylamino) benzimidazoles. J. Comb. Chem. 4: 359–361.

(143) M.-J. Lin and C.-M. Sun. 2004. Focused microwave-assisted parallel synthesis of bis-benzimidazoles. Synlett 4: 663–666.

(144) Y.-S. Su and M.-C. Sun. 2005. Microwave-assisted benzimidazole cyclization by bismuth chloride. Synlett 8: 1243–1246.

(145) M.-J. Lee and C.-M. Sun. 2004. Traceless synthesis of hydantoin by focused microwave irradiation. Tetrahedron Lett. 45: 437–440.

(146) M.-J. Lin and C.-M. Sun. 2003. Microwave-assisted traceless synthesis of thiohydantoin. Tetrahedron Lett. 44: 8739–8742.

(147) W.-B. Yeh, M.-J. Lin, M.-J. Lee and C.-M. Sun. 2003. Microwave-enhanced liquid-phase synthesis of thiohydantoins and thioxotetrahydropyrimidinones. Mol. Divers. 7: 185–198.

(148) S. Paul, M. Gupta, R. Gupta and A. Loupy. 2002. Microwave-assisted synthesis of 1,5-disubstituted hydantoins and thiohydantoins in solvent-free conditions. Synthesis 1: 75–78.

(149) G.G. Muccioli, J.H. Poupaert, J. Wouters, B. Norberg, W. Poppitz, G.K.E. Scriba and D.M. Lambert. 2003. A rapid and efficient microwave-assisted synthesis of hydantoins and thiohydantoins. Tetrahedron 59: 1301–1307.

(150) D. Kini, H. Kumar and M.Ghate. 2009. Microwave-assisted liquid-phase synthesis of benzimidazolo benzothiophenes for anti-microbial activity. E-J. Chem. 6: S25–S32.

(151) M. Zendehdel, A. Mobinikhaledi and F.H. Jamshidi. 2007. Conversion of acids to benzimidazoles with transition metal/zeolites. J. Incl. Phenom Macrocycl. Chem. 59: 41–44.

(152) R.R. Nagawade and D.B. Shinde. 2006. Zirconyl(IV) chloride-promoted synthesis of benzimidazole derivatives. Russ. J. Org. Chem. 42: 453–454.

(153) V.D. Patil, G. Medha and S. Mhatre. 2010. A mild and efficient synthesis of benzimidazole by using zinc acetate under solvent-free condition. Der Chemica Sinica 1: 125–129.

(154) S. Rostamizadeh, A.M. Amani, R. Aryan, H.R. Ghaieni and L. Norouzi. 2009. Very fast and efficient synthesis of some novel substituted 2-arylbenzimidazoles in water using ZrOCl$_2$ nH$_2$O on montmorillonite K10 as catalyst. Monatsh Chem. 140: 547–552.

(155) Z.H. Zhang, L. Yin and Y.M. Wang. 2007. An expeditious synthesis of benzimidazole derivatives catalyzed by Lewis acids. Catal. Commun. 8: 1126–1131.

(156) C.S. Reddy and A. Nagraj. 2008. A mild, efficient and one-pot synthesis of 2-substituted benzimidazoles by ZrOCl$_2$.8H$_2$O-catalyzed ring-closure reaction. Indian J. Chem. 47B: 1154–1159.

(157) A. Chawla, G. Kaur and A.K. Sharma. 2012. Green chemistry as a versatile technique for the synthesis of benzimidazole derivatives: review. Int. J. Pharm. Phytopharmacol. Res. 2: 148–159.

Five-Membered Fused
N,N-Heterocycles

7.1 Introduction

Heterocycles are an important class of compounds, accounting for over half of all known organic compounds [1a–e]. They are present in a wide variety of biologically active compounds, including anti-biotic, anti-tumor, anti-depressant, anti-inflammatory, anti-HIV, anti-malarial, anti-bacterial, anti-microbial, anti-viral, anti-fungal, anti-diabetic, fungicidal, herbicidal and insecticidal agents [2a–b]. They have been generally found as a key structural unit in synthetic agrochemicals and pharmaceuticals. Some of these compounds exhibit noteworthy photochromic, solvatochromic and biochemiluminescence properties [3a–b]. Most heterocycles have important applications in material science, such as in fluorescent sensors, dyestuff, information storage, brightening agents, plastics and analytical reagents [4a–b]. Moreover, they are also used in supramolecular and polymer chemistry, especially in conjugated polymers. Heterocycles also act as semiconductors, organic conductors, photovoltaic cells, molecular wires, organic light-emitting diodes, optical data carriers, light harvesting systems, chemically controllable switches, and liquid crystalline compounds [5a–b]. They are also of considerable interest because of their synthetic utility as protecting groups, synthetic intermediates, organic catalysts, chiral auxiliaries, and synthetic utility as metal ligands in asymmetric catalysts inorganic synthesis. Therefore, extensive attention has been paid by researchers to develop efficient new methods to synthesize heterocycles [6a–b].

Almost all the compounds we know of as drugs and vitamins, and many other natural products, are heterocycles. Among heterocycles, nitrogen-containing heterocyclic compounds have constantly intrigued researchers over decades of historical development of organic synthesis [7]. Nitrogen-containing heterocycles have been used as medicinal compounds for centuries and form the base of many common drugs such as morphine, captopril and vincristine (cancer chemotherapy). They are present in diverse natural compounds and drugs, and are of great significance in a wide variety of applications. Among drugs containing aromatic five-membered nitrogen heterocycles are cholesterol-reducing atorvastatin, anti-inflammatory celecoxib, anti-ulcerative cimetidine, anti-fungal fluconazole, and anti-hypertensive losartan.

Imidazole is found in the theophylline molecule found in tea leaves and coffee beans, which stimulates the central nervous system. Apart from its use for pharmaceutical purposes, it also has applications in industries. For instance, imidazole is extensively used as a corrosion inhibitor on certain transition metals like copper. The thermostable polybenzimidazole (PBI) contains imidazole fused to a benzene ring acts as a fire retardant. Imidazole can also be found in various compounds which are used for photography and electronics. Its derivatives have a wide range of pharmacological applications. Imidazole and its derivatives have been reported to have anti-inflammatory, analgesic, anti-neoplastic, cardiovascular activity, enzyme inhibiting, anti-fungal, anti-filarial, anti-helmintic, anti-ulcer and anti-viral properties [8–15].

Benzamidazoles, another category of heterocycles, are also well known for their pharmacological properties. In particular, they are widely used as anti-helmintic agents. It is interesting to note that a number of substituted pyridylsulfinylbenzimidazole molecules like omeprazole possess gastric anti-secretary and consequently anti-ulcerative properties [16].

7.2 Metal- and non-metal-assisted synthesis of five-membered fused heterocycles with two nitrogen atoms

Barium-assisted synthesis

Carpenter et al. [17] reported a reaction of *o*-aryl-isothiocyanate esters and *o*-phenylenediamines for the synthesis of benzimidazo[2,1-*b*]quinazolin-12(5*H*)-ones in 91–98% overall yields, *via* tandem *N,N'*-diisopropylcarbodiimide-assisted benzimidazole cyclization and MW-assisted benzimidazoquinazolinone cyclization with barium hydroxide **(Scheme 1)**. Commonly employed approaches generally produce benzimidazo[2,1-*b*]quinazolin-12(5*H*)-ones in low yields, sometimes along with by-products, and need temperatures as high as 200 °C.

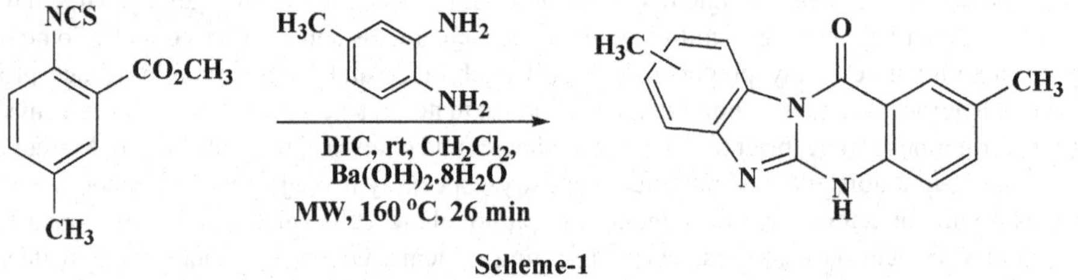

Scheme-1

Bismuth-assisted synthesis

Mohammadpoor-Baltork et al. [18] observed that a number of benzimidazoles and other five-membered heterocyclic compounds fused to one pyridine ring could be synthesized efficiently when *o*-phenylenediamines reacted with *o*-esters. These reactions were catalyzed by Bi(OTf)$_3$·xH$_2$O, Bi(TFA)$_3$, or BiOClO$_4$·xH$_2$O under solvent-free conditions **(Scheme 2)** [19].

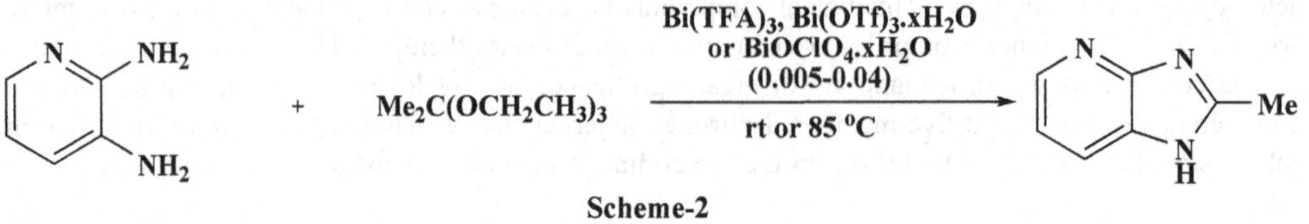

Scheme-2

Copper-assisted synthesis

A mixed copper(I)-copper(II) system can be generated *in situ* by partial reduction of copper sulfate with glucose. Using this, an eco-friendly and efficient multi-component cascade reaction of A^3-coupling of heterocyclic amidine with alkyne and aldehyde *via* 5-*exo-dig* cycloisomerization and prototropic shift provides therapeutically versatile *N*-fused imidazoles **(Scheme 3)** [20].

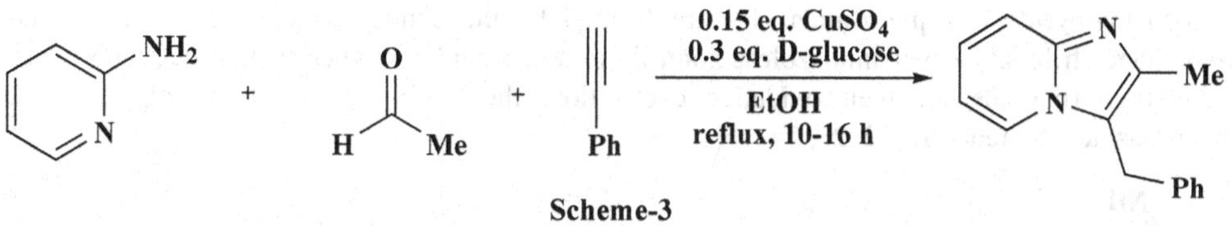

Scheme-3

A straightforward Cu-catalyzed three-component coupling reported for the synthesis of imidazopyridines [21] involved the condensation of 2-amino-5-methylpyridine with an aldehyde, providing pyrazolepyridine. A terminal alkyne was added to the intermediate imine in the presence of CuCl; Cu(II) triflate promoted the Lewis acid-assisted 5-*exo-dig* heteroannulation for the synthesis of bicyclic structure in good yields after chromatography (**Scheme 4**). However, despite being convergent and short, this method has limitations as the entire process requires a glove box arrangement and has thus been reported to take place only on small scales [22].

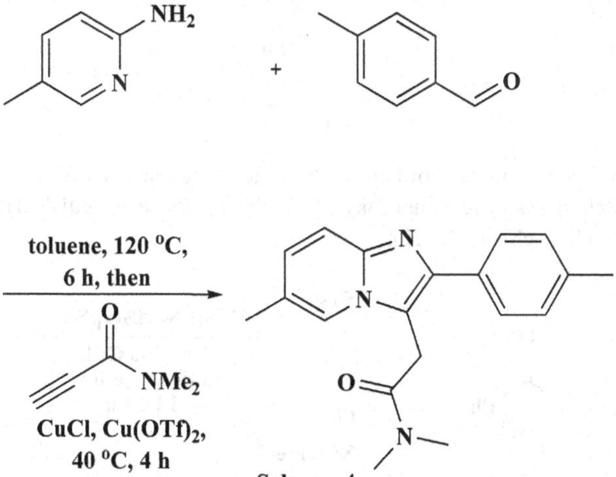

Scheme-4

Another efficient approach developed for the generation of imidazoquinoline, imidazopyridine and imidazoisoquinoline cores involved Cu-catalyzed three-component coupling of heteroaryl-, aryl- and alkyl aldehydes with terminal alkynes and 2-aminoheterocycles. The synthetic utility of this protocol was evident from the highly efficient one-pot synthesis of drugs zolpidem and alpidem (**Scheme 5**) [22].

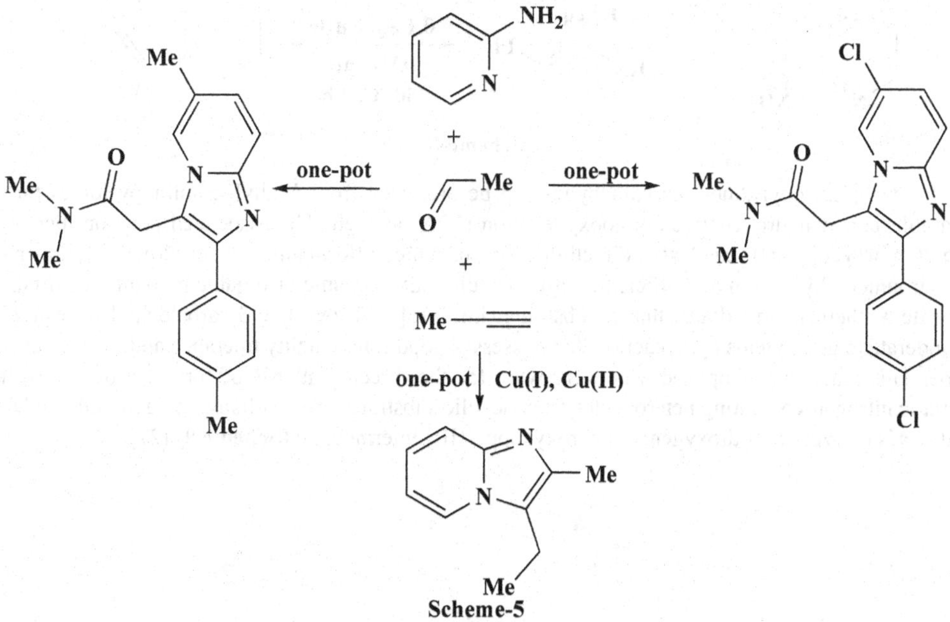

Scheme-5

3-Methylpyrido[1,2-*a*]benzimidazoles are formed by the condensation of 2-(arylhydrazono)-3-iminobutanenitrile and 2-benzimidazoleacetonitrile in acetic acid and when treated with cupric acetate in dimethylformamide for their oxidative cyclization, they afford *S*-triazolo[4,5-*b*]pyrido[1',2'-*a*] benzimidazoles (**Scheme 6**) [22–23, 24a–b].

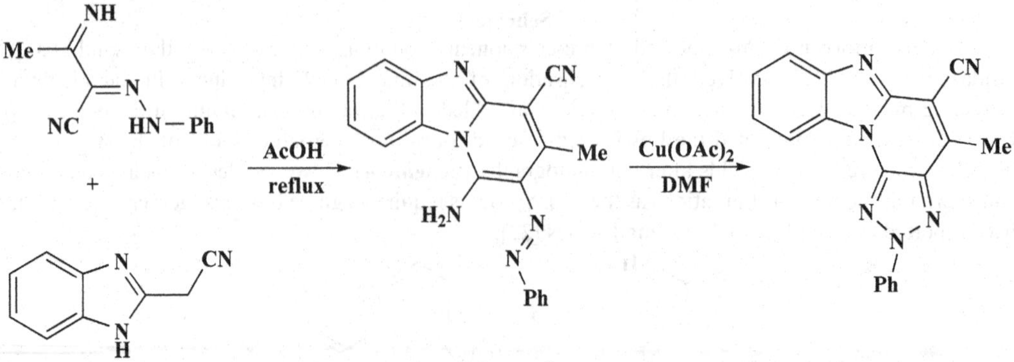

Scheme-6

Imidazo[1,2-*a*]pyridines can be formed in high to excellent yields by a one-pot reaction of 2-aminopyridines, aldehydes and terminal alkynes in the presence of CuI-NaHSO$_4$.SiO$_2$ combination catalyst in refluxing PhMe (**Scheme 7**) [25].

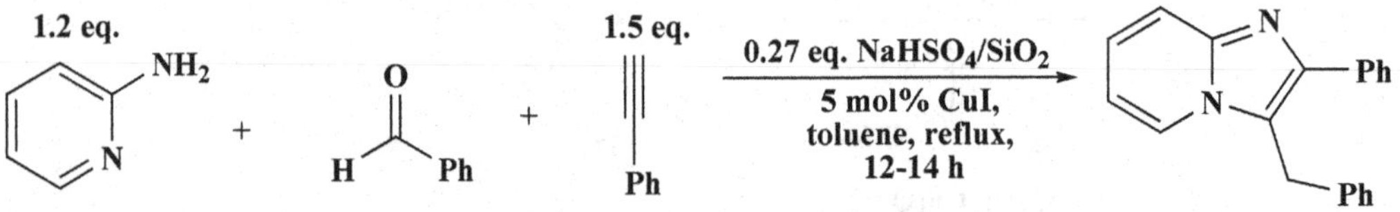

Scheme-7

Imidazo[1,2-*a*]pyridines can also be synthesized by a Cu-catalyzed one-pot reaction of aminopyridines and nitroolefins, using air as an oxidant. This reaction has been found to be very suitable for the preparation of several imidazo[1,2-*a*]pyridines (**Scheme 8**) [26].

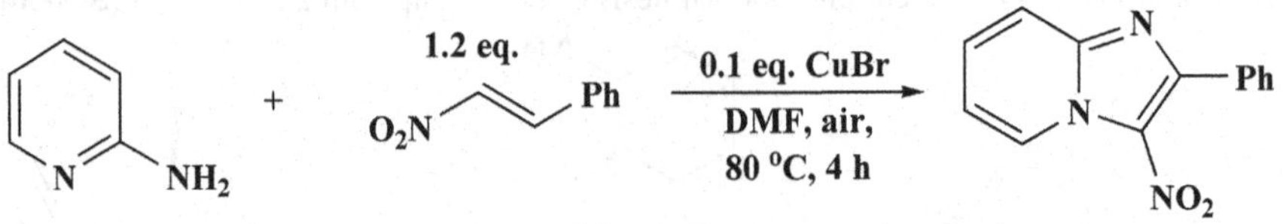

Scheme-8

Imidazo[1,2-*a*]pyridine-3-carbaldehydes can be produced from *N*-allyl–2-aminopyridines through intramolecular dehydrogenative aminooxygenation (IDA) as well. The associated reaction, performed with 20 mol% copper(II) catalyst in dimethylaniline or dimethylformamide under dioxygen atmosphere, is environmentally benign and efficient, with no additional inorganic or organic oxidants required. The versatile synthetic intermediates, that is, substituted imidazo[1,2-*a*]pyridine-3-carbaldehydes are prepared in moderate to good yields by a reaction that possesses good functionality tolerance and a wide substrate scope. This reaction has opened a new pathway for the direct synthesis of formyl group-substituted aromatic nitrogen-containing heterocycles from acyclic substrates. The carbonyl oxygen in the aldehyde products is derived from dioxygen, *via* peroxycopper(III) intermediate (**Scheme 9**) [27].

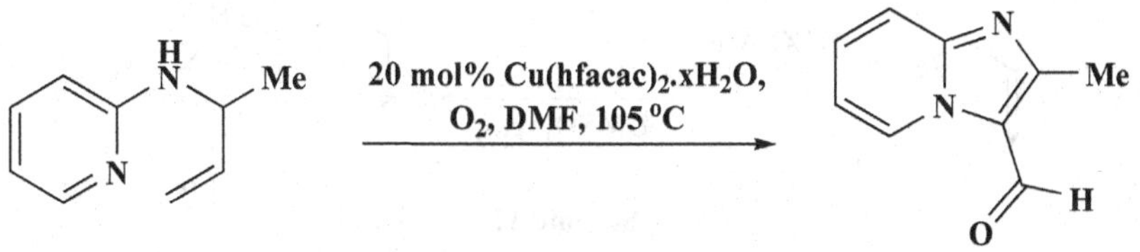

Scheme-9

Pyrido[3',2':5,6]pyrimido[1,2-*a*]benzimidazol-5(6*H*)-one (X = O) is synthesized by copper-catalyzed cyclocondensation of 2-bromobenzoic acid and 2-aminobenzimidazole in dimethylformamide and refluxing potassium carbonate. Upon treatment with sulfur and phosphorus in refluxing pyridine, it yields the 5-thione analog—pyrido[3',2':5,6]pyrimido[1,2-*a*]benzimidazol-5(6*H*)-thione. Consequently, the fused hexaazapentacyclic system is produced by treating this thione with hydrazine in refluxing ethanol followed by nitrosation (**Scheme 10**) [24, 28].

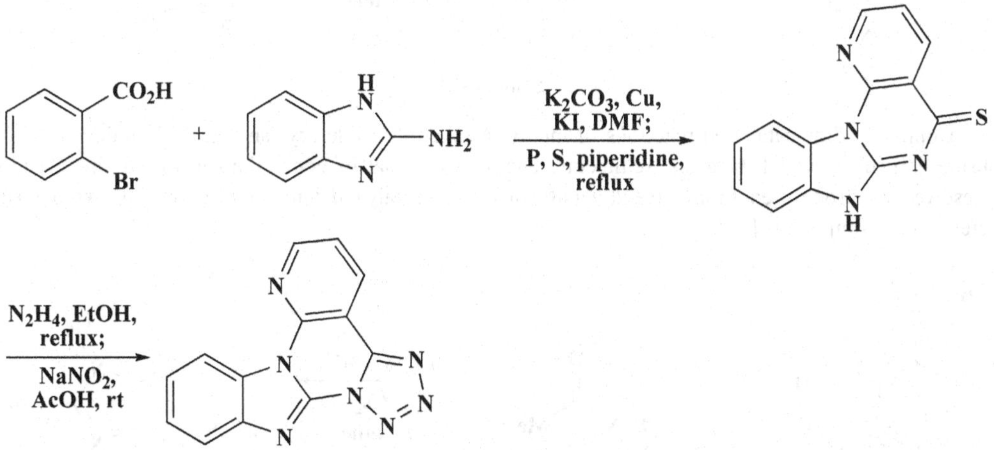

Scheme-10

O-Iodide functionalized *N*-acyl aromatic amines undergo different *N*-amination-based tandem reactions to form different heterocyclic compounds in the presence of L-proline, cuprous iodide and a base. Zhou et al. [29] reported a one-pot synthetic protocol for the synthesis of benzimidazoles, wherein substitution variation in both 1- and 2-positions of benzimidazoles was allowed. The *ortho*-substituent effect of the NHCOR group was observed to be an important factor for the transformation (**Scheme 11**) [30].

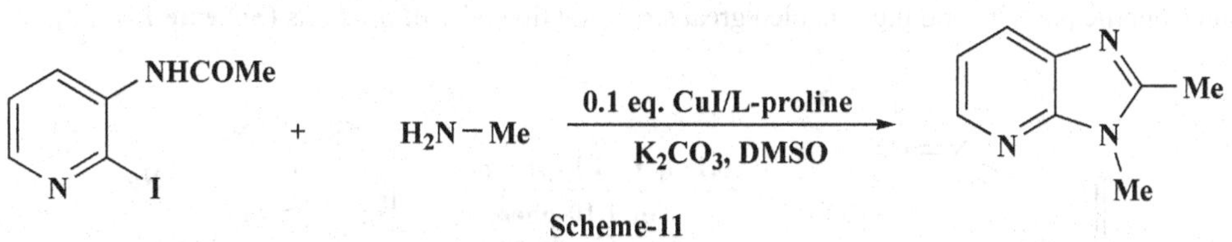

Scheme-11

Amination and subsequent intramolecular amidation of carbamate fragment yields heterocyclic products. Both iodide- and bromide-substituted substrates are good reactants, while a slight change in the reaction conditions like the base and the ligand are necessary to guarantee satisfactory yields for different aryl halides (**Scheme 12**) [30–31].

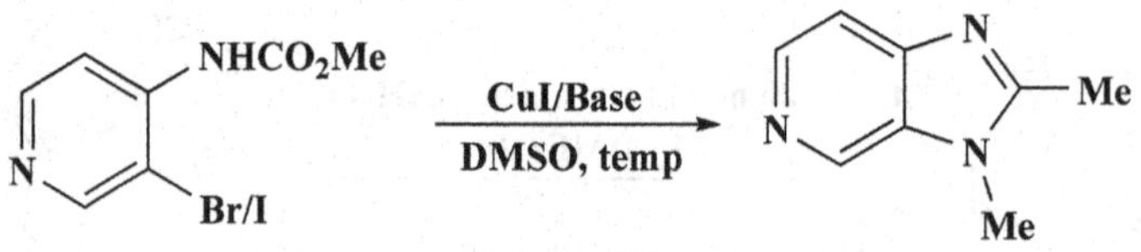

Scheme-12

Zhong and Sun [32] reported a tandem cascade carbon-chloride bond amination and intramolecular carbamate amidation for the synthesis of analogous pyrimidine-fused heterocycles from substrates, using *trans*-4-hydroxyl-L-proline ligand and cuprous chloride catalyst (**Scheme 13**) [30].

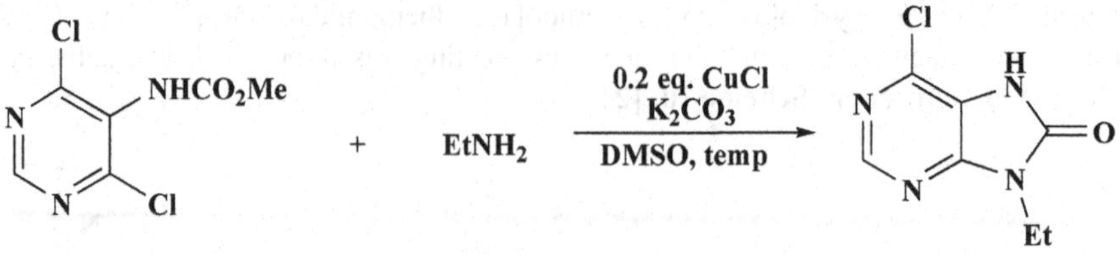

Scheme-13

Cuprous iodide-catalyzed reactions of methyl masked 2-iodide aryl amines with amides afforded purine derivatives which bear an architecture analogous to that of benzimidazoles (**Scheme 14**) [33]. These results strongly exemplify the diversity of copper-catalyzed tandem reactions for synthesizing heterocyclic compounds [30].

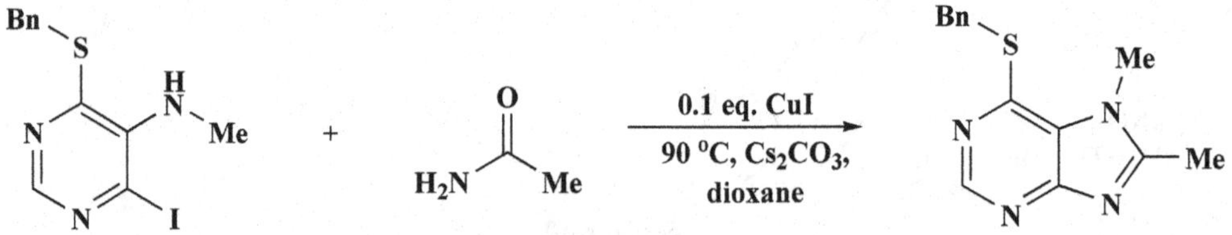

Scheme-14

Lv and Bao [34], during their research on new copper-catalyzed domino reactions, observed that diimine moiety was a versatile building block in designing copper-catalyzed cascade reactions. 1,2-Disubstituted benzimidazoles could be synthesized through subsequent intramolecular carbon-nitrogen coupling of imidamide intermediate through the attack of a nucleophile on the electron-deficient carbon in diimide. The advantage of this method was that imidazoles, amines and phenols were all used as nucleophilic partners and thus enabled great structural diversity of products (**Scheme 15**) [30].

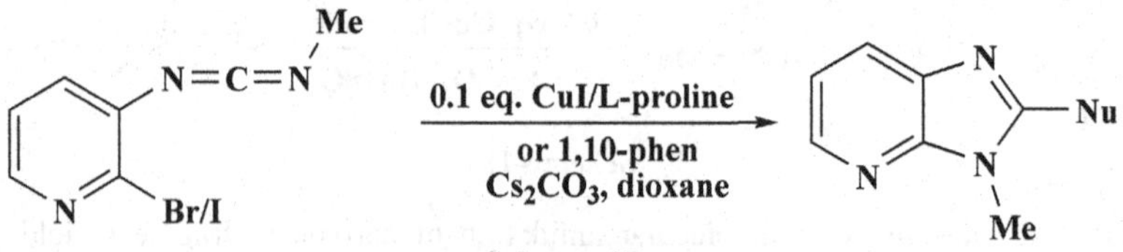

Scheme-15

The cascade synthesis of analogous heterocyclic compounds is also feasible when the positions of the nucleophile and the diimine species are exchanged. For instance, the reactions of *o*-haloaniline and *N,N*-disubstituted dimines produce 2,3-dihydro-1*H*-[*d*]imidazoles and imidazoles (**Scheme 16**) [30, 35].

Similar tandem reactions using amidines or guanidines as the coupling partner of *o*-dihaloarenes result in the formation of several benzimidazoles. Deng et al. [36–37] studied copper-catalyzed tandem reactions of amidines/guanidines and dihaloarenes; the results provided a novel pathway for the

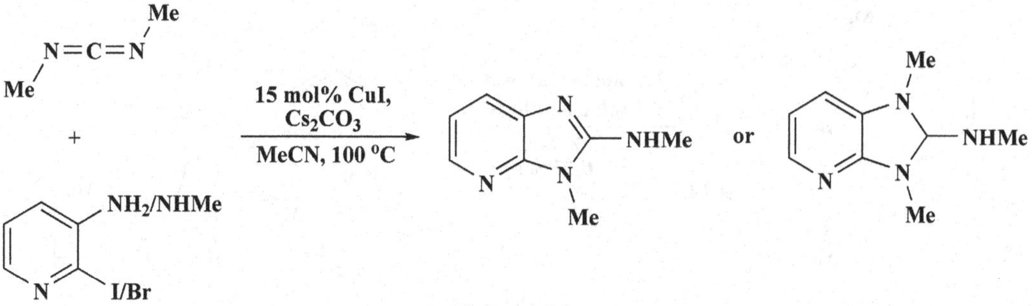

Scheme-16

preparation of benzimidazoles which were formed regioselectively when substituted *o*-dihaloarenes were used. As shown in **Scheme 17**, the regioselectivity in the synthesis of benzimidazoles was determined by the relative reactivity of haloatoms Br and Cl as well as the relative steric hindrance on the nucleophilic *N*-atoms in amidines or guanidines. When the substituents were the same as benzimidazoles, no evident selectivity was observed and both the isomers were formed in similar yields [30].

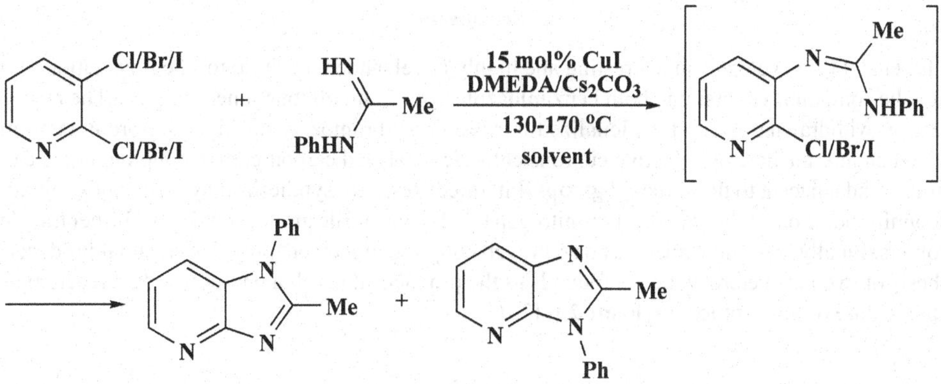

Scheme-17

Researchers working at Johnson & Johnson reported a protocol for the synthesis of benzimidazoles from amidines and 1,2-dihaloarenes **(Scheme 18)** [37–38].

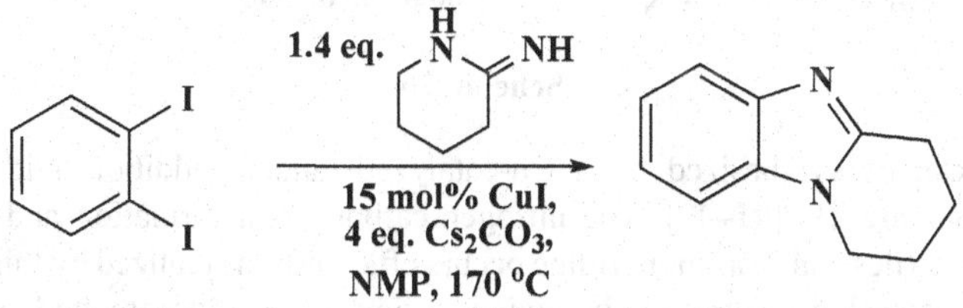

Scheme-18

2-Aminoindoles are structural components of various alkaloids. The synthesis of 2-aminoindoles is difficult by traditional methods like Fischer indole synthesis [30]. Yuen et al. [39] reported tandem coupling reactions of chiral *o-gem*-dibromovinylanilines for a one-pot synthesis of imidazoindolones. The tandem intramolecular coupling reaction occurred at 120 °C with cuprous iodide/racemic *trans*-1,2-cyclohexyldiamine/potassium carbonate as the catalyst in PhMe. The final products were possibly furnished through the formation of coupled indole intermediates. One of the rationalizations accounting for the epimerization was that the intermediates were not stable in terms of their chiral properties during the reaction process **(Scheme 19)**. With respect to the stereochemistry of the chiral center, good enantiomeric excess values were obtained in some entries by retaining the configuration of the starting materials, while epimerization occurred to a different extent in certain other reactions.

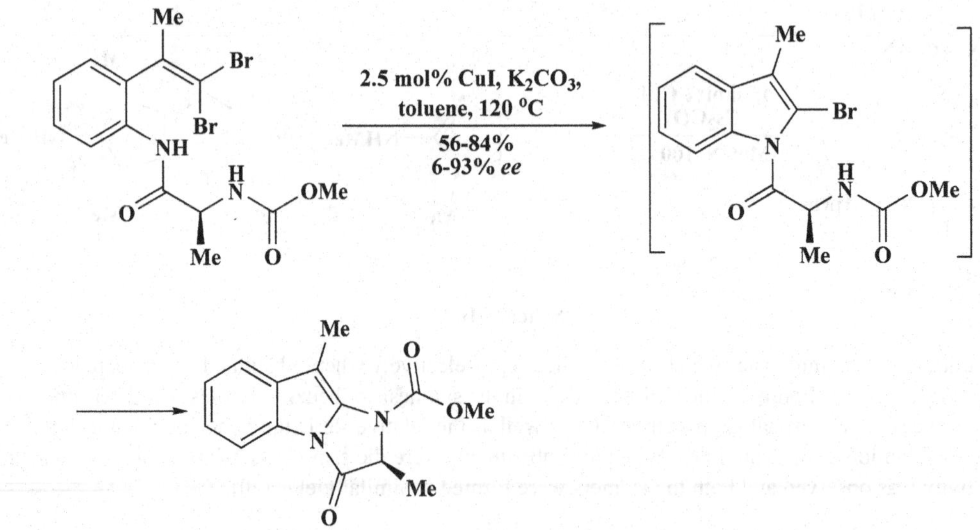

Scheme-19

Lu et al. [40] developed an interesting and highly novel tandem reaction for the preparation of fused tetracyclic compounds containing both benzimidazole and isoquinolinone functionalities. The reaction of β-electron-withdrawing group functionalized nitriles and N-bromoaryl-substituted 2-bromobenzamides provided an intermediate by selective cuprous chloride-catalyzed carbon-carbon coupling at the carbon-bromide bond adjacent to the carbonyl groups. Intermediates were synthesized by cyclization initiated by nucleophilic addition on the triple carbon-nitrogen bond of the nitrile group, which was further transferred to isomers. Finally, intramolecular carbon-nitrogen coupling of isomers provided the final products. The synthesis of two new heterocyclic compounds as the products, through a one-pot tandem reaction, was a notable feature of this protocol (**Scheme 20**) [30].

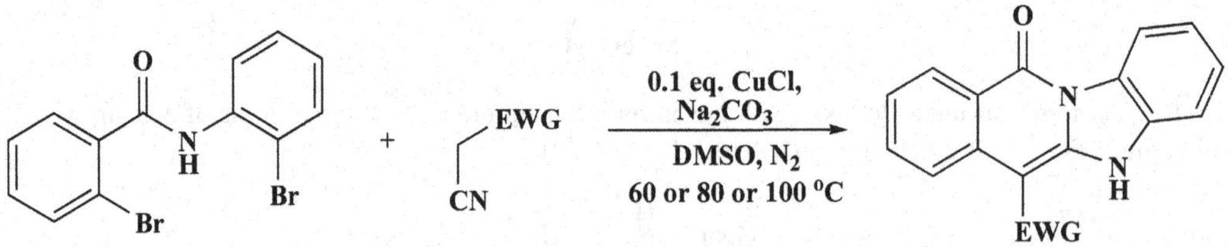

Scheme-20

1,2,4-Trizoles can be synthesized by a Cu-catalyzed tandem addition-oxidative cyclization of activated nitriles (**Scheme 21**) [41–42]. The nitrogen-carbon bond formation and oxidative nitrogen-nitrogen coupling of nitriles and 2-aminopyridine occurs efficiently, as realized by this reaction. Amidines and benzonitriles (possessing electron-withdrawing groups like trifluoromethyl or halogen) produce triazolopyridines in good yields. Compared to reactions with electron-deficient benzonitriles, the reactions with methoxy-substituted benzonitriles are slower and furnish moderate yields of triazolopyridines. This catalytic oxidative reaction occurs through the formation of amidine intermediate using molecular oxygen as the oxidant [43].

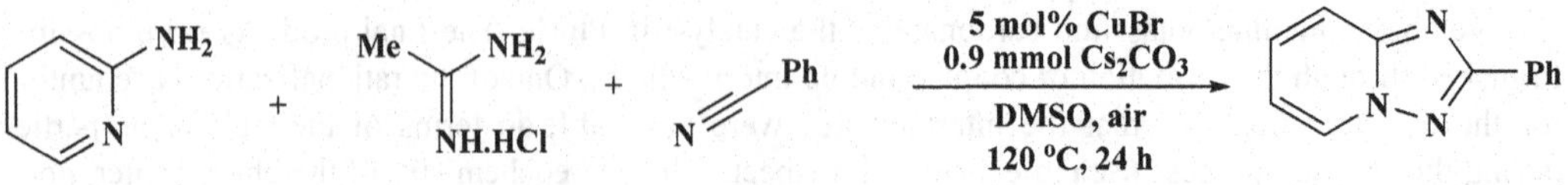

Scheme-21

Wang et al. [44] reported that this reaction occurred either through intramolecular hydroamination followed by dehydrogenative aromatization, or by direct amination of the vinyl carbon-hydrogen bond in *N*-(1-phenylallyl)-2-aminopyridine and subsequent double bond migration. Imidazo[1,2-*a*] pyridine-3-carbaldehydes (which are versatile synthetic intermediates) were synthesized instead of the desired products by using 20 mol% cuprous iodide(I) as the catalyst under a dioxygen atmosphere in dimethylformamide **(Scheme 22).** Both electron-withdrawing and electron-donating groups were tolerated in this reaction. A concise pathway for the synthesis of necopidem (anxiolytic drug) was achieved by this method. Mechanistic studies have directly shown that the carbonyl oxygen in the aldehyde products was derived from dioxygen by a reaction *via* peroxy-copper(III) intermediate [43].

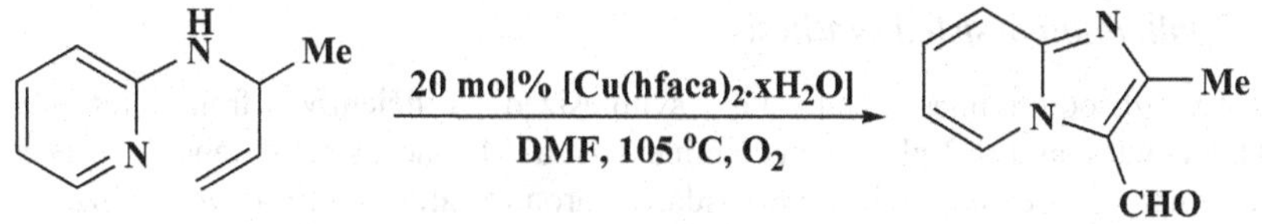
H
N
Me
N
20 mol% [Cu(hfaca)2.xH2O]
DMF, 105 °C, O2
N
N
Me
CHO

Scheme-22

In 2006, Martin et al. [45] reported a domino copper-catalyzed amidation/hydroamidation sequence of haloenynes for the synthesis of pyrroles. Pyrazoles were also synthesized by this protocol by using bis-Boc hydrazine as the nucleophile **(Scheme 23).** Preliminary mechanistic studies indicated that the hydroamidation step was both base-assisted and copper-catalyzed. Subsequently, Ackermann et al. [46] in 2009 reported an allied approach for the preparation of indoles.

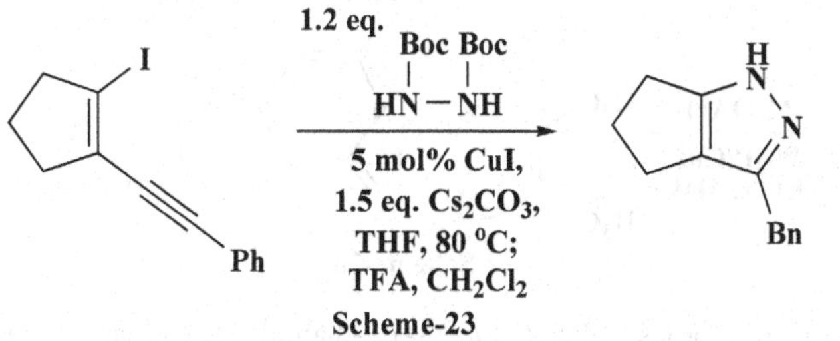
I
Ph
1.2 eq.
Boc Boc
HN—NH
5 mol% CuI,
1.5 eq. Cs2CO3,
THF, 80 °C;
TFA, CH2Cl2
H
N
N
Bn

Scheme-23

Pyrazole undergoes intramolecular copper-catalyzed arylation for the synthesis of novel tricyclic full agonists for the G-protein-coupled niacin receptor 109A which is helpful in the treatment of hyperlipidaemia **(Scheme 24)** [38, 47].

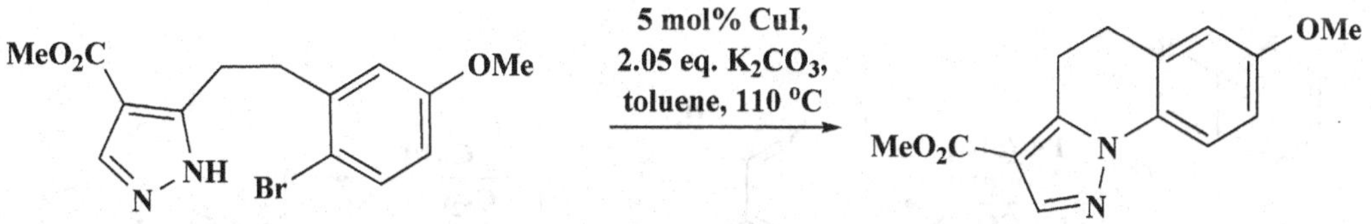
MeO2C
OMe
NH
N
Br
5 mol% CuI,
2.05 eq. K2CO3,
toluene, 110 °C
OMe
MeO2C
N
N

Scheme-24

Pabba et al. [48] synthesized indazoles through a two-step, one-pot condensation-arylation sequence. The aryl hydrazines were assembled with 2-haloacetophenones under copper(I) catalysis to provide the target molecules in high yields after two short MWI periods **(Scheme 25)** [49–50].

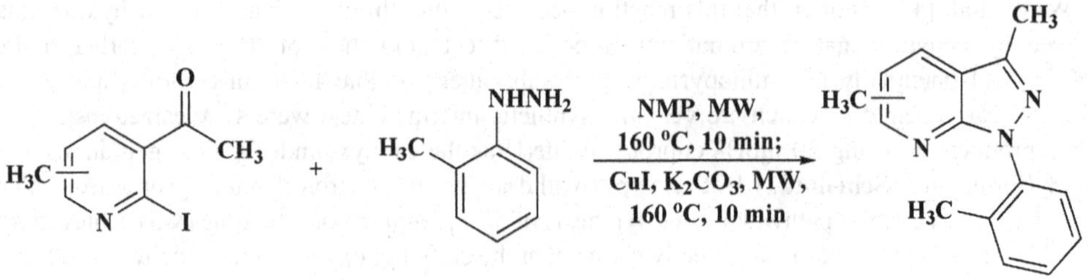

Scheme-25

Copper- and palladium-assisted synthesis

Benzimidazo[2,1-*a*]isoquinolines can be synthesized efficiently from terminal alkynes, 2-bromoarylaldehydes and 1,2-phenylenediamines by a MW-accelerated tandem process in which Sonogashira coupling, 5-*endo*-cyclization, oxidative aromatization and 6-*endo*-cyclization occur in a single synthetic operation (**Scheme 26**) [24, 51–52].

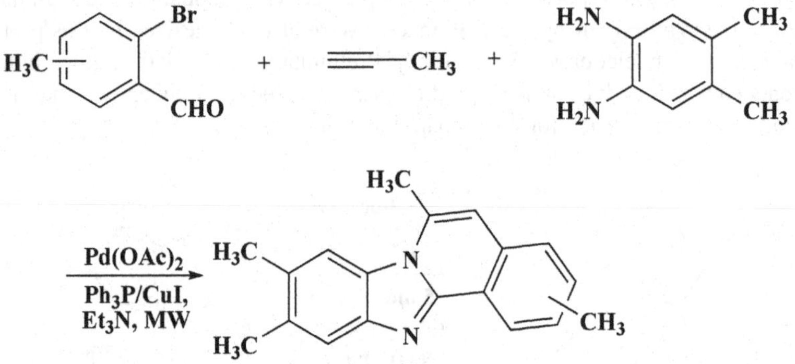

Scheme-26

Reactions based on cuprous iodide/(±)-*trans*-1,2-diaminocycloxehane catalytic system result in almost complete conversion after 2 hours to a target closed ring product which is suitable for the one-pot variant of the synthesis, without isolation and purification of the intermediate [53]. Intermediate thiazole compound, a model compound for optimization studies of cuprous iodide/L-catalyzed cyclization to a target cyclic molecule, yield [1,3]thiazolo[3',2':1,2]imidazo[4,5-*b*]pyridine as shown in **Scheme 27** [54].

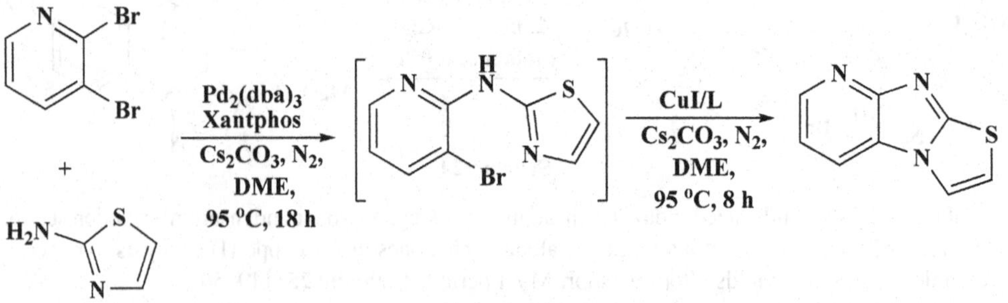

Scheme-27

Various cyclic guanidines are important medicinal targets and exhibit potent biological activity. Cyclic guanidines can be prepared using many of the techniques used for the synthesis of imidazolidin-2-ones. For instance, as depicted in **Scheme 28**, Zhao et al. [55] reported that acyclic guanidine undergoes rhodium-catalyzed carbon-hydrogen amination for synthesizing a cyclic guanidine. In another study, they investigated intra- and intermolecular diaminations to produce cyclic guanidines.

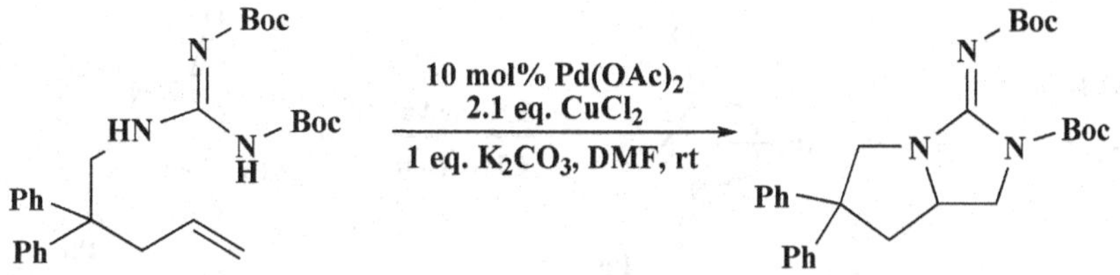

Scheme-28

When the conditions for a one-pot synthesis of [1,3]thiazolo[3′,2′:1,2]imidazo[4,5-*b*]pyridine are applied for preparing 9-thia-[1,4*b*,10]-triaza-indene[1,2-*a*]indene, the molecular peak detected for the ring-closure product 9-thia-[1,4*b*,10]-triaza-indene[1,2-*a*]indene after 2 hours of cuprous iodide/L-catalyzed second step is of a very low intensity. Even after longer reaction times, lower conversion rates are obtained **(Scheme 29)** [56].

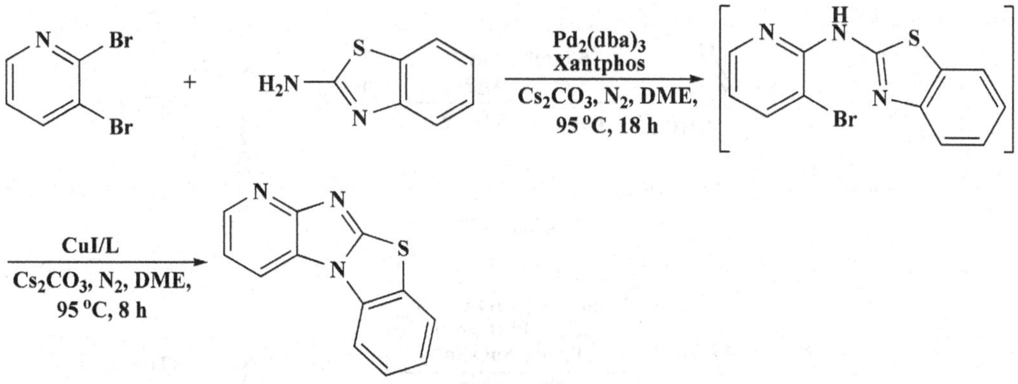

Scheme-29

Gold-assisted synthesis

Zhang et al. [57] reported a gold-catalyzed intramolecular [3+2]-annulation protocol for the synthesis of tricyclic indolines. The reaction of urea with gold catalyst and selectfluor oxidant promoted the oxidative cross-coupling of the aryl carbon-hydrogen and an alkyl gold complex formed *in situ* (derived from aminoauration of the allyl group) to provide the cyclic urea products in good yields **(Scheme 30)**.

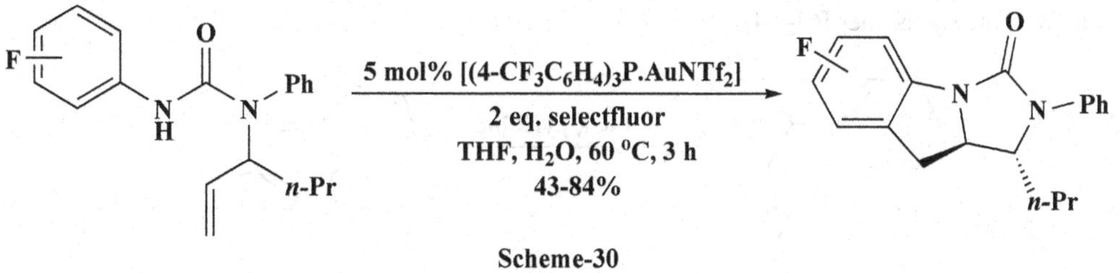

Scheme-30

Shapiro et al. [58] reported a novel [3+3]-annulation of azomethine imines with propargyl esters in the presence of Au(III) catalyst. The *β*-position of pyrazolidinone was substituted to provide the bicyclic product with high *cis* selectivity, which was determined during ring-closing rather than the synthesis of allyl-gold intermediate **(Scheme 31)** [59].

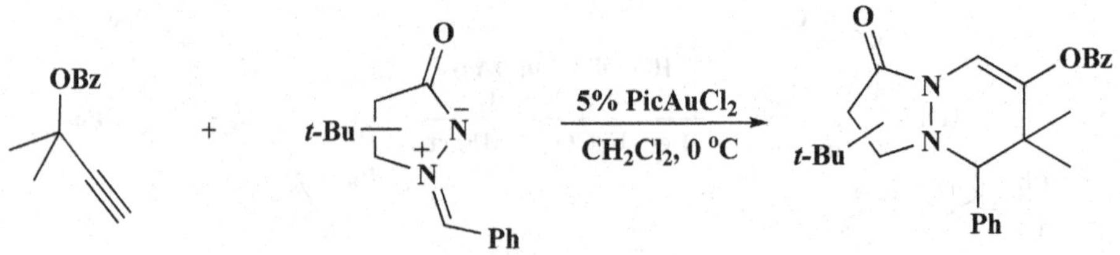

Scheme-31

A green and fast pathway to synthesize indole-1-carboxamides from *N'*-substituted *N*-(2-alkynylphenyl)ureas using AuPPh₃Cl/silver carbonate-catalyzed 5-*endo-dig* cyclization under MWI in water has also been developed by researchers [60]. Various functional groups such as *N'*-alkyl, aryl, heterocyclic, *N*-(2-ethynylpyridin-3-yl)ureas and *N*-substituted 2-ethynylphenyl can be tolerated and afford moderate to high yields of the desired products (**Scheme 32–33**) [59–60].

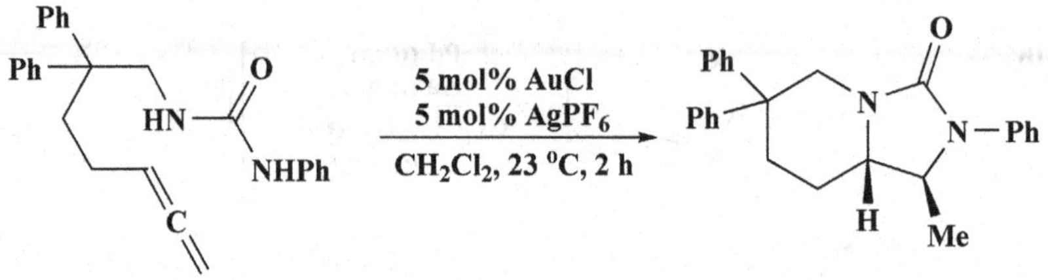

Scheme-32

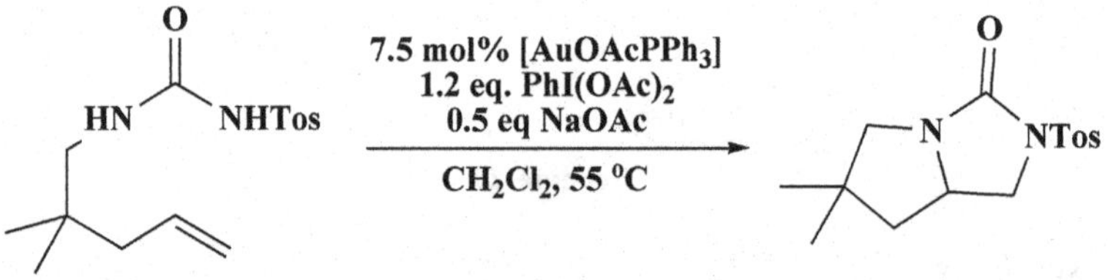

Scheme-33

Indium-assisted synthesis

Nitrogen-sulfur or nitrogen-nitrogen bond formation at the nitrogen atom of oxime has found ample mention in literature. For instance, a cyclic compound is formed from the *anti*-isomer of the oxime when an oxime bearing pyridyl group reacts with pyridine and tosyl chloride (**Scheme 34**), while no cyclization occurs with the *syn*-isomer [61–64].

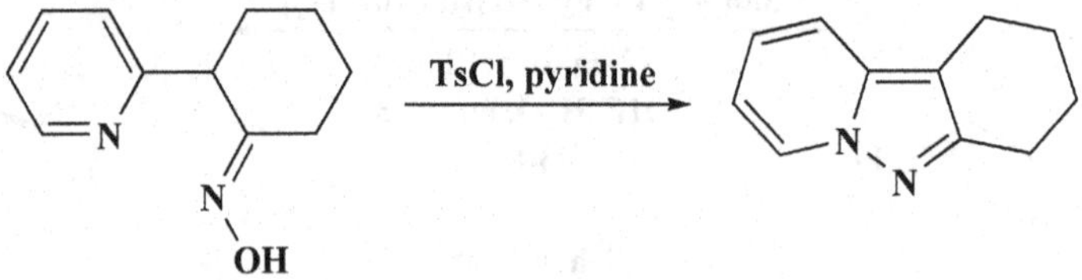

Scheme-34

Iodine-assisted synthesis

Quiclet-Sire and Zard [65] synthesized fused 4,5-dihydro-3*H*-pyrazoles in good to excellent yields through the reaction of hydrazones with iodine (**Scheme 35**). The diazo intermediate was formed by the iodination of hydrazones followed by expulsion of hydrogen iodide. The dihydropyrazoles were prepared by intramolecular cycloaddition of this diazo intermediate [66].

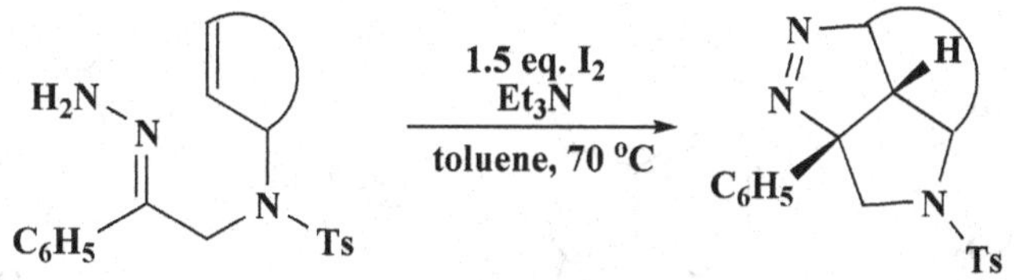

Scheme-35

The reaction of hydrazones with I$_2$ *via* intramolecular dipolar cycloaddition produces corresponding heterocycles **(Scheme 36)** [66].

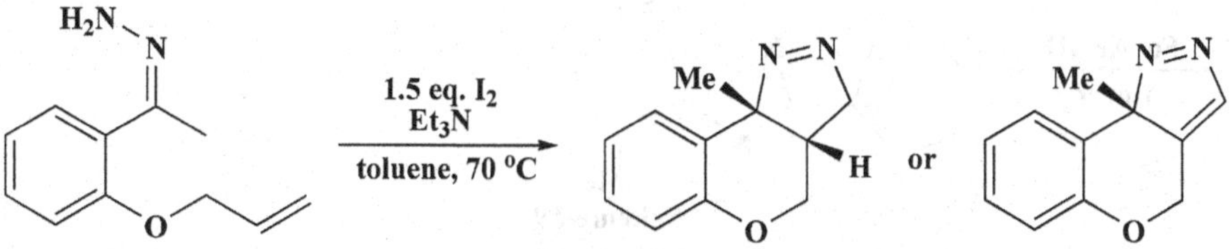

Scheme-36

Shibahara et al. [67] discovered that *N*-2-pyridylmethyl thioamides undergo an I$_2$-mediated oxidative desulfurization [68–69] to afford various 2-azaindolizines **(Scheme 37)**. An intermediate was obtained by the de-protonation of *N*-2-pyridylmethyl thioamides with pyridine, followed by double iodination at sulfur. 2-Aza-indolizines were formed upon intramolecular substitution of this intermediate with the pyridine nitrogen, and subsequent aromatization of the resultant intermediate [66].

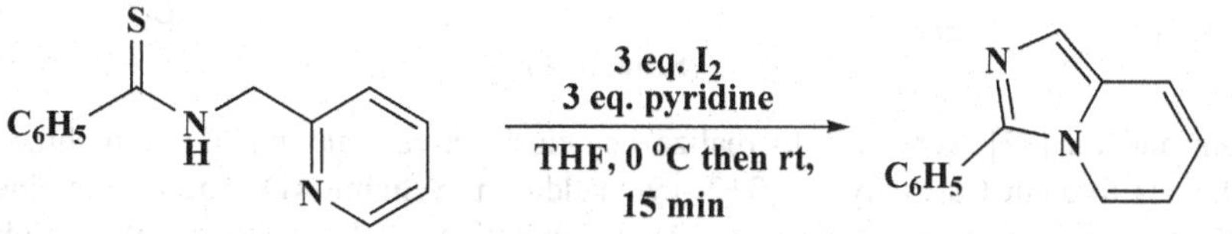

Scheme-37

Iron-assisted synthesis

Wang et al. [27] studied the direct intramolecular aromatic carbon-hydrogen amination of *N*-aryl-2-aminopyridines for the preparation of pyrido[1,2-*a*]benzimidazoles. This reaction was co-catalyzed by copper acetate and Fe(NO$_3$)$_3$.9H$_2$O under a dioxygen atmosphere in dimethylformamide. A variety of pyrido[1,2-*a*]benzimidazoles with various substitution patterns were prepared in moderate to excellent yields through this reaction **(Scheme 38)**. The mechanistic studies proposed that a copper(III)-catalyzed electrophilic aromatic substitution pathway was operative in this reaction. Fe(III) played a unique role and facilitated the synthesis of more electrophilic copper(III) species. A reversible and much less efficient copper(II)-assisted electrophilic aromatic substitution occurred in the absence of Fe(III) [43].

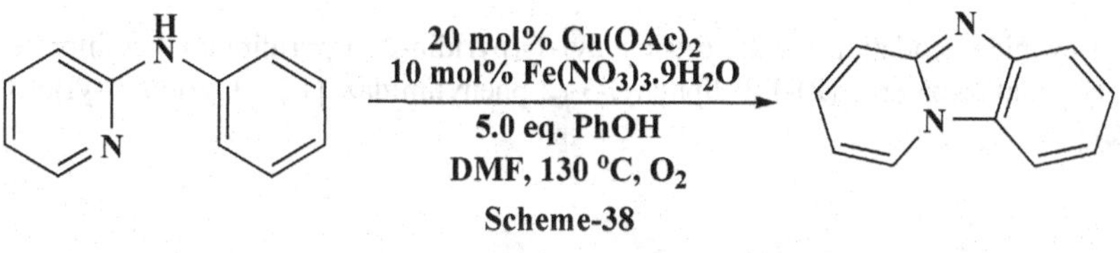

Scheme-38

2-Nitro-3-methylaniline heated with phthalic anhydride in *n*-amyl alcohol yields *N*-(2-nitro-3-methylphenyl)phthalimide which reacts with iron powder at 100 °C in 50% aqueous acetic acid to produce 6-methyl-11-oxoisoindolo[2,1-*a*]benzimidazole **(Scheme 39)** [24, 70].

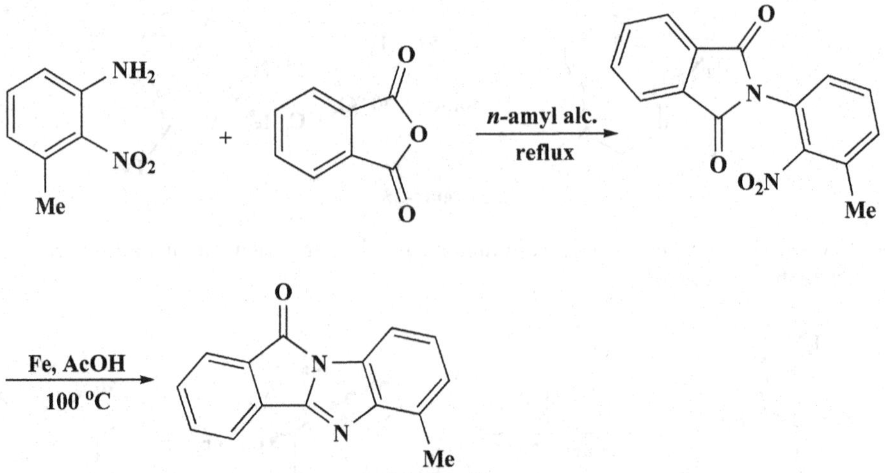

Scheme-39

The ring-expansion of *N*-cyclohexyl-*o*-nitroaniline forms hexahydroazepino-1',2'-1,2-benzimidazole through deoxygenation (**Scheme 40**) [71–72].

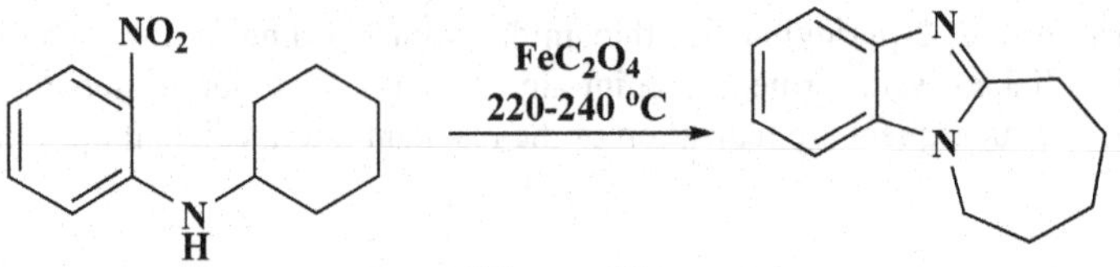

Scheme-40

Shintani and Fu [73] were able to realize enantioselective coupling of azomethine imines and terminal alkynes through Cu-catalyzed [3+2]-cycloaddition (**Scheme 41**). The azomethine imine and terminal alkynes underwent asymmetric [3+2]-cycloaddition in the presence of phosphaferrocene-oxazoline ligand and cuprous iodide catalyst to provide excellent yields of heterocyclic compounds with high stereoselectivity. Electron-deficient alkynes as well as un-activated alkynes were used as substrates in this method [74].

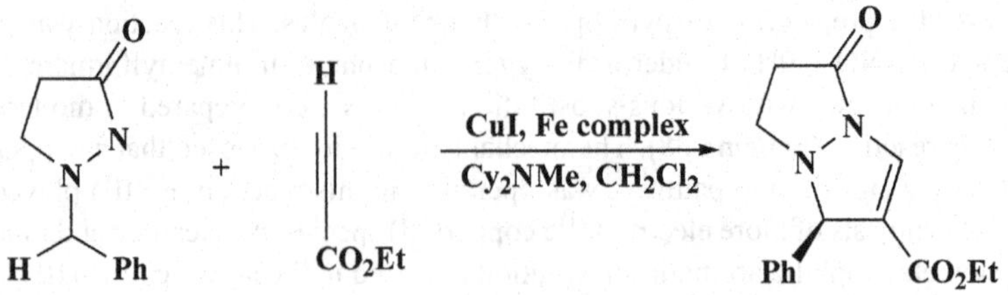

Scheme-41

The ferricyanide oxidation of 2,4,6-triphenyl-1-(pyridin-2-yl)pyridinium perchlorate is a facile method of synthesizing (*Z*)-1,3-diphenyl-3-(2-phenylimidazo[1,2-*a*]pyridin-3-yl)prop-2-en-1-on (**Scheme 42**) [75].

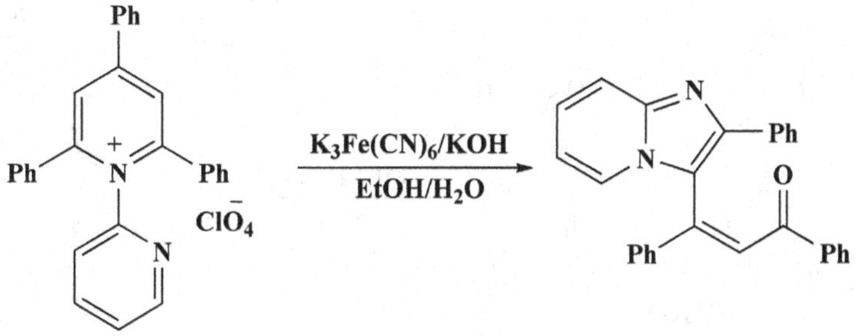

Scheme-42

5*H*-Benzimidazo[1,2-*d*][1,4]benzodiazepin-6(7*H*)-ones can be prepared from easily available 2-(2-aminophenyl)-1*H*-benzo[*d*]imidazoles and 2-bromoacetyl bromide under MWI. Benzimidazoles are formed in high yields by equimolar reaction of substituted *o*-phenylenediamines and 2-nitrobenzaldehyde in ethanol. The nitro group of nitro derivatives can be reduced efficiently using iron powder in a mixture of ethanol, concentrated hydrogen chloride and water (1:0.25:1) to produce 2-(1*H*-benzimidazol-2-yl)aniline derivatives in good yields. The condensation reaction of 2-bromoacetyl bromide and 2-(1*H*-benzimidazol-2-yl)aniline derivatives is performed in Na_2CO_3 and anhydrous tetrahydrofuran under MWI (300 W) to synthesize 5*H*-benzimidazo[1,2-*d*][1,4]benzodiazepin-6(7*H*)-ones (**Scheme 43**) [76].

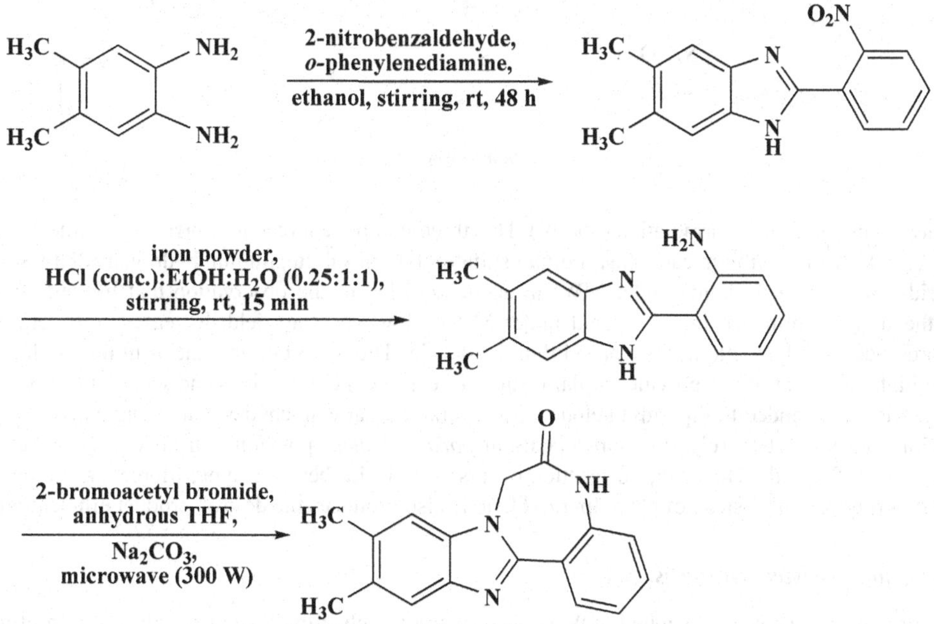

Scheme-43

Lithium-assisted synthesis

An efficient three-component one-pot condensation using various dipyridilketone and aromatic aldehydes with ammonium acetate and lithium chloride (as a mild Lewis acid catalyst) yields imidazo[1,5-*a*]pyridines under MWI (**Scheme 44–45**). Herein, dipyridylketone, *p*-methoxybenzaldehyde and ammonium acetate are used as model reactants for the synthesis of imidazo[1,5-*a*]pyridines during optimization. Of the different solvents like dimethylformamide, ethanol and acetic acid used in this condensation reaction, in the absence or the presence of lithium chloride, acetic acid has been found to be the best. However, the results of various durations of irradiation show that, in all solvents, the best yield is obtained in 3

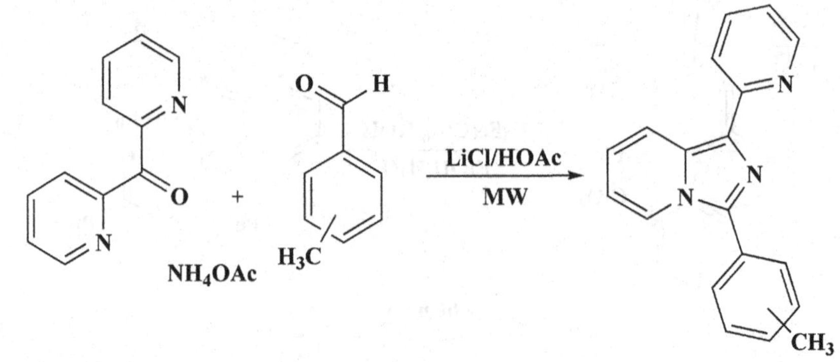

Scheme-44

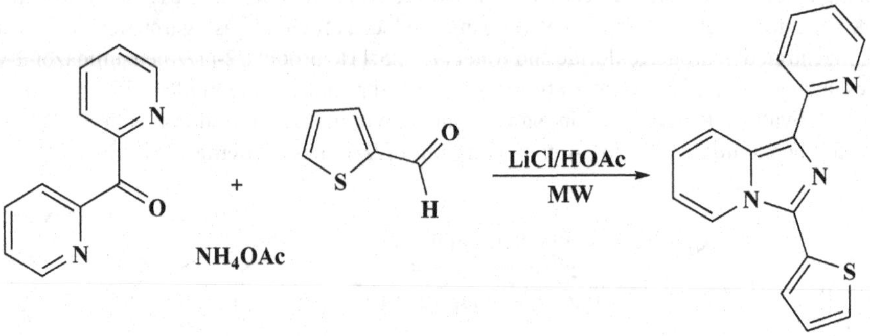

Scheme-45

minutes of medium power irradiation (300 W). This three-component reaction carried out in the presence of different salts like lithium carbonate, lithium sulfate, sodium sulfate, lithium hydrogensulfate, sodium chloride and sodium bromide under MWI in acetic acid led to the observation that lithium chloride has the best performance in acetic acid under MWI. However, the yield decreases considerably in the presence of only acetic acid or only lithium chloride. The optimum amount of lithium chloride is 2 eq., higher concentrations provide similar results. To explore the limitations and scope of this reaction, the reaction is extended to dipyridyl ketone with several aromatic aldehydes containing either electron-withdrawing or electron-releasing substituents in *para* and *meta* positions. In all cases, the reaction occurs very efficiently. However, no reaction occurs with 4-nitrobenzaldehyde. Moreover, the reaction with dipyridyl ketone instead of phenylpyridyl ketone also produces the desired product efficiently [77].

Magnesium-assisted synthesis

DiMauro and Kennedy [78] studied MW-assisted one-pot cyclization/Suzuki coupling for an efficient and rapid synthesis of several 2,6-disubstituted 3-amino-imidazopyridines. Diverse compound libraries were synthesized using 2-aminopyridine-5-boronic acid pinacol ester as a versatile and robust building block. The boronate functional group was tolerant towards Lewis acid-catalyzed cyclizations, and subsequent palladium(0)-catalyzed Suzuki coupling reactions occurred effectively with magnesium salts. This reaction highlighted the vast potential of metal-catalyzed MW-assisted multi-component reactions. The 2-aminopyridine-5-boronic acid pinacol ester also afforded intermediate Ugi condensation product which was converted into 3-amino-imidazopyridines by a four-component microwave-assisted coupling in a one-pot reaction **(Scheme 46)** [79].

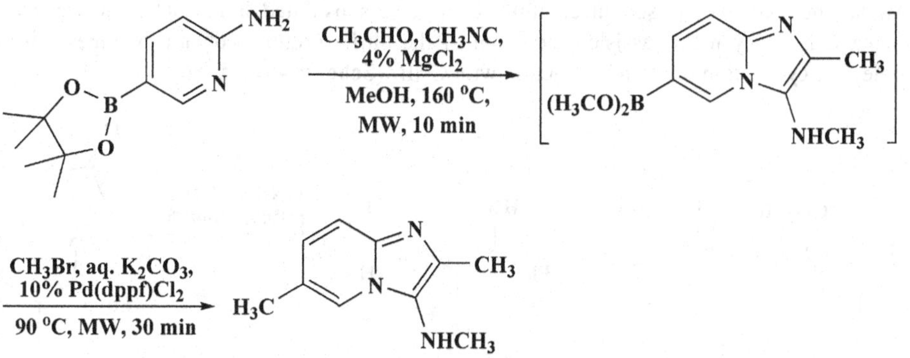

Scheme-46

Mercury-assisted synthesis

More complex polynitrogenated heterocyclic compounds can be prepared by intramolecular aminomercuration-demercuration. As shown in **Scheme 47**, bicyclic guanidine can be generated from an aniline derivative [80] and an intermediate is produced for the synthesis of asmarines [81–82].

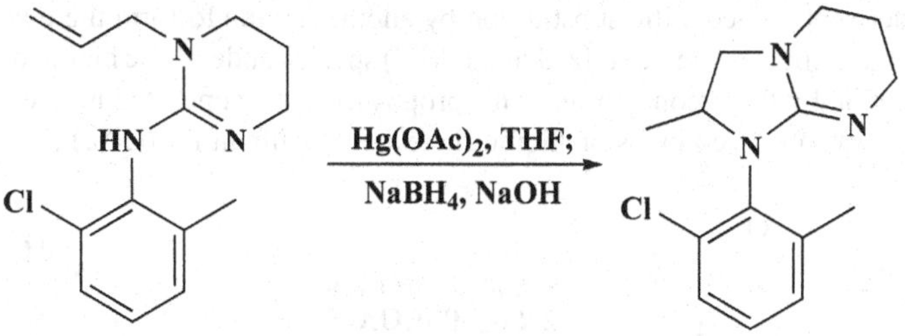

Scheme-47

Nickel-assisted synthesis

Spiro-pyrazolophthalazine-oxindoles can be synthesized in good to excellent yields by a three-component one-pot reaction of several malononitriles, isatins, and phthalhydrazide catalyzed by nickel chloride in PEG 600 (polyethylene glycol 600) **(Scheme 48)** [83]. The product is obtained by a sequence involving Knoevenagel reaction between nitriles and isatins, the aza-Michael addition of phthalhydrazide to the Knoevenagel-adducts, cycloaddition, and subsequent isomerization. Nickel chloride acts as a Lewis acid which activates the nitriles for their transformation into amines [84a–b].

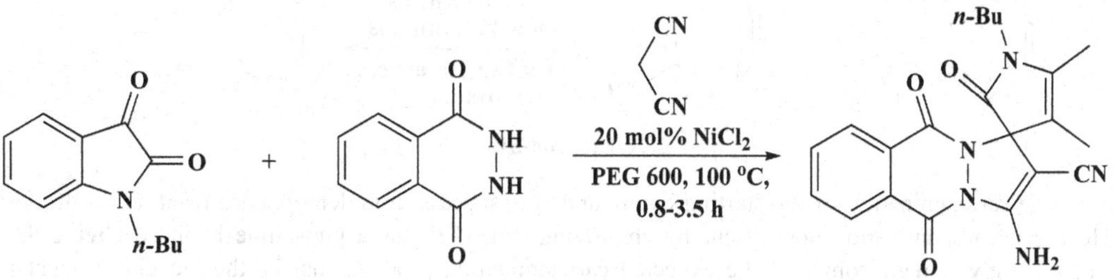

Scheme-48

Allylamines have been used successfully to provide substituted imidazo[1,2-*a*]pyrimidin-2-ones by a sequential Raney nickel-assisted reduction of the nitrile group to obtain diamines followed by intramolecular cyclization through its reaction with CNBr **(Scheme 49)** [85–86].

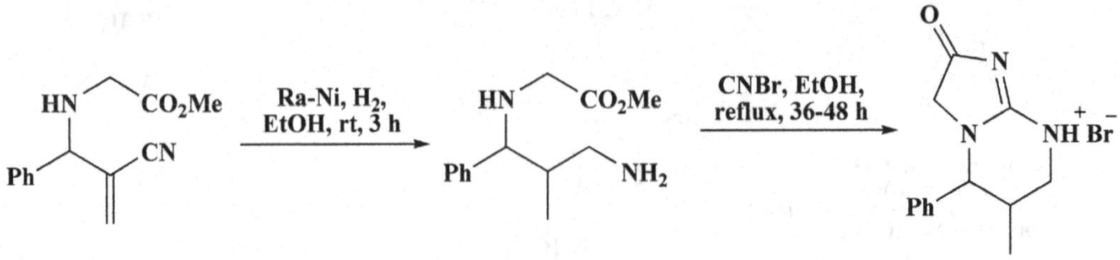

Scheme-49

Palladium-assisted synthesis

The mechanism of this reaction is a subject of much debate, although many reports have proposed that a PdII/PdIV catalytic pathway occurs under these oxidizing conditions. For example, Streuff et al. [87] developed an intramolecular diamination reaction of olefins under similar conditions as described for diacetoxylation **(Scheme 50)**. An initial intramolecular aminopalladation of the olefin provided Pd-C-species which underwent nucleophilic substitution by another amine to form the diaminated product. It was proposed that Pd-C species were oxidized into Pd(IV) species under these highly oxidizing conditions prior to the substitution by the second amine. This proposition was supported by the observation that no diaminated product was observed in a stoichiometric example without PhI(OAc)$_2$.

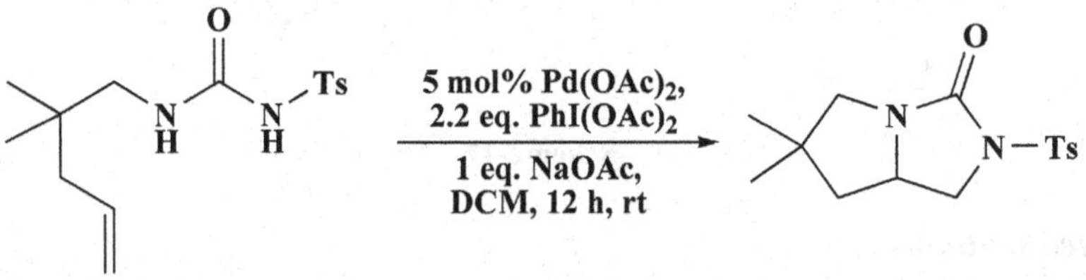

Scheme-50

Loones et al. [54] used an alternative for primary amines in double C-N bond forming reactions and reported that pyridines underwent consecutive palladium-catalyzed carbon-nitrogen bond formations with *o*-dihalopyridines **(Scheme 51)**. This reaction occurred efficiently with many amidine variants and resulted in a pathway for the production of polycyclic imidazoles.

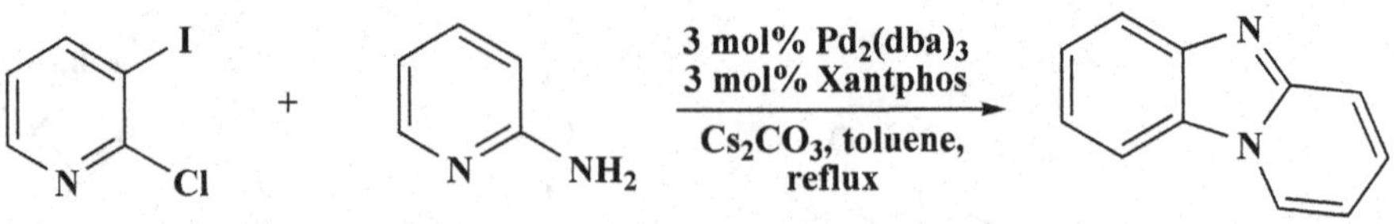

Scheme-51

The carboamination of the starting compound is unsuccessful under standard reaction conditions. However, carboamination does occur by employing PEt$_3$·HBF$_4$ as a phosphine ligand **(Scheme 52)**. Subsequently, a regioisomer of the expected carboamination product, that is, the bicyclic product is obtained [88].

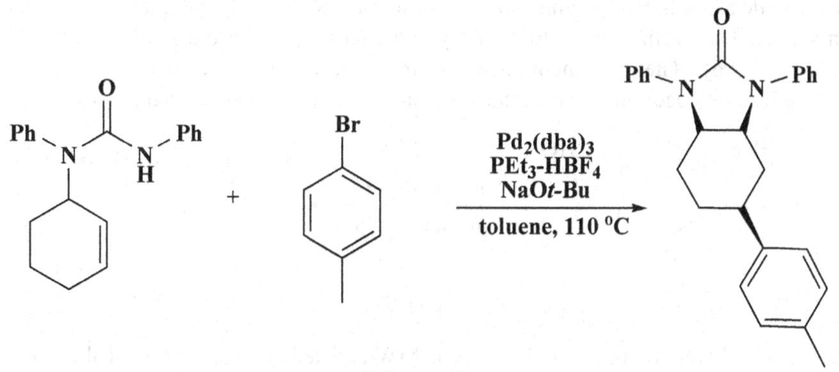

Scheme-52

The synthesis of allylpalladium complexes involves the oxidative addition of allylic electrophiles to palladium(0). This reaction has been explored by many groups and has been reviewed recently [89–90]. A representative example of this protocol is the total synthesis of (+)-Biotin [91], wherein the key step is an intramolecular amination of thiophene, resulting in 77% yield of bicyclic urea derivative. Palladium(0)-catalyzed reactions occur through an initial oxidative addition of the allylic acetate to form an intermediate *p*-allylpalladium complex which is captured by the pendant nucleophile in a formal reductive elimination process to generate the product and regenerate the Pd(0) catalyst. Both the reductive elimination and the oxidative addition steps occur with the inversion of configuration when soft nucleophiles are used. Hence, the overall configuration at the carbonate bearing the carbon stereocenter is retained. Propargylic electrophiles also undergo similar transformations **(Scheme 53)** [92].

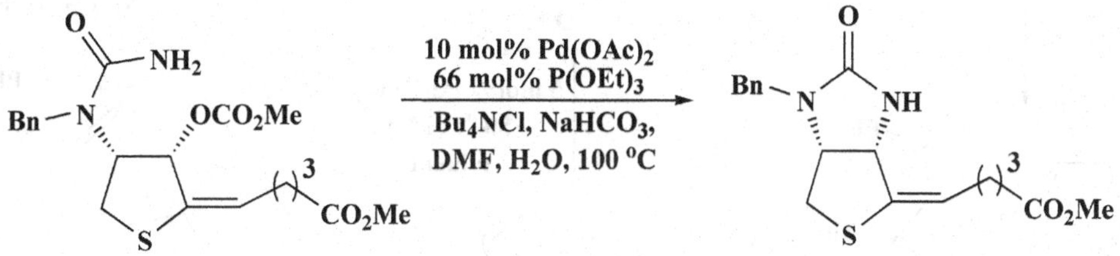

Scheme-53

N-Allylureas undergo carboamination with various alkenyl, heteroaryl and aryl bromides. *Trans*-4,5-disubstituted imidazolidin-2-ones are obtained with good to excellent diastereoselectivities from substrates containing an allylic substituent. *N*-Allylureas subjected to palladium-catalyzed carboamination with substrates possessing 1,1- or 1,2-disubstitued alkenes undergo reactions which are mechanistically analogous to pyrrolidine-forming carboamination reactions, with net *syn*-addition across the double bond. In general, the best yields of imidazolidin-2-one products are obtained when the N1-atom bears an aromatic group [93]. When the two nitrogen atoms are protected with a *p*-methoxyphenyl group and a benzyl group, these protecting groups undergo selective cleavage from the product with the help of ceric ammonium nitrate and lithium/ammonia respectively **(Scheme 54)** [94a–b].

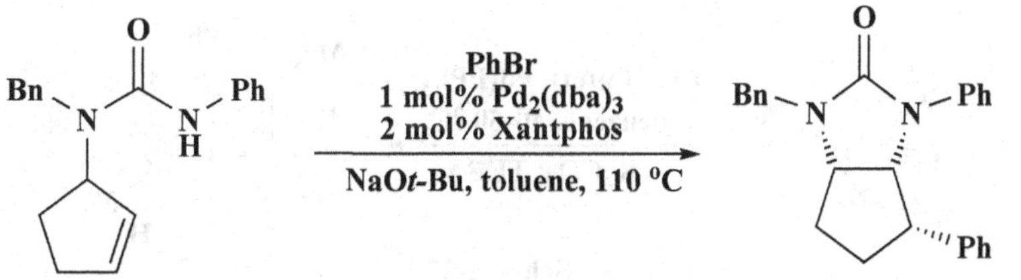

Scheme-54

Formaldehyde aminals can be synthesized from formaldehyde, *N*-Boc-protected allylic amines, and a nitrogen source. The cyclization of formaldehyde aminals results in the synthesis of imidazolidines (**Scheme 55**) [95–96]. The aforementioned substrates react under standard oxidative cyclization conditions to afford imidazolidines which can be easily transformed into vicinal diamines.

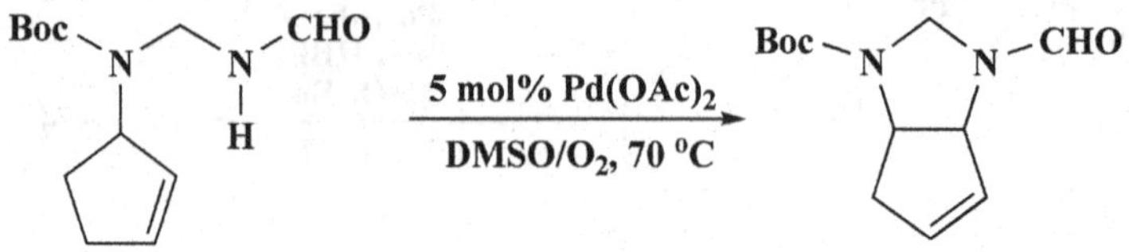

Scheme-55

Beccalli et al. [97] reported carboamination and MW-assisted hydroamination of allenamides bearing an indolyl unit (**Scheme 56**). The allenamides provided styryl-substituted indoloimidazoles through the formation of palladium π-allyl complex intermediate where trapping by the indole nitrogen occurred exclusively at the internal allenic carbon. This cyclization cascade synthesized indoloimidazoles in the presence of carbon monoxide. It was found that the use of MW was crucial for the hydroamination of allenamides. The palladium(II)-complex was produced by oxidative addition of the indolyl NH bond under MW activation. Subsequently hydro-palladation or hydride transfer occurred to provide the palladium π-allyl intermediate and steer the reaction towards vinyl indoloimidazoles.

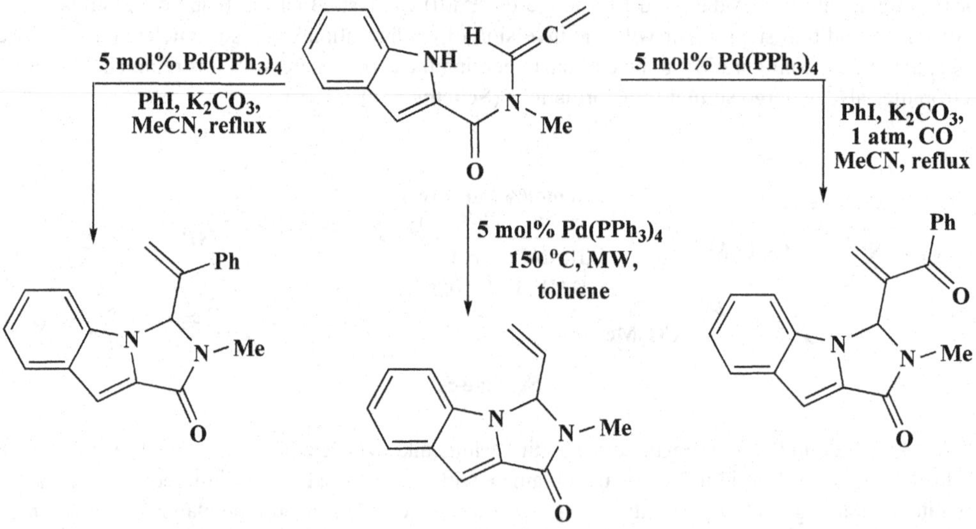

Scheme-56

Azirines react readily with carbon monoxide under homogeneous conditions in the presence of Pd(PPh$_3$)$_2$ [98]. The reaction involves dimerization followed by carbonylation, both catalyzed by palladium. In contrast, the reverse reaction sequence (carbonylation with subsequent cycloaddition of another molecule of un-reacted azirine) is also possible. The carbonylation reaction occurs through the formation of an aza (*n*-ally1)palladium intermediate (**Scheme 57**) [84, 99–100].

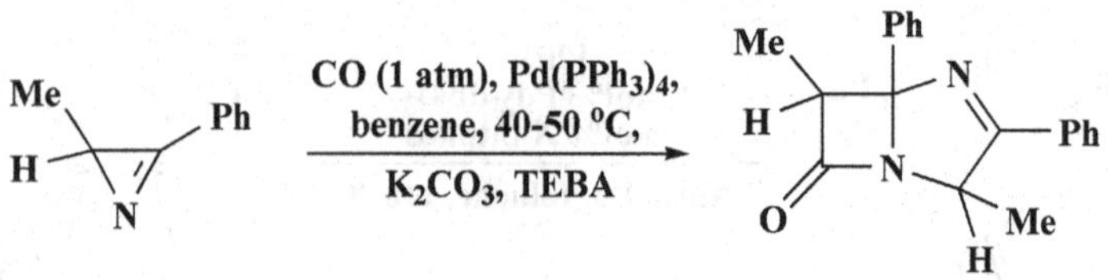

Scheme-57

Pd-catalyzed amide coupling produces imidazo[4,5-*b*]pyridines and pyrazines. Various substituted products can be accessed quickly by this reaction. A model system relevant to the natural product pentosidine has been demonstrated, as well as the total synthesis of the mutagen 1-Me-5-PhIP (**Scheme 58**) [101].

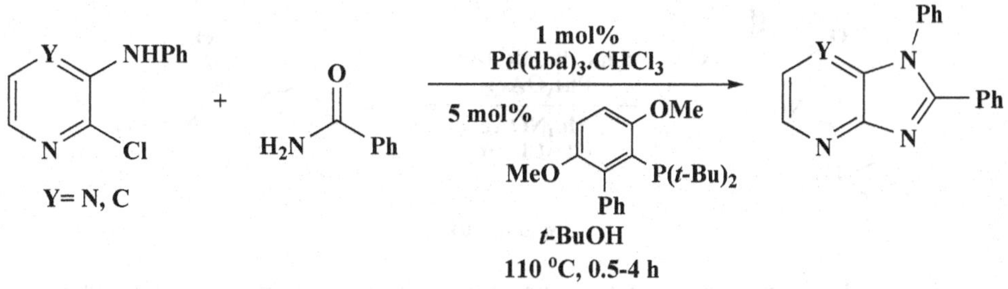

Scheme-58

Palladium-catalyzed annulation of 2-(4-methoxy-2-nitrophenyl)-2,3-dihydro-1*H*-isoindole by heating in dimethylformamide using 1,10-phenanthroline and bis(dibenzylideneacetone)palladium at 120 °C and subsequent saturation of the solution with carbon monoxide under pressure produces 7-methoxy-11*H*-isoindolo[2,1-*a*]benzimidazole in 90% yield (**Scheme 59**) [24, 102].

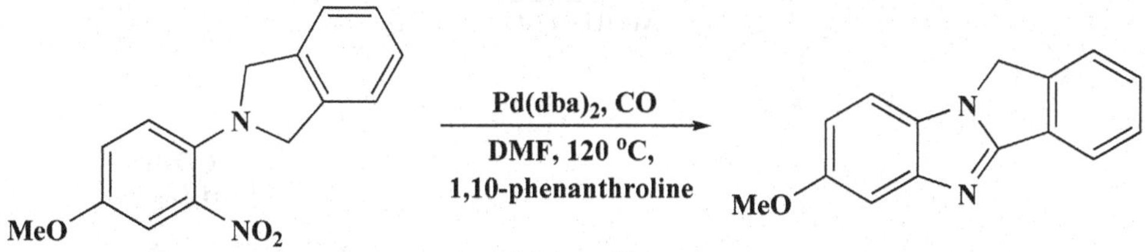

Scheme-59

2-Azidoaniline reacts with aromatic aldehydes to give *N*-(2-azidophenyl)imines which react with trimethylphosphine and then with diphenylketene to produce 6,11-dihydrobenzimidazo[1,2-*b*] isoquinolines in excellent yields by formal [4+2]-intramolecular cycloaddition of ketenimine with imine function of the intermediates. Benzimidazo[1,2-*b*]isoquinolines are obtained in good yields when 6,11-dihydrobenzimidazo[1,2-*b*]isoquinolines are refluxed with palladium/carbon in PhMe (**Scheme 60**) [24, 103–104].

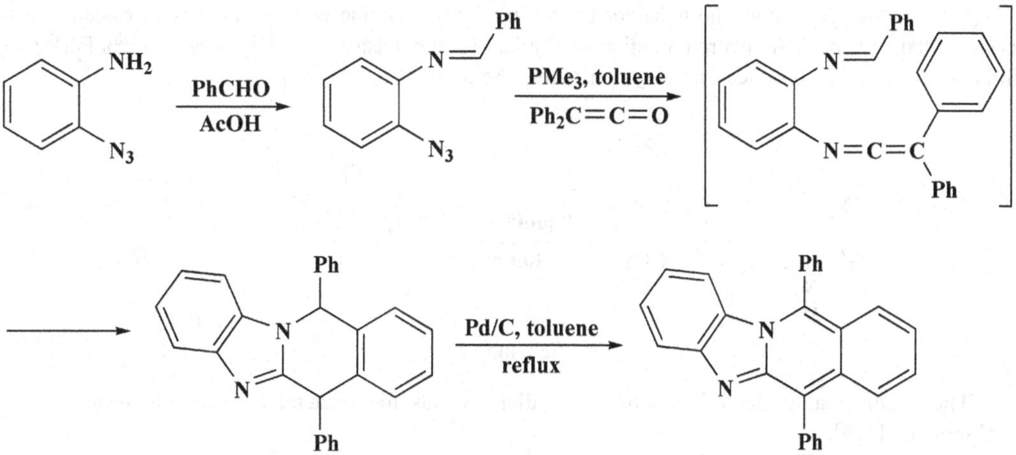

Scheme-60

Many valuable nitrogen heterocycles can be produced by intramolecular aminopalladation reactions. For example, a plethora of important nitrogen-containing structures are synthesized by intramolecular alkene diamination **(Scheme 61)** [105], chloroamination [106], and hetero-Heck-type transformations [107].

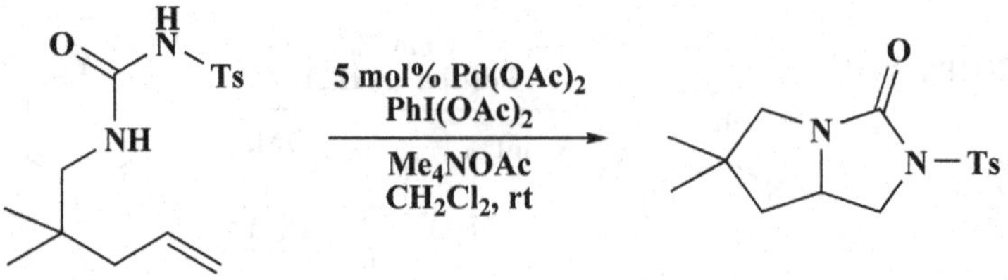

Scheme-61

1,5-Dibromo-2,4-dinitrobenzene, upon treatment with pyrrolidine, forms dinitrodipyrrolidine which is transformed into an intermediate by reduction followed by acylation. This intermediate is heated with formic acid and hydrogen peroxide at 70 °C for cyclization into 3*H*,7*H*-1,2,8,9-tetrahydropyrrolo[1,2-*a*]-pyrrolo[1`,2`:1,2]imidazo[4,5-*f*]benzimidazole in low yields **(Scheme 62)** [24, 108–110].

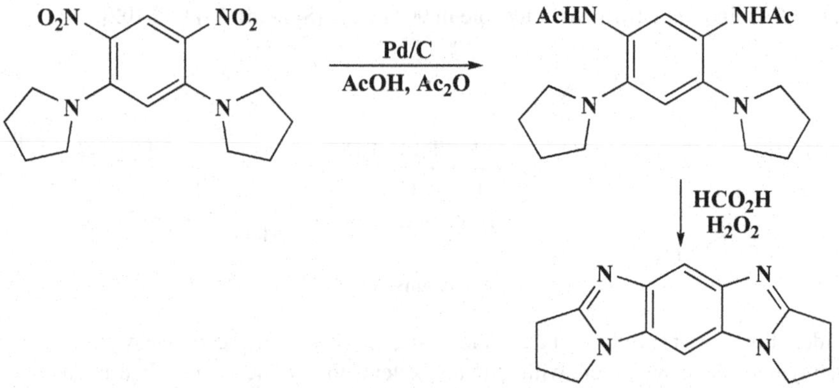

Scheme-62

Studies directed towards the reactivity of indole derivatives have attracted the use of *N*-allenyl 2-indolecarboxamides as substrates due to the presence of the allenyl group linked to the indole moiety where nitrogen atom acts as a nucleophile for intramolecular cyclization. Searching conditions to explore the protocols of cyclization, the behavior of the indolyl allenamide under microwave irradiation was investigated. 5-*Exo*-allylic hydroamination of *N*-allenyl 2-indolecarboxamides using 8 mol% Pd(PPh$_3$)$_4$ in PhMe as a catalyst produces the vinyl derivative **(Scheme 63)** [111–114].

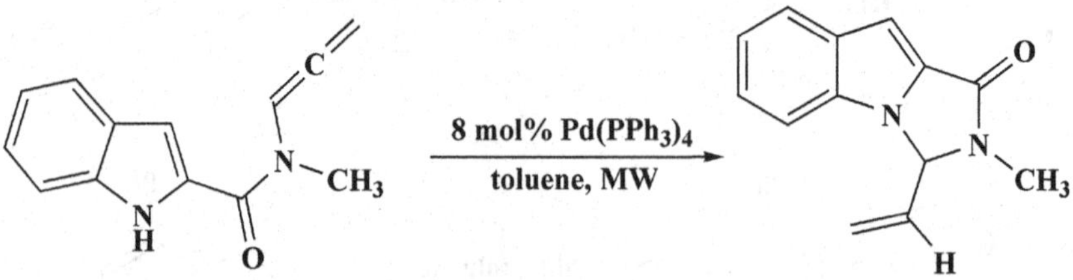

Scheme-63

The cyclization of deuterium-substituted allene yields the deuterated hydroamination product **(Scheme 64)** [115].

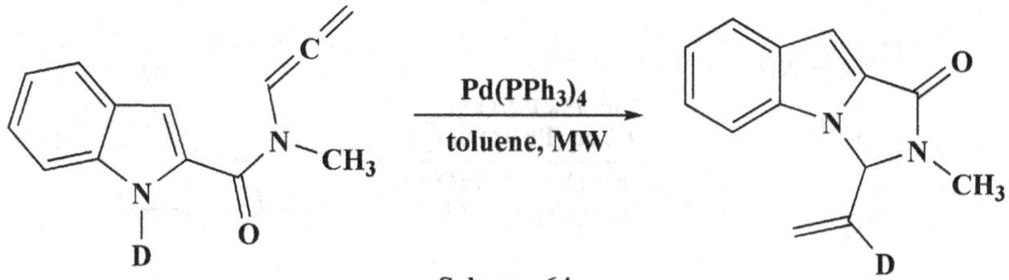

Scheme-64

The arylboronic acid reacted well with pyridine ring in Suzuki-Miyaura cross-coupling reactions [116–119]. The products of pyridine-bearing reactants are transformed into pyrazolo-3,4-pyridines upon heating with $NH_2NH_2 \cdot H_2O$ in PhMe for 4 hours at 80 °C. As the members of this class of heterocyclic compounds are kinase inhibitors, they exhibit anti-cancer properties [120]. The pyrazolo-3,4-pyridines are formed in 88% yield when pyridine-containing reactants are treated with $NH_2NH_2 \cdot H_2O$ at 80 °C in PhMe for 4 hours. The Suzuki-Miyaura cross-coupling and cyclization step can be combined in a one-pot protocol, starting from bromopyridine, which is subjected to palladium-catalyzed cross-coupling with arylboronic acid and subsequently reacts with $NH_2NH_2 \cdot H_2O$ to produce pyrazolo-3,4-pyridine in 77% yield **(Scheme 65–66)** [84b, 121].

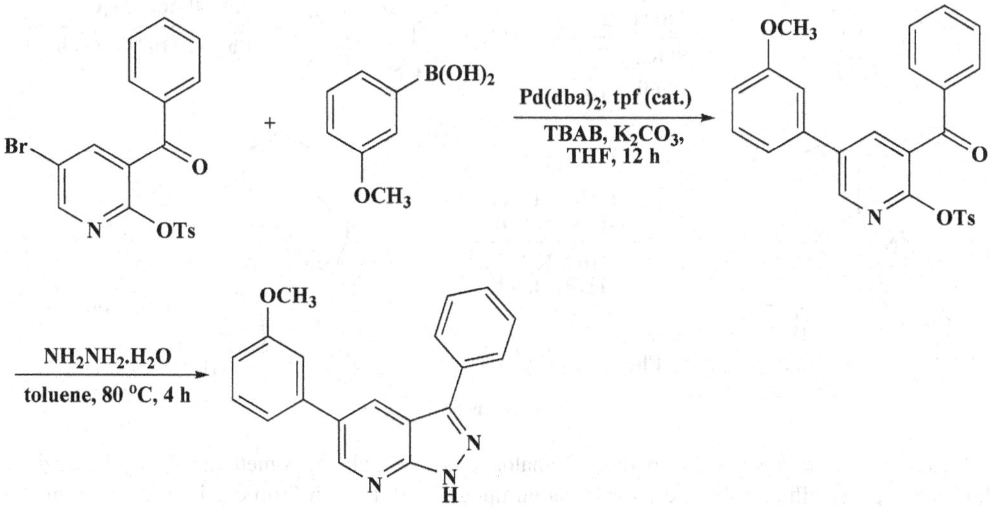

Scheme-65

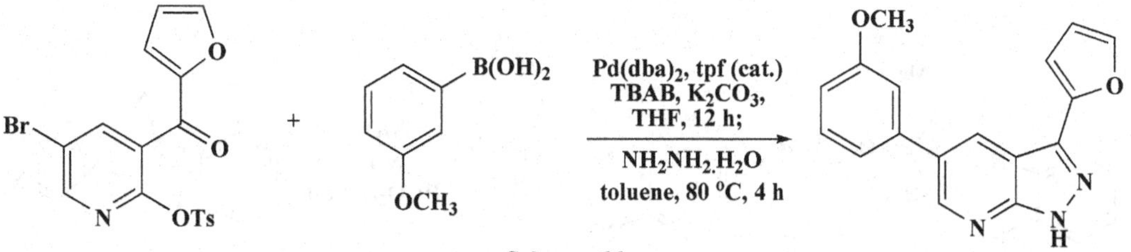

Scheme-66

An intramolecular vicinal alkene oxidation occur using robust tosyl ureas as nitrogen sources. Herein, hypervalent iodine oxidants like $PhI(OAc)_2$ have proved to be the most effective. Various five- and six-membered-ring annelation products of cyclic ureas can be prepared by this protocol. An example of diastereoselective alkene diamination is shown in **Scheme 67**. The proposed mechanism involves a *syn*-aminopalladation followed by *anti*-alkyl-nitrogen bond formation from a palladium(IV) intermediate. This mechanism is analogous to that suggested by Liu and Stahl [122] for related aminoacetoxylation reactions. The involvement of palladium(IV) catalyst state arising from the oxidation of *s*-alkyl palladium intermediate formed by aminopalladation has been confirmed by theoretical calculations [123].

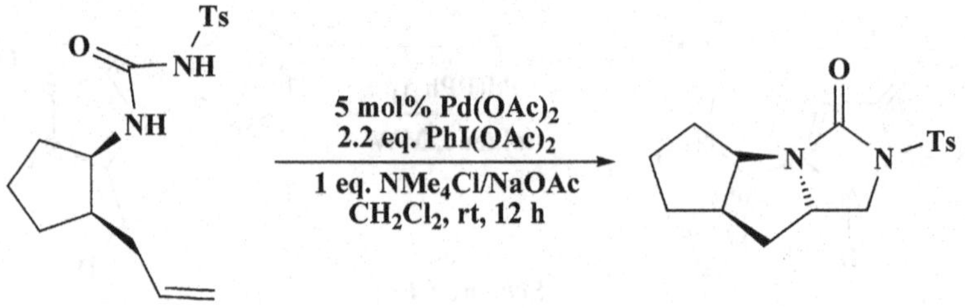

Scheme-67

Phosphorus-assisted synthesis

Primary allylamines (derived from acrylonitriles) can be utilized as substrates for the synthesis of isonitriles which undergo IMCR to yield substituted imidazo[1,2-*a*]pyridines (**Scheme 68**). Good yields of imidazoazepines are obtained upon reduction of the 2-nitro-group in substituted imidazo[1,2-*a*] pyridines followed by CNBr-mediated cyclization [86, 124].

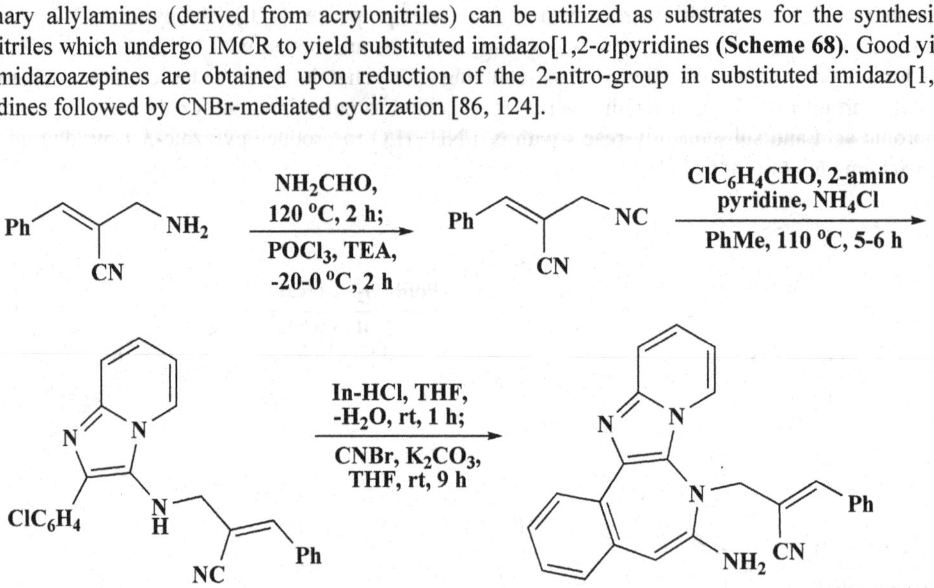

Scheme-68

Triazinone is transformed into its thio-analog with the help of *S*-methylated and hydrazinated phosphorus pentasulfide, followed by cyclization upon its reflux with formic acid to afford 4-methyl–1,2,4-triazolo[4',3':4,5][1,2,4]triazino[2,3-*a*]benzimidazole (**Scheme 69**) [24, 125–126].

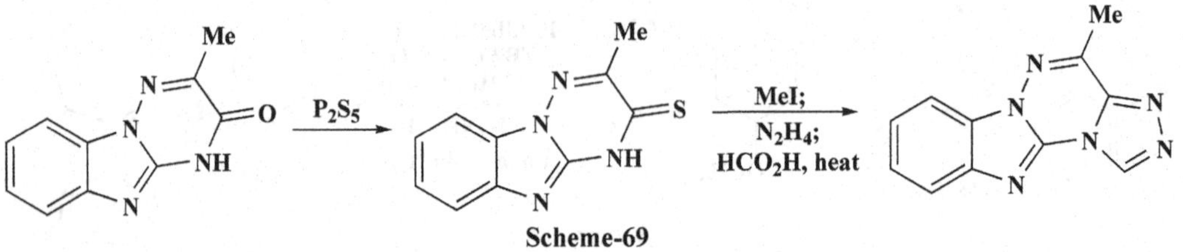

Scheme-69

Rhodium-assisted synthesis

Rech et al. [127] reported the condensation of substituted allylamines with 4-fluorophenyl tosylmethyl isonitrile in glyoxylic acid followed by rhodium-catalyzed intramolecular alkylation by carbon-hydrogen bond activation of enantiopure *N*-allyl-imidazoles to produce potent kinase inhibitors (**Scheme 70**) [86].

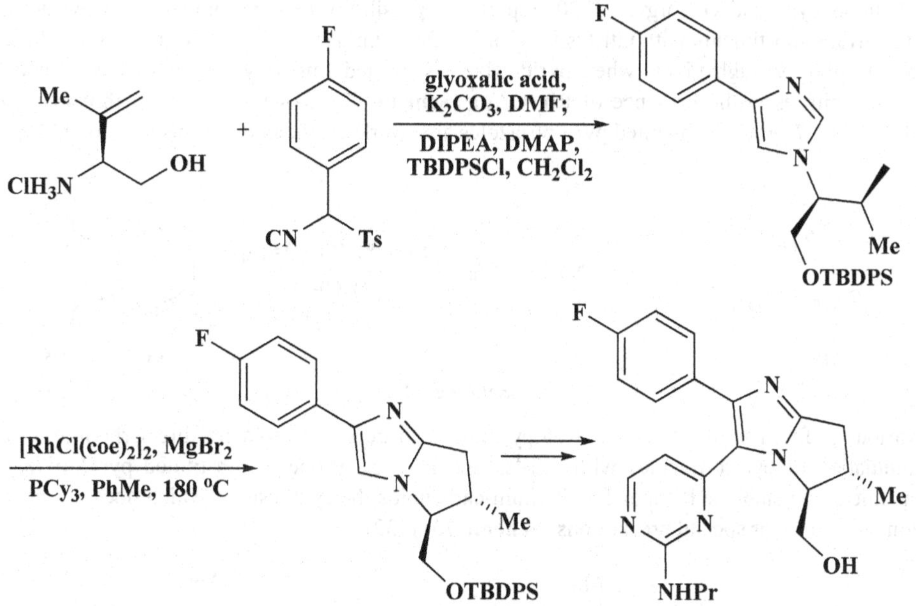

Scheme-70

The transannulation as illustrated in **Scheme 71** occurs through Rh carbenoid intermediate produced *in situ*. A direct nucleophilic attack [128] of the nitrile or alkyne affords an intermediate ylide which gets cyclized into cyclic zwitterion. Alternatively, [2+2]-cycloaddition of the Rh carbenoid with nitrile or alkyne provides metallacyclobutene which is also obtained from the cyclization of the intermediate ylide [86, 129–130].

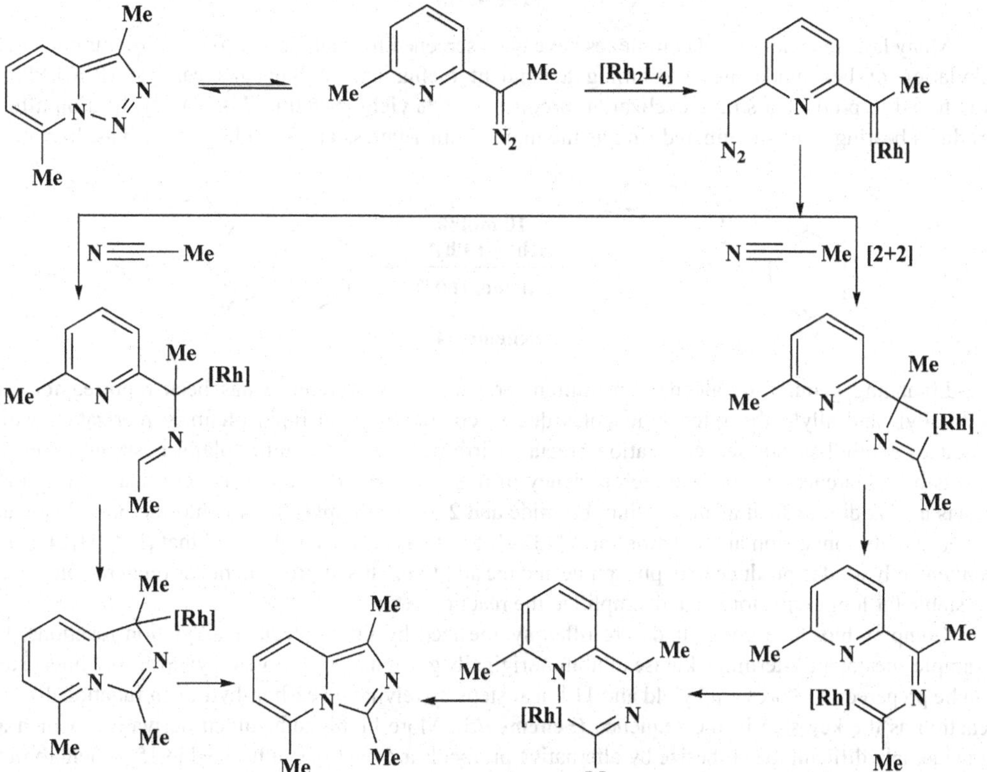

Scheme-71

Chattopadhyay and Gevorgyan [130] reported a possibility of transannulation of pyridotriazoles bearing various substituents with nitriles to afford *N*-fused imidazoles. *N*-Fused imidazopyridines were obtained in good to high yields when pyridotriazoles reacted smoothly with a variety of alkyl, aryl and alkenyl nitriles in the presence of Rh$_2$(OAc)$_4$ **(Scheme 72)**. 3-Aryl-, 3-carbomethoxy- as well as 7-methoxy- and 7-bromo-substitited pyridotriazoles also worked efficiently in this reaction [131].

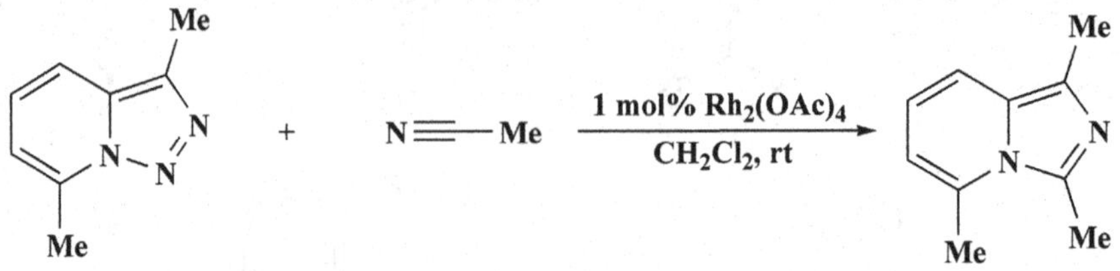

Scheme-72

Various *N*-fused imidazo- and pyrrolopyridines can be synthesized by direct rhodium-catalyzed transannulation of pyridotriazoles with nitriles and alkynes. Properly substituted pyridotriazoles act as convenient and stable substrates for rhodium-carbenoids the synthesis of which does not need slow addition techniques or special precautions **(Scheme 73)** [132].

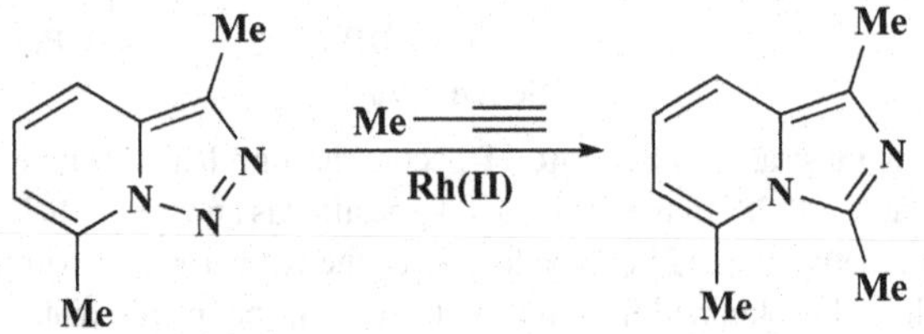

Scheme-73

Many late transition metal complexes have been screened for their ability to catalyze intramolecular alkylation of benzimidazoles containing a pendant olefin, and Wilkinson's catalyst (RhCl(PPh$_3$)$_3$) was found to produce a single cyclization product in 60% yield **(Scheme 74)**. The cyclization affords products bearing a five-membered ring as the major isomer unless an overriding steric bias, like allylic

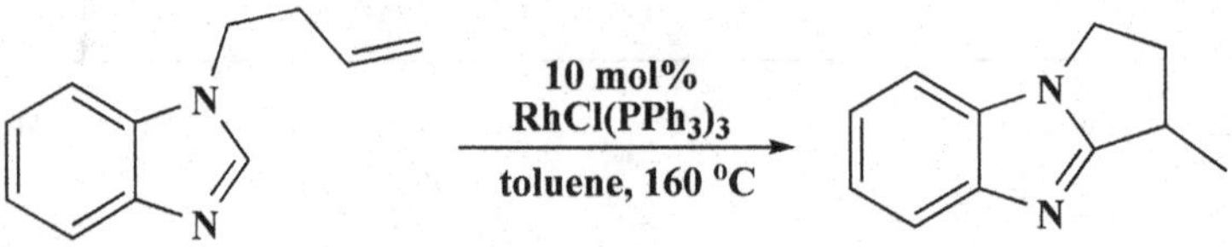

Scheme-74

α,α-dibranding geminal or alkene substitution, prevails. This preference has been reported for both homoallyl- and allyl-substituted imidazoles due to competitive and rapid olefin isomerization which produces an allyl-substituted cyclization precursor irrespective of the initial olefin position. Extensive efforts by researchers to improve the efficiency of this reaction led to the discovery that Brønsted and Lewis acid additives such as magnesium bromide and 2,6-dimethylpyridinium chloride provide notable increase in its conversion and reaction rates [133–135]. It has also been observed that [PCy$_3$H]Cl can be conveniently used to produce both phosphine and the additive. This improvement has rendered phosphine air stable for long-term storage and simplified the reaction setup.

Complex bioactive compounds are often synthesized by intramolecular alkylation reactions. For example, the potent *N*-terminal kinase inhibitor originally generated in 6% overall yield and 14 linear steps can be prepared in 13% overall yield and 11 linear steps by relying on carbon-hydrogen functionalization reaction as the key step in the sequence **(Scheme 75)**. More highly substituted derivatives which are considerably difficult to synthesize by alternative protocols are readily synthesized in 15% yield through this method and result in the identification of even more potent inhibitors [135–136].

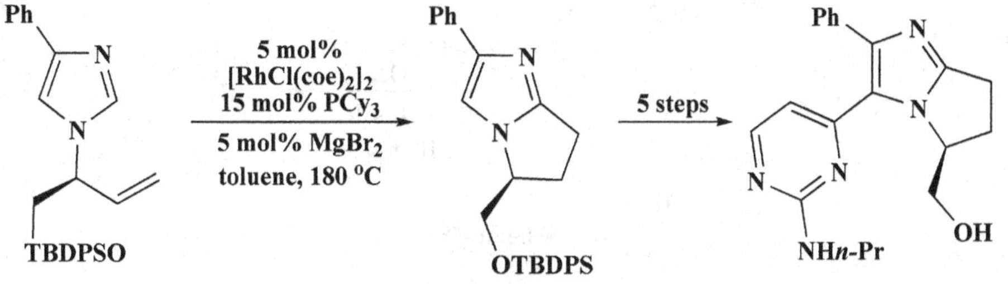

Scheme-75

Scandium-assisted synthesis

Blackburn et al. [137] reported a three-component condensation of 2-aminopyrazine, an aldehyde and an isonitrile in the presence of Sc(OTf)$_3$ catalyst to provide 3-aminoimidazo-[1,2-*a*]pyrazines **(Scheme 76)**. The reaction followed the initial stages of Ugi reaction wherein imine produced *in situ* was attacked by isonitrile to produce a nitrilium ion which was subsequently cyclized instead of being attacked by a carboxylic acid as in Ugi reaction. This three-component condensation can be performed under solid-phase conditions with any of the three reacting partners tethered to an amide resin [138–139].

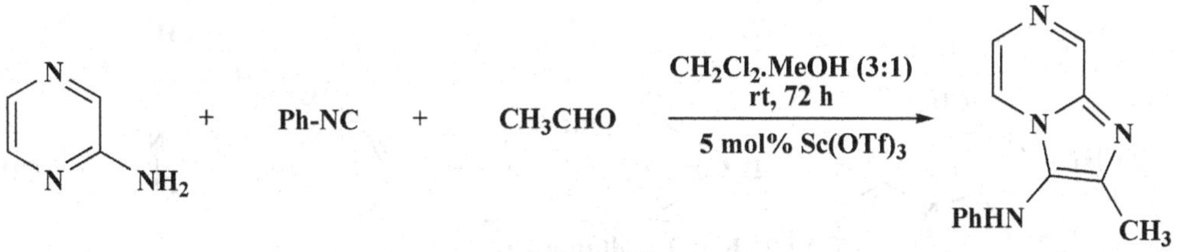

Scheme-76

The synthesis of 1-substituted 4-imidazolecarboxylates under MW dielectric heating conditions using Wang resin-bound 3-*N,N*-(dimethylamino)-isocyanoacrylate has been reported in literature [140]. The coupling of urea with aromatic and aliphatic diamines affords imidazolidine-2-ones by a MW-accelerated reaction in the presence of zinc oxide as the catalyst [141]. The synthesis of benzimidazoles from glycolic acid and *o*-phenylenediamine can be accelerated by 80 times, with reaction times reduced from 120 hours under standard thermal heating to 90 minutes under MWI [142]. Other MW-assisted protocols for the synthesis of benzimidazoles involve *in situ* reduction of *o*-nitro anilines to diamines [143] or a reaction of dianilines with aldehydes [144]. Intramolecular aryl-amination carried out under aqueous conditions gets completed after 20 minutes under MWI at 200 °C [145]. A facile method of synthesizing fused 3-aminoimidazoles involves a three-component Sc(OTf)$_3$-catalyzed cyclocondensation of aldehydes and heterocyclic amidines (2-aminopyridine) with isocyanides (Ugi MCR) **(Scheme 77)** [146a–b]. With trimethylsilylcyanide (TMSCN) **(Scheme 78)**, *N*-unsubstituted 3-aminoimidazo[1,2-*a*]pyridines are formed. A substantial acceleration of the reaction rate has been observed under MW dielectric heating of the methanolic reaction mixture in sealed reaction tubes at 140–160 °C. Consequently, the desired heterocyclic compounds can be formed within 10 minutes.

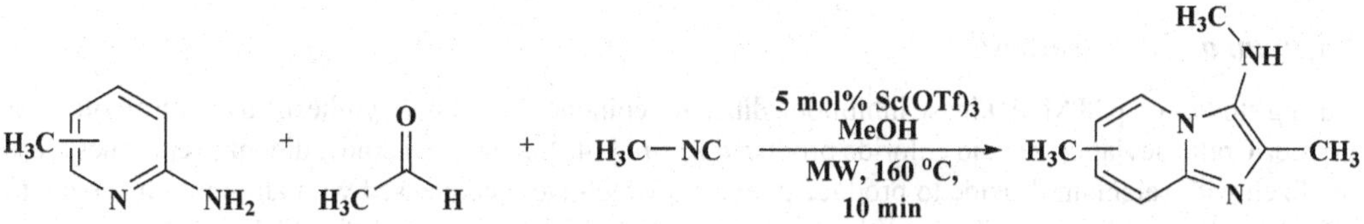

Scheme-77

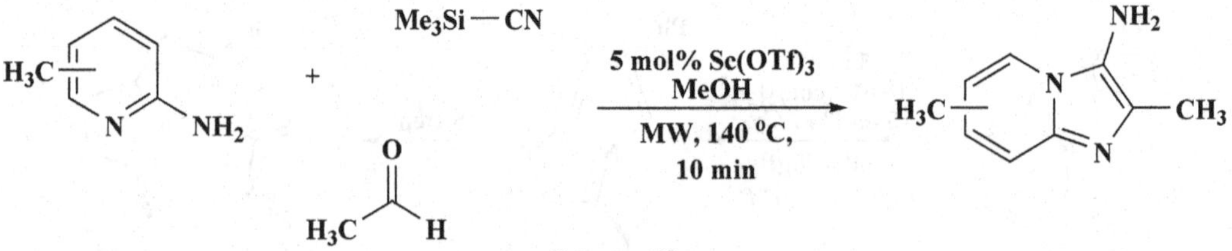

Scheme-78

The three-component condensation of perfluoroalkanesulfonyl-protected hydroxybenzaldehydes, isonitriles and 2-aminopyridines produces imidazo[1,2-*a*]pyridine ring system (**Scheme 79**) [147]. The Pd-catalyzed cross-coupling of these condensed products with boronic acids yields imidazo[1,2-*a*] pyridine.

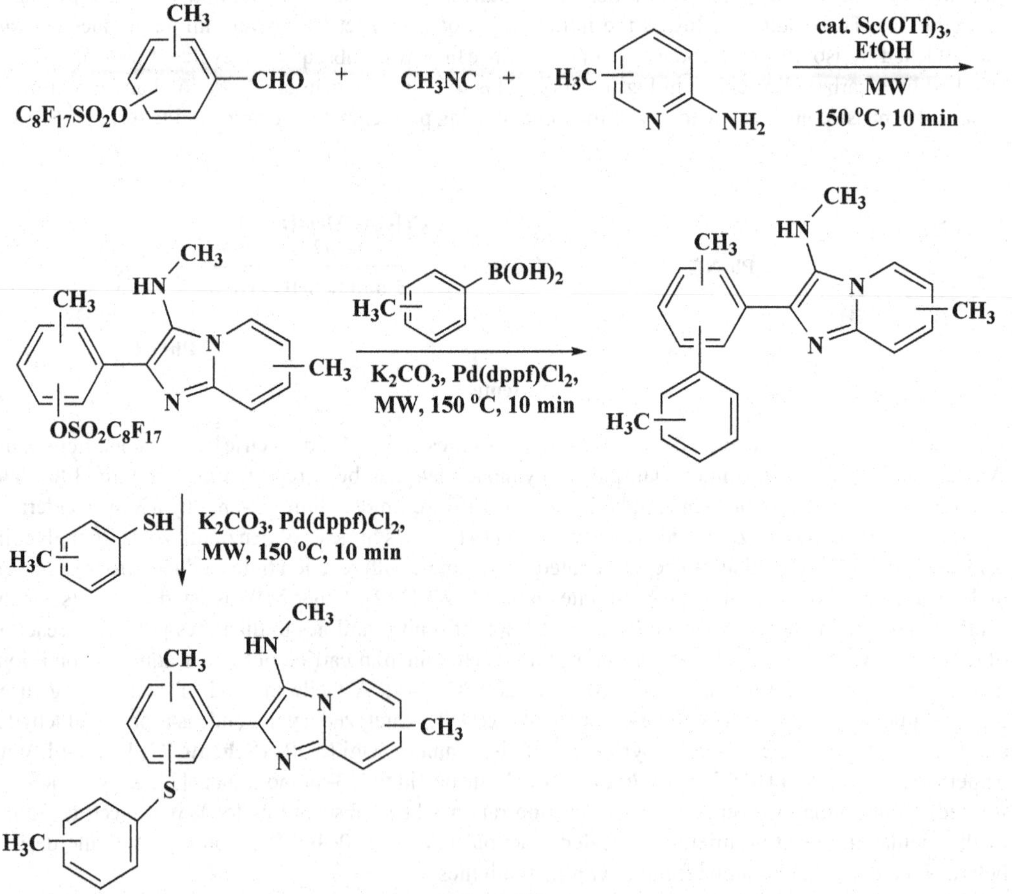

Scheme-79

Selenium-assisted synthesis

During studies on p38 MAP kinase inhibitors, dibenzoxepinones have been synthesized by intramolecular Friedel-Crafts acylation of acid chloride produced *in situ* [148]. In another study, dibenzoxepinones were oxidized with selenium dioxide to produce diketones which were condensed with ammonium acetate to afford imidazolodibenzoxepines [149]. The condensation of dimedone with furaldehyde also results in the synthesis of dipyrimidino-fused oxepines (**Scheme 80**) [150].

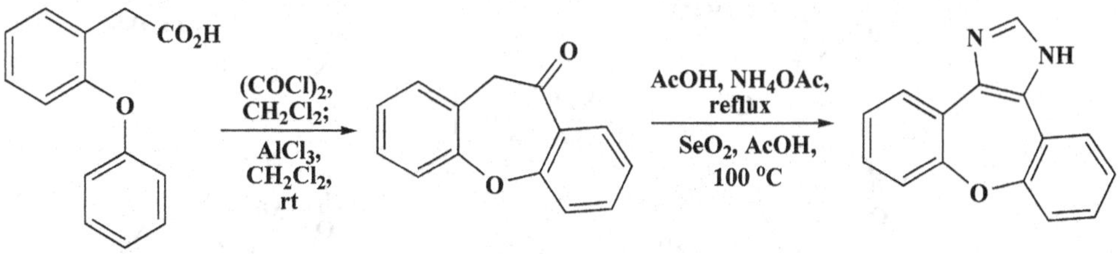

Scheme-80

Silver-assisted synthesis

Molteni [151] reported an intermolecular (but mainly intramolecular) 1,3-dipolar cycloaddition employing nitrilimines produced *in situ* from hydrazonoyl chlorides in the presence of silver carbonate **(Scheme 81)**. Herein, alkynes or alkenes served as dipolarophiles; large-and medium-ring heterocyclic systems were formed in this way. Stereoselective cycloadditions can also be performed as shown in **Scheme 82** [152], wherein nitrilimine is obtained from lactamic hydrazonoyl chloride with silver ion. Formation of azetopyrrolopyrazole with complete diastereo-, regio-, and enantioselectivity by intramolecular [3+2]-cycloaddition has also been reported [153].

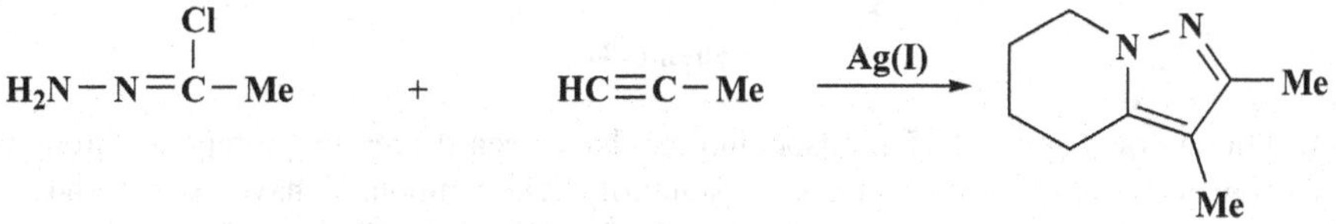

Scheme-81

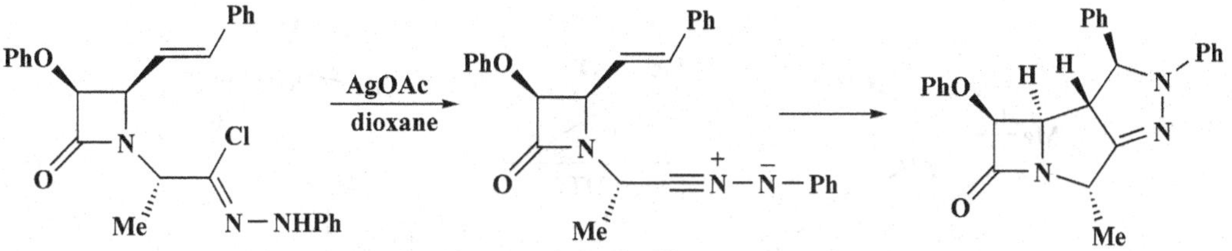

Scheme-82

Imidazolium triflate can be prepared from bis-oxazoline using silver triflate in combination with chloromethyl pivalate **(Scheme 83)** [153–154].

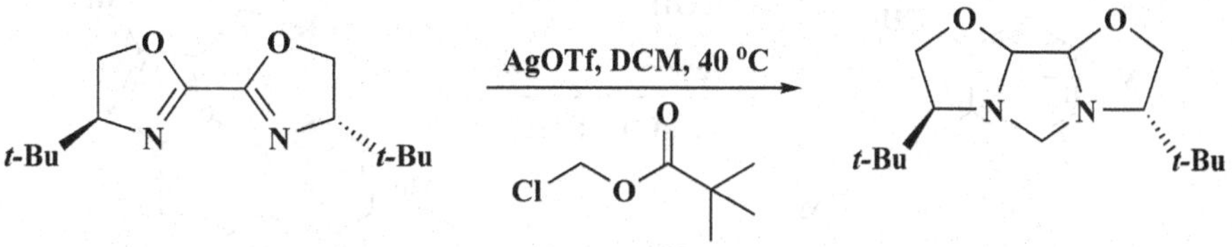

Scheme-83

Umesha et al. [155] synthesized pyrazoles through 1,3-dipolar cycloaddition of *in situ* produced nitrile imines and acetyl acetone by catalytic dehydrogenation of phenylhydrazones in the presence of chloramine-T oxidant. Consequently, regioselective cycloadducts were obtained in good yields. The nitrile imines were produced *in situ* by the reaction of aldehyde hydrazones with mercuric acetate [156]. The reaction of aldehyde hydrazones and mercuric acetate in the presence of olefins produced good yields of 1,3,5-trisubstituted 2-pyrazolines. Homochiral hydrazonoyl chlorides react with silver carbonate in dioxane to form nitrile imine which undergoes intramolecular cycloaddition without trapping agents to afford diastereoisomeric mixtures of 3,3-dihydro-pyrazolo[1,5-*a*][1,4]benzodiazepine-6(4*H*)-ones in enantiopure form **(Scheme 84)** [157–158].

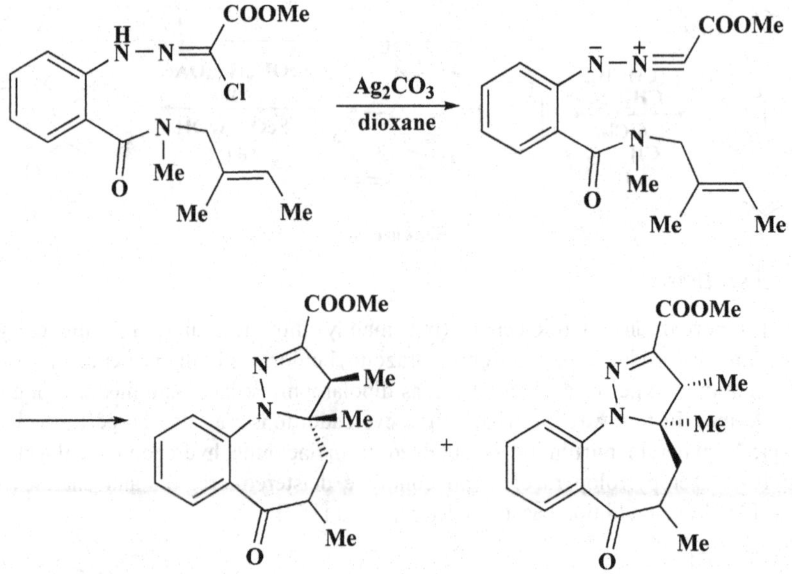

Scheme-84

A small library of *H*-pyrazolo[5,1-*a*]isoquinolines have been successfully prepared through cascade reactions. From preliminary biological assays, some of these compounds have been found to display promising inhibiting activities against TC-PTP, CDC25B and PTP1B (**Scheme 85–87**) [159–162].

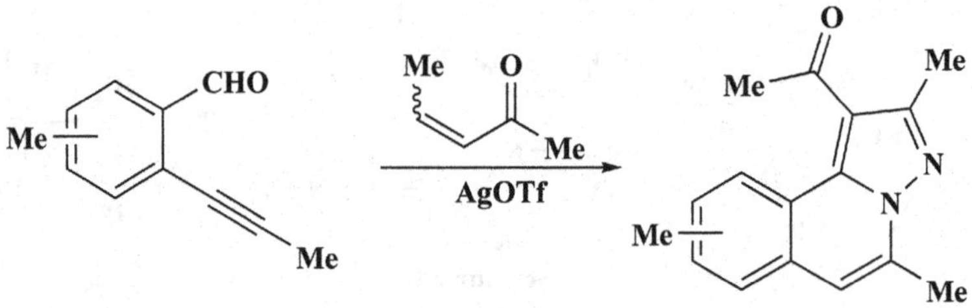

Scheme-85

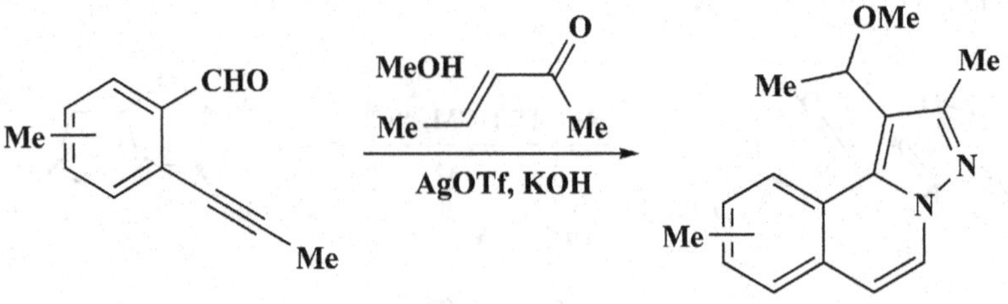

Scheme-86

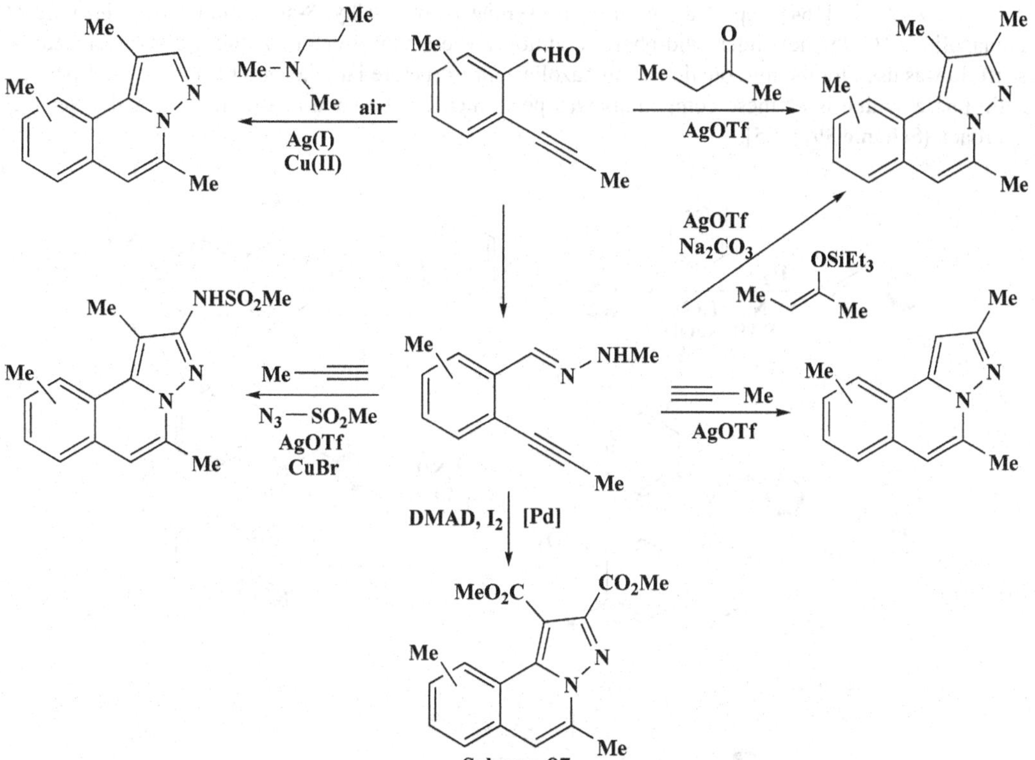

Scheme-87

Tin-assisted synthesis

5-Methyl-5,6-dihydrobenzimidazo[2,1-*a*]benzo[*f*]isoquinolines can be conveniently synthesized in three steps **(Scheme 88)**. In the first step, 1-bromo-2-naphthoic acid is heated with *o*-phenylenediamines in polyphosphoric acid (PPA) to yield 2-(1-bromo-2-naphthyl)-1*H*-benzimidazoles which subsequently undergo *N*-allylation with 3-bromoprop-1-ene and sodium hydride in tetrahydrofuran to produce 1-allyl-2-(1-bromo-2-naphthyl)benzimidazoles in 68–88% yields. Finally, the cyclization of 1-allyl-2-(1-bromo-2-naphthyl)benzimidazoles mediated in Bu₃SnH and refluxing PhMe forms 5-methyl-5,6-dihydrobenzimidazo[2,1-*a*]benzo[*f*]isoquinolines [24, 163].

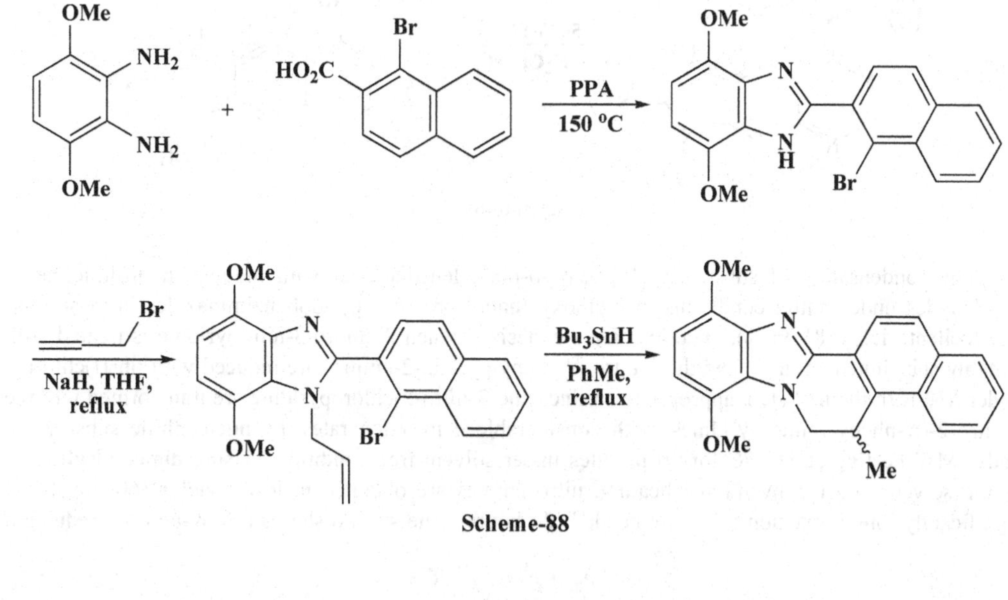

Scheme-88

Zhang et al. [164] reported an efficient synthesis of 3-alkyl-8-arylamino-1*H*-imidazo[4,5-*g*] quinazolin-2(3*H*)-thiones under solid-phase conditions, wherein the 4-chloro-7-fluoro-6-nitroquinazoline scaffold was used for the introduction of quinazoline core structure into the target compounds. A parallel solid-phase synthesis of these compounds was performed on the solid-phase employing the 'teabag' approach (**Scheme 89**) [165].

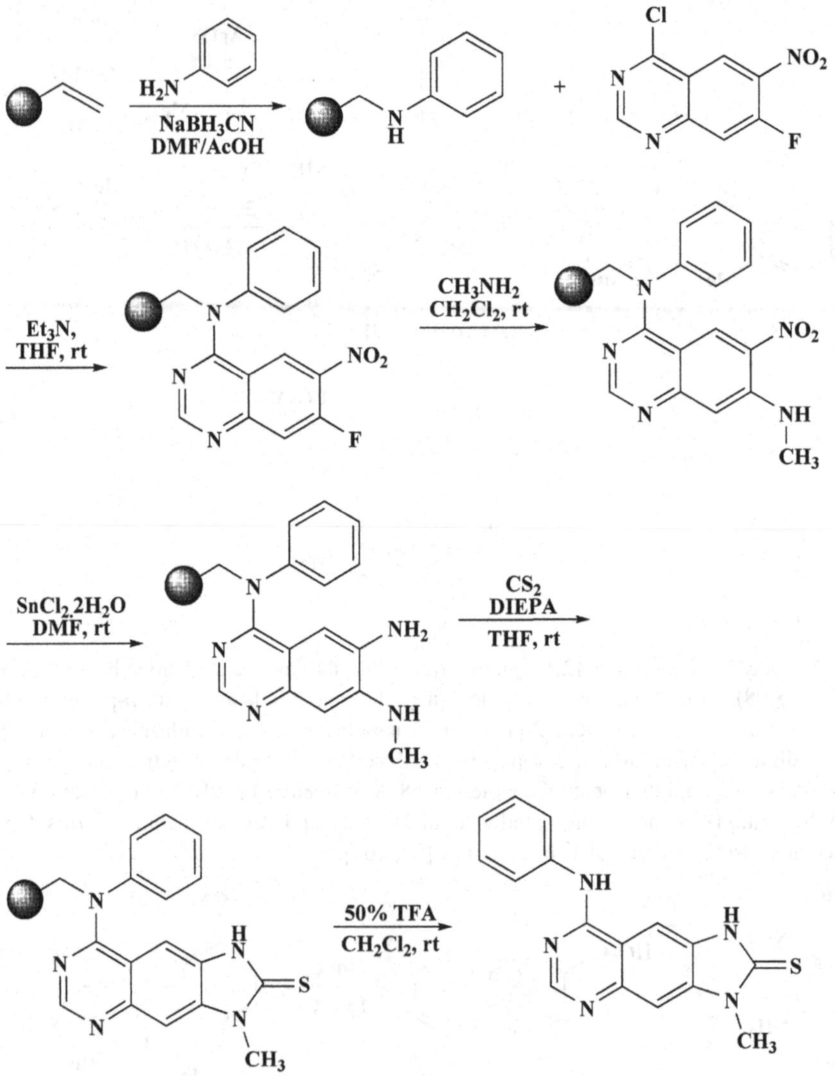

Scheme-89

The condensation of substituted *N*-phenyl-*o*-phenylenediamines with indole/benzo[*b*]thiophene-3-aldehydes under reflux conditions in methoxyethanol provides pyridobenzimidazoles in very good to excellent yields (81–96%). Diamines are synthesized when 2-chloro-3-nitropyridine is treated with suitably substituted anilines, and the 3-nitro-*N*-phenylpyridin-2-amines are reduced with tin(II)chloride under MWI. In the first step, appropriate anilines and 3-nitro-2-chloropyridine are transformed into the 3-nitro-2-(*N*-phenylamino)pyridines, with considerable conversion rates, by nucleophilic substitution under MW heating at 110 °C for 10 minutes under solvent-free conditions. Using dimethylsulfoxide as the solvent under conventional heating, nitro amines are obtained in lower yields (60–70%) and significantly longer reaction times are needed (24 hours). The second step is a MW-assisted reduction of 3-nitro-*N*-phenylpyridin-2-amines to 3-amino-2-*N*-arylaminopyridines. Most of these protocols use formate as the hydrogen transfer source. However, this MW method has the drawback of carbon dioxide formation which increases the pressure in the sealed MW tube to unsafe levels, thus making removal of the product tenuous at best. In the third step, pyridobenzimidazoles are formed in good to excellent yields (81–96%) when amines are condensed with substituted indole/benzo[*b*]thiophene-3-aldehydes at elevated temperatures (~125 °C) in the presence of methoxyethanol as the solvent **(Scheme 90)** [166].

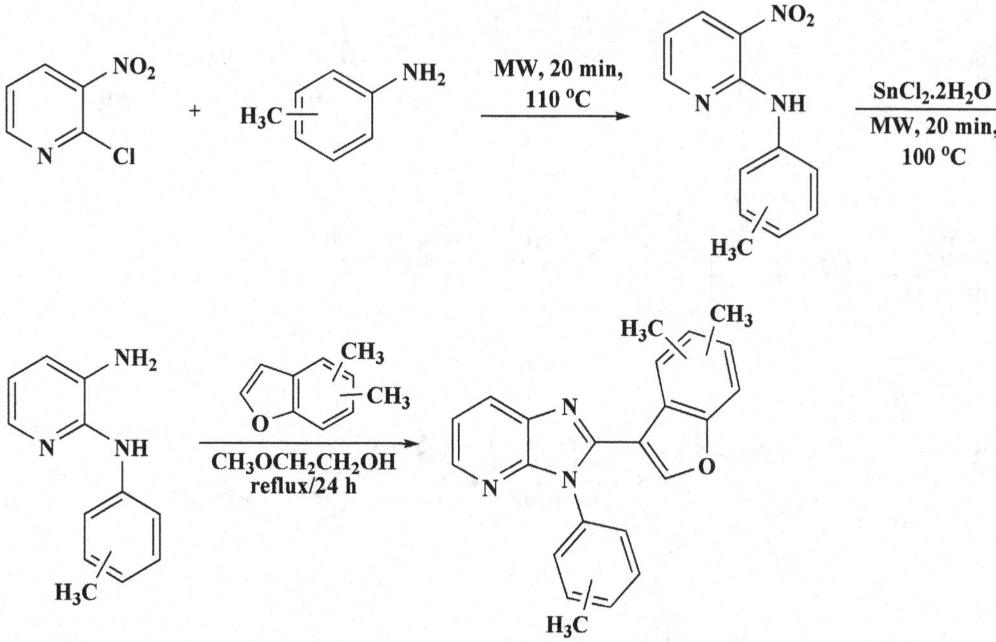

Scheme-90

7-Fluoro-4-methyl-6-nitro-2-oxo-2*H*-1-benzopyran-3-carboxylic acid, through its carboxyl group, couples with Rink amide resin. The resin-bound scaffold undergoes aromatic nucleophilic substitution with primary amines, and the reduction of the nitro group with $SnCl_2$ follows. The *o*-dianilino intermediates are cyclized with thiocarbonyldiimidazole to provide the resin-bound 1,3-dihydro-2-thioxo-6*H*-pyrano[2,3-*f*]benzimidazole-6-ones which undergo *S*-alkylation with alkyl halides and DIEA (*N,N*-diisopropylethylamine) **(Scheme 91)**. Ultimately, highly pure products are formed in good yields after trifluoroacetic acid cleavage [167].

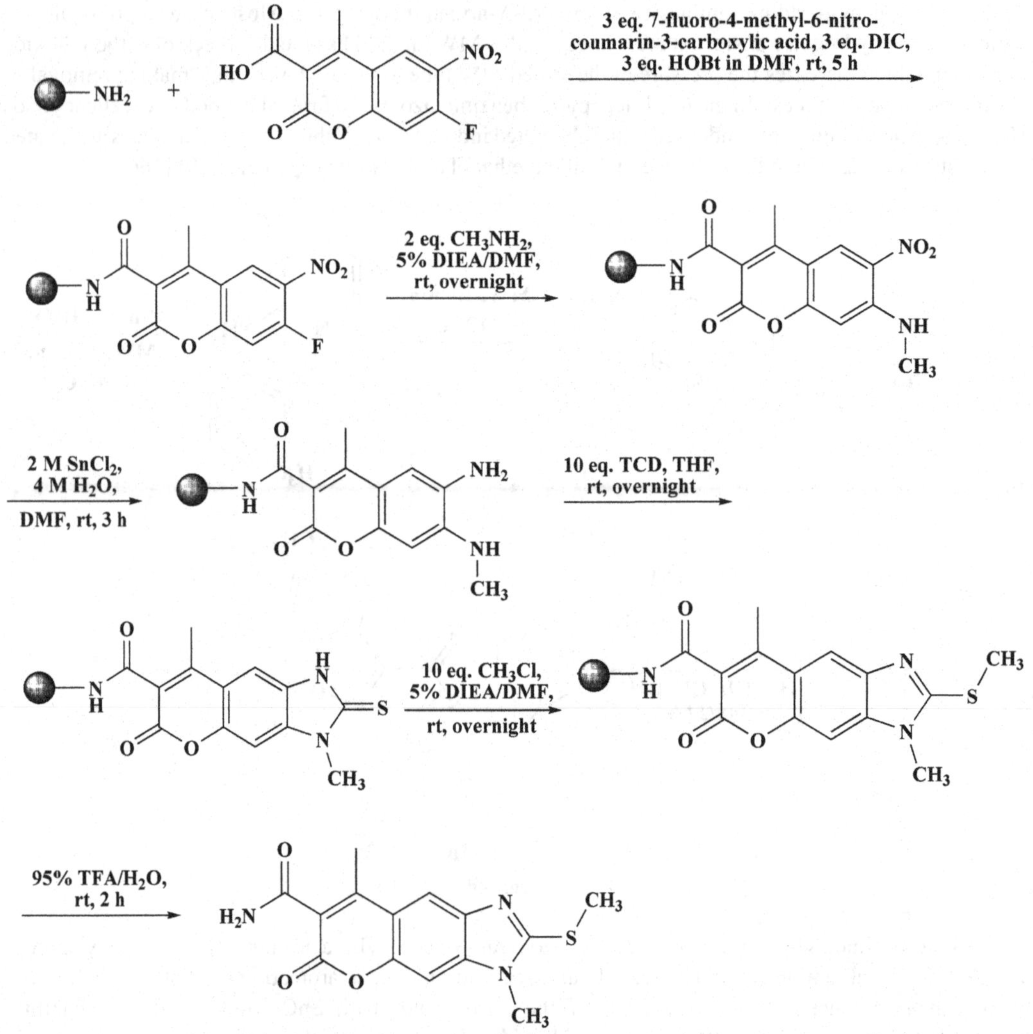

Scheme-91

Ytterbium-assisted synthesis

Certain substrates undergo 1,3-shifts aided by Lewis acid catalysts like Yb(OTf)$_3$ and La(OTf)$_3$ in CH$_2$Cl$_2$ to afford chromanediones, in addition to [3,3]-sigmatropic rearrangement of 3-allyl substituted flavone ethers. The reaction occurs *via* a dissociative intermediate and the outcome of the reaction is influenced by the ability of the aromatic group to stabilize the positive charge buildup. Several aromatic substituents like 2-thiophenyl, 2-furyl and 3,4-dimethoxyphenyl have been found to be compatible, while compounds with 3-methoxy and 3-furylphenyl are unreactive. A single-pot reaction sequence of chromanediones with benzaldehydes yields benzopyranoimidazoles, obviating further purification **(Scheme 92)** [168].

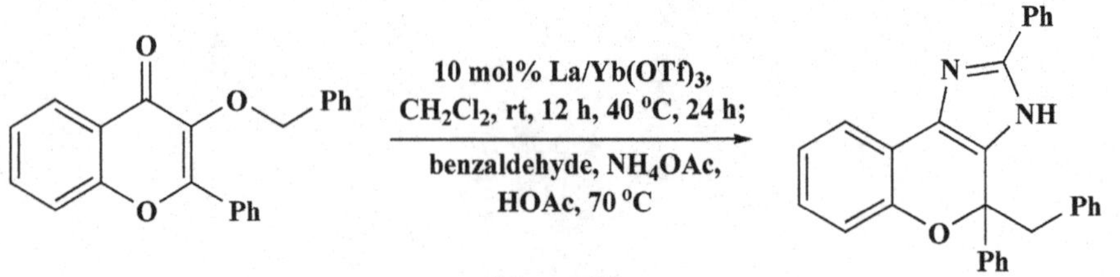

Scheme-92

Zinc-assisted synthesis

Mamada et al. [169] reported a solvent-free 'green' strategy for the synthesis of polycyclic benzimidazoles based on heating of arylene diamines and carboxylic acid anhydrides in the presence of $Zn(OAc)_2$ in solid state. Train sublimation of crude reaction mixtures aided in the isolation and purification of the products **(Scheme 93)** [170].

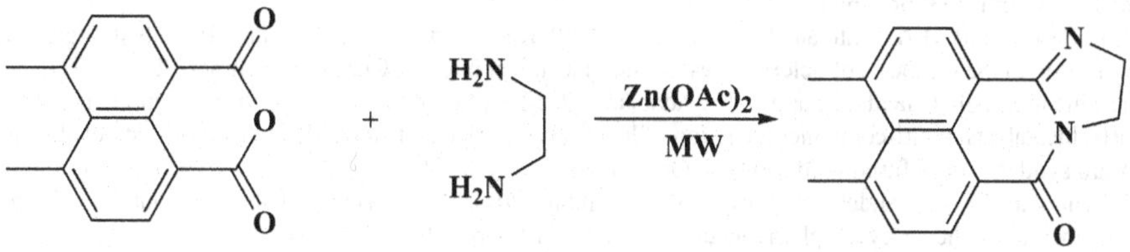

Scheme-93

References

(1)   (a) N. Kaur and D. Kishore. 2014. Nitrogen-containing six-membered heterocycles: solid-phase synthesis. Synth. Commun. 44: 1173–1211. (b) N. Kaur. 2015. Review of microwave-assisted synthesis of benzo-fused six-membered *N,N*-heterocycles. Synth. Commun. 45: 300–330. (c) N. Kaur. 2019. Application of silver-promoted reactions in the synthesis of five-membered *O*-heterocycles. Synth. Commun. 49: 743–789. (d) N. Kaur. 2019. Synthesis of seven and higher-membered heterocycles using ruthenium catalysts. Synth. Commun. 49: 617–661. (e) N. Kaur. 2018. Ruthenium catalyzed synthesis of five-membered *O*-heterocycles. Inorg. Chem. Commun. 99: 82–107.

(2)   (a) N. Kaur. 2014. Microwave-assisted synthesis of five-membered *O,N*-heterocycles. Synth. Commun. 44: 3509–3537. (b) N. Kaur. 2014. Microwave-assisted synthesis of five-membered *O,N,N*-heterocycles. Synth. Commun. 44: 3229–3247.

(3)   (a) N. Kaur. 2015. Six-membered *N*-heterocycles: microwave-assisted synthesis. Synth. Commun. 45: 1–34. (b) N. Kaur. 2015. Polycyclic six-membered *N*-heterocycles: microwave-assisted synthesis. Synth. Commun. 45: 35–69.

(4)   (a) N. Kaur and D. Kishore. 2014. Synthetic strategies applicable in the synthesis of privileged scaffold: 1,4-benzodiazepine. Synth. Commun. 44: 1375–1413. (b) N. Kaur. 2014. Microwave-assisted synthesis of five-membered *O*-heterocycles. Synth. Commun. 44: 3483–3508.

(5)   (a) N. Kaur. 2018. Ultrasound-assisted green synthesis of five-membered *O*- and *S*-heterocycles. Synth. Commun. 48: 1715–1738. (b) N. Kaur. 2018. Photochemical-mediated reactions in five-membered *O*-heterocycles synthesis. Synth. Commun. 48: 2119–2149.

(6)   (a) N. Kaur and D. Kishore. 2014. Solid-phase synthetic approach toward the synthesis of oxygen-containing heterocycles. Synth. Commun. 44: 1019–1042. (b) N. Kaur. 2015. Microwave-assisted synthesis of fused polycyclic six-membered *N*-heterocycles. Synth. Commun. 45: 273–299.

(7)   M. Garcia-Valverde and T. Torroba. 2005. Sulfur-nitrogen heterocycles. Molecules 10: 318–320.

(8)   G.L. Almajan, S.-F. Barbuceanu, G. Bancescu, I. Saramet, G. Saramet and C. Draghici. 2010. Synthesis and anti-microbial evaluation of some fused heterocyclic [1,2,4]triazolo[3,4-*b*][1,3,4]thiadiazole derivatives. Eur. J. Med. Chem. 45: 6139–6146.

(9)   A. Foroumadi, S. Emami, A. Hassanzadeh, M. Rajaee, K. Sokhanvar, M.H. Moshafi and A. Shafiee. 2005. Synthesis and anti-bacterial activity of *N*-(5-benzylthio–1,3,4-thiadiazol-2-yl) and *N*-(5-benzylsulfonyl-1,3,4-thiadiazol-2-yl) piperazinyl quinolone derivatives. Bioorg. Med. Chem. Lett. 15: 4488–4492.

(10)   B. Chandrakantha, A.M. Isloor, P. Shetty, H.K. Fun and G. Hegde. 2014. Synthesis and biological evaluation of novel substituted 1,3,4-thiadiazole and 2,6-di aryl substituted imidazo[2,1-*b*][1,3,4]thiadiazole derivatives. Eur. J. Med. Chem. 71: 316–323.

(11)   T. Karabasanagouda, A.V. Adhikari and N.S. Shetty. 2007. Synthesis and anti-microbial activities of some novel 1,2,4-triazolo[3,4-*b*]-1,3,4-thiadiazoles and 1,2,4-triazolo[3,4-*b*]-1,3,4-thiadiazines carrying thioalkyl and sulfonyl phenoxy moieties. Eur. J. Med. Chem. 42: 521–529.

(12)   D. Kumar, N.M. Kumar, K.-H. Chang, R. Gupta and K. Shah. 2011. Synthesis and *in vitro* anti-cancer activity of 3,5-bis(indolyl)-1,2,4-thiadiazoles. Bioorg. Med. Chem. Lett. 21: 5897–5900.

(13)   Y. Luo, S. Zhang, Z.-J. Liu, W. Chen, J. Fu, Q.-F. Zeng and H.-L. Zhu. 2013. Synthesis and anti-microbical evaluation of a novel class of 1,3,4-thiadiazole: derivatives bearing 1,2,4-triazolo[1,5-*a*]pyrimidine moiety. Eur. J. Med. Chem. 64: 54–61.

(14)   C.L. Rebolledo, P. Sotelo-Hitschfeld, S. Brauchi and M.Z. Olavarria. 2013. Design and synthesis of conformationally restricted capsaicin analogues based on the 1,3,4-thiadiazole heterocycle reveal a novel family of transient receptor potential vanilloid 1 (TRPV1) antagonists. Eur. J. Med. Chem. 66: 193–203.

(15)   B.A. Teicher, S.D. Liu, J.T. Liu, S.A. Holden and T.S. Herman. 1993. A carbonic anhydrase inhibitor as a potential modulator of cancer therapies. Anticancer Res. 13: 1549–1556.

(16)   E. Cereda, M. Turconi, A. Ezhaya, E. Bellora, A. Brambilla, F. Pagani and A. Donetti. 1987. Anti-secretary and anti-ulcer activities of some new 2-(2-pyridylmethyl-sulfinyl)-benzimidazoles. Eur. J. Med. Chem. 22: 527–537.

(17)   R.D. Carpenter, K.S. Lam and M.J. Kurth. 2007. Microwave-mediated heterocyclization to benzimidazo[2,1-*b*]quinazolin–12(5*H*)-ones. J. Org. Chem. 72: 284–287.

(18)   I. Mohammadpoor-Baltork, A.R. Khosropour and S.F. Hojati. 2007. Mild and efficient synthesis of benzoxazoles, benzothiazoles, benzimidazoles, and oxazolo[4,5-*b*]pyridines catalyzed by Bi(III) salts under solvent-free conditions. Montash. Chem. 138: 663–667.

(19)   J.A.R. Salvador, R.M.A. Pinto and S.M. Silvestre. 2009. Recent advances of bismuth(III) salts in organic chemistry: application to the synthesis of heterocycles of pharmaceutical interest. Curr. Org. Synth. 6: 426–470.

(20)   S.K. Guchhait, A.L. Chandgude and G. Priyadarshani. 2012. CuSO$_4$-Glucose for *in situ* generation of controlled Cu(I)-Cu(II) bicatalysts: multi-component reaction of heterocyclic azine and aldehyde with alkyne, and cycloisomerization toward synthesis of *N*-fused imidazoles. J. Org. Chem. 77: 4438–4444.

(21)   M. Baumann, I.R. Baxendale, S.V. Ley and N. Nikbin. 2011. An overview of the key routes to the best selling 5-membered ring heterocyclic pharmaceuticals. Beilstein J. Org. Chem. 7: 442–495.

(22)   N. Chernyak and V. Gevorgyan. 2010. General and efficient copper-catalyzed three-component coupling reaction towards imidazoheterocycles: one-pot synthesis of alpidem and zolpidem. Angew. Chem. Int. Ed. 49: 2743–2746.

(23)   S.V. Dhamnaskar and D.W. Rangnekar. 1988. Synthesis of triazoflo[4,5-*b*]pyrido[1',2'-*a*]benzimidazole derivatives as fluorescent disperse dyes and whiteners for polyester fibre. Dyes Pigm. 9: 467–473.

(24)   (a) K.M. Dawood and B.F. Abdel-Wahab. 2010. Synthetic routes to benzimidazole-based fused polyheterocycles. ARKIVOC (i): 333–389. (b) K.M. Dawood, N.M. Elwan, A.A. Farahat and B.F. Abdel-Wahab. 2010. 1*H*-Benzimidazole-2-acetonitriles as synthon in fused benzimidazole synthesis. J. Heterocycl. Chem. 47: 243–267.

(25)   S. Mishra and R. Ghosh. 2011. Mechanistic studies on a new catalyst system (CuI-NaHSO$_4$×SiO$_2$) leading to the one-pot synthesis of imidazo[1,2-*a*]pyridines from reactions of 2-aminopyridines, aldehydes, and terminal alkynes. Synthesis 21: 3463–3470.

(26)   R.-L. Yan, H. Yan, C. Ma, Z.-Y. Ren, X.-A. Gao, G.-S. Huang and Y.-M. Liang. 2012. Cu(I)-catalyzed synthesis of imidazo[1,2-*a*]pyridines from aminopyridines and nitroolefins using air as the oxidant. J. Org. Chem. 77: 2024–2028.

(27)   H. Wang, Y. Wang, C. Peng and J. Zhang. 2010. A direct intramolecular C-H amination reaction co-catalyzed by copper(II) and iron(III) as part of an efficient route for the synthesis of pyrido[1,2-*a*]benzimidazoles from *N*-aryl–2-aminopyridines. J. Am. Chem. Soc. 132: 13217–13219.

(28)   A. Da Settimo, G. Primofiore, F. Da Settimo, G. Pardi, F. Simorini and A.M. Marini. 2002. An approach to novel fused triazole or tetrazole derivatives starting from benzimidazo[1,2-*a*]quinazoline-5(7*H*)-one and 5,7-dihydro-5-oxopyrido[3',2':5,6]pyrimido[1,2-*a*]benzimidazole. J. Heterocycl. Chem. 39: 1007–1011.

(29)   B.L. Zhou, Q.L. Yuan and D.W. Ma. 2007. Synthesis of 1,2-disubstituted benzimidazoles by a Cu-catalyzed cascade aryl amination/condensation process. Angew. Chem. Int. Ed. 46: 2598–2601.

(30)   Y. Liu and J.-P. Wan. 2011. Tandem reactions initiated by copper-catalyzed cross-coupling: a new strategy towards heterocycle synthesis. Org. Biomol. Chem. 9: 6873–6894.

(31)   B.L. Zhou, Q.L. Yuan and D.W. Ma. 2007. Cascade coupling/cyclization process to *N*-substituted 1,3-dihydrobenzimidazol–2-ones. Org. Lett. 9: 4291–4294.

(32)   Q.-F. Zhong and L.-P. Sun. 2010. An efficient synthesis of 6,9-disubstituted purin-8-ones *via* copper-catalyzed coupling/cyclization. Tetrahedron 66: 5107–5111.

(33)   N. Ibrahim and M. Legraverend. 2009. Synthesis of 6,7,8-trisubstituted purines *via* a copper-catalyzed amidation reaction. J. Org. Chem. 74: 463–465.

(34)   X. Lv and W.L. Bao. 2009. Copper-catalyzed cascade addition/cyclization: an efficient and versatile synthesis of *N*-substituted 2-heterobenzimidazoles. J. Org. Chem. 74: 5618–5621.

(35)   G.D. Shen and W.L. Bao. 2010. Synthesis of benzoxazole and benzimidazole derivatives *via* ligand-free copper(I)-catalyzed cross-coupling reaction of *o*-halophenols or *o*-haloanilines with carbodiimides. Adv. Synth. Catal. 352: 981–986.

(36)   X.H. Deng and N.S. Mani. 2010. Reactivity-controlled regioselectivity: a regiospecific synthesis of 1,2-disubstituted benzimidazoles. Eur. J. Org. Chem. 4: 680–686.

(37)   X.H. Deng, H. McAllister and N.S. Mani. 2009. CuI-catalyzed amination of arylhalides with guanidines or amidines: a facile synthesis of 1*H*-2-substituted benzimidazoles. J. Org. Chem. 74: 5742–5745.

(38)   D.S. Surry and S.L. Buchwald. 2010. Diamine ligands in copper-catalyzed reactions. Chem. Sci. 1: 13–31.

(39)   J. Yuen, Y.Q. Fang and M. Lautens. 2006. CuI-catalyzed tandem intramolecular amidation using *gem*-dibromovinyl systems. Org. Lett. 8: 653–656.

(40)   J.Y. Lu, X.Y. Gong, H.J. Yang and H. Fu. 2010. Concise copper-catalyzed one-pot tandem synthesis of benzimidazo[1,2-*b*]isoquinolin-11-one derivatives. Chem. Commun. 46: 4172–4174.

(41)   A. Wang, H. Jiang and H. Chen. 2009. Palladium-catalyzed diacetoxylation of alkenes with molecular oxygen as sole oxidant. J. Am. Chem. Soc. 131: 3846–3847.

(42)   A. Wang and H. Jiang. 2010. Palladium-catalyzed direct oxidation of alkenes with molecular oxygen: general and practical methods for the preparation of 1,2-diols, aldehydes, and ketones. J. Org. Chem. 75: 2321–2326.

(43) Z. Shi, C. Zhang, C. Tanga and N. Jiao. 2012. Recent advances in transition metal-catalyzed reactions using molecular oxygen as the oxidant. Chem. Soc. Rev. 41: 3381–3430.

(44) H. Wang, Y. Wang, D. Liang, L. Liu, J. Zhang and Q. Zhu. 2011. Copper-catalyzed intramolecular dehydrogenative aminooxygenation: direct access to formyl-substituted aromatic *N*-heterocycles. Angew. Chem. Int. Ed. 50: 5678–5681.

(45) R. Martin, R.M. Rivero and S.L. Buchwald. 2006. Domino Cu-catalyzed C-N coupling/hydroamidation: a highly efficient synthesis of nitrogen heterocycles. Angew. Chem. Int. Ed. 45: 7079–7082.

(46) L. Ackermann, S. Barfuesser and H.K. Potukuchi. 2009. Copper-catalyzed *N*-arylation/hydroamin(d)ation domino synthesis of indoles and its application to the preparation of a chek1/KDR kinase inhibitor pharmacophore. Adv. Synth. Catal. 351: 1064–1072.

(47) H.C. Shen, F.X. Ding, Q.L. Deng, L.C. Wilsie, M.L. Krsmanovic, A.K. Taggart, E. Carballo-Jane, N. Ren, T.Q. Cai, T.J. Wu, K.K. Wu, K. Cheng, Q. Chen, M.S. Wolff, X.C. Tong, T.G. Holt, M.G. Waters, M.L. Hammond, J.R. Tata and S.L. Colletti. 2009. Discovery of novel tricyclic full agonists for the G-protein-coupled niacin receptor 109A with minimized flushing in rats. J. Med. Chem. 52: 2587–2602.

(48) C. Pabba, H.-J. Wang, S.R. Mulligan, Z.-J. Chen, T.M. Stark and B.T. Gregg. 2005. Microwave-assisted synthesis of 1-aryl-1*H*-indazoles *via* one-pot two-step Cu-catalyzed intramolecular-*N*-arylation of arylhydrazones. Tetrahedron. Lett. 46: 7553–7557.

(49) P. Nilsson, K. Olofsson and M. Larhed. 2006. Microwave-assisted and metal-catalyzed coupling reactions. Top. Curr. Chem. 266: 103-144.

(50) N. Kaur. 2014. Microwave-assisted synthesis of five-membered *O,N,N*-heterocycles. Synth. Commun. 44: 3229–3247.

(51) N. Okamoto, K. Sakurai, M. Ishikura, K. Takeda and R. Yanada. 2009. One-pot concise syntheses of benzimidazo[2,1-*a*] isoquinolines by a microwave-accelerated tandem process. Tetrahedron Lett. 50: 4167–4169.

(52) U. Halbes-Letinois, J.M. Weibel and P. Pale. 2007. The organic chemistry of silver acetylides. Chem. Soc. Rev. 36: 759–769.

(53) S.V. Ley and A.W. Thomas. 2003. Modern synthetic methods for copper-mediated C(aryl)[bond]O, C(aryl)[bond]N, and C(aryl)[bond]S bond formation. Angew. Chem. Int. Ed. 42: 5400–5449.

(54) K.T.J. Loones, B.U.W. Maes, R.A. Dommissse and G.L.F. Lemiere. 2004. The first tandem double palladium-catalyzed aminations: synthesis of dipyrido[1,2-*a*:3',2'-*d*]imidazole and its benzo- and aza-analogues. Chem. Commun. 21: 2466–2467.

(55) B. Zhao, H. Du and Y. Shi. 2008. Cu(I)-catalyzed cycloguanidination of olefins. Org. Lett. 10: 1087–1090.

(56) D.A. Bain. 2008. Chemical exchange. Annu. Rep. NMR Spectrosc. 63: 23–48.

(57) G. Zhang, Y. Luo, Y. Wang and L. Zhang. 2011. Combining gold(I)/gold(III) catalysis and C-H functionalization: a formal intramolecular [3+2]-annulation towards tricyclic indolines and mechanistic studies. Angew. Chem. Int. Ed. 50: 4450–4454.

(58) N.D. Shapiro, Y. Shi and F.D. Toste. 2009. Gold-catalyzed [3+3]-annulation of azomethine imines with propargyl esters. J. Am. Chem. Soc. 131: 11654–11655.

(59) H. Huang, Y. Zhou and H. Liu. 2011. Recent advances in the gold-catalyzed additions to C-C multiple bonds. Beilstein J. Org. Chem. 7: 897–936.

(60) D. Ye, J. Wang, X. Zhang, Y. Zhou, X. Ding, E. Feng, H. Sun, G. Liu, H. Jiang and H. Liu. 2009. Gold-catalyzed intramolecular hydroamination of terminal alkynes in aqueous media: efficient and regioselective synthesis of indole-1-carboxamides. Green Chem. 11: 1201–1208.

(61) F. Alexandre, L. Domon, S. Frere, A. Testard, V. Thiery and T. Besson. 2003. Microwaves in drug discovery and multistep synthesis. Mol. Divers. 7: 273–280.

(62) M. Hamana, H. Noda and J. Uchida. 1970. A novel cyclization reactions of 2-(2-quinolyl)- or 2-(2-pyridyl)-cyclohexanone oximes under conditions of the Beckmann rearrangement. Yakugaku Zasshi 90: 991–1000.

(63) D.H.R. Barton, W.B. Motherwell, E.S. Simon and S.Z.J. Zard. 1984. A mild and efficient method for the reduction of oximes to imines for further *in situ* reactions. Chem. Soc. Chem. Commun. 6: 337–338.

(64) Y. Ishida, S. Sasatani, K. Maruoka and H. Yamamoto. 1983. A new synthesis of imidoyl iodides *via* Beckmann rearrangement of oxime sulfonates. Tetrahedron Lett. 24: 3255–3258.

(65) B. Quiclet-Sire and S.Z. Zard. 2006. Observations on the reaction of hydrazones with iodine: interception of the diazo intermediates. Chem. Commun. 17: 1831–1832.

(66) P.T. Parvatkar, P.S. Parameswaran and S.G. Tilve. 2012. Recent developments in the synthesis of five- and six-membered heterocycles using molecular iodine. Chem. Eur. J. 18: 5460–5489.

(67) F. Shibahara, A. Kitagawa, E. Yamaguchi and T. Murai. 2006. Synthesis of 2-azaindolizines by using an iodine-mediated oxidative desulfurization promoted cyclization of *N*–2-pyridylmethyl thioamides and an investigation of their photophysical properties. Org. Lett. 8: 5621–5624.

(68) K.C. Nicolaou and H.J. Mitchell. 2001. Adventures in carbohydrate chemistry: new synthetic technologies, chemical synthesis, molecular design, and chemical biology. Angew. Chem. Int. Ed. 40: 1576–1624.

(69) M. Shimizu and T. Hiyama. 2005. Modern synthetic methods for fluorine-substituted target molecules. Angew. Chem. Int. Ed. 44: 214–231.

(70)   S.K. Meegalla, G.J. Stevens, C.A. McQueen, A.Y. Chen, C. Yu, L.F. Liu, L.R. Barrows and E.J. LaVoie. 1994. Synthesis and pharmacological evaluation of isoindolo[1,2-*b*]quinazolinone and isoindolo[2,1-*a*]benzimidazole derivatives related to the anti-tumor agent batracylin. J. Med. Chem. 37: 3434–3439.

(71)   G. Smolinsky and B.I. Feuer. 1966. Deoxygenation of nitro groups. The question of nitrene formation. J. Org. Chem. 31: 3882–3884.

(72)   A. Blank. 1891. Ueber carbazolsynthesen. Chem. Ber. 24: 306–306.

(73)   R. Shintani and G.C. Fu. 2003. A new copper-catalyzed [3+2]-cycloaddition: enantioselective coupling of terminal alkynes with azomethine imines to generate five-membered nitrogen heterocycles. J. Am. Chem. Soc. 125: 10778–10779.

(74)   Y.Y. Nakamura. 2004. Transition metal-catalyzed reactions in heterocyclic synthesis. Chem. Rev. 104: 2127–2198.

(75)   S. Bohm, R. Kubík, J. Novotny, J. Ondracek, B. Kratochvil and J. Kuthan. 1991. Reinvestigation of heterocyclic structures: oxidation products from 1-substituted 2,4,6-triphenylpyridinium salts. Collect. Czech. Chem. Commun. 56: 2326–2339.

(76)   D. Pessoa-Mahana, C. Espinosa-Bustos, J. Mella-Raipan, J. Canales-Pacheco and H. Pessoa-Mahana. 2009. Microwave-assisted synthesis and regioisomeric structural elucidation of novel benzimidazo[1,2-*d*][1,4] benzodiazepinone derivatives (09-3986AP). ARKIVOC (xii): 131–140.

(77)   A. Rahmati and Z. Khalesi. 2011. One-pot three-component synthesis of imidazo[1,5]pyridines. Int. J. Org. Chem. 1: 15–19.

(78)   E.F. DiMauro and J.M. Kennedy. 2007. Rapid synthesis of 3-amino-imidazopyridines by a microwave-assisted four-component coupling in one-pot. J. Org. Chem. 72: 1013–1016.

(79)   R.A. de Silva, S. Santra and P.R. Andreana. 2008. A tandem one-pot, microwave-assisted synthesis of regiochemically differentiated 1,2,4,5-tetrahydro-1,4-benzodiazepin-3-ones. Org. Lett. 10: 4541–4544.

(80)   F. Esser. 1987. Cyclic guanidines; I. Intramolecular mercury(II)-induced amination of alkenes as a convenient route to bicyclic guanidines. Synthesis 5: 460–466.

(81)   D. Pappo, S. Shimony and Y. Kashman. 2005. Synthesis of 9-substituted tetrahydrodiazepinopurines: studies toward the total synthesis of asmarines. J. Org. Chem. 70: 199–206.

(82)   T.E. Muller, K.C. Hultzsch, M. Yus, F. Foubelo and M. Tada. 2008. Hydroamination: direct addition of amines to alkenes and alkynes. Chem. Rev. 108: 3795–3892.

(83)   X.-N. Zhang, Y.-X. Li and Z.-H. Zhang. 2011. Nickel chloride-catalyzed one-pot three-component synthesis of pyrazolophthalazinyl spirooxindoles. Tetrahedron 67: 7426–7430.

(84)   (a) G.S. Singh and Z.Y. Desta. 2012. Isatins as privileged molecules in design and synthesis of spiro-fused cyclic frameworks. Chem. Rev. 112: 6104–6155. (b) H. Ren and P. Knochel. 2006. Regioselective functionalization of tri-substituted pyridines using a bromine-magnesium exchange. Chem. Commun. 7: 726–728.

(85)   R. Pathak and S. Batra. 2007. Expeditious synthesis of 5,6,7,8-tetrahydro-imidazo[1,2-*a*]pyrimidin–2-ones and 3,4,6,7,8,9-hexahydro-pyrimido[1,2-*a*]pyrimidin-2-ones. Tetrahedron 63: 9448–9455.

(86)   S. Nag and S. Batra. 2011. Applications of allylamines for the syntheses of aza-heterocycles. Tetrahedron 67: 8959–9061.

(87)   J. Streuff, C.H. Hoevelmann, M. Nieger and K. Muniz. 2005. Palladium(II)-catalyzed intramolecular diamination of un-functionalized alkenes. J. Am. Chem. Soc. 127: 14586–14587.

(88)   J. Ney and J.P. Wolfe. 2005. Selective synthesis of 5- or 6-aryl octahydrocyclopenta[*b*]pyrroles from a common precursor through control of competing pathways in a Pd-catalyzed reaction. J. Am. Chem. Soc. 127: 8644–8651.

(89)   C. Hyland. 2005. Cyclizations of allylic substrates *via* palladium catalysis. Tetrahedron 61: 3457–3471.

(90)   N.T. Patil and Y. Yamamoto. 2006. Palladium-catalyzed cascade reactions involving π-allyl palladium chemistry. Top. Organomet. Chem. 19: 91–113.

(91)   M. Seki, Y. Mori, M. Hatsuda and S. Yamada. 2002. A novel synthesis of (+)-biotin from L-cysteine. J. Org. Chem. 67: 5527–5536.

(92)   H. Ohno, A. Okano, S. Kosaka, K. Tsukamoto, M. Ohata, K. Ishihara, H. Maeda, T. Tanaka and N. Fujii. 2008. Direct construction of bicyclic heterocycles by palladium-catalyzed domino cyclization of propargyl bromides. Org. Lett. 10: 1171–1174.

(93)   N.J. Tom, W.M. Simon, H.N. Frost and M. Ewing. 2004. De-protection of a primary Boc group under basic conditions. Tetrahedron Lett. 45: 905–906.

(94)   (a) J.P. Wolfe. 2006. Stereoselective synthesis of saturated heterocycles *via* Pd-catalyzed alkene carboetherification and carboamination reactions. Synlett 4: 571–582. (b) J.P. Wolfe. 2008. Stereoselective synthesis of saturated heterocycles *via* Pd-catalyzed alkene carboetherification and carboamination reactions. Synlett 19: 2913–2937.

(95)   R.A.T.M. van Benthem, H. Hiemstra, G. Rodriguez Longarela and W.N. Speckamp. 1994. Formamide as a superior nitrogen nucleophile in palladium(II)-mediated synthesis of imidazolidines. Tetrahedron Lett. 35: 9281–9284.

(96)   E.M. Beccalli, G. Broggini, M. Martinelli and S. Sottocornola. 2007. C-C, C-O, C-N Bond formation on sp^2 carbon by Pd(II)-catalyzed reactions involving oxidant agents. Chem. Rev. 107: 5318–5365.

(97)   E.M. Beccalli, A. Bernasconi, E. Borsini, G. Broggini, M. Rigamonti and G. Zecchi. 2010. Tunable Pd-catalyzed cyclization of indole-2-carboxylic acid allenamides: carboamination vs microwave-assisted hydroamination. J. Org. Chem. 75: 6923–6932.

(98)   H. Alper, C.P. Perera and F.R. Ahmed. 1981. A novel synthesis of *β*-lactams. J. Am. Chem. Soc. 103: 1289–1291.

(99)   I. Paterson, H. Alper and C.P. Mahatantila. 1983. Palladium(0) and phase transfer catalyzed conversion of azirines to styrylindoles. Heterocycles 20: 2025–2028.

(100)   K.H. Howaradl. 1995. Transition metal-mediated carbonylative ring-expansion of heterocyclic compounds. Acc. Chem. Res. 28: 414–422.

(101)   A.J. Rosenberg, J. Zhao and D.A. Clark. 2012. Synthesis of imidazo[4,5-*b*]pyridines and imidazo[4,5-*b*]pyrazines by palladium-catalyzed amidation of 2-chloro-3-amino-heterocycles. Org. Lett. 14: 1761–1767.

(102)   J.W. Hubbard, A.M. Piegols and B.C.G. Soederberg. 2007. Palladium-catalyzed *N*-heteroannulation of *N*-allyl- or *N*-benzyl-2-nitrobenzenamines: synthesis of 2-substituted benzimidazoles. Tetrahedron 63: 7077–7085.

(103)   M. Alajarin, A. Vidal, F. Tovar and C. Conesa. 1999. Formal [4+2]-intramolecular cycloaddition ketenimine-imine. Synthesis of benzimidazo[1,2-*b*]isoquinolines. Tetrahedron Lett. 40: 6127–6130.

(104)   M. Alajarin, A. Vidal and F. Tovar. 2000. Periselective intramolecular [4+2]-cycloadditions of ketenimines: synthesis of pyrido[1,2-*a*]benzimidazoles. Tetrahedron Lett. 41: 7029–7032.

(105)   K. Muniz, C.H. Hoevelmann and J. Streuff. 2008. Oxidative diamination of alkenes with ureas as nitrogen sources: mechanistic pathways in the presence of a high oxidation state palladium catalyst. J. Am. Chem. Soc. 130: 763–773.

(106)   J. Helaja and R. Gottlich. 2002. A new catalytic hetero-Heck type reaction. Chem. Commun. 7: 720–721.

(107)   H. Tsutsui and K. Narasaka. 1999. Synthesis of pyrrole derivatives by the Heck-type cyclization of γ,δ-unsaturated ketone *O*-pentafluorobenzoyloximes. Chem. Lett. 28: 45–46.

(108)   I. Islam and E.B. Skibo. 1990. Synthesis and physical studies of azamitosene and iminoazamitosene reductive alkylating agents. Iminoquinone hydrolytic stability, *syn/anti* isomerization, and electrochemistry. J. Org. Chem. 55: 3195–3205.

(109)   E.B. Skibo, I. Islam, W.G. Schulz, R. Zhou, L. Bess and R. Boruah. 1996. The organic chemistry of the pyrrolo[1,2-*a*] benzimidazole anti-tumor agents. An example of rational drug design. Synlett 4: 297–309.

(110)   W.G. Schulz and E.B. Skibo. 2000. Inhibitors of topoisomerase II based on the benzodiimidazole and dipyrroloimidazobenzimidazole ring systems: controlling DT-diaphorase reductive inactivation with steric bulk. J. Med. Chem. 43: 629–638.

(111)   L. Besson, J. Gore and B. Cazes. 1995. Synthesis of allylic amines through the palladium-catalyzed hydroamination of allenes. Tetrahedron Lett. 36: 3857–3860.

(112)   M. Al-Masum, M. Meguro and Y. Yamamoto. 1997. The two-component palladium catalyst system for intermolecular hydroamination of allenes. Tetrahedron Lett. 38: 6071–6074.

(113)   M. Meguro and Y. Yamamoto. 1998. A new method for the synthesis of nitrogen heterocycles *via* palladium-catalyzed intramolecular hydroamination of allenes. Tetrahedron Lett. 39: 5421–5424.

(114)   S. Qiu, Y. Wei and G. Liu. 2009. Palladium-catalyzed intramolecular hydroamination of allenes coupled to aerobic alcohol oxidation. Chem. Eur. J. 15: 2751–2754.

(115)   A. Arcadi, G. Bianchi and F. Marinelli. 2004. Gold(III)-catalyzed annulation of 2-alkynylanilines: a mild and efficient synthesis of indoles and 3-haloindoles. Synthesis 4: 610–618.

(116)   N. Miyaura and A. Suzuki. 1995. Palladium-catalyzed cross-coupling reactions of organoboron compounds. Chem. Rev. 95: 2457–2483.

(117)   G.A. Molander and B. Biolatto. 2003. Palladium-catalyzed Suzuki-Miyaura cross-coupling reactions of potassium aryl- and heteroaryltrifluoroborates. J. Org. Chem. 68: 4302–4314.

(118)   W. Yang, Y. Wang and J.R. Corte. 2003. Efficient synthesis of 2-aryl-6-chloronicotinamides *via* PXPd$_2$-catalyzed regioselective Suzuki coupling. Org. Lett. 5: 3131–3134.

(119)   J. Witherington, V. Bordas, S.L. Garland, D.M.B. Hickey, R.J. Ife, J. Liddle, M. Saunders, D.G. Smith and R.W. Ward. 2003. 5-Aryl-pyrazolo[3,4-*b*]pyridines: potent inhibitors of glycogen synthase kinase-3 (GSK-3). Bioorg. Med. Chem. Lett. 13: 1577–1580.

(120)   R.N. Misra, H. Xiao, D.B. Rawlins, W. Shan, K.A. Kellar, J.G. Mulheron, J.S. Sack, J.S. Tokarski, S.D. Kimball and K.R. Webster. 2003. 1*H*-Pyrazolo[3,4-*b*]pyridine inhibitors of cyclin-dependent kinases: highly potent 2,6-difluorophenacyl analogues. Bioorg. Med. Chem. Lett. 13: 2405–2408.

(121)   H. Yu, Y. Fao, Q. Guo and Z. Lin. 2009. Theoretical investigations on mechanisms of Pd(OAc)$_2$-catalyzed intramolecular diaminations in the presence of bases and oxidants. Organometallics 28: 4507–4512.

(122)   G. Liu and S.S. Stahl. 2006. Highly regioselective Pd-catalyzed intermolecular aminoacetoxylation of alkenes and evidence for *cis*-aminopalladation and S$_{N}$2 C-O bond formation. J. Am. Chem. Soc. 128: 7179–7181.

(123)   K. Muniz. 2009. High-oxidation-state palladium catalysis: new reactivity for organic synthesis. Angew. Chem. Int. Ed. 48: 2–14.

(124)   M. Nayak, S. Kanojiya and S. Batra. 2009. The first synthesis of allyl isonitriles from Baylis-Hillman adducts, and their application in the synthesis of substituted imidazo[1,2-*a*]pyridines and tetraazadibenzoazulenes. Synthesis 3: 431–437.

(125)   P. Bilek and J. Slouka. 2002. Cyclocondensation reactions of heterocyclic carbonyl compounds VII synthesis of some substituted benzo-[1,2,4]triazino[2,3-*a*]benzimidazoles. Heterocycl. Commun. 8: 123–128.

(126)   V.P. Kruglenko, V.P. Gnidets and M.V. Povstyanoi. 2000. Synthesis of 2-methyl-1,2,4-triazolo[4,3-*d*]-1,2,4-triazino[2,3-*a*]-benzimidazole and 2-methyl-9-phenylimidazo[1,2-*b*]-1,2,4-triazolo[4,3-*d*]-1,2,4-triazine. Chem. Heterocycl. Compd. 36: 103–104.

(127) J.C. Rech, M. Yato, D. Duckett, B. Ember, P.V. LoGrasso, R.G. Bergman and J.A. Ellman. 2007. Synthesis of potent bicyclic bis-arylimidazole c-Jun *N*-terminal kinase inhibitors by catalytic C-H bond activation. J. Am. Chem. Soc. 129: 490–491.

(128) A. Padwa, D.J. Austin, A.T. Price, M.A. Semones, M.P. Doyle, M.N. Protopopova, W.R. Winchester and A. Tran. 1993. Ligand effects on dirhodium(II) carbene reactivities. Highly effective switching between competitive carbenoid transformations. J. Am. Chem. Soc. 115: 8669–8680.

(129) T.R. Hoye, C.J. Dinsmore, D.S. Johnson and P.F. Korkowski. 1990. Alkyne insertion reactions of metal-carbenes derived from enynyl-α-diazoketones [R′CN$_2$COCR$_2$CH$_2$CC(CH$_2$)$_{n-2}$CH:CH$_2$]. J. Org. Chem. 55: 4518–4520.

(130) B. Chattopadhyay and V. Gevorgyan. 2012. Transition metal-catalyzed denitrogenative transannulation: converting triazoles into other heterocyclic systems. Angew. Chem. Int. Ed. 51: 862–872.

(131) V. Bagheri, M.P. Doyle, J. Taunton and E.E. Claxton. 1988. A new and general synthesis of α-silyl carbonyl compounds by silicon-hydrogen insertion from transition metal-catalyzed reactions of diazo esters and diazo ketones. J. Org. Chem. 53: 6158–6160.

(132) S. Chuprakov, F. Hwang and V. Gevorgyan. 2007. Rh-catalyzed transannulation of pyridotriazoles with alkynes and nitriles. Angew. Chem. Int. Ed. 46: 4757–4759.

(133) K.L. Tan, R.G. Bergman and J.A. Ellman. 2002. Intermolecular coupling of isomerizable alkenes to heterocycles *via* rhodium-catalyzed C-H bond activation. J. Am. Chem. Soc. 124: 13964–13965.

(134) K.L. Tan, S. Park, J.A. Ellman and R.G. Bergman. 2004. Intermolecular coupling of alkenes to heterocycles *via* C-H bond activation. J. Org. Chem. 69: 7329–7335.

(135) J.C. Lewis, R.G. Bergman and J.A. Ellman. 2008. Direct functionalization of nitrogen heterocycles *via* Rh-catalyzed C-H bond activation. Acc. Chem. Res. 41: 1013–1025.

(136) P.P. Graczyk, A. Khan, G.S. Bhatia, V. Palmer, D. Medland, H. Numata, H. Oinuma, J. Catchick, A. Dunne, M. Ellis, C. Smales, J. Whitfield, S.J. Neame, B. Shah, D. Wilton, L. Morgan, T. Patel, R. Chung, H. Desmond, J.M. Staddon, N. Sato and A. Inoue. 2005. The neuroprotective action of JNK3 inhibitors based on the 6,7-dihydro-5*H*-pyrrolo[1,2-*a*]imidazole scaffold. Bioorg. Med. Chem. Lett. 15: 4666–4670.

(137) C. Blackburn, B. Guan, P. Fleming, K. Shiosaki and S. Tsai. 1998. Parallel synthesis of 3-aminoimidazo[1,2-*a*] pyridines and pyrazines by a new three-component condensation. Tetrahedron Lett. 39: 3635–3638.

(138) C. Blackburn. 1998. A three-component solid-phase synthesis of 3-aminoimidazo[1,2-*a*]azines. Tetrahedron Lett. 39: 5469–5472.

(139) S. Kobayashi, M. Sugiura, H. Kitagawa and W.W.-L. Lam. 2002. Rare-earth metal triflates in organic synthesis. Chem. Rev. 102: 2227–2302.

(140) B. Henkel. 2004. Synthesis of imidazole-4-carboxylic acids *via* solid-phase bound 3-*N*,*N*-(dimethylamino)–2-isocyanoacrylate. Tetrahedron Lett. 45: 2219–2221.

(141) Y.J. Kim and R.S. Varma. 2004. Microwave-assisted preparation of cyclic ureas from diamines in the presence of ZnO. Tetrahedron Lett. 45: 7205–7208.

(142) N. Boufatah, A. Gellis, J. Maldonado and P. Vanelle. 2004. Efficient microwave-assisted synthesis of new sulfonylbenzimidazole-4,7-diones: heterocyclic quinones with potential anti-tumor activity. Tetrahedron 60: 9131–9137.

(143) D.S. VanVliet, P. Gillespie and J.J. Scicinski. 2005. Rapid one-pot preparation of 2-substituted benzimidazoles from 2-nitroanilines using microwave conditions. Tetrahedron Lett. 46: 6741–6743.

(144) S. Perumal, S. Mariappan and S. Selvaraj. 2004. A microwave-assisted synthesis of 2-aryl–1-arylmethyl-1*H*-1,3-benzimidazoles in the presences of K-10. ARKIVOC (viii): 46–51.

(145) C.T. Brain and J.T. Steer. 2003. An improved procedure for the synthesis of benzimidazoles, using palladium-catalyzed aryl-amination chemistry. J. Org. Chem. 68: 6814–6816.

(146) (a) S.M. Ireland, H. Tye and M. Whittaker. 2003. Microwave-assisted multi-component synthesis of fused 3-aminoimidazoles. Tetrahedron Lett. 44: 4369–4371. (b) E. Suna and I. Mutule. 2006. Microwave-assisted heterocyclic chemistry. Top. Curr. Chem. 266: 49–101.

(147) Y. Lu and W. Zhang. 2004. Microwave-assisted synthesis of a 3-aminoimidazo[1,2-*a*]-pyridine/pyrazine library by fluorous multi-component reactions and subsequent cross-coupling reactions. QSAR Comb. Sci. 23: 827–835.

(148) A. Dorn, V. Schattel and S. Laufer. 2010. Design, synthesis and SAR of phenylamino-substituted 5,11-dihydro-dibenzo[*a*,*d*]cyclohepten–10-ones and 11*H*-dibenzo[*b*,*f*]oxepin-10-ones as p38 MAP kinase inhibitors. Bioorg. Med. Chem. Lett. 20: 3074–3077.

(149) R. Rupcic, M. Modric, A. Hutinec, A. Cikos, B. Stanic, M. Mesic, D. Pesic and M. Mercep. 2010. Novel tetracyclic imidazole derivatives: synthesis, dynamic NMR study, and anti-inflammatory evaluation. J. Heterocycl. Chem. 47: 640–656.

(150) A. Sachar, P. Gupta, S. Gupta and R.L. Sharma. 2010. A novel approach towards the synthesis of tricyclic systems based on pyridine, pyran, thiopyran, azepine, oxepin, thiepin, and pyrimidine rings under different solvent conditions. Can. J. Chem. 88: 478–484.

(151) G. Molteni. 2007. Silver(I) salts as useful reagents in pyrazole synthesis. ARKIVOC (ii): 224–246.

(152) P. del Buttero and G. Molteni. 2006. Stereoselective synthesis of highly functionalized tricyclic β-lactams *via* intramolecular nitrilimine cycloaddition. Tetrahedron: Asymmetry 17: 1319–1321.

(153) M. Alvarez-Corral, M. Munoz-Dorado and I. Rodriguez-Garcia. 2008. Silver-mediated synthesis of heterocycles. Chem. Rev. 108: 3174–3198.

(154) F. Glorius, G. Altenhoff, R. Goddard and C. Lehmann. 2002. Oxazolines as chiral building blocks for imidazolium salts and *N*-heterocyclic carbene ligands. Chem. Commun. 22: 2704–2705.

(155) K.B. Umesha, K.M. Lokanatha Rai and K. Ajay Kumar. 2002. A new approach to the synthesis of pyrazoles *via* 1,3-dipolar cycloaddition of nitrile imines with acetyl acetone. Indian J. Chem. 41B: 1450–1453.

(156) K.M. Lokanatha Rai and N. Linganna. 1997. Mercuric acetate in organic synthesis: a simple procedure for the synthesis of pyrazolines. Synth. Commun. 27: 3737–3744.

(157) G. Broggini, L. Garantu, G. Molteni, T. Pilati, A. Ponti and G. Zecchi. 1999. Stereoselective intramolecular cycloadditions of homochiral nitrile imines: synthesis of enantiomerically pure 3,3-dihydro-pyrazolo[1,5-*a*][1,4] benzodiazepine-6(4*H*)-ones. Tetrahedron: Asymmetry 10: 2203–2212.

(158) K. Ajay Kumar, M. Govindaraju and G. Vasanth Kumar. 2013. Nitrile imines: versatile intermediates in the synthesis of five-membered heterocycles. Int. J. Res. Pharm. Chem. 3: 140–152.

(159) D.P. Walsh and Y.-T. Chang. 2006. Chemical genetics. Chem. Rev. 106: 2476–2530.

(160) P. Arya, D.T.H. Chou and M.-G. Baek. 2001. Diversity based organic synthesis in the era of genomics and proteomics. Angew. Chem. Int. Ed. 40: 339–346.

(161) S.L. Schreiber. 2000. Target-oriented and diversity-oriented organic synthesis in drug discovery. Science 287: 1964–1969.

(162) S. Li, Y. Luo and J. Wu. 2011. Three-component reaction of *N*'-(2-alkynylbenzylidene)hydrazide, alkyne, with sulfonyl azide *via* a multi-catalytic process: a novel and concise approach to 2-amino-*H*-pyrazolo[5,1-*a*]isoquinolines. Org. Lett. 13: 4312–4315.

(163) E. Moriarty and F. Aldabbagh. 2009. Synthesis of aryl ring-fused benzimidazolequinones using 6-*exo-trig* radical cyclizations. Tetrahedron Lett. 50: 5251–5253.

(164) Y. Zhang, Z. Chen, Y. Lou and Y. Yu. 2009. 2,3-Disubstituted 8-arylamino-3*H*-imidazo[4,5-*g*]quinazolines: a novel class of anti-tumor agents. Eur. J. Med. Chem. 44: 448–452.

(165) N. Uchiyama, K. Saisho, R. Kikura-Hanajiri, Y. Haishima and Y. Goda. 2008. Determination of a new type of phosphodiesterase-5 inhibitor, thioquinapiperifil, in a dietary supplement promoted for sexual enhancement. Chem. Pharm. Bull. 56: 1331–1334.

(166) S. Kamila, H. Ankati, K. Mendoza and E.R. Biehl. 2011. Synthesis of novel pyridobenzimidazoles bonded to indoleorbenzo[*b*]thiophenestructures. Open Org. Chem. J. 5: 127–134.

(167) A. Song, J. Zhang, B.C. Lebrilla and S.K. Lam. 2004. Solid-phase synthesis and spectral properties of 2-alkylthio-6*H*-pyrano[2,3-*f*]benzimidazole-6-ones: a combinatorial approach for 2-alkylthioimidazocoumarins. J. Comb. Chem. 6: 604–610.

(168) K.C. Majumdar, R.N. De, A.T. Khan, S.K. Chattopadhyay, K. Dey and A. Patra. 1988. Studies of [3,3]sigmatropic rearrangements: rearrangement of 3-(4-*p*-tolyloxybut–2-ynyloxy)[1]benzopyran-2-one. J. Chem. Soc. Chem. Commun. 12: 777–779.

(169) M. Mamada, P. Anzenbacher and C.P. Bolivar. 2011. Green synthesis of polycyclic benzimidazole derivatives and organic semiconductors. Org. Lett. 13: 4882–4885.

(170) A. Chawla, G. Kaur and A.K. Sharma. 2012. Green chemistry as a versatile technique for the synthesis of benzimidazole derivatives: review. Int. J. Pharm. Phytopharmacol. Res. 2: 148–159.

Five-Membered *N,N,N*-Heterocycles

8.1 Introduction

Heterocyclic chemistry is constantly expanding and evolving as a result of the magnitude of research efforts directed towards this area. The majority of known molecules are heterocycles; they dominate the field of biochemistry, medicinal chemistry, and photographic sciences, and are of increasing importance in many other domains like polymers, adhesives and molecular engineering as well [1–4].

1,2,4-Triazole and its derivatives are an important class of compounds which possess diverse agricultural, industrial and biological properties- anti-microbial, sedative, anti-convulsant, anti-cancer, anti-inflammatory, anti-bacterial, hypoglycemic, anti-tubercular and anti-fungal [5–12].

Some of the present day drugs such as ribavirin (an anti-viral), rizatriptan (an anti-migraine agent), alprazolam (an anxiolytic), fluconazole and itraconazole (anti-fungal agents) are the best examples of potent molecules possessing triazole nuclei. Anti-fungal activity exhibited by many effective anti-fungal agents is attributed to the presence of the triazole ring system. Major examples of triazole-containing anti-fungal agents include fluconazole [13] (used for treating vaginal, oral and esophageal fungal infections caused by *Candida*; also effective in treating urinary tract infections, peritonitis, pneumonia and disseminated infections caused by *Candida*), voriconazole [14] (used in the treatment of certain fungal infections by blocking fungal cell wall growth, resulting in the death of the fungus) and itraconazole [15] (used for the treatment of fungal infections in both HIV- and non-HIV infected individuals, and active against fungal infections such as aspergillosis, blastomycosis, histoplasmosis and candidiasis, as well as those localized to toenails and fingernails).

Some derivatives of 1*H*-1,2,4-triazole are also found to be useful as strong anti-estrogens. For example, anastrozole [16] is used to treat breast cancer in women who have gone through menopause, by lowering the estrogen levels to help shrink tumors and slow their growth.

8.2 Metal- and non-metal-assisted synthesis of five-membered heterocycles with three nitrogen atoms

Aluminum-assisted synthesis

Kidwai and Mohan [17a] studied the reaction of substituted hydrazide, using various solid supports under MWI, for the preparation of new anti-fungal azoles including 1,2,4-triazole derivatives (Scheme 1) [17b].

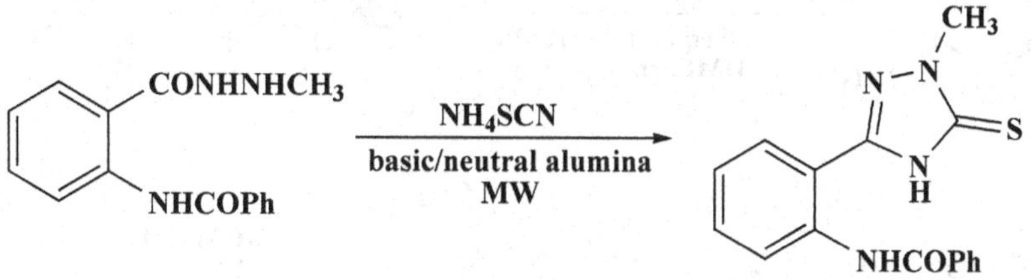

Scheme-1

Bismuth-assisted synthesis

Bicyclic triazoloheterocyclic derivative is obtained in 73% yield by tandem azidation/1,3-dipolar cycloaddition of trimethylsilyl azide and alkynyl acetal in the presence of Bi(OTf)$_3$·xH$_2$O catalyst (Scheme 2) [18–19].

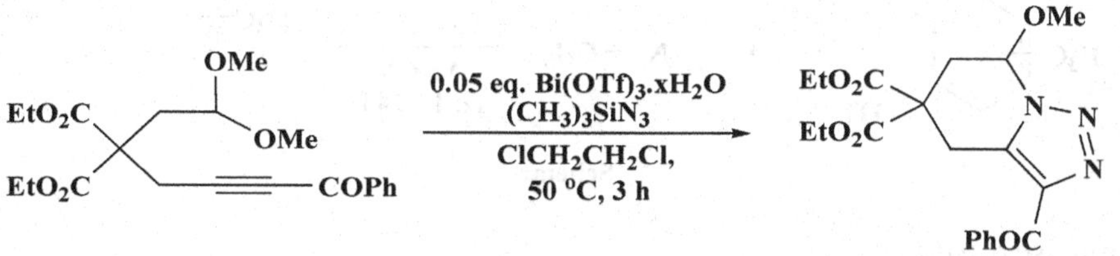

Scheme-2

Cerium-assisted synthesis

Oxidation of ketazine, obtained from 2-acetylpyridine, affords a mixture of condensed triazole and ketone [20]. The use of cerium ammonium nitrate reagent also provides same result. This suggests that the reaction involves cation radicals (Scheme 3) [21–22].

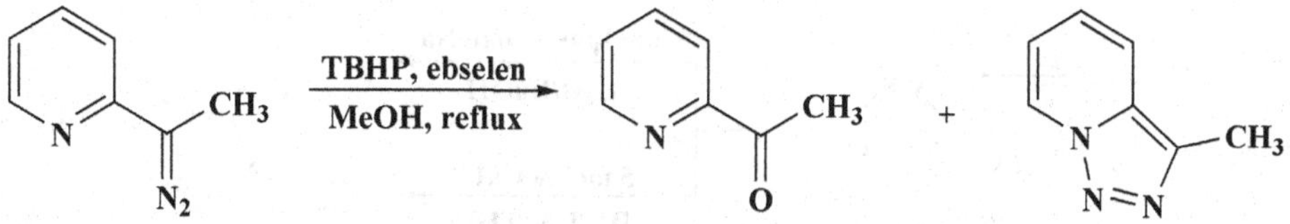

Scheme-3

Cesium-assisted synthesis

3-Amino-1,2,4-triazoles can be synthesized under solid-phase conditions. Resin-bound *S*-methylisothiourea reacts with carboxylic acids to yield resin-bound *S*-methyl-*N*-acylisothioureas which, with subsequent reaction with hydrazines under mild conditions, forms resin-bound 3-amino-1,2,4-triazoles regioselectively. The desired products are ultimately formed after cleavage (Scheme 4) [23].

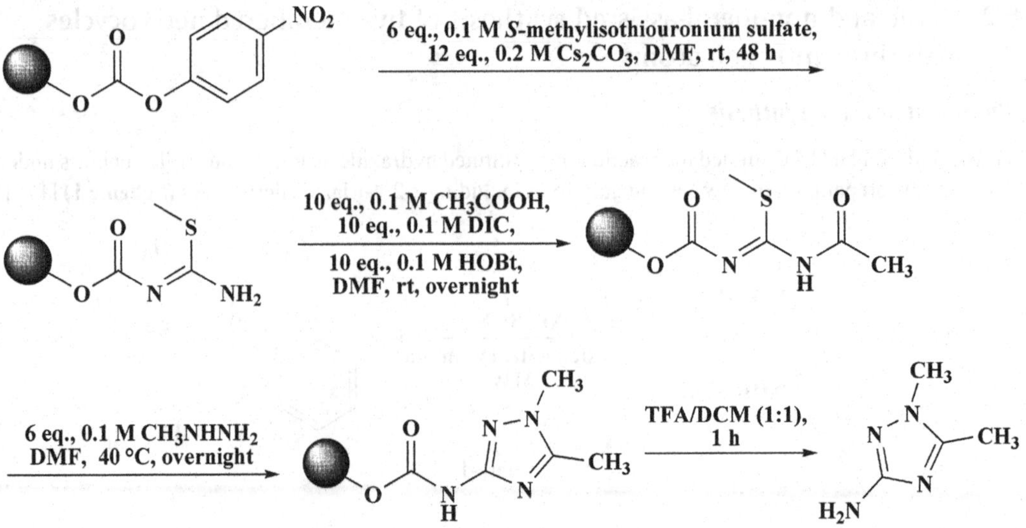

Scheme-4

A variety of substituted and functionalized benzotriazoles are synthesized by [3+2]-cycloaddition of azides to benzynes under mild reaction conditions **(Scheme 5)** [24].

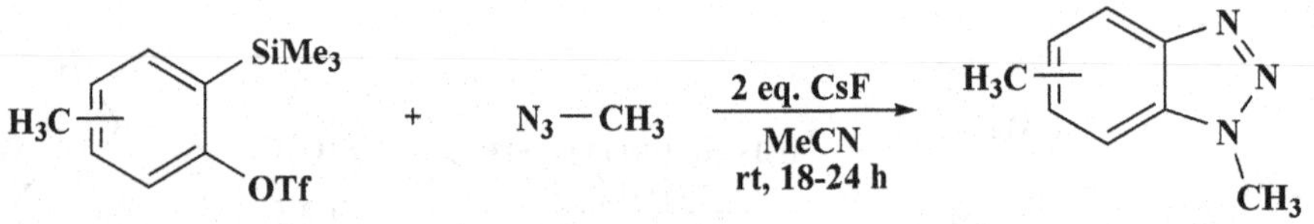

Scheme-5

Copper-assisted synthesis

Synthetic non-proteinogenic α-amino acids play an important role in modern drug discovery process. Such amino acids (especially α,α-disubstituted) incorporated into key positions of biologically active peptides, as a rule, lead to significant enhancement of proteolytic and conformational stability, increased selectivity, and improve the pharmaco-kinetic properties of the potential drugs [25]. α-CF$_3$-Substituted azahistidines are synthesized by 1,3-dipolar cycloaddition of different organic azides to propargyl-containing α-CF$_3$-α-amino acids in the presence of copper(I) **(Scheme 6)**.

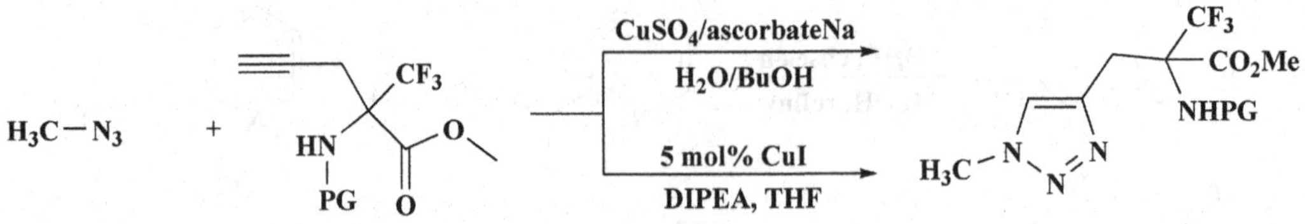

Scheme-6

1-Monosubstituted aryl 1,2,3-triazoles can be prepared in good yields by employing calcium carbide as a source of acetylene. The 1,3-dipolar cycloaddition reactions are performed in the presence of Cu catalyst in an acetonitrile-water mixture, without nitrogen protection **(Scheme 7)** [25].

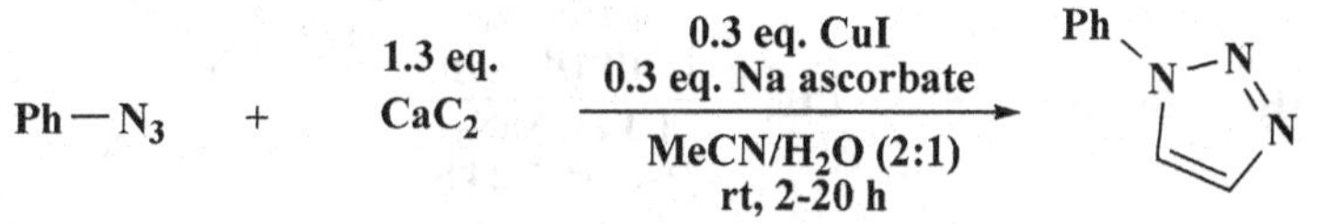

$$\text{Ph}-\text{N}_3 \quad + \quad \underset{\text{CaC}_2}{\overset{\text{1.3 eq.}}{}} \quad \xrightarrow[\substack{\text{MeCN/H}_2\text{O (2:1)} \\ \text{rt, 2-20 h}}]{\substack{\text{0.3 eq. CuI} \\ \text{0.3 eq. Na ascorbate}}}$$

Scheme-7

A tandem catalysis method based on decarboxylative coupling of alkynoic acids and 1,3-dipolar cycloaddition of azides reduces the handling of potentially explosive and unstable azides to a minimum, avoids the use of highly volatile or gaseous terminal alkynes, and provides several functionalized 1,2,3-triazoles with high purity and in excellent yields, hence eliminating the need for any additional purification **(Scheme 8)** [26].

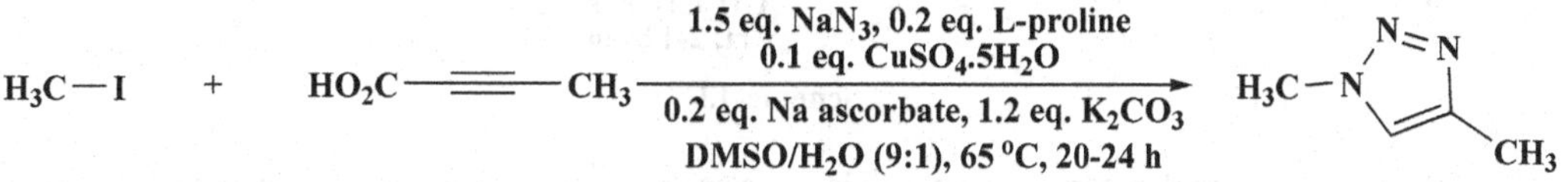

$$\text{H}_3\text{C}-\text{I} \quad + \quad \text{HO}_2\text{C}\!\!\equiv\!\!\text{CH}_3 \quad \xrightarrow[\substack{\text{0.2 eq. Na ascorbate, 1.2 eq. K}_2\text{CO}_3 \\ \text{DMSO/H}_2\text{O (9:1), 65 °C, 20-24 h}}]{\substack{\text{1.5 eq. NaN}_3\text{, 0.2 eq. L-proline} \\ \text{0.1 eq. CuSO}_4.5\text{H}_2\text{O}}}$$

Scheme-8

Anti-3-aryl-2,3-dibromopropanoic acids and sodium azide react to form 4-aryl-1*H*-1,2,3-triazoles, using inexpensive CuI as a catalyst, in dimethylsulfoxide (solvent) and Cs_2CO_3 (base) **(Scheme 9)** [27].

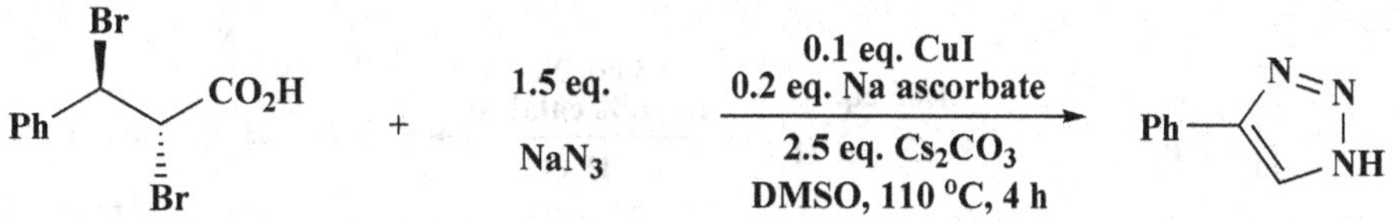

$$\quad + \quad \underset{\text{NaN}_3}{\overset{\text{1.5 eq.}}{}} \quad \xrightarrow[\substack{\text{2.5 eq. Cs}_2\text{CO}_3 \\ \text{DMSO, 110 °C, 4 h}}]{\substack{\text{0.1 eq. CuI} \\ \text{0.2 eq. Na ascorbate}}}$$

Scheme-9

Anti-3-aryl-2,3-dibromopropanoic acids and organic azides react by a one-pot protocol in dimethylsulfoxide, in the presence of CuI, to provide a series of 1,4-disubstituted 1,2,3-triazoles **(Scheme 10)** [28].

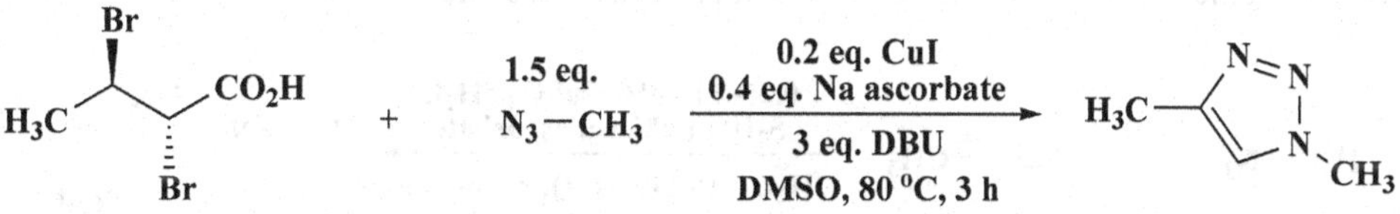

$$\quad + \quad \underset{\text{N}_3-\text{CH}_3}{\overset{\text{1.5 eq.}}{}} \quad \xrightarrow[\substack{\text{3 eq. DBU} \\ \text{DMSO, 80 °C, 3 h}}]{\substack{\text{0.2 eq. CuI} \\ \text{0.4 eq. Na ascorbate}}}$$

Scheme-10

The reaction between aliphatic and aromatic azides, using mild, Cu(I)-catalyzed 'click chemistry' in the presence of acetylene gas yields 1-substituted 1,2,3-triazoles conveniently **(Scheme 11)** [29].

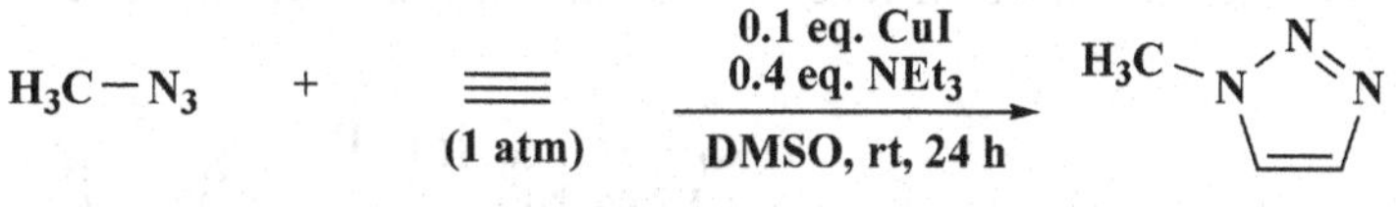

$$\text{H}_3\text{C}-\text{N}_3 \quad + \quad \underset{\text{(1 atm)}}{\equiv} \quad \xrightarrow[\substack{\text{DMSO, rt, 24 h}}]{\substack{\text{0.1 eq. CuI} \\ \text{0.4 eq. NEt}_3}}$$

Scheme-11

Commercially available [CuBr(PPh$_3$)$_3$] constitutes a true click catalytic system. This system, with 0.5 mol% or lower catalyst loadings, has been found to be active at room temperature without any additive. Moreover, it does not require any purification step for the isolation of pure triazoles **(Scheme 12)** [30].

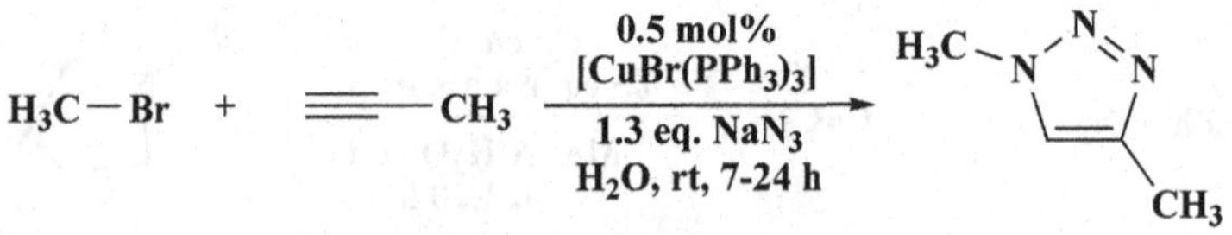

$$H_3C-Br \ + \ \equiv\!\!-CH_3 \ \xrightarrow[\substack{1.3 \text{ eq. NaN}_3 \\ H_2O, \text{ rt, } 7\text{-}24 \text{ h}}]{\substack{0.5 \text{ mol\%} \\ [CuBr(PPh_3)_3]}}$$

Scheme-12

It has been reported that Cu(I)-catalyzed azide-alkyne cycloaddition is promoted with the joint use of acid-base (**Scheme 13**) [31].

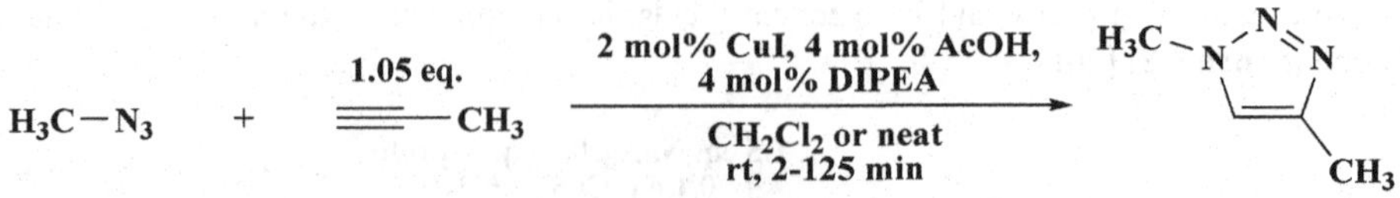

$$H_3C-N_3 \ + \ \overset{1.05 \text{ eq.}}{\equiv\!\!-CH_3} \ \xrightarrow[\substack{CH_2Cl_2 \text{ or neat} \\ \text{rt, } 2\text{-}125 \text{ min}}]{\substack{2 \text{ mol\% CuI, } 4 \text{ mol\% AcOH,} \\ 4 \text{ mol\% DIPEA}}}$$

Scheme-13

Cu(I) isonitrile complex serves as an efficient, heterogeneous catalyst for azide-alkyne 1,3-dipolar cycloaddition and three-component reactions of sodium azide, halides, and alkynes under mild conditions in H_2O to produce 1,4-disubstituted 1,2,3-triazoles in high yields. The complex can be recycled for at least five runs by simple filtration and precipitation, without any significant loss in activity (**Scheme 14**) [32].

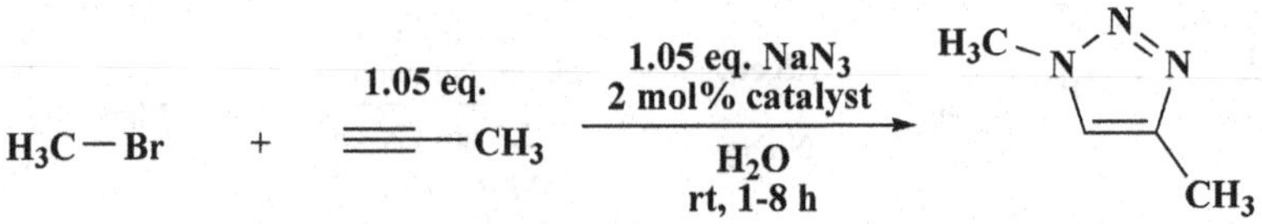

$$H_3C-Br \ + \ \overset{1.05 \text{ eq.}}{\equiv\!\!-CH_3} \ \xrightarrow[\substack{H_2O \\ \text{rt, } 1\text{-}8 \text{ h}}]{\substack{1.05 \text{ eq. NaN}_3 \\ 2 \text{ mol\% catalyst}}}$$

Scheme-14

1,4-Disubstituted 1,2,3-triazoles are synthesized by cycloaddition of Cu(I) acetylides and azides. This reaction has been found to be highly reliable and exhibits a wide scope with respect to both the components. Computational studies also suggest that a non-concerted mechanism involving unprecedented metallacycle intermediates works in this case (**Scheme 15**) [33].

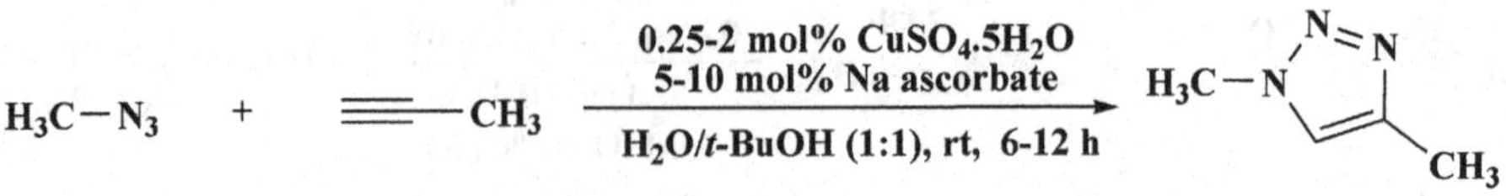

$$H_3C-N_3 \ + \ \equiv\!\!-CH_3 \ \xrightarrow[\substack{H_2O/t\text{-BuOH (1:1), rt, } 6\text{-}12 \text{ h}}]{\substack{0.25\text{-}2 \text{ mol\% CuSO}_4.5H_2O \\ 5\text{-}10 \text{ mol\% Na ascorbate}}}$$

Scheme-15

1,4,5-Trisubstituted 1,2,3-triazole can be synthesized regiospecifically by a CuI-catalyzed reaction. This is the first reported example of regiospecific synthesis of 5-iodo-1,4-disubstituted 1,2,3-triazole, which has been elaborated to a variety of 1,4,5-trisubstituted 1,2,3-triazoles (**Scheme 16**) [34].

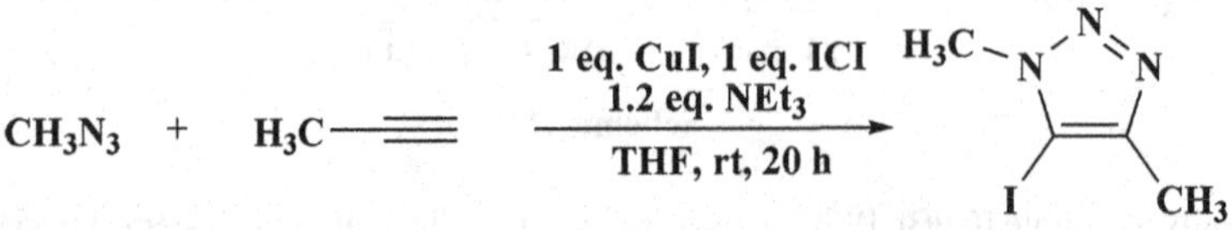

$$CH_3N_3 \ + \ H_3C-\!\!\equiv \ \xrightarrow[\substack{THF, \text{ rt, } 20 \text{ h}}]{\substack{1 \text{ eq. CuI, } 1 \text{ eq. ICI} \\ 1.2 \text{ eq. NEt}_3}}$$

Scheme-16

'Click' reactions of di- and tetraazides have been explored during the development of high-load nitric oxide donors. A mixture of bis-tetrazolo-1,4-diazepines and bis-triazoles is obtained by 1,3-dipolar cycloaddition of diazide and alkynes. Using copper(II) sulfate-sodium ascorbate, the bis-triazoles predominated; however, with copper(I) iodide-diisopropylethylamine the coupled products bis-tetrazolo-1,4-diazepines predominated, the latter resulting from copper(I)-catalyzed oxidative coupling **(Scheme 17)** [35].

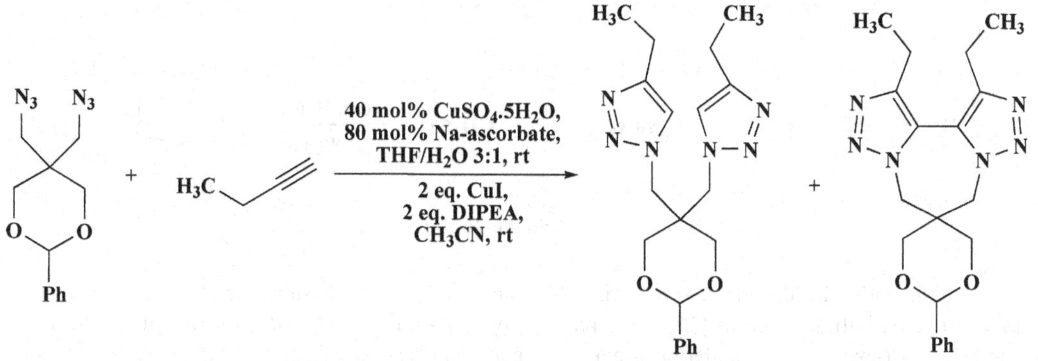

Scheme-17

Pyrido[3',2':5,6]pyrimido[1,2-*a*]benzimidazol-5(6*H*)-one (X = O) is formed by copper-catalyzed cyclocondensation of 2-aminobenzimidazole and 2-bromobenzoic acid in dimethylformamide and K$_2$CO$_3$ under reflux. When treated with sulfur and phosphorus in refluxing pyridine, it yields the 5-thione analog (X = S) which reacts with hydrazine in refluxing ethanol, followed by nitrosation, to afford a fused hexaazapentacyclic system **(Scheme 18)** [36, 37a–b].

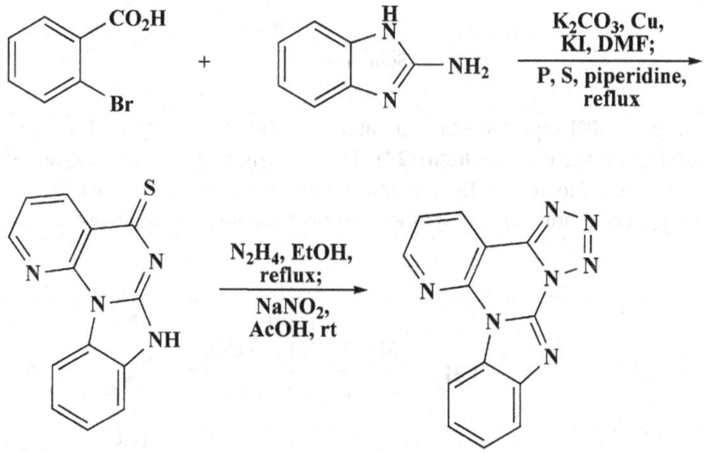

Scheme-18

Ruthenium-catalyzed direct arylations of 1,2,3-triazoles have proved to be complementary to palladium-based protocols with respect to site-selectivity [38]. In contrast to this, a modular one-pot multi-component reaction in the presence of inexpensive Cu catalysts produces fully decorated triazoles through a sustainable 'click' reaction or direct arylation sequence **(Scheme 19 and 20)** [38].

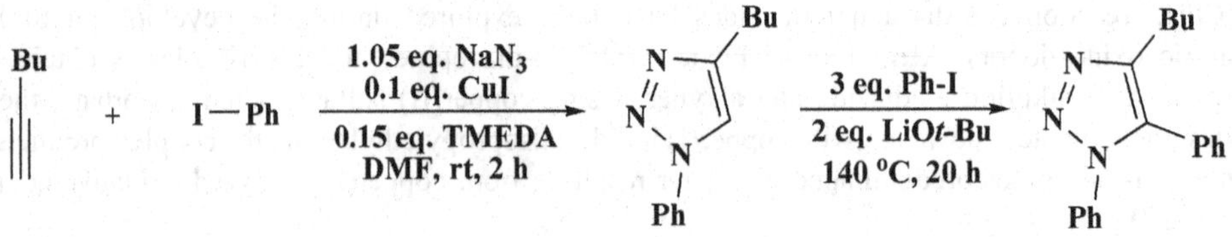

Scheme-19

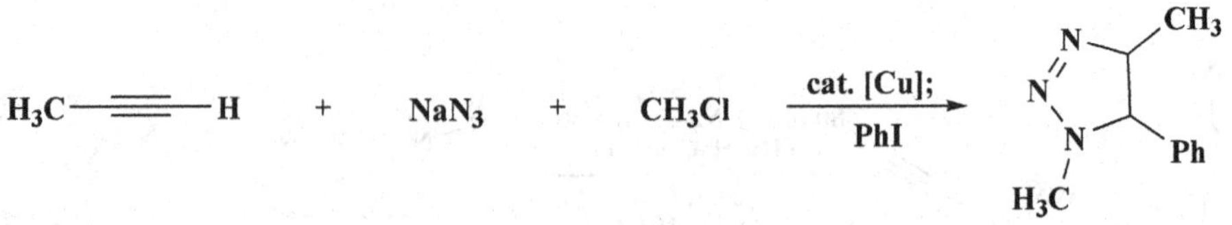

Scheme-20

The iodination of terminal alkynes with *N*-iodomorpholine affords substrates for the copper azide alkyne cycloaddition reaction [39]. This reaction occurs efficiently to afford differently substituted 5-iodo-1,2,3-triazoles which undergo carbon-hydrogen bond functionalization or Heck reaction to yield the desired fused triazole (**Scheme 21**) [40–48].

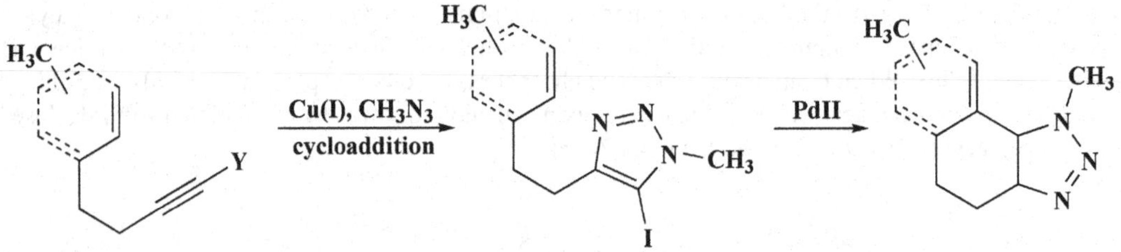

Scheme-21

Angell and Burgess [49] reported that Cu-catalyzed Huisgen reactions for an efficient synthesis of bis-triazoles depends on the base (**Scheme 22**). This chemistry affected the synthesis of ligands and the development of libraries to screen for pharmaceutical compounds [50], while also influencing the methods used to design conditions for the synthesis of bis-triazole or triazole products.

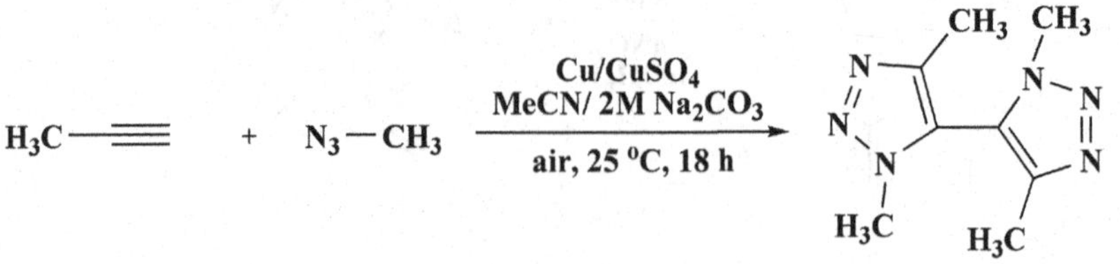

Scheme-22

The analog bearing triazole ring is prepared from methyl azide synthesized from methyliodide. However, methyl azide is not purified due to its instability and is, therefore, directly utilized in copper-assisted 'click' coupling (**Scheme 23**) [51].

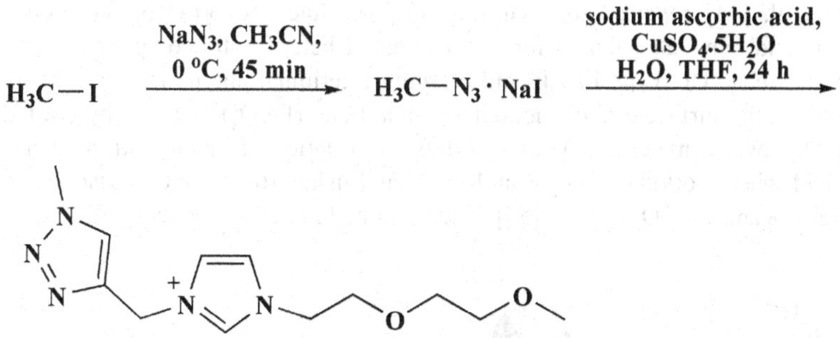

Scheme-23

The third monomeric analog bears a monomethoxy-poly(ethylene glycol) chain adjacent to the triazole ring. This is done for the purpose of studying the effect of a polymeric backbone on the activity and the stability of the resulting catalyst. For investigating the effect of hyper-branched backbone, based on the comparison of a linear analog with the hyper-branched one, PEG-azide was prepared from PEG-Ts **(Scheme 24)** [52].

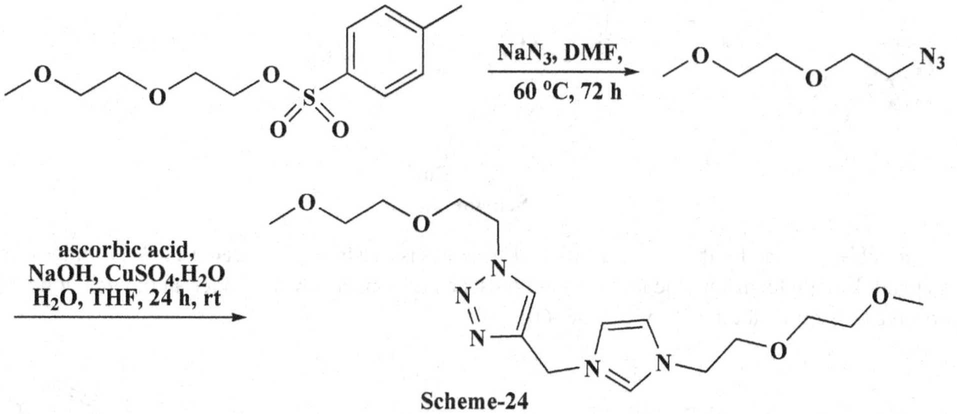

Scheme-24

3-Methylpyrido[1,2-*a*]benzimidazoles can be synthesized by the condensation of 2-(arylhydrazono)-3-iminobutanenitrile with 2-benzimidazoleacetonitrile in acetic acid. The former, when treated with cupric acetate in dimethylformamide for oxidative cyclization, yields *S*-triazolo[4,5-*b*]pyrido[1`,2`-*a*] benzimidazoles **(Scheme 25)** [37, 53].

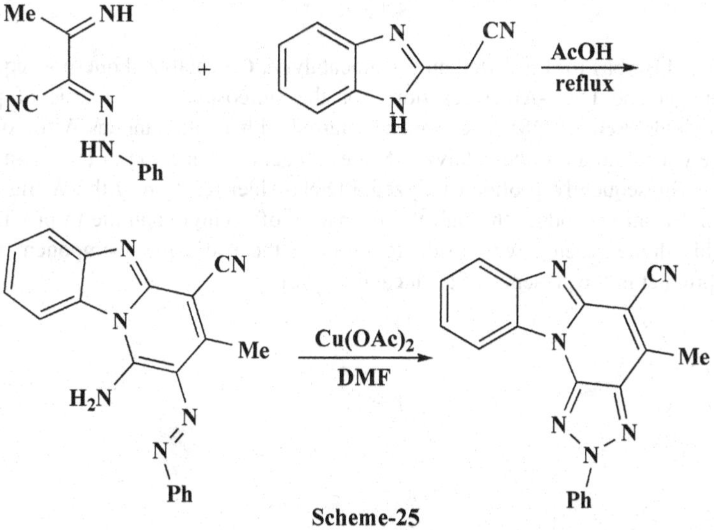

Scheme-25

Chouhan et al. [54] reported the synthesis of L-proline supported on magnetic nanoparticles as a recoverable and recyclable ligand for *N*-arylation of heterocyclic compounds in the presence of cuprous iodide catalyst. The L-proline ligand carrying a terminal alkyne moiety was immobilized by 'click chemistry' on the surface of the magnetite nanoparticles (Fe_3O_4) modified by grafted azido groups (**Scheme 26**). This hybrid material was used in the N1-arylation of indole with 4-bromoacetophenone giving 85% yield when working in DMF at 110 °C and using 10 mol% CuI associated to a "ligand" loading of 20 mol% and Cs_2CO_3 as base [55]. The separated material was recycled in the absence of Cu iodide.

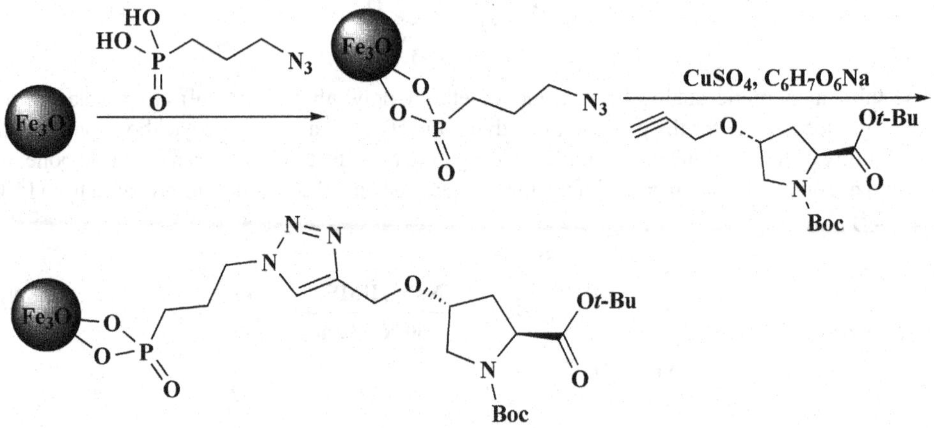

Scheme-26

A possible pathway for the synthesis of 1,2,4-triazoles is illustrated in **Scheme 27**. *N*-[Amino(*m*-tolyl) methylene]-4-methylbenzamidine reacts in the presence of copper powder under an oxygen atmosphere to form the target product in 68% yield [56–60].

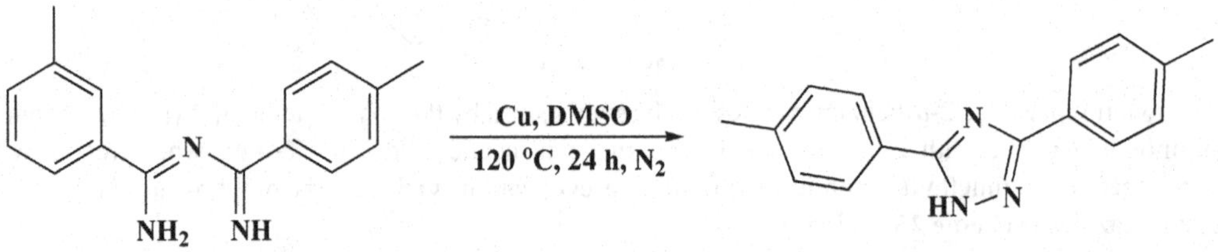

Scheme-27

Ramachary and Barbas [61] reported an organocatalytic, Cu-catalyzed one-pot sequence involving Wittig, Knoevenagel and Diels-Alder reactions for the stereospecific synthesis of poly-substituted triazoles in high yields (**Scheme 28**). The sequence started with a simultaneous Wittig olefination of the aldehyde and the ylide, and a proline-catalyzed Knoevenagel condensation of the CH-acidic compound and the aldehyde. Subsequently, proline catalyzed a Diels-Alder reaction of the Wittig product and the Knoevenagel condensation product through the formation of a vinyl enamine to provide a spirocyclic intermediate. This diyne reacted with azide to produce the pseudo-six-component poly-substituted triazoles as the product in the presence of a Cu catalyst [62].

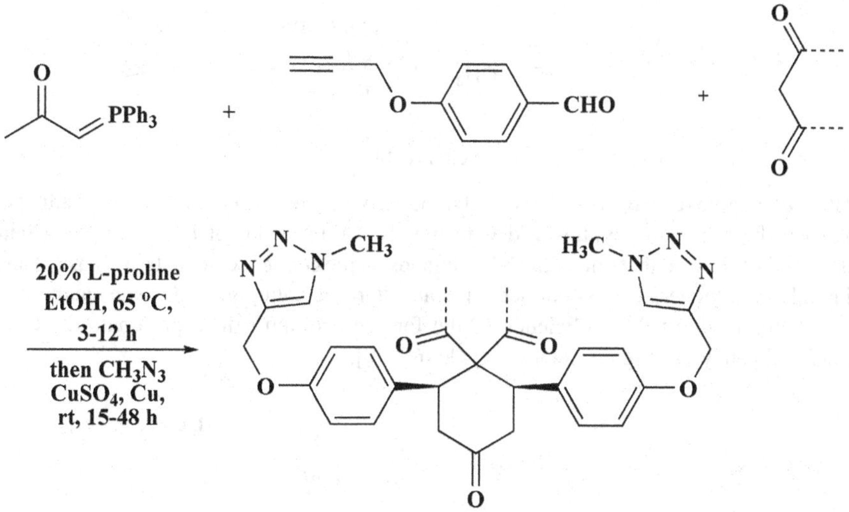

Scheme-28

In Cu-catalyzed Huisgen reaction, the catalytic cycle starts with the formation of a copper acetylide intermediate which precludes internal alkynes as cycloaddition partners. Moreover, a hypothetical activation towards cycloaddition *via* π coordination of Cu(I) to the alkyne (without de-protonation) is also ruled out since the calculated activation barrier for this process exceeds that of the un-catalyzed process. However, 4,5-disubstituted triazoles are formed in fair to good yields in the presence of [(SIMes) CuBr] after heating the reactants for 48 hours at 70 °C [63]. Optimization studies have shown that both the *N*-heterocyclic carbene ligand and the Cu salt are essential for this transformation. Although the Cu ion is considered as a poor π-back-donating ion, the ancillary ligands on the metal center play an essential role in its coordination with alkynes (**Scheme 29**) [64]. It is worth noting that the widely accepted reaction pathway for terminal alkynes is still applicable to this system. This proposition has been strongly supported by the isolation of an intermediate Cu(I) triazolide complex bearing a SIPr ligand [65–66].

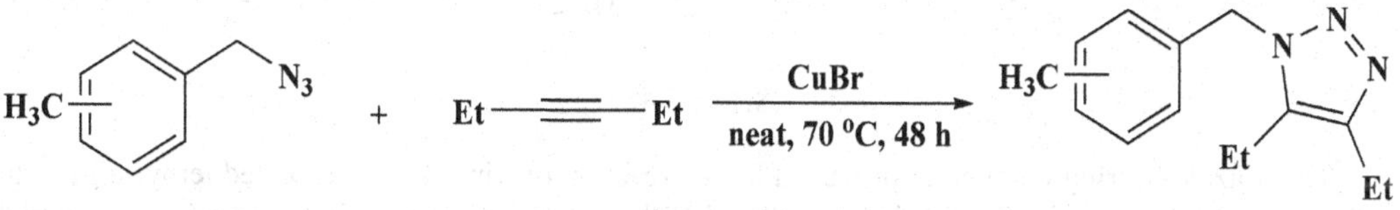

Scheme-29

The reaction between organic azides and terminal alkynes proceeds through a Cu-catalyzed pathway to afford 1,2,3-triazoles and is considered as the best 'click' reaction [66]. The screening of [(NHC)CuX] complexes for this reaction allows it to occur with only 0.8 mol% Cu catalyst under neat conditions [67]. Furthermore, this catalytic system facilitates an unprecedented use of internal alkynes in this context. These complexes have been reported to be efficient catalysts under simple reaction conditions [68]. Whereas a number of azides and alkynes do not react in the presence of [(SIPr)CuCl] in dimethylsulfoxide under ambient conditions, the reactions under consideration occur smoothly upon addition of H_2O at 60 °C. These studies enhance the use of this transformation, especially extending it into material science and biology (**Scheme 30**) [69].

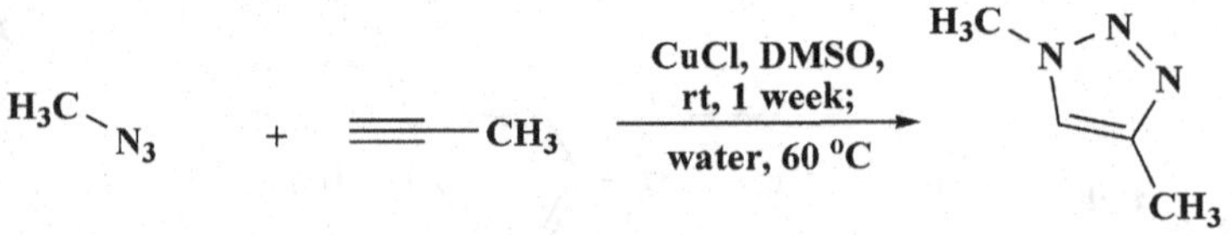

Scheme-30

[(NHC)Cu] complexes have been observed to be active even at very low catalyst loadings. An array of triazoles can be efficiently synthesized with merely 40–100 ppm of [(ICy)$_2$Cu]PF$_6$ (**Scheme 31**). Mechanistic studies showed that one of the NHC ligands on the copper center acts as a base, deprotonating the starting alkyne to generate a copper acetylide and start the catalytic cycle. The catalytic activity of this system increases due to the efficiency of the formed azolium salt to protonate the Cu triazolide intermediate, ultimately closing the catalytic cycle [69–70].

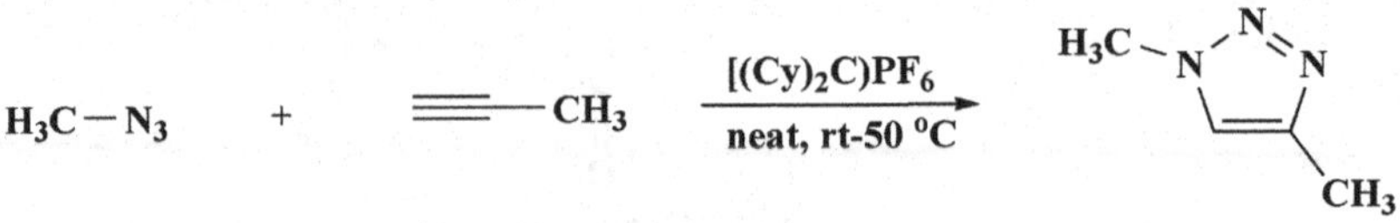

Scheme-31

Rostovtsev et al. [71a] reported a copper(I)-catalyzed regioselective Huisgen cycloaddition reaction (**Scheme 32**) wherein the 1,4-disubstituted product was obtained in high yields (91%) and 100% regioselectivity using Cu(I) catalyst (produced *in situ*) which completed the reaction in 8 hours at room temperature [71b].

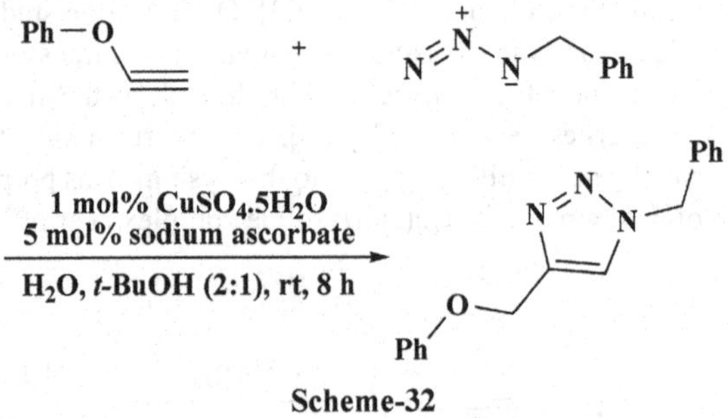

Scheme-32

The triazole-bearing compound produced by the reaction of triple-bond decorated terpyridine with a simple azide (N$_3$-CH$_2$-COOt-Bu) demonstrates that the metal ion complexing properties of a model terpyridinetriazole compound are similar to those of terpyridine (**Scheme 33**) [71a–b].

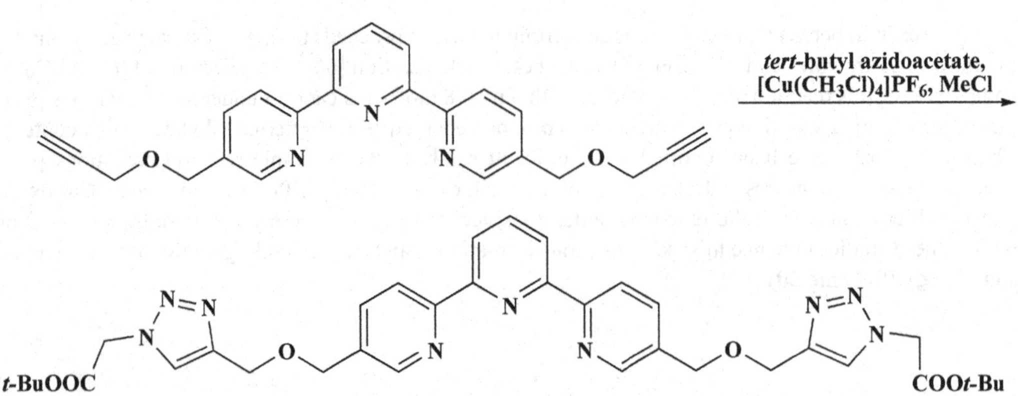

Scheme-33

Cu(I) acts as a regiospecific and efficient catalyst for the reaction of alkynes and azides to provide 1,2,3-triazoles [72–73]. For this reaction, the catalytic system consists of a Cu(II) salt and a reducing agent due to the inherent instability of cuprous salts. A wide range of ligands have also been shown to protect Cu(I) centers during this reaction [74–76]. Many [(NHC)CuX] complexes have been screened under standard cycloaddition conditions, wih the conclusion that [(SIMes)CuBr] is the best catalyst for this transformation. Whereas poor conversions have been reported in organic solvents, a strong acceleration is observed in H_2O. Furthermore, neat reactions occur smoothly with no detectable formation of by-products and the catalyst loading can be lowered to 0.8 mol%, without any loss of activity, which ensures a straightforward reaction work-up. The triazoles produced can be isolated in high purity and excellent yields through simple extraction or filtration. The azides produced *in situ* from NaN_3 and halides also react at room temperature in H_2O to yield triazoles efficiently. 1,3-Dipolar cycloaddition reactions result in the direct synthesis of a variety of heterocyclic systems and the emergence of 'click chemistry, or, more specifically, copper(I)-catalyzed azidealkyne cycloaddition (CuAAC). Copper(I) complexes have been used as catalysts in certain reactions such as conjugate reductions of cyclic enones [77], hydrosilylation [78–79], and carbene transfer [80]. However, Cu(I) catalysts are most known for CuAAC [81] a type of 'click' chemistry. Thermal cycloaddition of alkynes and azides usually needs prolonged heating and results in mixtures of both 1,5- and 1,4-regioisomers, whereas copper azide alkyne cycloaddition provides only 1,4-disubstituted product in excellent yields, with mild heating or at room temperature **(Scheme 34)** [66].

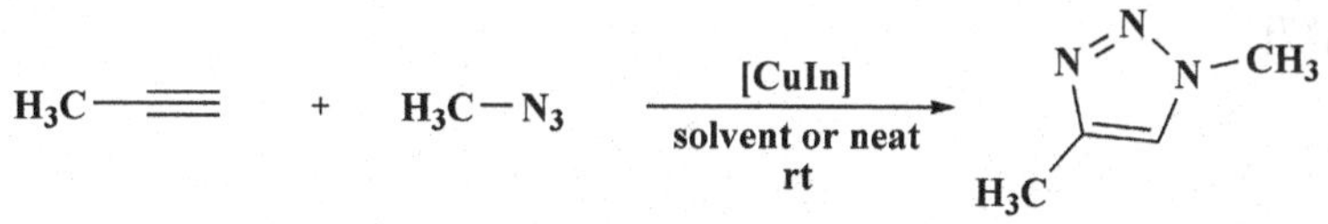

Scheme-34

Several quinolino-, pyrido-, qunoxalino- and pyrazinotetrazoles have been utilized efficiently as azide components in copper-catalyzed 'click' reaction with alkynes. This protocol allows an efficient preparation of a wide range of *N*-heterocyclic derivatives of 1,2,3-triazoles **(Scheme 35)** [82].

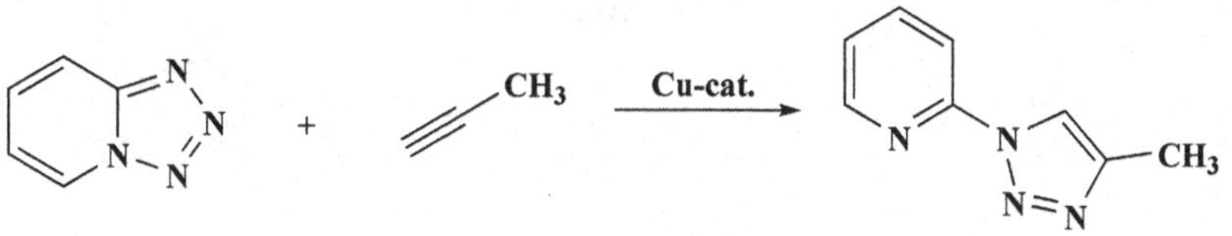

Scheme-35

An operationally simple and reliable one-pot protocol for one-carbon homologation of several aldehydes followed by copper-catalyzed azide-alkyne 'click' chemistry affords good yields of 1,4-disubstituted 1,2,3-triazoles without the isolation of alkyne intermediates **(Scheme 36)** [83].

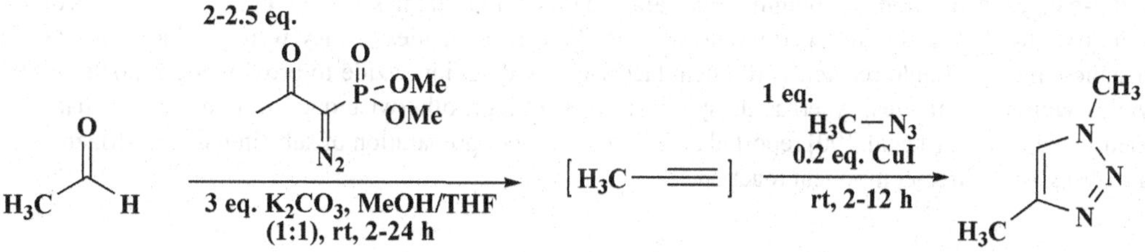

Scheme-36

Li et al. [84] synthesized pyrano[2,3-*d*]pyrimidines *via* a one-pot reaction of malononitrile, aromatic aldehydes and pyrimidine-4,6-diol at 80 °C in an ionic liquid medium, [bmim]BF$_4$. Zhao et al. [85] reported the reaction of sodium azides, halides and alkynes in a mixture of an ionic liquid, that is, [bmim]BF$_4$, and H$_2$O, using Cu(I) catalyst for the one-pot synthesis of 1,4-disubstituted 1,2,3-triazoles in good to high yields. Triazoles bearing a ferrocene substituent were prepared using ferrocenylacetylene as an alkyne component **(Scheme 37)**. The ionic liquid was recycled and reused five times without substantial loss in activity. Water played a positive role in these reactions due to the good solubility of sodium azide in H$_2$O [86].

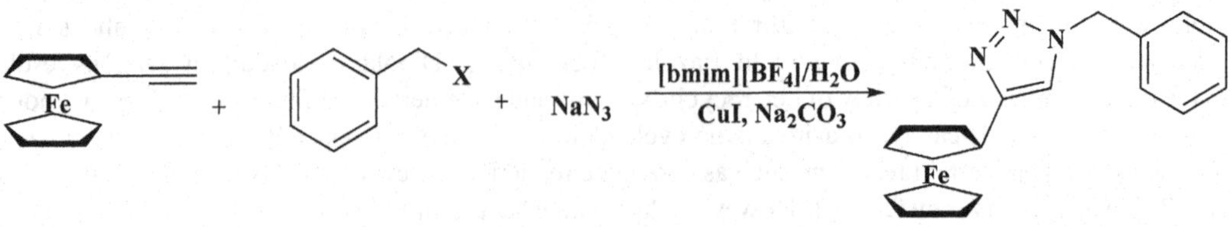

Scheme-37

1-{2-Benzoyl-[(1-benzyl-1*H*-1,2,3-triazol-4-yl)methoxy]-3-methylbenzofuran-5-yl}-1-ethanone can be synthesized when a mixture of benzylazide, 1-(2-benzoyl-3-methyl-6-(prop-2-ynloxy)benzofuran-5-yl), diisopropyl ethylamine and CuI in dimethylformamide react for 2 minutes under MWI at 320 W **(Scheme 38)** [87].

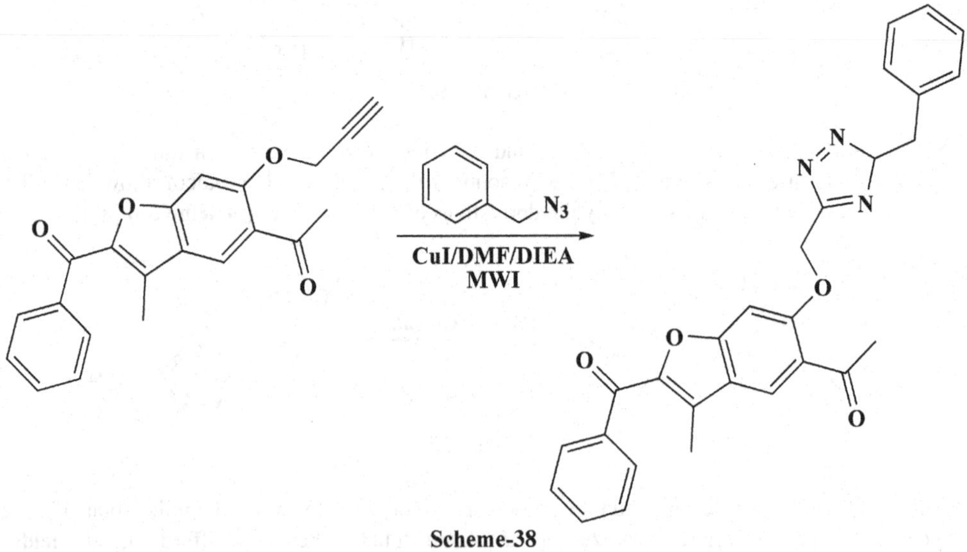

Scheme-38

Appukkuttan et al. [88] reported a MW-assisted preparation of 1,4-disubstituted 1,2,3-triazoles *via* Cu(I)-catalyzed reaction of sodium azide, alkyl halides and alkynes **(Scheme 39)**. This protocol was safe, user-friendly and eliminated the need to handle organic azides as they were produced *in situ*. For instance, methyl iodide reacted with phenylacetylene and sodium azide to provide the triaozle in 89% yields within 10 minutes, whereas its synthesis was difficult otherwise due to the hazardous nature of methyl azide [89]. Li et al. [90] reported an efficient one-pot preparation of substituted 1,2,4-triazoles *via* a MW-assisted three-component reaction.

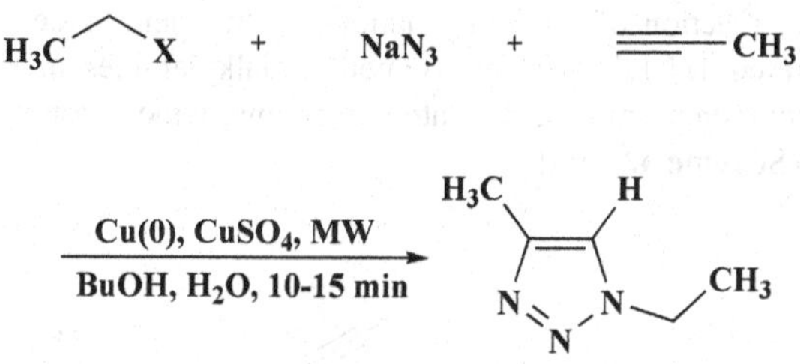

Scheme-39

Chtchigrovsky et al. [91] studied the use of copper(I) species anchored onto functionalized chitosan microspheres **(Scheme 40)**. To obtain the 1,4-substitued triazoles in quantitative yields, only a 0.1 mol% catalyst loading was needed for 15 minutes at 150 °C. All heterogeneous Cu catalysts used in this reaction contained Cu in the form of copper(I).

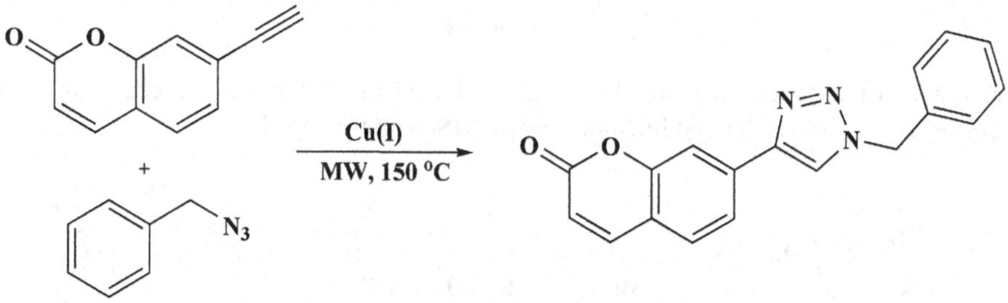

Scheme-40

A Cu(I)-catalyzed 1,3-dipolar cycloaddition reaction in triethylamine and H_2O-MeCN at room temperature occurs where the terminal triple bonds of propargylated nucleobases are ligated onto the azide residue of pseudo sugar to afford the 1,4-disubstituted regioisomer in 84 and 86% yields within 3 hours. The MW-assisted protocol results in the formation of triazole acyclonucleosides, with shorter reaction times and higher yields under mild reaction conditions. Herein, a facile, rapid and practical method was reported for the formation of triazole ring. The cycloaddition of azide and propargylated nucleobases under MW and solvent-free conditions, with cuprous iodide catalyst, provides the desired products in a reaction time of 1 minute and with quantitative yields **(Scheme 41)**. Consequently, the MWI protocol was selected as the best practical method for the preparation of triazole acyclonucleosides as it afforded the best results [92].

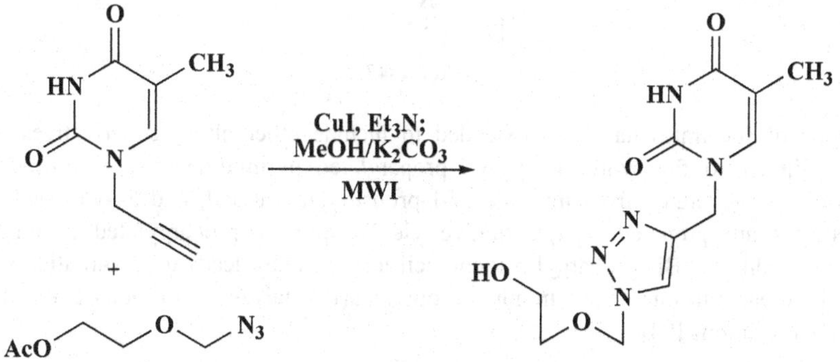

Scheme-41

The three-component reaction of propargyl halides, amines and azides in the presence of Cu(I) catalyst affords 1-substituted 1*H*-1,2,3-triazol-4-ylmethyl-dialkylamines in H$_2$O. The benefits of this synthesis were low environmental impact, high atom economy, wide substrate scope, good yields, and mild reaction conditions (**Scheme 42**) [93].

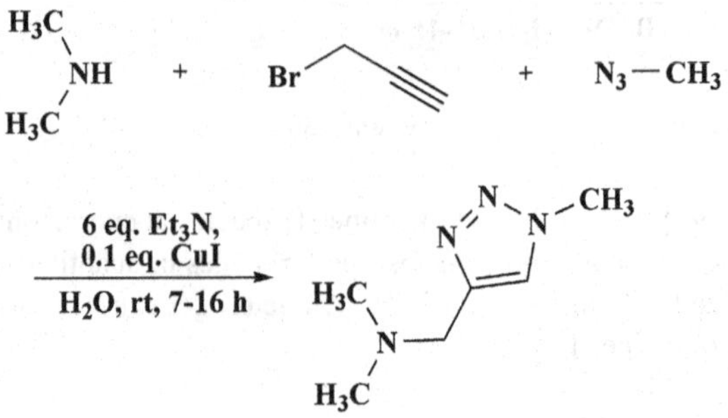

Scheme-42

The MW-assisted 'click' reaction has a general applicability for the preparation of triazole acyclonucleosides of purines and modified nucleobases (**Scheme 43**) [94].

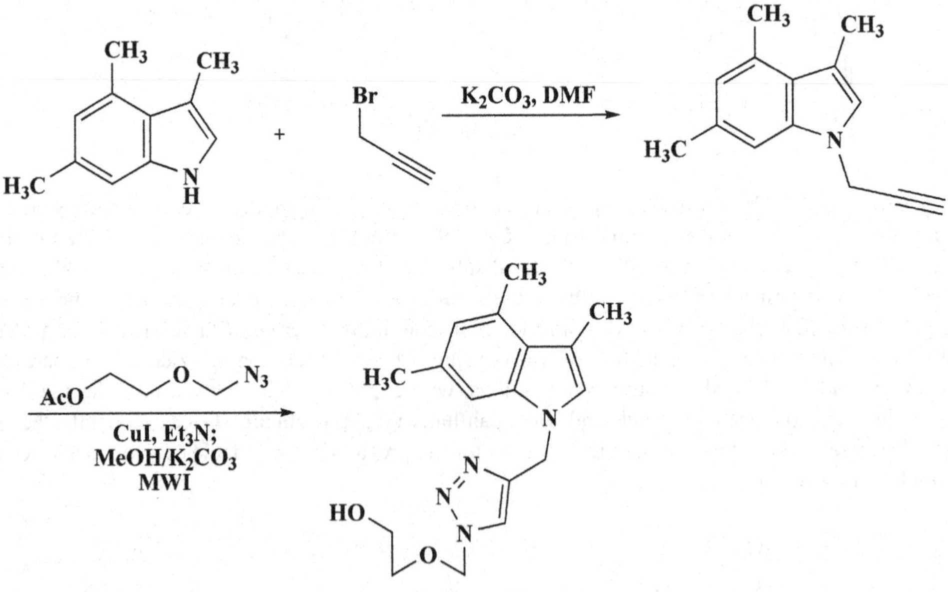

Scheme-43

The scope of substrates has been extended to include other alkinyle derivatives as shown in **Scheme 44**. Similar to the synthesis of *N*-1-proparglated pyrimidine, *N*-1, *N*-3-bis-propargylated pyrimidines can be generated from thymine, *N*-1-propargylated uracil, and 5-iodouracil. Bis-triazole acyclonucleosides are prepared in quantitative yields from bis-propargylated pyrimidines, using same reaction conditions. Bis-propargyl-5-iodouracil unexpectedly leads to quantitative yields of bis-triazoluracil in a reaction time of one minute, through Cu(I)-catalyzed 1,3-dipolar cycloaddition under 'click' reaction conditions [94].

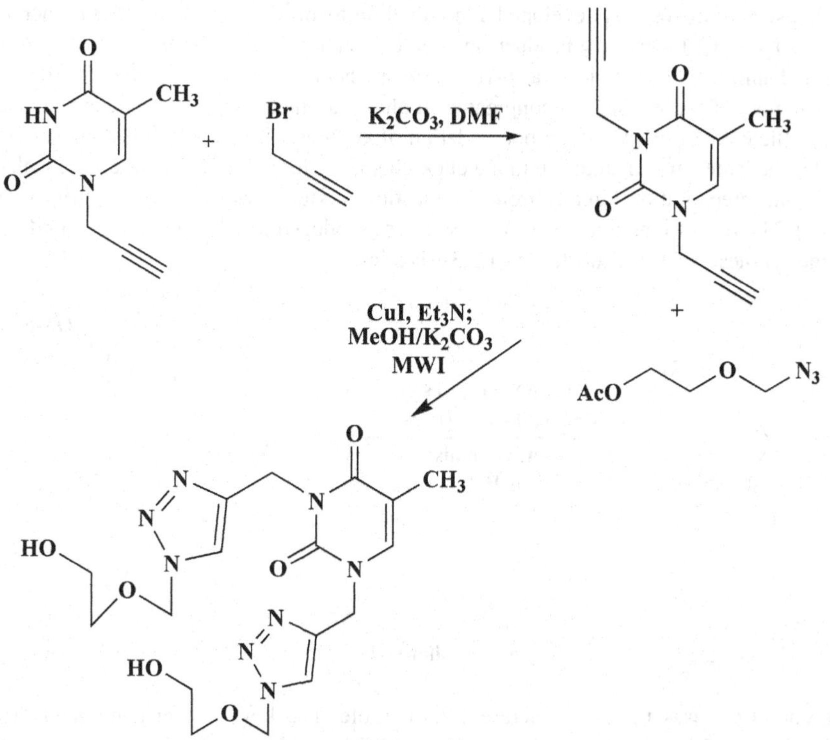

Scheme-44

Ultrasonic activation of metal catalysts due to mechanical depassivation has been extensively used in organic synthesis. Cravotto et al. [95] developed a Cu-catalyzed azide-alkyne cycloaddition reaction for the synthesis of 1,2,3-triazoles using ultrasound or ultrasound and MW simultaneously. When other activation methods were tested, the best results were obtained when terminal alkynes and azides were sonicated in dimethylformamide at 100 °C or dioxane/water at 70 °C in the presence of copper turnings **(Scheme 45)**. Water was substituted by dimethylformamide to prevent the synthesis of Cu complexes which made the purification of products difficult when 6-monoazido-cyclodextrins were used as the starting substrates. No significant differences were reported between the efficiencies of the reactions conducted employing US/MWI or only ultrasound.

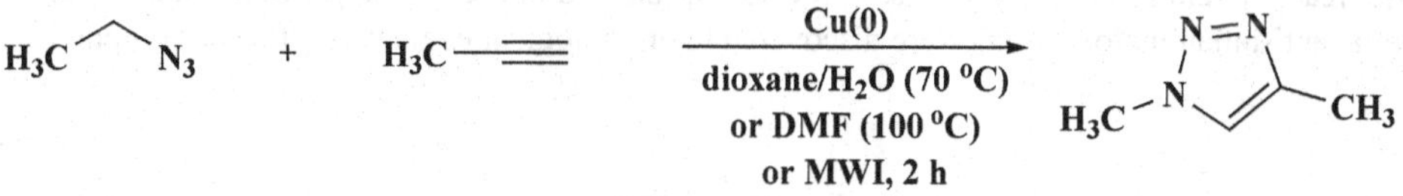

Scheme-45

Yang et al. [96] reported an efficient and convenient one-pot preparation of C-carbamoyl-1,2,3-triazoles in 72–93% yields under MWI **(Scheme 46)**, whereas the conventional approach needed longer reaction times and purification of intermediates.

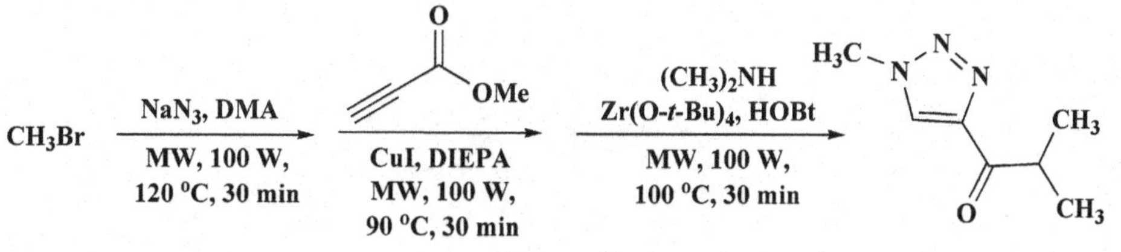

Scheme-46

Moorhouse and Moses [97] developed a modified protocol using MW radiation to increase the rate of formation of various 1,4-triazole products in good to excellent yields (80–99%) from readily available acetylenes and anilines. The reaction was particularly amenable to electron-deficient anilines and worked well with a number of alkynes such as conjugated, aliphatic, aromatic, electron-deficient and electron-rich species. An efficient and practical one-pot azidation of anilines with the combined reagent $TMSN_3$ and *t*-BuONO has become a useful addition to the click-chemistry toolbox. The products are isolated rapidly by filtration and precipitation directly from the reaction mixture, with no further purification required (**Scheme 47**). The rate of reaction of alkynes and azides produced *in situ* is enhanced significantly under MWI for the synthesis of 1,4-disubstituted 1,2,3-triazoles.

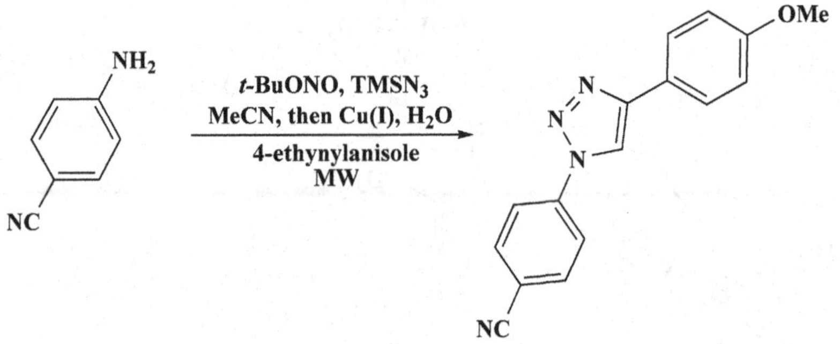

Scheme-47

Click chemistry uses the key structure 1,2,3-triazole. The key starting compound of the 'click' cycloaddition, ethyl 2-hydroxybut-3-ynoate, is prepared by adding ethynylmagnesium bromide to an aldehyde. Since 2-hydroxyl acid ethyl ester turns yellow at room temperature, it is kept under refrigeration. Many difficulties are faced when the product formed is transformed into an alkoxy compound due to the high activity of alkynyl. So, the alkoxy compound is obtained in the last step. Till date, many reactions have been reported for the synthesis of amides. The synthetic protocols for amides involve the reaction of a carboxylic acid with an amine in dicyclohexyl carbodiimide or that of an amine with an acyl chloride at reflux temperature. However, traditional protocols for the preparation of amides suffer from disadvantages such as inconvenient handling and long reaction times. Amides can be synthesized by the reaction as shown in **Scheme 48** [98].

In a study by DeSimone et al. [99], 1,3-dipolar cycloaddition and subsequent Suzuki cross-coupling were explored for the synthesis of 15 compounds after rational docking calculations (**Scheme 49**). For a faster reaction and satisfactory yields, MWI was introduced in the two steps. Consequently, three new potential anti-inflammatory drugs were discovered upon biological evaluation of these compounds.

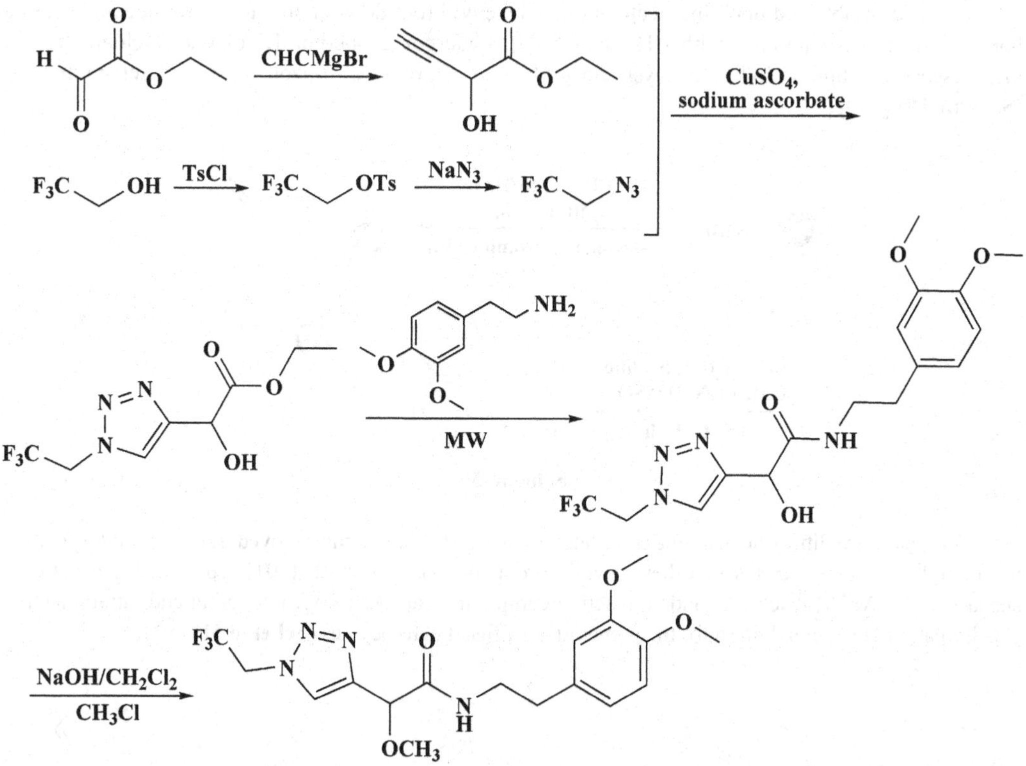

Scheme-48

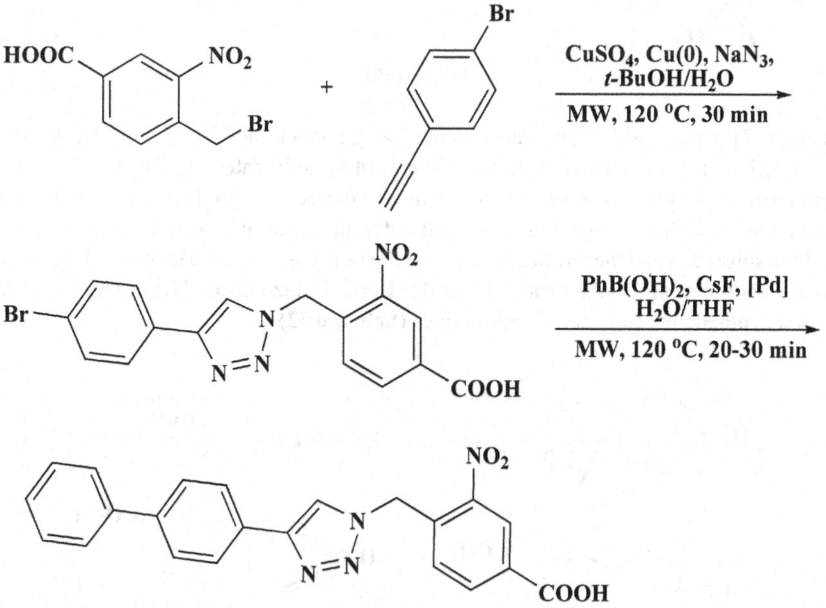

Scheme-49

White resin obtained upon the treatment of selenenyl bromide with propargyl bromide and sodium borohydride reacts smoothly with CH$_3$I and NaN$_3$, undergoing one-pot 1,3-dipolar cycloaddition in the presence of cuprous iodide catalyst and proline to afford 1,2,3-triazole-supported selenium resin (**Scheme 50**) [100].

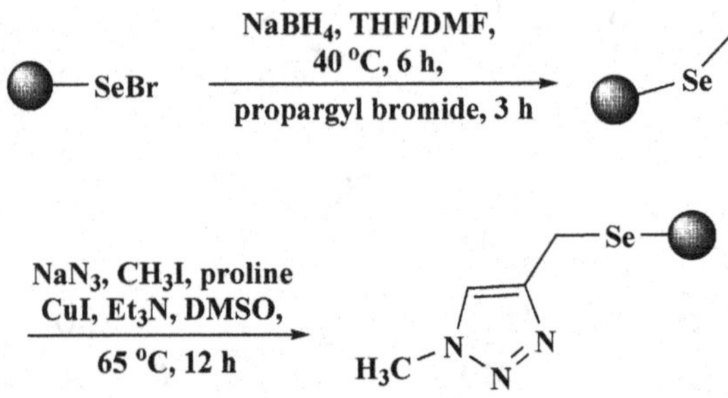

Scheme-50

Continuous modification of amine components wherein the haloatom is moved from their *ortho* position to ethynylbenzenes generates new heterocyclic products. Barange et al. [101] reported that a sulfonyl substrate and TMSN$_3$ react in a tandem reaction comprising carbon-nitrogen coupling and intramolecular 1,3-dipolar 'click' cycloaddition for the preparation of fused heterocycles (**Scheme 51**) [102].

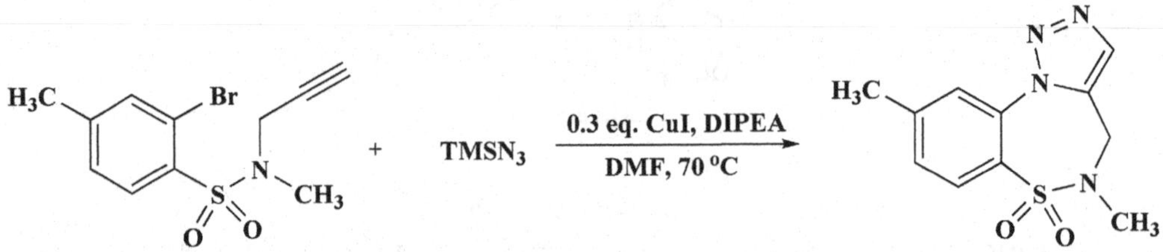

Scheme-51

Hu et al. [103] developed tandem reactions involving copper-catalyzed Sonogashira carbon-carbon coupling and 'click' 1,3-dipolar cycloaddition. The starting substrates bearing vinyl azide fragments reacted with terminal alkynes to produce triazole-fused heterocyclic products in the presence of cupric chloride and a base. Mechanistic studies propose that the intermediates were formed *via* carbon-carbon couplings of the substrates and the terminal alkynes in the first stage of the reaction, and the intramolecular 1,3-dipolar cycloaddition of the alkyne and the azide afforded triazole-fused heterocyclic products which served as good synthetic precursors of isoquinolines (**Scheme 52**).

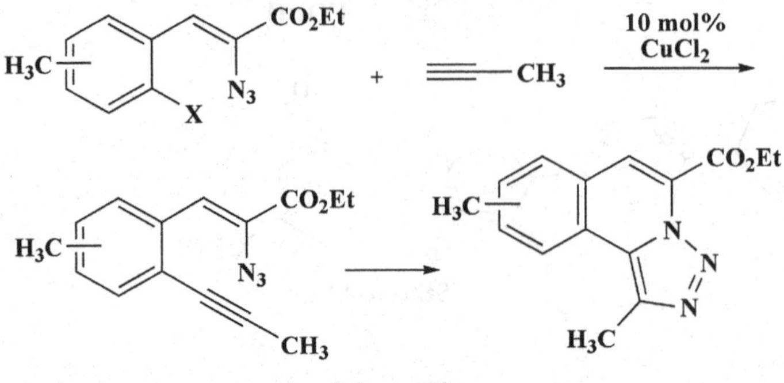

Scheme-52

Copper- and palladium-assisted synthesis

Iodo *N*-Boc NH-heterocycles react with TMSA under standard Sonogashira coupling conditions (PdCl$_2$(PPh$_3$)$_2$/cuprous iodide/triethylamine). The trimethylsilyl alkynes are de-protected with the help of tetrabutylammonium fluoride without isolation and subsequently reacted with 1 eq. of stable and commercially available benzyl azide to form *N*-Boc 3-triazolyl (aza)indoles and azoles in a one-pot protocol **(Scheme 53)**. The yields are very similar for pyrrole and (aza)indoles irrespective of the position and the number of nitrogen atoms. No further addition of cuprous iodide is required in the copper azide alkyne cycloaddition step. The steps proceed as 'spot-to-spot' reactions without noticeable amounts of by-products and the progress of the reaction can be monitored conveniently by thin layer chromatography. Because the copper azide alkyne cycloaddition reaction is performed under an argon atmosphere, no Glaser-type homodimerization products are detected. The electron-withdrawing Boc protective group makes the (aza)indolyl iodides stable for storage, whereas unprotected iodides are considerably sensitive to temperature and light and, therefore, inconvenient to handle [104]. Moreover, Sonogashira coupling is facilitated greatly, and even made feasible, by the diminished electron density of these heterocyclic compounds.

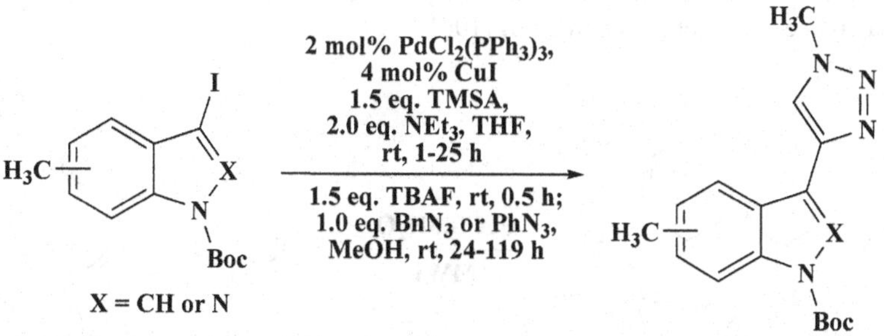

Scheme-53

The sequence can be extended to a four-component reaction of *N*-Boc protected 3-iodo-7-azaindole for the preparation of triazoles with different substituents on the N-1 atom. Additionally, this sequence includes the *in situ* formation of azide *via* nucleophilic substitution of a halide with cesium azide **(Scheme 54)**. Hence, electronically diverse benzyl substituents, *α*-phenylethyl substituents as well as homobenzyl groups can be introduced with a comparable yield of the desired products [104].

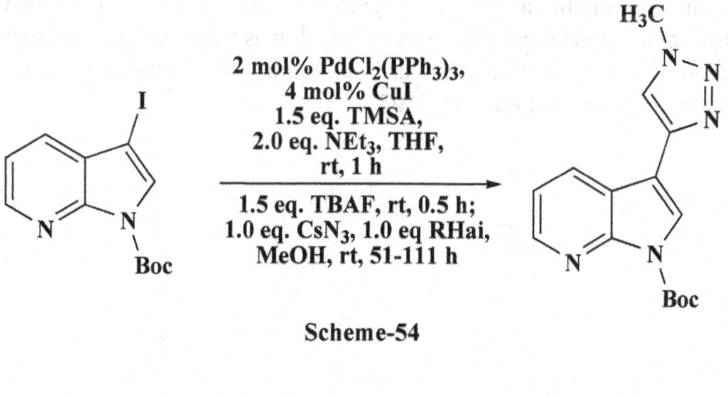

Scheme-54

Stable and commercially available 4- and 5-bromo-7-azaindoles produce *N*-Boc protected 4- and 5-triazolylazaindoles in very good yields through a four-component Boc-protection-Sonogashira coupling-TMS-de-protection-copper azide alkyne cycloaddition sequence (**Scheme 55**) [104].

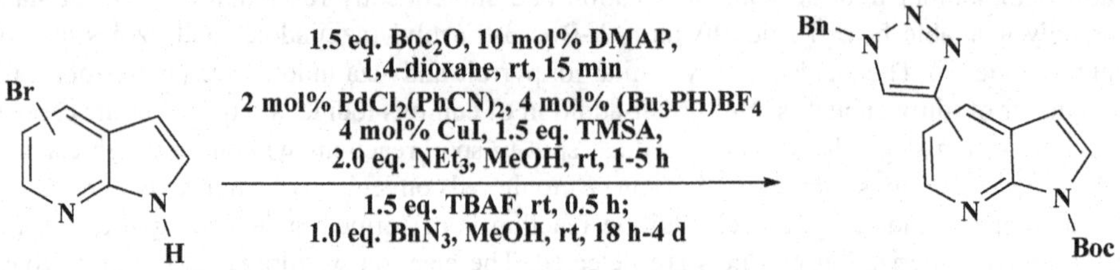

Scheme-55

Iron-assisted synthesis

Mogilaiah et al. [105] used arylaldehyde-3-(3-fluorophenyl)-1,8-naphthyridin-2-ylhydrazones for the synthesis of 9-aryl-6-(3-fluorophenyl)-1,2,4-triazolo[4,3-*a*][1,8]naphthyridines using FeCl$_3$.6H$_2$O under MWI and solvent-free conditions (**Scheme 56**) [106].

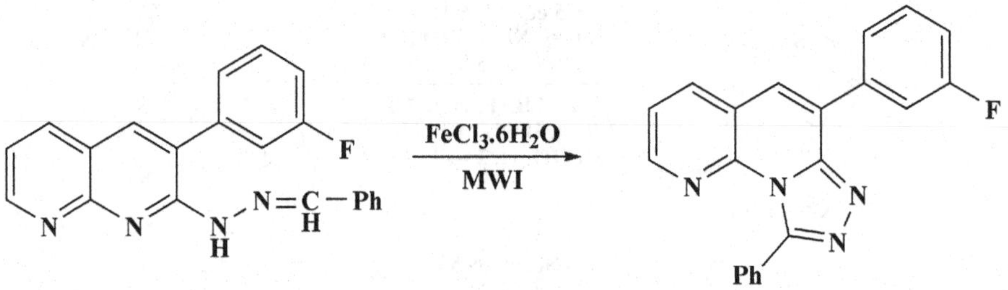

Scheme-56

Triazoles possessing anti-fungal compounds are a well-known group in pharmaceuticals. However, only a few members have found place in the list of best selling drugs such as itraconazole. All members of this class act as inhibitors in fungal cytochrome P-450 oxidase-mediated synthesis of ergosterol. Janssen pharmaceuticals marketed itraconazole consisting of a central diarylpiperazine unit as well as two triazole subunits. In one of the methods for synthesizing itraconazole [107], 1,2,4-triazole was introduced through direct nucleophilic substitution. In later stages of synthesis, the elaborated aniline derivative was trapped with phenyl chloroformate to synthesize carbamate which was transformed into triazolone by a double condensation reaction with formamidine and hydrazine. Finally, the synthesis was completed by a simple attachment of an isobutyl group (**Scheme 57**) [108].

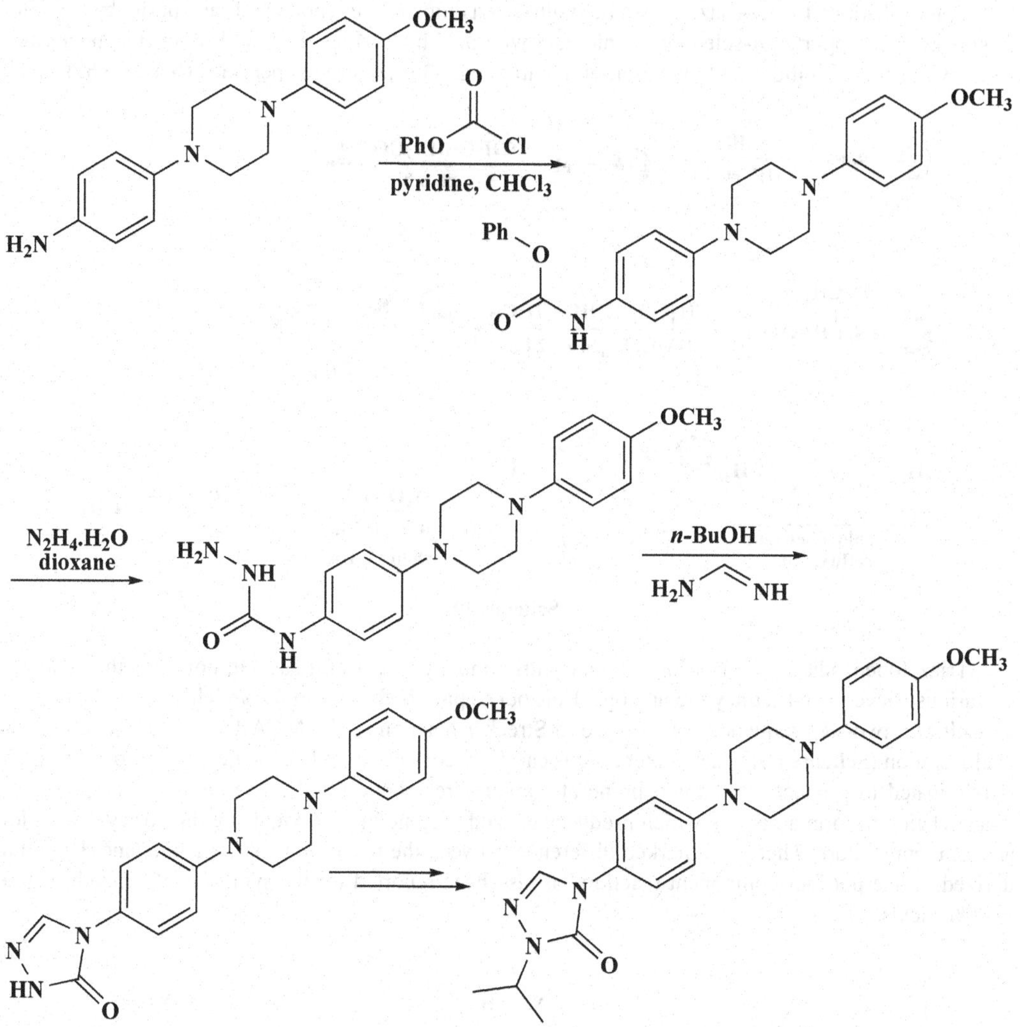

Scheme-57

Lithium-assisted synthesis

1,2,3-Benzotriazole-5-carboxylic acid yield benzotriazole resin by a two-step reaction sequence starting from Merrifield resin. Functionalization in good yield occurs with piperazine. The subsequent steps involve treatment of the formed resin with ethyl benzotriazolyl-5-carboxylate, bromination of the resin (by a standard protocol), lithiation, treatment with ethyl chloroformate, and heating with benzhydrazide at 120 °C in dimethylformamide for 36 hours. Finally, the desired polymer is formed by reflux in phosphorusoxy chloride and chloroform, followed by heating with ammonium acetate at 80 °C **(Scheme 58)** [109–110].

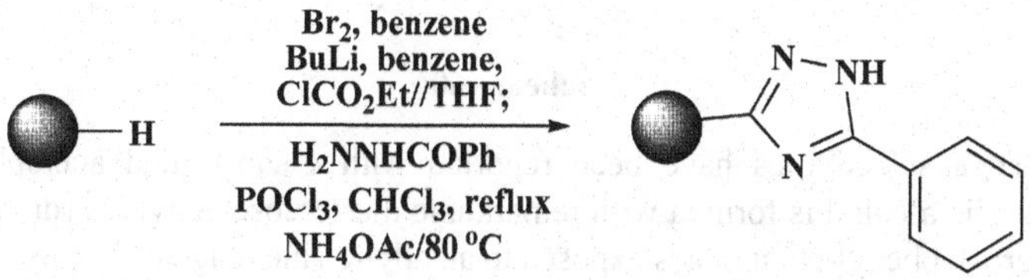

Scheme-58

Vinyl-substituted 1,2,4-triazoles can be synthesized efficiently in good yields and purity by acylation of polystyrene-supported α-selenopropionic acid with acid hydrazides, followed by cyclocondensation with arylphosphazoanilides and oxidation-elimination with 30% hydrogen peroxide **(Scheme 59)** [111].

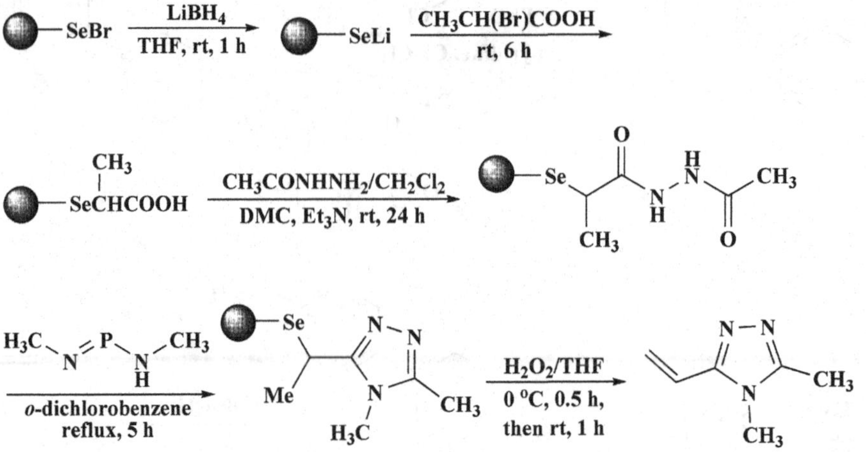

Scheme-59

Triazolobenzodiazepine bearing α-amino nitrile moiety is a useful starting point for the synthesis of various tricyclic and tetracyclic analogs. Benzodiazepine is formed in 69% yields from aldehyde as a result of a two-step sequence consisting of a Strecker reaction as the MCAP, and subsequent dipolar cycloaddition **(Scheme 60)**. The temperature is carefully controlled in this reaction as hydrogen cyanide is eliminated to produce the known imine at temperatures above 60 °C. The aminonitrile undergoes N-acetylation to form an amide which produces benzodiazepine in 96% yields by dipolar cycloaddition at room temperature. There is a marked difference between the reactivities of the amine and the amide derived. A one-pot four-component reaction has also been reported for the synthesis of benzodiazepine in 49% yields [112].

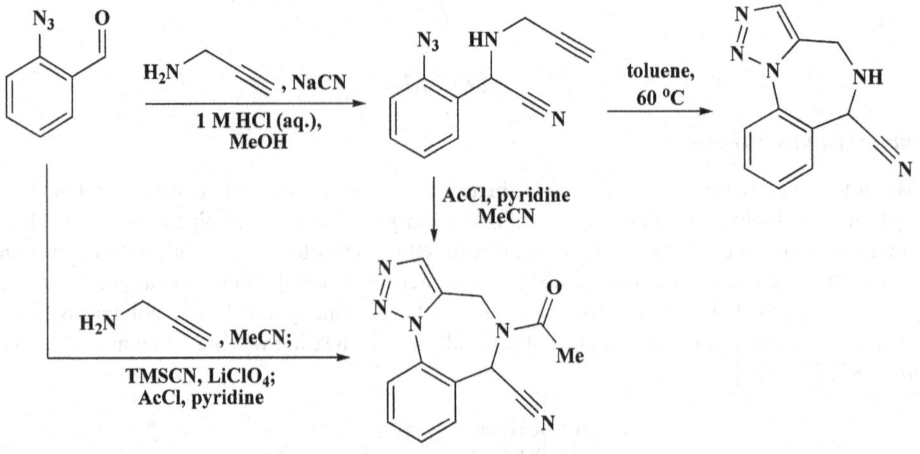

Scheme-60

Excellent diastereoselectivities have been reported with α-substituted acetophenone. Chloride-substituted homoallylic alcohol is formed with remarkable diastereoselectivities (dr > 98:2) and in 97% yields when 2-chloro-1-phenyl-ethanone is exposed to an allylic zinc reagent. The use of organoazides in 'click chemistry' has rendered azides as a very interesting class of structures [113]. Azides can be obtained in significant yields (93%) and diastereoselectivities (dr = 99:1) from 2-azido-1-phenyl-ethanone. A one-pot synthesis of 1,2,3-triazoles in 85–90% yields, with excellent regio- and diastereoselectivities, involves the use of CuCN·2LiCl (5 mol%) **(Scheme 61)**.

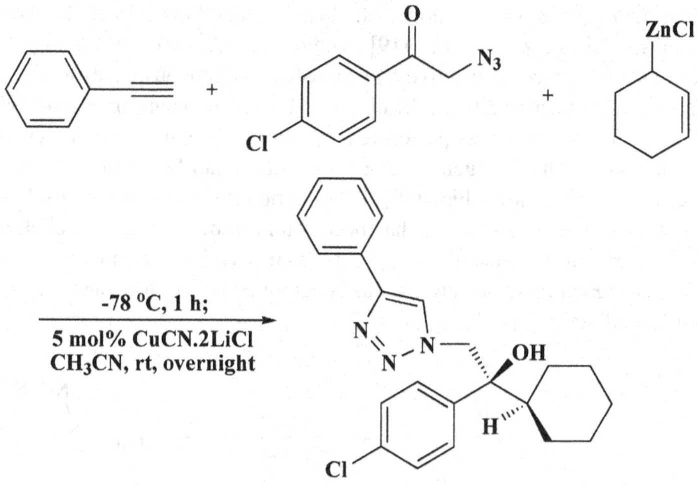

Scheme-61

Magnesium-assisted synthesis

Triazole-based monophosphine ligands can be efficiently synthesized through cycloaddition. Palladium complexes with these ligands are highly active catalysts for Suzuki-Miyaura coupling and amination of aryl chlorides **(Scheme 62)** [114].

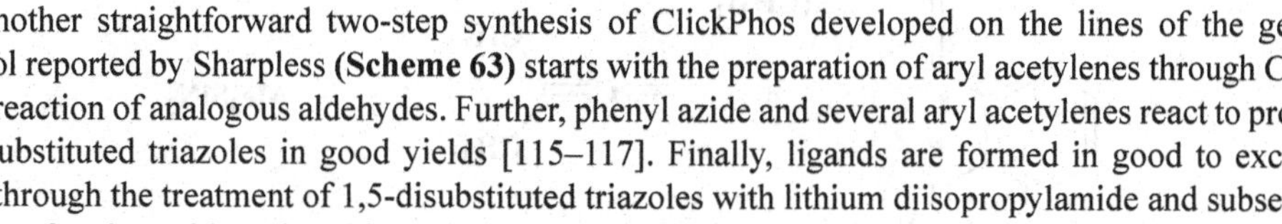

Scheme-62

Another straightforward two-step synthesis of ClickPhos developed on the lines of the general protocol reported by Sharpless **(Scheme 63)** starts with the preparation of aryl acetylenes through Corey-Fuchs reaction of analogous aldehydes. Further, phenyl azide and several aryl acetylenes react to produce 1,5-disubstituted triazoles in good yields [115–117]. Finally, ligands are formed in good to excellent yields through the treatment of 1,5-disubstituted triazoles with lithium diisopropylamide and subsequent addition of various chlorophosphines.

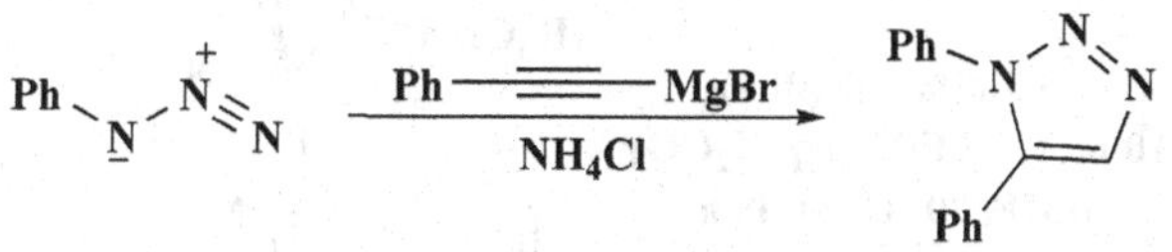

Scheme-63

Manganese-assisted synthesis

Triazoles can be synthesized efficiently by azide-alkyne Huisgen [3+2]-cycloaddition reaction, better known as the 'click' reaction [118]. These reactions catalyzed by copper(I) or copper(II) species and performed at slightly elevated or ambient temperatures take several hours for completion. Although limited in number, there have been reports of such cycloaddition reactions being catalyzed by heterogeneous catalysts. Lipshutz and Taft [119] synthesized 1,2,3-triazoles using charcoal-supported Cu (copper/carbon) as a highly regioselective and robust heterogeneous catalyst. The reaction required 1 eq. of base such as Et$_3$N under traditional heating conditions to ensure high reaction rates. However, under MW irradiation, the reaction was performed quantitatively within 3 minutes at 150 °C without any external base or ligand. The Huisgen [3+2]-cycloaddition can be carried out under MWI using a ligand/additive-free copper-manganese bimetallic heterogeneous catalyst as well (**Scheme 64**) [120]. For this, the optimum ratio of manganese:copper has been found to be 0.25:2. This click reaction produces 1,4-disubstituted 1,2,3-triazoles in quantitative yields from a variety of substrates under MW-assisted conditions and, after the reaction, the catalyst is easily removed by filtration and reused up to nine times without significant loss of activity [121].

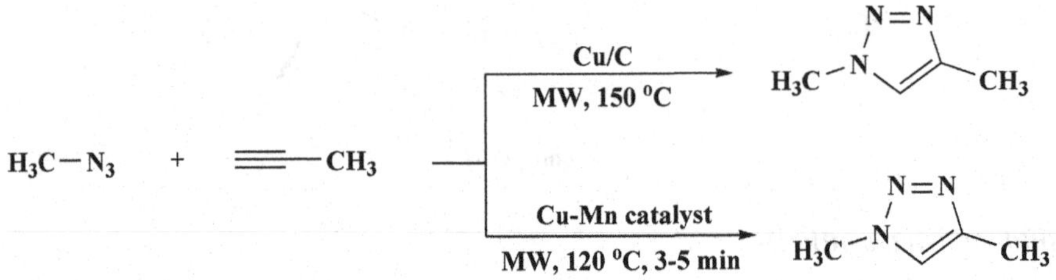

Scheme-64

Palladium-assisted synthesis

Azides are interesting trapping nucleophiles. They form propargyl type 1,3-dipoles upon nucleophilic displacement. As a consequence, annelated triazoles can be synthesized easily. For instance, triazoles can be prepared by a sequence of insertion of allenes, allylic substitution, and intramolecular 1,3-dipolar cycloaddition in a one-pot protocol (**Scheme 65**) [62, 122].

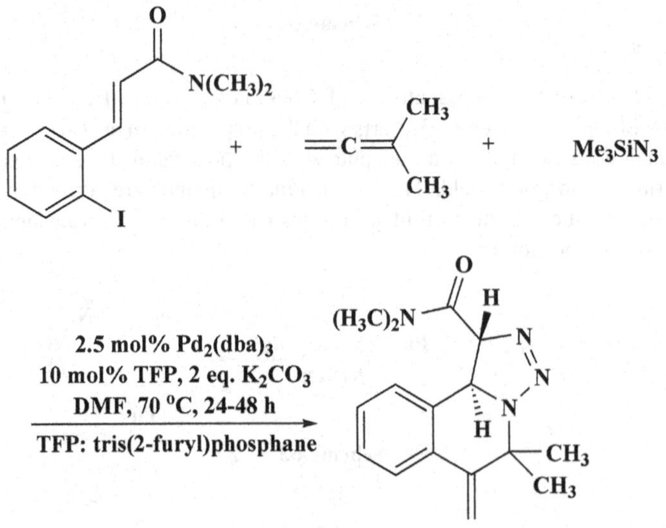

Scheme-65

Triazoles can be synthesized from non-activated terminal alkynes by coupling of trimethylsilyl azide and allyl methyl carbonate in the presence of palladium (0)-copper(I) bimetallic catalysts **(Scheme 66)** [123]; the desired product of this three-component reaction is not obtained in the absence of Cu(I). The allyltriazoles formed by coupling undergo deallylation. The reaction occurs through [3+2]-cycloaddition between an alkyne of the Cu acetylide (where Cu acts as an activating group for the alkyne) and the azide palladium complex to yield (3-allyl)(5-triazoyl)palladium as the intermediate. The triazoles are ultimately formed by subsequent reductive elimination of palladium(0) from this intermediate and protonolysis of the carbon-copper bond.

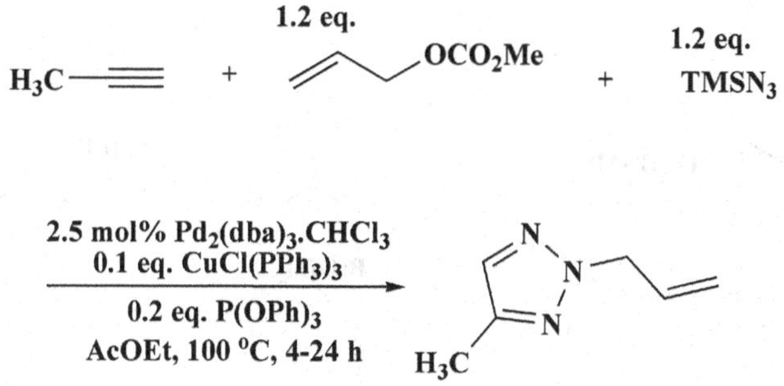

Scheme-66

A Pd-catalyzed reaction of alkenyl halides and sodium azide yields 1*H*-triazoles. This reaction represents a completely new reactivity pattern in palladium chemistry **(Scheme 67–68)** [124].

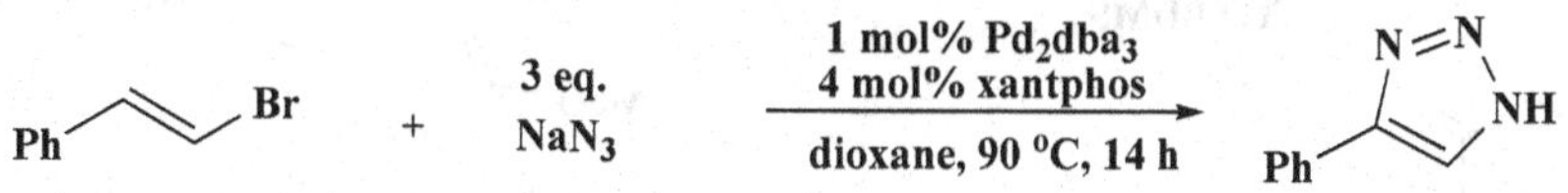

Scheme-67

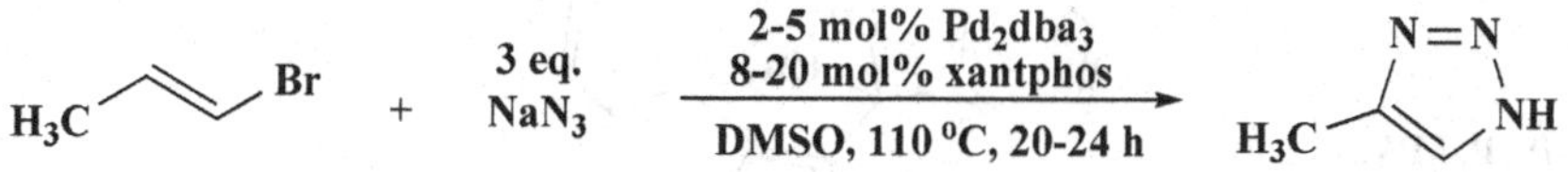

Scheme-68

Yamamoto et al. [125a–b] reported a general pathway for multi-component reactions involving azides in [3+2]-cycloaddition. The three-component palladium-catalyzed coupling of allyl methyl carbonate, alkynes and trimethylsilyl azide provided 2-allyl-1,2,3-tetrazoles **(Scheme 69)**. The reaction occurred *via* [3+2]-cycloaddition of allylpalladium azide to the alkyne, followed by the synthesis of (3-allyl) (5-triazoyl)-palladium [126]. This chemistry, similar to Cu-catalyzed alkyne/substituted azide coupling, was reported to occur *via* palladium-catalyzed oxidative addition of the allylic substrate, which allowed the subsequent preparation of a palladium-azide for cycloaddition. The latter product was eliminated by reduction to regenerate the palladium catalyst.

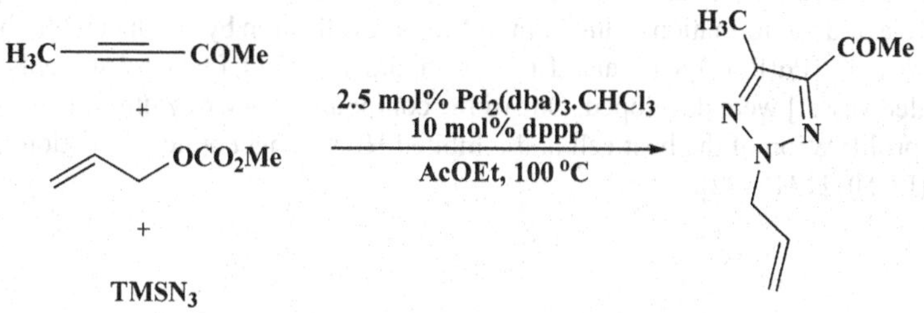

Scheme-69

Beryozkina et al. [127] reported a MW-assisted synthesis of 1,2,3-triazole-derived aza-analogs of naturally occurring lignan lactones (–)-steganone and (–)-steganacin (**Scheme 70**). The key biaryl intermediates were formed in high purity and yields by MW-assisted Suzuki-Miyaura cross-coupling reaction. The core biaryl skeleton of steganone and steganacin 7-aza-analogs containing the 1,2,3-triazole subunit was constructed in merely 15 minutes by MW-assisted Suzuki reaction, compared to 3 hours under conventional heating conditions resulting in a meager 42% yield. Finally, ring-closure produced the less constrained Steganacin analog by MW-assisted intramolecular Huisgen 1,3-dipolar cycloaddition at 120 °C in dimethylformamide for 15 minutes, while the highly constrained Steganone analog needed a higher temperature of 210 °C for 15 minutes in *o*-dichlorobenzene [128–129].

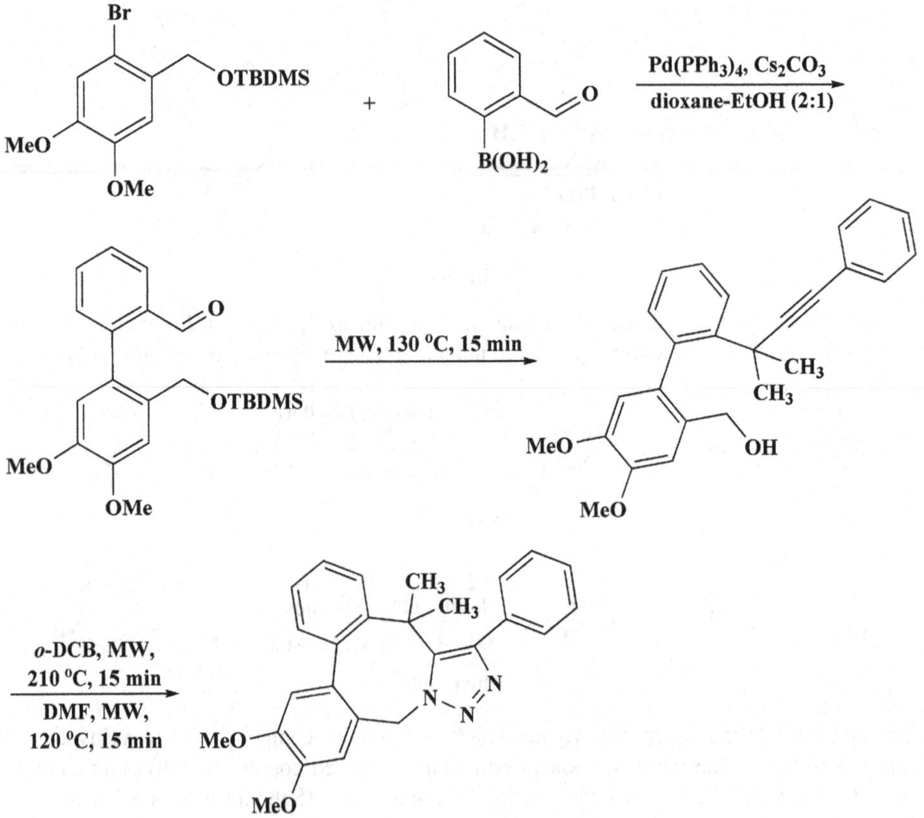

Scheme-70

Zhu et al. [130] designed novel acyclic 1,2,4-triazole nucleosides with various ethynyl moieties appended on triazole nucleobases. The preparation started with bromotriazole acyclonucleoside in 75–99% yields using an efficient one-step Sonogashira reaction under MWI in aqueous solution (**Scheme 71**). To avoid the formation of the intramolecular cyclization by-product under basic conditions, optimized conditions [Pd(PPh$_3$)$_4$/CuI and Li$_2$CO$_3$, in dioxane/H$_2$O (3:1) as solvents for 25 min at 100 °C in a sealed vessel] were developed. One of the compounds (having *p*-fluorphenyl substituent) did not inhibit the proliferation of the host cell and inhibited HCV subgenomic replication at 50% effective concentration (EC50) [131–132].

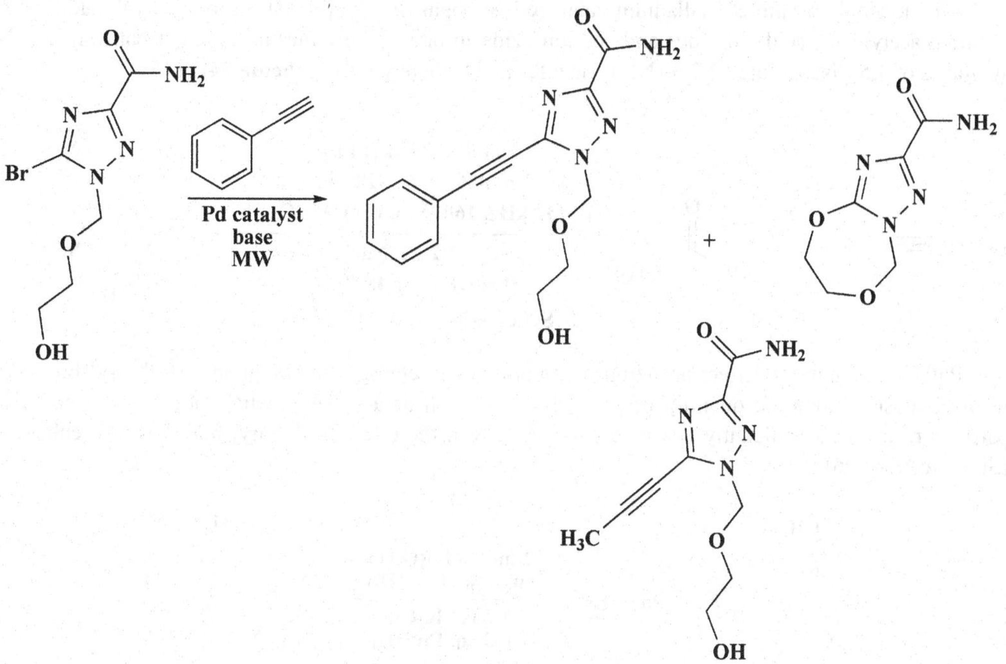

Scheme-71

A facile method of synthesizing [1,2,4]triazolo[4,3-a]pyridines is palladium-catalyzed addition of hydrazides to 2-chloropyridine. This reaction occurred chemoselectively at the terminal nitrogen atom of the hydrazide, followed by dehydration under MWI in acetic acid (**Scheme 72**) [133].

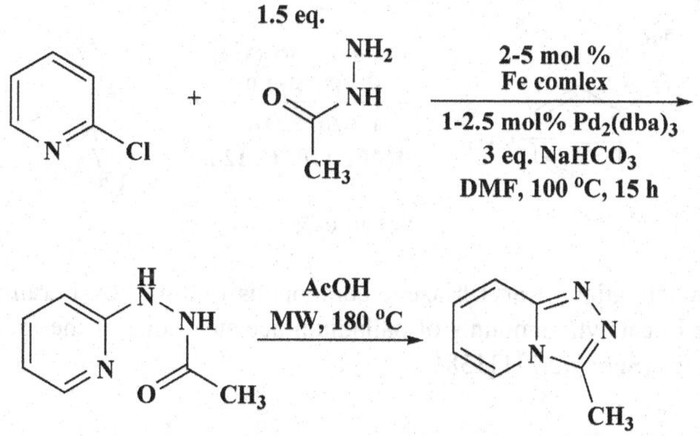

Scheme-72

Anti-3-aryl-2,3-dibromopropanoic acids and sodium azide react by a one-pot protocol using DMF as the solvent in the presence of Xantphos and Pd₂(dba)₃ to produce 4-aryl-1H-1,2,3-triazoles (**Scheme 73**) [134].

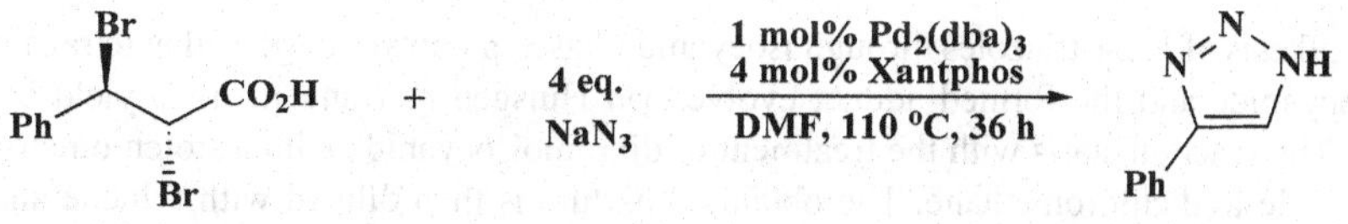

Scheme-73

An ultrasonic-promoted palladium-catalyzed Sonogashira coupling/1,3-dipolar cycloaddition of terminal acetylenes, acid chlorides and sodium azide in one-pot provides an efficient pathway for the synthesis of 4,5-disubstituted 1,2,3-(NH)-triazoles in excellent yields (**Scheme 74**) [135].

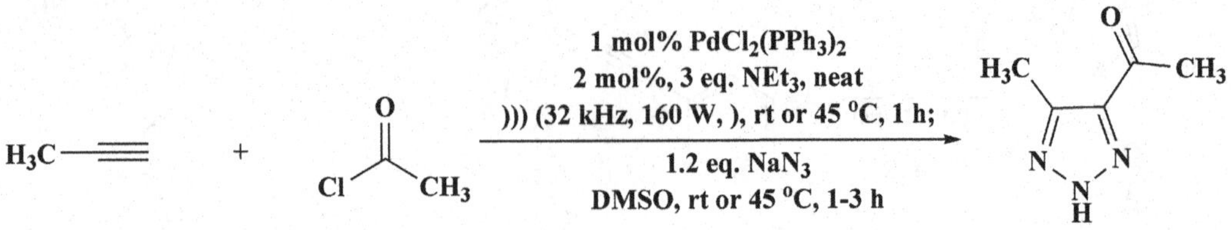

Scheme-74

Palladium-catalyzed direct arylation of a number of heterocyclic compounds with aryl bromides using a substoichiometric quantity of pivalic acid as well as a stoichiometric ratio of both coupling partners results in significantly fast reactions. Moreover, the effect of the aryl halide is also enhanced (**Scheme 75**) [136].

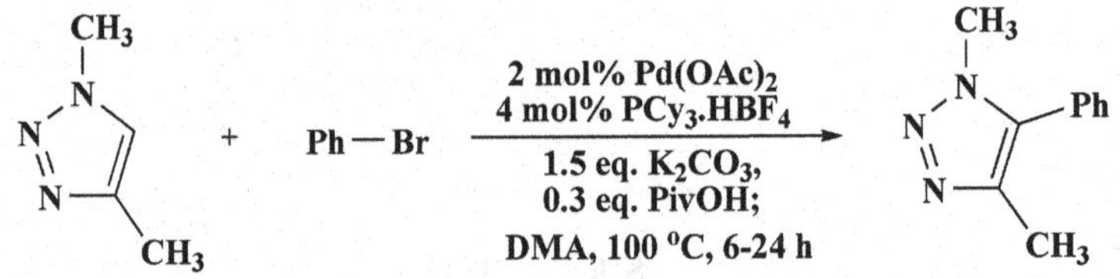

Scheme-75

The mechanism of synthesis of highly regioselective benzotriazoles in excellent yields by 1,7-palladium migration-cyclization-dealkylation sequence has also been investigated by researchers (**Scheme 76**) [137].

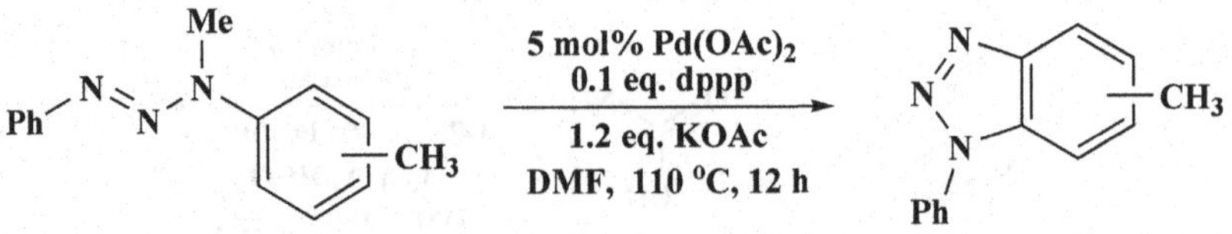

Scheme-76

Carbon-hydrogen activation of aryl triazene compounds followed by intramolecular amination at moderate temperature in catalytic amounts of palladium acetate is one of the methods of synthesizing 1-aryl-1*H*-benzotriazoles (**Scheme 77**) [138].

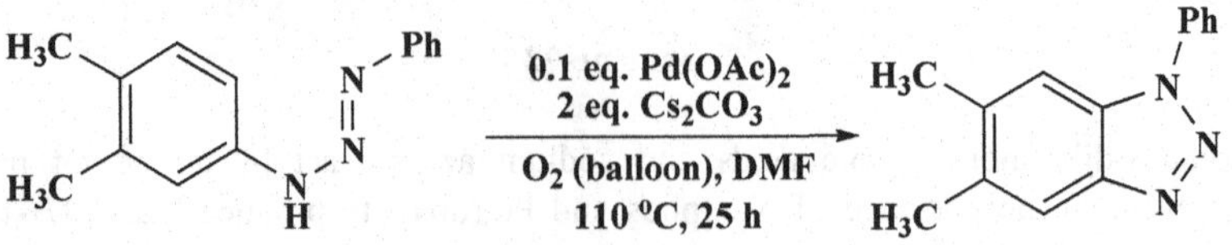

Scheme-77

The synthesis of 1,2,4-triazoles through isocyanide-based pathway suggests that tetrazoles react with dibromoisocyanide and the formed adduct evolves *via* Huisgen rearrangement to yield bromotriazole [139–144]. The reaction starts with the treatment of dibromoisocyanide with a stoichiometric amount of phenyl tetrazole in dichloromethane. The obtained mixture is then diluted with toluene and heated for 2.5 hours at 110 °C to yield bromotriazole which is subsequently subjected to Suzuki coupling.

Consequently, the desired 1,2,4-triazole is obtained in 40% yields by a one-pot four-step synthesis **(Scheme 78)** [145].

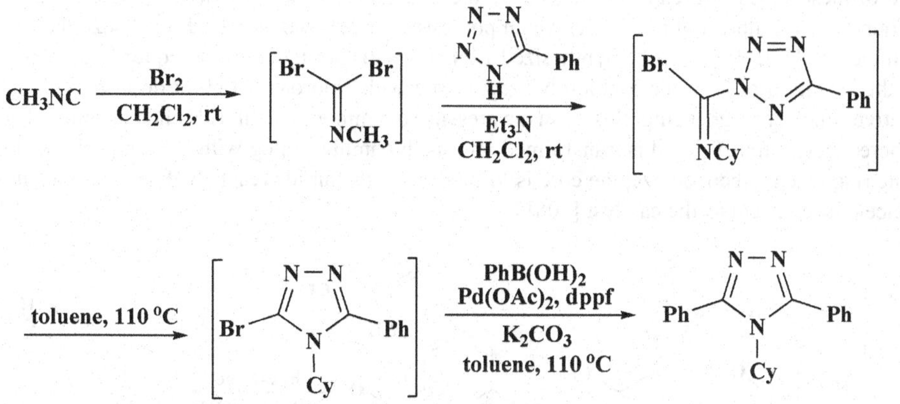

Scheme-78

The assessment of dichloromethane, 1,2-dicholoethylene and dimethylformamide as reaction solvents (in combinations or as individual solvents) has revealed that reactions are relatively cleaner in 1,2-dicholoethylene, although additional efforts are required to work out the appropriate reaction conditions for this one-pot Sonogashira/iodocyclization sequence. As reported earlier, various heterocyclic units are readily prepared by this iodocyclization protocol. This can be combined with other efficient transformations to broaden the scope of this strategy and allow easy access to heterocyclic compounds that are not accessible by iodocyclization yet. For example, alkyne subjected to click chemistry employing benzyl azide affords the desired triazole in good yields **(Scheme 79)** [146–147]. Iodotriazole is formed by iodo-desilylation, providing avenues for further introduction of heterocyclic compounds using Sonogashira coupling/iodocyclization, or other coupling reactions such as Suzuki-Miyaura coupling [148–151].

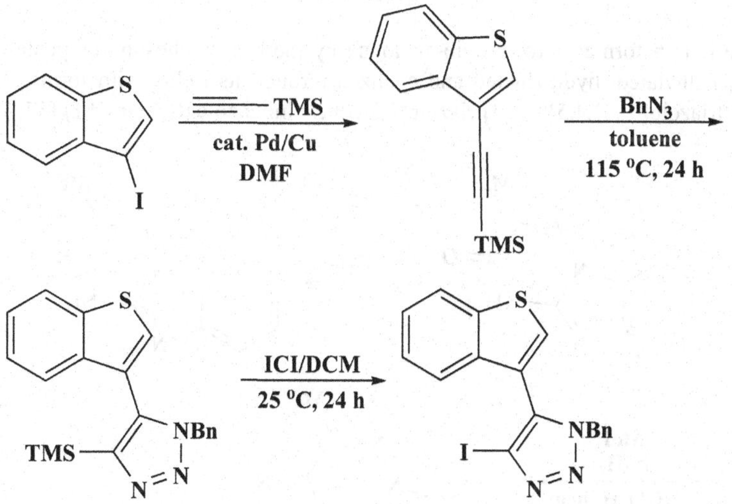

Scheme-79

Phosphorus-assisted synthesis

Benzodiazepines possess a variety of central nervous system related activities. The pharmacological profile of these drugs is greatly influenced by the attachment of a third ring. A representative of this compound class is alprazolam (Xanax) which possesses a 1,2,4-triazole fused to a benzodiazepine core. This molecule [152–153] can be synthesized (**Scheme 80**) through a short sequence of steps starting from the acylation of 2-amino-5-chlorobenzophenone with chloroacetyl chloride to provide an amide derivative which undergoes ring-closure in the presence of ammonium chloride and hexamine. The seven-membered lactam thus formed is transformed into its thioamide analog with P_2S_5 in pyridine. Finally, a triazole ring fused to benzodiazepine core is formed when thioamide reacts with acetyl hydrazide in the presence of acetic acid as the catalyst [108].

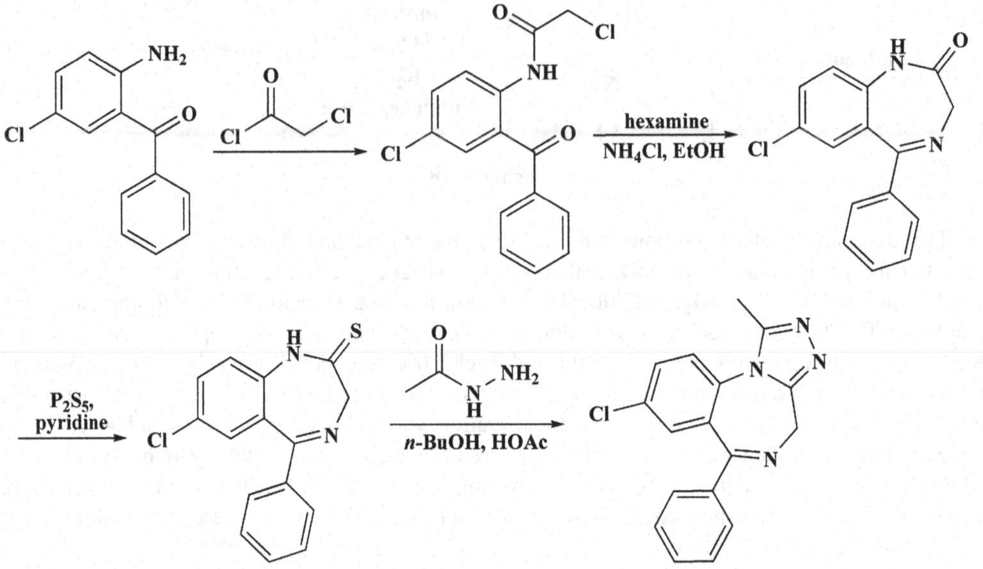

Scheme-80

Triazinone is transformed into its thio-analogue by the use of phosphorus pentasulfide which is successively S-methylated, hydrazinated and cyclized through its reflux with formic acid to produce 4-methyl-1,2,4-triazolo[4',3':4,5][1,2,4]triazino[2,3-a]benzimidazole (**Scheme 81**) [37, 154–155].

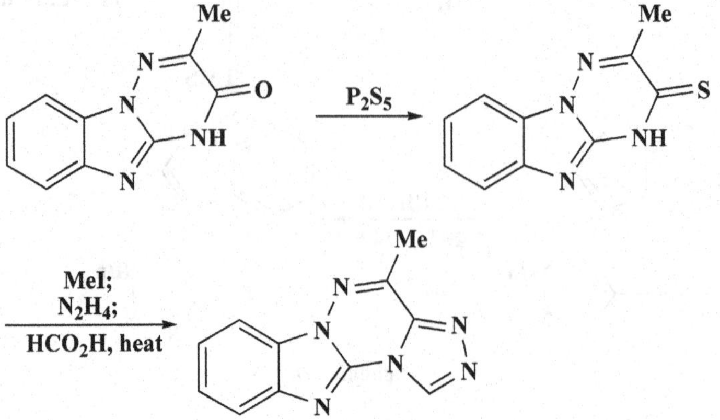

Scheme-81

Ruthenium-assisted synthesis

Primary and secondary azides react with a variety of terminal alkynes bearing a number of moieties in the presence Cp*RuCl(COD) or Cp*RuCl(PPh$_3$)$_2$ as the catalyst to form 1,5-disubstituted 1,2,3-triazoles selectively. Both of these complexes also promote the cycloaddition reactions of organic azides with internal alkynes to provide fully substituted 1,2,3-triazoles **(Scheme 82)** [156].

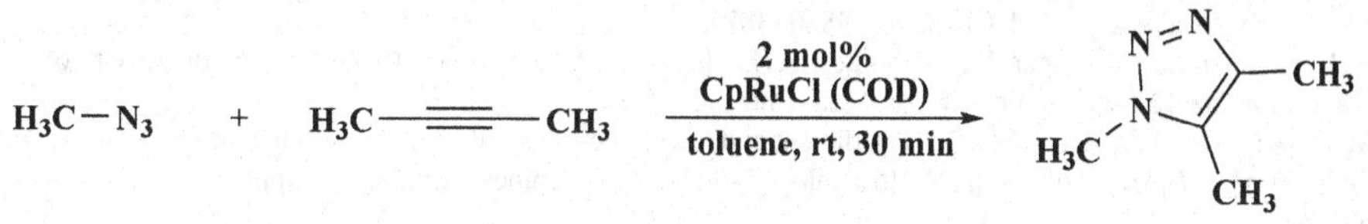

Scheme-82

Silicon-assisted synthesis

Acid hydrazides undergo a one-stage reaction with aryl or alkyl isothiocyanate in the presence of potassium hydroxide solution (10%) on the surface of montmorillonite K10 as well as on the surface of silica gel under MWI to produce 4,5-disubstituted 1,2,4-triazole-3-thiones. Even 4-substituted 1-aroyl-thiosemicarbazides react with potassium hydroxide solution (10%) on the surface of silica gel under MWI to afford these triazoles **(Scheme 83)** [157–158].

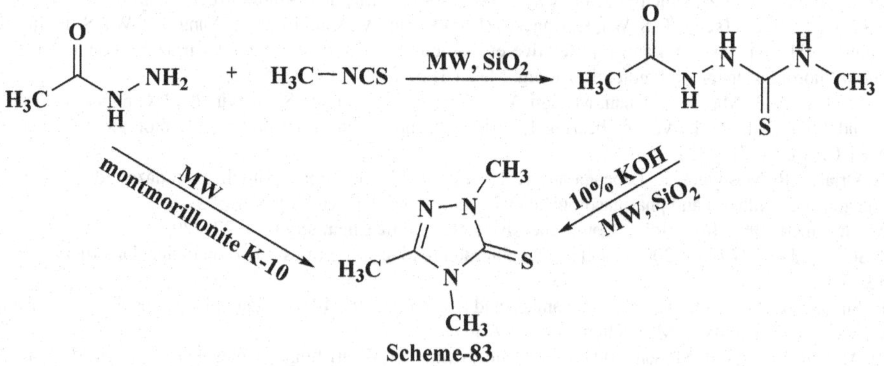

Scheme-83

Silver-assisted synthesis

1,1'-Bibenzotriazole can be synthesized in two steps through a method as depicted in **Scheme 84**. Carboni et al. [159] and Ortiz et al. [160] reported oxidation of 1,2-diaminobenzene to 2,2'-diaminoazobenzene—this is the first step of the synthesis. Both of these work-ups can also be modified into that described by Skrabal and Hohl-Blummer [161]. The second step follows diazotization of 2,2'-diaminoazobenzene with sodium nitrite in dilute hydrochloric acid, followed by reductive cyclization using sulfur dioxide to form 1,1'-bibenzotriazole in 40% yields, as reported by Harder et al. [162].

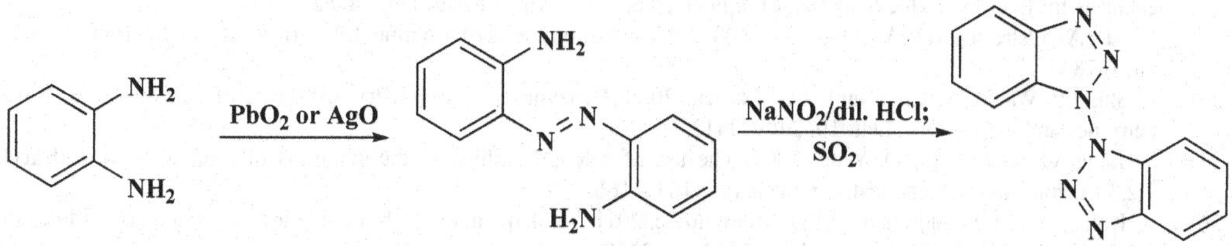

Scheme-84

316   *Metals and Non-Metals: Five-Membered N-Heterocycle Synthesis*

References

(1)   A. Thakur, P.S. Gupta, P.K. Shukla, A. Verma and P. Pathak. 2016. 1,2,4-Triazole scaffolds: recent advances and pharmacological applications. Int. J. Curr. Res. Aca. Rev. 4: 277–296.

(2)   K. Munstedt, I. Wunderlich, E. Blauth-Eckmeyer, M. Zygmunt and H. Varhson. 1998. Does dexamethasone enhance the efficacy of alizapride in *cis*-platinum induced delayed vomiting and nausea? Oncology 55: 293–299.

(3)   Q. Chao, L. Deng, H. Shih, L.M. Leoni, D. Genini, D.A. Carson and H.B. Cottam. 1999. Substituted isoquinolines and quinazolines as potential anti-inflammatory agents. Synthesis and biological evaluation of inhibitors of tumor necrosis factor alpha. J. Med. Chem. 42: 3860–3873.

(4)   W.J. Houlihan, P.G. Munder, D.A. Handley, S.H. Cheon and V.A. Parrino. 1995. Anti-tumor activity of 5-aryl–2,3-dihydroimidazo[2,1-*a*]isoquinolines. J. Med. Chem. 38: 234–240.

(5)   M. Gupta, S. Paul and R. Gupta. 2011. Efficient and novel one-pot synthesis of antifungal active 1-substituted-8-aryl-3-alkyl/aryl-4*H*-pyrazolo[4,5-*f*][1,2,4]triazolo[4,3-*b*][1,2,4]triazepines using solid support. Eur. J. Med. Chem. 46: 631–635.

(6)   T. Plech, M. Wujec, U. Kosikowska, A. Malm, B. Rajtar and M. Polz-Dacewicz. 2013. Synthesis and *in vitro* activity of 1,2,4-triazole-ciprofloxacin hybrids against drug-susceptible and drug-resistant bacteria. Eur. J. Med. Chem. 60: 128–134.

(7)   Q.J. Zhao, Y. Song, H.G. Hu, S.C. Yu and Q.Y. Wu. 2007. Design, synthesis and antifungal activity of novel triazole derivatives. Chin. Chem. Lett. 18: 670–672.

(8)   O. Prakash, D.K. Aneja, K. Hussain, P. Lohan, P. Ranjan, S. Arora, C. Sharma and K.R. Aneja. 2011. Synthesis and biological evaluation of dihydroindeno and indeno[1,2-*e*][1,2,4]triazolo[3,4-*b*][1,3,4]thiadiazines as antimicrobial agents. Eur. J. Med. Chem. 46: 5065–5073.

(9)   S. Eswaran, A.V. Adhikari and N.S. Shetty. 2009. Synthesis and antimicrobial activities of novel quinoline derivatives carrying 1,2,4-triazole moiety. Eur. J. Med. Chem. 44: 4637–4647.

(10)   S.N. Swamy, S. Naveen, B. Prabhuswamy, M.A. Sridhar, J.S. Prasad and K.S. Rangappa. 2006. Synthesis and crystal structure analysis of 2-(4-methyl-2′-biphenyl)–4-amino-1,2,4-triazole-3-thiol. Struct. Chem. 17: 91–95.

(11)   L.-Y. Wang, W.-C. Tseng, T.-S. Wu, K. Kaneko, H. Takayama, M. Kimura, W.-C. Yang, J.B. Wu, S.-H. Juang and F.F. Wong. 2011. Synthesis and anti-proliferative evaluation of 3,5-disubstituted 1,2,4-triazoles containing flurophenyl and trifluoromethanephenyl moieties. Bioorg. Med. Chem. Lett. 21: 5358–5362.

(12)   P.-L. Zhao, W.-F. Ma, A.-N. Duan, M. Zou, Y.-C. Yan, W.-W. You and S.-G. Wu. 2012. One-pot synthesis of novel isoindoline-1,3-dione derivatives bearing 1,2,4-triazole moiety and their preliminary biological evaluation. Eur. J. Med. Chem. 54: 813–822.

(13)   A. Orjales, R. Mosquera, L. Labeaga and R. Rodes. 1997. New 2-piperazinylbenzimidazole derivatives as 5-HT3 antagonists. Synthesis and pharmacological evaluation. J. Med. Chem. 40: 586–593.

(14)   M. Grimmett. 1996. Imidazoles. Comprehensive Heterocyclic Chemistry II, 3: 77–220.

(15)   J. Stacy and M.D. Childs. 2000. Safety of the fluoroquinolone anti-biotics: focus on molecular structure. Infect. Urol. 13: 1–3.

(16)   H. Shinaki, T. Ito, T. Iida, Y. Kitao, H. Yamada and I. Vchida. 2000. 4-Aminoquinolines: novel nociceptin antagonists with analgesic activity. J. Med. Chem. 43: 4667–4677.

(17)   (a) M. Kidwai and R.J. Mohan. 2004. Eco-friendly synthesis of anti-fungal azoles. Korean Chem. Soc. 48: 177–181. (b) I. Al-Masoudi, Y. Al-Soud, N. Al-Salihi and N. Al-Masoudi. 2006. 1,2,4-Triazoles: synthetic approaches and pharmacological importance. Chem. Heterocycl. Compd. 42: 1377–1403.

(18)   H. Yanai and T. Taguchi. 2005. Indium(III) triflate-catalyzed tandem azidation/1,3-dipolar cycloaddition reaction of ω,ω-dialkoxyalkyne derivatives with trimethylsilyl azide. Tetrahedron Lett. 46: 8639–8643.

(19)   J.A.R. Salvador, R.M.A. Pinto and S.M. Silvestre. 2009. Recent advances of bismuth(III) salts in organic chemistry: application to the synthesis of heterocycles of pharmaceutical interest. Curr. Org. Synth. 6: 426–470.

(20)   M. Giurg, H. Wójtowicz and J. Mochowski. 2002. Hydroperoxide oxidation of azomethines and alkylarenes catalyzed by ebselen. Polish J. Chem. 76: 537–542.

(21)   J. Mochowski, K. Kloc, R. Lisiak, P. Potaczek and H. Wójtowicz. 2007. Developments in the chemistry of selenaheterocyclic compounds of practical importance in synthesis and medicinal biology. ARKIVOC (vi): 14–46.

(22)   J. Mochowski, W. Peczyska-Czoch, M. Pitka-Ottlik and H. Wojtowicz-Mochowska. 2011. Non-metal and enzymatic catalysts for hydroperoxide oxidation of organic compounds. Open Catal. J. 4: 54–82.

(23)   Y. Yu, J.M. Ostresh and R.A. Houghten. 2003. Solid-phase synthesis of 3-amino-1,2,4-triazoles. Tetrahedron Lett. 44: 7841–7843.

(24)   F. Shi, J.P. Waldo, Y. Chen and R.C. Larock. 2008. Benzyne click chemistry: synthesis of benzotriazoles from benzynes and azides. Org. Lett. 10: 2409–2412.

(25)   Y. Jiang, C. Kuang and Q. Yang. 2009. The use of calcium carbide in the synthesis of 1-monosubstituted aryl 1,2,3-triazole *via* click chemistry. Synlett 19: 3163–3166.

(26)   A. Kolarovic, M. Schnurch and M.D. Mihovilovic. 2011. Tandem catalysis: from alkynoic acids and aryl iodides to 1,2,3-triazoles in one-pot. J. Org. Chem. 76: 2613–2618.

(27)   Y. Jiang, C. Kuang and Q. Yang. 2010. Copper(I) iodide-catalyzed synthesis of 4-aryl-1*H*-1,2,3-triazoles from *anti*-3-aryl-2,3-dibromopropanoic acids and sodium azide. Synthesis 24: 4256–4260.

(28)   X. Chen, Y. Yang, C. Kuang and Q. Yang. 2011. Copper(I) iodide-catalyzed synthesis of 1,4-disubstituted 1,2,3-triazoles from *anti*-3-aryl-2,3-dibromopropanoic acids and organic azides. Synthesis 18: 2907–2912.

(29)   L.Y. Wu, Y.X. Xie, Z.S. Chen, Y.N. Niu and Y.M. Liang. 2009. A convenient synthesis of 1-substituted 1,2,3-triazoles *via* CuI/Et$_3$N-catalyzed 'click chemistry' from azides and acetylene gas. Synlett 9: 1453–1456.

(30)   S. Lal and S. Diez-Gonzalez. 2011. [CuBr(PPh$_3$)$_3$] for azide-alkyne cycloaddition reactions under strict click conditions. J. Org. Chem. 76: 2367–2373.

(31)   C. Shao, X. Wang, Q. Zhang, S. Luo, J. Zhao and Y. Hu. 2011. Acid-base jointly promoted copper(I)-catalyzed azide-alkyne cycloaddition. J. Org. Chem. 76: 6832–6836.

(32)   M. Liu and O. Reiser. 2011. A copper(I) isonitrile complex as a heterogeneous catalyst for azide-alkyne cycloaddition in water. Org. Lett. 13: 1102–1105.

(33)   F. Himo, T. Lovell, R. Hilgraf, V.V. Rostovtsev, L. Noodleman, K.B. Sharpless and V.V. Fokin. 2005. Copper(I)-catalyzed synthesis of azoles. DFT study predicts unprecedented reactivity and intermediates. J. Am. Chem. Soc. 127: 210–216.

(34)   Y.M. Wu, J. Deng, Y.L. Li and Q.Y. Chen. 2005. Regiospecific synthesis of 1,4,5-trisubstituted 1,2,3-triazole *via* one-pot reaction promoted by copper(I) salt. Synthesis 8: 1314–1318.

(35)   O.A. Oladeinde, S.Y. Hong, R.J. Holland, A.E. Maciag, L.K. Keefer, J.E. Saavedra and R.S. Nandurdikar. 2010. "Click" reaction in conjunction with diazeniumdiolate chemistry: developing high-load nitric oxide donors. Org. Lett. 12: 4256–4259.

(36)   A. Da Settimo, G. Primofiore, F. Da Settimo, G. Pardi, F. Simorini and A.M. Marini. 2002. An approach to novel fused triazole or tetrazole derivatives starting from benzimidazo[1,2-*a*]quinazoline-5(7*H*)-one and 5,7-dihydro-5-oxopyrido[3′,2′:5,6]pyrimido[1,2-*a*]benzimidazole. J. Heterocycl. Chem. 39: 1007–1011.

(37)   (a) K.M. Dawood and B.F. Abdel-Wahab. 2010. Synthetic routes to benzimidazole-based fused polyheterocycles. ARKIVOC (i): 333–389. (b) K.M. Dawood, N.M. Elwan, A.A. Farahat and B.F. Abdel-Wahab. 2010. 1*H*-Benzimidazole-2-acetonitriles as synthon in fused benzimidazole synthesis. J. Heterocycl. Chem. 47: 243–267.

(38)   L. Ackermann, H.K. Potukuchi, D. Landsberg and R. Vicente. 2008. Copper-catalyzed "click" reaction/direct arylation sequence: modular syntheses of 1,2,3-triazoles. Org. Lett. 10: 3081–3084.

(39)   M. Koyama, N. Ohtani, F. Kai, I. Moriguchi and S. Inouye. 1987. Synthesis and quantitative structure-activity relationship analysis of *N*-triiodoallyl- and *N*-iodopropargylazoles. New anti-fungal agents. J. Med. Chem. 30: 552–562.

(40)   D. Alberico, M.E. Scott and M. Lautens. 2007. Aryl bond formation by transition metal-catalyzed direct arylation. Chem. Rev. 107: 174–238.

(41)   P. Thansandote and M. Lautens. 2009. Construction of nitrogen-containing heterocycles by C-H bond functionalization. Chem. Eur. J. 15: 5874–5883.

(42)   G.P. McGlacken and L.M. Bateman. 2009. Recent advances in aryl-aryl bond formation by direct arylation. Chem. Soc. Rev. 38: 2447–2464.

(43)   L. Ackermann, R. Vicente and A.R. Kapdi. 2009. Transition metal-catalyzed direct arylation of (hetero)arenes by C-H bond cleavage. Angew. Chem. Int. Ed. 48: 9792–9826.

(44)   R.F. Heck. 1982. Palladium-catalyzed vinylation of organic halides. Org. React. 27: 345–390.

(45)   A. Meijere and F.E. Meyer. 1995. Fine feathers make fine birds: the Heck reaction in modern garb. Angew. Chem. Int. Ed. Engl. 33: 2379–2411.

(46)   E. Negishi, C. Coperet, S.M. Ma, S.Y. Liou and F. Liu. 1996. Cyclic carbopalladation. A versatile synthetic methodology for the construction of cyclic organic compounds. Chem. Rev. 96: 365–394.

(47)   I.P. Beletskaya and A.V. Cheprakov. 2000. The Heck reaction as a sharpening stone of palladium catalysis. Chem. Rev. 100: 3009–3066.

(48)   J.M. Schulman, A.A. Friedman, J. Panteleev and M. Lautens. 2012. Synthesis of 1,2,3-triazole-fused heterocycles *via* Pd-catalyzed cyclization of 5-iodotriazoles. Chem. Commun. 48: 55–57.

(49)   Y. Angell and K. Burgess. 2007. Base dependence in copper-catalyzed Huisgen reactions: efficient formation of bis-triazoles. Angew. Chem. Int. Ed. 46: 3649–3651.

(50)   Z. Shi, C. Zhang, C. Tanga and N. Jiao. 2012. Recent advances in transition metal-catalyzed reactions using molecular oxygen as the oxidant. Chem. Soc. Rev. 41: 3381–3430.

(51)   S. Mori, T. Mori and Y. Mukoyama. 1993. Elution behavior of polyethylene glycols on a hydrophilic polymer gel column used for size exclusion chromatography. J. Liquid Chromatography 16: 2269–2279.

(52)   C.L. Cioffi, W.T. Spencer, J.J. Richards and R.J. Herr. 2004. Generation of 3-pyridyl biaryl systems *via* palladium-catalyzed Suzuki cross-couplings of aryl halides with 3-pyridylboroxin. J. Org. Chem. 69: 2210–2212.

(53)   S.V. Dhamnaskar and D.W. Rangnekar. 1988. Synthesis of triazoflo[4,5-*b*]pyrido[1′,2′-*a*]benzimidazole derivatives as fluorescent disperse dyes and whiteners for polyester fibre. Dyes Pigm. 9: 467–473.

(54)   G. Chouhan, D. Wang and H. Alper. 2007. Magnetic nanoparticle-supported proline as a recyclable and recoverable ligand for the CuI-catalyzed arylation of nitrogen nucleophiles. Chem. Commun. 45: 4809–4811.

(55)   L. Djakovitch, N. Batail and M. Genelot. 2011. Recent advances in the synthesis of *N*-containing heteroaromatics *via* heterogeneously transition metal-catalyzed cross-coupling reactions. Molecules 16: 5241–5267.

(56)   S. Ueda and H. Nagasawa. 2009. Facile synthesis of 1,2,4-triazoles *via* a copper-catalyzed tandem addition-oxidative cyclization. J. Am. Chem. Soc. 131: 15080–15081.

(57) I. Hager, R. Frohlich and E.U. Wurthwein. 2009. Synthesis of secondary, tertiary and quaternary 1,3,5-triazapenta-1,3-dienes and their CoII, ZnII, PdII, CuII and BF_2 coordination compounds. Eur. J. Inorg. Chem. 16: 2415–2428.

(58) J.P. Wikstrom, A.S. Filatov and E.V. Rybak-Akimova. 2010. Condensation of nitriles with amides promoted by coordinatively unsaturated bis-nickel(II)-hydroxy complex: a new route to alkyl- and aryl-imidoylamidines. Chem. Commun. 46: 424–426.

(59) M.N. Kopylovich, A.J.L. Pombeiro, A. Fischer, L. Kloo and V.Y. Kukushkin. 2003. Facile Ni(II)/ketoxime-mediated conversion of organonitriles into imidoylamidine ligands. Synthesis of imidoylamidines and acetyl amides. Inorg. Chem. 42: 7239–7248.

(60) S. Xu, Y. Jiang and H. Fu. 2013. Copper-catalyzed synthesis of 1,2,4-triazoles *via* sequential coupling and aerobic oxidative dehydrogenation of amidines. Synlett 24: 125–129.

(61) D.B. Ramachary and C.F. Barbas. 2004. Towards organo-click chemistry: development of organocatalytic multi-component reactions through combinations of aldol, Wittig, Knoevenagel, Michael, Diels-Alder and Huisgen cycloaddition reactions. Chem. Eur. J. 10: 5323–5331.

(62) D.M. D'Souza and T.J.J. Muller. 2007. Multi-component syntheses of heterocycles by transition metal catalysis. Chem. Soc. Rev. 36: 1095–1108.

(63) N. Candelon, D. Lastecoueres, A.K. Diallo, A.J. Ruiz, D. Astruc and J.M. Vincent. 2008. A highly active and reusable copper(I)-tren catalyst for the "click" 1,3-dipolar cycloaddition of azides and alkynes. Chem. Commun. 6: 741–743.

(64) J.S. Thompson, A.Z. Bradley, K.H. Park, K.D. Dobbs and W. Marshall. 2006. Copper(I) complexes with bis(trimethylsilyl)acetylene: role of ancillary ligands in determining π back-bonding interactions. Organometallics 25: 2712–2714.

(65) C. Nolte, P. Mayer and B.F. Straub. 2007. Isolation of a copper(I) triazolide: a "click" intermediate. Angew. Chem. Int. Ed. 46: 2101–2103.

(66) S. Diez-Gonzalez, A. Correa, L. Cavallo and S.P. Nolan. 2006. (NHC)Copper(I)-catalyzed [3+2]-cycloaddition of azides and mono- or di-substituted alkynes. Chem. Eur. J. 12: 7558–7564.

(67) P. Li, L. Wang and Y. Zhang. 2008. SiO_2-NHC-Cu(I): an efficient and reusable catalyst for [3+2]-cycloaddition of organic azides and terminal alkynes under solvent-free reaction conditions at room temperature. Tetrahedron 64: 10825–10830.

(68) S. Diez-Gonzalez, E.D. Stevens and S.P. Nolan. 2008. A [(NHC)CuCl] complex as a latent click catalyst. Chem. Commun. 39: 4747–4749.

(69) S. Diez-Gonzalez, N. Marion and S.P. Nolan. 2009. N-Heterocyclic carbenes in late transition metal catalysis. Chem. Rev. 109: 3612–3676.

(70) S. Diez-Gonzalez and S.P. Nolan. 2008. [(NHC)$_2$Cu]X complexes as efficient catalysts for azide-alkyne click chemistry at low catalyst loadings. Angew. Chem. Int. Ed. 47: 8881–8884.

(71) (a) V.V. Rostovtsev, L.G. Green, V.V. Fokin and K.B. Sharpless. 2002. A stepwise Huisgen cycloaddition process: copper(I)-catalyzed regioselective "ligation" of azides and terminal alkynes. Angew. Chem. Int. Ed. 41: 2596–2599.
(b) N.D. Bogdan, M. Matache, V.M. Meier, C. Dobrota, I. Dumitru, G.D. Roiban and D.P. Funeriu. 2010. Protein-inorganic array construction: design and synthesis of the building blocks. Chem. Eur. J. 16: 2170–2180.

(72) C.W. Tornøe, C. Christensen and M. Meldal. 2002. Peptidotriazoles on solid-phase: [1,2,3]-triazoles by regiospecific copper(I)-catalyzed 1,3-dipolar cycloadditions of terminal alkynes to azides. J. Org. Chem. 67: 3057–3064.

(73) R. Huisgen. 1989. Kinetics and reaction mechanisms: selected examples from the experience of forty years. Pure Appl. Chem. 61: 613–628.

(74) F. Perez-Balderas, M. Ortega-Munoz, J. Morales-Sanfrutos, F. Hernandez-Mateo, F.G. Calvo-Flores, J.A. Calvo-Asin, J. Isac-Garcia and F. Santoyo-Gonzalez. 2003. Multivalent neoglycoconjugates by regiospecific cycloaddition of alkynes and azides using organic-soluble copper catalysts. Org. Lett. 5: 1951–1954.

(75) T.R. Chan, R. Hilgraf, K.B. Sharpless and V.V. Fokin. 2004. Polytriazoles as copper(I)-stabilizing ligands in catalysis. Org. Lett. 6: 2853–2855.

(76) B. Gerard, J. Ryan, A.B. Beeler and J.A. Porco. 2006. Synthesis of 1,4,5-trisubstituted 1,2,3-triazoles by copper-catalyzed cycloaddition-coupling of azides and terminal alkynes. Tetrahedron 62: 6405–6411.

(77) V. Jurkauskas, J. Sadighi and S. Buchwald. 2003. Conjugate reduction of α,β-unsaturated carbonyl compounds catalyzed by a copper carbene complex. Org. Lett. 5: 2417–2420.

(78) H. Kaur, F.K. Zinn, E.D. Stevens and S.P. Nolan. 2004. (NHC)CuI (NHC = N-heterocyclic carbene) complexes as efficient catalysts for the reduction of carbonyl compounds. Organometallics 23: 1157–1160.

(79) S. Diez-Gonzalez and S.P. Nolan. 2008. Copper, silver, and gold complexes in hydrosilylation reactions. Acc. Chem. Res. 41: 349–358.

(80) M.R. Fructos, T.R. Belderrain, M.C. Nicasio, S.P. Nolan, H. Kaur, M. Mar Díaz-Requejo and P.J. Pérez. 2004. Complete control of the chemoselectivity in catalytic carbene transfer reactions from ethyl diazoacetate: an N-heterocyclic carbene-Cu system that suppresses diazo coupling. J. Am. Chem. Soc. 126: 10846–10847.

(81) S. Diez-Gonzalez and S.P. Nolan. 2007. N-Heterocyclic carbene-copper(I) complexes in homogeneous catalysis. Synlett 14: 2158–2167.

(82) B. Chattopadhyay, C. Rivera Vera, S. Chuprakov and V. Gevorgyan. 2010. Fused tetrazoles as azide surrogates in click reaction: efficient synthesis of N-heterocycle substituted 1,2,3-triazoles. Org. Lett. 12: 2166–2169.

(83) D. Luvino, C. Amalric, M. Smietana and J.J. Vasseur. 2007. Sequential Seyferth-Gilbert/CuAAC reactions: application to the one-pot synthesis of triazoles from aldehydes. Synlett 19: 3037–3041.

(84) Y.L. Li, B.X. Du, X.S. Wang, D.Q. Shi and S.J. Tu. 2006. One-pot synthesis of pyrano[2,3-*d*]pyrimidine derivatives in ionic liquid medium. J. Chem. Res. 3: 157–159.

(85) Y.B. Zhao, Z.Y. Yan and Y.M. Liang. 2006. Efficient synthesis of 1,4-disubstituted 1,2,3-triazoles in ionic liquid/water system. Tetrahedron Lett. 47: 1545–1549.

(86) M.C. Bagley, C. Brace, J.W. Dale, M. Ohnesorge, N.G. Philips, X. Xiong and J. Bower. 2002. Synthesis of tetra-substituted pyridines by the acid-catalyzed Bohlmann-Rahtz reaction. J. Chem. Soc. Perkin Trans. 1 14: 1663–1671.

(87) B. Das, C.R. Reddy, D.N. Kumar, M. Krishnaiah and R. Narender. 2010. A simple, advantageous synthesis of 5-substituted 1*H*-tetrazoles. Synlett 3: 391–394.

(88) P. Appukkuttan, W. Dahaen, V.V. Fokin and E. van der Eycken. 2004. A microwave-assisted click chemistry synthesis of 1,4-disubstituted 1,2,3-triazoles *via* a copper(I)-catalyzed three-component reaction. Org. Lett. 6: 4223–4225.

(89) M. Syamala. 2009. Recent progress in three-component reactions: an update. Org. Prep. Proced. Int. 41: 1–68.

(90) D. Li, H. Bao and T. You. 2005. Microwave-assisted and efficient one-pot synthesis of substituted 1,2,4-triazoles. Heterocycles 65: 1957–1962.

(91) M. Chtchigrovsky, A. Primo, P. Gonzalez, K. Molvinger, M. Robitzer, F. Quignard and F. Taran. 2009. Functionalized chitosan as a green, recyclable, biopolymer-supported catalyst for the [3+2]-Huisgen cycloaddition. Angew. Chem. Int. Ed. 48: 5916–5920.

(92) K. El Akri, K. Bougrin, B.J. Balzarini, A. Faraj and R. Benhida. 2007. Efficient synthesis and *in vitro* cytostatic activity of 4-substituted triazolyl-nucleosides. Bioorg. Med. Chem. Lett. 17: 6656–6659.

(93) Z.Y. Yan, Y.B. Zhao, M.J. Fan, W.M. Liu and Y.M. Liang. 2005. General synthesis of (1-substituted 1*H*-1,2,3-triazol-4-ylmethyl)-dialkylamines *via* a copper(I)-catalyzed three-component reaction in water. Tetrahedron 61: 9331–9337.

(94) J. Krim, B. Sillahi, M. Taourirte, E.M. Rakib and J.W. Engels. 2009. Microwave-assisted click chemistry: synthesis of mono and bis-1,2,3-triazole acyclonucleoside analogues of ACV *via* copper(I)-catalyzed cycloaddition (09-4115AP). ARKIVOC (xiii): 142-152.

(95) G. Cravotto, V.V. Fokin, D. Garella, A. Binello, L. Boffa and A. Barge. 2010. Ultrasound-promoted copper-catalyzed azide-alkyne cycloaddition. J. Comb. Chem. 12: 13–15.

(96) D. Yang, M. Kwon, Y. Jang and H.B. Jeon. 2010. A convenient and efficient synthesis of C-carbamoyl-1,2,3-triazoles from alkyl bromide by a one-pot sequential addition: conversion of ester to amide using Zr(O*t*-Bu)$_4$. Tetrahedron Lett. 51: 3691–3695.

(97) A.D. Moorhouse and J.E. Moses. 2008. Microwave enhancement of a 'one-pot' tandem azidation-'click' cycloaddition of anilines. Synlett 14: 2089–2092.

(98) P.G. Alsabeh, R.J. Lundgren, L.E. Longobardi and M. Stradiotto. 2011. Palladium-catalyzed synthesis of indoles *via* ammonia cross-coupling-alkyne cyclization. Chem. Commun. 47: 6936–6938.

(99) R. de Simone, M.G. Chini, I. Bruno, R. Riccio, D. Mueller, O. Werz and G. Bifulco. 2011. Structure based discovery of inhibitors of microsomal prostaglandin E2 synthase-1,5-lipoxygenase and 5-lipoxygenase-activating protein: promising hits for the development of new anti-inflammatory agents. J. Med. Chem. 54: 1565–1575.

(100) T. Pei, C.-Y. Chen, P.G. Dormer and I.W. Davies. 2008. Expanding the [1,2] aryl migration to the synthesis of substituted indoles. Angew. Chem. Int. Ed. 120: 4299–4301; 47: 4231–4233.

(101) D.K. Barange, Y.C. Tu, V. Kavala, C.-W. Kuo and C.F. Yao. 2011. One-pot synthesis of triazolothiadiazepine 1,1-dioxide derivatives *via* copper-catalyzed tandem [3+2]-cycloaddition/*N*-arylation. Adv. Synth. Catal. 353: 41–48.

(102) Y. Liu and J.-P. Wan. 2011. Tandem reactions initiated by copper-catalyzed cross-coupling: a new strategy towards heterocycle synthesis. Org. Biomol. Chem. 9: 6873–6894.

(103) Y.Y. Hu, J. Hu, X.C. Wang, L.N. Guo, X.Z. Shu, Y.N. Niu and Y.M. Liang. 2010. Copper-catalyzed tandem synthesis of [1,2,3]triazolo[5,1-*a*]isoquinolines and their transformation to 1,3-disubstituted isoquinolines. Tetrahedron 66: 80–86.

(104) E. Merkul, F. Klukas, D. Dorsch, U. Gradler, H.E. Greiner and T.J.J. Muller. 2011. Rapid preparation of triazolyl substituted NH-heterocyclic kinase inhibitors *via* one-pot Sonogashira coupling-TMS-de-protection-CuAAC sequence. Org. Biomol. Chem. 9: 5129–5136.

(105) K. Mogilaiah, R.S. Prasad and J.K. Swamy. 2010. Microwave-assisted synthesis of 1,2,4-triazolo[4,3-*a*][1,8] naphthyridines using FeCl$_3$.6H$_2$O under solvent-free conditions. Indian J. Chem. 49B: 335–339.

(106) V. Sharma, B. Shrivastava, R. Bhatia, M. Bachwani, R. Khandelwal and J. Ameta. 2011. Exploring potential of 1,2,4-triazole: a brief review. Pharmacologyonline 1: 1192–1222.

(107) J. Heeres, L.J.J. Backx and J. van Cutsem. 1984. Anti-mycotic azoles. 7. Synthesis and anti-fungal properties of a series of novel triazol-3-ones. J. Med. Chem. 27: 894–900.

(108) M. Baumann, I.R. Baxendale, S.V. Ley and N. Nikbin. 2011. An overview of the key routes to the best selling 5-membered ring heterocyclic pharmaceuticals. Beilstein J. Org. Chem. 7: 442–495.

(109) A.R. Katritzky, M. Qi, D. Feng, G. Zhang, M.C. Griffith and K. Watson. 1999. Synthesis of 1,2,4-triazole-functionalized solid support and its use in the solid-phase synthesis of tri-substituted 1,2,4-triazoles. Org. Lett. 1: 1189–1191.

(110) A.R. Katritzky, S. El-Zemity, H. Lang, E.A. Kadous and A.M. El-Shazly. 1996. A novel and convenient preparation of 1-(*γ*-aminoalkyl)-substituted 1,2,4-triazoles. Synth. Commun. 26: 357–365.

(111) G. Fu, L. Guo, X. Mao, S. Sheng, S. Fei and M. Cai. 2008. Solid-phase organic synthesis of vinyl-substituted 1,2,4-trizoles based on polymer-supported α-selenopropionic acid (08–3007CP). ARKIVOC (ii): 287–293.

(112) J.R. Donald and S.F. Martin. 2011. Synthesis and diversification of 1,2,3-triazole-fused 1,4-benzodiazepine scaffolds. Org. Lett. 13: 852–855.

(113) H.C. Kolb, M.G. Finn and K.B. Sharpless. 2001. Click chemistry: diverse chemical function from a few good reactions. Angew. Chem. Int. Ed. 40: 2004–2021.

(114) D. Liu, W. Gao, Q. Dai and X. Zhang. 2005. Triazole-based monophosphines for Suzuki-Miyaura coupling and amination reactions of aryl chlorides. Org. Lett. 7: 4907–4910.

(115) A. Krasinski, V.V. Fokin and K.B. Sharpless. 2004. Direct synthesis of 1,5-disubstituted 4-magnesio-1,2,3-triazoles, revisited. Org. Lett. 6: 1237–1240.

(116) A.K. Feldman, B. Colasson and V.V. Fokin. 2004. One-pot synthesis of 1,4-disubstituted 1,2,3-triazoles from *in situ* generated azides. Org. Lett. 6: 3897–3899.

(117) J.-G. Roveda, C. Clavette, A.D. Hunt, S.I. Gorelsky, C.J. Whipp and A.M. Beauchemin. 2009. Hydrazides as tunable reagents for alkene hydroamination and aminocarbonylation. J. Am. Chem. Soc. 131: 8740–8741.

(118) H.C. Kolb and K.B. Sharpless. 2003. The growing impact of click chemistry on drug discovery. Drug Discovery Today 8: 1128–1137.

(119) B.H. Lipshutz and B.R. Taft. 2006. Heterogeneous copper-in-charcoal-catalyzed click chemistry. Angew. Chem. Int. Ed. 45: 8235–8238.

(120) S.K. Yousuf, D. Mukherjee, B. Singh, S. Maity and S.C. Taneja. 2010. Cu-Mn bimetallic catalyst for Huisgen [3+2]-cycloaddition. Green. Chem. 12: 1568–1572.

(121) A. Das, A. Kulkarni and B. Torok. 2012. Environmentally benign synthesis of heterocyclic compounds by combined microwave-assisted heterogeneous catalytic approaches. Green Chem. 14: 17–34.

(122) X. Gai, R. Grigg, S. Rajviroongit, S. Songarsa and V. Sridharan. 2005. Synthesis of triazolo- and tetrazolo-tetrahydroisoquinolines and isoquinolines *via* temperature controlled palladium-catalyzed allene/azide incorporation/intramolecular 1,3-dipolar cycloaddition cascades. Tetrahedron Lett. 46: 5899–5902.

(123) S. Kamijo, T. Jin, Z. Huo and Y. Yamamoto. 2003. Synthesis of triazoles from non-activated terminal alkynes *via* the three-component coupling reaction using a Pd(0)-Cu(I) bimetallic catalyst. J. Am. Chem. Soc. 125: 7786–7787.

(124) J. Barluenga, C. Valdés, G. Beltrán, M. Escribano and F. Aznar. 2006. Developments in Pd catalysis: synthesis of 1*H*-1,2,3-triazoles from sodium azide and alkenyl bromides. Angew. Chem. Int. Ed. 45: 6893–6896.

(125) (a) S. Kamijo, T. Jin, Z. Huo and Y. Yamamoto. 2002. Regiospecific synthesis of 2-allyl-1,2,3-triazoles by palladium-catalyzed 1,3-dipolar cycloaddition. Tetrahedron Lett. 43: 9707–9710. (b) S. Kamijo, T. Jin and Y. Yamamoto. 2004. Four-component coupling reactions of silylacetylenes, allyl carbonates, and trimethylsilyl azide catalyzed by a Pd(0)-Cu(I) bimetallic catalyst. Fully substituted triazole synthesis from seemingly internal alkynes. Tetrahedron Lett. 45: 689–691.

(126) B. Yin, X. Zhang, X. Zhang, H. Peng, W. Zhou, B. Liu and H. Jiang. 2015. Access to poly-substituted indoles or benzothiophenes *via* palladium-catalyzed cross-coupling of furfural tosylhydrazones with 2-iodoanilines or 2-iodothiophenols. Chem. Commun. 51: 6126–6129.

(127) T. Beryozkina, P. Appukkuttan, N. Mont and E. van der Eycken. 2006. Microwave-enhanced synthesis of new (–)-steganacin and (–)-steganone aza analogues. Org. Lett. 8: 487–490.

(128) U. Weiss. 2005. Hepatitis C. Nature 436: 929–929.

(129) J.J. Feld and J.H. Hoofnagle. 2005. Mechanism of action of interferon and ribavirin in treatment of hepatitis C. Nature 436: 967–972.

(130) R. Zhu, M. Wang, Y. Xia, F. Qu, J. Neyts and L. Peng. 2008. Arylethynyltriazole acyclonucleosides inhibit hepatitis C virus replication. Bioorg. Med. Chem. Lett. 18: 3321–3327.

(131) R. Zhu, F. Qu, G. Quelever and L. Peng. 2007. Direct synthesis of 5-aryltriazole acyclonucleosides *via* Suzuki coupling in aqueous solution. Tetrahedron Lett. 48: 2389–2393.

(132) N.M. Nascimento-Junior, A.E. Kummerle, E.J. Barreiro and C.A.M. Fraga. 2011. MAOS and medicinal chemistry: some important examples from the last years. Molecules 16: 9274–9297.

(133) A. Reichelt, J.R. Falsey, R.M. Rzasa, O.R. Thiel, M.M. Achmatowicz, R.D. Larsen and D. Zhang. 2010. Palladium-catalyzed chemoselective monoarylation of hydrazides for the synthesis of [1,2,4]triazolo[4,3-*a*]pyridines. Org. Lett. 12: 792–795.

(134) W. Zhang, C. Kuang and Q. Yang. 2010. Palladium-catalyzed one-pot synthesis of 4-aryl-1*H*-1,2,3-triazoles from *anti*-3-aryl-2,3-dibromopropanoic acids and sodium azide. Synthesis 2: 283–287.

(135) J. Li, D. Wang, Y. Zhang, J. Li and B. Chen. 2009. Facile one-pot synthesis of 4,5-disubstituted 1,2,3-(NH)-triazoles through Sonogashira coupling/1,3-dipolar cycloaddition of acid chlorides, terminal acetylenes, and sodium azide. Org. Lett. 11: 3024–3027.

(136) B. Liegault, D. Lapointe, L. Caron, A. Vlassova and K. Fagnou. 2009. Establishment of broadly applicable reaction conditions for the palladium-catalyzed direct arylation of heteroatom containing aromatic compounds. J. Org. Chem. 74: 1826–1834.

(137) J. Zhou, J. He, B. Wang, W. Yang and H. Ren. 2011. 1,7-Palladium migration *via* C-H activation, followed by intramolecular amination: regioselective synthesis of benzotriazoles. J. Am. Chem. Soc. 133: 6868–6870.

(138) R.K. Kumar, M.A. Ali and T. Punniyamurthy. 2011. Pd-catalyzed C-H activation/C-N bond formation: a new route to 1-aryl-1*H*-benzotriazoles. Org. Lett. 13: 2102–2105.

(139) L. El Kaim, L. Grimaud and A. Schiltz. 2009. Isocyanide-based multi-component reaction 'without' isocyanides. Synlett 9: 1401–1404.

(140) R. Huisgen, J. Sauer and H.J. Sturm. 1958. Acylierung 5-substitutierter tetrazole zu 1.3.4-oxdiazolen. Angew. Chem. 70: 272–273.

(141) R. Huisgen, J. Sauer and M. Seidel. 1960. Ringöffnungen der azole, IV. Die synthese von 1.2.4-triazolen aus 5-substituierten tetrazolen und carbonsäure-imidchloriden. Chem. Ber. 93: 2885–2891.

(142) R. Huisgen, H.J. Sturm and M. Seidel. 1961. Ringöffnungen der azole, V. weitere reaktionen der tetrazole mit elektrophilen agenzien. Chem. Ber. 94: 1555–1562.

(143) R. Huisgen. 1980. 1,5-Electrocyclizations- an important principle of heterocyclic chemistry. Angew. Chem. Int. Ed. Engl. 19: 947–973.

(144) G.V. Boyd, J. Cobb, P.F. Lindley, J.C. Mitchell and G.A. Nicolaou. 1987. 3*H*-1,3,4-Benzotriazepines from (*N*-arylbenzimidoyl)-5-dimethylaminotetrazoles: 1,7- vs. 1,5- cyclization of extended dipolar nitrile imines. J. Chem. Soc. Chem. Commun. 2: 99–101.

(145) L.E. Kaim, L. Grimaud and P. Patil. 2011. Three-component strategy toward 5-membered heterocycles from isocyanide dibromides. Org. Lett. 13: 1261–1263.

(146) D.J. Hlasta and J.H. Ackerman. 1994. Steric effects on the regioselectivity of an azide-alkyne dipolar cycloaddition reaction: the synthesis of human leukocyte elastase inhibitors. J. Org. Chem. 59: 6184–6189.

(147) S.J. Coats, J.S. Link, D. Gauthier and D.J. Hlasta. 2005. Trimethylsilyl-directed 1,3-dipolar cycloaddition reactions in the solid-phase synthesis of 1,2,3-triazoles. Org. Lett. 7: 1469–1472.

(148) V.A. Rassadin, A.A. Tomashevskiy, V.V. Sokolov, A. Ringe, J. Magull and A. de Meijere. 2009. Facile access to bicyclic sultams with methyl 1-sulfonylcyclopropane-1-carboxylate moieties. Eur. J. Org. Chem. 16: 2635–2641.

(149) F. Bellina, A. Carpita and R. Rossi. 2004. Palladium catalysts for the Suzuki cross-coupling reaction: an overview of recent advances. Synthesis 15: 2419–2440.

(150) N. Miyaura and A. Suzuki. 1995. Palladium-catalyzed cross-coupling reactions of organoboron compounds. Chem. Rev. 95: 2457–2483.

(151) S. Mehta and R.C. Larock. 2010. Iodine/palladium approaches to the synthesis of polyheterocyclic compounds. J. Org. Chem. 75: 1652–1658.

(152) J.P. Hester, A.D. Rudzik and B.V. Kamdar. 1971. 6-Phenyl-4*H*-*s*-triazolo[4,3-*a*][1,4]benzodiazepines which have central nervous system depressant activity. J. Med. Chem. 14: 1078–1081.

(153) L.H. Sternbach, R.I. Fryer, W. Metlesics, E. Reeder, G. Sach, G. Saucy and A. Stempel. 1962. Quinazolines and 1,4-benzodiazepines. VI.1a Halo-, methyl-, and methoxy-substituted 1,3-dihydro-5-phenyl-2*H*-1,4-benzodiazepin-2-ones. J. Org. Chem. 27: 3788–3796.

(154) P. Bilek and J. Slouka. 2002. Cyclocondensation reactions of heterocyclic carbonyl compounds VII. Synthesis of some substituted benzo-[1,2,4]triazino[2,3-*a*]benzimidazoles. Heterocycl. Commun. 8: 123–128.

(155) V.P. Kruglenko, V.P. Gnidets and M.V. Povstyanoi. 2000. Synthesis of 2-methyl-1,2,4-triazolo[4,3-*d*]-1,2,4-triazino[2,3-*a*]-benzimidazole and 2-methyl-9-phenylimidazo[1,2-*b*]-1,2,4-triazolo[4,3-*d*]-1,2,4-triazine. Chem. Heterocycl. Compd. 36: 103–104.

(156) B.C. Boren, S. Narayan, L.K. Rasmussen, L. Zhang, H. Zhao, Z. Lin, G. Jia and V.V. Fokin. 2008. Ruthenium-catalyzed azide-alkyne cycloaddition: scope and mechanism. J. Am. Chem. Soc. 130: 8923–8930.

(157) S. Rostamizadeh, K. Mollahoseini and S. Moghadasi. 2006. A one-pot synthesis of 4,5-disubstituted 1,2,4-triazole-3-thiones on solid support under microwave irradiation. Phosphorus, Sulfur and Silicon and the Related Elements 181: 1839–1845.

(158) T. Beresneva, A. Mishnev, E. Jaschenko, I. Shestakova, A. Gulbe and E. Abele. 2012. Palladium-catalyzed synthesis of novel tetra- and penta-cyclic biologically active benzopyran- and pyridopyran containing heterocyclic systems. ARKIVOC (ix): 185–194.

(159) R.A. Carboni, J.C. Krauer, J.E. Castle and H.E. Simmons. 1967. Aromatic azapentalenes. I. Dibenzo-1,3a,4,6a-tetraazapentalene and dibenzo-1,3a,6,6a-tetraazapentalene. New heteroaromatic systems. J. Am. Chem. Soc. 89: 2618–2625.

(160) B. Ortiz, P. Villanueva and F. Walls. 1972. Silver(II) oxide as a reagent. Reactions with aromatic amines and miscellaneous related compounds. J. Org. Chem. 37: 2748–2750.

(161) P. Skrabal and M. Hohl-Blummer. 1976. Aspects of cyclization reactions of 2,2-diaminoazobenzene and 1,2-bis(2-aminophenylazo)benzene macrocyclic aza compounds II. Helv. Chim. Acta 59: 2906–2914.

(162) R.J. Harder, R.A. Carboni and J.E. Castle. 1967. Aromatic azapentalenes. V. 1,1'-and 1,2'-dibenzo-triazoles and their conversion to dibenzotetraazapentalenes. J. Am. Chem. Soc. 89: 2643–2647.

Index

Author's Biography

Dr. Navjeet Kaur received her B.Sc. from Punjab University Chandigarh (Punjab, India) in 2008. In 2010, she completed her M.Sc. in Chemistry from Banasthali Vidyapith. She was awarded with PhD in 2014 by the same university, under the supervision of **Prof. D. Kishore**. Presently, she is working as an Assistant Professor in Department of Chemistry, Banasthali Vidyapith and has entered into a specialized research career focused on the synthesis of 1,4-benzodiazepine based heterocyclic compounds (Organic Synthetic and Medicinal Chemistry). With 8 years of teaching experience, she has published over 137 scientific research papers, review articles, book chapters, and monographs in the field of organic synthesis in national and international reputed journals. She has published a book "Palladium Assisted Synthesis of Heterocycles" with CRC Press, Taylor & Francis group. She was presented the Prof. G. L. Telesara Award in 2011 by Indian Council of Chemists (Agra, Uttar Pradesh) at Osmania University (Hyderabad), and the Best Paper Presentation Award in National Conference on "Emerging Trends in Chemical and Pharmaceutical Sciences" (Banasthali Vidyapith, Rajasthan). She has attended about 40 conferences, workshops and seminars. Apart from all this, she has been working as NSS Program Officer since 2016, member of UBA (Unnat Bharat Abhiyan) since 2018 and has delivered numerous radio talks. **Dr. Navjeet** finds interest in Sikh literature and has completed a two-year Sikh Missionary course from Sikh Missionary College (Ludhiana, Punjab).

Dr. Navjeet Kaur is currently guiding 5 research scholars—**Meenu Devi, Yamini Verma, Pooja Grewal Pranshu Bhardwaj** and **Neha Ahlawat**—as their PhD supervisor.

Printed in the United States
by Baker & Taylor Publisher Services

Printed in the United States
by Baker & Taylor Publisher Services